5th International Conference on Marine Technology (SENTA 2020)

IOP Conference Series: Materials Science and Engineering
Volume 1052

Surabaya, Indonesia
8 December 2020

ISBN: 978-1-7138-4568-3
ISSN: 1757-8981

Printed from e-media with permission by:

Curran Associates, Inc.
57 Morehouse Lane
Red Hook, NY 12571

Some format issues inherent in the e-media version may also appear in this print version.

TABLE OF CONTENTS

IOP Conf. Series: Materials Science and Engineering 1052 (2021) 011001 doi:10.1088/1757-899X/1052/1/011001

Preface of SENTA 2020 Proceedings

The International Conference on Marine Technology, which is more popular known as SENTA, is an annual conference organized by the Faculty of Marine Technology, Institut Teknologi Sepuluh Nopember (ITS) Surabaya, Indonesia. Since 2001, the Faculty of Marine Technology – ITS has organized the annual SENTA conference, formerly known as Seminar Nasional Teknologi dan Aplikasi Kelautan (National Conference on Marine Technology and Application), which provided a regular forum for researchers and engineers in Marine Technology in Indonesia. Then in 2016, SENTA conference has turned into an international conference in order to expand its research network.

Due to the Covid-19 pandemic situation, The 5th International Conference on Marine Technology (SENTA 2020) held virtually on 8 December 2020. All honourable authors contributed to the conference through submissions of their research papers on all topics related to "Transformation from Archipelagic into Maritime States for Prosperous Indonesia", including, but not limited to: Naval Architecture, Ship Production and Material Technology, Marine Robotics and Digital Ship Technology, Marine and Safety Engineering, Marine Renewable Energy, Offshore and Coastal Engineering, Shipping, Port and Maritime Logistic, also Maritime Education Training. All of the authors presented their manuscript in two plenary sessions divided into eight rooms according to their topics through zoom platform. Each author delivered their presentation for about 12 minutes.

On behalf of SENTA committee, I would like to extend gratitude to all the authors who contribute to the submission of the result of their latest research. The committee of SENTA received 113 submissions. From those submissions, there were 78 papers that were presented in SENTA 2020. All the accepted paper has passed the criteria and has been reviewed by the reviewer and then the authors has revised it according to the reviews. 73 papers were accepted for publication. Also, I would like to express my grateful to the Technical Program Committee and the reviewers for their support in the reviewing process to keep the quality of manuscripts.

SENTA 2020 is a high quality conference with a competitive submission process. The committee received submissions from 9 countries, they are Indonesia, Japan, Malaysia, Nigeria, Saudi Arabia, South Korea, Sri Lanka, United Kingdom and United States. SENTA 2020 becomes a great opportunity for researchers, students and practitioners to participate in exchanging information and experiences to keep up to date the research, technology development and innovation related to maritime technology.

There are four prominent keynotes speakers to whom I would like to say thank for their support to SENTA 2020, they are:

1. Prof. Satryo Soemantri Brodjonegoro, President of Indonesia Academy of Science (AIPI), Indonesia
2. Prof. Mehmet Atlar, Professor Associate HOD Research & KE, University of Strathclyde, United Kingdom
3. Prof. Joem Kee-Paik, Professor of Safety Design and Engineering, Pusan National University, South Korea
4. Prof. Jason Monty, Head of Department of Mechanical Engineering, University of Melbourne, Australia

In addition, I sincerely appreciate Dr. Rafet Emek Kurt from University of Strathclyde - United Kingdom for SAFEMODE Project Presentation. SAFEMODE is a research project on the human factor element in marine and aviation safety. Conducted by more than 30 institutions worldwide and financed by Euro-Horizon 2020.

I would also appreciate the IOP conference publisher so that we could make an agreement to publish the SENTA proceeding to the IOP Conference Series: Materials Science and Engineering (MSE). We also would like to deliver the warmest greeting for our colleges, partnership universities, industry and government institutions.

Once again, thank you for your support, I wish that the aims of SENTA 2020 still can be achieved and gives great benefits for all authors though we could not conduct the conference as usual. I also hope that the SENTA 2020 proceeding can be useful for the readers in supporting their research related in marine technology as well as further contributing to develop the technology in the maritime field.

General Chair of SENTA 2020

Professor I Ketut Aria Pria Utama

Faculty of Marine Technology

Institut Teknologi Sepuluh Nopember, Surabaya, Indonesia

December, 2020

IOP Conf. Series: Materials Science and Engineering 1052 (2021) 011001 doi:10.1088/1757-899X/1052/1/011001

Program Committees:

1. Dr. Eng. Trika Pitana
2. Dr. Eng. Rudi Waluyo
3. Wasis Dwi Aryawan, Ph.D.
4. Dr-ing. Setyo Nugroho
5. Herman Pratikno, Ph.D.
6. Benny Cahyono, Ph.D.

Organizing Committees

1. Prof. I Ketut Aria Pria Utama
2. Dr. Eng. Yuda Apri Hermawan
3. Dr. Eng. Dhimas Widhi Handani
4. Dr. Achmad Baidowi
5. Dr. Nurhadi Siswantoro
6. A.A. BGS. Dinariyana Dwi Putranta, Ph.D.
7. Dr. Dendy Satrio
8. Dr. Eng. M. Badrus Zaman
9. Sufian Imam Wahidi, M.Sc.
10. Triwilaswandio Wuruk Pribadi, M.Sc.
11. Sholikan Arif, M.T.
12. Febriani Rohma Dhana, M.T
13. Danu Utama, M.T.
14. Erzad Iskandar Putra, M.T.
15. Ardi Nugroho Yulianto, M.T.
16. Teguh Putranto, M.T.
17. Raditya Danu Riyanto, M.T.
18. Dini Adni Navastara, M.Sc.
19. Hasan Iqbal Nur, M.T.

List of reviewers:

1. Prof. Achmad Zubaydi
2. Prof. Eko Budi Djatmiko
3. Prof. Mukhtasor
4. Prof. Ketut Buda Artana
5. Prof. Heri Supomo
6. Prof. Aris Sulisetyono
7. Prof. Aguk Zuhdi M. Fathallah
8. Prof. Semin
9. Dr. Ketut Suastika
10. Dr. Made Ariana
11. Dr.Eng. I Gusti Ngurah Sumanta Buana
12. Suntoyo, Ph.D
13. Dr. Eng. Shade Rahmawati

14. Dr. Emmy Pratiwi
15. Sutopo Purwono Fitri, Ph.D.
16. Nur Syahroni, Ph.D.
17. Dr. Eng. Yeyes Mulyadi
18. R. Haryo Dwito A, Ph.D.

Organized by

Supported by

IOP Conf. Series: Materials Science and Engineering 1052 (2021) 011002 doi:10.1088/1757-899X/1052/1/011002

Peer review declaration

All papers published in this volume of *IOP Conference Series: Materials Science and Engineering* have been peer reviewed through processes administered by the Editors. Reviews were conducted by expert referees to the professional and scientific standards expected of a proceedings journal published by IOP Publishing.

- **Type of peer review: Single-blind**
  The "blind" in single blind review refers to what information authors can see. In a single blind peer review, reviewers' identities are kept hidden from authors.
- **Conference submission management system: Easychair**
- **Number of submissions received: 113 papers**
- **Number of submissions sent for review: 102 papers**
- **Number of submissions accepted: 78 papers**
- **Acceptance Rate (Number of Submissions Accepted / Number of Submissions Received X 100): 69 %**
- **Average number of reviews per paper: 2-3 reviews**
- **Total number of reviewers involved: 21 reviewers**
- **Any additional info on review process:**

  - After revision has been submitted by author, the committee recheck the paper to ensure if the review of reviewer has been accommodated in the paper. Further correspondence with author is done if necessary, to improve quality of paper based on contents and format.
- **Contact person for queries:**
  I Ketut Aria Pria Utama : kutama@na.its.ac.id
  Sufian Imam Wahidi : sufian@its.ac.id

IOP Conf. Series: Materials Science and Engineering 1052 (2021) 012001 doi:10.1088/1757-899X/1052/1/012001

Preliminary Study of Ship Maneuvering Prediction of Container Ship

A N Yulianto[1,2], W D Aryawan[1], T Putranto[1,3], S H Sujiatanti[1,4], G M Ahadyanti[1,5], D B Purwanto[1]

[1]Department of Naval Architecture, Institut Teknologi Sepuluh Nopember, Indonesia
[2]Department of Naval Architecture and Marine Engineering, Mokpo National University, South Korea
[3]Department of Civil Engineering and Architecture, Tallinn University of Technology, Estonia
[4]Department of Transportation and Environmental Systems, Hiroshima University, Japan
[5]Department of Naval Architecture, Ocean and Marine Engineering, University of Strathclyde, United Kingdom

Abstract. Ship maneuvering is defined as the inherent capability of a vessel to turn its path based on the hydrodynamics principles. The maneuvering performance becomes important to the safety of navigation when a ship operates. In case of vessel turning motion, International Maritime Organization (IMO) requires the maximum tactical diameter is five times of the ship lengths. Tactical diameter is the distance travelled by the ship's center of gravity in a parallel direction to the ship's initial path. In this paper, semi-empirical approach proposed by Clarke (1983) was used to obtain the first derivative of hydrodynamics equation. After that, the maneuvering performance was calculated using Son and Nomoto's mathematical model (1981). Based on IMO regulations, the calculation of tactical diameter was conducted with 35° of rudder position. The calculation and analysis of tactical diameter was carried out according to relevant procedure of the ship maneuvering provided by IMO. The calculation result shows satisfactory result based on IMO regulations. The maximum tactical diameter does not exceed the requirements of IMO maneuvering standards. A further investigation was performed to see whether there was any impact on ship's main dimensions to the maneuverability performance. In the future study, the development of open source software program to analyze the ship maneuvering performance is considered. It is expected to be able to contribute for facilitating in the preliminary ship design process.

1. Introduction

In the last years, cargo ships have been built to a larger size to meet the increasing load capacity leading to the transport efficiency. However, a larger size of ship has an impact on the limitation of maneuverability performance, especially in channels or crowded port, hence the likelihood of accidents caused by collision and contact is high [1]. With a good maneuver, a ship can turn easily to avoid a collision. Therefore, the ship maneuverability is one of the important factors that needs to be ensured for the passenger and cargo safety.

Analysis of ship maneuvering is typically related to the seakeeping performance. For example, side forces induced by turning circle maneuver effect to the transverse movement, i.e. sway motion. This is a concern of static transverse stability problem that should be convinced when a ship maneuver [2]. In addition, the hull form of the ship is also a factor that effects the magnitude of hydrodynamics forces naturally dominating in maneuvering performance. In numerical simulation, estimation of this hydrodynamics coefficient is indispensable to complete the calculation of maneuvering equation.

Moreover, the ship motions have also significantly contributed to its maneuverability, especially for the calm water [3] and wave conditions [4]. It is well known that the strategy to optimize such design stage is using parametric study with respect to hull sizing [5]. Therefore in this paper, parametric study involving 10 main dimensions variation had been carried out for various L/B, L/D, and B/D ratios. Such main dimension variations were derived based on the existing and available data in BKI.

Prediction of ship trajectory is usually carried out to estimate the future position in the vessel maneuvers. Free-running model for maneuvering was carried out to perform the turning circle test by varying the propeller and rudder types and sizes [6]. Experimental results were satisfactory; however, it is costly in time to prepare the model compared to numerical method. Some papers have analyzed the maneuvering performance of ship and its hydrodynamics performances. Hydrodynamics derivatives for ship are calculated to confirm the maneuvering equation using analytical approach [7] and computational fluid dynamics (CFD) [8] which the solution of mathematical model is complicated. In another paper, Clarke et al. [9] proposed the development of semi-empirical formula to calculate hydrodynamics coefficient of a ship.

The aim of this paper is to analyze the turning ability of containerships based on the International Maritime Organization (IMO) regulation. IMO criteria for turning circle maneuver provides that the tactical diameter should not exceed five ship lengths, see in Figure 1. In order to obtain the prediction of tactical diameter when a ship operates, a simplification formula introduced by Son and Nomoto [10] is used. Hydrodynamics coefficients in this formula were calculated using Clarke approach [9]. In addition, the principal dimensions associated with the ship were required. Finally, the estimation of tactical diameter for turning circle test was obtained and compared to the IMO criteria.

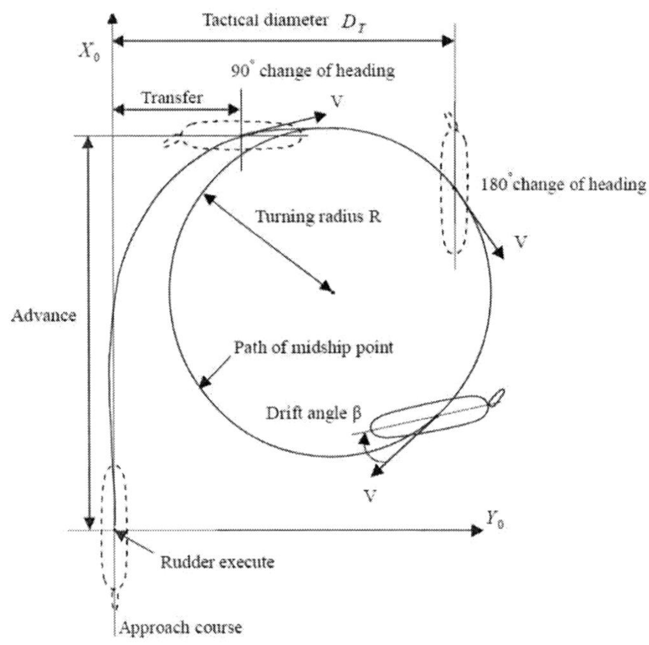

Figure 1. Turning Circle Diagram.

2. Method

Ship maneuvering is one of important parts in the area of ship performance. Many researchers are competing to find ways to solve maneuver problems in both testing and numerical approaches. International Towing Tank Conference (ITTC) has proposed the standard of ship maneuvers for sea trials such turning circle, zig-zag maneuver, pull-out maneuver, spiral maneuver, reverse spiral maneuver and stopping trial. Figure 2 shows the coordinate systems in the investigating of ship maneuverability. It shows the earth-fixed coordinate system O_0-$x_0y_0z_0$ and the body-fixed coordinate system O-xyz that moves together with the ship. The heading angle (ψ) can be defined as the angle between the direction of x_0 axis and x axis. In the earth-fixed coordinate system, the ship center of gravity (x_{0G} and y_{0G}) can be assigned as the position of ship and the heading angle (ψ) is determined as the orientation of ship [11].

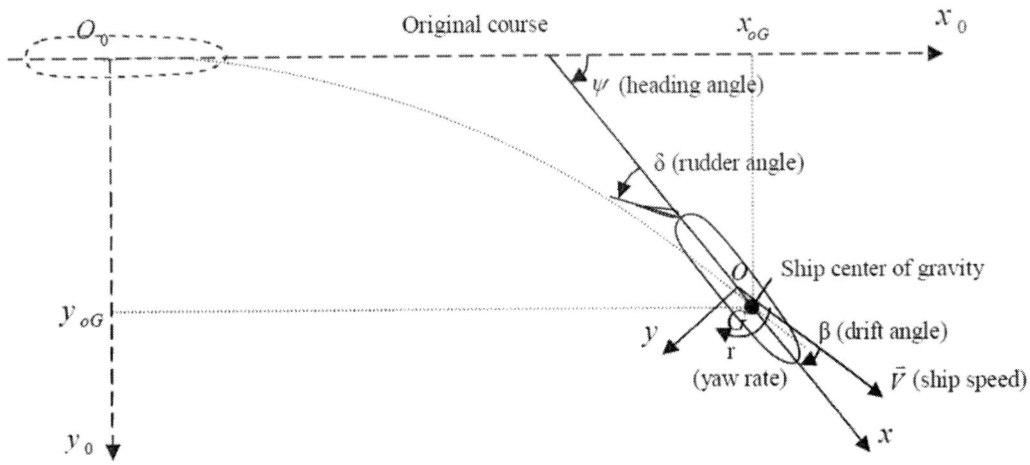

Figure 2. Coordinate System

In this paper, the analysis will focus for turning circle analysis. Turning circle is carried out to calculate the ship's steady turning radius. It also performs the steering ability of ship under course-changing maneuvers. The steering machine and rudder control have important role in the turning circle performance [12]. Tactical diameter will be important thing in the analysis part. The turning radius can be expressed by [12]:

$$R = \frac{Y'_v(N'_r - m'x'_G) - N'_v(Y'_r - m')}{(Y'_vN'_\delta - N'_vY'_\delta)}\frac{1}{\delta}L \tag{1}$$

Where,
m: mass of ship
L: length of ship
Y: hydrodynamic force
N: moment
x_g: center of gravity

The ship speed factor does not give effect in the turning radius formula. It can be said that ship speed has independent role in terms of analysis [12]. The turning radius formula is based on linear theory which assume that δ (rudder angle) is small. Y'_v will be used to denote the y-component of the hydrodynamic force. N'_v is termed the virtual moment of inertia coefficient. The linearized y-component of the force created by rudder deflection acting at the center of gravity of the ship is Y'_δ and the linearized component of the moment created by rudder deflection about the z-axis of the ship is N'_δ [13]. The hydrodynamic derivatives are developed using Clarke approach to generate the turning radius of container ship [9]. Table 1 shows the ship data for maneuvering analysis [14]. There are ten variations of container ship that will be executed for turning circle. The ship data was obtained from the ship register of Biro Klasifikasi Indonesia (BKI). The prediction of mass and center of gravity is an important part of ship maneuvering prediction. In this case, the mass and the center of gravity are derived based on Schneekluth's method [15].

In general, the relationship of ship main dimensions, ship length (L), breadth (B), and depth (D), can be shown as in Table 1. The ratio of main dimension has a significant role in the ship preliminary design. It can be seen the range of L/B is in between 4.02 and 6.74, L/D is in between 11.79 and 15.97, and B/D is in between 2.03 and 3.03. The acceptance criteria of maneuvering follow the regulation of IMO Resolution MSC 137 (76) annex. 6 [16]. It stated that the tactical diameter should be less than five times ship lengths.

Table 1. Ship Data for Maneuvering Analysis.

No.	Ship Name	Loa *(m)*	Lbp *(m)*	B *(m)*	D *(m)*	L/B	L/D	B/D
1	Kendhaga Nusantara 10	74.05	69.20	17.20	4.90	4.02	14.12	3.51
2	Akashia	95.90	89.00	15.20	7.20	5.86	12.36	2.11
3	Caraka Jaya III	98.00	92.00	16.50	7.80	5.58	11.79	2.12
4	Estuari Mas	119.90	115.00	21.80	7.20	5.28	15.97	3.03
5	Icon Perdana	84.60	78.00	16.00	6.50	4.88	12.00	2.46
6	Pulau Hoki	120.97	114.00	20.80	8.00	5.48	14.25	2.60
7	Sendang Mas	112.23	109.17	16.20	8.00	6.74	13.65	2.03
8	Spil Hana	135.70	133.00	22.50	10.20	5.91	13.04	2.21
9	Umbul Mas	119.90	115.03	21.80	7.30	5.28	15.76	2.99
10	Teluk Bintuni	114.30	107.60	16.00	7.80	6.73	13.79	2.05

3. Discussion and Result

The prediction of the turning circle for the selected containerships is presented in Table 2. According to the criterion for tactical diameter required by IMO, it can be seen from the table that all containerships have passed this criterion. As shown in the table, the vast majority of the prediction of the tactical diameter averaging at around 3 in ship lengths.

Table 2. Prediction of Turning Circle for the Containerships.

No.	Ship Name	Tactical diameter (in ship lengths) TD/L
1	Kendhaga Nusantara 10	2.67
2	Akashia	3.22
3	Caraka Jaya III	3.54
4	Estuari Mas	3.00
5	Icon Perdana	3.20
6	Pulau Hoki	3.43
7	Sendang Mas	3.05
8	Spil Hana	4.61
9	Umbul Mas	3.04
10	Teluk Bintuni	3.07

A further analysis has been carried out to observe whether there was any influence on the ship's main particulars to the maneuverability capability. The tactical diameter predicted then was plotted against the containerships' L/B, L/D and B/D ratios as shown in Figure 3, Figure 4, and Figure 5 respectively.

Figure 3 shows the relationship between the containerships' L/B ratio and the tactical diamater over the ship's length. The scatter diagram shows a low degree of correlation between the two variables. It can be seen from the graph that along with increasing price ratio L/B, the tactical diameter in ship lengths are fluctuating. However, if we look closely from L/B 4.02 up to 5.58, there was an upward trend despite the decreasing 2 figures at value of L/B 5.28 (which represents Estuari Mas and Umbul Mas).

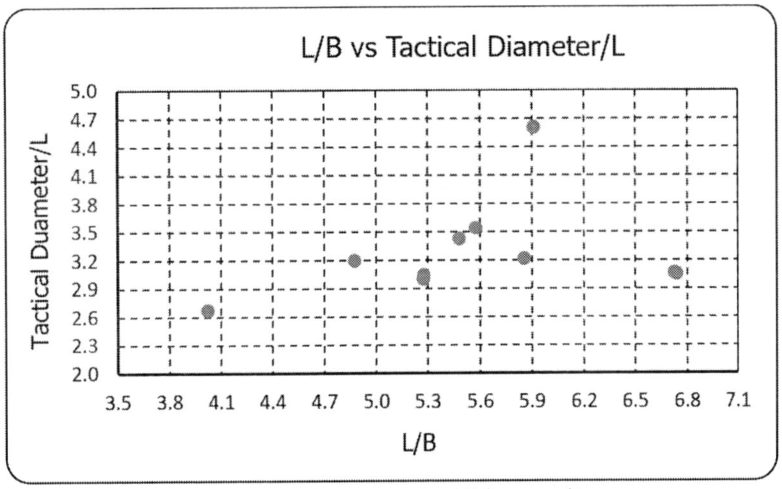

Figure 3. L/B vs. Tactical Diameter/L.

Figure 4 depicts the containerships L/D ratios against the predicted tactical diameter over ship lengths. Similar with Figure 3, the graph also shows a fluctuation over the value of L/D ratios, though it can be seen there was a downward trend in the overall. Because the data point spread in this figure is so random that we cannot draw a line through them, therefore, it is fair to say that the L/D and tactical diameter variables have no correlation.

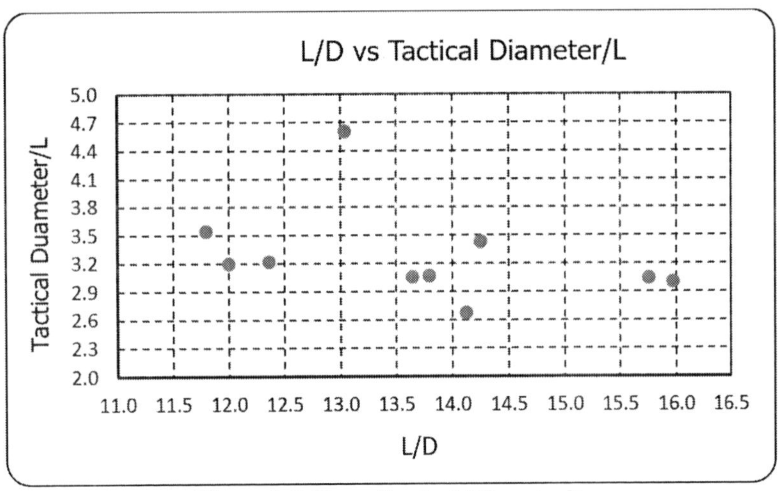

Figure 4. L/D vs. Tactical Diameter/L.

Lastly, the observation was made by plotting the tactical diameter against B/D ratios as seen in Figure 5. Unlike the previous figures, in this diagram, data points are close to each other and a line can be drawn by following their pattern. It was clear that there was an upward trend from the value of B/D 2.03 to the value of B/D 2.21, the scatter diagram presents a strong positive correlation between the two variables. Afterwards, the tactical diameter shows a downward trend along with the increasing price ratio B/D excluding the increasing value at B/D 2.6, the scatter diagram presents a strong negative correlation between the two variables with a presence of an outlier.

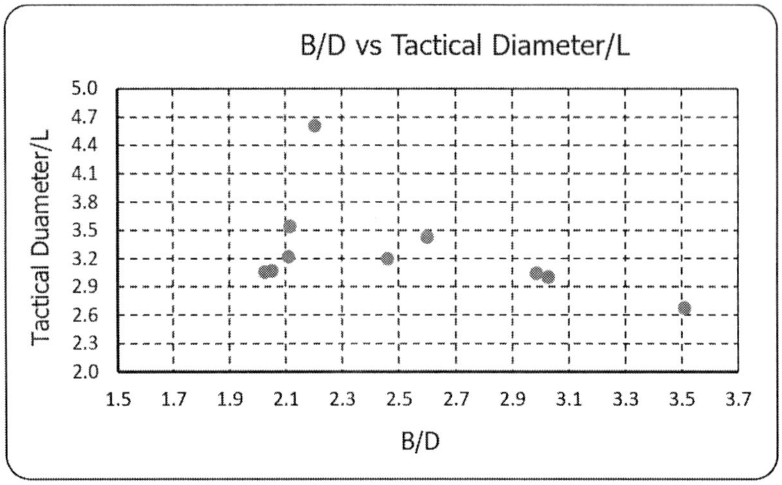

Figure 5. B/D vs. Tactical Diameter/L.

Based on the discussion from Figure 3 until Figure 5, the correlation between the ship's main particulars and tactical diameter are varied depends on the main dimensions ratios. In Figure 3 the scatter diagram has a moderate correlation with a few potential outliers, while in Figure 4 the scatter diagram shows no correlation between the two variables. However, in Figure 5, the scatter diagram has a high degree of correlation, with a strong positive correlation from the value of B/D 2.03 to the value of B/D 2.21, and a strong negative correlation from the value of B/D 2.21 to the value of B/D 3.51.

4. Conclusion

In this study, ship maneuvering prediction was carried out for several selected container ship. The calculation of turning circle radius has been conducted and obtained satisfactory result based on IMO Resolution MSC 137. According to ship data from Biro Klasifikassi Indonesia (BKI), the tactical diameter of ship maneuvering does not exceed than five times of ship length (L). The difference trend from the result might be caused by the limitation of the ship's sample that were taken. The development of open source code of ship maneuvering performance will be initiated in the future research. It will make easier for a naval architect in the ship design process, especially in the preliminary ship design.

Acknowledgement

The authors would like to thank all colleagues at Department of Naval Architecture, Institut Teknologi Sepuluh Nopember for their support and valuable comments on this research.

References

[1] J. Xue, P. H. A. J. M. Van Gelder, G. Reniers, E. Papadimitriou and C. Wu, "Multi-attribute decision-making method for prioritizing maritime traffic safety influencing factors of autonomous ships' maneuvering decisions using grey and fuzzy theories," *Safety Science,* vol. 120, p. 323–340, 1 12 2019.

[2] T. Putranto, W. D. Aryawan, H. A. Kurniawati, D. Setyawan and S. R. W. Pribadi, "Resistance and Stability Analysis for Catamaran Fishing Vessel with Solar Cell in Calm Water," in *MATEC Web of Conferences*, 2018.

[3] Y. Sanada, K. Tanimoto, K. Takagi, L. Gui, Y. Toda and F. Stern, "Trajectories for ONR Tumblehome maneuvering in calm water and waves," *Ocean Engineering,* vol. 72, p. 45–65, 1 11 2013.

[4] G. Chillcce and O. el Moctar, "A numerical method for manoeuvring simulation in regular waves," *Ocean Engineering,* vol. 170, p. 434–444, 15 12 2018.

[5] W. D. Aryawan and G. M. Ahadyanti, "Response-Based Metocean Criteria for the Optimization of Floating Production Facility for Marginal Oil Field at Java Sea," in *International Conference on Ship and Offshore Technology*, 2015.

[6] R. Suzuki, Y. Tsukada and M. Ueno, "Estimation of full-scale ship manoeuvring motions from free-running model test with consideration of the operational limit of an engine," *Ocean Engineering,* vol. 172, p. 697–711, 15 1 2019.

[7] R. Skejic and O. M. Faltinsen, "A unified seakeeping and maneuvering analysis of ships in regular waves," *Journal of Marine Science and Technology,* vol. 13, p. 371–394, 28 11 2008.

[8] H. p. Guo and Z. j. Zou, "System-based investigation on 4-DOF ship maneuvering with hydrodynamic derivatives determined by RANS simulation of captive model tests," *Applied Ocean Research,* vol. 68, p. 11–25, 1 10 2017.

[9] D. Clarke, P. Gedling and G. Hine, "The Application of Manoeuvring Criteria in Hull Design Using Linear Theory," 1982.

[10] K. Son and K. Nomoto, "On the Coupled Motion of Steering and Rolling of a High Speed Container Ship," *Journal of the Society of Naval Architects of Japan,* vol. 1981, p. 232–244, 1981.

[11] Z. Zou, Ship Manoeuvring and Seakeeping, Shanghai: Shanghai Jiao Tong University, 2006.

[12] T. Fossen, "Guidance and control of ocean vehicles," New York, 1994.

[13] E. V. Lewis, Principle of Naval Architecture Second Revision, Jersey: SNAME, 1989.

[14] B. K. Indonesia, "Ship Register," [Online]. Available: https://bki.co.id/shipregister.html.

[15] H. Schneekluth and H. Bertram, Ship Design for Efficiency and Economy, boston: Butterworth Heinemann, 1998.

[16] IMO, Standard for Ship Manoeuvrability (MSC.137(76), International Maritime Organization, 2002.

An Estimation of hourly average solar radiation using artificial neural network in the city of Surabaya

A Kurniawan[1]*, E S Koenhardono[1], I R Kusuma[1], J Prananda[1], S Sarwito[1] and A A Masroeri[1]

[1]Department of Marine Engineering, Institut Teknologi Sepuluh Nopember, Surabaya 60111 Indonesia

*e-mail: adi.kurniawan@ne.its.ac.id

Abstract. To maximize the effectiveness of solar energy systems installed on the ship, the panels need to be positioned on their optimal inclination angle. To determine this angle, information on hourly average solar radiation is required in addition to position and direction data. However, the amount of hourly solar radiation is not continuously measured in Indonesia. This study proposed a method to estimate the hourly solar radiation for any month. This is based on Artificial Neural Network (ANN) with the month and the hour required for the estimate and the previous daily average solar radiation as the input. The proposed ANN method was validated by comparing the estimation results with the measured hourly average solar radiation in Surabaya from May to July 2020. Its effectiveness was affirmed with the coefficient of determination (R^2) of 0.983 or higher for each of the three observed months.

1. Introduction

Indonesia is a maritime state located on the equator, therefore a solar-assisted ship is very suitable to be used as a major mode of transportation [1], [2]. However, limited available space to install the photovoltaic (PV) panels on the ship forced the engineers to maximize the efficiency of the PV system [3], [4]. One possible method to maximize the efficiency of this system is by installing the panels on their optimal inclination angle and not on a horizontal position [5].

To determine this angle, the information of hourly solar radiation is required while the ship is sailing. In contrast to the daily average solar radiation, the measurement of hourly solar radiation is not continuously performed in Indonesia [6], which reduces the possibility of determining the optimal inclination angle of PV panel on ship sailing in the Indonesia area.

Nevertheless, with the recent technologies, the use of estimation techniques helped to overcome the problem of limited information on solar radiation. Several methods were validated to accurately predict the amount of hourly solar radiation, either based on physical or historical statistical models [7]–[12]. However, each of them requires information on weather parameters that are difficult to access in Indonesia, such as the hourly clearness index among others.

This study proposed a new estimator for the hourly average solar radiation based on artificial neural network. The ANN was utilized either as an estimator of the daily average [13], [14], or hourly solar radiation [15], [16] and this is more suitable to be used in this region since it only requires three inputs which are easily accessible in Indonesia. These inputs include the month and the hour when the estimation is required and also the previous amount of the daily average solar radiation. Furthermore, this estimation technique enables the determination of the PV panel inclination angle installed on the ship sailing in Indonesia, which in turn reduces the operational fuel cost and greenhouse gas emissions by cutting the usage of diesel generator.

2. Material and Method

The structure of the proposed ANN method to estimate the hourly average solar radiation is presented in Figure 1. As mentioned in section 1, there are three inputs for the ANN and the data of daily average solar radiation of the previous month is easily accessible from NASA Power database.

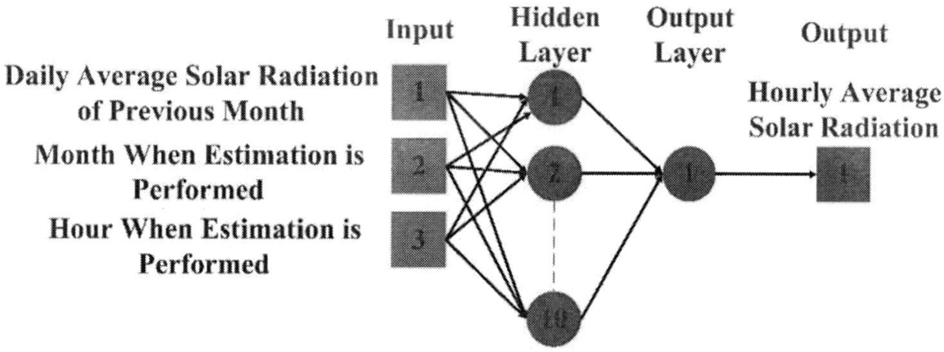

Figure 1. Structure of the proposed ANN.

The ANN was trained using the data of the daily average solar radiation in the city of Surabaya from January to December 2019. The number of hours is selected from the earliest sunrise (5 AM) to the latest sunset (5 PM), resulting in a sample size of 156 for each input-output parameter. In addition, the Levenberg-Marquardt backpropagation method was adopted to train the ANN, which represents a fast and stable convergence [17]. Meanwhile, the stopping criteria of this method is the minimization of mean squared error (MSE) defined

$$\text{MSE} = \frac{1}{n}\sum_{k=1}^{n} e_k^2 = \frac{1}{n}\sum_{k=1}^{n}\left(t_k - y_k\right)^2 \tag{1}$$

with n as the total number of samples, k is the index, e is the error of the output, t is the target value of the output taken from the data, and y is the estimated output value obtained from the ANN [18].

The minimum MSE during training was reached after 27 iterations as shown in Figure 2. The MSE itself reaches the value of 1000. It is understandable since the daily solar radiation is in the order of thousands Wh/m^2. The training of ANN is stopped when the error is stop to decrease, indicated by the gradient. Furthermore, the good agreement between the estimated and target value of the output from the training data was confirmed by 0.996 regression value as presented in Figure 3. The white dots are the representation of plotting between the actual target value in horizontal axis and the output value of the ANN for the same input in vertical axis. The straight diagonal black line is the regression line that approach the distribution of data plot. If the output of the ANN can be exactly same with the target data, then the line will touch all the dots and the value of R will be 1.

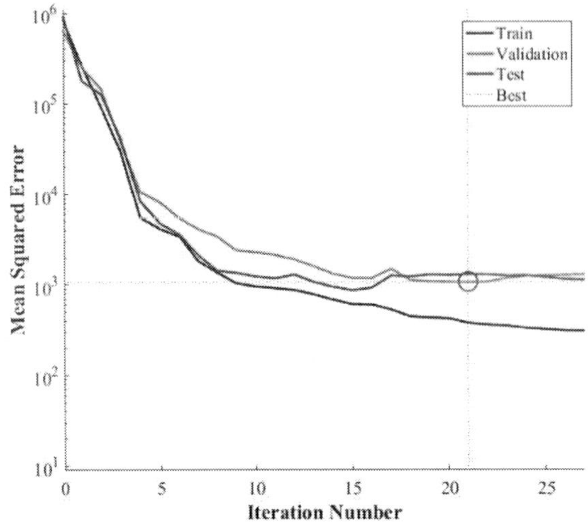

Figure 2. Error performance of ANN training.

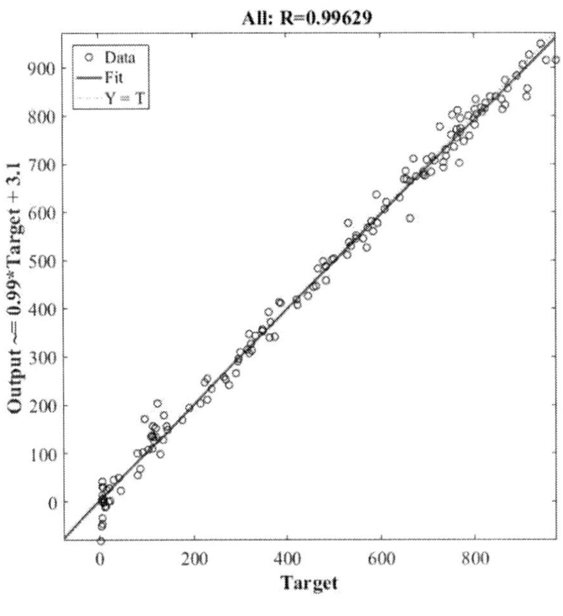

Figure 3. Regression line of ANN training.

3. Result and Discussion

The proposed method was validated by comparing the estimation results with the measured hourly average solar radiation for May, June, and July 2020 in Surabaya. Also, the graphical comparisons for each month are shown in Figure 4 to 6. It showed a high similarity between predicted and measured values, especially for July. For May and June, although there is gap between the estimated and measured value, but the curves of the estimated value have same pattern with the measured value. Furthermore, the predicted values tend to be higher than the measured, especially during the peak sunlight between 10 AM and 1 PM.

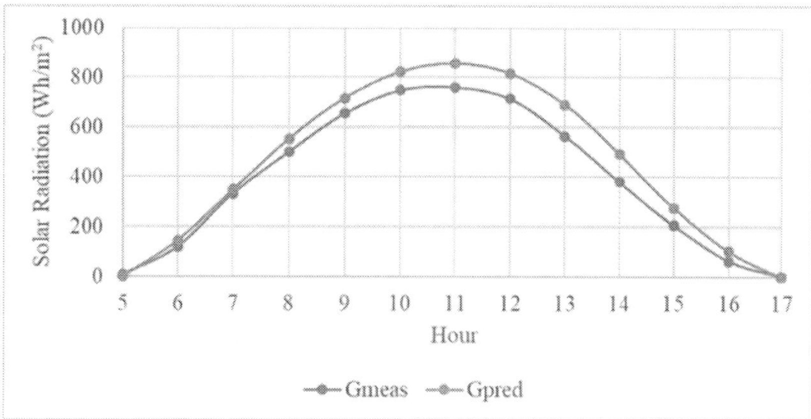

Figure 4. Comparison of predicted and measured hourly solar radiation in Surabaya for the month of May.

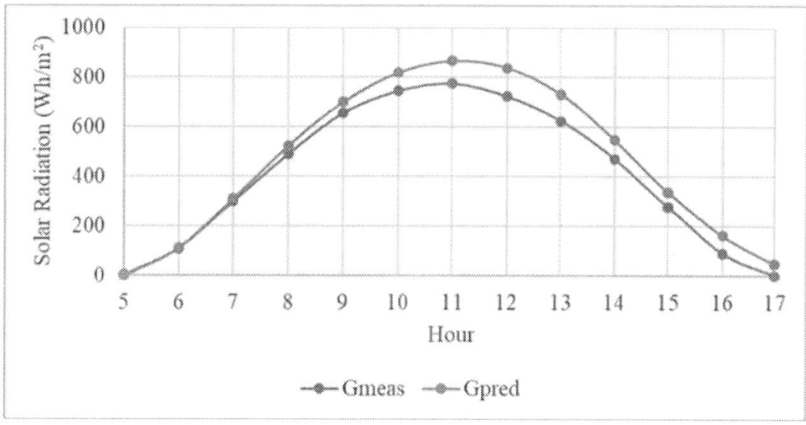

Figure 5. Comparison of predicted and measured hourly solar radiation in Surabaya for the month of June.

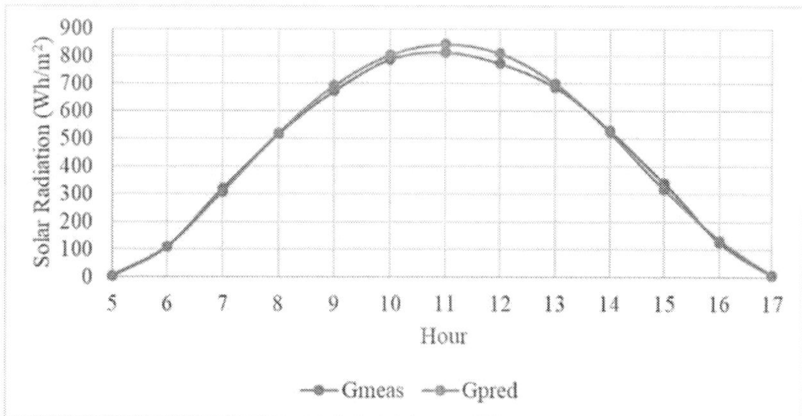

Figure 6. Comparison of predicted and measured hourly solar radiation in Surabaya for the month of July.

Further validation of the proposed method was done by calculating the statistical parameter in the form of the coefficient of determination (R^2) for each month as follow:

$$R^2 = 1 - \frac{\sum_{h=5}^{20} (G_{meas(h)} - G_{pred(h)})^2}{\sum_{h=5}^{20} G_{pred}^{\ 2}} \quad (2)$$

where G_{meas} is the measured value of hourly solar radiation, G_{pred} is the predicted value of hourly solar radiation, and h is the index of hour.

The R^2 of each month is listed in Table 1 and the lowest R^2 is 0.983 for the estimation of May. The similar result is obtained for the estimation of June, but more accurate result is obtained for the estimation of July, with R^2 of 0.999. For comparison, the MARS method generates an average R^2 of 0.912 [8] while the best R^2 for various regression methods is 0.93 [11].

Table 1. Statistical results of hourly solar radiation estimation in Surabaya for the year of 2020.

Month	R^2
May	0.983
June	0.985
July	0.999

4. Conclusion

This study aims to develop an estimation method of hourly solar radiation, which may be utilized for various purposes such as determining the optimal inclination angle of PV panels installed on ship. The determination of the optimal inclination angle on ship based on the estimation of solar radiation using land data has been proposed by the research in Japan. Therefore, it is possible to apply the similar method using the data of solar radiation in Indonesia. The proposed method is applicable for use in Indonesia since all the required inputs are accessible. Furthermore, the accuracy of the proposed method was validated by comparing the predicted with the actual measured value of hourly average solar radiation. Meanwhile, the calculation of the statistical parameters also indicates a good performance of the estimation method.

Acknowledgment

This study is financially supported by Institut Teknologi Sepuluh Nopember by the scheme of "Penelitian Dana Departemen" no. 1732/PKS/ITS/2020.

References

[1] Putri D.P. and E. S. Koenhardono, "Perencanaan Sistem Pembangkit Listrik Hybrid (Sel Surya dan Diesel Generator) Pada Kapal Tanker," *J. Tek. ITS*, vol. 5, no. 2, pp. 394–399, 2016.

[2] A. Kurniawan and E. Shintaku, "Control of photovoltaic system connected to DC bus in all-electric ship," in *Proceeding - ICAMIMIA 2017: International Conference on Advanced Mechatronics, Intelligent Manufacture, and Industrial Automation*, 2018, pp. 110–115.

[3] K. Mahmud, S. Morsalin, and M. I. Khan, "Design and Fabrication of an Automated Solar Boat," *Int. J. Adv. Sci. Technol.*, vol. 64, pp. 31–42, Mar. 2014.

[4] A. Kurniawan, E. Haryanto, and A. A. Masroeri, "A neural network based maximum power

point tracker with KY converter for photovoltaic system on a moving vehicle," in *The 1st International Conference on Advanced Mechatronics, Intelligent Manufacture, and Industrial Automation*, 2015, no. Icamimia, pp. 117–120.

[5] A. Kurniawan and E. Shintaku, "Determining the optimal inclination and orientation angles of solar panels installed on ship," *Int. J. Renew. Energy Res.*, vol. 10, no. 1, pp. 45–53, 2020.

[6] Nasruddin, M. A. Budiyanto, and M. H. Lubis, "Hourly solar radiation in Depok, West Java, Indonesia (106.7942 Longitude, -6.4025 Latitude)," *IOP Conf. Ser. Earth Environ. Sci.*, vol. 105, no. 1, 2018.

[7] A. Calcabrini, H. Ziar, O. Isabella, and M. Zeman, "A simplified skyline-based method for estimating the annual solar energy potential in urban environments," *Nat. Energy*, vol. 4, no. 3, pp. 206–215, 2019.

[8] D. H. W. Li, W. Chen, S. Li, and S. Lou, "Estimation of hourly global solar radiation using Multivariate Adaptive Regression Spline (MARS) – A case study of Hong Kong," *Energy*, vol. 186, p. 115857, 2019.

[9] E. Akarslan and F. O. Hocaoglu, "A novel method based on similarity for hourly solar irradiance forecasting," *Renew. Energy*, vol. 112, pp. 337–346, 2017.

[10] F. O. Hocaoglu and F. Serttas, "A novel hybrid (Mycielski-Markov) model for hourly solar radiation forecasting," *Renew. Energy*, vol. 108, pp. 635–643, 2017.

[11] E. Paulescu and R. Blaga, "Regression models for hourly diffuse solar radiation," *Sol. Energy*, vol. 125, pp. 111–124, 2016.

[12] P. F. Jiménez-Pérez and L. Mora-López, "Modeling and forecasting hourly global solar radiation using clustering and classification techniques," *Sol. Energy*, vol. 135, pp. 682–691, 2016.

[13] M. Marzouq, H. El Fadili, Z. Lakhliai, and K. Zenkouar, "A review of solar radiation prediction using artificial neural networks," in *International Conference on Wireless Technologies, Embedded and Intelligent Systems*, 2017.

[14] A. Kurniawan and E. Shintaku, "Estimation of the Monthly Global , Direct , and Diffuse Solar Radiation in Japan Using Artificial Neural Network," *Int. J. Mach. Learn. Comput.*, vol. 10, no. 1, 2020.

[15] G. Notton, C. Voyant, A. Fouilloy, J. L. Duchaud, and M. L. Nivet, "Some applications of ANN to solar radiation estimation and forecasting for energy applications," *Appl. Sci.*, vol. 9, no. 1, 2019.

[16] A. Khosravi, R. N. N. Koury, L. Machado, and J. J. G. Pabon, "Prediction of hourly solar radiation in Abu Musa Island using machine learning algorithms," *J. Clean. Prod.*, vol. 176, pp. 63–75, 2018.

[17] A. Kurniawan and E. Shintaku, "A neural network-based rapid maximum power point tracking method for photovoltaic systems in partial shading conditions," *Appl. Sol. Energy*, vol. 56, no. 3, pp. 156-67, 2020.

[18] D. Brkić and Ž. Ćojbašić, "Intelligent Flow Friction Estimation," *Comput. Intell. Neurosci.*, vol. 2016, pp. 1–10, 2016.

Modifying Single-Cylinder Diesel Engine Become Dual-fuel System to Reduce NOx Emission and Maintain Fuel Efficiency

A. Santoso[1], Semin[2] and B. Sampurno[3] and B. Cahyono[4]

[1] The Dept. of Marine Engineering, Institut Teknologi Sepuluh Nopember, Indonesia
[2] The Dept. of Marine Engineering, Institut Teknologi Sepuluh Nopember, Indonesia
[3] The Dept. of Industrial Machinery Engineering, Institut Teknologi Sepuluh Nopember, Indonesia
[4] The Dept. of Marine Engineering, Institut Teknologi Sepuluh Nopember, Indonesia

Abstract. Tier III of Annex VI from MARPOL 73/78 about NOx emission control soon may be implemented at a global level. There are three potential methods to fulfill the Tier III. First is derating power, second is using gas as fuel, and the third one is installing SCR. This research deals with using CNG as diesel engine fuel. This initial study was occupied a single-cylinder diesel engine. Gas cannot burn inside conventional diesel cycles. Dual fuel technology also tends to reduce fuel efficiency compared with the conventional one. Therefore, the single-cylinder engine was modified its piston crown part to overcome both problems. The modification of piston crown geometry affects the engine compression ratio. Modify the conventional diesel engine to become the dual-fuel type is concluded as a potential technique to be developed. The study results show NOx reduction up to 33.68% and the fuel efficiency increase up to 1.512% without any disruption to the engine operating conditions.

Keywords: Dual-fuel, CNG, Diesel Engine, Piston Crown, NOx Emission, Fuel Efficiency

1. Introduction

Air pollution has been growing day by day at a linear rate due to the fast-growing of industrialization in almost every sector such as manufacturing, power plants, automotive, and sea transportation. Therefore, many governments and professional institutions develop rules and regulations to control the situations and keep a better atmosphere. In the maritime sector, IMO has become a global regulator that set up a special board Marine Environmental Protection Committee (MEPC) and produce MARPOL 73/78 Regulations [1]. Technical Code of Annex VI Regulation 13 has an achievement to a comprehensive level of Tier III in detaining the NOx emission from ship propulsion and electrical power plant [2]. The NOx emission is the most important issue for diesel engines that is a dominant power plant used onboard a ship. To fulfill the Tier III requirement that is known very stringent compared with the previous Tier I at about a lower value of 80%. To achieve the Tier III level, there are three methods that can be adopted for the marine diesel engine. They are derating engine power [3], using gas as a fuel [4], and installing a more complicated SCR (Selective Catalytic Converter) unit [5]. This paper is deals with the second method, using natural gas CNG as the fuel of a diesel engine. Following the most common technology that was already settled in the marine power plant then gas will be burned on a diesel engine combustion chamber under dual-fuel technology called the diesel ignited gas engine [6].

In any methods available now, the effort in reducing emission always encounter decrement of engine efficiency and reversely, improving engine efficiency will affect in increasing emission level [7]. Moreover, gas fuel in any form and technology also faces a problem of the decreasing of overall engine efficiency. It is the most important parameter of the successful design and operation of an engine [1] due to the relation with the fuel consumption to certain power output. It is the main reason why the use of gas as fuel should be adjusted by modifying a certain parts of the diesel engine. This paper deal with the modification of piston crown [8]. Based on the engine model Yanmar TF85MH, its piston crown was modified in seven alternatives. Basic modifications refer to the changed of the piston dimension but the results show a trendline of the changing of compression ratio. The seven models after validated are used to simulate in-cylinder parameters such as maximum pressure and temperature, gas velocity, and heat transfer flux to perform further investigations of the engine overall efficiency and NOx emission. The best modifications are a new piston structure that gives as lower NOx emission as possible and at the same time capable of maintaining overall efficiency. Later, in the comparative study, the eligible method to evaluate the efficiency is not comparing dual-fuel with a diesel engines but between dual-fuel standards with the modified dual-fuel engines [9].

2. Literatures Review

MARPOL 73/78 Annex VI Regulation 13 is used to control the NOx emission from the global maritime fleet [2]. Due to good developments of marine diesel engine technology that supported well by the inventions of any methods and products to control emissions then the MEPC regularly fit the regulation updated with the technological achievements. Tier I was ratified based on the 1997 Protocol then it was improved as Tier II that globally come into force in 2011 and finally Tier III comes into force in 2016 for a specific area called NOx Emission Control Area (ECA)[10].

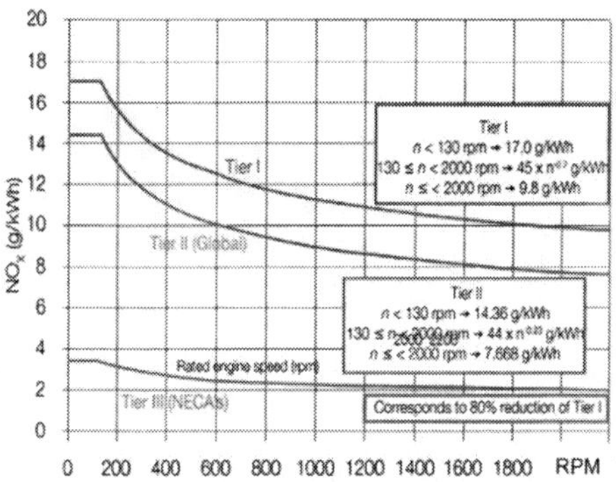

Figure 1. The Technical Code of Annex VI Regulation 13 [2].

Typically, natural gas in the form of liquid natural gas (LNG) or compressed natural gas (CNG) is contains a big portion of methane gas. Fully use of natural gas as a diesel engine fuel (is called a gas engine) has been established well in the land industry such as automotive and electric power plants. But under IGF Regulation the gas engine cannot use in the ships due to redundancy reason [11]. A ship insisted to bring a certain type of fuel oil instead of gas. It is the reason why dual-fuel technology becomes the best choice for a specific application on board a ship. Technically speaking, the natural gas cannot be burned directly by in-cylinder diesel engine pressure.

There are three methods to burn fuel gas under dual-fuel technology. The first is direct injection gas engine (DIGE) technology, where a certain amount of diesel oil functioned as a pilot burner injected directly into the combustion chamber [6]. When the fresh air flowing toward the combustion chamber then a specific portion of gas is injected into the intake manifold and entering together to intake port at the expansion step. As common diesel cycle, before end of the compression stroke and just about before top dead centre, the start of injection (SOI) of fuel oil is done till performs initial flame and the start of combustion (SOC) will burn the mixing of air, gas, and fuel oil. Figure 2 shows the DIGE process.

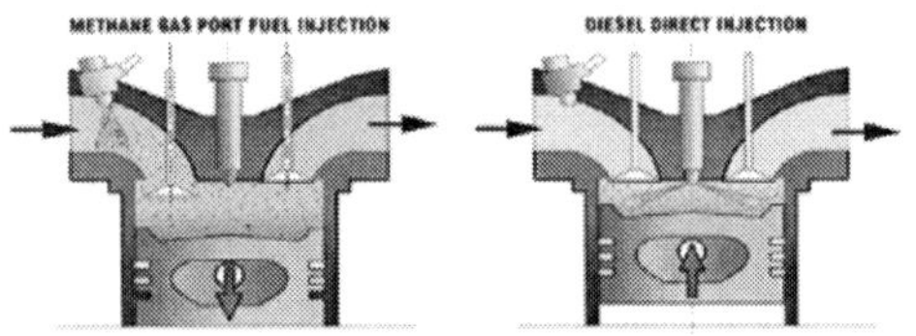

Figure 2. The principle process of The DIGE method [6].

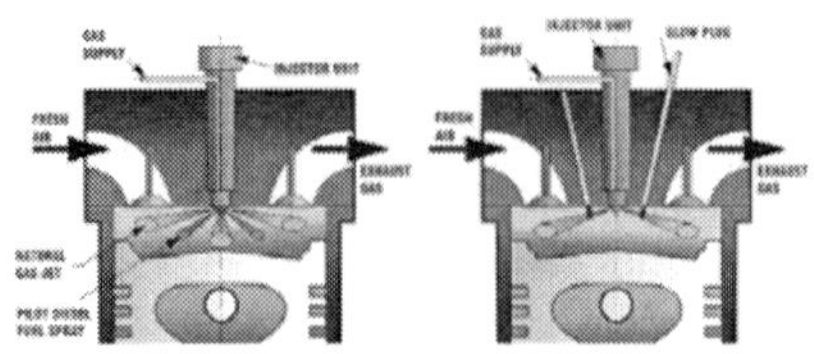

Figure 3. The principle process of the HPDI and HSCI method [12].

Figure 3 (left) shows the principle operation of high-pressure direct injection (HPDI) where the engine has a multi-nozzle fuel injector unit with two channels. In this second method, fresh air is inserted into the chamber as a conventional diesel engine. At the end of the compression step, at SOI the fuel oil is injected into the chamber and perform flames at SOC then followed by fuel gas injection to perform the advance combustion process. Figure 3 (right) shows the principle operation of the third method called hot surface assisted compression ignition (HSCI) where initial flame resulted by assisted a heater unit called a glow plug. The gas injected continuously by using an injector unit contained two channels [12].

3. Modeling and Simulation

As stated before, there are seven modifications of the piston crown structure will be studied here [7]. All of the original and seven modified models will be shown in Figure 4 below.

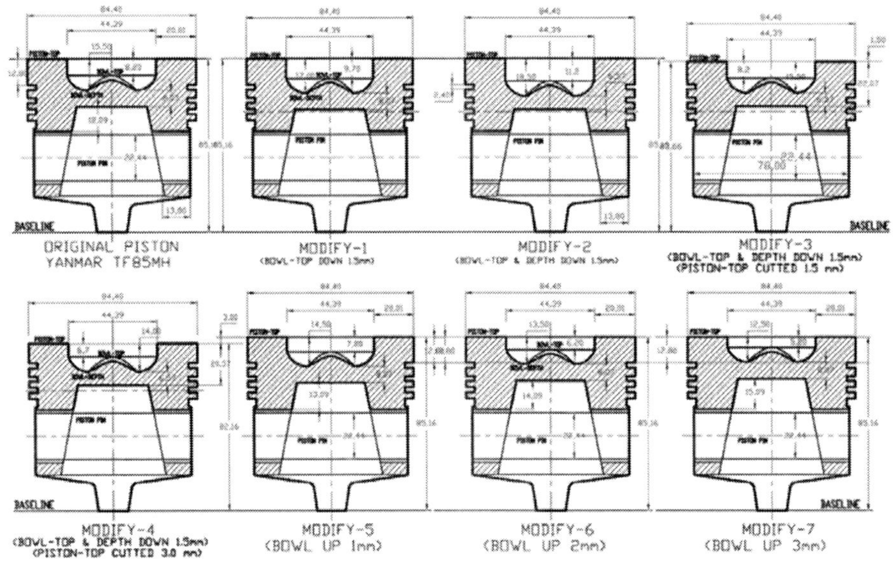

Figure 4. The Seven Modified Models of Piston Crown [8].

After the validation step then all of the modification models are simulated to investigate the combustion results based on the parameters such as; maximum pressure and temperature combustion, gas velocity or gas flow rate, and heat transfer flux that affect to the cylinder liner, cylinder head, and piston based on the CFD analysis as shown in Figure 5. From those initial simulation can be stated that the modify-3 and modify-4 give a significant drop on the four parameters studied then both modifications are not eligible for further analysis [8]. The remaining models were taken into a further investigation to calculate the overall engine efficiency and the potential NOx emission level.

IOP Conf. Series: Materials Science and Engineering 1052 (2021) 012003 doi:10.1088/1757-899X/1052/1/012003

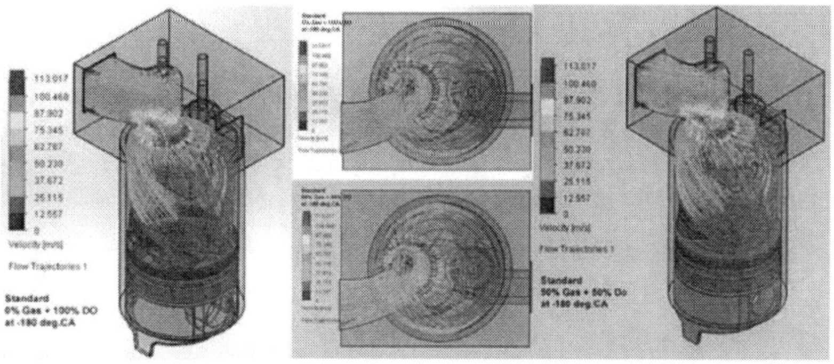

Figure 5. A sample of the simulation result for in-cylinder combustion by using CFD.

4. Analysis and Discussion

Based on the data results generated from the initial simulation then the overall engine efficiency (%) and the NOx emission (g/kWh) can be calculated as shown in Figure 6 and 7. The value of 0% gas means it works at a diesel engine conventional. And the value of 25%, 50%, and 75% gas means it works at a dual fuel engine. Working with gas more than 50% portion technically is a doubt effort because of the high risk of engine knocking phenomenon that may occur [13]. It should be a clear answer to why our modification works are needed to improve the gas portion [7]. Because using more gas in practically gives benefit in terms of economic and technical aspects [14]. It is a challenge to test our modified piston to work up to 75% gas portion so that the data of conditions between 50% to 75% also can be recorded.

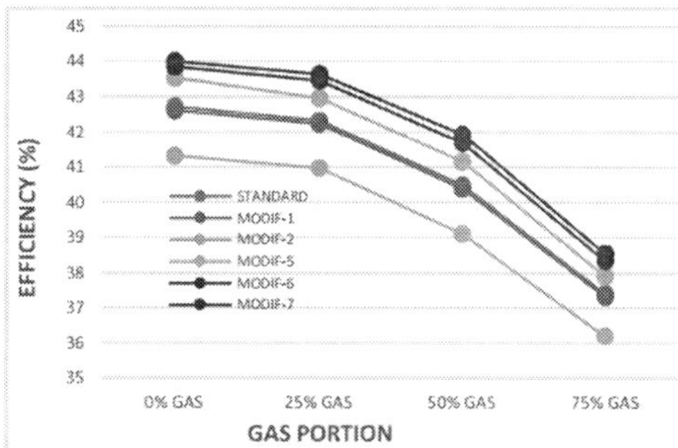

Figure 6. The Trend Graphic of the Calculated Overall Engine Efficiency.

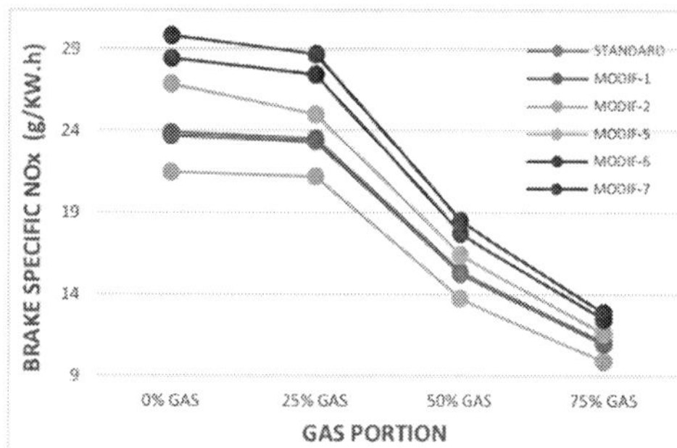

Figure 7. The Trend Graphic of Calculated NOx Emission Level.

The modification 1 to 7 gives a higher engine compression ratio that contributes to higher engine thermal efficiency during combustion processes. The higher efficiency contributes an efficient fuel combustion processes that tend to produce higher in-cylinder maximum pressure and temperature which leads to higher NOx emission formation. On the other hand, Figure 6 and Figure 7 also show that adding the gas portion will reduce both the efficiency and NOx emission. Both the data is a contradiction situation when the lowering efficiency is a disadvantage result and the lowering NOx emission is advantage progress for the modified dual-fuel engine operations.

Understanding the trend from the above graphics then it is an interesting fact about how to evaluate the way to get the best modification. If the above graphics have been merged then new graphical data can be generated to get more valuable analysis based on the increment and decrement values only as shown in Figure 8 below. Following the previous statement before, the comparative study was only done by

IOP Conf. Series: Materials Science and Engineering 1052 (2021) 012003 doi:10.1088/1757-899X/1052/1/012003

comparing dual-fuel only. It means the modified piston when running with a 50% gas portion is compared to the 25% dual fuel to evaluate the increment or decrement values. Figure 8 shows that the only modify-5 can give increasing overall efficiency and reducing the NOx emission simultaneously when the modified dual-fuel engine running under a 25% gas portion. Burning 50% gas portion to modify-5 will result in lower efficiency and contribute to much lower NOx emission. In normal consequences, the modify-6 and modify-7 even give better efficiency, but they produce higher NOx emission value, so it is not acceptable to the basic criterion of this study objective. Finally, modify-5 can be chosen as the best one.

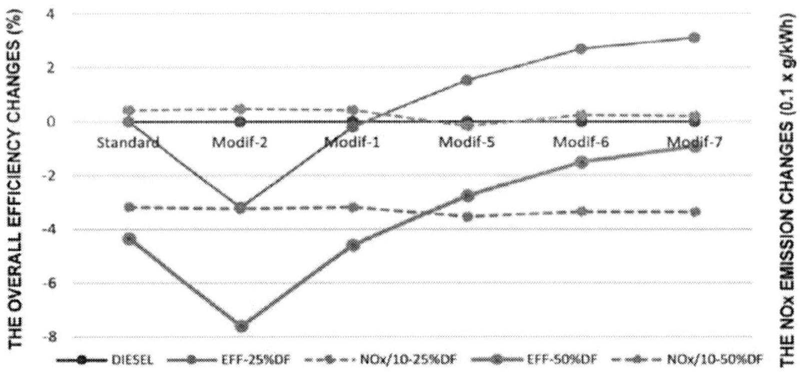

Figure 8. The Resume of Simulation Results of Overall Efficiency and NOx Emission.

The modify-5 can reduce NOx emission of 1,63% while possibly maintain the overall efficiency of about 1,512 % compared with the usage of the standard piston when burning 25% gas portion. By the interpolation method, the new piston may be operated safely up to the gas portion of 33.8%, where there is no increasing the NOx emission level, but the engine can run under the higher efficiency condition. In the 50% gas portion, the modify-5 can reduce of 35.43% NOx emission but cause a 2.76% efficiency drop. The modify-6 causes the NOx emission increase of 2.25% but the efficiency can increase 2,7% when 25% gas is burned, and NOx emission can reduce 33.61%, and efficiency drop about 1.51% when burning 50% gas portion. Finally, the modify-7 makes the NOx emission increase by 2.02% but the efficiency can increase by 3.12% when the engine burns 25% gas portion. In the same way, when the modify-7 burns of 50% gas portion then the NOx emission can be reduced by about 33.68% while the efficiency also drops in about 0.92%. For a 75% dual fuel system all of the modified piston can reduce NOx emission at least 53.78% but the overall efficiency cannot be maintained and drop significantly at least 8.9%.

5. Conclusion

The study results show that the NOx emission can be reduced up to 1.63% and in simultaneously the fuel efficiency can be maintained positively up to 1.51% without any disruption to the engine operating conditions when the engine uses the mode of 25% dual-fuel. While let the efficiency drop about 2.76%, the engine can run at 50% dual-fuel mode and capable of reducing NOx emission up to 33.61%. Finally, the modifying piston at a single-cylinder diesel engine can perform a proper modified dual-fuel diesel engine that offers reducing NOx emission and a slight chance to increase the overall efficiency indeed. Further investigation is needed to evaluate the fuel efficiency in economic aspects such as fuel prices, supply, and bunkering so it can be imagined the consequences of the efficiency drop in the dual-fuel mode compared with the diesel engine conventional that use the higher price of diesel oil.

6. References

[1] Ismail MM, Zulkifli FH, Fawzi M, and Osman SA 2016 Conversion Method of a Diesel Engine to a CNG-Diesel Dual Fuel Engine and Its Financial Savings, ARPN J. Engineering and Applied Sciences, 11(8), April 2016, 5078-5083.

[2] Campara L, Hasanspahic N and Vujicic S 2018 Overview of MARPOL ANNEX VI Regulations for Prevention of Air Pollution from Marine Diesel Engine, *SHS Web of Conferences* 58, 01004, GLOBMAR2018, EDP Science.

[3] Maiwada B, Muaz N and Musa GH 2016 Diesel Engine Modifications Techniques to Minimize its Exhaust Emission (Theoretical Survey), *Int. Journal on Theoretical and Applied Research in Mechanical Eng.* (IJTARME), Vol-5 (1), https://www.researchgate.net/publication/300452481

[4] Zaglinskis J, Rapalis P, and Lazareva N 2017 An Overview of Natural Gas Use in Ships: Necessity and Engine Supply, *Periodica Polytechnica Transportation Engineering*, February 2018, pp. 1-9.

[5] Azzara A, Rutherfod D, and Wang H 2014 Feasibility of IMO Annex VI Tier III Implementation using Selective Catalytic Reduction, Working Paper 2014-4, March 2014, The International Council on Clean Transportation, London.

[6] Hutter R et al 2017 Low-Load Limit in a Diesel-Ignited Gas Engine, *Energies*, 10, 1450, pp.1-27.

[7] Herdzik J 2011 LNG as a Marine Fuel – Possibilities and Problems, *Journal of KONES Powertrain and Transport* Vol18 No2.

[8] Santoso A, Semin, Cahyono B, and Sampurno B 2020 New Development of Piston Crown for Dual Fuel Diesel Engines to Improve Efficiency and Reduce NOx Emissions: A Review, *International Journal on Engineering Application (IREA)*, No1 Vol8, pp.1-9, Rome.

[9] Santoso A 2020 Developing Piston Crown for Dual-fuel Diesel Engine to Increase Efficiency and Reduce Exhaust Gas Emission, Doctorate Dissertation in the Ocean Technology Postgraduate Programme ITS, Surabaya.

[10] Djadjev I 2015 Complying with MARPOL 73/78, University of Groningen, Netherlands, April 2015, publications at: https://www.researchgate.net/publication/277708775.

[11] IMO, The International Code of Safety for Ships Using Gases or Other Low-Flashpoint Fuels (IGF Code) (IMO Publishing Co., London, 2016).

[12] Wei L and Geng P 2016 A Review on Natural Gas/Diesel Dual Fuel Combustion, Emissions and Performance", *Fuel Processing Technology*, 142, Elsevier, pp. 264-278.

[13] Semin, Zaman MB, and Santoso A 2019 The Effect of Compression Ratio Improvement on Engine Performance of Yanmar TF 85 MH-DI Fueled using Diesel Oil – CNG, *International Review of Mechanical Engineering (IREME)*, Vol13(3), Rome.

[14] Elgohary MM, Seddiek IS and Salem AM 2014 Overview of alternative fuels with emphasis on the potential of liquefied natural gas as future marine fuel, *Proc IMechE Part M: J Engineering for the Maritime Environment* p 1–11.

IOP Conf. Series: Materials Science and Engineering 1052 (2021) 012004 doi:10.1088/1757-899X/1052/1/012004

Designing 69 Meters Littoral Mission Ship Based on the 60 Meters Fast Patrol Boat Platform to Improve Multi-Task Warfare

A. Santoso
Department of Marine Engineering, Institut Tenologi Sepuluh Nopember, Surabaya, Indonesia

Abstract. 60 meters fast patrol boat class is well known as one of the successful programs in the build of local warship in the national shipyards. Due to the higher tension in the South China Sea recently, therefore, it is urgently to have some better ships both quantitatively and qualitatively. To reduce procurements costs and shorten the delivery time then an LMS type with a length of 69 meters would be proposed based on the proven design of 60 meters FPB. Improvement of the platform not only about its size and capacity but also multi-task missions. The LMS will support missions of patrolling and intercepting as a fast patrol craft, moreover additional tasks including anti-surface warfare, mines-warfare, hydrography and intelligence, surveillance, and reconnaissance (ISR), also bio & chemical warfare duties. LMS will have a low radar cross-section by improving its stealth design and the addition of a helideck to accommodate a medium-sized helicopter. Ship able carries up to three standard ISO-containers as complementary logistics when the deck not being occupied by helicopter. LMS needs only 20% more engine power when its draught is 0.1 m higher and 21.5% more displacement compare with the FPB at a similar speed of 24 knots.

Keywords: Littoral Mission Ship, LMS, Fast Patrol Boat, FPB, Warship, Warfare

1. Introduction

Recently there is higher tension in the South China Sea when it involves two big countries China and the USA. Due to the geographical position, the conflict affects many countries such as Indonesia, Malaysia, Taiwan, etc. In this crucial situation, a strong military power is high in demand. Basic philosophy in defense is how to show that the country is not in the weak condition in military assets. This idea can prevent potential disturbances from other countries. Indonesian MoD do actual acts by considering buying several military weapons includes a naval ship, main battle tank, jet-fighter, combat heli, etc. The capabilities of the national industry also take into account building specific war machines [1]. In the maritime sector, many local shipyards have been proven to build many types of a fast patrol boats and a special shipyard PT. PAL Indonesia excellently capable of the built corvette, LPD, and even submarine. This potential resource becomes a vital backbone to support national military defense and security. Therefore, build a combat ship such as LMS or LMV will be a good challenge for national shipyards. Building strategy and engineering planning must be setup effectively based on yard capacity, facilities, experience, human resources and expertise, and finances [2].

2. Literatures Review

2.1. *LMS Definition and Duties*

Littoral Mission Ship (LMS) also named as Littoral Mission Vessel (LMV) for the bigger size. There are two types of LMS developed by certain countries. Firstly, it is a monohull model and secondly is a trimaran model called Littoral Combat Ship (LCS) [3]. The LMS is a type of small surface combatant vessel considered to perform littoral or coastal mission and security tasks in which high-speed maneuverability, agility, and sprint speed are required [4]. The ship can be equipped with a helicopter flight deck and missiles. This class of warship insists on fully electronics support to control the weapons system known as Combat Management System (CMS). The ship can also be provided with a drone launcher facility.

2.2. *Arrangement and Systems of LMS*

Typical arrangement and system of the LMS not so different from any type of patrol boat [5]. The unique system is additional equipment and weapon to support the multi-mission of the ship. The relatively small size of this combatant ship looks likely complicated and robust. The body of the ship and exterior silhouette similar to a common patrol boats. It is a key point to develop an LMS from the basic hull of the FPB type. Extensive length can be developed in the sectional area called the parallel middle body.

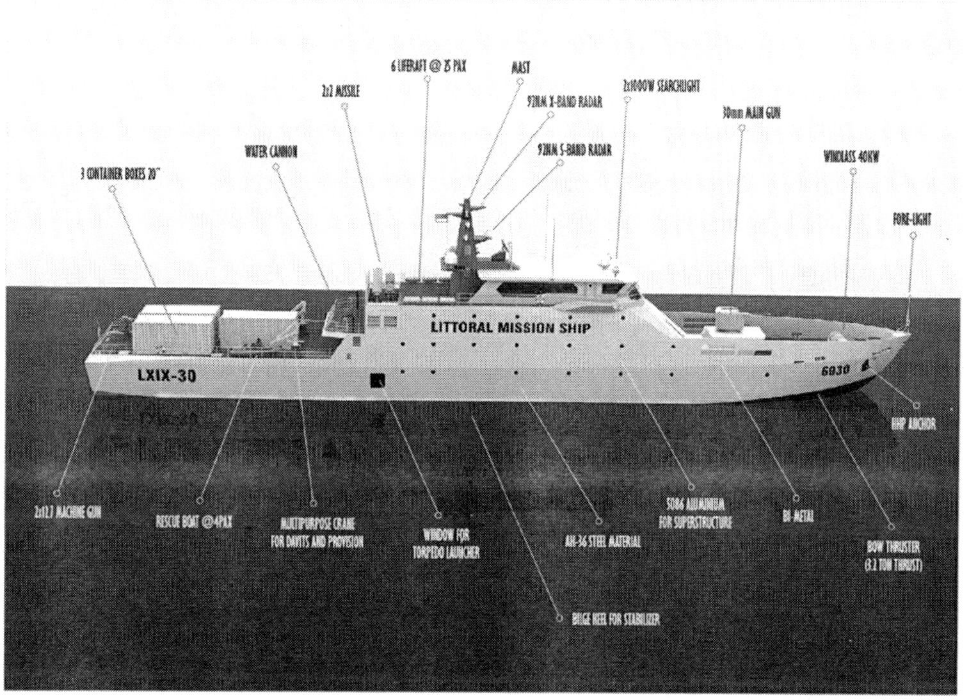

Figure 1. Technical Specification of the Developed 69 Meter Littoral Mission Ship.

3. Development Model

Figure 2 shows the design of 60 meters fast patrol boat. This one has been built by several shipyards for Indonesian Navy. Main hull uses high tension AH-36 and the superstructure use aluminium 5086.

Figure 2. Profile View of the Existing 60 Meter Fast Patrol Boat.

Figure 3. 3D View of the Developed 69 Meter Littoral Mission Ship.

Figure 3 shows the developed design of the LMS. There are two decks above the main deck compared with FPB that only has one deck. LMS needs more internal space to accommodate more crews and passengers [6], electronic equipment and command room, etc. Similar to FPB, the LMS uses high tension steel AH-36 for the main hull and aluminium 5086 for superstructure and welded together via bi-metal technology. Light aluminium as a superstructure gives better stability and easier shaped [7].

4. Analysis and Discussion

Upgrading the FPB to be an LMS mostly simple in terms of the hull construction. Based on the steel weight calculation, there is a relatively small addition of the hull materials at about 29 tonnes of aluminium and 66 tonnes of AH-36. The additional steel cost is about IDR 8.44 billion or it can say less than 1% from the LMS platform price. More details of the technical specification data of the ships can be resumed in Table 1 below. The hull outfitting may improve higher even it is not significant values, also for machinery and electrical outfitting for the basic platform. The electronic combatant system takes much bigger budgeting value due to the more complicated warfare's mission of the LMS [8] compared with the FPB.

Table 1. Comparison of the Technical Specification Data

Specification Parameter	Unit	60 Meter Fast Patrol Boat	69 Meter Littoral Mission Ship
Length Over All (LOA)	m	60.00	69.20
Length of Water Line (LWL)	m	56.96	64.14
Breadth (mld) (B)	m	8.20	9.00
Height (D)	m	4.90	5.60
Draught (T)	m	2.30	2.40
Displacement	Ton	506	679
Lightweight (LWT)	Ton	342	459
Deadweight (DWT)	Ton	164	220
Coefficient Block		0.483	0.499
Cruising Speed (Vs)	Knots	24	24
Engine Power	HP	2 x 4130	2 x 4830
Crews	person	46	49
Guest + cadets + detention	person	10	27
FOT	KL	130	160
FWT	KL	30	20
Endurance (cruising)	NM	2500	2500

The main propulsion system commonly takes bigger budgeting at almost 40% for military high-speed platforms [9]. The higher the speed needs the bigger power of engines that is correlated with the more expensive engine prices. Theoretically, the power demand of a ship is rank three of the ship's speed as shown in Equation 1 below.

$$HP \approx (V_s)^3 \quad \text{......} \quad (1)$$

For a fully combatant ships that occupied an advanced CMS, the prices of platforms commonly less than the price of the electronic combatant system, moreover to the main propulsion system. Upgrading the size of the ships from 60 meters FPB to the 69 meters LMS needs 20% more powerful engines to overcome the increasing ship draught at about 0.1 meters and 21.5% of bigger displacement to achieve a similar speed at 24 knots. Figure 4 resumes all of the designed speeds that was simulated by software where the power demand is relatively increasing at the average value of about 1500 HP.

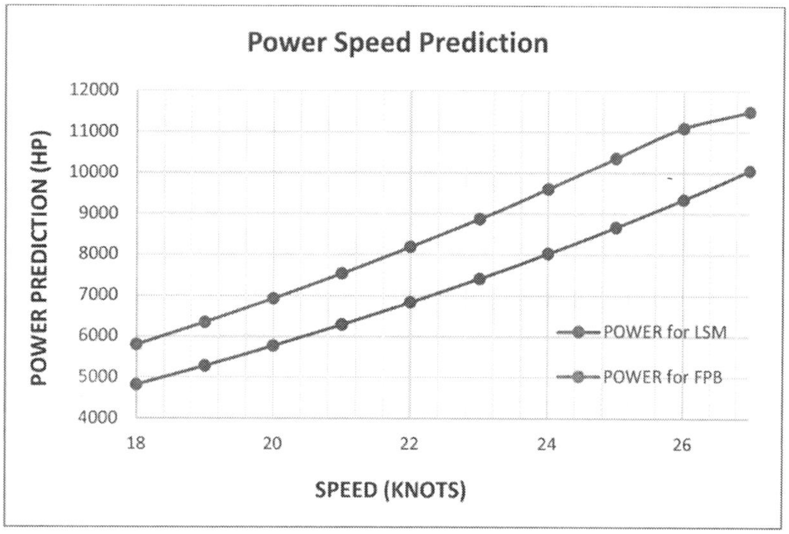

Figure 4. Power Speed Prediction for the Developed 69 Meter Littoral Mission Ship.

Based on the calculation of the speed power prediction above, then the suitable engine maker can be selected and determined. Both of the ships can be served by using MTU engines range power. FPB proposes to use MTU type 16V4000M93 and the LMS uses MTU type 20V4000M73L. Based on the basic requirement as a fast ship, the LMS still possible to upgrade up to 27 knots by using MTU type 20V4000M93L without any much penalty to the engines's weight and dimension which are primary constraints in the development of the fast patrol ships. The LMS can also be equipped by using diesel-electric propulsion technology integrated with the latest automation system such as PMS and IPMS that already installed in the predecessor PKR Class and SIGMA Class [10].

5. Conclusion

LMS needs only 20% more engine power when its draught is increase at about 0.1 meters and 21.5% more displacement compare with the FPB at a similar speed of 24 knots. Bigger displacement is a worthy value to make improvements of the interior functioning to fulfill the more complex duty of the LMS. It is still possible to be upgraded up to 27 knots by using a power engine of 2 x 5765 HP. From the shipbuilder's point of view, the capability to construct LMS gives more benefits in the financial aspect. The LMS budget is about two times more than the FPB Class based on the platform only exclude the CMS and electronic warfare supporting system.

IOP Conf. Series: Materials Science and Engineering 1052 (2021) 012004 doi:10.1088/1757-899X/1052/1/012004

References

[1] Santoso A, and Semin 2009 Improving the Warfare Performances of OPV90 to Become Corvette Class, National Conference of Marine Technology (SENTA), 17 December 2009, Surabaya

[2] Washington J, and Gaddin G Systems Engineering Planning, Article on the Leading Edge – Combat Systems Engineering & Integration, February 2013, p.50-57

[3] Murphy MN 2010 Littoral Combat Ship - an Examination of Its Possible Concepts of Operation, Center for Strategic and Budgetary Assessments, CSBA, Washington DC, USA.

[4] Alkire B, et al 2007 Littoral Combat Ships Relating Performance to Mission Package Inventories, Homeports, and Installation Sites, RAND - National Defense Research Institute, Santa Monica, USA.

[5] Esbati S 2018 Design for Support in the Initial Design of Naval Combatants, PhD Thesis in Naval Architecture and Marine Engineering, Department of Mechanical Engineering, University College London.

[6] Ong BSA, Chan CWJ, Koh ETD, and Huang Y 2017 Designing and Validating the Optimised Manning Concept for Littoral Mission Vessel, DSTA Horizons, Singapore.

[7] Mateer R, Scott SA, Owen I, and White MD 2018 Superstructure Aerodynamics of the Type 26 Global Combat Ship, https://www.researchgate.net/publication/328383281

[8] Pedatzur O 2004 Feasibility Study for a Littoral Corvette Ship, Conference paper February 2004, https://www.researchgate.net/publication/278410288

[9] Kertmen A 2006 Evaluation of the Littoral Combat Ship (LCS) Potential for the Turkish Navy, Thesis of Naval Postgraduate School, Monterey, California, The publication article September 2006 at https://www.researchgate.net/publication/235068481

[10] Santoso A, Kusuma IR, and Lin J 2018 The Application of Vessel Integrated Automation Systems to Improve Ship Safety and Security, *Maritime Safety International Conference* (MASTIC), 9-11 July 2018, Denpasar, Bali.

Extending the Capability of Vessel Integrated Automation System for Damaged Control Flooding System in 80 Meter Offshore Patrol Vessel

M.I. Saihilmi[1], A. Santoso[2], and I.R. Kusuma[3]

[1] Master Student of Double Degree in the Hochschule Wismar Germany – Marine Engineering ITS
[2] The Department of Marine Engineering ITS Surabaya, Indonesia
[3] The Department of Marine Engineering ITS Surabaya, Indonesia

Abstract. Indonesia is an archipelago country that owns thousands of islands. The worthy treasures must be protected properly from any maritime disruption such as foreign invasion, pirate, hijack, terrorist, illegal fishing, illegal logging, etc. The internal ship also needs to be protected from any disaster at sea caused by natural and enemy. There is a high demand for many numbers of modern OPVs that have the capability of a warship in terms of safety and security system. One of the features is the DCFS (Damage Control Flooding System). The system assures that the ship could intact from hull damages on any recoverable scale. In this research, DCFS was built for working automatically under the integration of the installed VIAS (Vessel Integrated Automation System). A sub-program was developed so that the DCFS capable of automatically switch on/off all ship alarms and sensors systems, fire and bilge pumps systems, even main engine and auxiliary engine when the emergency call situation declared. Integrated DCFS under ship automation system will work faster than any human response to the sources of disaster can be detected and override much earlier. All integrated DCFS can be monitored in a progressive LCD multi-screen by interactive windows and menus as the results of this research.

Keywords: PLC Module, Control System, Integrated Automation, VIAS, DCFS, OPV

1. Introduction

Modern ship design has been grown linear with the development of an electronic system. The ship's electronics fill almost all the equipment, apparatus, and system inside the vessel. The existences make many benefits of easiness operation, improving safety and security [1], less energy, fewer crews number [2], and full maintenance program [3]. The advanced electronic and information technology brings the ship to a higher level of automation from remotely operated till the future autonomous vessel.

Ship accidents become a catastrophic event that changes the rule of the game in the shipping industry from the ancient age to this modern era [4]. Ship accidents always happens and repeat all the time in shipping history. This disaster always happens as look likely unavoidable situation as simple thinking about the unlucky or lucky matters for the ship. The milestone of the technology development to avoid ship accidents came when the famous Titanic sank a hundred years ago. Then learning from this deadliest disaster during peacetime, then the subsequent evaluation recommended changes the maritime regulations, leading to the establishment of the International Convention for the Safety of Life at Sea (SOLAS) in 1914. The regulation has always been improved and upgraded refer to the marine disaster occurred in the yearly events. But revised SOLAS regulations and modernized ships look likely to only to reduce the number but not eliminate the accidents in totally. Some maritime accidents such as sunk, capsize, fire [5], explosion, crashed or collision, grounded, etc. still happens every year [1]. Moreover, in the military vessels, they face risk damage caused by mine, torpedo, bomb, and missile that possibly cause flooding and fire [6]. Ridiculously, this might be happened due to human error during peacetime.

IOP Conf. Series: Materials Science and Engineering 1052 (2021) 012005 doi:10.1088/1757-899X/1052/1/012005

2. Literatures Review

In the first step of the development, DCFS is carried out by evaluating the principle dimension and the general arrangement of the 80 Meter OPV as shown in Table 1 and Figure 1.

Table 1. Ship's general arrangement data

PRINCIPLE DIMENSION 80 METERS OPV	
Length Over All (LOA)	80.00 meter
Length Water Line (LWL)	72.25 meter
Breadth (mld)	14.00 meter
Height (mld)	7.95 meter
Maximum Draught	2.50 meter
Maximum Speed	22 knots
Crews	60 persons
Medical Crews	10 persons
Passengers	26 persons
Water Cannon @ 300 m^3/hr	4 units
Helicopter medium size	1 unit
Fuel Oil Capacity	280 KL
Fresh Water Capacity	60 KL
Endurance (at cruising speed)	3000 NM

Figure 1. General Arrangement of 80 Meter Offshore Patrol Vessel (OPV).

The main duty of this 80 meters Offshore Patrol Vessel is to a floating base that has the capability of unlimited patroling tasks for surveillance and security acts to protect Indonesian sea-territorial from any illegal activities and securing national sovereignty. 80 meters OPV is designed by using a combination of high tensile steel AH-36 as the main hull and marine aluminum 5086 as the superstructure. The designed ship powered by 2 (two) units medium speed marine diesel engines with a rating of continuous heavy duty which is capable of operated remotely from the ship's wheelhouse.

Damaged Control Flooding System (DCFS)

DCFS is a standard system for naval vessel or warship of any type and size [7]. Even the system is very well known capable of protecting the ship from flooding due to damage in any ship hull but there is no common implementation in a commercial vessel also in any service vessel [8]. Operational costs always a problem even in warships. For example, Canadian Forces have priority to reduce the operating costs, approaches to fewer crew levels without obstructing operational capabilities and safety [9]. DCFS can reduce crews because this automation facility needs less people to overcome the ship's shell damage and fire control. This project has several objectives including a state-of-the-art review of damage and fire control technologies, modeling and simulation of damage control activities and the evaluation of how automation will affect the crewing levels required to maintain the ship's damage and fire control

IOP Conf. Series: Materials Science and Engineering 1052 (2021) 012005 doi:10.1088/1757-899X/1052/1/012005

capabilities, material identification to resist damage and fire, and the choose of right technology such as a wireless condition monitoring systems [10].

Varela [11] explain how the importance of a ship control system to support emergency planning decision when flooding been occurred. Flooding times and stability parameters are measured, allowing for the crew to take the adequate measures, such as isolate or counter-flood all of the rooms, before the flooding takes incontrollable. The simulation is supported by a virtual environment in real-time, which provides all the functionalities to evaluate the seriousness event and any consequences of the situation, as well as to test, monitor, and carry out final emergency actions. The flooding simulation combined with virtual monitoring may require a large computational power to ensure the accuracy of the simulation results [12]. Therefore, the availability of automation systems onboard a ship not only functioning for any real damage and fire accidents but also can be used for training and exercise situations. The functionality will approach the crew's awareness in many scenarios of flooding and fire.

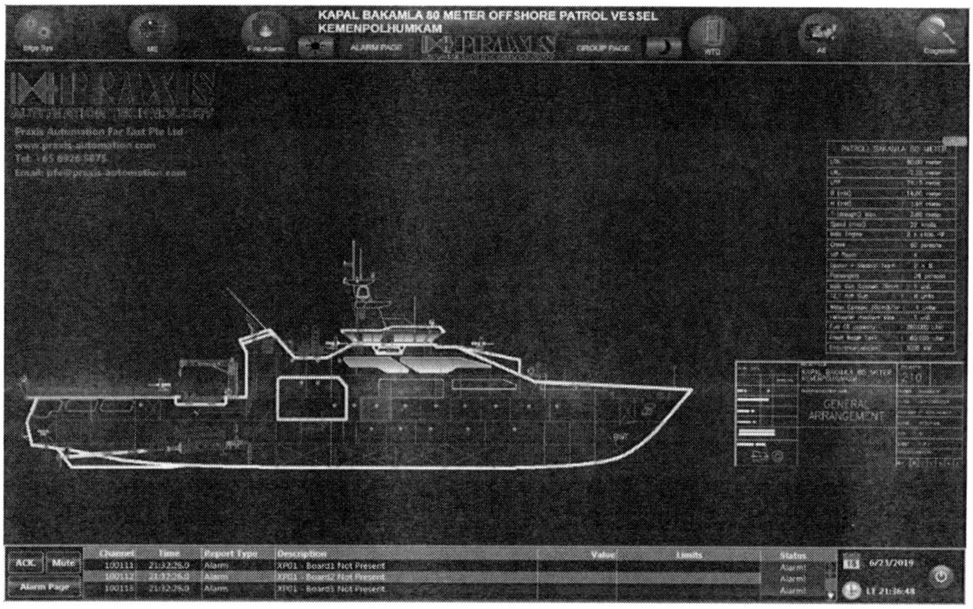

Figure 2. Monitoring Screen to handle DCFS-VIAS Operational System.

Detect of flooding in the ship, by required sensor or transducer will automatically send a notification to the system to take an action. Progressive flooding inside a damaged ship can seriously endanger the safety. Level sensors can be used to detect the flooding, then based on the data finally the breach acts can be estimated. For decision support, the prediction of flooding extent and the intermediate phases is necessary. For this purpose, a new simplified, but still reasonably accurate, flooding prediction method called DCFS has been developed as shown in Figure 2.

Concerning the ship flooding simulation, *Varela* and *Soares* [13] developed a *Virtual Reality based Decision Support System* to assist the coordination of damage control teams and to take the appropriate countermeasures in case of flooding, fire, and possible chemical/biological contamination onboard the military vessels. For the specific case of progressive flooding, simulation allows the crew to quickly capable of check which compartments have been flooded, the order by which they will flood, and how much time they will take to be flooded. Ideally, the simulation must complete in time for the operator to take the adequate countermeasures before the ship enters into an irreversible capsize or sinking situation. Within this scope, the simulation ends when the stability is recovered or when the ship capsizes. The current paper presents an onboard decision support system for ship flooding emergency response. The system runs a simulation of the progressive flooding of ship compartments given an initial condition, which includes the current load and damage conditions of the ship. As the way system decide to improve or control when the flooding happens. The controller or micro-controller will contribute here.

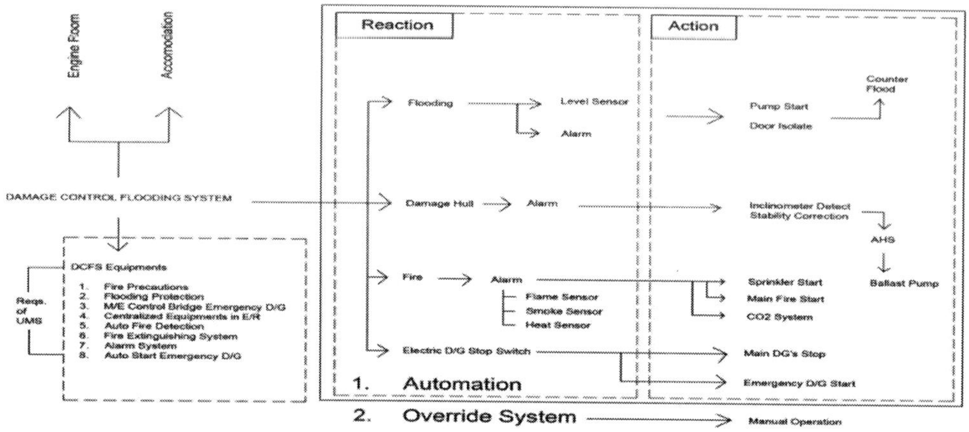

Figure 3. Schematic Diagram of the Developed Damage Control Flooding System.

DCF System will take an action to overcome the flooding, by implementing the Data-acquisition (DAQ) using PLC [14] to make the real sensors will automatically notify the real situation. The actuator will work to turn off the valves on and to turn on or off the pump needed. Relationship between equipment, when the ship's hull broken, for example, caused by hit missile from the enemy during duty, it may make leak damages to the construction and seawater will come inside the ship, then low-level sensor detector triggered and the emergency alarm will on. Signal then could be sent to operate the pump automatically, some of the pumps can operate, then crews will reconstruct those damaged hulls immediately. During the emergency conditions, it can be arranged that the system will run another seawater pump such as a ballast or GS pump to support the effort to remove leak water overboard.

Figure 3. shows the scheme of the ship DCFS. DCFS planned mainly to overcome flooding in the engine room and accommodation room where placed in the lower part of the main deck. There is two options provide for the DCFS, series of reaction components group to determine the situation and series of action components group to overcome the situation.

3. Research Methodology

DCFS consists of hardware in the form of modules, software to processing data, and electronic control. PLC-Modules are used for data acquisition (DAQ) and interface for all input-output systems. DCFS as a part of the military type vessel therefore will be designed and applied for the 80 meters OPV as the requirements code to achieve UMS Notation. Therefore, the main idea of this research work is how the most efficient of the pump arrangement which is integrated into automated DCFS to fulfill the minimum requirements of the UMS Notation. DCFS is integrated with well's established VIAS technology.

The other purpose of this study is to maximize the use of the ship's existing systems and minimizing any new installation when adopting DCFS and then proposing UMS Notation by designing the optimal use of integration automation for the DCFS and finally when the UMS's requirements are achieved, the successful system will be built in a complete and effective built-in MIMIC.

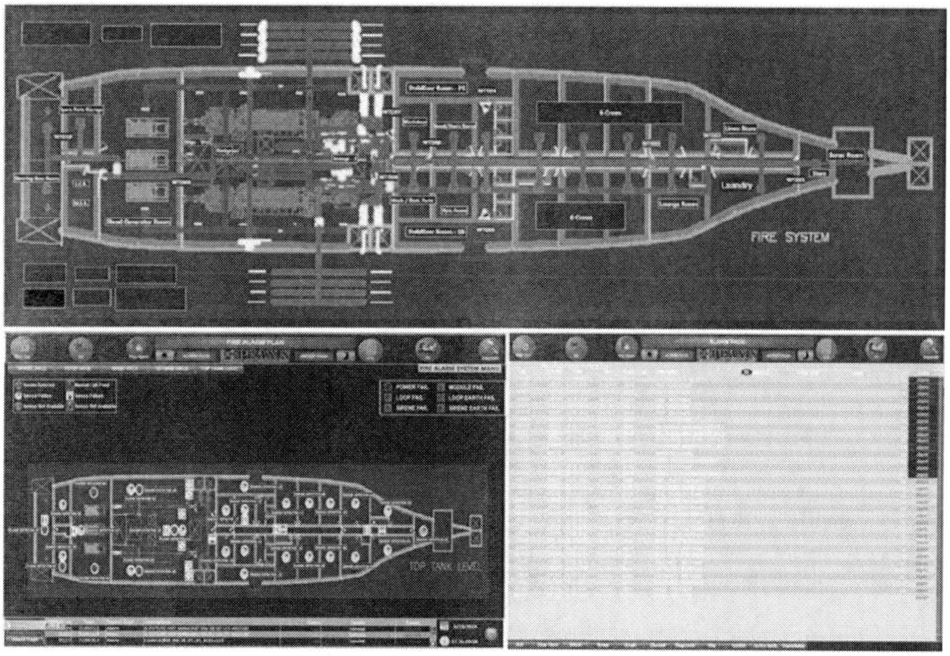

Figure 4. Acts of situation Alerting system under DCFS-VIAS.

4. Technical Analysis

The availability of data will be a key performance to build a sophisticated modern engineering model [15]. Operational data of the equipment on-board the ship and the sequence operational information will be collected and re-arranged to perform DCFS under the mode of VIAS. VIAS become efficient tools to handle the engineering solution for effective protection from the risk of fire and flooding. The existence of a sub-program will be developed specifically for DCFS. See Figure 4 as the proposed system. The final model that is built based on the form of GUI (Graphical User Interface) can be used to simulate any scenario of fire disaster on a ship and overtaking flooding that may come simultaneously [16]. The existence of water from the fire extinguisher system may perform flooding in some ship's compartments. The emergency bilge system acts to pump out the water so that ship's stability and safe draught always maintain good conditions. When the fire and/or flooding occurred inside the engine room then the automation system will shut down the main and auxiliary engines and other machinery [17]. VIAS can also arrange the closing and opening of watertight doors automatically under sequences that are set up by logical programming. Even responsive detectors and the alarming system work well in alerting crews but they need a certain time to indicate the sources of alarms [18], then even wasting a lot of time to hurry toward the fire and/or flooding point. It is common sense in a conventional system. DCFS-VIAS offers an interactive MIMIC screen to monitor and show any single fire and/or flooding point progressively. Based on the logical control as shown in Figure 3 for example, then, this VIAS can do execution to handle many acts as stated above.

VIAS also have has an *Internet View System* that capable of sending emergency messages to the Headquarter or any SAR institution in an earlier time if necessary [19]. Earlier acts usually produce a successive ending for any emergencies situation [20] instead of other political issues. Integrating the 80 meters OPV may also give economic benefits in the form of fuel-saving, fewer crews, less maintenance by adopting a power management system (PMS) and an integrated power management system (IPMS), etc. Ship Owner should not consider that additional budget to adopt DCFS-VIAS as a cost but it is an investment to keep the ship always safe and get a possible longer lifetime, off course the most important is to avoid any fatalistic accident during ship operational times.

5. Conclusion

The new module DCFS has been proven to work under DAQ that connecting the sources of data to the automated control all of the operational fire and bilge system and finally they can be displayed under the MIMIC screen for crew monitoring. Inserting a lot of water on-board a ship when fire extinguishing works make the ship heeling may be changed. Instead of progressive working of emergency bilge pumps then using a tank stabilizer can make it easier to adjust the angle of healing in any situation. All features possibly to be integrated into one automation act.

From all of the development works in building a new DCFS module under VIAS technology, then this project concludes that the following goals have been achieved:

1. New DCFS module capable of using the existing fire and bilge system on-board the 80 Meters OPV. It means that there is not any additional pumps and sensor & extra alarm system, but certain minor installations still in demands for improving system compatibilities.
2. The complete system on the new DCFS module has been built effectively in the form of a virtual MIMIC display. Input and output data of the new module DCFS-VIAS can work progressively in taking over any potential disaster especially which are caused by fire and flooding.

6. References

[1] Zaman MB, Santoso A, Kobayashi E, Wakabayashi N, and Maimun A 2015 Formal Safety Assessment (FSA) for Analysis of Ship Collision Using AIS Data, *The Int. J. on Marine Navigation and Safety on Sea Transportation*, pp67-72, Vol9, No1, March 2015, Gydnia Poland.

[2] Saihilmi MI 2019 Integration of Machinery System to Support Damage Control Flooding System (DCFS) in Order to Fulfill Unattended Machinery Space (UMS) Notation Applied to 80 Meter Offshore Patrol Vessel, Final Project report in the Joint Degree Program Department of Marine Engineering ITS and Hochschule Wismar Germany, Surabaya, Indonesia.

[3] Zaman MB, Santoso A, Hasanuddin, and Busse W 2020 Risk Evaluation of Ferry in the Bali Straits Using FMEA Method, *IOP Conf. Series: Earth and Environmental Science* 557, 012045.

[4] Shtay AD and Gharib W 2009 An Intelligent Control System for Ship Collision Avoidance, *Int. Journal of Engineering & Technology* IJET-IJENS Vol:09 No:10, December 2009.

[5] Bistrovic M and Tomas V 2014 Application of New Techniques and Information Technology for Early Fire Detection on ships, *Journal Nase Mone* 61 (5-6), 2014, pp.87-95.

[6] Hiltz JA and Daniels J 2006 Damage Control Technology – A Literature Review, *Technical Memorandum DRPC* Atlantic TM 2006-045, March 2006, Canada.

[7] Calabresse F et all 2012 Damage Control System: an application for Ship Safety and Security, 9th *IFAC Conference on Manouvering and Control of Marine Craft*, Sept 19-21, Arenzona, Italy.

[8] Pitana T, Prastowo H, Zaman MB, Nurwahyudi A and Gunawan R 2017 Flooding Causes Analysis in The Engine Room of KM. Nusantara Akbar, *International Journal of Marine Engineering Innovation and Research*, Vol. 1(2), Mar. 2017. 88-97, Surabaya, Indonesia.

[9] Santoso A, Kusuma IR, and Lin J 2018 The Application of Vessel Integrated Automation Systems to Improve Ship Safety and Security, *Maritime Safety International Conference* (MASTIC), 9-11 July 2018, Denpasar, Bali.

[10] Zhang L and Wang G 2009 Design and Implementation of Automatic Fire Alarm System based on Wireless Sensor Networks, *Proceedings of the 2009 Int. Symposium on Information Processing* (ISIP'2009), August 21-23, pp. 410-413, Huangshan, China.

[11] Varela JM, Rodriques JM and Soares CG 2014 On-Board Decision Support System for Ship Flooding Emergency Response, *Procedia Computer Science*, Vol29, Pp.1688-1700, ICCS 2014, Elsevier.

[12] Bulitko VV and Wilkins DC 1999 Automated Instructor Assistant for Ship Damage Control, From: *IAAI-99 Proceedings*. Copyright © 1999, *American Association for Artificial Intelligence*. (www.aaai.org). USA.

[13] Varela JM and Soares CG 2007 A virtual environment for decision support in ship damage control, *Journal of Computer Graphics and Applications*, IEEE, 27(4): 58-69.

[14] Sheena PA, Suneetha MN and Prasanna KH 2016 PLC Based Control and Monitoring in Ship, *Int. Journal of Electrical and Electronics Research* ISSN 2348-6988 (online) Vol. 4, Issue 3, pp.74-81, July - September 2016, Available at: www.researchpublish.com.

[15] Sen K et all 2015 Automated Fire Detection and Controlling System, *International Advanced Research Journal in Science, Engineering and Technology* Vol. 2, Issue 5, May 2015.

[16] Ruponen P, Larmela M and Pennanen P 2012 Flooding Prediction Onboard a Damaged Ship, Presented by Napa Ltd, *Proceedings of the 11th International Conference on the Stability of Ships and Ocean Vehicles* 23-28 September 2012, Athens, Greece.

[17] Santoso A, Kusuma IR, and Saihilmi MI 2019 Pengembangan Sistem Otomasi Terintegrasi Dengan Alat Safety Darurat Untuk Mengatur Start/Stop Mesin Diesel Di Kapal Yang Menggunakan Bahan Bakar Gas, *National Conference* - Implementasi Hasil Riset Sumber Daya Laut dan Pesisir Dalam Peningkatan Daya Saing Indonesia, Fakultas Teknik dan Ilmu Kelautan Universitas Hang Tuah, Surabaya 11 Juli 2019.

[18] Dey P et all 2013 Design and Implementation of an Automatic Fire Extinguishing System Based on Fault Secure Multi-Detectors, *Proceedings of the International Conference in Mechanical Engineering and Renewable Energy* ICMERE2013, 1-3 May, Chittagong, Bangladesh

[19] Lin J 2017 PRAXIS Ship Automation, Guest Lecturer *slide presentation* for Department of Marine Engineering ITS, 22 November 2017, Surabaya Indonesia.

[20] Zaman MB, Priyanta D, and Afianza I 2016 Evacuation Analysis of 1200 GT Passenger Ship in Case of Fire using Agent-Based Modeling Approach, *Int. J. of Marine Engineering Innovation and Research*, IJMEIR, Vol. 1(1), Dec. 2016. 9-11 (ISSN: 2541-5972), Surabaya, Indonesia.

The potential study of fishing area and its relationship to marine security in Natuna island

R D Putra[1], A V Sitohang[2], M P Suhana[2,*], E Prayetno[1], A H Yunianto[1], D Nusyirwan[1], S Nugraha[3] and I K Bachtiar[3]

[1]Department of Naval Architecture, Faculty of Engineering, Raja Ali Haji Maritime University, Kampus UMRAH Senggarang 29111, Tanjungppinang, Indonesia
[2]Marine Science Department, Faculty of Marine Science and Fisheries, Raja Ali Haji Maritime University, Kampus UMRAH Senggarang 29111, Tanjungppinang, Indonesia
[3]Department of Electrical Engineering, Faculty of Engineering, Raja Ali Haji Maritime University, Kampus UMRAH Senggarang 29111, Tanjungppinang, Indonesia

Abstract. Natuna Island is one of the small outer islands in Indonesia bordering Malaysia, Singapore, Thailand, Vietnam, and directly facing the South China Sea. Natuna seawaters have enormous potential fishery resources and one of the fishing areas used as a fishing activity for Indonesian and foreign fishermen. The potential also made illegal fishing practice rampant in Natuna seawaters and led to the Indonesian maritime security patrol being directed to Natuna seawaters to stop the illegal fishing process. The marine security patrol activities require high operational costs, and This can be minimized in several ways, particularly by knowing the points that can contain extensive fish resources, and the security patrols can be carried out directly at these points of location. To obtain information about these locations, map the potential fishing areas in Natuna seawaters for each period or monsoon. This study aims to apply remote sensing analysis to collect comprehensive information regarding strategic fishing areas prone to violations and security threats related to fishing activities based on monthly analyzes. The data used in this study were chlorophyll-a concentration and SST from MODIS level-3 with a resolution of 4 km from January to December 2019. SST and chlorophyll-a concentration data used pre-process analysis with SeaDas Software to ensure the data projection in the World Geodetic System (WGS) 84 format. The data processing stage uses ER Mapper software to project the distribution of SST and chlorophyll-a concentration into a contour. Post-processing data used ArcGIS software to determine crosses' results on the SST contour and the combined chlorophyll-a concentration. Image data processing results show that the lowest fishing point on Natuna Island in August with 51 fishing points, and the highest fishing point on Natuna Island in February with 584 fishing points. The results showed that fishing points vulnerable to illegal fishing are mostly located in the northern part of Natuna Island. The fishing zone is mostly far from Natuna island's coastal line. The results of this study indicate that the fishing area in the outer regions of Natuna Island needs to be increased in security and protection, especially in regions with a high level of fishing activity. This information is expected to become a reference for the authorities in the marine security sector in Natuna waters to take certain policies related to marine security patrols' operational activities in Natuna waters, particularly in minimizing operational costs of marine security patrols in Natuna seawaters.

Keywords: Fishes, Coral, Small Outer Island, Lutjanidae, Serranidae

1. Introduction

Indonesia's marine resources' potential consists of renewable resources such as fishery resources (fishing and aquaculture) [1,2] and non-renewable resources such as oil and natural gas and several types of minerals [3,4]. The large potential of fisheries causes most Indonesian people who live in coastal areas are fishermen. In general, most fishermen in determining the fishing area are only based on experience and direct observation. As a result, fishing operation time is not effective and efficient in determining fishing grounds [5]. Fish resources are differentiated based on their habitat, including pelagic fish [6]. Pelagic fish live in the middle layer of the water surface and generally live in groups either with the same group species or other fish species [7]. Pelagic fish are phototaxis positive and attracted to floating objects and streamline body shape with fast swimmers [8]. Small pelagic fish is one of the abundant fisheries resources and is widely caught for Indonesians' consumption compared to other pelagic fish groups. This condition also encourages rampant illegal fishing activities to meet small pelagic fish' consumption needs, both domestically and abroad. Small pelagic fish groups including Anchovies (*Encrasicholina spp., Stolephorus spp., Thryssa spp., Setipinna spp., Papuaengraulis spp.*), Sardines (*Sardinella spp., Amblygaster spp.*), Herrings (*Herklotsichthys spp., Pelona spp.*), Round herrings (*Dussumieria spp.*), Sprats (*Spratelloides spp.*), Mackerels (*Rastrelliger spp.*), Scads (*Decapterus spp., Selar spp., Selaroides spp., Atule spp.*), Fusiliers (*Caesio spp., Pterocaesio spp., Gymnocaesio spp.*), Flying-fish (*Hirundichthys spp., Cypselurus spp., Cheilopogon spp., Paraexocoetus spp.*), Half-beaks (*Hemiramphus spp., Hyporamphus spp., Oxyporamphus spp.*), Silversides (*Atherinomorus spp., Hypoatherina spp., Stenatherina spp.*) [9].

Natuna Island is one of Indonesia's outermost areas bordering Malaysia, Singapore, Thailand, and Vietnam and directly adjacent to the South China Sea. Natuna has potential coastal and marine resources [10], especially fishery resources that can trigger illegal fishing [11]. Illegal fishing is a criminal act of fishing in a country's sea area, carried out illegally [12]. Illegal fishing efforts cause losses or reduce the income obtained to increase economic fishing management yields and actions. Illegal fishing is a problem frequently faced by countries with a huge coastal area because this problem has existed for a long time. However, illegal fishing has not been eradicated until now due to the difficulty of monitoring multiple marine regions simultaneously. Even countries with advanced technology in the defense and security sectors have also been hit by illegal fishing and reduce their national fishery production and fish resources [13]. An unknown fishing ground area and a lack of strategy to carry out marine protection activities cause the Natuna Sea less effective in protecting fishery resources. The fishing ground area is an area for fishing [14]. The fisherman usually determines the fishing ground area by observing bubbles or ripples on the sea surface and flying birds. The abundance of fish in a region can also be predicted based on oceanographic conditions [15]. Oceanographic conditions greatly affect the fish abundance, particularly the distribution of chlorophyll-a and temperature [16,17]. The vastness of the Natuna Sea also causes finding a fishing location to take a relatively long time.

On the other hand, foreign fishers who carry out illegal fishing with advanced technology can effectively identify fishing ground locations. Knowing the problem of fishing grounds every month in the Natuna Sea, local fishermen's fishing activities become more effective and increase income. This maritime security strategy focuses on securing area points where illegal fishing is likely to occur. Information on fishing ground points in the Natuna Sea will minimize illegal fishing incidents on Natuna Island in the long term.

2. Research Methodology

2.1. Study Sites Location

The selection of research locations is based on local fishermen's productive fishing locations [18,19]. This research was conducted in January 2018 - December 2018 in the Natuna Sea as a research location. The Natuna Sea is a region that stretches from the Natuna Islands to the Lingga Islands in the Riau Archipelago province, Indonesia. The South China Sea borders the northern part of the Natuna Sea. The Karimata Strait also bounds the Natuna Sea in the southeast and the Singapore Strait in the west.

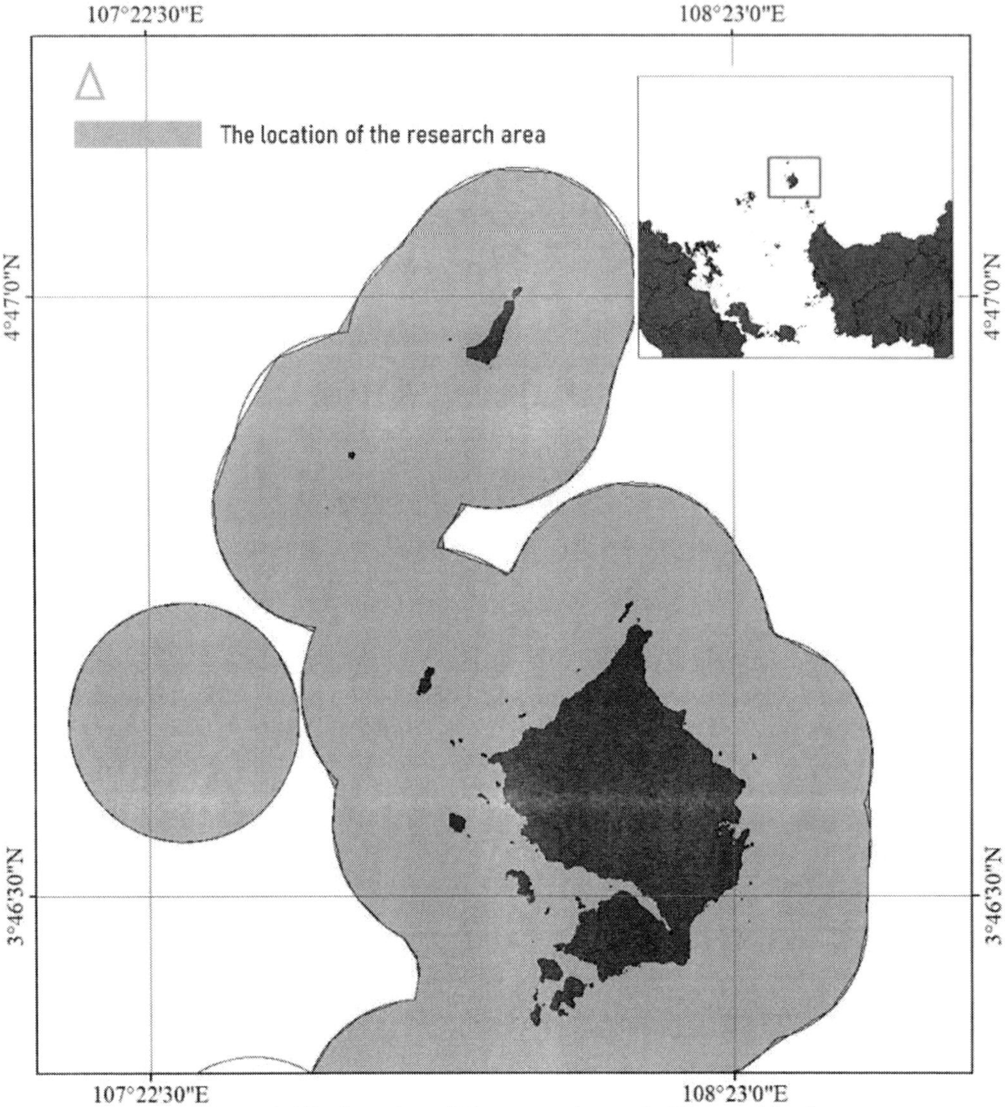

Figure 1. The location of the research area in Natuna Island.

2.2. Data Collected

The data used in this study are primary data from the extraction of chlorophyll-a and SST through remote sensing technology information obtained from MODIS sensor recordings processed and provided by NASA. Image downloaded from NASA's Web Marine Biological Processing Group http://oceancolor.gsfc.nasa.gov. The data is in the form of images from January 2018 to December 2018, at level 3, with 4 km of resolution. The image data from MODIS products for seawater include three characteristics: the color of seawater, SST, and primary seawater production through the detection of chlorophyll content. MODIS level 3 imagery is processed product data with an atmosphere corrected image quality, eliminating the very high light scattering caused by atmospheric components. The corrected components were Rayleigh scattering and aerosol scattering. Also, MODIS level 3 imagery is used for climatological data and ozone data, which are environmental data to improve image output. The advantages of Aqua MODIS imagery are its spectral wavelength (radiometric resolution), more precise land cover (spatial resolution), and more frequent observations (temporal resolution). Research on SST dynamics and primary chlorophyll-a productivity using Aqua MODIS satellite data provides better results describing field conditions. It is evident from research using SST that satellite image data has a strong relationship with in situ data.

2.3. Data Analysis
download SST and chlorophyll data at http://oceancolor.gsfc.nasa.gov. The downloaded data is then processed first in SeaDas, where the data is reprojected, the function of which must be ensured that the data projection is the World Geodetic System (WGS) 84 [20]. WGS 84 is the Global Positioning System, the standard used in cartography [21]. After confirming that the projection is WGS 84, it will be saved in GeoTiff form for further processing in ER Mapper. Data reprojected on SeaDas in image form of GeoTiff format and entered into ER Mapper to crop zoom data to the Natuna region in Er Mapper software and select appropriate Formula Editor to edit the image selecting variable ranges for SST and Chlorophyll. Change variable 1 to 25, and variable 2 to 31 and the variable change in chlorophyll 1 to 0.1 and variable 2 to 1.2 are the minima and maximum SST thresholds and chlorophyll for pelagic fish [22,23]. The variable modifier is used to project the contours of the distribution of SST and chlorophyll, and the data is stored in the raster data format (.ers).

The data resulting from changing formulas from SST and Chlorophyll data are opened in the ArcGIS software. The data was made into SST and Chlorophyll contours, using a tool available in ArcGIS, specifically Arc Toolbox, then searched for surface click contours, input temperatures, and chlorophyll data. In the raster form, the data input process is carried out alternately with temperature and chlorophyll. The raster input menu is filled with contour intervals for the content temperature of 0.5 and chlorophyll 0.2, and the basic contour is filled with the minimum chlorophyll and temperature ranges. Chlorophyll temperature and contours are mapped and appear in ArcGIS. SST contours and chlorophyll were analyzed by looking at crosses' presence at the combined temperature and chlorophyll contours. The results of these contour crosses can be seen using the tools contained in ArcGIS by combining each temperature and chlorophyll first. After the merging process is successful, the two temperature contours intersect, and the combination's to chlorophyll-a concentration. This intersection menu appears with the output changes to a point, then the area suspected of being a fishing area is adjacent. Warm temperatures and high chlorophyll content meet the contours of SST and chlorophyll [24].

3. Result and Discussion

3.1. Chlorophyll-a conditions from January to December in Natuna Island
The chlorophyll analysis results in January were in the range of 0.8 - 2.79 mg/m^3, spread over the seawaters of Natuna, wherein the areas close to the island had more chlorophyll concentration seawaters received additional nutrients from the seawaters. The chlorophyll concentration generally occurs on land due to rainfall that brings nutrients to the seawaters. January is still the western (rainy) season, where there is still a lot of rainfall [25]. In February, chlorophyll-a conditions ranged from 1.84 - 5.73 mg/m^3, and the chlorophyll-a concentration in February was higher than in January. Figure 4 shows that the farther away from the island, the chlorophyll-a values concentration is lower because the largest nutrients in the waters mostly come from land [26]. In March, the analysis of chlorophyll data showed that the Natuna chlorophyll conditions ranged from 1.22 - 4.62 mg/m^3. The chlorophyll concentration decreased slightly due to reduced rainfall, which reduced the mainland's supply of nutrients.

The factor chlorophyll can develop properly, and lack of nutrients prevents chlorophyll development from being inhibited [27]. In April, the chlorophyll conditions in Natuna conditions ranged from 0.823 - 4.46 mg/m^3, with the highest range in Natuna seawaters 0.823 - 1.44 mg/m^3. Figure 4 shows that the closer to the island, the increasing chlorophyll concentration conditions [28]. May is the beginning of the eastern (dry) season, and this affects oceanographic conditions and causes chlorophyll conditions to be in the range of 1.33- 3.46mg / m3. The chlorophyll concentration near the island was lower than the previous month. The dry season causes plankton development with photosynthesis due to the high intensity of sunlight, but chlorophyll requires sufficient nutrients to thrive. Nutrient intake from the mainland has decreased due to reduced rainfall intensity this month [29]. The chlorophyll analysis results in June tended to decrease in the range from 0.127-2.7 mg / m3. The decrease in chlorophyll concentration this month is due to the lack of nutritional intake from the mainland. Besides, improper temperature conditions will inhibit the development of chlorophyll. The decrease of chlorophyll concentration shown in Figure 8, that the largest nutrient intake comes from the land where the sea waters near the island have a higher chlorophyll concentration, while the farther from the island, the chlorophyll concentration tends to decrease [30].

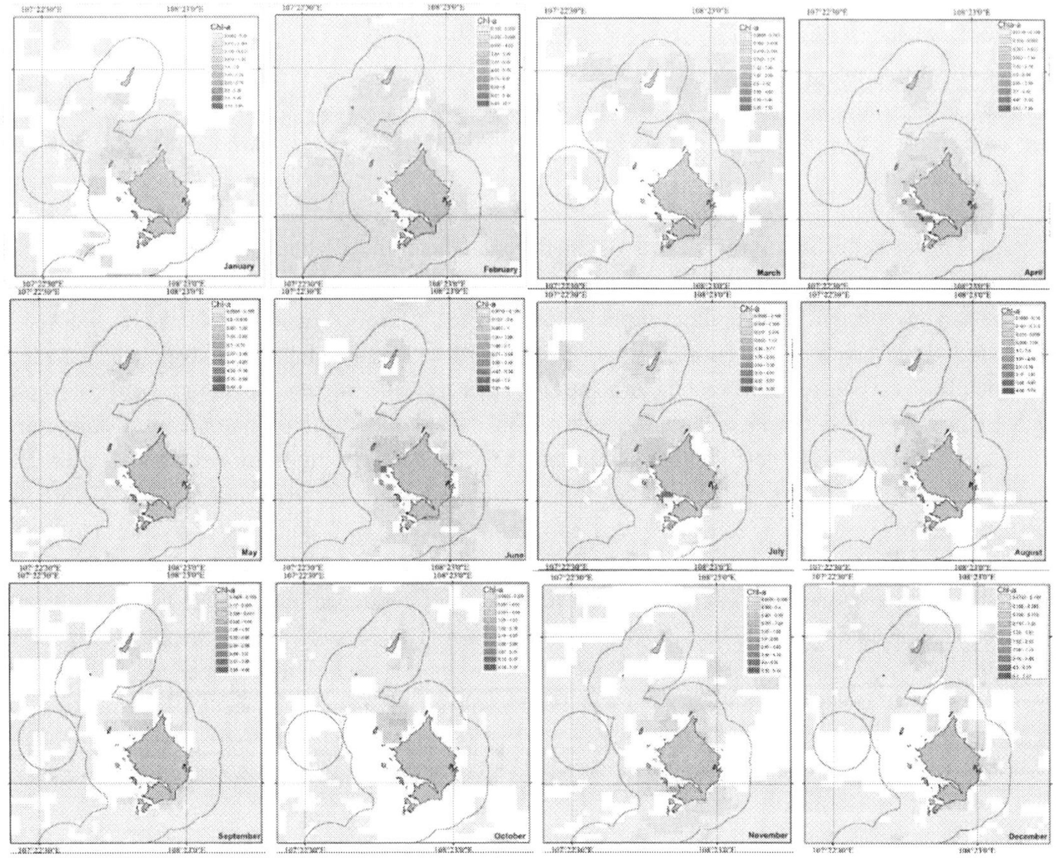

Figure 2. The distribution of *Chlorophyll-a* concentration from January to December.

The distribution of chlorophyll in Natuna waters in July ranges from 0.512 - 2.63 mg/m^3. The northern part of Natuna has a higher concentration, where the chlorophyll concentration in these waters slightly increases because it is directly adjacent to the South China Sea. The current brings higher chlorophyll concentrations in the northern part to the southern Natuna seawaters. It produces the chlorophyll concentration in July slightly increase due to increased intake from the South China Sea and the presence of intake from the mainland, causing chlorophyll-a concentrations near the Natuna island to be more abundant. August is the end of the eastern (dry) season, and this affects oceanographic factors, which cause the distribution of chlorophyll to range from 0.559 - 3.83 mg/m^3. There is a slight increase in chlorophyll because this month usually starts to rain, which transports nutrient intake from land to seawater (Putra et al., 2017). The yield of chlorophyll distribution in September ranged from 0.642 - 3.2 mg/m^3. The chlorophyll distribution in this month has a non-significant difference due to the lack of rainfall and the intensity of sunlight, which is an inhibiting factor for the growth of plankton (chlorophyll) in the seawater [31]. The results of the chlorophyll analysis in October ranged from 1.05 - 3.96 mg/m^3. In October, the distribution of chlorophyll increases along with the increase in rainfall.

The condition of the increase in chlorophyll is directly proportional to the intensity of rainfall, but this is different in October when there was a decrease in chlorophyll distribution due to the El Nino phenomenon [32]. Chlorophyll conditions in November ranged from 1.25 to 4.19 mg/m^3, and there was an increase in chlorophyll concentration due to increased rainfall, which caused the chlorophyll concentration to increase [33]. The results of the chlorophyll analysis in December ranged from 1.25 - 4.89 mg/m^3. December is the western season where the intensity of rain has started highly increase and made the distribution of chlorophyll increase significantly, as indicated by the high concentration of chlorophyll found around Natuna Island.

3.2. Sea Surface Temperature (SST) conditions from January to December in Natuna Island

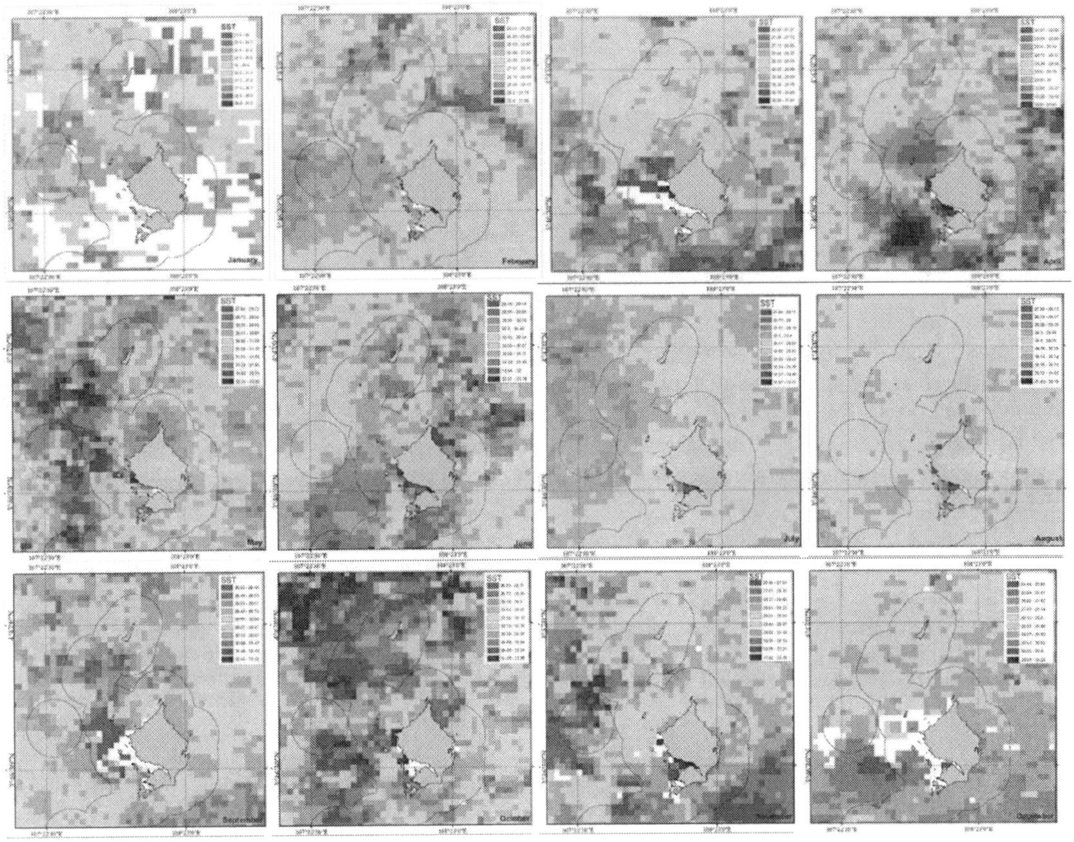

Figure 3. The distribution of sea surface temperature conditions from January to December.

SST (SST) conditions in January range between 21.4 - 29.8 °C in the with SST in January were still warm, current movements and high rainfall cause the SST of seawater conditions in Natuna to become cooler. In February shoe the SST shows in Natuna tend to be warm, ranging from 26.63-29.79 °C, with some points reaching 31 °C in the area around the island. February is the end of the western season, where rainfall is still relatively high but compared to the previous month, February tends to be warmer. In March, the Natuna island's SST range from 26.36 - 31.41 °C, with the seawater conditions in the southern part warmer than the northern part because it entered the transitional season. The western (rainy) season with SST this month tending to be colder, and the east (dry) season with the SST conditions are warmer. In March, the southern part of Natuna shows that SST is warmer due to the influence of currents and winds that transport dry winds so that the intensity of the rain decreases, producing the water conditions tend to be warmer [34]. The SST in April ranges from 27.47-31.41 °C and temperature increases due to the eastern monsoon's influence where this month, the intensity of the rain decreases. The eastern monsoon's effect with a slightly warmer SST distribution in the south to the north of Natuna. The warmer temperature because the wind from the east had the dry condition and produced the seawaters warmer. Currents and more sunlight intensity also influence warm waters. However, this month, because it is a transitional month, this month is still affected by the western season (rainy), making some parts of the water cooler or has a lower temperature [35]. The east (dry) season tends to be dry and warm [36], and Natuna seawaters are predominantly yellow to red color condition with an SST range of 29.73 °C for the lowest temperature and the highest temperature 33.25 °C. The SST in June is predominantly orange, which indicates that the seawaters of the Natuna are warm. The SST range in June ranges from 29.55-33.18 °C with a dominant temperature range of 30.88 - 31.11 °C where this condition tends to increase this month due to the east monsoon's influence, which tends to be warmer [37].

The SST condition of the Natuna ranges from 28.72 - 30.96 °C. A change in the northern Natuna island's SST tends to be lower because it is directly adjacent to the South China Sea. The low temperature of the South China Sea impact Natuna seawater by distribution from currents. The southern part region in Natuna island was not affected by this change due to the East monsoon (dry) tends to be warmer. SST conditions in August range from 29.08 - 30.71 °C. This month, there was a decrease in seawater temperature due to the intensity of rain and changes in the current direction from the South China Sea with low SST. the condition was a factor in the decreasing SST conditions in Natuna during August [38]. September has entered transitional season two and is influenced by the east (dry) season to the west (rainy) season. In the transition season, the SST tends to decrease because the wind blowing from the Asian continent tends to be lower and influences the northern part of Natuna seawaters, which tends to be lower than the southern part [25] with showing that SST value range from 26.82 - 30.87 °C. October is the second transitional season where conditions are supposed to be lower, but in October, the temperature conditions of Natuna waters can be categorized as warm, ranging from 26.63 - 32.95 °C. The northern region of Natuna was warmer than the southern region. This event occurred due to an El Nino event (increase in water temperature) in the South China Sea waters, which impacted Natuna waters [26]. The results of the analysis of SST in November ranged from 27.65 - 33.38 °C. The SST has decreased due to increased rain intensity, and indicated by there are several locations that had the lower SST. Although there are still many areas that tend to be hot, this happens because the El Nino phenomenon influences it in the northern part of Natuna [39]. In December, SST was 25.84 - 30.6 °C, with the northern part lower than the southern region with warmer conditions. December was the west season with an increase in rain intensity and a cause to reduce SST in Natuna seawater [40]. This condition causes the SST in the southern part of Natuna to be warmer due to the warmer temperature on land carried by the wind.

3.3. Potential fishing ground areas on Natuna Island from January to December

In January, the image data processing found 307 fishing ground areas found mostly in the northern part of Natuna Island. More fishing ground areas from the island to the South China Sea / offshore sea with larger fishing vessels presenting the fishing catch production of small local fishers in Natuna island have decreased. Apart from the inadequate ship facilities, the bad weather discourages local fishers from fishing activities. In February, based on data processing analysis, 584 fishing grounds were scattered throughout the Natuna seawaters. The number of fishing grounds found the impact on the more significant the fish catch production [41]. This month was estimated that the fish catch is more abundant and allows local fishers to get the maximum number of fish with abundant fishing ground areas. In March, there were 136 points of fishing grounds. The number of fishing grounds in this month significantly reduced compared to February. Several factors have reduced fishing ground points, including oceanography conditions that impact fish with less prefer temperature and chlorophyll. The southern part of Natuna Island's fishing grounds tends to be closer to the mainland than the northern part of Natuna Island within the 12-mile fishing ground point. In February, the distant fishing ground causes small local fishers to conduct fishing activities in the southern part of Natuna island to preserve the effectiveness and be more protected than fishing activities on the open seas with inadequate ships and fishing gear. There were 187 fishing ground points in April, and there was an increase in the fishing ground area located close to the Natuna island.

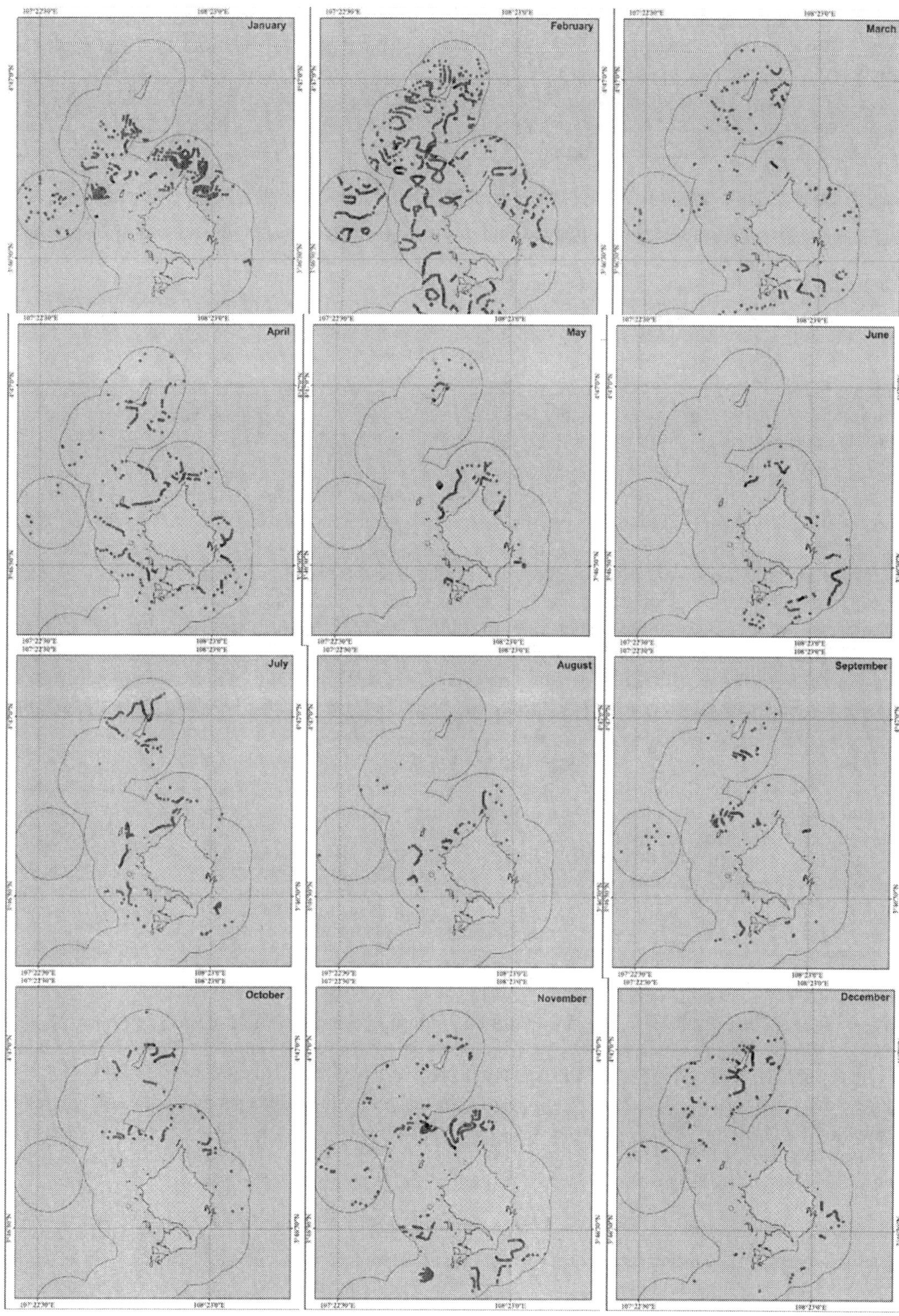

Figure 4. Potential fishing ground areas on Natuna Island from January to December.

The increase in the number of fishing grounds due to several factors, including the waters' fertility, impacts the abundance of chlorophyll-a. Sufficient rainfall intensity is one factor that makes chlorophyll-a fertile because it receives nutrients from the land. Nutrients from land flow to the sea with rain processes, but the currents also affect rivers' flow and bring abundant nutrients to the seawaters [42]. The higher nutrient in seawater impact the number of fishing ground points were located around the island. It provided increased fishing production of local fisher due to many fishing grounds [43]. Based on the data analysis results, the fishing ground area in May was found 106 fishing ground points. Compared to the previous month, the number of fishing ground points had decreased. This month, the fishing ground points were almost all around the northern part of the island, and local fishers can produce fishing activities. Oceanographic factors strongly influence the decrease in the number of fishing areas. The fishing ground areas around the island are due to river currents that carry nutrients to sea waters so that parts far from the island are inadequate chlorophyll-a conditions for fish.

The results of the analysis of temperature data and chlorophyll-a in July found 105 fishing points. In July, the fishing grounds were mostly located close to the northern mainland of Natuna Island. Several fishing areas were found due to the temperature, and chlorophyll-a conditions were adequate and suitable for the fish growth [44]. Based on the data analysis results, August is the month where fishing grounds were found. There are 51 fishing points because the SST this month was relatively warm for fish. The warmer temperatures more inhibit the growth of chlorophyll, and proceed fishing grounds less common. the seasonal factors also influence the fishing ground this month because the end of the eastern season changed in the current pattern and affects the number of fishing ground points. In September, 103 fishing grounds were found and more expected far from the mainland. There was a small fishing area located in the southern part of Natuna Island close to the mainland. The fishing area tends to be dense and accumulated in one area located in the northern part of the Natuna Islands due to the influence of the South China Sea conditions with more abundant nutrients and relatively lower temperature conditions. October is the month when the transitional season in Natuna and affecting the number of fishing ground points. Based on the analysis results in October, the fishing ground was found more than 100 points. The number of fishing ground not-significant, different from the previous month, and still in a transitional season. The differences were seen in the fishing areas further away from the mainland of Natuna Island. The currents changes cause the distance of the fishing ground point far away from the mainland due to the current pattern that brings chlorophyll-a away from the mainland towards the north location. The significant changes in SST due to oceanographic phenomena also present the fishing ground point was lower in quantity [45]. November is the starting month for the western (rainy) season, where rainfall intensity increases.

Increasing rainfall can affect fishing areas. The number of fishing grounds this month increased to 289 fishing ground points. Increased rainfall causes nutrients from the land to flow into the seawaters, and the increased intensity of rain can reduce the warm SST, provide suitable for fish to grow. The fishing ground points were densely populated in the north and south, while in the eastern part, there is none fishing ground pint because warm temperatures still influence it, and the effect of the east (dry) season was still happening. In December, the image data processing found 112 fishing ground points where the number of fishing points decreased this month. December had high rain intensity, and SST turns low, resulting in the temperature less suitable for fish. Figure 5 shows a relatively small fishing ground point in the eastern part of the mainland because the eastern seawaters of Natuna cool down more slowly. After all, the east monsoon is still influenced by the high seas carried with currents and winds when changing seasons. In the northern part of the Natuna Islands seawaters, the fishing grounds are very dense because the currents carry nutrients and the warm South China Sea. The Natuna seawaters also become warm and making the area's conditions suitable for fish life.

3.4. Linking Potential Fishing Areas and Marine Security

Illegal fishing is a significant problem in most coastal countries. In this case, the authorities do not supervise rampant illegal fishing practices, especially in Natuna seawaters by foreign fishing vessels. In neighboring countries, illegal fishing activities are carried out using several destructive fishing gears, including the Trawl Pair, the Squid Fishing Line, the Hand Line, the Gillnet, and the Hand Fishing Rod. However, most illegal fishing is carried out using Trawl Pair as fishing gear, with the number of fish caught reaching 1 ton per boat. The fisherman's capacity in conducting surveillance operations is still limited in terms of facilities, human resources (HR), and operational funds. It becomes an obstacle to optimally carrying out its duties and functions, especially with an extensive capture fishery area coverage and requires strong institutional capacity. It is also known that coordination between related agencies has not been optimal, causing many gaps in the occurrence of violations at sea, both in terms of quantity and quality [46]. Also, the monitoring or controlling illegal fishing fleets in the Natuna Sea is ineffective and inefficient because of the vast waters of the Natuna Sea. Routine monitoring activities in Natuna marine waters are insufficient and efficient because of the high costs required for monitoring activities and a long time to reach the maritime boundary area [47]. The security forces need infrastructure and resources to patrol Natuna seawaters. The detection of foreign vessels' presence carrying out illegal fishing activities in the Natuna sea border area is not optimal where foreign vessels will depart immediately before BAKAMLA, Coastguard, or the Indonesian Navy arrive at the location.

The results of the feeding ground mapping show that Natuna has a significant potential for fishing areas. The result of our studies can help local communities or local fishers know and take advantage of the potential of marine fisheries in Natuna every month. Most of the local Natuna fishermen still use small boats with limited technology in their fishing activities. Most of the local fishing boats are also not equipped with sonar technology to detect the presence of high potential fishing locations. The fishing zone map will help local fishers find out the potential fisheries in the Natuna Sea to carry out fishing activities in the Natuna Sea optimally. The fishing zone map results provide information on extensive fishing locations in Natuna waters north of the Natuna Sea. This location is the location where foreign fishers found illegal fishing. The utilization of the Fishery Zone Map provides clear and precise information for local fishermen. It is hoped that local fishermen can carry out fishing activities in that location. On another positive side, the local fisherman's fishing activities at that Natuna seawater location will decrease illegal fishing cases because the illegal foreign vessels will reconsider their illegal activities because they consider crossed the border route. The inactivity or non-activity of local fisherman causes most illegal fishing cases by foreign vessels. The optimal local fishing fleet doing the North Natuna Sea fishing activity indirectly impacts Indonesia's National Maritime Security. The local fishing ship has a key role in protecting the sovereignty and showing that the location of these seawater areas in Indonesia, and several foreign vessels consider not crossing national borders and not carrying out illegal activities.

Another benefit for fishermen uses a fishing ground location map, particularly for the local fishermen to improve their personal and state economy. Local fishermen can also provide information if an illegal case occurs in the vast Natuna seawaters. The local fishermen can become informants and provide valid information to the coast guard, Navy, and BAKAMLA regarding the certainty that several fishing fleets with illegal fishing activities on Natuna seawaters. The illegal foreign fishing fleet will enter illegal foreign vessels fishing areas with high fish production. This study's results are very useful in helping marine security activities become more optimal, effective, and efficient.

4. Conclusion

The AQUA MODIS satellite can determine potential fishing ground points using SST imagery data and chlorophyll-a. There are many fishing grounds in the waters of the Natuna district. The higher fishing ground in Natuna seawater was still gradually influenced by seasonal winds. Most fishing ground points were in November-December and January-April, with more than 110 fishing potential points. The May-October period is a period with fewer potential fishing areas because it tends to decline. The most potential fishing ground points were in February with 584 points, and the least is in August with 51 points. Therefore, Natuna seawaters had potential fishing ground points. The number of fishing grounds can provide information that supports marine security activities and functions to optimize marine security activities from violations and threats to maritime security related to fishing activities. The mapping fishing ground location in Natuna seawaters also minimized spending budgets to carry out marine security activity by the Navy or BAKAMLA. Information about the potential of fishing grounds in Natuna waters obtained from this study can be used as a reference. From the results of this study, it is known that the areas that have the potential to catch fish. With this information, the marine security patrol can directly patrol the area without always surrounding all Natuna waters to find illegal fishing activities in Natuna. That way, marine security patrol activities can be more efficient in terms of time and operational costs. This departs from the fishing potential location search system used by foreign fishermen based on satellite information systems to find potential fishing locations by looking at SST conditions and chlorophyll-a concentrations in the waters using satellite data. Therefore, by applying the same working principles, marine security patrols in Natuna waters will be much more efficient in time and operational costs. It is very useful in supporting marine security activities to become more optimal, effective, and efficient.

Acknowledgment
We gratefully thank Raja Ali Haji Maritime University for financial support. We thank the Faculty of Engineering, Maritime Raja Ali Haji University, for supporting this project. We also thank the reviewer for the constructive comments and suggestions that substantially improved the paper

References

[1] Gorris P 2016 Deconstructing the reality of community-based management of marine resources in a small island context in Indonesia *Front. Mar. Sci.* **3**
[2] Sanchirico J N and Wilen J E 2007 Global marine fisheries resources: Status and prospects *Int. J. Glob. Environ. Issues* **7** 106–18
[3] Hidayat T, Fitrianingrum L, Suharto and Sudibyo 2020 Ecology and sociology analysis for mineral resources management in Indonesia *IOP Conf. Ser. Earth Environ. Sci.* **483**
[4] Putrasari Y, Praptijanto A, Santoso W B and Lim O 2016 Resources, policy, and research activities of biofuel in Indonesia: A review *Energy Reports* **2** 237–45
[5] McCluskey S M and Lewison R L 2008 Quantifying fishing effort: A synthesis of current methods and their applications *Fish Fish.* **9** 188–200
[6] Edwards S F 2005 *6 Ownership of multi-attribute fishery resources in large marine ecosystems* vol 13 (Elsevier B.V.)
[7] Ekau W, Auel H, Hagen W, Koppelmann R, Wasmund N, Bohata K, Buchholz F, Geist S, Martin B, Schukat A, Verheye H M and Werner T 2018 Pelagic key species and mechanisms driving energy flows in the northern Benguela upwelling ecosystem and their feedback into biogeochemical cycles *J. Mar. Syst.* **188** 49–62
[8] Orue B, Lopez J, Moreno G, Santiago J, Soto M and Murua H 2019 Aggregation process of drifting fish aggregating devices (DFADs) in the Western Indian Ocean: Who arrives first, tuna or non-tuna species? *PLoS One* **14**
[9] Dalzell P J 1993 Small Pelagic Fishes *Nearshore Mar. Resour. South Pacific* 97–113
[10] Simangunsong F and Hutasoit I 2018 A study of the development of natuna regency as a key site on Indonesia's outer border with particular regard to national defense and security issues in the South China Sea *J. Mar. Isl. Cult.* **7** 63–73
[11] Kurniaty R, Ikaningtyas and Ruslijanto P A 2018 Analysis on traditional fishing grounds in Indonesia's Natuna waters under International Law *IOP Conf. Ser. Earth Environ. Sci.* **137**
[12] Pramod G, Nakamura K, Pitcher T J and Delagran L 2014 Estimates of illegal and unreported fish in seafood imports to the USA *Mar. Policy* **48** 102–13
[13] Cabral R B, Mayorga J, Clemence M, Lynham J, Koeshendrajana S, Muawanah U, Nugroho D, Anna Z, Mira, Ghofar A, Zulbainarni N, Gaines S D and Costello C 2018 Rapid and lasting gains from solving illegal fishing *Nat. Ecol. Evol.* **2** 650–8
[14] Amorim P, Sousa P, Jardim E, Azevedo M and Menezes G M 2020 Length-frequency data approaches to evaluate snapper and grouper fisheries in the Java Sea, Indonesia *Fish. Res.* **229** 105576
[15] Oozeki Y, Inagake D, Saito T, Okazaki M, Fusejima I, Hotai M, Watanabe T, Sugisaki H and Miyahara M 2018 Reliable estimation of IUU fishing catch amounts in the northwestern Pacific adjacent to the Japanese EEZ: Potential for usage of satellite remote sensing images *Mar. Policy* **88** 64–74
[16] Nurdin S, Mustapha M A and Lihan T 2013 The relationship between sea surface temperature

and chlorophyll-a concentration in fisheries aggregation area in the archipelagic waters of spermonde using satellite images *AIP Conf. Proc.* **1571** 466–72

[17] Lan K W, Shimada T, Lee M A, Su N J and Chang Y 2017 Using remote-sensing environmental and fishery data to map potential yellowfin tuna habitats in the Tropical Pacific Ocean *Remote Sens.* **9** 1–14

[18] Solanki H U, Mankodi P C, Nayak S R and Somvanshi V S 2005 Evaluation of remote-sensing-based potential fishing zones (PFZs) forecast methodology *Cont. Shelf Res.* **25** 2163–73

[19] Saitoh S, Chassot E, Dwivedi R, Fonteneau A, Kiyofuji H, Kumari B, Kuno M, Matsumura S, Platt T, Raman M, Sathyendranath S, Solanki H and Takahashi F 2009 Remote sensing applications to fish harvesting *Remote Sens. Fish. Aquac.* 57–76

[20] Martinis S, Twele A, Strobl C, Kersten J and Stein E 2013 A multi-scale flood monitoring system based on fully automatic MODIS and terraSAR-X processing chains *Remote Sens.* **5** 5598–619

[21] Boucher C and Altamimi Z 2001 ITRS, PZ-90 and WGS 84: Current realizations and the related transformation parameters *J. Geod.* **75** 613–9

[22] Putra R D, Suryanti A, Kurniawan D, Pratomo A, Irawan H, Said Raja'I T, Kurniawan R, Pratama G and Jumsurizal 2018 Responses of herbivorous fishes on coral reef cover in outer island Indonesia (Study Case: Natuna Island) *E3S Web Conf.* **47** 1–18

[23] Putra R D, Suhana M P, Kurniawn D, Abrar M, Siringoringo R M, Sari N W P, Irawan H, Prayetno E, Apriadi T and Suryanti A 2019 Detection of reef scale thermal stress with Aqua and Terra MODIS satellite for coral bleaching phenomena *AIP Conf. Proc.* **2094**

[24] Santos A M P 2000 Fisheries oceanography using satellite and airborne remote sensing methods: A review *Fish. Res.* **49** 1–20

[25] Nababan B and Simamora K 2012 Variability of Chlorophyll-a Concentration and Sea Surface Temperature of Natuna Waters *J. Ilmu dan Teknol. Kelaut. Trop.* **4** 121–34

[26] Hendiarti N, Suwarso, Aldrian E, Amri K, Andiastuti R, Sachoemar S I and Wahyono I B 2005 Seasonal variation of pelagic fish catch around java *Oceanography* **18** 114–23

[27] Aarti P D, Tanaka R and Tanaka A 2006 Effects of oxidative stress on chlorophyll biosynthesis in cucumber (Cucumis sativus) cotyledons *Physiol. Plant.* **128** 186–97

[28] Yu Y, Xing X, Liu H, Yuan Y, Wang Y and Chai F 2019 The variability of chlorophyll-a and its relationship with dynamic factors in the basin of the South China Sea *J. Mar. Syst.* **200**

[29] Teuku Mahlil, Inoue T, Matsumoto Y, Aoki S, Kato S, Yokota K, Rasul E and Saga M 2018 Effect of Nutrient Inputs on Water Quality Change and Phytoplankton Growth in Atsumi Bay *J. Eng. Technol. Sci.* **50** 548–65

[30] Maisyarah S, Wirasatriya A, Marwoto J, Subardjo P and Prasetyawan I B 2019 The Effect of the ENSO on the Variability of SST and Chlorophyll-a in the South China Sea *IOP Conf. Ser. Earth Environ. Sci.* **246**

[31] Harris G P and Lott J N A 1973 Light intensity and photosynthetic rates in phytoplankton *J. Fish. Res. Board Canada* **30** 1771–8

[32] Trenggono M, Amron A, Avia Pasha W and Lazuardy Rolian D 2018 Effects of el nino on distribution of chlorophyll-a and sea surface temperature in northen to southern sunda strait *E3S Web Conf.* **47**

[33] Moura M E P de, Rocha L D S and Nabout J C 2017 Effects of global climate change on chlorophyll-a concentrations in a tropical aquatic system during a cyanobacterial bloom: a microcosm study *Ambient. e Agua - An Interdiscip. J. Appl. Sci.* **12** 390

[34] Xie S P, Deser C, Vecchi G A, Ma J, Teng H and Wittenberg A T 2010 Global warming pattern formation: Sea surface temperature and rainfall *J. Clim.* **23** 966–86

[35] Kuswanto T D, Syamsuddin M L and Sunarto 2017 Relationship between Sea Surface Temperature and Chlorophyll-a in Tuna Fishing at Lampung Bay *J. Perikan. dan Kelaut.* **VIII** 90–102

[36] Rukminasari Nita N & K A and Program 1995 The Effect of Acidic Level of Media on Calcium Concentration and Growth of Halimeda sp *J. Ilmu Kelaut. dan Perikan.* **24** 28–34

[37] Yuniarti A, Maslukah L and Helmi M 2013 Study of Sea Surface Temperature Variability Based on Aqua MODIS Satellite Imagery 2007-2011 in the Bali Strait *J. Oceanogr.* **2** 416–21

[38] Nahdyah N, Zainuddin M and Farhum A 2017 The Mapping of potential fishing gorund areas for mackerel (Rastrelliger kanagurta) in Bantaeng Regency seawaters, South Sulawesi *J. Sains Teknol.* **17** 172–8

[39] Hadi S and Ningsih N S 2005 Study of Upwelling Location for Determining Potential of Tuna Fishing Ground *Ilmu Kelaut. - Indones. J. Mar. Sci.* **10** 61–7

[40] Nurman A 2010 Utilization of MODIS Data to Detect Fishing Ground Areas of East and West Coast of North Sumatra *J. Geogr.* **2** 17–30

[41] Gephart J A, Deutsch L, Pace M L, Troell M and Seekell D A 2017 Shocks to fish production: Identification, trends, and consequences *Glob. Environ. Chang.* **42** 24–32

[42] Gillanders B and Kingsford M 2002 *Impact of Changes in Flow of Freshwater on Estuarine and Open Coastal Habitats and the Associated Organisms*

[43] Garcia S M and Rosenberg A A 2010 Food security and marine capture fisheries: Characteristics, trends, drivers and future perspectives *Philos. Trans. R. Soc. B Biol. Sci.* **365** 2869–80

[44] Zainuddin M 2011 Skipjack Tuna in Relation To Sea Surface Temperature and Chlorophyll-a Concentration of Bone Bay Using Remotely Sensed Satellite Data *J. Ilmu dan Teknol. Kelaut. Trop.* **3**

[45] Munarsih E and Rini P 2019 Relationship between chlorophyll-a and sea surface temperature to tuna catch in the Southern Water of Java *J. Penelit. Sains* **21** 163–7

[46] Putranto G N D, Zid M and Miarsyah M 2020 Maritim Limit Conflict, Ilegal Fishing and Enformance of Effort in Natuna Sea Region between Indonesia and China *urnal Perikan. dan Kelaut.* **9** 185–93

[47] Winarso G and Kurniawan E 2014 Remote Sensing Application for Supporting Maritime Security Operations at Arafuru Sea *J. Penginderaan Jauh* **11** 128–41

Heat Transfer Characteristics of Latent Heat Thermal Energy Storage for Floating Concentrate Solar Power Systems

M Shibahara[1] and T Hinoki[2]

[1] Graduate School of Maritime Sciences, Kobe University, 5-1-1, Fukaeminami, Higashinada, Kobe, Hyogo, 658-0022, Japan
[2] Institute of Advanced Energy, Kyoto University, Gokasho, Uji, Kyoto, 611-0011, Japan

Abstract. The heat transfer characteristics of the latent heat thermal energy storage (LHTES) for floating concentrated solar power (CSP) systems were investigated experimentally and numerically. The understanding of natural convection heat transfer between the heat transfer fluid (HTF) and a tube bundle in the heat exchanger is important for the thermal design of the LHTES. As a primitive study, the analytical model of natural convection heat transfer was developed between the HTF and the tube bundle in the heat exchanger of LHTES. In this study, D-mannitol was used as the HTF for the LHTES. To simplify the tube bundle of the heat exchanger, a heated cylinder was installed in the liquid D-mannitol. In the experiment, the average temperature of the cylinder and the heat flux were measured, and the heat transfer coefficient under the natural convection was obtained. Meanwhile, the numerical analysis of natural convection heat transfer was conducted using the commercial computational fluid dynamics (CFD), PHOENICS. The temperature and velocity profiles around the heated cylinder were obtained in the numerical simulation. The heat transfer coefficient was compared with the author's experimental data. The numerical result was in good agreement with the measured heat transfer coefficient.

1. Introduction

The technology of concentrated solar power (CSP) systems have been developed [1,2]. Diendorfer et al. [3] proposed the floating offshore parabolic trough collector system. By applying the vertical collector system in the offshore area, the thermodynamic efficiency would increase due to the huge cooling seawater. Since their work was focused on the floating model and the optical performance as a function of time and location without the thermal energy storage (TES), the fundamental research of the TES was scarce for the CSP system.

The TES is categorized as sensible heat systems and latent heat systems [4]. In general, the molten salts are applied for the TES of CSP systems. Since the temperature of molten salts is high, it is difficult to operate the TES. For example, the rate of mixture is 40 wt.% of KNO_3 to 60 wt.% of $NaNO_3$ as solar salts. The temperature of solar salts is ranged from 563 K to 823 K. Since the waste heat recovery of the CSP is needed below the melting temperature of molten salts (approximately 511K), the latent heat system using phase change material (PCM) is considered [5].

The latent heat thermal energy storage (LHTES) using PCMs is also implemented in various waste heat resources such as power plants, EVs and marine transportations, and industrial applications. The PCM depends on the fusion temperature and latent heat. High-temperature PCMs are utilized as energy conservation technologies. PCMs can be categorized as alkali hydroxide, alkali nitrate, hydrate salt,

IOP Conf. Series: Materials Science and Engineering 1052 (2021) 012007 doi:10.1088/1757-899X/1052/1/012007

organic material, and sugar alcohol [6-7]. As the PCM candidates for the LHTES of CSP, D-mannitol is an attractive material since the latent heat is relatively higher than the other materials in medium temperature ranges. Therefore, it is important to understand the heat transfer process of liquid D-mannitol for the design of the LHTES recovering waste heat from the CSP system.

Shibahara et al. [8] investigated the heat transfer characteristics of the PCM such as a hydrate salt and sugar alcohol. As the PCM candidates for waste heat recovery of ship, the sodium acetate trihydrate and the D-mannitol were used since these materials have high latent heat. They clarified that the effect of heat inputs on the melting process of the PCM. Moreover, they researched the natural convection heat transfer of D-mannitol which is one of sugar alcohol for medium TES [9]. They measured the transient heat transfer coefficient of liquid D-mannitol at various liquid temperatures and exponential periods of heat generation rate. Then, the boiling incipience of liquid D-mannitol around the heated cylinder was investigated experimentally [10]. Although the effects of liquid temperatures and exponential periods on the natural convection and boiling heat transfer of D-mannitol were revealed in previous studies [9-10], the temperature and velocity profiles around the heated cylinder were not clarified numerically because the thermal properties of D-mannitol were not enough to conduct the numerical simulation.

In this study, the numerical simulation of natural convection heat transfer for liquid D-mannitol was conducted using the measured thermal properties of D-mannitol. Moreover, the numerical result was compared with the experimental data. The purposes of this study are to clarify the heat transfer characteristics of D-mannitol numerically and to develop the analytical model of D-mannitol.

2. Experimental method

2.1. Experimental setup

Figure 1 shows the schematic of the experimental setup. The experimental apparatus is composed of a D.C power source, a standard resistance (R_s), a double bridge circuit, a silicon oil bath, a glass vessel, a platinum cylinder, a thermocouple, and a stirrer. In the experiment, the D-mannitol was heated in the silicon oil bath, and the temperature of D-mannitol was measured by the thermocouple. For visualization of the experiment, the video camera was set outside of the oil bath. The platinum cylinder (heater) was connected to the copper electrode, and it was annealed before using it in the experiment. Moreover, the electrical resistance (R_T) vs. the averaged temperature of the heater (T_a) was calibrated as follows:

$$R_T = R_0(1+c_1T_a - c_2T_a^2), \tag{1}$$

where R_0, c_1 and c_2 are the electric resistance at 0 °C, coefficient of Eq. (1), respectively.

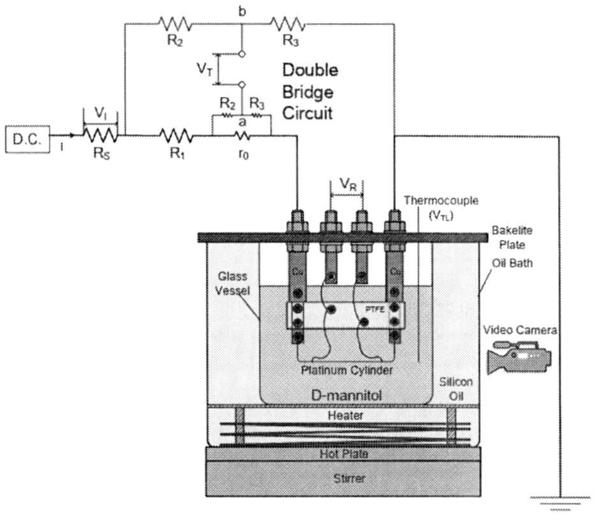

Figure1. Experimental apparatus.

2.2. Measurement system and experimental conditions

The averaged temperature of the cylinder was calculated from V_T and I ($=V_I/R_s$), based on the principle of Kelvin double bridge, as follows:

$$R_T = \frac{R_2+R_3}{R_2 I}V_T + \frac{R_1 R_3}{R_2}. \tag{2}$$

where V_T and I are the unbalanced voltage and the direct current, respectively. In the experiment, the heater was heated by an electric current from the D.C. power source ($Q=V_R I$) after the liquid temperature was controlled by the silicon oil bath.

The heat flux can be expressed by the energy balance as follows:

$$q = \frac{d}{4}\left(\dot{Q} - \rho_h c_h \frac{dT_a}{dt}\right), \tag{3}$$

where d, $\dot{Q}$, ρ_h, c_h, and t denote the diameter of the cylinder, the heat generation rate of the heater, density of the heater, the specific heat of the heater, and time, respectively. The surface temperature can be calculated from following the unsteady heat conduction equation with boundary conditions as follows:

$$\frac{\partial T}{\partial t} = a\left(\frac{\partial^2 T}{\partial r^2} + \frac{1}{r}\frac{\partial T}{\partial r}\right) + \frac{\dot{Q}}{\rho_h c_h}, \tag{4}$$

$$\left.\frac{\partial T}{\partial r}\right|_{r=0} = 0, \tag{5}$$

$$-\lambda\left.\frac{\partial T}{\partial r}\right|_{r=d/2} = q, \tag{6}$$

$$T_a = \frac{8}{d^2}\int_0^{d/2} Trdr. \tag{7}$$

The density of liquid D-mannitol was measured by the Archimedes' principle as follows:

$$\rho = \frac{|m_{air} - m_{liquid}|}{V} + \rho_{air}, \tag{8}$$

where m and V are a weight of float and volume of float, respectively. ρ_{air} is a density of air. The measurement result of density is shown in Figure 2. The density of liquid D-mannitol decreased in an increase of temperature. Based on the measurement result, the empirical correlation of density was obtained by the least squared method as follows:

$$\rho = 1.478 \times 10^3 + 4.415 \times 10^{-1}T - 1.618 \times 10^{-3}T^2. \tag{9}$$

The thermal expansion coefficient can be calculated by the following equation:

$$\beta = -\frac{1}{\rho}\frac{\partial \rho}{\partial T}. \tag{10}$$

The thermal conductivity of the D-mannitol was measured by the laser flash method (ULVAC-RIKO, TC-7000). To measure the specific heat of D-mannitol, the adiabatic scanning calorimetry was used (SHINKU-RIKO, SH-3000M). The experimental condition and the thermal properties of D-mannitol are summarized, as shown in Table 1. The experiment was carried out with various liquid temperature and exponential periods.

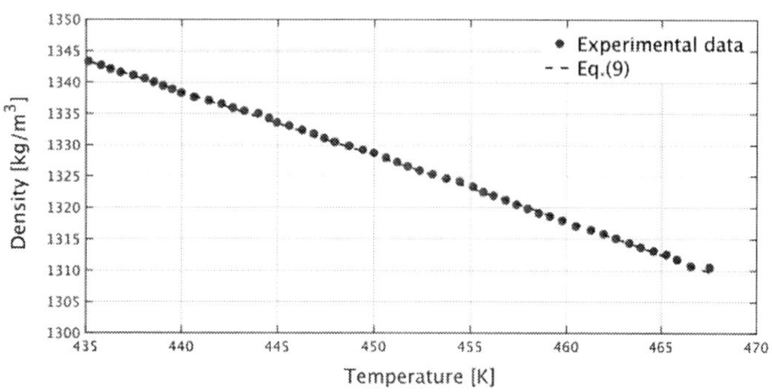

Figure 2. Relationship between the density and the liquid temperature.

IOP Conf. Series: Materials Science and Engineering 1052 (2021) 012007 doi:10.1088/1757-899X/1052/1/012007

Table 1. The experimental conditions and the thermal properties of D-mannitol.

Experimental conditions		Thermal properties	
Heater configuration	Cylinder	Test fluid	D-mannitol
Material	Platinum	Density (ρ) at 470 K	1325 [kg/m^3]
Diameter (d)	1.0 [mm]	Specific heat (c_p) at 460 K	2690 [J/kgK]
Effective length (L)	25.9 [mm]	Thermal conductivity (λ) at 420 K	0.347 [W/mK]
Exponential periods (τ)	11.52 [s]	Kinetic viscosity (ν) at 460 K [11]	2.84×10^{-5} [m^2/s]
Liquid Temperature (T_L)	477.15 [K]	Thermal expansion coefficient (β)	6.68×10^{-4} [1/K]

3. Numerical method

3.1. Analytical model and boundary conditions

The analytical model is shown in Figure 3. The heated cylinder was arranged in the liquid D-mannitol. The diameter of the cylinder was 1.0 mm, and the effective length of the cylinder was 25.9 mm. The heat flux from the cylinder was set to a uniform heat flux with $q=q_0\exp(t/\tau)$ as the boundary condition. Based on the comparison of sensitivity analysis for the computational grid, the number of the computational grid for r, θ, and z was 24, 20, and 25, respectively. In order to treat the near-wall condition between the solid region (cylinder) and fluid region (D-mannitol), the first mesh from the cylinder, Δr, was arranged in Δr < 144 μm.

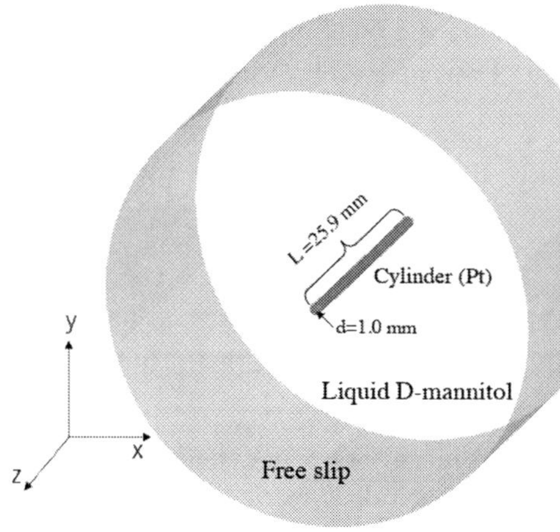

Figure 3. Analytical model in the numerical simulation.

3.2. Numerical method

The numerical analysis of natural convection heat transfer was conducted using the commercial CFD code, PHOENICS ver.2013. To solve the pressure and velocity fields, the SIMPLE algorithm was applied in the CFD code. Governing equations of mass, momentum, and energy were discretized by finite volume (FVM) method in cylindrical geometry. Since this analytical model was a natural convection problem, the momentum equation was taken into account for the buoyancy term using the Boussinesq approximation. The time step was set to 1 s and the initial temperature was the same at the experiment for the transient analysis.

The surface temperature can be expressed by solving Fourier's law as follows:

$$T_s = \frac{q\Delta r}{2\lambda} + T_{cell}, \tag{11}$$

where T_{cell} is the cell temperature of the first mesh on the cylinder. Since T_s is influenced by Δr in Eq. (11), the size of Δr was selected comparing the numerical result with the experimental data. It was 0.15 mm, approximately.

4. Experimental and numerical results

Figure 4 shows the relationship between heat flux and the surface temperature differences (T_s-T_L) at the liquid temperature of 477.15 K. The symbol and dashed line were the experimental data and the numerical result. As shown in Figure 4, the numerical result is in good agreement with the experimental data. The surface temperature differences increased with an increase in heat flux.

Figure 5 shows the relationship between the heat transfer coefficient and the heat flux. The heat transfer coefficient was defined as $h = q/(T_s-T_L)$. The heat transfer coefficient slightly increased as the heat flux increased.

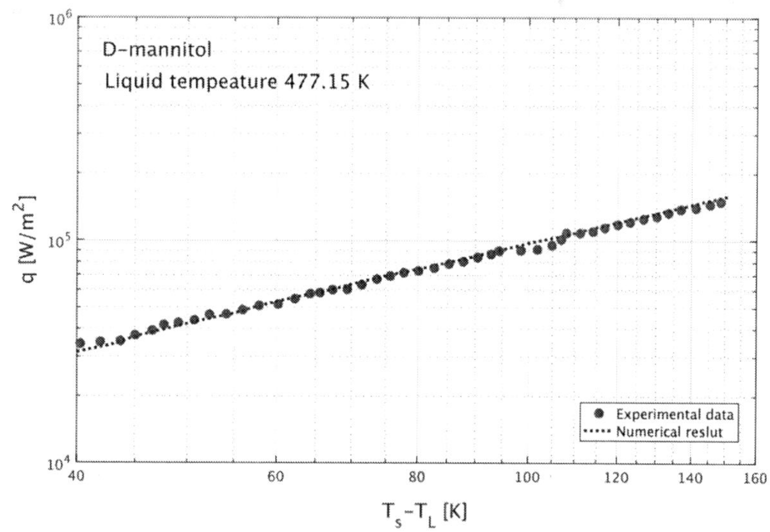

Figure 4. Relationship between the heat flux and the surface temperature differences.

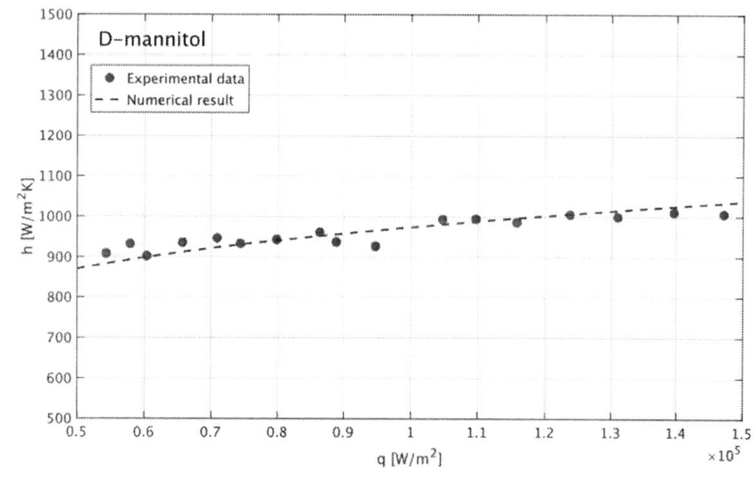

Figure 5. Relationship between heat transfer coefficient and heat flux.

Figure 6 displays the temperature profiles in the liquid D-mannitol. The bulk temperature of the liquid D-mannitol was 204 K in the numerical simulation. As shown in Figure 6, the liquid D-mannitol was heated by the cylinder with the uniform heat flux, and the temperature increased with time. Since this heat transfer process was natural convection, the plum of liquid D-manntiol was formed as the heated time proceeded.

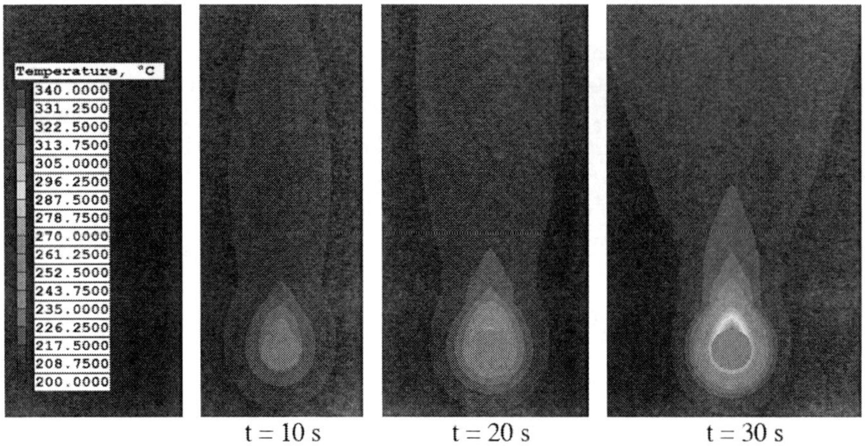

Figure 6. Temperature profiles in the liquid D-mannitol.

In order to understand the heat transfer characteristics between the heated cylinder and the liquid D-mannitol, the temperature distribution around the heated cylinder was illustrated in Figure 7. The analytical time was ranged from 10 to 30 s. $\theta = 0°$ and $\theta = 180°$ are the top and the bottom of the cylinder, respectively. As shown in Figure 7, it was understood that the temperature decreased as θ increased from 0° to 90°. On the other hand, the temperature increased when θ was ranged from 270° to 360° due to the natural convection as mentioned above. Since the plum of natural convection was developed with the heated time, the temperature gradient of 30 s is higher than that of 10s. Moreover, each temperature was uniform at the range of θ from 90° to 270°.

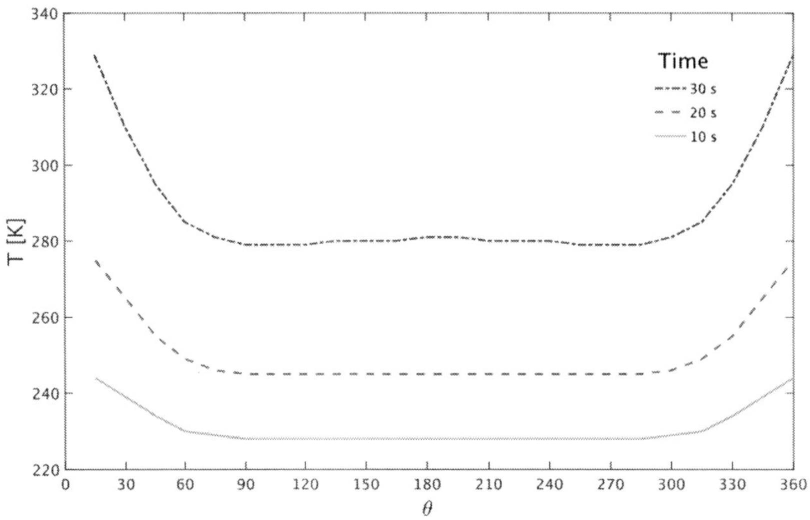

Figure 7. Temperature distribution around heated cylinder with various times.

Figure 8 shows the velocity profiles at various times. The velocity derived from natural convection exists on the heated cylinder. As the heated time proceeded, the velocity increased around the cylinder. Since the velocity at the bottom of the cylinder was lower than that of the top, it was understood that the temperature distribution was uniform at the range of θ from 90° to 270° as shown in Figure 7.

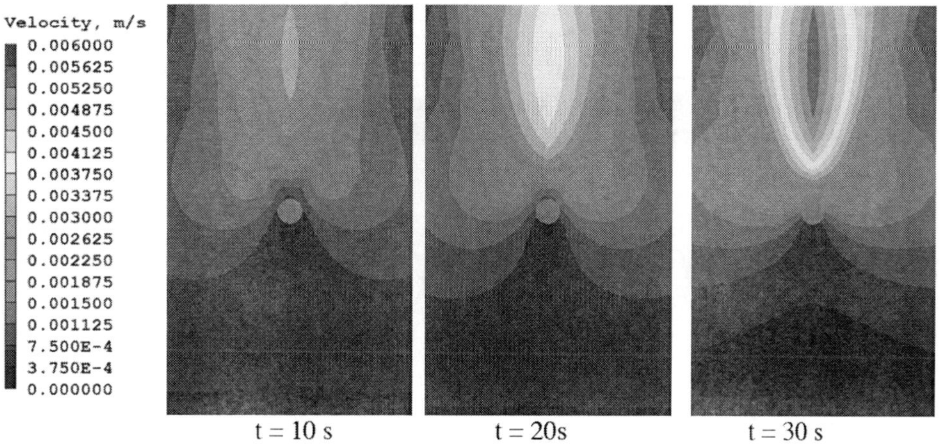

Figure 8. Temperature profiles in the liquid D-mannitol.

5. Conclusions

The numerical simulation of natural convection heat transfer for liquid D-mannitol was conducted using the measured thermal properties of D-mannitol. Moreover, the numerical result was compared with the experimental data. The temperature and velocity profiles around the heated cylinder were obtained in the numerical simulation. The heat transfer coefficient was compared with the author's experimental data. The numerical result was in good agreement with the measured heat transfer coefficient. The natural convection around the cylinder in liquid D-mannitol was clarified.

Acknowledgment

This work is supported by the "Joint Usage/Research Program on Zero-Emission Energy Research, Institute of Advanced Energy, Kyoto University, ZE2020A-31.

References

[1] Itiki R., Santo S. G. D., Itiki C., Manjrekar M., Chowdhury B. H., 2019, A comprehensive review and proposed architecture for offshore power system, Electrical Power and Energy Systems, 111, pp.79-92.
[2] Sutopo P. F., Muhammad B.Z., Fahri A. A., 2019, Design of organic rankine cycle (ORC) power plant systems by using flat-plate solar collector, International Journal of Marine Engineering Innovation and Research, 4(3), pp. 199-207.
[3] Diendorfer C., Haider M., Lauermann M., 2014, Performance analysis of offshore solar power plants, Energy Procedia, 49, pp.2462-2471.
[4] Robak, C. W., Bergman T. L., Faghri A., 2011, Economic evaluation of latent heat thermal energy storage using embedded thermosyphons for concentrating solar power applications, Solar Energy, 85, pp.2461-2473.
[5] Chirino H., Xu B., Xu X., 2019, Parametric study of cascade latent heat energy storage (CLHTES) system in concentrated solar power (CSP) plants, Journal of the Enegy Institute, 92, pp.653-664.
[6] Sharma A., Tyagi V.V., Chen C.R., Buddhi D., 2009, Review on thermal energy storage with phase change materials and applications, Renewable and Sustainable Energy Reviews, 13, pp.318-345.
[7] Shibahara M., Liu Q.S., Fukuda K., 2015, Heat transfer characteristics of D-mannitol as a phase change material for a medium thermal energy system, Heat and Mass Transfer, 52, pp.1993-2004.
[8] Shibahara M., Liu Q.S., Fukuda K., 2015, Thermal characteristics of phase change materials for waste heat recovery system, Journal of Japan Institute Marine Engineering, 50(6), pp.63-70.

[9] Shibahara M., Liu Q.S., Fukuda K., 2016, Transient natural convection heat transfer of liquid D-mannitol on a horizontal cylinder, Renewable Energy, 99, pp.971-977.
[10] Shibahara M., Fukuda K., Liu Q.S., 2018, Transient nucleate boiling on a horizontal heated cylinder in liquid D-mannitol, Applied Thermal Engineering, 144, pp.342-348.
[11] Hoshi A., Saito S. T., 2003, Study of phase change steam accumulator and application to solar Rankine cycle system (in Japanese), Trans JSME Ser. B, 69, pp.697-704.

Port Connectivity Model in The Perspective of Multimodal Transport: A Conceptual Framework

F Indriastiwi, S P Hadiwardoyo and Nahry
Civil Engineering Department, Universitas Indonesia, Depok, Indonesia

Abstract. As an archipelagic country, the use of several modes or multimodal freight transport in Indonesia is indispensable. As a place of "interchanging ," Port is an essential element in a transportation system that supports a region's economy and becomes more in the region island where incoming and outgoing goods rely on it. Port connectivity is not just about how "good" the Port is connected to the maritime transportation network and reached by regular lines services but also contributes to the entire transport network. Generally, the higher port connectivity is, the more attractive it is to facilitate cargo transportation and reduce transportation costs and time. Port connectivity is very complex, and it will be more useful if analysed from more perspective by using comprehensive analysis. This paper proposes the conceptual framework port connectivity in the multimodal perspective that sees from three aspects: maritime connectivity, connectivity of port with the hinterland, and connectivity at the Port and using the same measurement such as generalized transport cost. Analysis on various aspects will give a better understanding of port connectivity. Conceptual framework consists of three stages, the first stage is defining the definition of connectivity, seconds stage is connectivity analysis, and the last stage is synthetic of results. The importance of conceptualizing port connectivity is because there is still a lack of research on port connectivity. It is mostly analysed only on one aspect, as we know that port connectivity is very complicated. It will be more useful if analysed from a more different point of view with a comprehensive analysis.

Keywords: Maritime, network, port, connectivity, multimodal

1. Introduction

The large volume of cargo transported indicates by sea transport to entire world make the world economy development closely related to maritime transport quality such as efficiency, quality of shipping, and port's activities [1]. A port as an "interchange" is an essential element from transportation system, has a role as an interface for various modes of transportation that supports a region's economy and becomes more crucial to support an island region where incoming and outgoing cargo rely on it [1, 2].

One of the attributes that indicate the quality of maritime transport system is the port connectivity. Port connectivity can be described as how "good" the port connects to the maritime transportation network. Generally, the higher port connectivity is the more attractive it is in terms of facilitating cargo transportation and reducing transportation costs and time [3].

The high-level connectivity of ports has many transportation options through the port. Carriers can make choices for mode transportation, not just related to the shipping services but also inland transport that involves at least two-mode (multimodal transport) to reduce time and increase the capacity of the transportation network. The port with high-level connectivity will contribute to the transport network;

IOP Conf. Series: Materials Science and Engineering 1052 (2021) 012008 doi:10.1088/1757-899X/1052/1/012008

on the other hand, ports with low connectivity and less available options will give a low contribution to the transport network [3].

Understanding port connectivity can help port operators and shipping companies develop strategies that can give benefits to them. For instance, port operators who want their port's position as a transshipment hub port can be used as a benchmarking to improve services. In contrast, ocean liners can use it to select hub ports or redesign the service network to follow the hub and spoke system [4].

Recently, research related to port connectivity has increased due to the desire to understand better the importance of port connectivity and its role in supply chains. Its strategy also changes related to logistics. Port connectivity is an essential factor that is related to port competitiveness.

Mostly port connectivity define connectivity from shipping or maritime side such as developed container port connectivity index, based on economic, network topology, and the position of port within global shipping network system [5].

Measuring port connectivity also consider the transportation time and capacity maritime network [3]. The other method is trying to combine connectivity index, attribute of port key operating dimensions. The connectivity index is measure based on number of ports that can be directly reached from port and the number of ports that can be jointly through both while attribute of port key operating dimensions consist from number of port calls, draught. Ship turn round time, port cargo traffic, annual operating hours, port charges, availability of intermodal transports [6].

Connectivity of port is also useful to define competitiveness of port, by measuring the average ship capacity of liner shipping services that deployed between port and annual frequencies liner shipping between port [7]. Connectivity also can be measured using liner shipping connectivity index, that is derived from five components: Number of container ships on the liner services; capacity of container ships (TEUs Unit); number of services; maximum vessel size; and number companies that deploying ships. All components are related to services to and from country's port. Mostly analysis of port connectivity and especially that trying to derive the index are measure at the maritime segment.

Therefore, the role of ports is vital in moving goods from the origin and destination of goods. Port functions are not only related to the maritime network, but they are also function in the transportation and logistics chain, so the connection between maritime and port with its hinterland becomes essential [8]. Some researchers consider that the port hinterland connectivity and distance seem like one of the variables that impact the cargo flow from the port to its hinterland [9]. The dry port also becomes the issues that arise when discussing port hinterland connectivity, dry port as an inland node that establishes between seaport and hinterland, supposed to have a purpose as cargo unitization and has a strong link to seaport using reliable, high capacity, and frequent inland mode such as inland waterway, rail or road [10]. Another issue is enhancing the resilience network between port and hinterland containers [11]. Furthermore, research that discusses network integration not just between port and hinterland but between maritime or sea and the landside network is still lack.

There is the interaction between inland and maritime transportation, for example, the choice of load and discharge port and origin-destination transit time considerations, which is the evidence that holistic optimization of intermodal transportation has implications for a liner shipping [12]. The quality of inland and maritime transport connection also significantly changes the hinterland[13] Landside and sea integration become vital in logistics. These also develop along with the transportation network and need to incorporate to measure port connectivity [14]

Since there are the function or port as an "integrator" for sea mode and land mode, multimodal transport is an important consideration when building the port connectivity model. The origin and destination cargo locate at inland locations, so it is essential to consider the inland transport network when arranging the shipping network[12].

The port itself as an infrastructure transport has to meet the performance required to support trade connectivity, such as capacity and availability related to container and transshipment volume and port productivity[15].

Therefore, it is essential to develop port connectivity from the multimodal perspective that consists of maritime connectivity network, connectivity at port, and hinterland port connectivity.

Connectivity is an attribute of the network that indicates the quality and cost to move freight from origin to destination [16, 17]. The cost to move freight in the transport model is transport costs. Besides the cost, several attributes indicate the network's quality, such as time, capacity, and external attributes such as environmental variables e.g., pollution, accident, noise, congestion. These attributes or non-monetary costs then translate into the monetary using the value of time. The cost that consists of monetary elements and non-monetary elements is generalized costs [18]. Since the port connectivity analysis from three different aspects (maritime connectivity network, connectivity at port, and hinterland port connectivity) has different measurements, the cost is one of the variable measurements that can integrate those aspects. The transport cost or the generalized cost will be one of the useful variables that have fair representation as a measurement of port connectivity.

Moreover, there is a lack of a robust conceptual framework related to port connectivity that consider maritime segment, port, and hinterland side, and using the same measurement such as generalized transport cost and this paper is trying to fill that gap. This paper formulates the conceptual framework to measure port connectivity especially based on the generalized transport cost. Port connectivity has been compared to how the port can reduce the overall generalized transport cost. This paper aims two-fold: first, construct a theoretical framework for port connectivity, and second, propose conceptual framework

This paper will be structured as follows: the second section proposed a literature review concerning the former studies related to multimodal transport, port connectivity, maritime/shipping connectivity and connectivity to hinterland, and connectivity at port,, section three will propose the conceptual framework port connectivity. Finally, the last section will be concluded and give some directions for further research.

2. Literature Review

2.1. Multimodal Transport

The freight transportation chain consists of three-segment: pre haul (first mile), long haul (door to door transit of containers), and end-haul (last mile). Mostly pre haul, and end haul carried out with land modes, and for the long haul, transportation can use the road, rail, air, and water modes [19].

Multimodal freight transportation is freight transport using of at least two different modes sequentially can be a road or rail vehicle or vessel and the unit of transportation can be a box, a container, a swap body [19].

Multimodal freight transport has a complex structure and involves many stakeholders, using at least two modes of transport such as railways, Maritime routes and the roads to deliver cargo door-to-door [20].

Planning in multimodal transport consist of multiple objects and involves a multi-criteria decision making process, the objectives might such of the minimization of cost, time, and carbon emissions as well as improvement of service levels and utilization [21]. The purpose of multimodal network design is to identify network structures (number and location and terminals) to minimize network costs and transport demand. [22].

Some studies related to multimodal transport network have been done and related with the sea transport and other modes using several approaches. One stream of the studies is related to network design for multimodal strategic planning to determine suitable action planning to improving transport infrastructure[23] that incorporated multiple stakeholder, multi type container [24] or optimal intermodal freight routing for export and import container cargo [25]. The other one is network design for shipping service of large scale intermodal or intermodal liner shipping service [12] incorporated inland and maritime transportation [26], inventory, emission [27].

Beside the network design there also some studies related to routing problem Routing problem for container multimodal transport that incorporating inland/feeder transport [28, 29], inventory, emission [27].

The objective of the studies is related to minimizes cost such as total cost [12, 27], total operating cost Meng et al., 2012), transportation cost of carriers and operational cost of hub operators [24], minimization of travel time and of travel cost [25], minimizing the expected total cost [29], and minimum generalized transport costs [28].

2.2. Port Connectivity
Connectivity is an attribute of the network that indicates the possibility to reach all nodes from all nodes and refers to the quality and costs to move freight or persons from an origin to a destination at a specific time and using various modes [16, 17].

Connectivity's itself can act as the degree to which nodes in a network connection to each other [30]. According to various authors [31-33], there are two primary perspectives on connectivity: (1) the accessibility or (in)direct connectivity perspective; and (2) the centrality or hub connectivity perspective.

There at least three main perspectives on connectivity, namely a narrow perspective focused on the availability and characteristics of infrastructure and transport services, a broader perspective that includes trade facilitation procedures, and a supply chain management perspective [34].

Port as a node in a supply chain that is serving logistic services and information transfers, it functions as an interface that links sea and land transport, so there is a new perspective to understand port connectivity related to dynamics liner shipping network inter-port relationships in the supply chains. [35].

The various objectives of analyzing port connectivity are also important for consideration. There are several objectives in analyzing port connectivity, including understanding more about networks, connectivity related to port and hinterland connectivity, and port competitiveness.

Some studies derived the connectivity index to measures the liner shipping connectivity [36], port container accessibility [3, 5-7] or integrated port connectivity that derived from hinterland connectivity, inner bay connectivity and international connectivity [14].

Several methods use for measure index connectivity that derived from five component such as number of container ships on the liner services; capacity of container ships (TEUs Unit); number of services; maximum vessel size; number companies that deploying ships. All components are related to services to and from country's port [36].

Model that develop connectivity is network-based hub port assessment (NHPA) model consider the connectivity and cooperatives, that can be used as a platform analysis of port which can be develop into a global or regional hub port and the sustainability of hub status for existing hub ports [37]

Models for port connectivity also derive from a global container liner shipping network perspective by determine transportation time and capacity. [3].

A comprehensive model was developed with an index that integrated port connectivity using Multi criteria decision (TOPSIS method) with variable that related to hinterland connectivity, inner bay connectivity and international connectivity [14]).

2.3. Maritime/ Shipping connectivity
The maritime container liner networks have some different characteristics compared with other transportation, such as: (1) Mostly, the routes of container liners are circular or usually it's called pendulum route. The ships call is not running bi-directionally but visit a series of port and return to origin port, while bus and railway mostly running bi-directionally on routes. (2) The circular route makes the network is directed and asymmetric (3) Lines consisting into main lines and branch lines, mostly for main line involves a serial port calls a cross the ocean, are long haul lines [38].

The maritime network is also influenced by the commodity. Commodity diversity plays a vital role in maritime traffic distribution among ports, [39] applied global maritime flow in which ports and links are differentiated into the simple basis of five major commodity groups, coupling from single to multiple commodities with several measurements.

Mostly studies about liner/ shipping/ maritime connectivity using network analysis such as using graph theory [2, 4, 40-42] or gravity methods [43, 44]. Some studies also analyses using statistical analysis to identify important node (port) based on different centrality measures [38] or to identify the impacts of liner shipping connectivity and port infrastructure on freight rates using the case study Caribbean freight rates [45].

Several attributes and variables to analyses shipping/ maritime connectivity used in the connectivity studies, namely frequency and vessel size [42, 43], maximum and minimum vessel capacity [40, 43], number of link between port [40], flow of containers [2, 5, 7], Flow commodities [39], Distance between port [2, 41].

The one best approach to analyze ports positions in the network including all carriers, services, and ports connected is using classical methods of network analysis. [4]. The main attribute from connectivity refers to quality and cost, so considering the cost to move cargo from origin port to destination port is one of the critical factors, but in recent studies cost variable relatively does not applied much to measure shipping/ maritime connectivity The shipping/ maritime connectivity is related to maritime transport cost and economic [46-48].

How good the maritime/ shipping connectivity also can be translated to how "good" the quality and the cost to move cargo from origin and destination, so to improve the connectivity can also be by reducing the cost. It is related to design and optimize the shipping network. Many studies have been done related to design shipping network and they have the objective to minimize cost and are involved with many variables such operating cost, sailing cost, transit cost, inventory, penalty, handling cost [49-53], speed [52, 54, 55], bunker cost [55, 56], time [56-58], transit time restriction [52], emission [51, 59], uncertain demand [26].

2.4. Port connectivity to hinterland

Hinterland is an inland area of a port where most of its business activities originate from that area. It is the catchment area of the port as the points of origin/destination of the cargo that movement through a particular port. The traditional concept of the hinterland is an area that has a continuous line boundary which covers the economic influence of the port in a land area [60].

Port hinterland is a geographical area where most of the area is served by ports where the ports have a substantial part of their business (most of the traffic flow is generated). Port's Hinterland is an area that can be reached at a lower cost and faster time compared to other ports [61-63].

The boundary of the hinterland area is difficult to determine because it depends on various things such as commodities (for i.e. bulk or containers), time (examples relating to season, technological change, port policy, economy, etc.) and modes of transportation, as well as the dynamic market. Port hinterland is not static but dynamic [64, 65].

The ability port to attract traffic is strongly influencing port hinterland because the origin and destination of the movement of goods is at landside [61].

Two main factors that affect to the shape of the hinterland are: 1) the effectiveness of infrastructure networks (such as toll roads and railways), it can determine the direction of hinterland development and (2) the location of the inland terminal (intermodal platform). The inland terminal plays an important role in expanding the port market area, the strategic location of the terminal is an attractive factor attractive gravitational factor), and reduces the "frictions" caused by distance [60].

The problem of seaports' accessibility to the hinterland can be approached from infrastructural and market perspectives. From an infrastructural perspective to resulting in the seamless operating multimodal, the critical consideration is a good network of roads, railways, and rivers/canals, together with efficient interconnecting systems. From the market perspective, efficient and effective companies that provide hinterland transport services to deliver to customers. [66].

Reduce transit and transport time, and lower transport costs is essential for better port connectivity with inland container transport networks therefore ensuring product quality and decreasing risk product damage also important.

the gap of accessibility among regions can reduce by providing better connection by implementation of intermodal networks reduces the gap of accessibility among regions by providing better connections to container ports. The benefit of intermodals in terms of freight cost savings can be greater in remote places to sea ports, given low enough shipping cost decay effects and proximity to intermodal infrastructures.[67].

The function of the port as a distribution strongly involves the storage of goods at hinterland transport. Therefore, hinterland transportation capacity should be necessary to correspond with the volume of loading and unloading cargo at its port. Therefore, it does not always consider because loading and unloading a vessel not always correspond to the moment of loading of the hinterland mode. It is essential to consider the capacity of hinterland transport because one of the competitive factors in the cost of the logistic chain is transport cost, so essential to ascertain the sufficient capacity of the hinterland of transport.

In terms of the relation of container network, mainly the research about containership routing and scheduling in liner shipping focuses on the ocean side, and there are a few studies on the inland transportation containers directed at the optimization of both maritime and inland transportation. The choice of load and discharge port and transit time considerations are some of example interaction between inland and maritime transportation [12].

It is evident to have a better result of integrated multimodal/intermodal transport optimization, which incorporated maritime and inland transportation with several variables such as network shipping and inland network Transport cost such as shipping cost, inventory cost, transshipment cost [12, 27, 29], inland cost [27], uncertainty travel times [29], emission [27], repositioning container [26].

2.5. Connectivity at Port

Connectivity at port basically related to the port capacity and availability also multimodal facility, for example a container port terminal provides facilities to transfer container between sea vessels and land transportation modes, such as truck and rail. Port terminal has three areas. The sea-side area includes the quays, crane and other loading and unloading equipment to facilitate the loading and unloading of containers into and from ships. The land-side area is an interface between the land and sea transportation systems where are locate the truck and train receiving gates. This area is the place for rail cars to load and unload cargos. The yard area is the place for Loading and unloading cargos to the truck and stacking loaded and empty containers.

Capacity port as a transhipment port becomes essential to determine, in line with increasing the function seaport as the hubs. Imported goods to supply to hinterland and cargo from hinterland are grouped and loaded into the ships that take place at the seaport [3].

The facilities of port also the important to analyses as variable such as number of cranes, storage area, berth and water depth, and inter-modal connections, number of port calls; distance between ports [35, 37, 46], port throughput [43]. However, the capacity and availability can translate it into the cost and time; the port with sufficient capacity and availability service has good performance. The impact is at a low cost and reasonable time because of the port services' effectiveness.

2.6. Generalized transport cost

One indicator to be developed as an indicator to measure connectivity is a transport cost. Transport cost consists at least three part i.e., shipping transport cost, port cost, and inland transport cost.

Total costs are ships costs, port costs, inland/ feeder transport cost, inventory costs and external costs is CO2 cost [27]. Component of ship cost are capital cost at sea, operating cost at sea, bunker cost at sea, capital cost in port, operating cost in port, bunker cost in port. Port cost has component port due and terminal handling charge. Inventory cost has component inventory cost at sea and inventory cost in port (during ship operation and dwell time), inventory cost during inland transport. Cost of inland transport in this paper consider using road capital cost, operating cost, overhead cost, infrastructural cost (e.g. toll fee).

Generalized transport costs are costs to move one unit for each commodity, calculated based on the cost structure for each mode of transportation and transhipment, including external costs such as the cost of CO2 emissions if they are incorporating in the transportation model.

As seen on figure 1, the components of generalized cost can vary depending on the costs incorporated into the transport modeling. Generalized cost consists of maritime transport components with ship size and maritime access (time and cost), port performance and costs at the port, and generalized hinterland costs

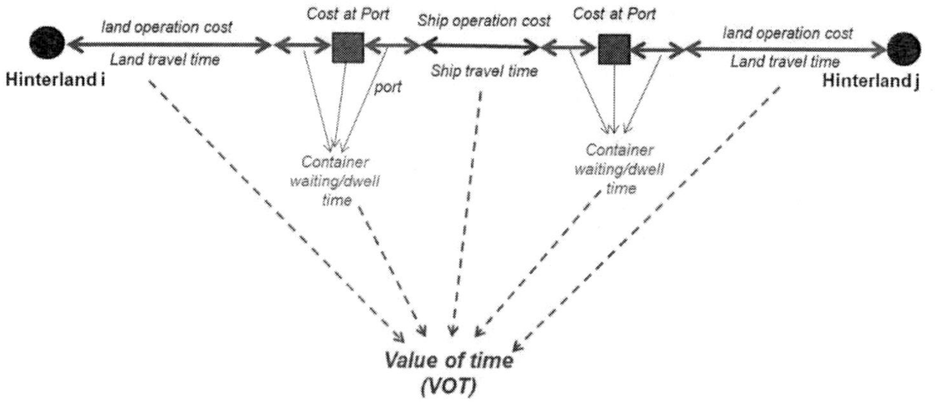

Figure 1. Illustration Generalized Transport Cost Component.

3. Proposed Conceptual Framework

The conceptual framework is formulating the port connectivity that integrated the maritime, port and hinterland segment and the assessment of connectivity is using relative accessibility with generalized transport cost as a measurement variable. As seen on figure 2, There are three stages of the process. The first is definition of connectivity, in this part, it is proceeded to define the variables of measurement; the second process is an analysis that consists of the essential elements, which are an assessment of the connectivity and policy scenario to improve the port connectivity. The last part is the synthesis of the results.

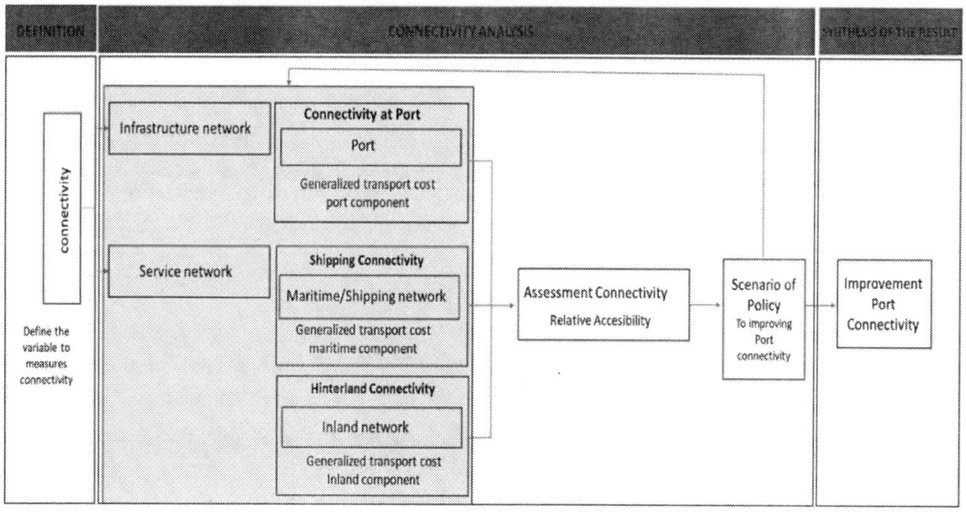

Figure 2. Proposed Conceptual Framework.

3.1. *Definition of Connectivity*

The conceptual framework begins with definition of connectivity. Connectivity indicates the quality and costs of the network and the possibility to reach all nodes from all nodes.

As seen on figure 3, connectivity can be seen from a narrow perspective, which consists of an infrastructure network and service network. The broader perspective is more related to supply chain management and trade facilitation. This paper will see from the narrow perspective that the infrastructure network focuses on the port as a node that links sea and land transport. The service network consists of the shipping network and hinterland network, involving at least two modes or as a multimodal network. Port connectivity seen from a multimodal perspective means that the ocean side or seaside and the landside are involved in the multimodal network.

The narrow perspective can be seen from three aspects of connectivity: maritime/ shipping connectivity representing the connectivity at the seaside. It gives a view of how good port connect with other port. Hinterland connectivity describes how "good" port connects with the hinterland, and connectivity at port represents port availability and capacity. It can also be one of the indicators of port performance.

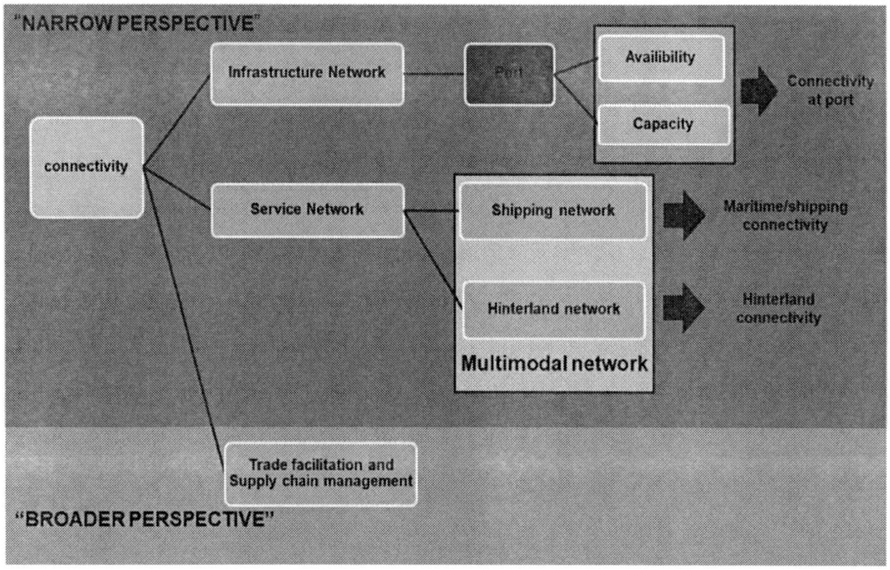

Figure 3. Connectivity Framework.

Source: author elaboration

Network's effectiveness or feasibility of the network could obtain from overall cost or unit cost when travel within it [30]. This paper trying to develop port connectivity model based on the generalized transport cost.

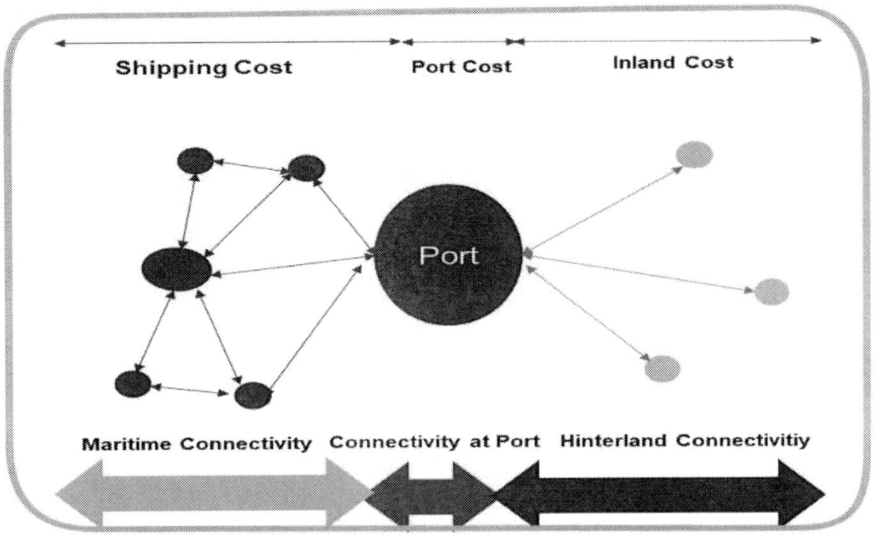

Figure 4. Connectivity and transport cost.

Source: author elaboration

As seen on figure 1 and 4, every segment from the cargo movement, it translates into the generalized transport cost. Each connectivity aspect, like maritime connectivity, connectivity at the port, and hinterland connectivity measure in generalized transport costs, indicates that better connectivity has lesser generalized transport costs.

The use of generalized cost because the model is also trying to capture the non-monetized attribute and translate it into a monetized attribute. This model incorporates the time that spends to move from origin to destination and translate it into monetized through the value of time.

3.2. Connectivity analysis

The second step is the connectivity analysis. This step is building the model port connectivity. As the measurement already decided, which is generalized transport cost, the analysis that suitable is adapted from the accessibility analysis based on the generalized transport cost. Generalized transport costs derive from each segment of the transport chain movement that consists of the maritime, port, and hinterland segment. Each part has activities and translates it into the generalized cost. After defining the generalized cost, its use to assess the port connectivity by accessibility analysis.

Using equation (6), the port connectivity can assess from the value of relative accessibility. The port that has minimum value of relative accessibility is the most accessible port.

Government or port operator may have several scenarios of policy regarding the transport network including port, such as development of port, development of rail or road, increasing the ships frequency, increasing the number of ships. All the scenario has an impact to generalized transport. The change on the generalized transport will affect to the value of the accessibility which mean also have an impact to the port connectivity.

The mostly connectivity model using graph theory, and the model doesn't integrate maritime and inland sides. A model that incorporated both maritime, port and inland side developed using Multi-criteria decision, with several variables and then assets using an index [14]. Using the same measurement, such as time and capacity, the model has conducted, but the analysis only at the maritime side or port to port connectivity [3]. A model that will build in this paper using generalized transport cost as a measurement of the node is port to port and port to the hinterland, and using the same measure is generalized transport cost. the used of exact size make it more comfortable to identify which part from maritime or land part which has low connectivity then the scenario of policy will be more focus on the domain that needs more improvement.

The study area divides into the zone. Each zone is associated with a centroid and represents origin or destination. The origin or destination can be a zone or terminal transport (port) and connectivity modeled using a transport network. The suitable model adapts from [68] modeled the generalized cost that consists of generalized cost at maritime, port, and inland.

As connectivity measure using generalized cost, the simple measure is using accessibility that represented through impedance matrix. Impedance represent through generalized transport cost.
Generalized transport cost that incorporated to the model adapted [68] from as follows:

$$GC_{odi} = GC_{oi}^{h} + GC_{i} + GC_{ij}^{m} \text{ or } GC_{ji}^{m} + GC_{i} + GC_{id}^{h} \quad (1)$$

GC_{odi} is Generalized transport cost between hinterland o to hinterland d via port i. m is maritime link (port origin i to loading and port j to unloading in the destination d) and h is hinterland link (port i to loading/unloading from/to hinterland).
Generalized cost Maritime component :

$$GC_{oi}^{m} = \sum_{s=1,\dots,s} MS_{si}(\, C_{si} + t_{si} * vot_{m}) \quad (2)$$

where:
MS : market share for ship category s for each loading/unloading port (i)
C_{si} : cost shipping
t_{si} : shipping time for each shipping category
vot_{m} : value of time at shipping.
Generalized cost Port component adapted from [68]:

$$GC_{i} = PC_{i} + CC_{i} \quad (3)$$

$$PC_{i} = v_{i} + r_{i} + d_{i} * vot_{di} + z_{i} * vot_{zi} \quad (4)$$

PC_{i} : Generalized port and terminal cost
CC_{i} : Penalty for congestion port
v : port cost
r : terminal cost
d : handling time (days)
vot(d) : value of time for handling (days)
vot(z) : value of time for storage (days)
Generalized cost hinterland component consist from origin/ destination with road or rail from loading/unloading to hinterland.

$$GC_{id}^{h} = \sum_{m} MS_{dmi}(C_{dmi} + (t_{dmi} * vot_{t})) \quad (5)$$

MS mode share for each mode (rail, road) to hinterland *d, C* is cost for each mode (m), t is inland travel time and vot_{t} is value of time at inland.

the model use accessibility matrix, the relative accessibility for transport terminal/ port a_{odi}consists origin o and destination d via port i, transportation cost from o to port i, c_{oi}^{m}, m is maritime link, using mode m and transport terminal cost p_{i} , and transport cost from port i to hinterland d, h is maritime link, using mode h the function as follow adapted from [69]:

$$a_{odi} = \beta_{1} c_{oi}^{h} + \beta_{2} p_{i} + \beta_{3} C_{ij}^{m} \text{ , or } a_{odi} = \beta_{1} c_{ji}^{m} + \beta_{2} p_{i} + \beta_{3} C_{id}^{h} \quad (6)$$

From the equation, the accessibility of the transport terminal/port i, from origin o to port i, and also same for the reverse, from port i to zone o has resulted. In standard application of accessibility measurement, the smallest distance, cost, time will imply the higher accessibility and it means also has a good connectivity.

3.3. Synthesis of the Results

The last stage is to synthetise of the results, this part is trying to formulate the scenario of policy that has most impact to the port connectivity. The assessment on connectivity will find the scenario of policy that has the highest impact to the port connectivity. Further the connectivity model will result in the robust scenarios of policy related to the improvement of port connectivity.

4. CONCLUSION AND FUTURE WORK

Connectivity is one indicator that the network has a good connection; in terms of transport, it means that the network can facilitate freight or passenger movement effectively and efficiently. In terms of port connectivity, the Port should be having a good connection between other ports and the hinterland. The port is also functioned as an interchange between mode, maritime mode, and rail or road mode, and it is essential to capture the connectivity from the multimodal perspective.

This paper offered the conceptual framework to formulate port connectivity using generalized transport cost as a variable and measures using relative accessibility. The generalized cost is built with considering the whole cargo movement segment, starting from the hinterland, maritime segment, and port itself. This paper will give contribution significantly to the connectivity model using the accessibility measurement.

Future work will discuss and elaborate further on the measurement of generalized cost. The question that may arise is about the detailing of the generalized transport cost variable because there are so many variables that might be involved in it; besides, the conceptual framework's applicability to the real case should be challenging.

Further, the conceptual framework will be applied to assess the connectivity of port in Indonesia. As an archipelagic country, using more than two-mode is inevitable, the port will be an interchange between maritime transport and land transport. Port connectivity be an essential variable to consider in the scenario of policy.

ACKNOWLEDGMENT

The authors would like to thank the Indonesia Endowment Fund for education, Ministry of Finance of the Republic Indonesia, for the scholarship support to the first author of this paper.

REFERENCES

[1]. Lee, P.T.-W., et al., *Maritime logistics and port connectivity in the globalised economy.* Transportation research. Part E, Logistics and transportation review, 2016.

[2]. Tovar, B., R. Hernández, and H. Rodríguez-Déniz, *Container port competitiveness and connectivity: The Canary Islands main ports case.* Transport Policy, 2015. **38**: p. 40-51.

[3]. Jiang, J., et al., *Port connectivity study: An analysis framework from a global container liner shipping network perspective.* Transportation research part E: Logistics and transportation review, 2015. **73**: p. 47-64.

[4]. Ducruet, C., S.-W. Lee, and A.K. Ng, *Centrality and vulnerability in liner shipping networks: revisiting the Northeast Asian port hierarchy.* Maritime Policy & Management, 2010. **37**(1): p. 17-36.

[5]. Bartholdi, J.J., P. Jarumaneeroj, and A. Ramudhin, *A new connectivity index for container ports.* Maritime Economics & Logistics, 2016. **18**(3): p. 231-249.

[6]. Tang, L.C., J.M. Low, and S.W. Lam, *Understanding port choice behavior—a network perspective.* Networks and Spatial Economics, 2011. **11**(1): p. 65-82.

[7]. Cullinane, K. and Y. Wang, *A capacity-based measure of container port accessibility.* International Journal of Logistics: Research and Applications, 2009. **12**(2): p. 103-117.

[8]. Acciaro, M., et al., *Contested port hinterlands: An empirical survey on Adriatic seaports.* Case

studies on transport policy, 2017. **5**(2): p. 342-350.

[9]. Wang, L., A. Goodchild, and Y. Wang, *The effect of distance on cargo flows: a case study of Chinese imports and their hinterland destinations.* Maritime Economics & Logistics, 2018. **20**(3): p. 456-475.

[10]. Nguyen, L.C. and T. Notteboom, *The relations between dry port characteristics and regional port-hinterland settings: findings for a global sample of dry ports.* Maritime Policy & Management, 2019. **46**(1): p. 24-42.

[11]. Chen, H., J.S.L. Lam, and N. Liu, *Strategic investment in enhancing port–hinterland container transportation network resilience: A network game theory approach.* Transportation Research Part B: Methodological, 2018. **111**: p. 83-112.

[12]. Liu, Z., et al., *Global intermodal liner shipping network design.* Transportation Research Part E: Logistics and Transportation Review, 2014. **61**: p. 28-39.

[13]. Guerrero, D., *Impacts of transport connections on port hinterlands.* Regional Studies, 2019. **53**(4): p. 540-549.

[14]. Wang, G.W., et al., *Port connectivity in a logistic network: The case of Bohai Bay, China.* Transportation Research Part E: Logistics and Transportation Review, 2016. **95**: p. 341-354.

[15]. Arvis, J.-F., et al., *Maritime Networks, Port Efficiency, and Hinterland Connectivity in the Mediterranean.* 2018: World Bank Publications.

[16]. Rietveld, P., *Some notes on interconnectivity in transport networks*, in *Overcoming isolation*. 1995, Springer. p. 18-28.

[17]. De Langen, P.W. and K. Sharypova, *Intermodal connectivity as a port performance indicator.* Research in Transportation Business & Management, 2013. **8**: p. 97-102.

[18]. Grosso, M., *Variables Influencing Transport Mode Choice: A Generalized Cost Approach.* Proceedings XIII Riunione Scientifica Società Italiana dei Trasporti e della Logistica, Messina, 2011.

[19]. SteadieSeifi, M., et al., *Multimodal freight transportation planning: A literature review.* European journal of operational research, 2014. **233**(1): p. 1-15.

[20]. Mutlu, A., Y. Kayıkçı, and B. Çatay, *Planning multimodal freight transport operations: A literature review.* 2017.

[21]. Chang, T.-S., *Best routes selection in international intermodal networks.* Computers & operations research, 2008. **35**(9): p. 2877-2891.

[22]. Zhang, M., B. Wiegmans, and L. Tavasszy, *Optimization of multimodal networks including environmental costs: a model and findings for transport policy.* Computers in industry, 2013. **64**(2): p. 136-145.

[23]. Yamada, T., et al., *Designing multimodal freight transport networks: A heuristic approach and applications.* Transportation science, 2009. **43**(2): p. 129-143.

[24]. Meng, Q. and X. Wang, *Intermodal hub-and-spoke network design: incorporating multiple stakeholders and multi-type containers.* Transportation Research Part B: Methodological, 2011. **45**(4): p. 724-742.

[25]. Cho, J.H., H.S. Kim, and H.R. Choi, *An intermodal transport network planning algorithm using dynamic programming—a case study: from Busan to Rotterdam in intermodal freight routing.* Applied Intelligence, 2012. **36**(3): p. 529-541.

[26]. Meng, Q., T. Wang, and S. Wang, *Short-term liner ship fleet planning with container transshipment and uncertain container shipment demand.* European Journal of Operational Research, 2012. **223**(1): p. 96-105.

[27]. Tran, N.K., H.-D. Haasis, and T. Buer, *Container shipping route design incorporating the costs of shipping, inland/feeder transport, inventory and CO 2 emission.* Maritime Economics & Logistics, 2017. **19**(4): p. 667-694.

[28]. Hao, C. and Y. Yue, *Optimization on combination of transport routes and modes on dynamic programming for a container multimodal transport system.* Procedia Engineering, 2016. **137**(1): p. 382-390.

[29]. Zhao, Y., et al., *A chance-constrained stochastic approach to intermodal container routing problems.* PloS one, 2018. **13**(2): p. e0192275.
[30]. Paleari, S., R. Redondi, and P. Malighetti, *A comparative study of airport connectivity in China, Europe and US: which network provides the best service to passengers?* Transportation Research Part E: Logistics and Transportation Review, 2010. **46**(2): p. 198-210.
[31]. Veldhuis, J., *The competitive position of airline networks.* Journal of Air Transport Management, 1997. **3**(4): p. 181-188.
[32]. Burghouwt, G. and R. Redondi, *Connectivity in air transport networks: an assessment of models and applications.* Journal of Transport Economics and Policy (JTEP), 2013. **47**(1): p. 35-53.
[33]. Malighetti, P., S. Paleari, and R. Redondi, *Connectivity of the European airport network: "Self-help hubbing" and business implications.* Journal of Air Transport Management, 2008. **14**(2): p. 53-65.
[34]. Calatayud, A., et al., *Understanding connectivity to international markets: a systematic review.* Transport Reviews, 2016. **36**(6): p. 713-736.
[35]. Lam, J.S.L. and W.Y. Yap, *Dynamics of liner shipping network and port connectivity in supply chain systems: analysis on East Asia.* Journal of Transport Geography, 2011. **19**(6): p. 1272-1281.
[36]. Hoffman, J., *Liner shipping connectivity.* UNCTAD Transport Newsletter, 2005. **27**(1): p. 4-12.
[37]. Low, J.M., S.W. Lam, and L.C. Tang, *Assessment of hub status among Asian ports from a network perspective.* Transportation Research Part A: Policy and Practice, 2009. **43**(6): p. 593-606.
[38]. Hu, Y. and D. Zhu, *Empirical analysis of the worldwide maritime transportation network.* Physica A: Statistical Mechanics and its Applications, 2009. **388**(10): p. 2061-2071.
[39]. Ducruet, C., *Network diversity and maritime flows.* Journal of Transport Geography, 2013. **30**: p. 77-88.
[40]. Laxe, F.G., M.J.F. Seoane, and C.P. Montes, *Maritime degree, centrality and vulnerability: port hierarchies and emerging areas in containerized transport (2008–2010).* Journal of Transport Geography, 2012. **24**: p. 33-44.
[41]. Ducruet, C. and F. Zaidi, *Maritime constellations: a complex network approach to shipping and ports.* Maritime Policy & Management, 2012. **39**(2): p. 151-168.
[42]. Mohamed-Chérif, F. and C. Ducruet, *Regional integration and maritime connectivity across the Maghreb seaport system.* Journal of Transport Geography, 2016. **51**: p. 280-293.
[43]. Montes, C.P., M.J.F. Seoane, and F.G. Laxe, *General cargo and containership emergent routes: A complex networks description.* Transport Policy, 2012. **24**: p. 126-140.
[44]. Kaluza, P., et al., *The complex network of global cargo ship movements.* Journal of the Royal Society Interface, 2010. **7**(48): p. 1093-1103.
[45]. Wilmsmeier, G. and J. Hoffmann, *Liner shipping connectivity and port infrastructure as determinants of freight rates in the Caribbean.* Maritime Economics & Logistics, 2008. **10**(1-2): p. 130-151.
[46]. Márquez-Ramos, L., et al., *"Special issue on Latin-American research" maritime networks, services structure and maritime trade.* Networks and Spatial Economics, 2011. **11**(3): p. 555-576.
[47]. Duval, Y. and C. Utoktham, *Trade facilitation in Asia and the Pacific: Which policies and measures affect trade costs the most?* 2011, ARTNeT Working Paper Series.
[48]. Wilmsmeier, G., J. Hoffmann, and R.J. Sanchez, *The impact of port characteristics on international maritime transport costs.* Research in Transportation Economics, 2006. **16**: p. 117-140.
[49]. Giovannini, M. and H.N. Psaraftis, *The profit maximizing liner shipping problem with flexible frequencies: logistical and environmental considerations.* Flexible Services and Manufacturing Journal, 2019. **31**(3): p. 567-597.
[50]. Tu, N., et al., *Shipping network design in a growth market: The case of Indonesia.*

Transportation Research Part E: Logistics and Transportation Review, 2018. **117**: p. 108-125.

[51]. Dulebenets, M.A., *A comprehensive multi-objective optimization model for the vessel scheduling problem in liner shipping.* International Journal of Production Economics, 2018. **196**: p. 293-318.

[52]. Karsten, C.V., B.D. Brouer, and D. Pisinger, *Competitive liner shipping network design.* Computers & Operations Research, 2017. **87**: p. 125-136.

[53]. Song, D.-P. and J.-X. Dong, *Cargo routing and empty container repositioning in multiple shipping service routes.* Transportation Research Part B: Methodological, 2012. **46**(10): p. 1556-1575.

[54]. Cariou, P., et al., *Liner shipping network design with emission control areas: A genetic algorithm-based approach.* Transportation Research Part D: Transport and Environment, 2018. **63**: p. 604-621.

[55]. Cheraghchi, F., et al., *Modeling the speed-based vessel schedule recovery problem using evolutionary multiobjective optimization.* Information Sciences, 2018. **448**: p. 53-74.

[56]. Zhen, L., et al., *Models on ship scheduling in transshipment hubs with considering bunker cost.* International Journal of Production Economics, 2016. **173**: p. 111-121.

[57]. Wang, S., Q. Meng, and Z. Liu, *Containership scheduling with transit-time-sensitive container shipment demand.* Transportation Research Part B: Methodological, 2013. **54**: p. 68-83.

[58]. Gelareh, S. and D. Pisinger, *Fleet deployment, network design and hub location of liner shipping companies.* Transportation Research Part E: Logistics and Transportation Review, 2011. **47**(6): p. 947-964.

[59]. De, A., et al., *Composite particle algorithm for sustainable integrated dynamic ship routing and scheduling optimization.* Computers & Industrial Engineering, 2016. **96**: p. 201-215.

[60]. Ferrari, C., F. Parola, and E. Gattorna, *Measuring the quality of port hinterland accessibility: The Ligurian case.* Transport Policy, 2011. **18**(2): p. 382-391.

[61]. Garcia-Alonso, L., A. Martinez-Pardo, and J.A. Vallejo-Pinto, *Analysis of the spatial development of the hinterland of ports: a case study.* International Journal of Shipping and Transport Logistics, 2016. **8**(2): p. 111-128.

[62]. Gouvernal, E., J.-P. Rodrigue, and B. Slack. *The divergence of regionalization: the challenges of the Mediterranean ports of Europe.* in *IAME Conference*. 2012.

[63]. Wilmsmeier, G., J. Monios, and B. Lambert, *The directional development of intermodal freight corridors in relation to inland terminals.* Journal of Transport Geography, 2011. **19**(6): p. 1379-1386.

[64]. Notteboom, T., *The Relationship between Seaports and the Inter-Modal Hinterland in Light of Global Supply Chains.* 2008.

[65]. Guerrero, D., *Deep-sea hinterlands: Some empirical evidence of the spatial impact of containerization.* Journal of transport geography, 2014. **35**: p. 84-94.

[66]. Van der Horst, M.R. and L.M. Van der Lugt, *Coordination mechanisms in improving hinterland accessibility: empirical analysis in the port of Rotterdam.* Maritime Policy & Management, 2011. **38**(4): p. 415-435.

[67]. Thill, J.-C. and H. Lim, *Intermodal containerized shipping in foreign trade and regional accessibility advantages.* Journal of Transport Geography, 2010. **18**(4): p. 530-547.

[68]. Zondag, B., et al., *Port competition modeling including maritime, port, and hinterland characteristics.* Maritime Policy & Management, 2010. **37**(3): p. 179-194.

[69]. Cartenì, A., *Accessibility indicators for freight transport terminals.* Arabian Journal for Science and Engineering, 2014. **39**(11): p. 7647-7660.

Investigation of Buoy Hydrodynamic Damping Based on Model Testing Data Series of Indonesia Tsunami Buoy

W.H. Nugroho[1], Arifin[1], N.J.H. Purnomo[1], and B. Ali[2]

[1]Pusat Teknologi Rekayasa Maritim, BPPT
[2]Balai Teknologi Hidrodinamika, BPPT

Abstract. Due to the recent Tsunami event in Indonesia in 2018, the Ina-TEWS program has been started again by the Indonesian Government. A new design has been manufactured and already deployed to the Indonesian water. This design is named Buoy G – 3. The tsunami early warning system (TEWS) consists of a tsunami buoy, an ocean bottom unit (OBU), a satellite and a ground station. The changes of the water pressure due to the seismic movement prior to the tsunami event will be acquired by OBU and sent the reading to the tsunami buoy by acoustic signal. Then, tsunami buoy transmits the signal to the ground station via satellite. The continuously link between OBU – Buoy – Satellite makes the buoy motions especially roll/ pitch has to be restricted. Because of these the hydrodynamic motion damping assessment of this tsunami buoy becomes so important. This paper describes a hydrodynamic damping analysis for roll/pitch of the tsunami buoy that based on model testing data series of Indonesian Tsunami Buoy G -1 to G -3. The data analysis will be presented on graphics that show the aspect ratio of buoy diameter over draft (D/T) and form ratio of buoy displacement over weight of the cylindrical submerged form (Δ/W) against damping ratio (β). This study finds that by using the linear damping approximation the higher the aspect ratio D/T the larger roll/pitch damping ratio β, but also the higher ratio of Δ/W the smaller roll/pitch damping ratio β.

Keywords: tsunami buoy, roll/pitch damping coefficients, motion decay test

IOP Conf. Series: Materials Science and Engineering 1052 (2021) 012009 doi:10.1088/1757-899X/1052/1/012009

1. Introduction

Due to the recent Tsunami event in Indonesia in 2018, as well as the large potential for Tsunamis in Indonesian region, it is necessary for the Government of Indonesia to revitalize the buoy-based early warning system for tsunami disasters developed by BPPT nationally. The Tsunami Buoy system is one of the tsunami early warning systems that works based on the sea level elevation anomaly that passes above the sensor placed on the seabed in the form of the Ocean Bottom Unit (OBU). This system is placed in the open seas, far from the coast.

The concept for designing this Indonesian Tsunami Early Warning System (Ina - TEWS) buoy is to design and build a buoy that has a good buoyancy, stability and sea-keeping capability as well as a structural integrity in where electronic devices and a capable communication system to be installed inside the buoy. The buoy will be moored with a taut mooring system in deep water. In terms of function, the surface buoy subsystem is designed to be able to meet the following functional requirements such as a good stability to receive and send signals to satellites also to make sure sensors except the acoustic modem are always above sea level and be able to position the antenna at a specified minimum height.

The tsunami buoy system consists of an Ocean Bottom Unit (OBU) and a Buoy (SB) unit at sea surface. OBU actively and continuously communicates with the Surface Buoy through an underwater acoustic modem. The surface buoy acts as a data receiver from the OBU and transmits the data via satellite to the tsunami monitoring centre at the read down station (RDS). A telemetry system is needed for sending tsunami signals to a data processing centre at the ground station in real-time. These tsunami buoys will be placed on the open sea with more than 2000m water depth on 14 (four teen) locations of Indonesian water that shown in Figure 1. Henceforth, the continuously link between OBU – Buoy – Satellite makes the buoy motion especially rolling/pitching has to be restricted and has ability to return to its equilibrium position as soon as possible. These conditions can only be predicted if the motion damping of the buoy is characterized. Because of these the hydrodynamic damping assessment of this tsunami buoy becomes so important. Some references have been found to support the importance of the roll damping analysis. The Averaging method by [3] assumes the profile of the roll decay is sinusoidal and the rate of change of the amplitude and phase is constant at their average values for each cycle, that make this method to be more applicable for light damped systems [7] has been studied this motion damping analysis to apply on the Tsunami buoy G – 1 and continue the study on Tsunami Buoy of G – 2 [8], [2] made the roll damping characterization program (RDCP) to characterize and quantify the roll damping characteristic of a free-floating maritime platform. [6] conducted Experimental Determination of Non-Linear Roll Damping of an FPSO Pure Roll Coupled with Liquid Sloshing in Two-Row Tanks.

This paper describes a hydrodynamic damping analysis for roll/pitch of the tsunami buoy that based on model testing data series of Indonesian Tsunami Buoy G -1 to G -3 (shown in Figure 2) in Maneuvering Ocean Engineering (MOB) tank at Balai Teknologi Hidrodinamika (BTH), BPPT. The data analysis will be presented on graphics that show the aspect ratio of buoy diameter over draft (D/T) and form ratio of buoy displacement over weight of the cylindrical submerged form (Δ/W) against damping ratio (β).

Figure 1. Tsunami Buoy deployment plan in Indonesia water [10].

Figure 2. Indonesia Tsunami Buoy G – 1, G – 2 and G- 3.

2. Theoritical basis

In this section a governing equation for the buoy motion decay is presented. When the energy of the buoy motion is gradually dissipated by friction and other resistances, the motion are said to be decayed. The motion gradually reduces or changes in frequency or intensity or cease and the buoy rests in its equilibrium position. When the system behaviour is almost linear (equivalent) damping can be derived as follows: The linear rotational motion (t) during a free extinction test can be described by, assuming a linear system:

$$(I + I_A).\frac{d^2\phi}{dt^2} + B_\phi.\frac{d\phi}{dt} + C_\phi.\phi = 0 \tag{1}$$

Where: I = vessel inertia for rotation
and I_A = added inertia for rotation
$B\phi$ = angular damping coefficient
and $C\phi$ = angular spring coefficient.

The spring coefficients are a hydrostatic spring. The non-damped natural period of this system can be calculated as:

$$T_\phi = 2\pi\sqrt{\frac{I + I_A}{C_\phi}} \tag{2}$$

For such a system the critical damping $B\phi_{crit}$ is defined as:

$$B_{\phi_{crit}} = 2\sqrt{(I + I_A).C_\phi} \tag{3}$$

If the damping is equal to, or larger than, the critical damping, no overshoot of dynamic amplification occurs in the system. To determine the degree of damping in a system, the damping is sometimes expressed as a ratio β of the linear damping coefficient B and the critical damping B_{crit}:

$$\beta = \frac{B}{B_{crit}} \tag{4}$$

The method to determine a damping coefficient of the buoy above can be explained in this section. From the recorded decay curve which is presented on Figure 3 the motion decay or free extinction natural periods can be derived.

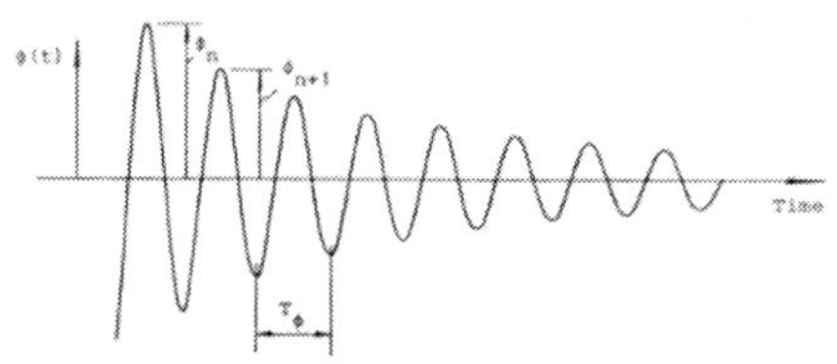

Figure 3. Time history of a motion decay test.

In Figure 3, where ϕ(t) is a time trace of motion ϕ, ϕ_n is a motion amplitude of n-th oscillation and Tϕ is a natural period of motion ϕ. The damping coefficients were derived from the decrease of motion amplitude for two successive oscillations.

The linear damping can be obtained from the decay tests as follows.

- First, a plot is made from the decay tests on logarithmic paper to determine the mean logarithmic decrement δ of the amplitudes. The logarithmic decrement δ is defined as:

$$\delta = \frac{\ln(\phi_1) - \ln(\phi_n)}{n} \tag{5}$$

  Where n is the number of oscillations.

- Second the logarithmic decrement is used to determine the damping B or the dimensionless damping β as a percentage of the critical damping:

$$B_\phi = \frac{2.(I + I_A).\delta}{T_\phi} \tag{6}$$

$$\beta = \frac{\delta}{2\pi} \tag{7}$$

3. Motion decay test

The tsunami buoy model used in this study was made of fiber glass, before testing the model, its radius of gyration was measured by hanging the buoy model using a crane and adjusting the mass distribution of the buoy model. The motion decay test the measurement was performed by using an Optical wireless motion tracking system that can determine accurately the angle of buoy motion decay. During the decay test in the water tank has to be in a calm state. An extensive series of decay tests were performed on the buoy model. Each decay test was performed by giving the model an initial offset in one direction of freedom and, then, releasing it which presented on Figure 4. The resulting vessel motions were recorded in one degree of freedom which is roll or pitch. Some of the decay tests were repeated several times. Of these repeats the best record was analyzed to determine the natural periods and the damping ratio. All measured data that shown on Figure 5 was recorded and stored on computer for further processing. For that purpose the analogue signals were sampled at a rate of 50 Hz. The notations of the motions were taken in relation to their directions with respect to the model system of axes (see Figure 6). Each rotation was considered positive when, from zero position, it rotated clockwise about the respective axis of the local system when looking into the positive direction of that axis.

Figure 4. Roll decay test of Buoy G-3.

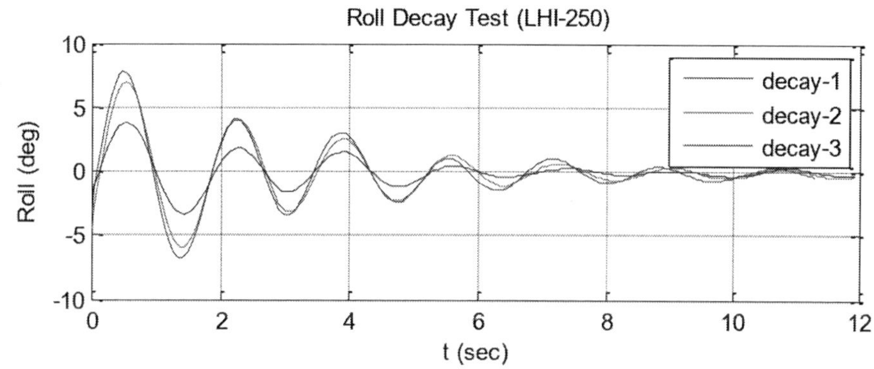

Figure 5. An example of Time traces roll decay test of Buoy G 3.

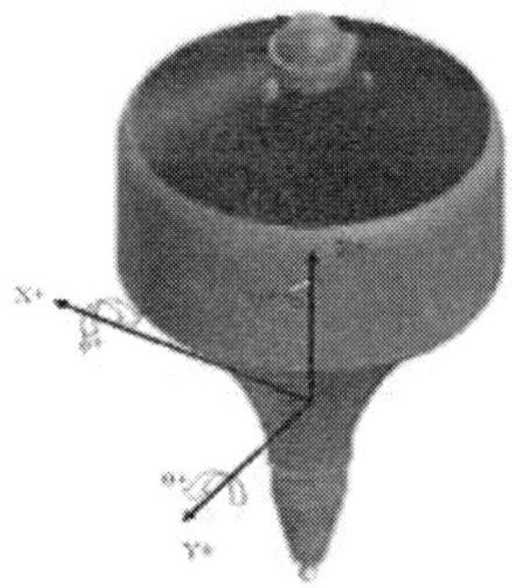

Figure 6. The buoy model system of axes.

ϕ : roll starboard down
θ : pitch bow down
ψ : yaw bow to portside

4. Results & discussions

A number of decay tests have been performed in MOB to determine damping ratio β as a percentage of the critical damping of the buoy models derived from these decay tests by the methods described above.

There are 6 test buoy conditions obtained in this study. These data are the result of the motion decay test on the G1 - G2 - G3 buoy models. Two plots were generated to present the damping coefficient as a function, namely the function of the ratio of diameter to draft of buoy (D/T) and the ratio of displacement of the buoy to weight of the submerged cylindrical form (Δ/W). The function of the ratio of the diameter to draft of buoy (D/T) on the damping coefficient β is shown in Figure 7. The function of the ratio of displacement of the buoy to the weight of the submerged cylindrical form (Δ/W) on the damping coefficient β is shown in Figure 8.

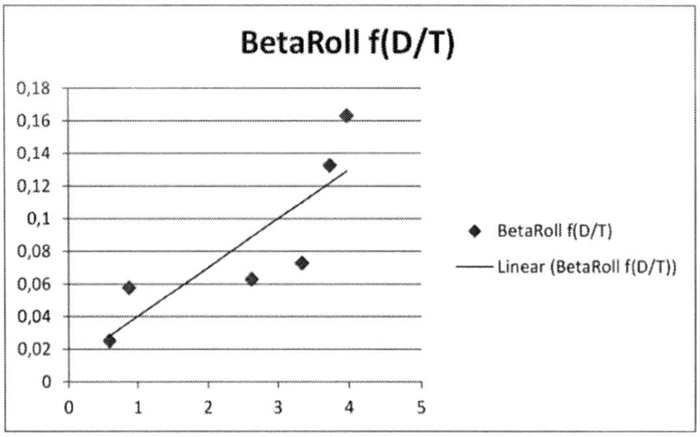

Figure 7. The Diameter over Draught (D/T) vs Damping ratio β.

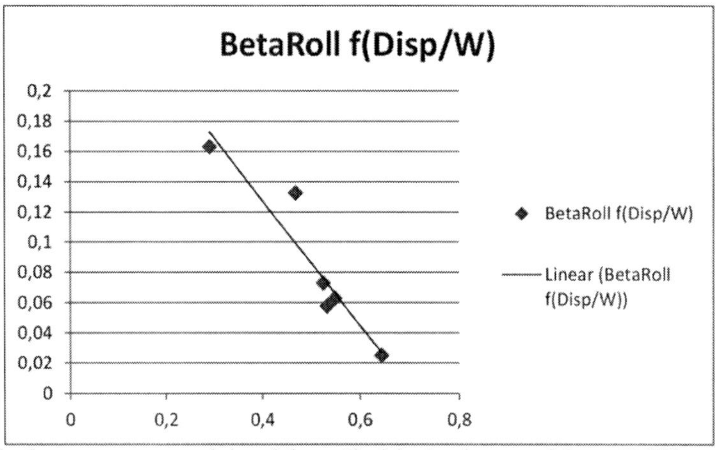

Figure 8. The Displacement over weight of the cylindrical submerged form (Δ/W) vs damping ratio β.

As in the theoretical basis explained when the energy of the buoy motion is gradually dissipated by friction and other resistances, the motion are said to be decayed. The motion gradually reduces or changes in frequency or intensity or cease and the buoy rests in its equilibrium position. A lightly damped harmonic motion moves with almost the same frequency, but it loses amplitude and velocity and energy as times goes on. The timescale over which the amplitude decays is related to the time constant. As the resistive force increases or damping force increases, the decay happens more quickly. If the system gradually increased the amount of damping, the period and frequency begin to be affected, because damping opposes and hence slows the back and forth motion. If there is very large damping, the system does not even oscillate and it slowly moves toward equilibrium. In Figure 7, it can be seen that there is a tendency for the hydrodynamic damping coefficient value of the buoy to increase with the increase in the ratio value of the buoy diameter to its draft. It can also be said that in order to obtain a large hydrodynamic damping value to quickly move to the original equilibrium position a designer can choose a buoy diameter size that exceeds its draft, which seems to apply to a ratio (D / T) equal to 1 to 4, according to the results obtained in this study. However, in Figure 8 it can be seen that the damping coefficient value tends to decrease with increasing the ratio (Δ / W) 0.25 to 0.65, which means that the ability to roll the buoy to its balance position takes longer time. In case of the tsunami buoy the high damping coefficient is preferable because of the continuously link between OBU – Buoy – Satellite to meet a good stability to receive and send signals to satellites.

Concluding remarks

This study finds that by using the linear damping approximation the higher the aspect ratio D/T equal to 1 to 4 the larger roll/pitch damping ratio β, but also the higher ratio of Δ/W of approximately 0.25 to 0.65 the smaller roll/pitch damping ratio β. Results of this study clearly can be useful for a naval architect to design a buoy with a suitable damping performance to operate in the open sea.

Future Works

Some possible future works can be proposed; (1) Since the study based on the linear damping approximation based on series of model tests, to predict the motion of the buoy to be more accurate a quadratic damping approximation can be studied (2) In some studies found that a damping coefficient is also function of the excitation frequency and current speed, therefore a study of the damping coefficient of the buoy in various wave condition and current speed is also an interesting subject to be conducted.

Acknowledgements

This research was financially supported by the Agency of Assessment and Application oftechnology, Ministry of Research Technology, Republic Indonesia through Program of Rancang Bangun Buoy dan OBU (Inovasi Sistem dan Teknologi INA-TEWS), PTRIM, TIRBR – BPPT FY2020

References

[1] Bhattacharya, R, "Dynamics of Marine Vehicles", John Wiley & Sons Inc (1978).
[2] Dawson Edward and Riding Murray, "Roll Damping Characterisation Program: User Guide Maritime Division", Defence Science and Technology Organisation, Australia, 2014.
[3] Flower, J.; Sabti Aljaff, W," Kryloff-Bogoliuboff's solution to decaying nonlinear oscillations in marine systems", Int. Shipbuild. Prog. 1980.
[4] Lewis E," PNA Vol III," Motions in Waves and Controlability", SNAME 1988.
[5] Reddy D.V, Arockiasamy M, "Offshore Structures, Vol 1", Krieger Publising Company, Malabar, Florida, 1991.
[6] Igbadumhe Jane-Frances, Sallam Omar , Fürth Mirjam , and Feng Rihui, "Experimental Determination of Non-Linear Roll Damping of an FPSO Pure Roll Coupled with LiquidSloshing in Two-Row Tanks",J. Mar. Sci. Eng. 2020, 8, 582.
[7] Wibowo HN, Samudro, Sahlan RB, "Aplikasi Teknologi Kelautan untuk Merekayasa Buoy Seawatch sebagai Buoy TEWS – Indonesia", SENTA 2007, ITS, Surabaya-Indonesia, 2007.
[8] Wibowo HN, Samudro, Arifin, "On The Analysis Design Of Indonesian Tsunami Buoyhull Based On The Hydrodynamic Test", 6TH Marine Technology Conference, UI, Jakarta, Indonesia, 2008.
[9] Wibowo HN, Arifin, Buddin Hakim, "PROGRAM MANUAL 2020, Rancang Bangun Buoy dan OBU (Inovasi Sistem dan Teknologi INA-TEWS)", (PTRIM), TIRBR – BPPT, Januari 2020.
[10] Wibowo H Nugroho, "Konsep dan Desain Rancang Bangun INA-Buoy BPPT", Webinar Inovasi Teknologi INA – BUOY BPPT dalam Perspektif Penanggulangan Bencana Tsunami di Indonesia, 29 September 2020.

IOP Conf. Series: Materials Science and Engineering 1052 (2021) 012010 doi:10.1088/1757-899X/1052/1/012010

Naval Architectural Aspects in The Design of a Hybrid Hydrofoil-Submarine Craft.

W Wardhana[1], E M Wardhana[2], M Soetardjo[3] and O F Nichita[4]
[1] Department of Ocean Engineering, ITS Surabaya, Indonesia
[2] Department of Marine Engineering, ITS Surabaya, Indonesia
[3] Indonesia Hydrodynamics Laboratory, BPPT, Indonesia
[4] Department of Marine Engineering, ITS Surabaya, Indonesia

Abstract. Indonesia is the largest archipelago country in the world. Territorial disputes and illegal fishing are some main threat of maritime defense. One of the solutions is to provide a multi-mission vehicle. Submarine craft has the advantage of undetectable to ambush the threats. Hydrofoil craft has the advantage of higher speed than the conventional high-speed craft. The purpose of this research is to combine two distinctive and one conventional concept into one craft. This craft has to has a light structure to support hydrofoil mode and able to submerge. The material of the hull uses aluminum and for wing and propulsion system structure use carbon steel. This craft uses hydraulic propulsion that the main engine moves the hydraulic pump. The hydraulic pump flows oil to the hydraulic motor where located in the nacelle. The hydraulic motor rotates the shaft and propeller. This propulsion system requires the propeller to submerge in the water, therefore the strut which supports the wing and propulsion system has to design retractable. All important parameters are included in this paper. *(Keywords: hydrofoil, submarine, conventional, hydraulic propulsion, Aluminium)*

1. Introduction

The hybrid-submarine hydrofoil craft, the crocodile craft, built to operate in three modes such as conventional mode, hydrofoil mode, and submarine mode. This craft is a result of Ristek-Dikti research program designed to secure Indonesia's seawater region, especially the western part, which relatively shallow. As we know that pirating, smuggling, and territorial violating in Indonesia's seawater are endanger national sovereignty that has to tackle. This craft has a length 12 meter and a width 3 meter designed in high-speed in hydrofoil mode and undetected in submarine mode, as a result, it increases shock effects because those illegal fishing vessels do not recognize it. It is propelled by 2x350 Hp diesel engine with 40 knots in hydrofoil mode, 15 knots in submarine mode, and approximately 25 knots in the conventional craft. This craft uses Aluminium as the main hull construction material and carbon steel in the strut and wing system. The design is the result of optimized space and craft's maximum speed. The struts are designed to be able to lift the craft in hydrofoil mode and able to rotate in certain degrees and the craft uses a hydraulic propulsion system because it should be able to deliver thrust in every mode as the picture is shown below[1][2][3][4][5][6].

Figure 1. The Croc Design Concept.

IOP Conf. Series: Materials Science and Engineering 1052 (2021) 012010 doi:10.1088/1757-899X/1052/1/012010

Philosophically, hydrofoil or submarine craft is illustrated as two circles below. In the design process, moving those circles close to each other makes a feasible area of this craft that relatively has a narrow area. Hydrofoil is a light structure as this craft has to move upward above free surface and wave whereas submarine is a heavy structure as it has to submerge below free surface. Hence, hydrofoil can move faster but still detected by radar because it is still above free surface. Submarine, on the other hand, can move slowly but it is undetected because it is under water, This is the factor try to maintain so that this type of craft has two importance factor in defense, it is fast and at the same time undetected[1][2][3][7].

2. Methods

The design thinking method illustrated in the following flowchart below. At the design step, the craft's form and weight obtained at the hydrofoil design, and the size of the ballast tank in submarine mode also should be considered in this step. This condition is also applied to estimate engine power, strut and wing specification, hydraulic pump, and hydraulic motor. The size and location of the ballast tank obtained from the craft's static stability. After this step, submarine and conventional mode craft can be considered. The towing tank test is taken as a reference for resistance calculation and craft performance for hydrofoil mode, submarine mode, and conventional mode. The method can then be describe in the following figure[1][2].

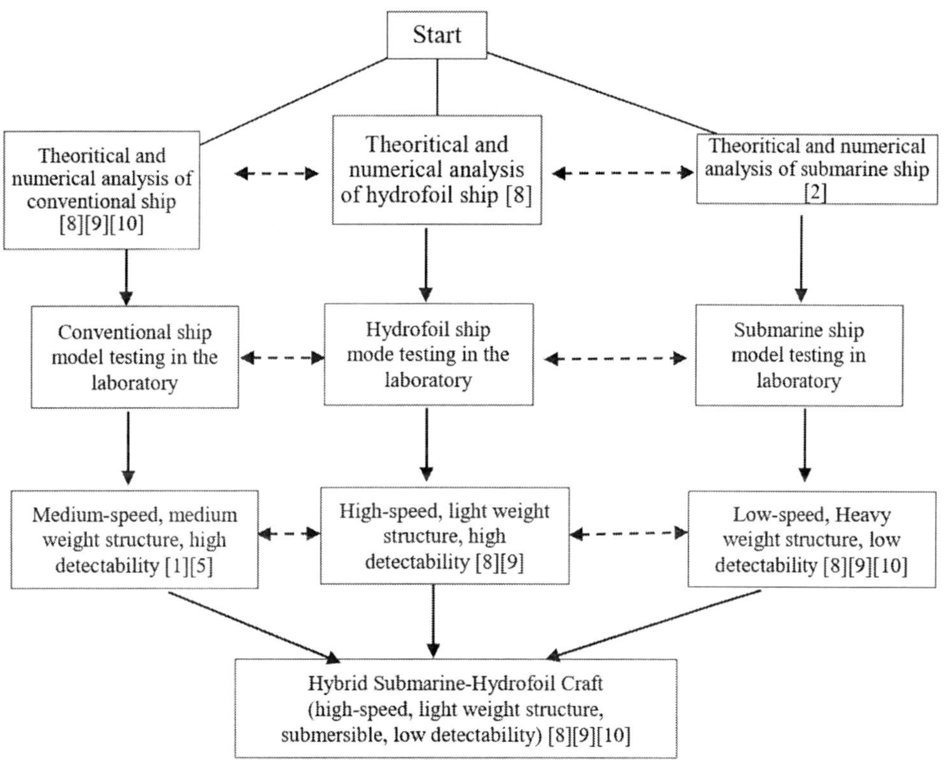

Figure 2. Design Thinking Flowchart.

This flow chart shows that the design should start from design in hydrofoil mode in which the equilibrium of

$$\sum F = 0, \sum M = 0 \tag{1}$$

IOP Conf. Series: Materials Science and Engineering 1052 (2021) 012010 doi:10.1088/1757-899X/1052/1/012010

should be satisfied during this mode[1][2]. The first consists of the resistance of all components such as the resistance of wing, nacelle, propeller etc and must be equal to the propulsion of the craft. The dimension of the wing, including chord and type of the wing section, is set up so that its resistance is found. The dimension is designed such that if the Lift and Drag force is not enough, it is then re-designed until it is able to support the craft to get its equilibrium. It is found that the chord size of the aft wing is 65cm and the fore chord is 40cm. The propeller is chose to be B series 3 blades with diameter 50 cm. It is also found that the engine power is 2 x 350 HP and is designed using hydraulic propulsion syatem.

As the craft has to be designed submerge then it is need to design the hull to have the equality of weight and volume displacement which result in the specific form. The main ballast tank is placed in the middle and in the fore and after part of the hull. The fore and after part is used as trim and compensating tank to arrange the trim of the hull.

3. Hull Configuration

Hydrofoil craft is designed to be light structure. However, the submarine structure requires heavy structure. To achieve these conditions and space requirements, therefore the hull is designed to accommodate two main engines and ballast tank space requirements. Two fixed wings at a half-length to the fore of the hull have to build to increase lift force when take-off in hydrofoil mode. At the top of the craft, there is a sail as an envelope of the piping system that connects installation in the engine room and accommodation room. Therefore, the craft able to control in accommodation room located in the fore of the craft. This craft designed for four crews consists of a pilot, a co-pilot, and two researchers. The main ballast tank is located in mid-ship to not affect the trim of the craft, fore and stern ballast tank use when necessary to control the trim. The engine room is located in the stern, but the fuel tank is located in mid-ship, in which the same room as the main ballast tank.

The propulsion system of this craft uses hydraulic propulsion which the main engine propels a hydraulic pump. The hydraulic pump flows oil to the hydraulic motor in the nacelle uses two hose or flexible pipe. The hydraulic motor connects with the shaft system to propel the propeller. The result of the hydrofoil calculation obtains the length of the stern wing chord is 65 cm and the fore wing system is 40 cm. The esssential characteristics of hydrofoil craft is that the craft be supported by buoyancy force when floating on the water and by dynamic lift of the foil while flying above the water. The lift of the foil is the function of foil area, lift coefficient, and dynamic pressure ($\frac{1}{2}\rho v^2$). The basis formula of foil area to support its weight is therefore [11],

$$S = \frac{W}{0.5\,\rho\, v^2\, c_L} = \frac{790\Delta}{c_L v_k^2} \tag{2}$$

Where,

S = *Total area of foil = ft2*
W = *Weight = lb*
ρ $= density\ of\ water\ (lb\frac{sec^2}{ft^4})$
v $= speed = ft/sec$
v_k $= speed = knots$
c_L $= lift\ coefficient$
Δ $= weight = tons$

The configuration of the craft is as shown below[5][6].

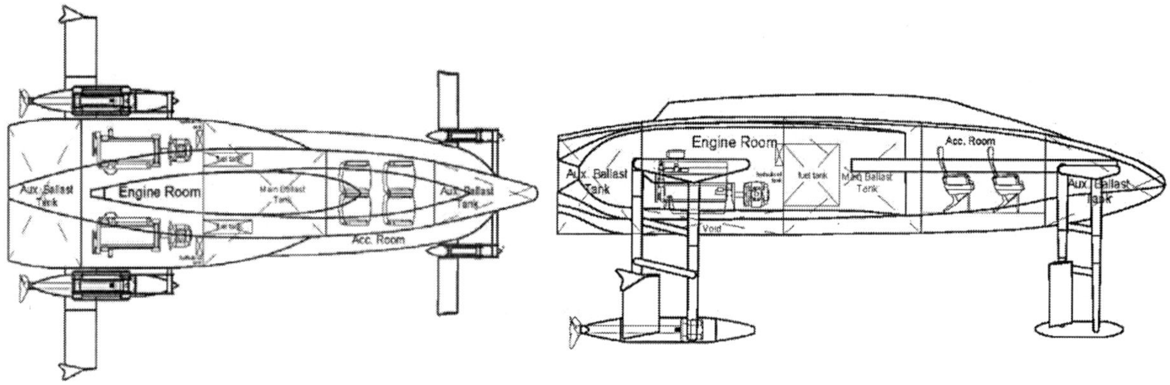

Figure 3. General Arrangement of The Crocodile.

4. Static Calculation of Hydrofoil Craft

Hydrofoil craft calculation obtains by configuring various chord geometry of wing. The best configuration obtains at a trim angle of 4 degrees [2]. The dihedral angle ohe fore and stern wing are the same approximately 40 degrees. The hydrostatics can be seen as follows,

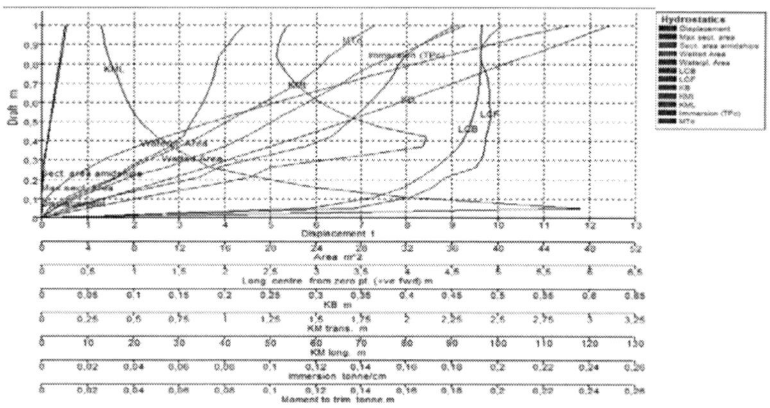

Figure 4. Hydrostatic Curve of The Crocodile.

This graph is calculated for draft until 1 meter which is the draft for surface mode. This draft is about half of the depth of the craft. This graph shows the static characteristics of the craft. This graph is produced using Maxsurf software [10].

5. Polygon Diagram

Some of critical conditions are obtained by running the craft model in Maxsurf software describes below:

Table 1. Various Static Stability Position of Crocodile Ship with Reference Point at Keel Plate

No.	Position	Draft FP [m]	Draft AP [m]
(1)	Hydrofoilcraft condition with the fore and the stern strut both in perpendicular position	1,028	1,052
(2)	Hydrofoil craft condition with the fore and the stern strut both in perpendicular position with fully ballast water	1,725	1,325
(3)	Submarine craft condition with the fore strut rotates in maximum position and the stern strut rotates 90 degrees without ballast water	0,693	1,346
(4)	Submarine condition with the fore strut rotates in maximum position and the stern strut rotates 90 degrees with ballast water	1,572	1,517
(5)	Conventional craft condition with the strut is in maximum-rotated position and the stern strut rotates 45 degrees.	0,761	1,250

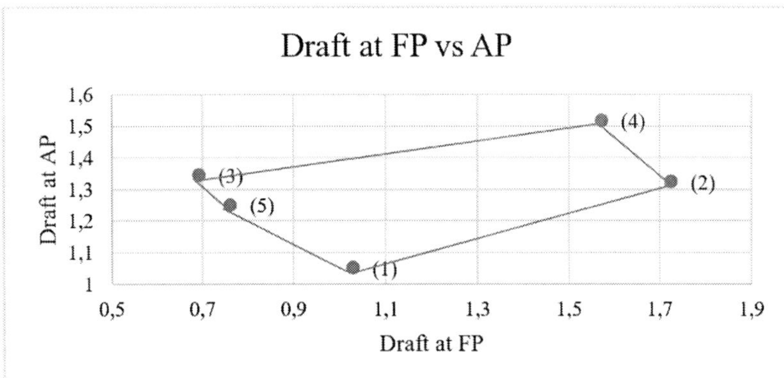

Figure 5. Polygon Diagram shown as respect to position number at Table 1.

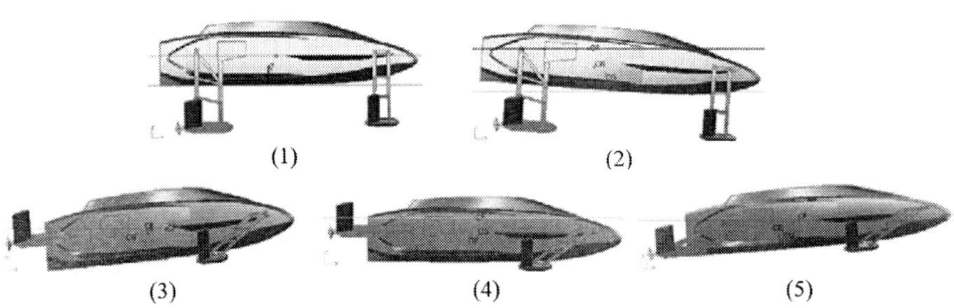

Figure 6. The Crocodile Static Stability Condition Respect to Position Number at Table 1.

5. Resistance Calculation of Hydrofoil Craft

The resistance calculation of hydrofoil craft that affected by force and moment and has 3 factors:

1. Hydrodynamics forces consist of [8][9][10]:
   a. Wave making resistance
   b. Flows around the foil and its interaction
   c. Current
2. Aerodynamics forces that caused by upper surface craft's body interact with airflows
3. Propulsion forces that caused by the propeller and its propulsion system

Equilibrium condition could be achieved if:

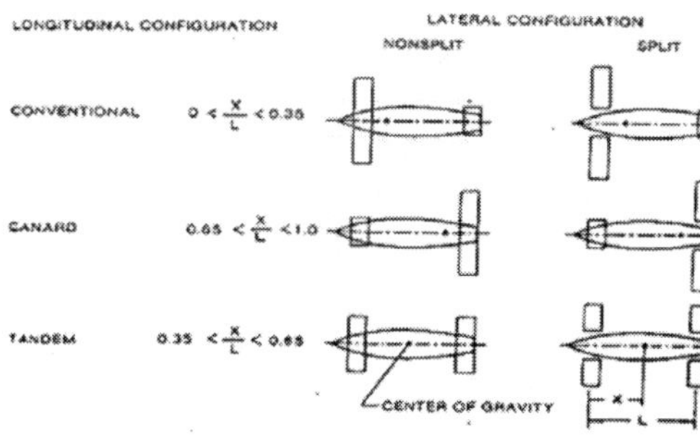

Figure 7. Foil Configuration of Hydrofoil Craft [3].

$$\sum FH\left(DH, \tau, {}^{\delta}/_{\alpha}, T\right) = 0, Horizontal\ forces\ equilibrium \quad (3)$$

$$\sum FV\left(DH, \tau, {}^{\delta}/_{\alpha}, T\right) = 0, Vertical\ forces\ equilibrium \quad (4)$$

$$\sum M\left(DH, \tau, {}^{\delta}/_{\alpha}, T\right) = 0, Equilibrium\ Moment \quad (5)$$

The design of the Crocodile has the biggest weight distribution at the stern of the craft. Hence the most suitable foil configuration is canard, as shown in the figure above. The canard configuration has a bigger stern wing dimension than the fore wing.

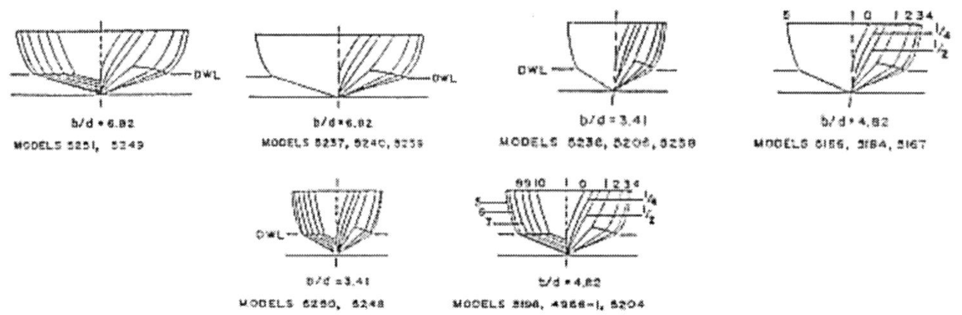

Figure 8. Types of high-speed hull form [3].

the hull form of the Crocodile is not simply adopted conventional high-speed as illustrated above but it has to has knuckle in the mid-section which a combination of conventional high-speed planing craft and hydrofoil craft.

The resistance of the hydrofoil ship is formulated below[8][9][10]:

$$RT = RH + DFL + DST + RAP + RAA + DW \quad (6)$$

Which,
RH = Resistance of Hull Form
DFL = Resistance of the Foil
DST = Resistance of the Struts
RAP = Appendages Resistance
RAA = Air Resistance
DW = Wave Resistance

6. Dynamics Ship's Motion

The equation of hydrofoil craft is expressed below [8][9][10]:

$$m(\dot{u} + qw - rv) = X + mg\sin\theta \quad (7)$$
$$m(\dot{v} + ru - pw) = Y - mg\cos\theta\sin\phi \quad (8)$$
$$m(\dot{v} + ru - pw) = Z - mg\cos\theta\sin\phi \quad (9)$$
$$\dot{p}I_{xx} + qr(I_{zz} - I_{yy}) = K \quad (10)$$
$$\dot{q}I_{yy} + rp(I_{xx} - I_{zz}) = M \quad (11)$$
$$\dot{r}I_{zz} + pq(I_{yy} - I_{xx}) = N \quad (12)$$

These equations are used to check whether the craft is stable in hydrofoil mode under a specific wave type. The equations are run in 200 seconds, and the motion of the craft is recorded.

7. The Software

Using all equations above, the resistance calculation of hydrofoil craft requires input and output as shown below [8][9][10]:

Table 2. General Input Data.

General Input Data	
Foil area distribution type	CANARD
Total craft weight	15000.00 [kg]
X-coordinate C.O.G.	3.40 [m]
Z-coordinate C.O.G	0.95 [m]
Projected chine length of hull	10.00 [m]
Average chine beam	2.50 [m]
Mid-ship deadrise angle	20.00 [deg]
Zero speed trim angle	2.30 [deg]
Design speed trim angle	0.00 [deg]
Hull clearance at design speed	1.00 [m]
Density of water	1025.00 [kg/m3]
Temperature of water	15.00 [deg-C]
Foil roughness allowance	0.00040
Hull correlation allowance	0.00040
Ambient pressure	99600.00 [N/m2]

Table 3. Propulsion Properties Input Data.

Propulsion Properties	
Propulsion type/ IPROP	3
Thrust vector angle	0.00 [deg]
X-coord. thrust vector	0.90 [m]
Y-coord. thrust vector	2.09 [m]
Z-coord. thrust vector	-1.68 [m]
Number of Propulsion units	2

Table 4. Foil System Properties Input Data.

Panel No.	X-Coord. 1 [m]	X-Coord. 2 [m]	Y-Coord. 1 [m]	Y-Coord.2 [m]	Z-Coord. 1[m]	Z-Coord. 2 [m]
1	9.67	9.67	-2.14	-0.38	0.00	-1.48
2	9.67	9.67	-0.38	0.00	-1.48	-1.48
3	9.67	9.67	0.00	0.38	-1.48	-1.48
4	9.67	9.67	0.38	2.14	-1.48	0.00
5	9.67	9.67	-0.38	-2.14	0.31	0.00
6	9.67	9.67	-0.38	-0.388	0.31	-1.48
7	9.67	9.67	0.38	0.38	-1.48	0.31
8	9.67	9.67	-0.63	0.38	0.00	0.31
9	1.61	1.61	0.00	-0.63	0.00	-1.68
10	1.61	1.61	-0.63	0.00	-1.68	-1.68
11	1.61	1.61	0.00	0.63	-1.68	-1.68
12	1.61	1.61	0.63	2.93	-1.68	0.00
13	1.61	1.61	-1.25	-2.93	0.41	0.00
14	1.61	1.61	-0.63	-0.63	0.20	-1.68
15	1.61	1.61	0.63	0.63	-1.68	0.20
16	1.61	1.61	2.93	1.25	0.00	0.41

Using that input for a certain configuration, the hydrofoil mode can be ensured to be stable, Hence, at the end of the iteration, results are as follows

Table 5. Propeller design output data.

Propeller Design Data	
Pitch-diameter ratio	1.4
Blade area ratio	1.000
Open Water Efficiency	0.691
Number of revolutions	2720.3 [1/min]
Delivered thrust	9.2 [kN]
Absorbed torque	0.9 [kNm]
Delivered power	253.5 [kW]

8. Towing Tank Experiment

Three experiments have been carried out in the towing tank under the conventional mode, hydrofoil mode, and submarine mode as shown in the following figure, with the following results [9][10]

Table 6. Hydrofoil Mode in Towing Tank Test.

No	V Ship (Knot)	V Ship (m/s)	V model (m/s)	Strip Chart (Kotak)	Resistance Model (Kg)
1	4	2,0576	0,6507		0,0000
2	6	3,0864	0,9760		0,0000
3	8	4,1152	1,3013		0,0000
4	10	5,1440	1,6267	5,00	0,1124
5	12	6,1728	1,9520	10,00	0,2247
6	14	7,2016	2,2773	17,00	0,3820
7	16	8,2304	2,6027	31,50	0,7079
8	18	9,2592	2,9280	70,00	1,5730
9	20	10,2880	3,2534	84,00	1,8876
10	22	11,3168	3,5787		0,0000

IOP Conf. Series: Materials Science and Engineering 1052 (2021) 012010 doi:10.1088/1757-899X/1052/1/012010

Table 7. Convention Ship Mode in Towing Tank Test.

No	V Ship (Knot)	V Ship (m/s)	V model (m/s)	Strip Chart (Kotak)	Resistance Model (Kg)
1	4	2,0576	0,6507		0,0000
2	6	3,0864	0,9760	5,50	0,3667
3	8	4,1152	1,3013	14,50	0,9667
4	10	5,1440	1,6267	28,00	1,8667
5	12	6,1728	1,9520	48,00	3,2000
6	14	7,2016	2,2773	6,50	3,2500
7	16	8,2304	2,6027	5,50	3,7500
8	18	9,2592	2,9280	8,00	4,0000
9	0	0,0000	0,0000		0,0000
10	0	0,0000	0,0000		0,0000

Table 8. Submarine Mode in Towing Tank Test.

No	V Ship (Knot)	V Ship (m/s)	V model (m/s)	Strip Chart (Kotak)	Resistance Model (Kg)
1	4	2,0576	0,6507		0,0000
2	6	3,0864	0,9760	1,50	0,4054
3	8	4,1152	1,3013	5,50	1,4865
4	10	5,1440	1,6267	9,00	2,4324
5	12	6,1728	1,9520	17,00	4,5946
6	14	7,2016	2,2773		0,0000
7	16	8,2304	2,6027		0,0000
8	18	9,2592	2,9280		0,0000
9	0	0,0000	0,0000		0,0000
10	0	0,0000	0,0000		0,0000

Figure 9. Experiment conducted in towing tank.

A wooden model experiment with length 1.1 meter, the experimental result is then extrapolated to find the resistance of full-scaled using ITTC extrapolation skin [2].

9. Manufacturing Process

The making of the Croc prototype is underway and now is about 90%. Hull is almost finished, and only the propulsion system and the control engine is not yet finished. The strut and wing are already installed but left only the piston and its equipment.

Figure 10. The Crocodile production process in the workshop.

10. Conclusion

Naval architectural aspects of Crocodile-Hydrofoil craft is presented. The craft is designed using Aluminium material with the wing is made of carbon steel. The craft uses NACA 16 type Canard with surface piercing wing system. The struts are designed retractable to make certain angle needed for every mode. The principle dimension of the craft is L = 12 m, B = 3 m, H = 3m, with engine power 2 x 350 HP with design speed 35 knots for hydrofoil mode, 25 knots for surface speed and around 15 knots for submerge mode. This carft is designed for 4 crews, 1 pilot, 1 co-pilot and 2 researchers.

Acknowledgement

The author would like to thank to DRPM Research DIKTI for the fund and Tita, Diar, and Frea for their patient and help during the research.

References

[1] Djatmiko, E. B. 1995, *"Kajian Numerik dan Eksperimental Kapal SWATH"*, RUT Ristek.
[2] Jones, B 2008, *"Technological Aspects of Submarine"*, Jornal of Underwater Tech, Canberra
[3] Lewis, E.V. 2002, *"Principle of Naval Architecture"*, London
[4] Rencana Induk Penelitian RIP ITS 2014-2018
[5] Suntoyo 2010, *"Hybrid Trimaran-Hydrofoil Bertenaga Surya"*, Riset Insentif, Ristek
[6] Sarwito, S. 2015, *"Pengembangan Sistem Utama Kapal Perang Crocodile-Hydrofoil untuk Teknologi Pertahanan dan Keamanan Laut Yang Efisien dan Efektif"*, PUPT Kemenristek-dikti
[7] Sulisetyono, A. 2015, *"Kajian Kapal Trimaran Bertenaga Listrik"*
[8] Walree, van, F. 2007, *"Hydrofoils, Model Tests and Computations", Hydrodynamics: Computations, Model Tests and Reality*
[9] Wardhana, W. 1998, *"Kajian Numerik dan Eksperimental Interaksi Sayap Hidrofoil"*, RUT IV
[10] Wardhana, W. 2004, *"Kajian Numerik dan Eksperimental Manned Hydrofoil Model"*, RUT X
[11] Hydrofoil Research Project for Office of Naval Research Navy Department, Washington DC, 1995, *"Hydrofoil Handbook Volume I Design of Hydrofoil Craft."*

Method for Estimating Price of Second hand Ship with Multi Method

A Azhar
Naval Architecture and Ship Building Engineering, Hang Tuah University of Surabaya, Jl Arif Rahman Hakim 150, Surabaya, Indonesia.
Email: ali.azhar@hangtuah.ac.id

Abstract. Demand fleet as a means of marine transport could be met by building new ships and buying second-hand vessels. The current constraints in building a new ship are the cost relatively expensive and take a long time, while the profit for the procurement of second hand ships time relative faster and cheaper price. This study aimed to identify the price of second hand vessels and to develop a cost estimation model second hand vessels. Collecting data with survey vessel prices do with field surveys and market surveys via the internet on websites selling used shipbroker. Data survey include the price, type of vessel, the main dimension, tonnage, machines, materials and speed. Multi-methods data processing and analysis consisting of market price approach, comparative vessel and physical pricing approach. Estimated price or appraisal vessel obtained from the average value of the three methods.
Keywords: prices market, ship comparison, physical price, average value

1. Introduction

The demand for fleets as a means of sea transportation can be met by building new ships and buying second-hand ships. The current constraints in building new ships are relatively expensive and take a long time, while the advantages for procurement of second hand vessels are relatively faster and prices

Determination or estimation of second hand ship prices can be done by econometric analysis method and modelling second hand ship prices, and obtained used prices with different ship sizes/segments will give different reactions to changes in market variables [1]. The financing structure of second hand vessels both technically and economically is different from new vessels, and depends on several main factors by the varied practices between one country and another [2]. In Indonesia the factors considered are as follows: remaining life-time vessels, new ship market prices, market prices of similar used vessels (national and international), technical assessment of ships per item, scrapping cost levels. But in addition to the benchmark price, which is influenced by the above factors, the price of second hand vessels is strongly influenced by non-technical factors that are negotiable. These factors are; fees or fee agencies (brokers), the rules of a region or country, and the level of needs of the buyer. The process of estimating the value of used ships, nowadays many are carried out by experienced people and usually work as brokers.

Another estimation model with econometrics to analyse used container prices, by applying hedonic regression models, depreciation costs and age costs [3], and nonlinear prediction models to estimate the cyclonic movements of the Panamax ex-ship market [4]. The two models of estimated ship prices are specifically container ships and Panamax so it is difficult to apply to other types of vessels.

The logistics network model for ship dismantling activities, ranging from the ship procurement process to the redistribution process to steel manufacturing companies as the final consumer of this business [5]. This study uses single ship entry and scrap metal components as analysis and evaluation materials. In the next research can be used multi-ship, the sale of reusable components other than scrap metal and the determination of second hand vessel prices are more realistic.

Designing a second hand ship condition survey model-based online via the web that can be used to store the data of survey results and this program produces output in the form of current condition and price of second hand vessels [6]. The program also requires thoroughness and experience for surveyors to determine the estimated condition and price of ship components surveyed.

Modelling the price of used tankers using static methods with variable ship life, dwt, shipyard builders, Class and the country of origin of shipbuilders [7]. The results obtained used tanker price modelling in the form of the regression model and did not conduct a direct survey of the ship model so that the objectivity of the estimation is less holistic.

The purpose of the preparation of second hand ship price estimation with multi-methods is to identify the price of second hand vessels with field surveys and market surveys, as well as to compile second hand ship estimates with multi-methods that are self-established methods of market price approach, ship comparison and physical price of the ship so that the accuracy of the estimate is more guaranteed.

2. Methods

The research was conducted with a community survey study to obtain data on the offer price or sale of ships that have specifications comparable or almost comparable to the vessels to be estimated. Ship price surveys will be conducted with field surveys and market surveys over the internet on the websites of ship sales brokers both domestically and abroad. From survey data covering price, ship type, main size, GRT, engine, material and ship speed. Primary and secondary data have been identified and modelled with three approaches: market price approach with regression analysis method, comparative ship and physical price of the ship. On each of these approaches will be obtained ship prices, so there will be three different ship prices. Each price of the three approaches will be summed up and taken the average value so that it can be summed up the estimated price of the ship.

3. Results

3.1.Identification of Secondhand Ship Prices

Identify primary and secondary data related to second hand ship prices with field surveys and market surveys. Ship price survey will be conducted by surveying the market through the internet on the websites of ship sales brokers both domestically and abroad. The brokerage websites accessed are http://www.boatsource.ca/, http://maritimesales.com/, http://www.ship-trader.com/, http://www.shipbroker.net/ (figure 1). From survey data covering price, ship type, the main size, GRT, engine, material and ship speed. Field surveys aim to observe, provide qualitative and quantitative assessments related to the condition, price, ship type, main size, GRT, engine, material and ship speed. Examples of market survey results and field surveys as follows (figure 1 and figure 2):

Figure 1. Example display of second hand ship brokerage sites offered in http://www.shipbroker.net/

Figure 2. Main engine (Example of field survey documentation)

3.2.Estimated Price of Second hand Ships

3.2.1.Second hand Ship Market Price Approach

The results of the market survey against several second hand ship sales broker sites on the internet then obtained ship data that has similar so that it can be used as a comparison of the price of secondhand ships. Some similarities in the ship's data include the year of manufacture, ship body type, type and number of payloads, type of construction material, propulsion system, and ship condition. The ship's data can be expressed on a ship's offer price chart which is a function of GRT or DWT or other dominant variables. The linear regression approach will be able to estimate the offer value of the secondhand ship in graphical form (figure 3), and that value will be reduced by 15% selling price factors and a 5% correction to estimate the secondhand ship price.

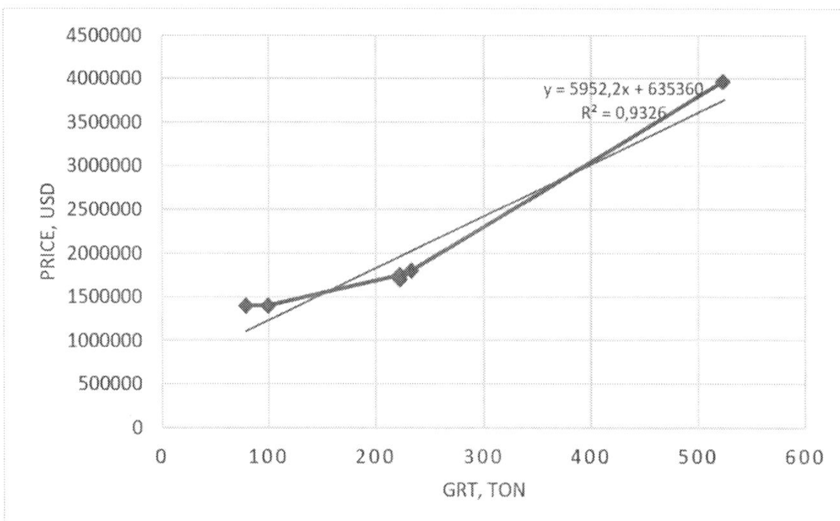

Figure 3. Example of linear graph of second hand ship price

3.2.2.Comparative Ship Approach

In this approach, the factor of the selection of comparative vessels is to determine the accuracy of the estimated price of the ship assessed. The specifications of the comparative ship should have similarities to the ship's estimated price in terms of ship type, main size, material, main engine and drive system. The estimated price of secondhand vessels using the price of the comparative ship, first calculating the price per GRT of the comparative ship, then the price per GRT is multiplied by the size of the used ship GRT, it will be obtained the estimated price of the secondhand ship.

The price of the secondhand ship needs to be corrected if there is a difference in the speed of service between the comparative ship and the secondhand ship, as this illustrates the difference in the performance of the ship's main engine.

The calculation is corrected if the difference in the year of manufacture between the comparative ship and the secondhand ship. Depreciation of the price due to the increasing age on the comparative ship can be calculated by assuming that the economic age of the ship is 35 years so that linearly the amount of depreciation per year can be calculated. So the final price of the secondhand ship after making correction due to the difference in speed and years of manufacture.

The difference in the number of cargo or passenger capacity between the comparative ship and the secondhand vessel needs to be taken into account as this concerns the number of safety equipment and passenger accommodation facilities. The cost of this difference can be estimated by first assuming the cost of providing safety equipment and passenger accommodation facilities taken 20% of the sale price of the comparison ship. These costs are divided by the passenger capacity of the comparative ship so that there are costs for safety facilities and accommodation per person. Furthermore, this cost is multiplied by the difference in passenger capacity between the two ships, and the depreciation value is obtained. Thus the final price of the secondhand ship is obtained after being corrected due to differences in passenger capacity and a reduced correction factor of 5%.

3.2.3. Ship Physical Pricing Approach

The estimated price of a new building can be explained as follows:

- Assumption : hull built year, hull shipyard, m/e built year, main engine;
- Principal particulars : length overall, length water line, breadth moulded, depth moulded, service speed

Example of shipbuilding cost calculation sees table 1.

To determine what the current value of the ship is is to calculate the amount of depreciation value which includes: depreciation due to the life of the ship, depreciation due to reduced ship function and depreciation due to the physical condition of the ship.

3.2.3.1. Depreciation due to ship life.

- It is assumed the economic lifespan of the ship's body is 35 years.
- The lifespan of the ship up to this time is n years.
- Depreciation of ship value per year is calculated linearly from the price of a new building in the year of construction until it reaches a value of 0 after the age of 35 years and the depreciation of the value per year of (1/35 x 100%)
- The depreciation value of the ship due to the life of the ship can be calculated with the following approach formula:

$$DO = 1/(1+k)^n \qquad (1)$$

where : k = depreciation per year
n = age of the ship in years

3.2.3.2. Depreciation due to reduced ship function

- Planning the function of the secondhand ship is the speed of service (knot), the size of the GRT, and the passenger capacity of n people.
- Used vessels experienced a decrease in service speed based on sea trial documents. Depreciation of values due to this decrease in speed can be calculated as follows:

$$DV = \frac{\text{Current Capacity}}{\text{Initial capacity}} \times \frac{\text{Current Speed}}{\text{Initial speed}} \qquad (2)$$

Table 1. Example of shipbuilding cost calculation

No	Item	Volume		Price Unit (USD)	Amount Unit (USD)
A	Hull				
1	Aluminum Plate	60	Ton	8,500	510,000
2	Aluminum Profile	20	Ton	10,000	200,000
3	Electrode, Oxygen, Acetylene, Bolt etc.	1	Shipset	90,000	90,000
4	Paint and cathodic protection	5000	Liter	10	50,000
5	Others	1	Shipset	10,000	10,000
	Sub Total A				860,000
B	Hull Fitting				
1	Hull Outfit & Deck Machinery	1	Shipset	285,000	285,000
2	Furniture, Air Conditioner, Window etc.	1	Shipset	450,000	450,000
3	Life Saving Appliances	1	Shipset	80,000	80,000
	Sub Total B				815,000
C	Machinery				
1	Main engine, Gearbox, Stern Arrangement etc.	2	Sets	582,000	1,164,000
2	Water Jet and electric controller	2	Sets	145,500	291,000
3	Aux. Engine and Generator incl. acc Standard maker	2	Shipset	98,300	196,600
4	Pump and others	1	Sets	23,800	23,800
5	Pipe, elbow, valve and cock	1	Sets	90,800	90,800
	Sub Total C				1,766,200
D	Electric Equipment				
1	Main switchboard and panel	1	Shipset	25,000	25,000
2	Transformer, Battery, Charger, Lighting etc.	1	Shipset	50,000	50,000
3	Communication, Navigation & Automatic Equip.	1	Shipset	120,000	120,000
4	Others	1	Shipset	10,000	10,000
	Sub Total D				205,000
E	Labor Cost				
1	Shop drawing / drafter	50,000	Man-hour	8	400,000
	Sub Total E				400,000
F	Third Parties Cost				
1	Certificate, classification and authority fee	1	Shipset	100,000	100,000
	Sub Total F				100,000
G	General & Service Expenses				
1	Design, Administration, Selling, Familiarization	1	Shipset	175,000	175,000
	Sub Total G				175,000
	Total Amount				4,321,200

3.2.3.3. Depreciation due to the physical condition of the ship

- The physical condition of the ship in question is the construction condition, main engine and auxiliary engine, propulsion system, a navigation system, electrical installation, safety equipment, deck equipment and accommodation equipment whether or not it is still functioning properly at the time of purchase.
- From the results of the field survey, it can be explained the physical condition of the ship (table 2).

Table 2. Example physical condition of second hand vessels

No	Item	Condition (%)
1.	Construction and shell of the ship	70
2.	Main engine and auxiliary machine	65
3.	Propulsion system	70
4.	Navigation system	70
5.	Electrical installation	70
6.	Safety equipment	70
7.	Deck fittings	70
8.	Accommodation supplies	75
	Average condition (DP)	70

The conclusion of the estimated price of the second hand ship at the time of purchase after depreciation of the price of the new building is the price of the new ship x DO x DV x DP. Based on three approaches, namely the approach of ship market data, comparative vessels and the physical price of the ship, and after being taken on average it can be concluded that the value of the second hand ship is USD.

3.3. The Compare Estimated Price of Second hand Ships

The comparison of ship price estimates aims to verify the multi methods that have resulted from the study. Comparison is done with the method of approaching the age of the ship and country of origin [2] and the data of calculation results from multi-method passenger ship type ferry catamaran in 2006. Based on table 3 presented the main size of the ship to be estimated, and in table 4 shows the percentage difference in the estimated price of second hand ships between the two methods by 24.92%. A significant difference indicates that the estimated price of used vessels requires thoroughness and accuracy and the ship is an order product whose specifications and complexity are very varied. So that multi-method for estimating the price of second hand ships consisting of world ship market data, comparative vessels and physical prices of ships, is quite comprehensive and holistic.

Table 3. Technical specifications

No	Item	Description
1	Year of Manufacture	1990
2	Material	Aluminium
3	Length (LOA)	34 m
4	Width (B)	8 m
5	Draught (D)	1,2 m
6	Service speed	34 knots
7	Maximum speed	38 knots
8	GRT	258 tons
9	Net Tonnage	89 tons
10	Main engine	MTU (2 unit)
11	Propulsion	Water jets (2 unit)
12	Number of passangers	200 orang

Table 4. The Compare Estimated Price of Second hand Ships

No	Estimation method	Price (USD)			
		Ship market data	Comparative vessels	Physical prices of ships	Results
1	Multi-method	1,769,566.6	1,650,085.31	1,587,702.2	1,669,118
2	Adji [2]				1,253,148
3	Price Difference				415,970
4	Price Difference (%)				24,92

4. Conclusion

Identification of second hand ship prices is carried out by field survey to ship object, internet market survey on ship sales broker sites in http://www.boatsource.ca/, http://maritimesales.com/, http://www.ship-trader.com/, http://www.shipbroker.net/. and the physical price approach of the ship is carried out by calculating the cost of building new ships minus various depreciations. Based on three methods (multi-method) namely approach of ship market data, comparative ship and physical price of the ship, then the estimated price or estimate of a second hand ship is taken the average value of the three methods.

Acknowledgments

Thank you to Hang Tuah University Surabaya and all parties who have helped with the publication activities in scientific forums.

Bibliography

[1] Tsolakis S, Cridland C, Haralambides H, 2003 Econometric Modelling of Secondhand Ship Prices *Marit Econ Logist* **5** pp 347–377. https://doi.org/10.1057/palgrave.mel.9100086

[2] Adji S W, 2004 *Industri Perkapalan Indonesia Menyongsong Masa Depan: Prospek Dunia Usaha Dan Potensi Pembiayaannya Oleh Perbankan* Workshop Bank Indonesia Jakarta 7 – 8 Juni Jurusan Teknik Sistem Perkapalan Fakultas Teknologi Kelautan Institut Teknologi Sepuluh Nopember Surabaya

[3] Wu W M, 2010 The Decomposition of Secondhand Ship Price: An application of Hedonic Regression Model *The 40th International Conference on Computers & Indutrial Engineering* Awaji pp 1-5 doi: 10.1109/ICCIE.2010.5668289

[4] Chang C C, Lai T C, 2011 Nonlinear model for Panamax Secondhand Ship *Applied Economics* 43:17, 2193-2198, doi: 10.1080/00036840903120670

[5] Ningdyah W K, Rusdiansyah A, Anityasari M, 2010 *Pemodelan Pengambilan Keputusan dalam Jaringan Reverse Logistics untuk Industri Pembongkaran Kapal Bekas* (Doctoral dissertation Thesis Jurusan Teknik Industri Institut Teknologi Sepuluh Nopember Surabaya).

[6] Prasetya H, 2014 *Perancangan Model Survei Kondisi Kapal Bekas Berbasis Online* (Doctoral Dissertation Institut Teknologi Sepuluh Nopember Surabaya)

[7] Supomo, H., & Baihaqi, I. 2019. Studi Pemodelan Harga Kapal Tanker Bekas Dengan Metode Statistik. *Kapal: Jurnal Ilmu Pengetahuan dan Teknologi Kelautan*, **16(**1), 23-32.

IOP Conf. Series: Materials Science and Engineering 1052 (2021) 012012 doi:10.1088/1757-899X/1052/1/012012

Human Error Assessment of Situation Awareness in Bridge Operations: A Case Study of Indonesian Maritime Accidents

L. P. Bowo[1], **R. E. Prilana**[1], **and M. Furusho**[1]

[1]Graduate School of Maritime Sciences, Kobe University, Kobe, Japan

Abstract. In the maritime industry, the role of human on-board operations is frequently cited in accidents report due to its relation to human error. Furthermore, according to the European Maritime Safety Agency, human error in bridge operations in the 2018 maritime accident cases due to lack of situational awareness has affected 65%. This study aims to assess the Indonesian maritime accidents concerning situational awareness in bridge operation in the collision, grounding, and sinking accidents by applying the new propose HRA method, Maritime Accidents Analysis and Reduction Technique (MAART). Finally, the forty-seven maritime accident data derived from the National Transportation and Safety Committee of Indonesia from 2008 to 2018 are assessed. In conclusion, Indonesian seafarers' situational awareness has to be given more concern because most of the analyzed accidents' causal factors show a lack of continuous monitoring, communication, and coordination on-board operation.

1. Introduction

Human and technology interactions in collaboration and interdependencies to achieve system purposes can be best described as complex socio-technical systems [1]. In the maritime industry, the role of human on-board operations is frequently cited in accidents report due to its relation to human error [2]. In recent technology development, the operator working on-board is the convenience with the introduction of automation. The automation helps seafarers process information and comprehend it before making critical decisions; however, it also unexpected consequences [3]. The convenience of automation frequently left the operator incapable of responding sufficiently to challenges posed by the system. This phenomenon is termed as "Out of the Loop" syndrome [4]. Situational Awareness (SA) definition, in general, is the operator's dynamic awareness of the progressing external situation [5]. The concept of SA was firstly brought by the aviation industry [6] and widely developed in the operation of a complex socio-technical system [7]. The maritime sector, the especially on-board process, is a complex socio-technical system consisting of procedures, technologies, and personnel. This dynamic environment may cause accidents [8]. It is proven by the research conducted by EMSA and Grech et al., that more than 60% of human errors can be assigned to SA-related matter [9][10].

In this study, Indonesia's maritime accidents with a strong relation to SA condition, such as collision, grounding, and sinking, will be analyzed. There are 47 vessels studied that occurred in the period from 2008 to 2018 to determine the role of human factors that might have been related to the SA in maritime accidents. The new development method for analyzing human factors in maritime accidents, named Maritime Accident Analysis and Reduction Technique (MAART), is applied. This methodology is based on Human Error Assessment, and Reduction Technique (HEART) developed and adapted to analyze the human error probability in the maritime domain, especially for maritime accident analysis.

This study aims to assess the Indonesian maritime accidents concerning situational awareness in bridge operation in the collision, grounding, and sinking accidents by applying the new propose HRA method, MAART method.

2. Situational Awareness in the maritime industry

The SA on-board operation is required not only for the vessel's crew but also for the pilot who helps the vessel's pilotage to the berth and vice versa. Sharma et al. explained three levels of information requirements for pilotage [11], [12]. Level one is constructing the initial framework, such as information about the vessel's status, equipment status, route, traffic, and weather conditions. The role of communication on the bridge is crucial. It is related to the information sharing about vessel conditions, equipment status, and the weather condition that will be faced during the voyage. This information has to be shared with all the seafarers and all crews that will work on-board, so all the personnel on-board have the same shared mental model to form an accurate picture. Level two comprehends the vessel and system states and learns the effect of traffic and weather compared to planned and expected conditions. In this level, route planning, traffic, and weather conditions are further discussed. It needs good communication and good coordination among seafarers to prepare the plan and mitigate the risk of the voyage.

Moreover, in level three is projecting the information available to different conditions. Here, continuous monitoring from seafarers is critical. Because to project the situation well, the seafarer must process the information received before the voyage and process it with the additional information that faces during the voyage. With these ways, seafarers can create an excellent judgment to create an action to mitigate the risk that might be occurred. Therefore, managing a go communication is an essential dimension that prevents it from the source of misinformation and creates an ambiguous manner that can affect the seafarers working conditions [13], [14]. With recent technological advances in the navigation deck, the technology also provides SA-related information for the seafarers; thus, man-machine interactions are essential things on-board operation [15]. Good planning enables seafarers to minimize the risk. Knowing the probabilities of every condition that might face during the voyage, the seafarers can create the mitigation plan and scenarios to avoid undesirable events.

The human factor is an essential element to keep the SA safe because the loss of SA on-board operation is one of the immediate causes of the maritime accident [16]. According to the EMSA annual overview of marine casualties and incidents report in 2019, 66% of maritime accidents from 2011 to 2018 were attributed to human actions, whereas 65% occurred in on-board operations [10]. They further breakdown the on-board operation working failures into other sub-factors of human activities, such as inadequate training programs, design errors, risk assessment, inefficient procedures, planning and coordination, lack of knowledge, ineffective work methods, and situational awareness. It turns out that inadequate work methods and situational awareness were the two sub-factors most reported causing maritime accidents [10]. Therefore, it is essential to research the SA condition of the seafarers' on-board operation, especially for the maritime accidents that are related to the SA, such as collision, grounding, and sinking.

3. Data and methodology

3.1. Data collection

There are two ways to gather the data that can be used for analyzing the accidents, first doing the investigation directly to the related parties in the accidents by conducting the interview, known as the primary data source. The data reports prepared by an official investigation board called the secondary data source. This secondary data source is reliable because it is created from primary data sources and analyzing first-hand information obtained by the accident investigator [17], [18]. Many kinds of research used the accident reports as the data sources for investigating the accidents in a particular maritime accident [19]–[21].

In this study, the maritime accident data reports were collected from the National Transportation and Safety Committee (NTSC) website. The data report that contained only from 2008 to 2018. Three maritime accident types stare strongly related to SA of the seafarers on-board that analyzed is study, collision, grounding, and sinking accidents. In the collision accidents, the vessels involved in the accident were analyzed separately. Therefore the total number of vessels is more than the total number of cases.

Table 1. Maritime Accidents Data Reports and Vessels Number.

No.	Accidents Type	Cases	Vessels
1.	Collision	14	24
2.	Grounding	4	4
3.	Sinking	19	19
	Total	**37 cases**	**47 vessels**

The accident reports that were thoroughly analyzed for this study were synopses, general information, analysis sections, and the conclusions. To minimize the subjectivity, the reviewers extracted the embedded information based only on the words that were written in the reports, avoiding further investigation and assumptions that could create subjective opinions. All the data were summarized and analyzed by using Microsoft Excel. The reports were all reviewed by researchers who are experts in the field of human factors.

3.2. MAART methodology

Maritime Accident Analysis and Reduction Technique (MAART) method is a methodological extension of the Human Error Assessment and Reduction Technique (HEART) as a qualitative and quantitative approach for human error probability calculation in maritime accident analysis which was introduced by[22]. In the MAART technique, the thirty-eight EPC that has been established by William [23] has been categorized into man, machine, media, and management factors because the thirty-eight EPC was still general, and every EPC are related to factors other than human. This categorization aims to make a better understanding that influences seafarers' conditions that lead to accidents. Furthermore, the MAART method also provides the quantification process to determine the value of the Affect Proportion Effect (APE) by utilizing the TOPSIS to minimize APE value's subjectivity.

Table 2 shows the generic task used in this study to determine the working condition on-board operation at the moment of the accidents. There are nine GT to describe every task on-board from the unexpected action due to unfamiliarity with the situation which has the highest nominal human unreliability (NHU) to activity that responds to the system correctly even when there is an augmented or automated supervisory system providing an accurate interpretation of the system stage which has the lowest NHU. The assessor will select one of the GT that describes the working condition in the accident report. This selection also strongly relates to the weather condition. If the working conditions are the same in the two or more reports, but the weather conditions are different, it can influence GT selection. The more severe the weather condition, the GT that selected will have a higher value of NHU.

In the generic task, it can differ as a challenging task and convenience task. The challenging tasks consist of generic tasks A, B, and C, which in the description in Table 2, the seafarers do the tasks that required a high level of comprehension and skill, do the job without supervision or procedures from the shipping company to facilitate the seafarers. Worse, the seafarer has to face the task which they do knowing how to handle the task. Those conditions required good handling from the seafarer, which had been taught in the training and education process before working on-board. Due to the high importance of seafarers ability to encounter challenging conditions on-board, the failure to do so will cause a high probability of the accident occurrence.

Besides those three generic tasks, others are considered the convenience task, which the seafarers routinely do. Still, due to the lack of focus and attention, the task was a failure. Moreover, if the condition was responded correctly and a familiar one, the accidents still can occur due to external factors. Nevertheless, the probability of this occurrence is lower. All of these conditions are explained and differed briefly in Table 2.

Table 2. Generic Tasks (GT).

Code	Type of work	Condition
A	Totally unfamiliar	Performing the work at speed with no real idea of likely consequences.
B	Restore the system to an original state on a single attempt	Doing work without supervision or procedures.
C	Complex task	It requires a high level of comprehension and skill.
D	A fairly simple task	Performing the work rapidly or given scant attention.
E	The routine, highly practiced, rapid task	Involving a relatively low level of skill.
F	Restore a system to original	An error occurred even though following procedures with some verification.
G	Entirely familiar, highly practiced, routine task occurring several times per hour, performed to the highest possible standards by a highly motivated, highly trained, and experienced person, totally aware of implications of failure, with time to correct the potential error	Without the benefit of significant job aids.
H	Respond correctly to the system command	Even when there is an augmented or automated supervisory system providing an accurate interpretation of the system stage.
M	Miscellaneous task for which no description can be found.	

Figure 1 shows the range of NHU value of every GT that has been defined before by the William in the HEART method first development. There are the minimum limitation which stated by the blue line, the maximum value stated by the green line, and the average value stated by the red line. In this study, the average value is used for the calculation process.

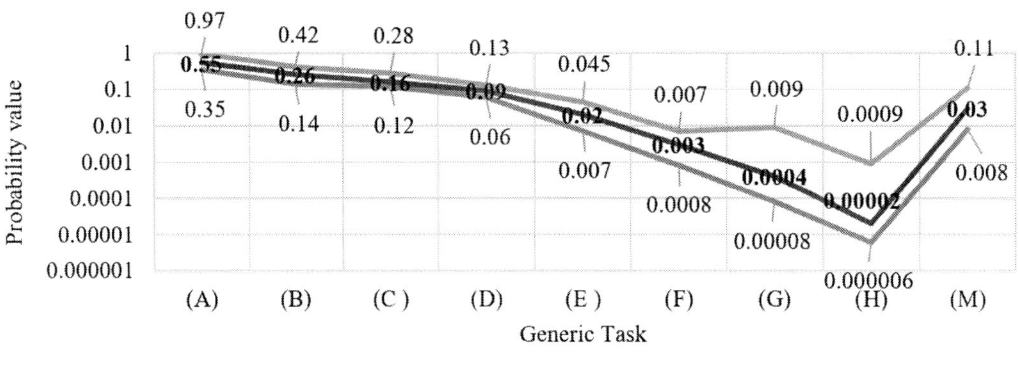

Figure 1. The value of NHU.

After the selection of GT and determine the value of NHU, the next procedure is selected as the EPC – 4M. There are thirty-eight EPC that have been categorized into 4M factors (man, machine, media, and management) [18], [24] as shown in Table 3. Four main factors categorize into five sub-factors for the man and four sub-factors for management factors. In the table also shown the multiplication factors for every EPC – 4M that will be used for the HEP calculation as shown in the formula (14).

The SA condition on-board operation can be described well with the EPC – 4M. Many factors can influence the SA condition on-board operation. Machinery problems, weather conditions, and important ones are human conditions. This human condition can differ as to the human element and human to human interactions, which can be mentioned as management factors. All EPC – 4M in communication, coordination, and monitoring subfactors are strongly related to the SA. As mentioned in the previous section, there are three preparation levels for the voyage to maintain good SA [11], [12]. The first level is constructing the initial framework. The EPC – 4M in the communication sub-factors can describe the first level of SA well. In the second level, the seafarers comprehend all the information received on the first level and create the best plan to safely and successfully achieve the goal. The EPC – 4M in the communication sub-factors and coordination sub-factors can represent the second level's causal factors. The third level of SA is projecting all the information in the first and second level. Communications, coordination, and monitoring are essential factors for keeping the SA effective.

Table 3. Error Producing Conditions – 4M (EPC – 4M).

Man Factors		×	**Media Factors**		×
Physical limitations			EPC 33	Poor environment	1.15
EPC 27	Physical capabilities	1.4	**Management Factors**		×
EPC 36	Task pacing	1.06	*Communication*		
EPC 38	Age	1.02	EPC 13	Poor feedback	4
Psychological limitations			EPC 14	Delayed/incomplete feedback	3
EPC21	Dangerous incentives	2	EPC 16	Impoverished information	3
EPC28	Low meaning	1.4	EPC 18	Objectives conflict	2.5
EPC 29	Emotional stress	1.3	EPC 19	No diversity of information	2.5
EPC 31	Low morale	1.2	*Coordination*		
EPC 34	Low mental workload	1.1	EPC 2	Time shortage	11
Experience			EPC 6	Model mismatch	8
EPC 1	Unfamiliarity	17	EPC 10	Knowledge transfer	5.5
EPC 12	Misperception of risk	4	EPC 24	Absolute judgments required	1.6
EPC 22	Lack of experience	1.8	EPC 25	Unclear allocation of function	1.6
Skill and Knowledge			EPC 37	Supernumeraries/ lack of human resources	1.03
EPC 7	Irreversibility	8	*Monitoring*		
EPC 9	Technique unlearning	6	EPC 17	Inadequate Checking	3
EPC 11	Performance ambiguity	5	EPC 26	Progress tracking lack	1.4
EPC 15	Operator inexperience	3	*Procedures*		
EPC 20	Educational mismatch	2	EPC 4	Features over-ride allowed	9
Health			EPC 5	Spatial and functional incompatibility	8
EPC 30	Ill-health	1.2	EPC 32	Inconsistency of displays	1.2
EPC 35	Sleep cycles disruption	1.1			
Machine Factors					
EPC 3	Low signal-noise ratio	10			
EPC 8	Channel overload	6			
EPC 23	Unreliable instruments	1.6			

Furthermore, to obtain the APE weight for every selected EPC – 4M it applied the TOPSIS calculation in the MAART method. This calculation tool is utilized to minimize the subjectivity in obtaining the APE weight. The Saaty's 1–9 linguistic relative importance scale is used [25], as shown in Table 4. These importance scales will be used by the accidents accessor to determine the importance among the selected EPC, to know about the interdependencies among it.

Table 4. Saaty's pair-wise comparison scale.

Importance scale	Definition	Importance scale	Definition
1	Equal importance	7	Extreme importance
3	Moderate importance	9	Absolute extreme importance
5	Strong importance	2, 4, 6, 8	Intermediate values

1. A pair-wise comparison matrix (**D**) can be established following Formula (1). In the formula, x_{ij} (i = 1, 2, ..., m, j = 1, 2, ..., n) has the relative importance of the *i*th element compared to the *j*th. In this study, every selected EPC will be compared to the other selected EPCs to determine the interdependencies of EPCs. By comparing these EPCs, it can be observed that every EPC is related to each other, and there will be a tendency for an EPC to be a major factor in an accident.

$$D = \begin{bmatrix} x_{11} & x_{12} & \cdots & x_{1n} \\ x_{21} & x_{22} & \cdots & x_{2n} \\ \vdots & \vdots & \ddots & \vdots \\ x_{m1} & x_{m2} & \cdots & x_{mn} \end{bmatrix} \tag{1}$$

$$x_{ii} = 1, \quad x_{ij} = 1/x_{ji}, \quad x_{ji} \neq 0$$

2. The normalized decision matrix is constructed and weighted.

a. Normalized decision matrix

To construct the normalized decision matrix, first, the attribute weight (w_i) for each EPCi must be obtained by utilizing Formula (2).

$$w_i = \sqrt{\sum_{i=1}^{m} x_{ij}^2} \tag{2}$$

After obtaining the attribute weight, the normalized decision matrix (r_{ij}) is constructed by dividing the value from the pair-wise comparison matrix to the attribute weight, as shown in Formula (3).

$$r_{ij} = \frac{x_{ij}}{w_i} \tag{3}$$

b. Weighted normalized decision matrix.

$$p_{ij} = r_{ij} \times x_{ij} \tag{4}$$

3. The ideal and negative ideal solutions are determined.

a. Ideal solution

$$d_{ij}^{+} = (p_{ij} - p_{i\,max})^2 \tag{5}$$

b. Negative ideal solution

$$d_{ij}^{-} = (p_{ij} - p_{i\,min})^2 \tag{6}$$

4. The separation from the ideal solution is determined.

$$d_i^{+} = \sqrt{\sum_{j=1}^{n} (d_{ij}^{+})^2} \tag{7}$$

5. The separation from the negative ideal solution is determined.

$$d_i^{-} = \sqrt{\sum_{j=1}^{n} (d_{ij}^{-})^2} \tag{8}$$

6. Relative closeness to the ideal solution.

$$\xi_i = \frac{d_i^{-}}{d_i^{+} + d_i^{-}} \tag{9}$$

7. Normalization.

The summation of all the EPC ideal solution values is not one; it is often more than one and sometimes even less than 1. Thus, it needs to be normalized before using this value for the HEP calculation. The last value used in the HEP calculation is the normalization value (N) to be the APE weight. This value shows which EPC has the highest value of the importance, which implicates this is the main factor of the accident because its particular EPC is the most important compared with other EPCs. If the weight is approved, then it can be used for the HEP calculation. Therefore, in this study, EPC's highest value was named the top of the EPC series. Formula (10) shows the calculation formula for the normalization value.

$$\mathrm{N} = \frac{\xi_i}{\Sigma \xi} \tag{10}$$

8. Consistency verification

The next step proves the consistency of the data. This step verifies whether the comparison pair-wise matrix is consistent or not. The consistency index (CI) can be calculated using the following formula:

$$\sum_{j=1}^{n} x_{ij} N = \lambda_{max} N_i \tag{11}$$

$$CI = \frac{\lambda_{max} - n}{n - 1} \tag{12}$$

A consistency verification calculation is needed to specify a reasonable consistency. The consistency ratio (CR) value was ≤ 0.10. Otherwise, expert judges will be revised to obtain consistent results.

$$CR = \frac{CI}{RI} \tag{13}$$

In the equation, RI stands for a random index. It is subjected to several items that are compared in the matrix. The RI values are provided in Table 5. The RI values will be used to calculate the CR value. This RI value shows the number of selected EPC in a case. The number of selected EPC in every case is varied, so the RI in every case is different.

Table 5. Random index values [26].

n	1	2	3	4	5	6	7	8	9	10
RI	0	0	0.58	0.90	1.12	1.24	1.32	1.41	1.45	1.49

The formula for HEP calculation is shown in the formula (14). The HEP calculation components in the MAART method are NHU, which can be obtained when selecting the generic task, the multiplication value in the EPC, and APE's weight.

$$HEP = NHU \times \{\textstyle\prod_i [(EPC_i - 1) \times APE_i + 1]\} \tag{14}$$

4. Results

The maritime accidents (collision, grounding, and sinking) in Indonesia will be different by three explanations. The working type of accidents, the EPC – 4M that involve, and the EPC series before accidents.

4.1. Working type

In these results of working type, it explains about the navigation condition and situation before accidents. There are five GT types obtained from all 47 vessels that were analyzed in this study. Table 6 shows GT's result for every kind of accident and stated in the number of cases. A detailed explanation of every selected GT can be found in the previous section. The GT for every type of accidents show a different result. In collision accidents, the accidents occurred in the C, D, E working type, where mostly it occurred in the E generic task, which occurs in the routine task.

Furthermore, only 4 cases occurred in the C working type, where it is stated as a complex task that requires a high level of skill and competency. In grounding accidents, the results for four cases analyzed are even in the four types of GT as follow, generic task A, C, D, and E. Moreover, the results for the sinking accidents is also different from the other two accidents. There are five types of generic tasks discovered in the sinking accidents: generic tasks A, B, C, D, and E. The most occurrence sinking accidents are found when the seafarers on-board faced a complex task that required a high level of skills.

In contrast, it also occurred in a fairly simple task but performing the work rapidly or given scant attention. From all cases, the highest percentage for working type condition that the accidents occurring is the routine, highly practiced, rapid task involving low level of skill. There are 22 vessels in the Indonesian maritime accidents that occurred due to GT E, following by generic task C, D, A and B.

Table 6. GT obtained in Indonesian maritime accidents.

Accidents	A	B	C	D	E	Total
Collision	0	0	4	2	18	24
Grounding	1	0	1	1	1	4
Sinking	2	2	6	6	3	19
All	3	2	11	9	22	47

4.2. Situational Awareness Factors

Table 7 shows the EPC – 4M found in 47 vessels analyzed by using the MAART method. In total, there are 215 EPC – 4M found, with the distribution as follows collision accidents have 104 EPC – 4M, grounding has 20 EPC – 4M, and sinking accidents has 91 EPC – 4M. In the table, the information is written from the most occurrence factors in all accidents. Besides, there are thirty-one EPC – 4M out of thirty-eight EPC – 4M, which distribution as follows, fourteen EPC – 4M in the management factors, fourteen EPC – 4M in man factors, two machine factors, and media factor. Every accident analyzed has different factors and differs from which EPC – 4M occurred in the accidents.

There are twenty-three of thirty-eight EPC-4M found for twenty-four vessels analyzed in the collision accident, where the most common factors are management factors in the monitoring sub-factors. EPC 17 and EPC 26 of EPC – 4M in the monitoring factors are the most common collision accidents. The machinery problem is not a factor causing collision accidents. There are several factors that only found in the collision accidents, such as EPC 13 for poor feedback, EPC 37 for lack of human resources, EPC 29 for emotional stress and EPC 35 for ill health. Furthermore, EPC 5, EPC 15, EPC 6, EPC 14, EPC 25, EPC 8, and EPC 31 were not found in the collision accidents unlike in the grounding and sinking accidents. Moreover, only one case has poor environment as the factor for accidents occurrence in the collision accidents. It means that the rest of the accidents occurred during fine day and good visibility.

There are fourteen of thirty-eight EPC-4M found in the four analyzed cases of grounding accidents in Indonesia. Moreover, management problem in total has the highest cases comparing with other factors. Lack of monitoring, poor coordination, and inadequate communication during working on board operation, causing the grounding accidents in Indonesia. Unlike with the collision accidents, poor environment in the grounding accidents is causing 50% of the total cases.

In sinking accidents, the management factors still dominated the causing factors of the accidents. Monitoring subfactors has the highest number of causal factors found in the cases, following by the inadequate communication and poor coordination. Unlike other maritime accidents, sinking accidents has the highest number of machine factors. Almost 50% of the sinking accidents were caused by the unreliable instrument and machinery on board. Furthermore, the ability of the seafarer for skill and knowledge in the sinking accidents were poorer than other accidents.

IOP Conf. Series: Materials Science and Engineering 1052 (2021) 012012 doi:10.1088/1757-899X/1052/1/012012

Table 7. EPC – 4M factors found in the analysis.

4M - Subfactors	EPC	Collision	Grounding	Sinking	Total
Management - Monitoring	EPC 17	14	2	7	23
Management - Monitoring	EPC 26	12	-	9	21
Machine	EPC 23	5	3	9	17
Man - Experience	EPC 12	9	1	3	13
Man - Experience	EPC 22	6	1	6	13
Management - Communication	EPC 13	9	-	-	9
Management - Communication	EPC 16	3	1	5	9
Management - Procedures	EPC 5	-	2	6	8
Man - Skill and knowledge	EPC 11	6	-	2	8
Man - Physical limitations	EPC 21	4	1	3	8
Management - Coordination	EPC 10	4	2	1	7
Man - Skill and knowledge	EPC 20	4	-	3	7
Man - Skill and knowledge	EPC 9	1	1	4	6
Man - Skill and knowledge	EPC 15	-	1	5	6
Management - Coordination	EPC 24	1	1	4	6
Media	EPC 33	1	2	3	6
Man - Physical limitations	EPC 34	5	-	1	6
Management - Coordination	EPC 2	4	-	1	5
Management - Communication	EPC 18	2	-	3	5
Management - Communication	EPC 19	1	1	3	5
Management - Coordination	EPC 37	5	-	-	5
Man - Experience	EPC 1	1	-	2	3
Management - Coordination	EPC 6	-	-	3	3
Management - Communication	EPC 14	-	-	3	3
Man - Psychological limitations	EPC 28	2	-	1	3
Man - Physical limitations	EPC 36	2	1	-	3
Management - Coordination	EPC 25	-	-	2	2
Man - Psychological limitations	EPC 29	2	-	-	2
Machine	EPC 8	-	-	1	1
Man - Psychological limitations	EPC 31	-	-	1	1
Man - Health	EPC 35	1	-	-	1
Total		**104**	**20**	**91**	**215**

4.3. Human Error Probability

Table 8 presents the calculation process of the HEP result for collision accident case number one. The GT that was selected for the condition before the accident is a complex task that requires a high level of comprehension and skill, which has an NHU of 0.16. The reason why this case has generic task C is because it enters the Madura strait without guidance from a local tug boat, whereas in the regulations, this ship must use a tug service to enter the Madura Strait shipping channel. Table 8 presents the EPC series in case one, which has EPC 21 as the top of EPC series, followed by EPC 12, EPC 11, EPC 29, and EPC 1.

Furthermore, in Table 8 it shows the multiplication factors that belongs to every EPC – 4M, the detail of this information can be derived in the previous chapter. Moreover, the value APE of every EPC – 4M is determined by using the TOPSIS calculation process, by determining the interdependencies among selected EPC – 4M factors by using the Saaty's importance scale and using all the calculation formula in the TOPSIS. The value of the APE weight calculation can be processed further if it passed the requirements, Consistency Ratio (CR) is below or equal to 0.1. if the number is exceed 0.1, then the pair-wise comparison matrix has to be re-analyze again. In the example below, the number of CR is 0.061, since the CR value is < 0.1, so the value can be used and it is stated in the table below.

Table 8. HEP Calculation.

TOP		BODY							
EPC 21		EPC 12		EPC 11		EPC 29		EPC 1	
×	APE	×	APE	×	APE	×	APE	×	APE
2	0.47	4	0.32	5	0.11	1.3	0.09	17	0.003
HEP Value						0.71			

In Table 9, there is the information about the year and date of the collision accident occurrences. As mentioned in the data section, two vessels involve several cases of collision accidents. It is stated with the same case number. The EPC series consists of TOP EPC, which has the highest weight of APE and followed by the rest of EPC – 4M, respectively. Th number of EPC – 4M in every case is different, it is because of the condition and complexity of information that provide in the case itself. The average of the HEP value for collision is 18%.

Table 9. Collision's HEP Value.

No.	Year	Date	Top EPC	EPC	EPC	EPC	EPC	EPC	EPC	EPC	HEP
1	2009	22-May	21	12	11	29	1				0.71
2	2010	19-May	13	17	26	2	11	37			0.14
2	2010	19-May	12	20	17	2					0.14
3	2010	2-Jun	29	17	26	24	28				0.04
3	2010	2-Jun	13	10	26	2	17	22			0.2
4	2010	4-Aug	20	22							0.04
4	2010	4-Aug	12	17	28	20	11	18	26	34	0.09
5	2011	18-Mar	16	13	12						0.1
5	2011	18-Mar	22	17	12	20	37	34	11	2	0.06
6	2011	26-Sep	9	22	12	33					0.52
6	2011	26-Sep	17	22	26	37	34				0.21
7	2012	26-Sep	11	13	12						0.16
7	2012	26-Sep	26	13	21	10	11				0.06
8	2012	11-Dec	34	10	21	19					0.57
9	2013	31-May	26	12	16	23					0.07
9	2013	31-May	36	26	12	17					0.04
10	2014	1-Apr	17	18	21	22	23	10	16		0.56
11	2016	19-Nov	26	17	36	23	37				0.29
11	2016	19-Nov	26	23	17						0.03
12	2017	7-Apr	26	13	17						0.06
12	2017	7-Apr	26	13	17						0.06
13	2018	22-May	37	35	34						0.02
14	2018	19-Jul	17	13							0.09
14	2018	19-Jul	13	23							0.07

Table 10 is the EPC series and HEP value for four cases of grounding accidents. The average of the HEP value for grounding is 48%. The number of EPC – 4M selected in every case varies; it depends on the cases' complexities. The case number 4 in the grounding accidents has the HEP value 1. It means that it definitely because of human factors. This value can occur if the selected generic task has a higher NHU value, which is had by the challenging task and needs a high level of comprehension and skill to do the task.

Table 10. Grounding's HEP Value.

No.	Year	Date	Top EPC	EPC	EPC	EPC	EPC	EPC	EPC	EPC	HEP
1	2016	22-Dec	19	16	10						0.07
2	2017	12-Jun	17	23	5	33					0.74
3	2018	20-Feb	24	5	36	12	10	17	22	23	0.63
4	2018	10-Aug	21	15	9	33	23				1

Table 11 shows the EPC series and HEP value of sinking accidents in Indonesia from 2008 to 2018. The average of the HEP value for sinking is 43%. In the sinking accidents EPC series, in some cases, it shows that the main factors of the accidents are not because of the lack of communication, monitoring, and coordination but machinery problems. Furthermore, machinery problems are also found in other sinking cases as one factor that leads to accidents.

Table 11. Sinking's HEP Value.

No.	Year	Date	Top EPC	EPC	EPC	EPC	EPC	EPC	EPC	EPC	HEP
1	2008	17 May	22	1							1
2	2009	11-Jan	15	20	22	2					1
3	2009	22-Nov	8	9	5	16	15	26	18	31	1
4	2010	6-Mar	22	16	10	14	15	5			0.12
5	2011	27-Aug	16	12	23	24					0.37
6	2013	24-Dec	23	24	19						0.14
7	2013	3-Jul	17	16	19	26	1				0.94
8	2014	26-Aug	22	14	26						0.22
9	2014	3-Jan	17	26	14	11	24				0.91
10	2016	14-Oct	26	12	17	23					0.27
11	2016	4-Mar	6	22	26	9	19	21	17	11	0.71
12	2016	13-Dec	17	21	6	25	26	18			0.13
13	2016	29-Dec	16	23	22	5	33				0.71
14	2017	20-Mar	28	17	23						0.38
15	2017	6-May	23	26	33						0.14
16	2017	17-Sep	21	23	34						0.31
17	2018	3-Jan	26	12	9	24	6	5			1
18	2018	27-Jan	9	15	20	23	25	5			0.12
19	2018	18-Jun	17	20	18	5	15	23	33		0.99

The HEP value of collision accidents is lower than the grounding and sinking accidents. The generic task selected for collision accidents is mostly E generic, which has lower NHU value than other selected generic tasks.

5. Discussion and Considerations

In Indonesia, maritime transportation plays a vital role in Indonesia's social-economic situation as an archipelagic country. However, the safety of maritime transportation in Indonesia is needed to be improved in specific ways. The analysis results show that most of Indonesia's accidents were routine jobs requiring low-level high skill. In the generic task that belongs to MAART, nine GT describes the working condition on-board, where three out of nine are considered the difficult task under challenging situations. The three GT includes in this explanation are GT A, B, and C. The working type is working in terrible weather and traffic condition and terrible bridge team condition on sailing.

Furthermore, the rest of the GT is considered a convenient working type because it occurred in the routine task and only needs a low level of skill. The seafarers have much training in the situations. However, in this study, most of the maritime accidents analyzed in this study show that accidents occurred in the convenience condition, which they should control adequately. Based on EPC – 4M, the required information regarding the causal factors affecting Indonesian maritime accidents can be found, especially the SA condition factors. Factors that lead the accidents in Indonesia are management, man, and machine problems. In the management factors, lack of monitoring, communication, and coordination while working on-board are the traits and increase the probability of accident occurrences. Furthermore, most of the accidents occurred in a conducive weather condition. It means that mostly the accidents occurred due to a lack of maintaining the progress voyage.

As explained by Endsley [12], the three levels of SA requirements on-board, according to the EPC – 4M factors, every level of it can be represented by the EP – 4M sub-factors. In the first level, where all the information has to be collected and shared with the rest of the crew on-board to have a shared mental model, communication is essential. For instance, the information about upcoming weather conditions during the voyage must be checked well before departure. The seafarers can plan the mitigation to avoid or minimize the bad weather's impact on the vessel. In the second level of SA condition, comprehends the vessel and system states and learns the effect of traffic and weather on planned and expected conditions requires good communication and good coordination. For this level, the seafarers on-board, whether it is the bridge team or the bridge team, and pilot, have actively checked the actual traffic by using all available means to prevent the accidents. The seafarers rely not only on the plan themselves but also on the actual situation that might sometimes be different. Active and effective communications while monitoring the voyage are essential to inform about the traffic situations and the vessel conditions; they help all the crew on the bridge coordinate well.

Furthermore, in the third level, projecting the information available to different conditions, continuous monitoring tasks, and coordination is crucial. Most of the collisions are occurred due to the failure to continuously monitoring the voyage, mainly when they sailed in the good weather. They are lulled by the conditions they think are safe, so they relax their attention to continue to monitor the situation. It also found out that the machinery problems had taken place in most of the accidents. From forty-seven vessels analyzed, seventeen vessels with unreliable machinery conditions, which occurred mostly in the sinking accidents, even became the main factor of the sinking accidents. The shipping company and the seafarers on-board have to give more concern to the maintenance schedule and its quality.

The MAART method is a proposed method in the extension of the HEART methodology that has been established by William in the 1980s. The HEART methodology is a robust method that helps assessors analyze the accidents based on qualitative and quantitative processes in a single method. However, due to its first development in the nuclear power plant analysis, it needs some adjustments to analyze other industries, especially the maritime industry. The application of MAART in Indonesian maritime accidents suggests the relation between EPC and the 4M factors, making a better understanding of which related factors cause accidents.

The difficulty in conducting this study is that the availability of maritime accident reports that can be accessed openly from the NTSC official website is limited. The larger number of reports that can be accessed will contribute to the more significant validation of the results. Regardless, these are the implications that can be drawn from these findings. First, the information about SA derived through to EPC – 4M can improve seafarer's workplace habits and shipping company procedures for the operation and machine maintenance. Second, the MAART method can be utilized well to analyze maritime accidents. Furthermore, it can provide information regarding the accidents' causal factors and the human error probability value. This value can be based on the authorities to set the mitigation action and judge for better policy. Understanding the maritime accidents' causal factors can derive more information that the maritime stakeholders can learn from the accident analysis [11].

6. Conclusion

The study explored Indonesian maritime accidents involving the SA theory and the seafarers working type in the accident's occurrences. This study aims to assess the Indonesian maritime accidents concerning situational awareness in bridge operation in the collision, grounding, and sinking accidents by applying the new propose HRA method, MAART method. In conclusion, Indonesian maritime society has to give more concern to seafarers' quality on-board by giving them more training to encounter the undesirable conditions and the quality of machinery, whether it is the main engine or the supporting machinery. Furthermore, the application of the MAART method to analyze maritime accidents concerning SA can perform well. The EPC – 4M that consists of the MAART method can breakdown the causal factors in the maritime accidents well and explain the categorization of the EPC to 4M and its sub-factors.

References

[1] G. H. Walker, N. A. Stanton, P. M. Salmon, and D. P. Jenkins, "A review of socio-technical systems theory: a classic concept for new command and control paradigms," *Theor. Issues Ergon. Sci.*, vol. 9, no. 6, pp. 479–499, Nov. 2008.

[2] M. R. Grech, T. Horberry, and A. Smith, "Human Error In Maritime Operations: Analyses Of Accident Reports Using The Leximancer Tool," in *Proceedings Of The Human Factors And Ergonomics Society 46th Annual Meeting*, 2002, vol. 46, no. 19, pp. 1718–1722.

[3] S. Nazir, A. Kluge, and D. Manca, "Automation in Process Industry: Cure or Curse? How can Training Improve Operator's Performance," in *24 European Symposium on Computer Aided Process Engineering*, vol. 33, J. J. Klemeš, P. S. Varbanov, and P. Y. B. T.-C. A. C. E. Liew, Eds. Elsevier, 2014, pp. 889–894.

[4] M. R. Endsley and E. O. Kiris, "The Out-of-the-Loop Performance Problem and Level of Control in Automation," *Hum. Factors*, vol. 37, no. 2, pp. 381–394, Jun. 1995.

[5] P. M. Salmon, N. A. Stanton, and D. P., *Distributed Situation Awareness: Theory, Measurement and Application to Teamwork*. Ashgate, Surrey., 2009.

[6] M. R. Endsley, "Toward a Theory of Situation Awareness in Dynamic Systems," *Hum. Factors*, vol. 37, no. 1, pp. 32–64, Mar. 1995.

[7] A. J. Masys, "A systemic perspective of situation awareness: An analysis of the 2002 mid-air collision over Überlingen, Germany," *Disaster Prev. Manag. An Int. J.*, vol. 14, no. 4, pp. 548–557, 2005.

[8] F. Sanfilippo, "A multi-sensor fusion framework for improving situational awareness in demanding maritime training," *Reliab. Eng. Syst. Saf.*, vol. 161, no. December 2016, pp. 12–24, 2017.

[9] M. R. Grech, T. Horberry, and A. Smith, "Human Error In Maritime Operations: Analyses Of Accident Reports Using The Leximancer Tool," in *Proceedings Of The Human Factors And Ergonomics Society 46th Annual Meeting -- 2002*, 2002, vol. 19, pp. 1718–1722.

[10] EMSA, "Annual Overview of Marine Casualties and Incidents 2019," 2019.

[11] A. Sharma, S. Nazir, and J. Ernstsen, "Situation awareness information requirements for maritime navigation: A goal directed task analysis," *Saf. Sci.*, vol. 120, no. October 2018, pp. 745–752, 2019.

[12] M. R. Endsley, "Situation awareness global assessment technique (SAGAT)," in *Proceedings of the IEEE 1988 National Aerospace and Electronics Conference*, 1988, pp. 789–795 vol.3.

[13] B. G. Kanki, "Chapter 4 - Communication and Crew Resource Management," B. G. Kanki, R. L. Helmreich, and J. B. T.-C. R. M. (Second E. Anca, Eds. San Diego: Academic Press, 2010, pp. 111–145.

[14] H. Sandhåland, H. Oltedal, and J. Eid, "Situation awareness in bridge operations - A study of collisions between attendant vessels and offshore facilities in the North Sea," *Saf. Sci.*, vol. 79, pp. 277–285, 2015.

[15] M. . Endsley, *Designing for Situation Awareness*. Boca Raton, Fla, US: CRC Press, 2012.

[16] C. Hetherington, R. Flin, and K. Mearns, "Safety in shipping: The human element," *J. Safety Res.*, vol. 37, no. 4, pp. 401–411, 2006.

[17] F. H. Hawkins, *Human Factors in Flight*. Aldershot, UK: Gower Technical Press, 1987.

[18] L. P. Bowo and M. Furusho, "Usability of Human Error Assessment and Reduction Technique with a 4M framework (HEART–4M) – A Case Study on Ship Grounding Accidents," *J. ETA*

Marit. Sci., vol. 7, no. October, pp. 266–279, 2019.
[19] A. Mazaheri, J. Montewka, J. Nisula, and P. Kujala, "Usability of accident and incident reports for evidence-based risk modeling – A case study on ship grounding reports," *Saf. Sci.*, vol. 76, pp. 202–214, 2015.
[20] W. Mutmainnah, L. P. Bowo, A. B. Sulistiyono, and M. Furusho, "Causative Chains that Leads to Ship Collisions in Japanese Maritime Traffic System (MTS) as Final Outcome of MOP Model," *Appl. Mech. Mater.*, vol. 874, pp. 221–227, 2018.
[21] L. P. Bowo and M. Furusho, "A Comparison of the Common Causes of Maritime Accidents in Canada, Indonesia, Japan, Australia, and England," vol. 589, pp. 256–267, 2018.
[22] L. P. Bowo, R. E. Prilana, and M. Furusho, "A Modified HEART – 4M Method with TOPSIS for Analyzing Indonesia Collision Accidents," vol. 14, no. 3, 2020.
[23] J. C. Williams, "HEART – a proposed method for assessing and reducing human error," in *9th Advances in Reliability Technology Symposium, University of Bradford*, 1986.
[24] L. P. Bowo, R. E. Prilana, and M. Furusho, "A Hybrid Methodology for Maritime Accident Analysis: The Case of Ship Collision." Jun-2019.
[25] T. L. Saaty, "Axiomatization of the Analytic Hierarchy Process." pp. 91–108, 1985.
[26] T. L. Saaty, "How to make a decision: the analytic hierarchy process," *Interfaces (Providence).*, vol. 26, no. 4, pp. 19–43, 1994.

Glossary

APE : Assessed Proportion Effect
EPC : Error Producing Conditions
HEP : Human Error Probability
GT : Generic Task
NHU : Nominal Human Unreliability
MAART : Maritime Accident Analysis and Reduction Technique

Acknowledgment

Financial support for this study was provided by Kobe University, Graduate School of Maritime Sciences. The authors wish to thank the reviewers for their work in reviewing the manuscript contents and all the suggestions.

IOP Conf. Series: Materials Science and Engineering 1052 (2021) 012013 doi:10.1088/1757-899X/1052/1/012013

Porpoising instability study of the floatplane during take off operation on calm water

M H N Aliffrananda[1] and A Sulisetyono[1]

[1]Department of Naval Architecture, Institut Teknologi Sepuluh Nopember, Surabaya, 60111, Indonesia

Abstract. Floatplane is a type of airplane that can take-off and land on the surface of the water, for that it is equipped with a float that is installed under the fuselage to serve the buoyancy force. An important phenomenon that often occurs and should be avoided during a take-off operation is porpoising because it causes longitudinal instability of the float which has an impact on passenger safety and comfort. This paper describes the porpoising mechanism and the variables that can influence the occurrence of this phenomenon. The most dominant variables include the location of the longitudinal center of gravity (LCG) and the deadrise angle which are optimized by considering the take-off speed. A numerical arrangement based on computational fluid dynamics (CFD) techniques is proposed to simulate a three-dimensional fluid flow around a free-floating float in calm water. The accuracy of the simulation results shows a good fit compared to the previously published experimental and numerical results.

1. Introduction

A floatplane is a type of seaplane that can take off and land both in water and on runways on land. It can be used for tourist transportation, passenger transportation, and emergencies such as rescue missions and disaster response actions. Therefore, floatplane is often looked at and developed, especially by countries that have a long coastline *[1]*. However, the floatplane has problems related to safety and passenger comfort, namely porpoising instability when taking off. At the time of takeoff, the floatplane has three conditions, namely displacement condition, plowing condition, and planing condition. The planing condition is a condition in which the entire weight of the floatplane is supported by the hydrodynamic lift from the floating body and generally occurs when traveling at high speed. In the planing condition, if the pitch angle is too high or too low, the plane will experience instability. In such a case it will be possible to cause porpoising and result in accidents if the pitch angle can't be controlled properly *[2]*.

Porpoising is a rhythmic pitching motion of amphibious aircraft, caused by dynamic instability in the force acting on the floater. The sensation of this movement is often equated with swinging. This phenomenon can occur if the plane exceeds a certain pitch angle range and the plane begins to experience a cyclic oscillation of the pitch angle which can increase and end with the plane plunging into the water and possibly turning over. It can also happen if the aircraft's longitudinal stance is too high during takeoff. It could cause the plane to stop and drop its front half into the water. Porpoising can also be caused by placing a transverse step too far forward or behind the center of gravity. Porpoising can be reduced by flattening the end of the forebody by 1.5 times the width. It will result in relatively uniform pressure at the bottom of the hull, where the curved section will exert a variable pressure resulting in a more dynamic response *[3]*.

IOP Conf. Series: Materials Science and Engineering 1052 (2021) 012013 doi:10.1088/1757-899X/1052/1/012013

The porpoising phenomenon that occurs on the floatplane during take-off must be avoided because it can cause problems with the operability of the floatplane. Some of the problems that can be caused by this phenomenon, namely:

1. Porpoising can cause problems in the use of time and length of the course used for taking off. The floatplane must be able to take off in less than 60 seconds, and the maximum length of the course used for take-off is less than 10000 feet or equivalent to 3.048 kilometers for a large floatplane and it would be better if the take-off path length is less than 5000 feet or equivalent to 1,524 kilometers *[3]*.
2. Porpoising can cause accidents during take-off. Take-off occurs at high speed as well as porpoising occurs at high speed. When porpoising occurs the hydrodynamic forces experienced by the float change fast enough so that there will be a large longitudinal moment that occur fast enough. This phenomenon can cause the floatplane to reverse if there is no equilibrium at the moment along the length of the ship and the correct pitch attitude is not corrected *[2]*. This is very dangerous for the safety of pilots and passengers.
3. Porpoising can cause inconvenience to both the pilot and the passenger. The large amplitude of the pitch and the acceleration value of the large pitch angle will cause the floatplane to swing at high speed. This can cause passengers and pilots to experience the same swing as the response from the floatplane due to hydrodynamic forces *[4]*. Excessive swinging can cause nausea and vomiting for passengers who are not used to extreme swing conditions.
4. Porpoising can cause defects in the construction of the float hull. The large amplitude of the pitch and the angular acceleration under pitch conditions can cause a large impact load on the hull of the float. This impact load does not only occur once but repeatedly. It can cause damage to the construction of the float hull so that the float must be repaired and the floatplane cannot be operated for a while. It is not desirable by the owner due to the expense of repairs for damaged construction.

From the several problems caused by the porpoising phenomenon. Porpoising is something that must be avoided and is a consideration in designing the float hull. One of the easiest ways to solve the porpoising problem on the floatplane is to adjust the position of the longitudinal center of gravity (LCG) *[5]*, by shifting the main body of the floatplane forward or backward, it will change the LCG position of the float and change the shape of the float hull deadrise. Therefore, this research will focus on determining the optimum location of the LCG and the effect of the deadrise angle on the float to avoid the porpoising that occurs on the floatplane during takeoff operations.

2. Floater model and computational grids

Simulation is carried out by only modeling the float of the floatplane, the main body of the plane is ignored and focuses on the part of the floatplane that is submerged in water. **Figure 1** shows the design of the float that will be simulated in this study. The float has a deadrise angle of 20 degrees. The deadrise angle on the float will be varied by 10 degrees and 20 degrees, to determine the effect of the deadrise angle on porpoising. The hull of the floater is divided into two parts, namely afterbody and forebody separated by a transverse step with a forebody length of 5.4 meters. The transverse step has a function to reduce water resistance when the float is in planing conditions. The float used in this study has a transverse step height of 150 mm. The overall length of the float model is 9.6 meters with a width of 1.25 meters and a height of 1.1 meters. The distance between the demi hulls is 3.95 meters, with a displacement of the float of 7,664 tons.

Figure 2 shows the comparison of the body plan between a floater with a deadrise angle of 10 degrees and a floater with a deadrise angle of 20 degrees. It can be seen in the figure that due to the change in the deadrise angle there is a change in the draft of the floater. Whereat the 20-degree deadrise angle, the submerged part of the ship was 0.8 meters, while due to the deadrise angle that changed to 10 degrees, the submerged part of the ship changed to 0.745 meters. There was a drop of 0.055 meters in the submerged part of the floater. This occurs because a change in the deadrise angle causes a change in the submerged volume of the floater. Due to the need for constant displacement and the increased volume

IOP Conf. Series: Materials Science and Engineering 1052 (2021) 012013 doi:10.1088/1757-899X/1052/1/012013

of floaters that are immersed in water so as to provide the same hydrostatic force, there is an arrangement of the submerged floater so that there is a change in the draft of the floater.

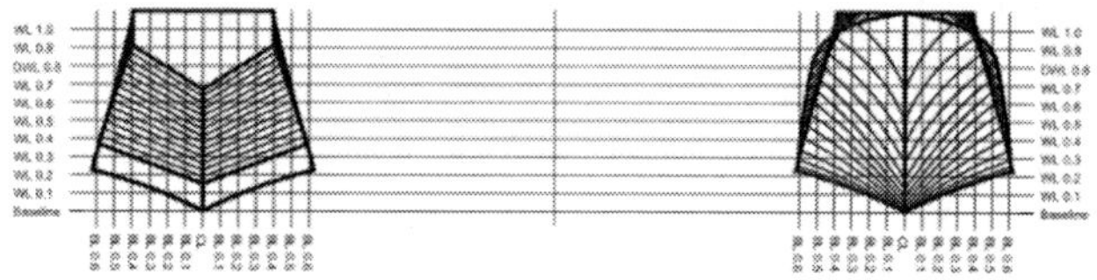

(a) Floater Body Plan

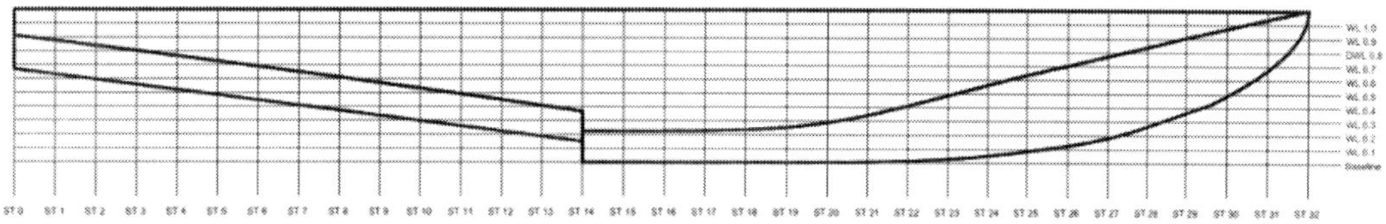

(b) Floater Side View

Figure 1 Linesplan Floater with 20° deadrise angle.

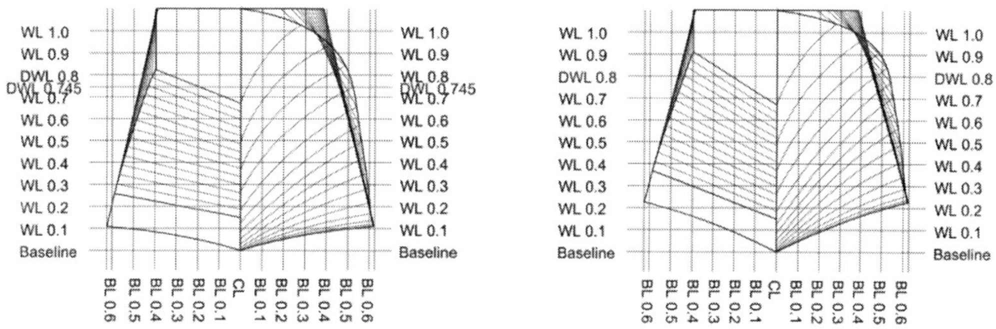

(a) 10° deadrise angle (b) 20° deadrise angle

Figure 2 Comparison between a body plan floater with a 10° deadrise angle and a floater with a 20° deadrise angle.

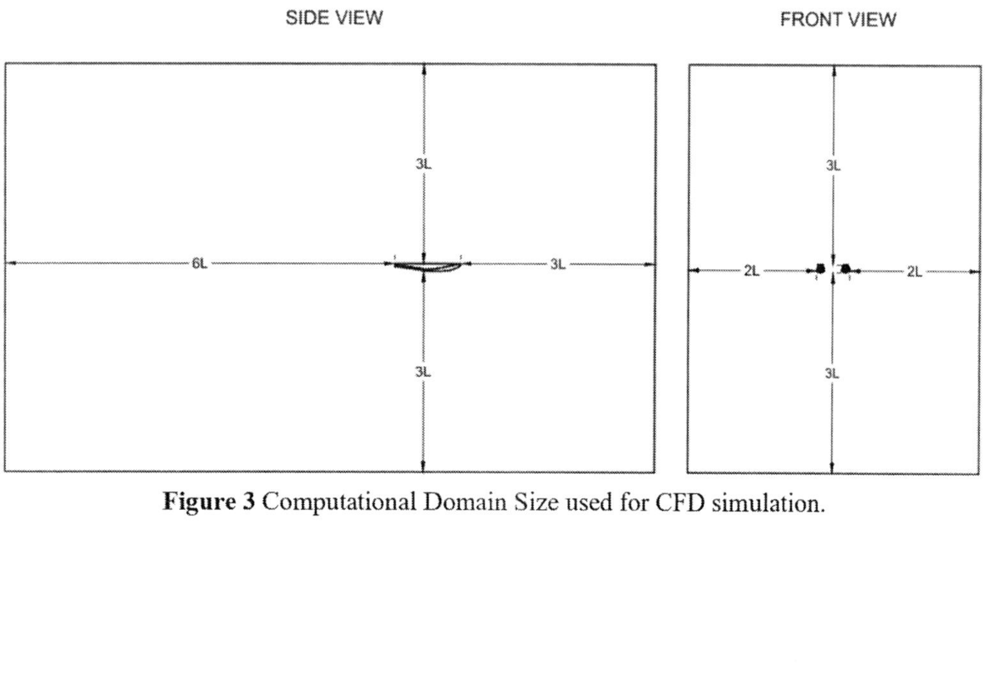

Figure 3 Computational Domain Size used for CFD simulation.

Figure 3 shows the size of the computational domain used in this study. The computational domain is rectangular shaped and represents the simulation as performed on a towing tank. The size of the domain that is on the back of the floater is made to be longer than the other parts in order to see the shape of the water surface on the back of the floater when it is moving. The computational domain is created using a multiple of the length of the object as a basis in order to make it easier to determine the size of the computational domain as a whole.

It can also be seen in the figure that the domain size of the length of the ship and the height of the ship is made quite large, this is intended to provide space for the ship to move in two degrees of freedom, namely, heave motion and pitch motion. In addition, by using the dynamic mesh method, the grid used will experience deformation during the simulation, causing the numerical calculation process to become unstable and potentially error in the middle of the simulation process due to the large deformation of the mesh. So, to avoid this problem, the domain size is enlarged to lengthen the ship and ascend the ship.

Figure 4 shows the mesh of a floater with 5.4 million cells. Since the hydrodynamic force is the main force component that supports the net force that occurs in the float in high-speed operation, getting an accurate picture of the overall fluid flow that occurs is very important. Therefore, the mesh in the float wall area is smoothed which serves to get a more accurate picture of the flow that occurs in the float wall area [6]. In addition, the mesh in the free surface area is also refined to obtain accurate fluid flow on the free surface and to get a better picture of the free surface that occurs on the back of the object. Heave and pitch simulations on the float are performed using the dynamic mesh method during the calculation process.

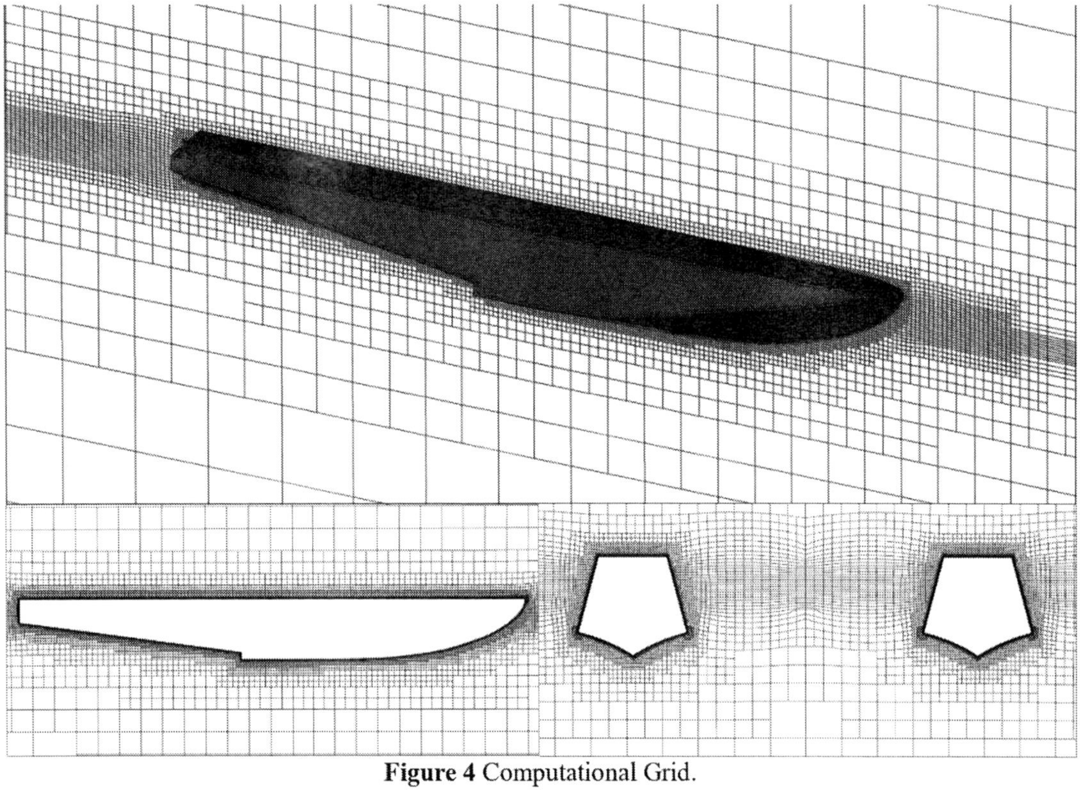

Figure 4 Computational Grid.

3. Mathematical model and computing method

The CFD flow solver, uses the incompressible unsteady Reynolds-averaged Navier Stokes equations (RANSE). The solver is based on the finite volume method to build the spatial discretization of the transport equations. The face-based method is generalized three-dimensional unstructured meshes for which non-overlapping control volumes are bounded by an arbitrary number of constitutive faces. The velocity field is obtained from the momentum conservation equations and the pressure field is extracted from the mass conservation constraint, or continuity equation, transformed into a pressure-equation. In the case of turbulent flows, additional transport equations for modeled variables are solved in a form similar to that of the momentum equations and they can be discretized and solved using the same principles. Incompressible flow phases are modeled through the use of conservation equations for each volume fraction of phase.

The flow solver can deal with multi-phase flows and moving grids. In the multi-phase continuum, considering incompressible flow of viscous fluid, mass, momentum and volume fraction conservation equations can be written as:

$$\frac{\partial}{\partial t}\int_V \rho dV + \int_S \rho(\vec{U} - \vec{U}_d)\cdot\vec{n}dS = 0 \tag{1}$$

$$\frac{\partial}{\partial t}\int_V \rho U_i dV + \int_S \rho U_i(\vec{U} - \vec{U}_d)\cdot\vec{n}dS = \int_S (\tau_{ij}I_j - pI_i)\cdot\vec{n}dS + \int_V \rho g_i dV \tag{2}$$

$$\frac{\partial}{\partial t}\int_V c_i dV + \int_S c_i(\vec{U} - \vec{U}_d)\cdot\vec{n}dS = 0 \tag{3}$$

Where V is the domain of interest, or control volume, bounded by the closed surface S with a unit normal vector $\vec{n}$ directed outward, where V is the domain of interest, or control volume, bounded by the closed surface S moving at the velocity $V_{ect}U_d$ with a unit normal vector $V_{ect}n$ directed outward. $\vec{U}$ and p represent, respectively, the velocity and pressure fields. τ_{ij} and g_i are the components of the viscous stress tensor and the gravity vector, whereas I_j is a vector whose components vanish, except for the component j which is equal to unity. c_i is the ith volume fraction for fluid i and is used to distinguish the presence $c_i = 1$ or the absence $(c_i = 0)$ of fluid i. Since volume fraction between 0 and 1 indicates the presence of a mixture, the value of $^1/_2$ is selected as a definition of the interface. The effective flow physical properties (viscosity and density) are obtained from each phase physical properties μ_i and ρ_i with the following constitutive relations:

$$\rho = \sum_i c_i \rho_i \tag{4}$$

$$\mu = \sum_i c_i \mu_i \tag{5}$$

$$1 = \sum_i c_i \tag{6}$$

When the grid is moving, the so-called "space conservation law" must also be satisfied:

$$\frac{\partial}{\partial t} = \int_V dV - \int_S \vec{U}_d\cdot\vec{n}dS = 0 \tag{7}$$

The $K - \omega$ turbulence model is used in this research to model the turbulence that occurs in this simulation. The $K - \omega$ model of Wilcox [7] has proven to be superior in numerical stability to the $K - \varepsilon$ model primarily in the viscous sublayer near the wall. This model does not require explicit wall-damping functions as do the $K - \varepsilon$ model and other two-equation models due to the large values of ω in the wall regions. The numerical wall boundary conditions require the specification of the distance from the wall to the first point off the wall. The two transport model equations for the K and ω scalar turbulence scales are defined below.

$$\frac{\partial \rho K}{\partial t}+\frac{\partial}{\partial x_j}\left(\rho U_j K-(\mu+\sigma^*\mu_t)\frac{\partial K}{\partial x_j}\right)=\tau_{tij}S_{ij}-\beta^*\rho\omega K \quad (8)$$

$$\frac{\partial \rho \omega}{\partial t}+\frac{\partial}{\partial x_j}\left(\rho U_j \omega-(\mu+\sigma\mu_t)\frac{\partial \omega}{\partial x_j}\right)=\alpha\frac{\omega}{K}\tau_{tij}S_{ij}-\beta\rho\omega^2 \quad (9)$$

Following Wilcox [8], the model constants are defined as $\alpha = {}^5/_9$, $\beta = {}^3/_{40}$, $\beta^* = {}^9/_{100}$, $\sigma = 0.5$, $\sigma^* = 0.5$.

4. Validation of the numerical method

Before performing the numerical simulation of the floater, the hydrodynamic force acting from the numerical simulation must be verified first. Some of the verifications carried out include determining the optimum number of grids that can be used in the simulation. However, due to the limitations of the testing facilities, the validation of the results from the pitch and heave within a certain time will be compared with the experiments conducted by other researcher regarding the porpoising motion on high-speed craft. Validation related to numerical simulations is carried out to obtain simulation results that are close to and following real events and to obtain reliable results.

4.1. Grid Independence Study

Grid independence study is performed to eliminate/reduce the influence of the number of grids/grid size on the computational results. It is always good practice to follow this for every individual geometry, which is tedious. In this study, the parameters used as a reference in carrying out the grid independence study are the values of drag force and lift force that occurs in the floater [9].

Figure 5 shows the relationship between the drag force and the lift force acting as the number of grids increases. As the number of grids increases, the value will be smaller and then the force value will not change significantly. When the force value has not changed significantly, the number of grids in the area will be used as the optimum number of grids to be used in the numerical simulation. It can be seen from the graph that the drag and lift force start not to change after passing through the number of grids of 5 M. Therefore, a grid number of 5.4 M is used in this study.

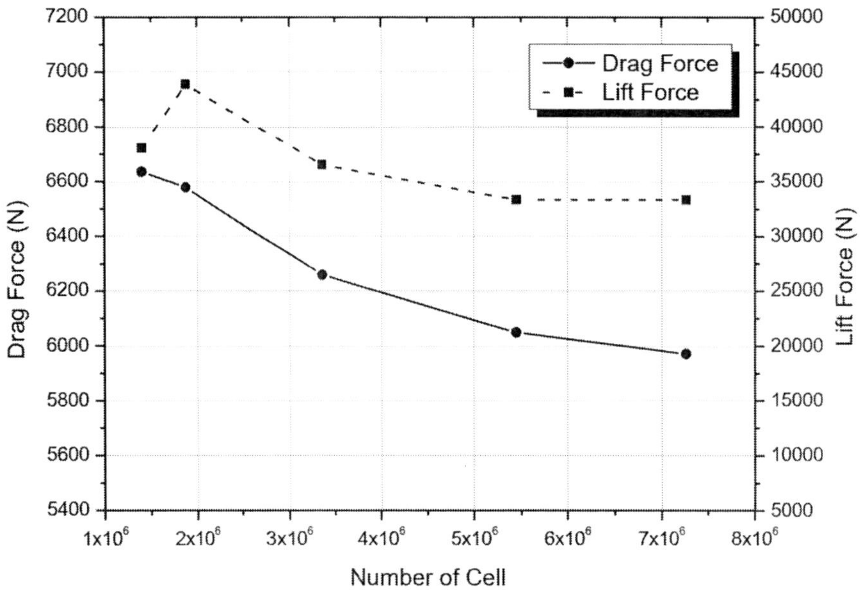

Figure 5 Grid Independence Study Graph.

4.2. Porpoising Motion Validation

Validation related to the porpoising motion was carried out using experimental data conducted by Sajedi and Ghadimi [10]. Sajedi and Ghadimi experimented with the longitudinal stability of high-speed craft, in this case, porpoising motion. The experiments were carried out using two different ship models, namely a ship without a stepped hull and a ship with a stepped hull. In this study, the experimental results from Sajedi and Ghadimi regarding a ship without a stepped hull will be used to validate the numerical simulation of porpoising motion [11].

Table 1 Comparison of the characteristics between the Sajedi and Ghadimi Ship Model and the CFD Ship Model that will be used as a validation tool in this study.

Parameter	Sajedi and Ghadimi Ship Model	CFD Ship Model	Error
L (mm)	2640	2640	0%
LCG (mm)	791	791	0%
VCG (mm)	184.6	184.6	0%
LBP (mm)	2368.18	2373.35	0.22%
m (kg)	86.024	86.350	0.38%
V (m^3)	0.08585	0.08424	1.87%
τ_s (degrees)	2.34	2.33	0.58%
B (mm)	551.9	551.9	0%

Table 1 shows the comparison between the experimental ship model and the CFD simulated ship model. It can be seen on the table that the two ship models are almost similar with the largest error in the volume of the ship with a value of 1.87%. However, the overall error of each parameter is below 2%, so that the two models have similarities and tolerable errors.

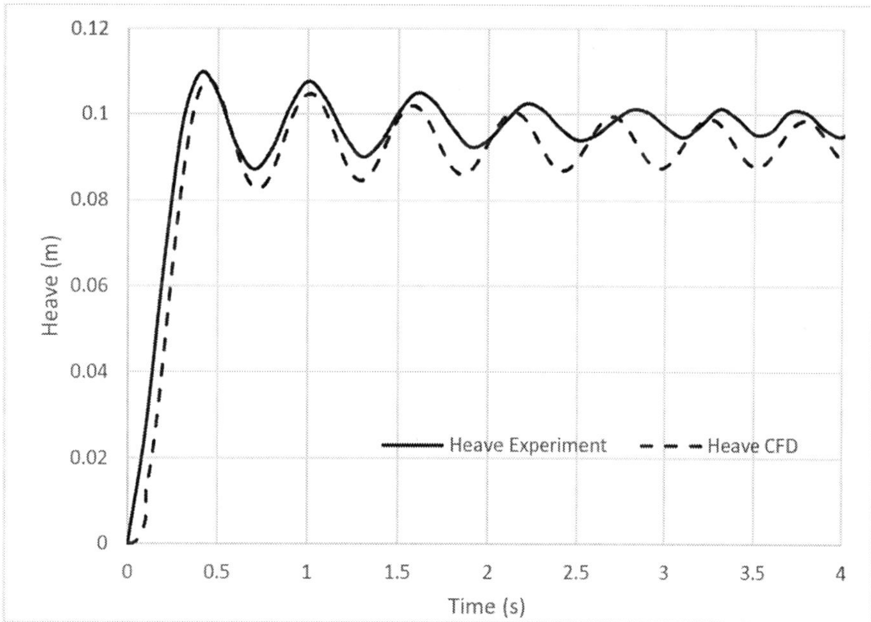

Figure 6 Time History Comparison of Heave Motion.

Figure 6 shows a comparison of the experimental and numerical simulations related to the motion of the high-speed craft model. It can be seen from the graph that the ship's heave movements are similar up to 1.5 seconds. Then in the next second the period of the heave porpoising movement increased and preceded the heave movement that occurred in the experiment. This happens because the numerical simulation is starting to become unstable due to the deformation of the grid. This can be corrected by increasing the size of the domain horizontally and vertically so that the deformation experienced by the computational grid can be reduced. However, the heave motion porpoising phenomenon can be fully captured by numerical simulations. Therefore, it can be said that numerical simulations can model the porpoising phenomena that occur on ships working at high speed.

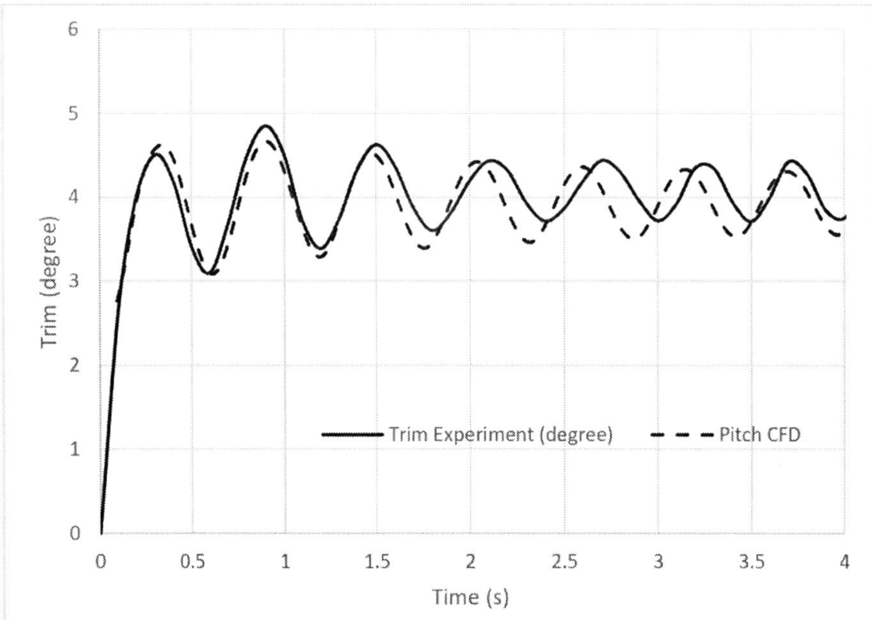

Figure 7 Time History Comparison of Pitch Motion.

It can also be seen in **Figure 7** that up to 1.5 seconds the phenomenon of pitch motion porpoising that occurs on ships has similarities between the experiment and CFD then in the next second the pitch motion in the numerical simulation changes in period. However, the phenomenon of pitch motion porpoising can be fully captured by numerical simulations. It can be concluded that numerical simulations can be used as a tool in modeling the porpoising phenomenon that occurs in ships operating at high speed.

5. Simulation results of porpoising motion

Porpoising motion is a problem most often experienced by ships operating at high speed. Thus, studies related to porpoising motion are important concerning the many losses that can be experienced due to this phenomenon. One of the things that are considered related to the porpoising phenomenon is the history of the movement of objects in a function of time. The movement of objects over time can be observed related to the porpoising phenomenon. When the porpoising phenomenon occurs, you will see oscillations from the heave and pitch on the object [12]. Besides, the characteristics of the heave and pitch amplitude also need to be observed. This research will show the results of the simulation of the heave and pitch movements of the floater over time.

One of the things reviewed is the movement of objects from time to time. The movement of the object in question is heave motion or vertical displacement of the center of gravity. The next object movement to be reviewed is pitch motion, which is the longitudinal angular position of the ship. **Figure 8** and **Figure 9** shows the comparison between the results of the porpoising motion on a floater with a speed of 15,492 m/s. It can be seen in the graph that there are two variations of the deadrise angle value and two variations of the LCG value, the LCG value is a multiple of the LOA (Length Over All) value.

The variation of the deadrise angle is 10 degrees and 20 degrees. While the LCG vary based on floater's LOA which is 0.50LOA and 0.55LOA.

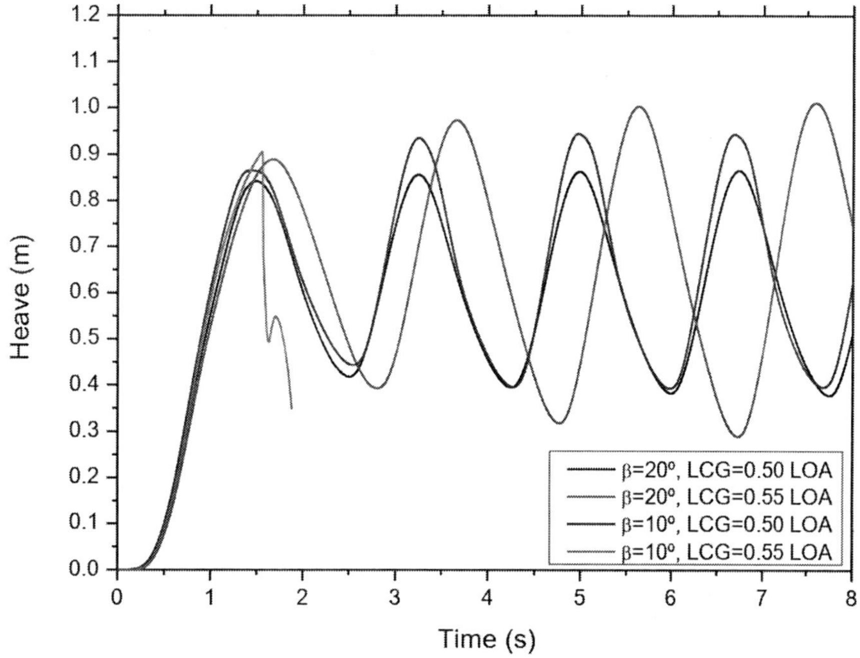

Figure 8 Comparison of heave motion time history of two variations of LCG and two variations of deadrise angles.

Figure 8 shows the heave motion of the floater. The graph shows that the largest heave value is generated from the deadrise angle of 20 degrees and LCG 0.55LOA configuration. From the graph it can also be seen that at a small deadrise angle a larger motion heave can be seen in the same LCG configuration. This is because the small deadrise angle will cause an increase in lift force on the object so that it can increase the displacement of the ship's heave motion [13]. The graph shows, that in the first second when the ship starts to get the lift, there is a similarity in the heave elevation up to 1.5 seconds, then the heave elevation of each configuration begins to experience differences.

Table 2 Average Heave Porpoising Period and Amplitude

	Average Heave Porpoising Period (s)	Average Heave Porpoising Amplitude (m)
β=20°, LCG=0.50LOA	1.75	0.23
β=20°, LCG=0.55LOA	1.97	0.32
β=10°, LCG=0.50LOA	1.73	0.26
β=10°, LCG=0.55LOA	-	-

It can also be seen that a configuration with an LCG value of 0.50LOA produces the same heave period, which is approximately 1.75 seconds, even though the deadrise angles have different sizes, period data can be seen in **Table 2**. However, the amplitude value of the float movement varies and increases with time. Besides, it can also be seen that in the configuration of the deadrise angle of 10 degrees and the LCG of 0.55LOA, the heave motion at 1.8 seconds underwent a very sharp change. The CG of the object drops rapidly in a very short time and may not be able to correct its position again.

Such conditions cause a large enough impact load in the bow float area and have the potential to damage the structure [14].

Table 2 shows the heave porpoising period and amplitude of the deadrise angle and LCG configurations. As previously explained, the LCG value of 0.50 LOA has almost the same period average, only 0.02 second difference. However, there is an increase in the average amplitude wherein the LCG 0.50 LOA configuration, at the deadrise angle of 20 degrees the average amplitude is 0.23 meters and at the deadrise angle of 10 degrees of 0.26 meters, there is an increase in the average amplitude of 3 centimeters, so it can be concluded that the change in deadrise angle. can increase the heave porpoising amplitude and shifting the LCG closer to the bow of the floater can increase the average period and amplitude of the heave porpoising.

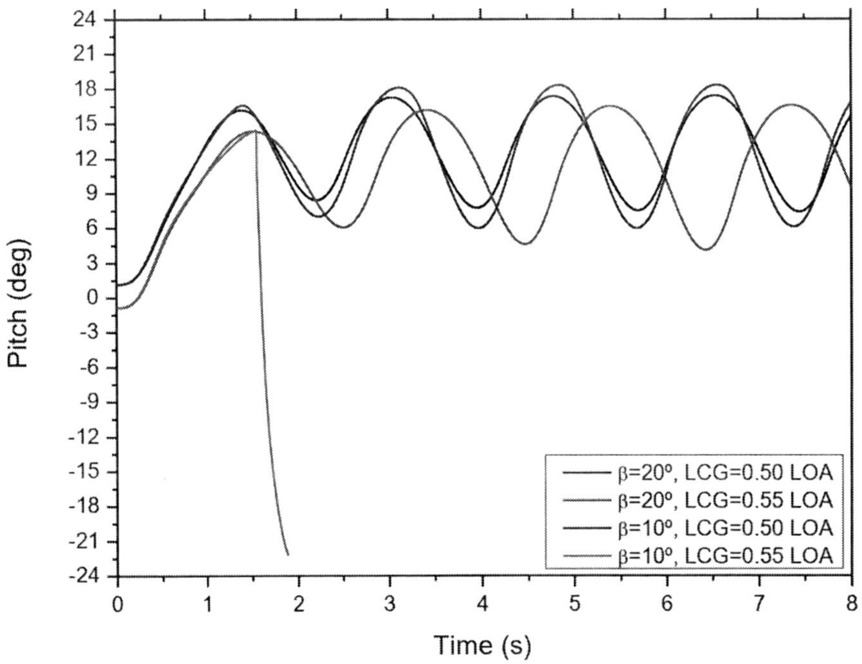

Figure 9 Comparison of pitch motion time history of two variations of LCG and two variations of deadrise angles.

Figure 9 shows the pitch motion that occurs in the floater. As a result of the location of the LCG, the initial condition of the floater is in a trim condition, where at 0.50LOA LCG, the floater has a stern trim of 1,174 degrees, while in the LCG configuration of 0.55LOA the floater has a bow trim of 0.874 degrees. Therefore, it can be seen on the graph, that at 0 seconds the pitch of the ship has a different value, where the negative value indicates that the floater is experiencing a bow trim. The graph shows that the largest pitch motion is generated by the deadrise angle configuration of 10 degrees and LCG of 0.50LOA which is shown at 3.1 seconds with a value of approximately 18 degrees. It can be seen from the graph the initial seconds to 1.5 seconds, the increase in the pitch angle has almost the same gradient in each configuration, then starts to change in the next second. Besides, it can also be seen that in the deadrise angle configuration of 10 degrees and LCG of 0.55LOA, the pitch of the motion at 1.8 seconds experiences a very drastic change. The float experienced a very extreme change in trim conditions in a very short time, reaching 22 degrees of the bow trim, causing the entire bow of the ship to enter the water. In this condition, the float is no longer possible to return to a stable condition and there is the possibility of an accident. **Figure 12** shows the extreme conditions of the ship with the bow of the ship completely submerged in water.

Table 3 Average Pitch Porpoising Period and Amplitude

	Average Pitch Porpoising Period (s)	Average Pitch Porpoising Amplitude (deg)
β=20°, LCG=0.50LOA	1.73	4.68
β=20°, LCG=0.55LOA	1.95	5.57
β=10°, LCG=0.50LOA	1.71	5.86
β=10°, LCG=0.55LOA	-	-

Table 3 shows the pitch porpoising period and amplitude that occurred in the simulation time interval of eight seconds. From the table, it is shown that the 0.50LOA LCG configuration has a porpoising period that is almost similar, with difference of 0.02 seconds, even though the deadrise angle values are different. Besides, it can also be seen that at the same deadrise angle configuration, namely the deadrise angle of 20 degrees, an increase in the amplitude of the porpoising pitch in the LCG configuration is 0.55LOA. This occurs because the center of gravity of the float is closer to the bow, making it easier for the bow to fall. When the bow falls, the bow gets more hydrodynamic force and raises the bow higher. At the same deadrise angle configuration, the shift of LCG to the bow of the floater will increase the period and amplitude of the pitch porpoising, this also occurs in the heave porpoising.

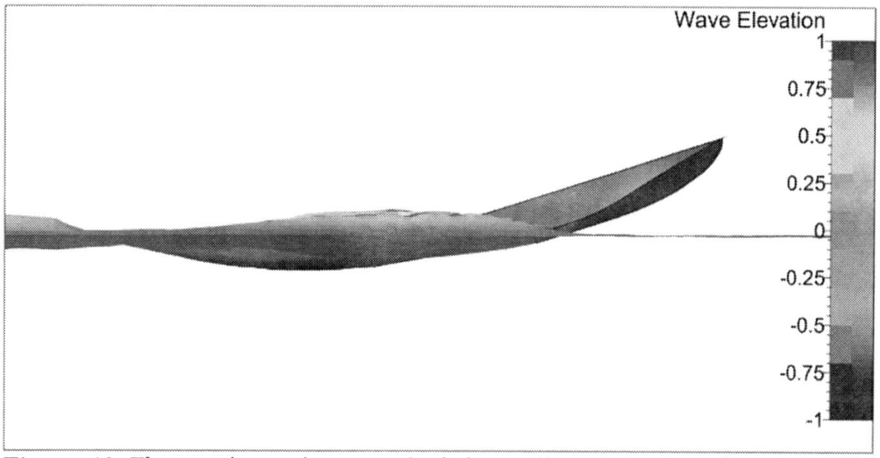

Figure 10 The maximum heave and pitch conditions occur at 6.5 seconds in configuration β = 20°, LCG 0.50 LOA.

Figure 10 shows the maximum heave and pitch conditions at a 20 degrees deadrise angle configuration with an LCG value of 0.50LOA. In this condition, the pitch is 16.26 degree and the heave elevation is 0.756 meters. It can also be seen that at the maximum heave and pitch conditions the water surface forms a wave with an amplitude of about 0.5 meters around the floater body. It can also be seen that the part of the floater which is immersed in water is very small, if the operating speed is increased, the floater body may come out of the water completely.

Figure 11 shows a series of side view images of the ship's motion cycle from the first second to the fourth second. These images are shown to provide a comprehensive picture of the movement and porpoising motion experienced by the floater. Unstable oscillations around the center of the object's center of gravity can be seen clearly. The floater moves and bends above the water surface. The amplitude of the porpoising pitching movement in the bow of the ship increases over time which can be seen from the 0 seconds to the 1.5 seconds. The floater experiences an extreme pitch angle with a maximum pitch of 16.2 degrees, in that condition the floater will throw its body forward and try to return to its position by raising the bow. In this condition, passengers who are not used to extreme conditions will experience nausea and vomiting.

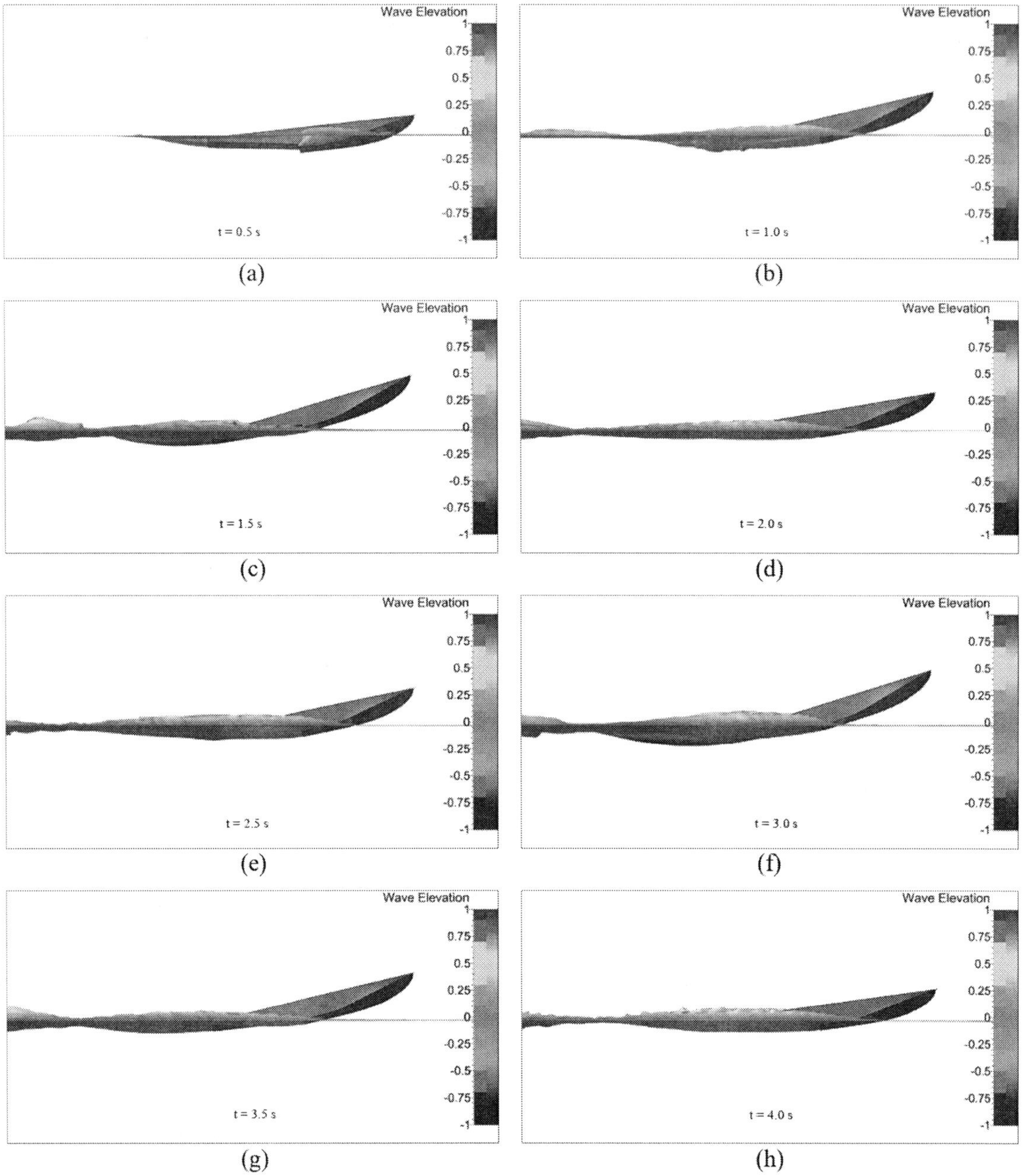

Figure 11 Series of movement showing the motion of the vessel with configuration β=20°, LCG=0.50LOA at intervals of 0.5 seconds to 4 seconds labeled from (a) to (h).

It can also be seen at the third second, the floater is unable to leave the water due to its high speed. However, the lift generated by the air has not been able to lift the weight of the entire floater so that the floater may hit the water very sharply in the next cycle. This porpoising motion occurs in all deadrise angle and LCG configuration conditions described in the previous section. One in four configurations underwent extreme movement in a very short period and caused extreme pitch conditions and was unable to regain its position and was likely to result in an accident. It can also be seen that the 1.5 second and the 3.0 second in Figure 12 are the peak of the pitch condition in one cycle, then in the next second the pitch condition decreases then increases again, this is called the heave and pitch rhythmic oscillatory motion of the ship. This condition will continue continuously if the pitch condition is not improved.

(a) Side View

(b) Perspective View

Figure 12 The occurrence of extreme conditions of heave and pitch motion at the deadrise angle configuration $\beta = 10^{\circ}$ and LCG = 0.50 LOA seen from the side view and perspective view.

Figure 12 shows the extreme conditions experienced by the floater. The floater experienced a bow trim of 22 degrees, so the floater's bow was completely submerged in the water. It is most likely that in this condition the floater will not be able to return to the displacement position or the planing position because the bow that is immersed in water will tend to be immersed and then the floater will reverse if the high-velocity fluid flows continuously. Therefore, in these conditions the best way to prevent the floater from turning over is to reduce the operating speed when extreme conditions occur. The speed reduction allows the floater to be more stable but also has the possibility that the floater will continue to spin and then reverse completely. It can also be seen that the free surface formed indicates that the water surface is subjected to an impact load from the hull of the floater due to a large change in pitch with a very fast period. The free surface experiences turbulence and a basin that reaches 1 meter below the water surface, and irregular water splashes occur in all directions.

6. Conclusion

The CFD simulation described in this study is used to study high-speed fluid flow. CFD simulations are able to capture the porpoising phenomenon on ships or floaters from seaplanes operating at high speed. Unsteady flow RANSE and K-ω turbulence models are used for the numerical simulation of high-speed fluid flows. The use of unstructured hexahedral mesh and dynamic mesh is used to model motion that occurs in objects due to fluid flow. In addition, the numerical problem solving has been validated using porpoising motion experiments on high-speed craft. Overall, the dynamic characteristics of the floater porpoising motion have been investigated in detail.

1. By decreasing the deadrise angle value, it can increase the lift force that occurs in the floater, causing the floater to experience an increase in heave motion faster in the same time compared to floaters with high deadrise angles.
2. The high deadrise angle makes it easier for the hull of the floater to break the surface of the water so that the reduction in impact loads that occur at the bottom of the floater's hull can be reduced. Besides, the small deadrise angle can increase the occurrence of accidents on seaplane floaters traveling at high speed due to the inability of the hull to break the surface of the water when the bow of the ship falls to the surface of the water quickly.
3. The position of the LCG that is close to the bow can increase the heave and pitch period that occurs on the ship when experiencing porpoising. It can also increase the heave and pitch amplitude over time if the floater pitch conditions are not immediately corrected.
4. During porpoising motion, pitching oscillations occur due to hydrodynamic forces acting on the object. At high velocities, the center of the hydrodynamic force moves forward from the object's center of gravity, which results in a continuous amplification of the unstable oscillations. Careful control of the trim angle is very important, especially when the floater is in the planning condition. When porpoising begins to occur, it is recommended to reduce the speed so that the floater will go to a displacement condition without further instability.

The author realizes that there are still many shortcomings in this paper. Therefore, the author tries to provide recommendations for the development of research related to floatplane motion behavior in the future. It would be better if the simulation time is longer, this is to provide an overview of changes in floater motion conditions as a whole phenomenon. Besides, it is also necessary to make variations related to the use of the dynamic mesh method used, where different methods can be used such as the overset grid method to provide an overview of the accuracy level of each dynamic mesh method. It is also necessary to simulate a floatplane when taking off in a rough water environment.

References

[1] L. Qiu and W. Song, "Efficient Decoupled Hydrodynamic and Aerodynamic Analysis of Amphibious Aircraft Water Takeoff Process," *Journal of Aircraft,* vol. 50, pp. 1369-1379, 2013.

[2] Federal Aviation Administration, Seaplane, Skiplane, and Float/Ski Equipped Helicopter Operations Handbook, Aviation Supplies & Academics Inc., 2004.

[3] S. Gudmundsson, General Aviation Aircraft Design, Oxford: Elsevier, 2014.

[4] O. M. Faltinsen, Hydrodynamics of High-Speed Marine Vehicles, Cambridge: Cambridge University Press, 2005.

[5] A. Sulisetyono, "Computing Porpoising Instability and Motions of a High Speed Boat in Waves," DalTech Dalhoisie Univeristy, Master Thesis, Nova Scotia, 1999.

[6] X. Duan, W. Sun, C. Chen, M. Wei and Y. Yang, "Numerical Investigation of The Porpoising Motion of a Seaplane Planing on Water With High Speeds," *Aerospace Science and Technology,* 2018.

[7] D. C. Wilcox, "Reassessment of The Scale-determining Equation for Advanced Turbulence Models," *AIAA Journal,* vol. 26, pp. 1299-1310, 1988.

[8] D. C. Wilcox, Turbulence Modeling for CFD, California: DCW Industries, Inc, 1993.

[9] I. Fadhlurrohman and A. Sulisetyono, "Resitance Analysis on Float of Floatplane Due to Trim Angle Variation Using CFD Method," Department of Naval Architecture, Faculty of Marine Technology, Sepuluh Nopember Institute of Technology, Surabaya, 2019.

[10] S. M. Sajedi and P. Ghadimi, "Experimental and Numerical Investigation of Stepped Planing Hulls in Finding an Optimized Step Location and Analysis of Its Porpoising Phenomenon," *Mathematical Problems in Engineering,* 2020.

[11] S. M. Sajedi and P. Ghadimi, "Experimental Investigation of The Effect of Two Steps on The Performance and Longitudinal Stability of a Mono-hull High-speed Craft," *Cogent Engineering,* 2020.

[12] T. Katayama, "Mechanism of Porpoising Instabilities for High-Speed Planing Craft," Department of Marine System Engineering, Graduate School of Engineering, Osaka Perfecture University, Sakai, Osaka, Japan, 2004.

[13] D. Savitsky, "Hydrodynamics Design of Planing Hulls," *Marine Technolgy Vol 1,* pp. 71-95, 1964.

[14] Z. A. Washoya and A. Sulisetyono, "Analysis of Impact Loads With Deadrise Angle Variations in Floats of Floatplane When Landing With CFD Simulation," Department of Naval Architecture, Faculty of Marine Technology, Sepeluh Nopember Institute of Technology, Surabaya, 2019.

Study application of the domestic freight logistics system in Indonesia

S Nugroho[1] and W A Larasati[1]

1. Department of Sea Transportation Engineering, Institut Teknologi Sepuluh Nopember Surabaya

Abstract. With the geographic conditions of Indonesia, the availability and need for commodities in each province are different, the fluctuations and disparities in the prices of basic goods between regions are quite high, those are driving factors for inter-island trade. Inter-island trade has an important role in the economic sectors of society. The geographic condition of Indonesia with 62.89% sea area, means that the modes of transportation that can be selected in inter-island trade are sea transportation and air transportation. The choice of transportation mode is closely related to the 9R of logistics principles, namely: Right Way, Right Product, Right Quantity, Right Quality, Right Place, Right Time, Right Customer, Right Cost. The development of the digital era that has penetrated all aspects of human life has also influenced the patterns of world trade which have begun to change since the presence of e-commerce. This study will try to potray how the current inter-island trade logistics system in Indonesia is conduct and how electronic commerce (e-commerce) affect the logistics business. This research aim to provide the right input for business actors in choosing the mode of inter-island trade transportation.

1. Introduction

Indonesia is the largest archipelagic country in the world which has 17,508 islands, with a land area of 1,922,570 km and an area of body water 3,257,483 km. The geographic conditions of Indonesia cause each region in Indonesia has the diversity of the face of the earth, this diversity makes each region have its own main product which gives rise to product specialties from each region. Referring to the 2018 data released in 2019 by Badan Pusat Statistik (BPS), of the ten cities with the largest economies of scale according to total GRDP, six of them are cities in Java [1]. So that it can be concluded that Java Island is still the center of the national economy, where many companies both in the production and trade sectors operate.

With the geographic conditions of Indonesia, the availability and need for commodities in each province are different, the fluctuations and disparities in the prices of basic goods between regions are quite high, those are driving factors for inter-island trade.

Inter-island trade is the activity of trading and / or distributing goods from one island to another within a province or between provinces, which is carried out by inter-island trading business actors by means of crossing the goods using sea or river transportation. Thus, trade between regions can be interpreted as the trade and distribution of goods from one province to another which is different within the same country [2].

The development of the digital era that has penetrated all aspects of human life has also influenced the patterns of world trade which have begun to change since the presence of e-commerce. Online transactions are an option for residents in areas where the availability of goods is limited and not diverse. This phenomenon is expected to continue to grow in Indonesia. Based on data released by Asosiasi Penyelenggara Jasa Internet Indonesia [3] the number of internet service users in Indonesia continues to increase from year to year as shown in figure 1. The blue graph representing the survey results shows that in the last ten years the number of internet users in Indonesia has increased by 86%. In the early 2000 the number of internet users looks stagnant, then it incrased gradualy. The highest increase in the number of Indonesian internet users occurred in 2015 and 2016, where the number of users in each year increased by 22 million people from the previous year. There were 143.26 million internet users in Indonesia in 2017, or around 54.68% of Indonesia's total population. The number increased by more than 10 million people compared to the previous year which was 132.7 million. It is estimated that the numbers of internet users in Indonesia will continue to rise, the green graph which represent the projection of users illustrates that by 2023 the number of internet service users in Indonesia will increase by 32%, breaking the 212.6 million user [4].

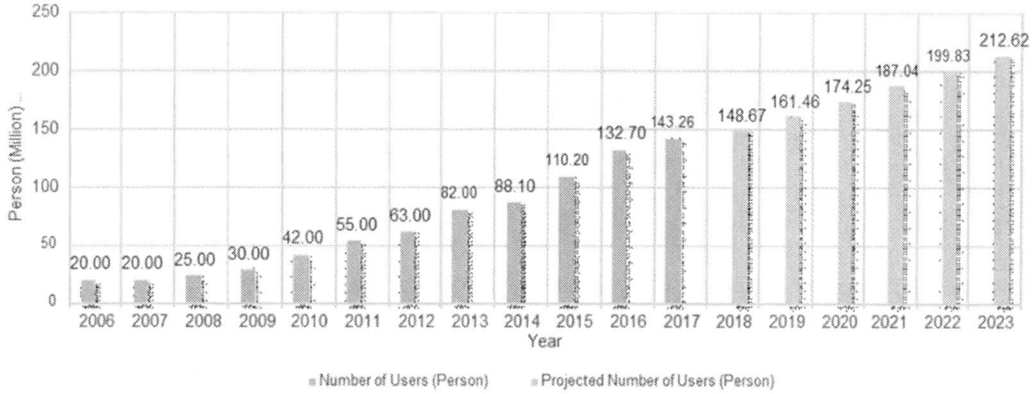

Figure 1. Development of Internet Users in Indonesia [4]. The blue graph representing the survey results while the green graph represent the projection of user. The graphic show that the number of internet users in Indonesia increased each year and will breaking the 212.6 million user by 2023.

This study will try to potray how the current inter-island trade logistics system in Indonesia is conduct and how electronic commerce (e-commerce) affect the logistics business. This research aim to provide the right input for business actors in choosing the mode of inter-island trade transportation.

2. Literature Review

2.1 Value Chain

A value chain is a business model that describes the full range of activities needed to create a product or service. For companies that produce goods, a value chain comprises the steps that involve bringing a product from conception to distribution, and everything in between—such as procuring raw materials, manufacturing functions, and marketing activities [5].

Value Chain Porter (discovered by Michael Porter) is a model used to help analyze specific activities that can create value and competitive advantage for the organization [6]. These activities are divided into 2 types, namely:

- Primary Activities which includes Inbound Logistics, Operations, Outbound Logistics, Marketing and Sales, and Service.
- Support Activities which includes Procurement, Technology Development, Human Resource Management, Firm Infrastructure.

Inbound Logistics are activities associated with the receiving input, storage and distribution of inputs / raw materials. Operations are activities associated with converting inputs or raw materials into final product forms. Outbound Logistics are activities associated with the collection, storage and distribution of products to buyers. Marketing and Sales are the activity of persuading or attracting buyers to buy, while Service are activities associated with providing services to increase and maintain product value.

Procurement, refers to the purchasing function, Technology Development, consists of various activities that can be grouped into efforts to improve products and processes. Technology development is essential for competitive advantage in all industries. Human Resource Management includes recruitment activities, training, human resource development. Lastly Firm Infrastructure are the company's infrastructure activities consist of a number of activities including general management, planning, finance, accounting and quality management.

2.2 Logistics Concepts

There are no correct terms or definitions regarding logistics due to differences in the goods produced, the companies that produce them, and the different systems applied. Although it has various definitions, one definition that is widely agreed upon around the world and which also touches the essence of logistics is that of Logistics = Material Management + Distribution [7]. Logistics focuses on the physical flow and information as well as storage (storage) from raw materials to the final distribution of finished products. Thus, material management represents storage and flow in the production process, which for distribution represents storage and flow from the point of the finished product to the end consumer or end user.

Logistics company is a company that provides transportation services for delivery of goods from the place of collection of goods to their destination. In addition, logistics companies also provide warehouse rental services for other companies to store their goods [7]. Logistics company can be divided into three categories:

- Logistics Service Provider: only provide one type of service.
- Third Party Logistics: provide goods delivery services and warehouse management.
- Lead Logistics Provider: the entire shipping process is handled by one company only.

In the Regulation of the Minister of Trade of the Republic of Indonesia, Number 29 / M-DAG / PER / 5/2017, it is stated that in inter-island trade the distribution of goods can be authorized to third parties, namely logistic service providers [2].

2.3 Supply Chain

Supply chain covers an even broader cope of the business area. This includes the supply of raw materials and components as well as the delivery of products to the final customer. Thus: Supply Chain = Suppliers + Logistics + Customers. In general, it can be said that: supply and materials management represents the storage and flows into and through the production process; while distribution represents the storage and flows from the final production point through to the customer or end user [7]. Supply Chain Management is an activity that involves coordinating the management of raw materials / materials, business information and financial flows in business relationships between participating organizations / companies with added value benefits, maximizing profitability through efficiency, and achieving customer satisfaction. Supply Chain Management is a cycle that runs continuously in line with a company's business processes which include [10]:

- Material flow: the physical flow of products from suppliers to customers, including product returns, services, recycling and disposal.
- Information flow: includes demand forecast, purchase transmission and delivery status reports
- Financial Flow: includes credit card information, credit terms, payment schedules.

The mode of transportation has an important role in the physical flow process of the product. Supply Chain Managers need to understand well the feasibility, advantages and disadvantages of each type of transportation in making product delivery / distribution decisions. In general, each mode of transportation has its advantages and disadvantages in terms of various considerations. The following is shown in Figure 2 to see the differences in the advantages and disadvantages of each mode of transportation commonly used to ship goods:

Table 1. Advantages and Disadvantages of Each Transportation Mode [4].

Modes of transportation	Truck	Train	Ship	Airplane
Volume	Average	Very Large	Very Large	Large
Time Flexibility	High	Low	Low	Low
Route Flexibility	High	Very Low	Very Large	Very Large
Speed of Delivery	Average	Average	Low	Very High
Cost	Average	Low	Low	High
Inventory (in-transit)	Low	High	Very High	Low

Like the table shown each modes of transportation have its advantage and disadvantage. There are some decision factor for both shipper and carrier to determine the best way to deliver their goods, for example, the use of expensive modes of transportation results in fast delivery and significant reductions in inventory.

2.4 E-commerce

According to Chaffey, define e-business as all information exchange and transactions through electronic media, both within an organization (for example ordering processing) and with external stakeholders who support various business processes. Meanwhile, what is meant by e-commerce is all transactions and exchanges of information through electronic media between organizations and external stakeholders [8]. Based on the subject, e-commerce divided into [9]:

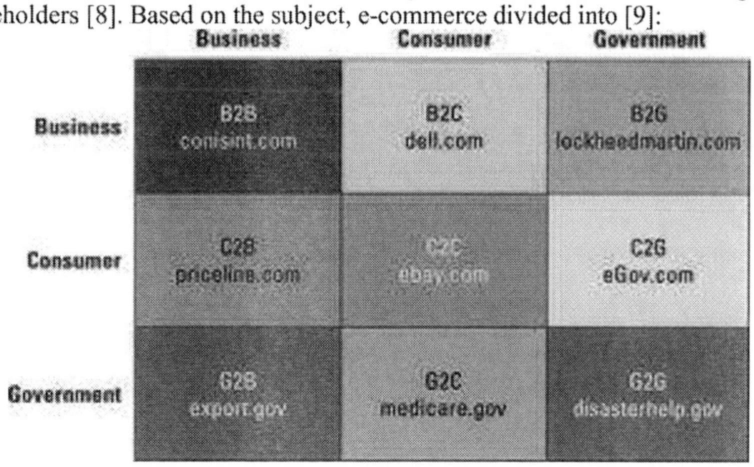

Figure 2. Types of e-commerce based on the subject involve and example of the platform [11].

- Business to Business (B2B) e-commerce: a form of buying and selling of products or services involving two or more companies and is carried out electronically.
- Retail (Business to Customer / B2C) e-commerce: B2C can be defined as a type of electronic commerce in which a company sells its goods directly to buyers.

- Customer to Business (C2B) e-commerce: is a form that is in contrast e-commerce in general, where consumers play an active role by informing the internet audience about their needs.
- Customer to Customer (C2C) e-commerce: users can interact with each other and the content provided by the users themselves.
- Business to Government (B2G) e-commerce: the government collaborates with businesses (private companies) in the form of provision of regulations or application media.
- Government to Business (G2B) e-commerce: The intermediary for the relationship between the government and the private sector is through the website, which is done online and on mobile.
- Government to Citizen (G2C) e-commerce: The general public in this case becomes the consumer and the government becomes the seller.

3. Method

This study was conducted in several industries (based in Surabaya, Gresik, Jogjakarta, and Bekasi) and Logistics Service Provider company (based in Surabaya). Data were collected by field observations and interviews to owners of industries to found out how the business process caried out, how the current logistics and shipment handled, current marketshare, supporting infrastructure owned, and their opinon about e-commerce. Meanwhile for the Logistics Service Provider company the interview will cover what is their current services, supporting infrastructure owned, how is business process of the services they offer, current marketshare, and their opinon about how the e-commerce will affect their bussines. The interview was required in order to capture an overview of the actual logistics process in the field, both from the industrial side and the logistics service provider. It would also be used in the analysis of how e-commerce will affect the future of the industries and Logistics Service Provider.

4. Discussion

To be able to find a real portrait of how the logistics process is carried out, a field review is being conduct on both the industry side as a service user and from the side of the logistics service provider company.

a. Overview of Industry - Fabric Business Field (UD. S)

UD.S is located in the Pengampon Square area, Surabaya. Operates in the textile sector, serving orders for various types of fabrics. Fabric is supplied by a textile industry in Jakarta. Suppliers send the latest fabric catalogs via Messenger Apps, orders are made via Messenger Apps or PO. Deliveries from Jakarta are made via expedition using trucks. Shipping costs are paid directly to the fabric supplier as the total bill paid.

Marketing is currently carried out with a sales system that makes offers to fabric stores. UD. S does not have a cloth display stand. Regarding e-commerce, UD.S does not have a plan to take part on online bussines. They reason that for fabrics, relying only on photos is considered less competitive in terms of price and quality. The average customer comes from Java, sometimes there are also orders from outside Java, for example Makassar.

IOP Conf. Series: Materials Science and Engineering 1052 (2021) 012014 doi:10.1088/1757-899X/1052/1/012014

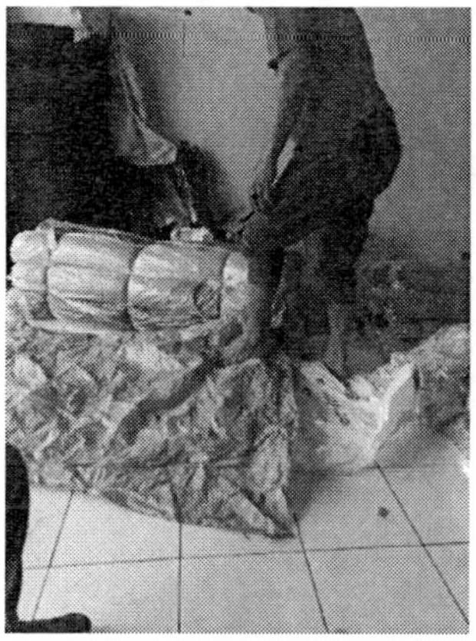

Figure 3.a

Figure 3.b

Figure 3.c Sequences process of how UD S performs packaging and delivery of orders to consumers.

For delivery to customer, the fabric is packed in plastic and then covered with 2 layers of sacks which are sewn so that it does not tear easily. The order will then delivered to the courier services that has been designated by the customer. The company has 1 truck fleet to support shipments to courier services. Shipping costs are borne by the buyer. The fee of shipping cost is a matter for the customer and the courier services, UD. S is only responsible for delivering goods to the courier services.

b. *Overview of Industry - Fabric Business Field (QBB)*

QBB is located in Bekasi. Operates in the textile sector, serving orders for various types of fabrics and woven products from all over Indonesia. Fabric is supplied by a textile industry in Bandung. Deliveries from Bandung are fully processed by fabric suppliers and made by trucks. Shipping costs are paid directly to the fabric supplier. Weaving products are obtained from the craftsmen all around Indonesia and usually sent via cargo services.

Marketing is currently carried out using offline (shop) and online systems (marketplace, social media, and web). Most customers come from Java, as well as several shops from Ternate and Kalimantan who regularly order fabrics. Shipping to customer usually uses courier services, while for bulk order are sent via cargo services. Shipping costs are borne by the buyer. The owner said that before the pandemic the offline store has the biggest sale, but after the pandemic most sale comes from marketplace.

c. *Overview of Industry - Convection and Screen Printing (TT)*
TT is located in Topaz, Gresik. Operates in the textile sector, Serving orders for making shirts, polo, jackets and others as well as screen printing. Production is carried out by customer order. Fabric supplier is a textile industry in Bandung. Suppliers send the latest fabric catalogs via Messenger Apps, orders are made via Messenger Apps. Deliveries from Bandung are made via courier services of choice by buyer using trucks. Shipping costs are paid directly to the courier services. Marketing is currently done offline (workshops) and online via social media.

Regarding e-commerce, TT does not have a plan to take part on marketplace. They reason that for fabrics, relying only on photos is considered less competitive in terms of price and quality.

The average customer comes from Java, there are also some orders from outside Java, for example Medan and Sintang. Orders are packed in plastic and then covered with a plastic bag with their brand logo. The order will then be delivered to the courier services that has been designated by the customer. For shipments outside Java, the average customer chooses Indonesia's goverment owned couriers service because the price is cheap and can reach remote areas such as Sintang. Shipping costs are borne by the buyer.

d. *Overview of Industry - Convection and Screen Printing (VS)*
VS is located in Benjeng, Gresik. Operates in the textile sector, Serving orders for making shirts, polo, jackets and others as well as screen printing. Production is carried out by customer order. Fabric supplier is a textile industry in Kapasan, Bandung and Jogja. Orders are made via Messenger Apps or PO. Delivery of raw materials is adjusted to the urgency of the order, for urgent matters fabric will be sent via train courier, if it is not urgent they use cargo services. But, the downside of cargo services is that we can't trace and track our shipment. Shipping costs are paid directly to the courier services.

Marketing is currently done offline (workshops) and online via social media. There are plans to explore and join marketplace of e-commerce, but they will only sell and offer a finished product, not a design and screen printing consulting service. They reason that the consulting service are very flexible and need a lot of communication, so offering services via marketplace is not a good idea. The plan to set up a marketplace store still has to be adjusted, there are many things to be considered like the additional employee who will take care of the marketplace bussines side, so as not to affect offline store performance.

The average customer comes from Java, sometimes there are also orders from outside Java. VS also won an order for the procurement of 6000 uniforms for the Papua region. Orders are packed in plastic and then covered with 2 layers of sacks which are sewn so that they do not tear easily.

The order will then be delivered to the courier services that has been designated by the customer. For outsides Java shipments in bulk, the average customer chooses to send by using containers LCL. Shipping costs are borne by the buyer.

e. *Overview of Industry – Sugar Factory (SF)*
SF is located in Jogjakarta. Nira suppliers are cooperatives in Kulonprogo, Cilacap and Cilongkok plantations. Delivery of raw materials to the production site uses pickups or 3-wheeled vehicles. 1kg of sugar requires 6000 liters of Nira which is carried out in 5 transportation times.

Marketing is currently carried out offline to companies that have entered into contracts, business actors, and trademark holders. They centered on serving export but also serve local shipments. For export shipments the costs are borne by the exporter, while for local shipments the costs are borne by the manufacturer. Local shipments use private transport from the manufacturer. Local consumers are still centered in the Java area, especially Central Java and the Special Region of Jogjakarta.

f. Overview of Logistics Service Provider – 3PL (CKL)

CKL's area of expertise is more focused on energy and manufacturing. However, CKL also provides services for other types of industries. Services offered by CKL: Transportation Management; Project Logistics; Warehousing Management; Bonded Logistics Center; Remote Site / Supply Base Services; Port Management; Cargo Shipping; Coal Barging; Loading and unloading for mining products. CKL has more than 40 network coverage throughout Indonesia.

They offered an end to end solution service, from inventory management (warehouses) to shipping services. Shipment can be conduct by Air Freight Services, Land/Road/Rail Freight Services, Ocean Freight Services. To support their bussines they have various infrastucture like truck fleets, vessel, warehouses. Currently they owned 130 truk fleets in various capacity and type. The fleets also empowered with GPS so customer can trace and tracks the update of their shipments. CKL own and operated several LCT and lease a few other type of vessel to accomodate their customer needs. Ten warehouses are being operated across Indonesia, six of them also certified as Bonded Logistics Center. As warehouse services they offer crossdocking, kitting, and distribution centre. The crossdocking is their top service, which shortening the warehouse process such as put away, refill dan picking. CKL also offering value added service like packing and repacking. These service take place for a shipment with a packaging that doesn't meet CKL standard or when the order from the distributor comes in a pieces not as a whole package.

Their bussines process are as follows: recieve inquiry, create operating plan, create project cost estimation, execution, reporting. For the example of cost component, here are the sea freight cost components: Freight Cost, Handling, Documentation, Trucking, LOLO, OPT/OPP, Stuffing/Unstuffing. The price for each cost component is different in each management area. The exact same service may have a different rate based on the site it carried out. For example trucking service to warehouse cost are different between these three city area as shown in tabel 2.

CKL priorities shipment based on urgent matters.

For now, CKL does not intend to expand its business by cooperating with e-business players. Especially looking at the areas of expertise of their customers who are mostly involved in energy and manufacture, its a special commodities which requires special handling. Their marketshare is customer-oriented with special care.

Table 2. CKL Rates.

City	LCL (IDR)	20" (IDR)	40" (IDR)
SUB	72,000.00	780,000.00	936,000.00
JKT	84,000.00	1,872,000.00	2,196,000.00
BPP	96,000.00	3,792,000.00	5,676,000.00

g. Overview of Logistics Service Provider – 3PL (PLL)

PLL's area of expertise is more focused on retail, consumer goods, and automotive, industrial. They offers custom brokerage, distribution, custom and distribution. Currenty PLL Indonesia is doing a trial run for their track and trace application which already use in their China and Vietnam branch for over two years.

Different from CKL, PLL doesn't have any assets. They rented all of their warehouses and fleets. Currently they operated eight warehouses across the country. Another differentiation with CKL, PLL open consolidation services. They offer a multi picks, multi drop, and multi customer consolidation. But the risk are higher for the LCL services, the source said that sometimes they need to process the shipment not in full container because the there aren't any goods to process, the waiting time for the LCL service can add to one month. PLL also offer multi modal and warehouses services too. Crossdocking and value added include. The value added services conduct when there are an order from distributor which comes in a pieces not as a whole package.

Their bussines process are as follows: recieve inquiry, create operating plan, create project cost estimation, execution, reporting. As in the LCL services, the goods will store at the PLL hub in container while waiting for another goods in the same destination. When the container arrive at the destination there are two choices, delivering goods based on how they place nearer the container door or the container unpack according to customer and delivered using a smaller fleet.

PLL priorities shipment based on the value of the goods, they have a list which cargo should be process first.

In line with CKL, PLL does not intend to expand its business by cooperating with e-business players. However, because PLL's marketshare is mostly engaged in FMCG, they think the presence of digital-based companies engaged in logistics services can be considered as their competitors in the future. They have the same market share, and digital-based companies able to offer user-oriented concepts as well as digitization that from customers point of view are added value in the services provided.

5. Conclusion

Based on the explanation that has been delivered, the following conclusions are obtained:

- The key factor for shipper to determine the best way to deliver their goods are the urgency of the matters (time), value of the goods, their destination, and shipping cost. So the decision factor are volume of the goods, speed of delivery, and cost.
- Logistics Service Provider priorities their order on the urgency of the matters (time) and value of the goods. So the decision factor are volume of the goods, time flexibility, route flexibility, cost, and inventory.
- Each logistics service provider has a different operating system depending on the type of customer served and company policy.
- Digital business is something big nowadays, but not all industry players are interested in getting into it. Especially competition in a marketplace where buyers are pegged on pictures and prices, without being able to directly evaluate goods.
- For industry to get into the marketplace requires a lot of preparation, for example from the human resources side.
- For logistics service providers with FMCG customers, the presence of digital-based companies in the field of logistics services have a possibility to become their competitors in the future.

6. Suggestion

For future works/study regarding these matter, we would like to suggest more comprehensive study regarding how the digital base business logistics affect the current logistics bussines, how is the future picture of the conventional logistics business competition with digital-based logistics and IT businesses.

References

[1] Badan Pusat Statistik 2019 *Publikasi Produk Domestik Regional Bruto (PDRB) Kabupaten/Kota di IndonesiaTahun 2014–2018* (Jakarta: Badan Pusat Statistik)

[2] Kementrian Perdagangan 2017 *Antar Pulau-Perdagangan* (Jakarta: Kementrian Perdagangan)
[3] Asosiasi Penyelenggara Jasa Internet Indonesia 2018 *Laporan Survey Infografis Penetrasi & Perilaku Pengguna Internet Indonesia 2017* (Jakarta: Asosiasi Penyelenggara Jasa Internet Indonesia dan Teknopreneur)
[4] Devira, F 2019 *Perancangan Sistem Logistik Ekspor Komoditas Kayu Manis* (Surabaya: Institut Teknologi Sepuluh Nopember)
[5] Carla T 2020 *Value Chain* https://www.investopedia.com/terms/v/valuechain.asp
[6] Setiawan A, Satyarini R, Hasan A, and Adriani R L 2014 *Identifikasi Value Chain pada Bidang Fashion sebagai Bagian dari Industri Kreatif Untuk Membuat Bandung Juara* (Bandung: Lembaga Penelitian dan Pengabdian kepada Masyarakat Universitas Katolik Parahyangan) p 9-10
[7] Rushton, A, Croucher, P, and Baker, P 2014 *The Handbook of Logistics and Distribution Management 5th Edition* (London: KoganPage)
[8] Pujawan I, and E R M 2017 *Supply Chain Management Edisi 3* (Surabaya: Penerbit ANDI)
[9] Fahrurroji A 2014 *Mengenal Model Bisnis E-Commerce* https://afahrurroji.net/mengenal-model-bisnis-e-commerce/
[10] Fahmi M 2013 *Supply Chain Management VS Value Chain* https://green4zone.wordpress.com/2013/11/23/supply-chain-management-vs-value-chain/
[11] Luhur M A 2020 *Studi Integrasi Layanan E-Ticketing Kapal Laut Penumpang Pada E-Dagang (E-Commerce)* (Surabaya: Institut Teknologi Sepuluh Nopember)

Comparative analysis of singlebeam and multibeam echosounder bathymetric data

Khomsin , D G Pratomo , and I Saputro

Department of Geomatics Engineering, Institut Teknologi Sepuluh NopemberKampus ITS, Keputih, Sukolilo, Surabaya, 60111, Indonesia

Abstract. Gresik Port, East Java, is a port that has a very strategic location because there are several large industries such as the PT. Petrokimia Gresik, PT. Semen Gresik, PT. Maspion, PT. Wilmar, PT. Pelindo III, and other factories. This condition makes the need to obtain accurate necessary water information such as bathymetry. Bathymetry is an essential parameter of port management. There are currently two types of equipment used frequently to measure bathymetry, namely Singlebeam and Multibeam Echosounder (SBES and MBES). However, these two bathymetric survey instrument tools may not provide the same depth. This study compared the bathymetry data derived from MBES and SBES, collected at the same time and place. The depth comparison test between the SBES and MBES shows the Z_0 of the statistical Z test is 0.018 m. This Z_0 value is smaller than the Z table for the 95% confidence level, which ranges $-1.96 < Z_0 < 1.96$. This result means that the data has a good quality within the standard 1a of the IHO SP-44 5th edition in 2008.[1] The comparison between the SBES and MBES surfaces reveals that the cut volume is greater than the fill volume, which is 21078.81 m^3. This result means the SBES data has shallower average depths than the MBES data. The dredge volume difference of the SBES and the MBES data with 15 m depth as a reference is 23617 m^3 inside the area 22.1 Ha. It means the dredge volume difference of 0.11 m^3 per m^2 or 11 cm difference depth in m^2.

Keywords: SBES, MBES, Data Comparison

1. Introduction

Indonesia has a geographic condition where two-thirds of its territory is the sea. The sea is a communication route for shipping traffic between islands, countries, and continents, both for passenger and goods transportation in Indonesia.[2] This condition makes the role of the sea essential for Indonesia. Gresik Port is a port that has a very strategic location because there are several large industries, such as the PT. Petrokimia Gresik, PT. Semen Gresik, PT. Pelindo III Gresik, PT. Maspion and other factories. Gresik Port (Pelindo III) has a dry bulk & log terminal dock and 78 wharves to serve the loading and unloading of coal, log logs, bag cargo, and general cargo. Port management requires information, namely a nautical chart. One of the components of a nautical chart is the depth of the sea (bathymetric map). [3] Bathymetric surveys usually require special equipment that functions to get depth values in water. Some methods can be used in a bathymetric survey to get seafloor topography, such as mechanical, acoustic, and remote sensing.[4]

Currently, SBES and MBES tools are frequently to obtain bathymetric data in the sea. Simultaneously, the bathymetric survey using these two types of equipment will produce different depths. MBES data will have denser than SBES data because the SBES survey has only one depth point solution, whereas the MBES has multiple depth solutions for each acoustic wave emission. This study will analyze the comparison between the SBES and the MBES data at the same location simultaneously. This study will also explore the difference in surface and dredging volume at the port between the SBES and the MBES data.

2. Data and methods

The location used for the case study in this research is the Gresik port (Pelindo III) (Figure 1). The port area of Gresik is coastal waters. In this study, PT. Pelabuhan Indonesia III (Pelindo III) supported all of SBES and MBES data in the XYZ coordinates. The SBES data quality test will be carried out, comparing the SBES and MBES data such as the bathymetric map, the difference between the surface, the cross-section, and the dredge volume calculation.

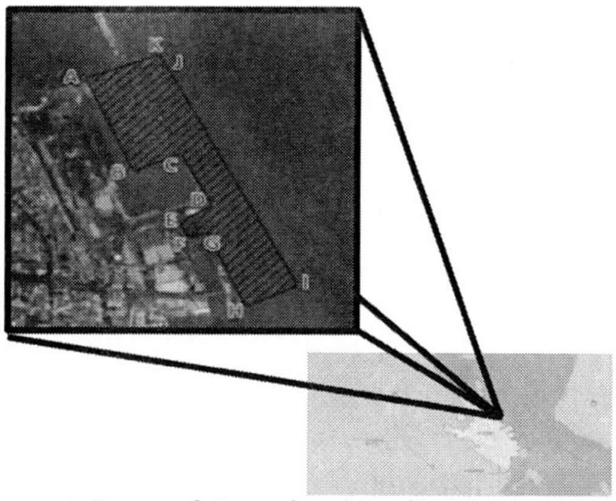

Figure 1. Research Location (Gresik Port Pelindo III).

The stages of the implementation of this research are as follows :

2.1 Identification of problems

This stage is intended to identify early on the case to be studied, in the form of problem formulation, objectives and benefits of Multibeam and Single-beam Echosounder Bathymetry Data Comparative Analysis in Gresik Port Waters.

2.2 Study of literature

The literature study is intended to collect references that will support the processing steps up to the analysis. Regarding this research, it is necessary to study literature regarding the bathymetric survey, single-beam echosounder, multibeam echosounder, germination strip, data quality test, data comparison test, volume calculation and previous research as a reference..

2.3 Data collection

Data collection aims to collect the data needed to conduct research, namely data using single-beam echosounder and multibeam echosounder data during the roaming process, as well as images of the research area.

2.4 Data processing

After all the data has been collected, the next step is data processing. At this stage, the SBES data quality test will be carried out, a comparison test of the SBES and MBES data and then a bathymetry map is made, the difference between the surface differences, the cross-section sample differences and the volume calculation.

2.4.1 calibration

Sound wave velocity correction for shallow water is carried out using a barcheck. Barcheck correction is a correction done to remove systematic errors from the tool. A barcheck is made of a circular or rectangular metal plate suspended from a rope or scale chain and placed under the transducer. A scale rope or chain is used as a comparison of the measurement results with the results read by the tool perum. Comparison of the results of depth measurements is carried out for each change in depth, from 0 meters to the maximum depth to be summarized at 1 meter intervals [6].

The barchek measurement data obtained is used to find the relationship between the actual depth and the measured depth using an echosounder in the form of a linear equation. The linear equation formed can be seen in equation 1:

$$d_c = [[(b_{ari} - b_{ari+1}) \div (r_{eci} - r_{eci+1})] . (d_0 - r_{eci})] + b_{ari} \ldots(1)$$

Information:

d_c	: true depth
d_0	: depth of observation results
b_{ari}	: barcheck depth at check point i
b_{ari+1}	: depth of reading tool at barchek point i + 1
r_{eci}	: depth of reading tool at barchek point i
r_{eci+1}	: depth of reading tool on barchek i + 1,
i+1	: order of point depth calibration and $r_{eci} < d_0 < r_{eci}+1$

In deep waters, correction with barcheck is not recommended because the influence of subsurface currents and surface waves causes the barcheck chain to curve so that the size of the barcheck data obtained does not match the actual. Therefore, to analyze the speed of sound waves using the physical properties of seawater with equation 2 as follows [7]:

$$V = 1410 + 4{,}21\ T - 0{,}0370\ D2 + 1{,}14\ S + 0{,}018 \ldots\ldots(2)$$

Information :
V : Speed of sound at temperature (T), salinity (S) and depth (D)
S : Salinity of seawater (%)
T : Temperature (^{0}C)
D : The depth of the sea (meter)

2.5 Analysis

Analyze the results obtained regarding the results of the SBES data quality test, comparison test of SBES and MBES data and bathymetry maps, differences in surface differences, differences in cross section samples and volume calculations.

2.5.1 Data Quality Test

Each depth measurement, often the depth value given is not the actual depth value but can be an outlier, so a quality test is needed to be able to eliminate it. The references used to determine the tolerance limit for the depth difference value are IHO SP 44 of 2008 [1] and SNI 7988-2014 [5] with a confidence level of 95%. The equation used to calculate the tolerance limit can be seen in equation 3 [8]:

$$\pm\sqrt{a^2 + (b \times d)^2} \ \ldots\ldots\ldots\ldots (3)$$

Information :
a : independent error
b : Depth dependent error factor
d : average depth
(bxd) : Depth dependent error (the sum of all dependent depth errors).

The a and b values in equation 3 are adjusted to the order of the survey carried out as listed in Table 1 according to IHO SP-44 of 2008 :

Tabel 1. Survey Orde According to IHO

Orde / Ketentuan	Spesial	1a	1b	2
Konstanta	a=0,025m b=0,075m	a=0,5m b=0,013m	a=0,5m b=0,013m	a=1,0m b=0,023m

The quality test is carried out by first assuming that the Hlu (main lane depth) is the same as the Hls (depth of cross lane) which overlaps with each other (at the same x and y positions). Further presented in equation (4) to equation (7) [9] as follows:

a. Hlu = Hls = H..............(4)
b. μU – μS = δ.................(5)
c. *Mean error* : $\overline{H}=\frac{1}{n}\Sigma(Hlu-Hls)$...........(6)
d. Standard deviation :

$$\sigma = \pm \frac{\sqrt{\sum_{i=1}^{B}(Hi-H)^2}}{n-1} \quad ....(7)$$

Information:
Hlu : main lane depth (m)
Hls : depth of cross lane (m)
H : True value (actual depth difference = 0) (m)
Hi : The difference between the main and cross lane depth values (m)
σ : Standard deviation
δ : Difference in depth

The quality test is continued by calculating the error value for the depth difference data with a confidence level of 95% which refers to the IHO SP-44 2008, which is 1.96 x σ. If the depth difference data error value is still within the depth tolerance limit obtained by equation 2.13, then the quality of the depth data sample falls within the tolerance referred to in IHO SP-44 in 2008. Whereas if the data error value for depth difference is outside the depth tolerance limit, then the sample quality depth data does not fall within tolerances.

2.5.2 Comparative Test

The comparison test between MBES and SBES data was carried out by statistical tests using the Z test. The choice of statistical tests using the Z test was because the number of research samples that could be categorized as large samples (> 30). In this comparison test stage, the data from the bathymetry measurement uses a multibeam echosounder with depth data from the singlebeam echosounder by looking at the mean error, standard deviation, and Z0 values of the entire sample depth data. The equations for calculating the value of the mean error, standard deviation, and Z0 value are in equations (6), (7), (8) and (9).

This depth data comparison test uses a large sampling statistical test with paired samples using Z table. Based on the Z0 results, it can be seen whether the depth data measurement results are within tolerance or not with a 95% confidence level (± 1.96).

$$Z_0 = \frac{X_1 - X_2}{S_{X_1 - X_2}} \quad \ldots\ldots\ldots\ldots\ldots\ldots\ldots\ldots (8)$$

Where,

$$S_{X_1 - X_2} = \sqrt{\frac{\sigma_1^2}{n_1} + \frac{\sigma_2^2}{n_2}} \quad \ldots\ldots\ldots\ldots\ldots\ldots\ldots\ldots (9)$$

Keterangan:
X_1 = MBES average (m)
X_2 = SBES average (m)
σ_1^2 = MBES Sample Variants (m)
σ_2^2 = SBES sample Variants (m)
n_1 = Amount of MBES Samples (m)
n_2 = Amount of SBES Samples (m)

Hypothesis :
a. Ho ; $\mu 1 = \mu 2$ atau $\mu 1 - \mu 2 = \delta$
b. HI ; $\mu 1 \neq \mu 2$ atau $\mu 1 - \mu 2 \neq \delta$

So that :
a. Ho rejected, if $Z_0 > +1{,}96$ atau $Z_0 < -1{,}96$
b. Ho accepted, if Z_0 $-1{,}96 < Z_0 < +1{,}96$

Information:
H : True value (difference in depth = 0)
Hi : The difference in the depth value of the multibeam echosounder measurement results and the singlebeam echosounder depth point data.
δ : Difference in depth

2.5.3 Volume calculation

Cut and fill is a method for determining the volume of excavated or landfill in a place. It can also be used to calculate solid materials. The principle of calculating the volume is one area multiplied by one representative of the height. If there are several areas or several highs, representatives are made, for example by averaging the area or leveling the height. The method that can be used to calculate volume is 1. Grid "Borrow Pit" method, this method of calculation is done by creating a grid / plot with elevation information at each corner. Volume is obtained by multiplying the average height of each plot area by the area of each plot. The total volume is the total volume of the total plot. [10]

2.6 Preparation of reports

Reports are compiled to report the results obtained from the research that has been done.Results and discussion.

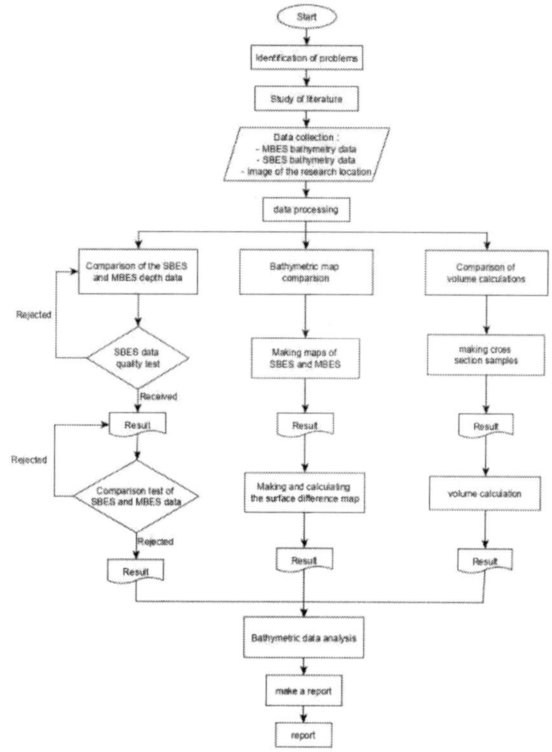

Figure 2. Research Flowchart.

3. Result

This study has four results, namely the SBES Data Quality Test, Comparison Test of Singlebeam and Multibeam Echosounder Data, Analysis of Bathymetric Map, Dredge Volume Difference Analysis.

3.1 SBES Data Quality Test

The singlebeam echosounder data quality test has been carried out at all points between the main and the cross lane. Testing the single-beam echosounder data is based on the IHO SP-44 5th edition in 2008 [1] standard for a 95% confidence level, 1.96 x σ. There should be no difference in the value of the depth between the main and the cross lane for all points that are close together in ideal conditions (Figure 3). However, due to various factors, this perfect condition is not always obtained. The number of samples from the data filtering process coincided with each other, and the different depth values can be seen in Figure 4. The standard deviation of the depth difference (σ_Z) between the main and cross lane is 0.127 meters, and the 95% confidence level (1.96 x σ) is 0.250 meters. The tolerance limit for IHO at the survey order-level 1a is 0.500 meters. Therefore, the 95% confidence level's standard deviation value is smaller than the maximum tolerance limit allowed in the IHO standard. The quality of this single-beam echosounder measurement can be accepted.

IOP Conf. Series: Materials Science and Engineering 1052 (2021) 012015 doi:10.1088/1757-899X/1052/1/012015

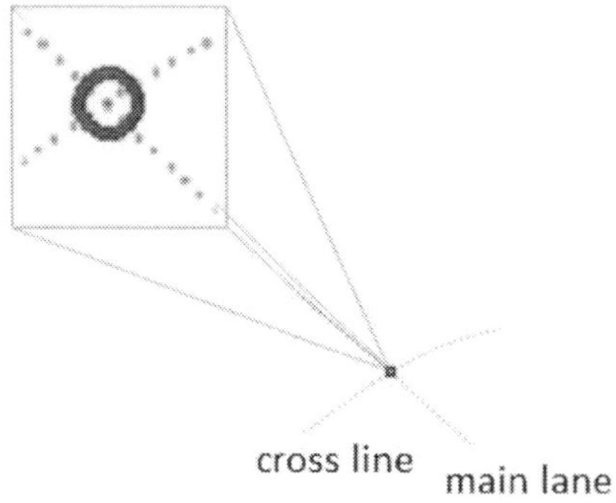

Figure 3. The point between the main and the cross survey line.

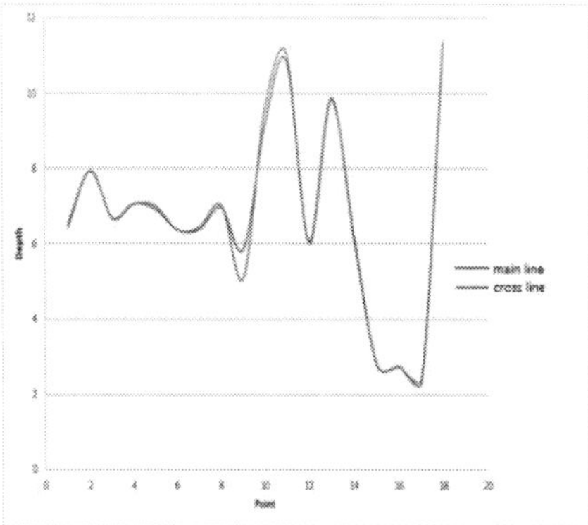

Figure 4. Coincided point comparison between the main survey lane and cross survey lane.

3.2 Comparison Test of Singlebeam and Multibeam Echosounder Data.

The comparison test of SBES and MBES data uses the Z test because the number of samples used in this statistical test is a large sample (> 30). This data comparison test is carried out for all points that overlap between the SBES and the MBES grid point. The number of samples obtained after previously carrying out the data filtering process through MBES data interpolation adjusting the SBES data was 6279 points. Ideally, there is no difference in the depth values between the SBES and MBES data for all the points that overlap. However, these ideal conditions are not always obtained. There is a difference between the SBES grid point depth value and the MBES. Furthermore, a statistical test was carried out for the 95% confidence level by calculating the Z_0 value. The depth data comparison test between the SBES and MBES data was conducted to determine the degree of difference in the two tools' measurement results. This difference value should not exceed the standard set in the 5th edition of the IHO SP-44 for a 95% confidence level. Based on the statistical test results using the Z test, the Z_0 was 0.018m. At the 95% confidence level, the Z table value is in the range $-1.96 < Z_0 < 1.96$. From the results of the Z_0 and Z tables, the $Z_0 <$ the Z table. So it can be concluded that the SBES and MBES data for the 95% confidence level are not significantly different, so it can also be said that the MBES data were by the IHO SP-44 standard 5th edition in 2008. [1]

3.3 Analysis of Bathymetric Map

The next step is making a bathymetry map between SBES and MBES processing (Figure 5). The research area has a maximum depth of about 17 meters, indicated in blue as the maximum color, and has a minimum depth between 0 - 1 meter shown in red.

SBES	MBES

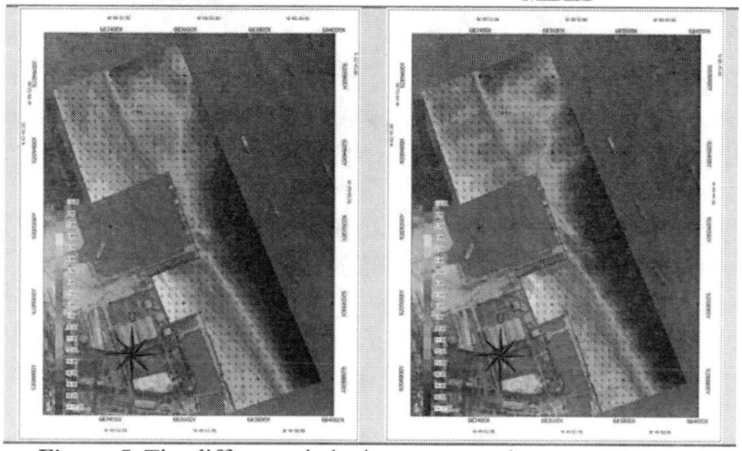

Figure 5. The difference in bathymetry map between SBES and MBES.

Both data processing, both SBES and MBES data, have little different surfaces. Figure 6 shows the SBES data has more average depth than the MBES data due to the software's interpolation process. It is also because the MBES data has more points than the SBES data so that the depth of the sea surface can be adequately visualized and match the existing surface. The MBES surface is more detailed than the SBES.

SBES	MBES

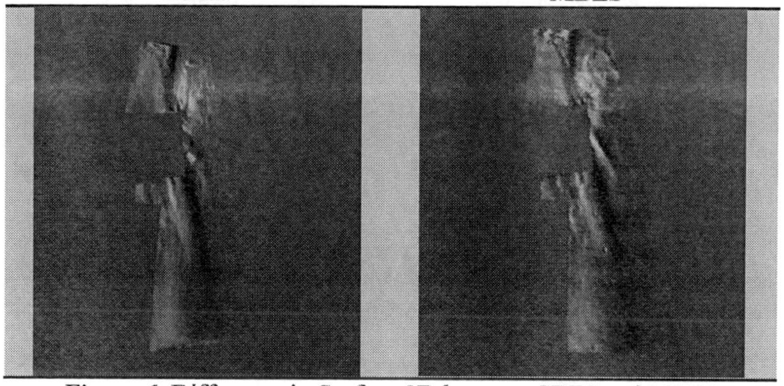

Figure 6. Difference in Surface 3D between SBES and MBES.

The difference surfaces in-depth shows that if the data is positive (+), the depth of the SBES is shallower than the MBES. On the contrary, if the data is negative (-), the depth of the MBES is shallower than the SBES. The depth difference in the SBES and MBES data can be calculated as the cut and fill volume where the MBES data is a reference for the existing surface with an area of 22.1 Ha. The difference in the bathymetric map and comparing the two surfaces (Figure 7), the MBES data as the reference, resulted in a cut value of 79793.82 m^3, which was higher than the fill 21078.81m^3. This volume shows that the SBES data majority has a shallower depth than the MBES data.

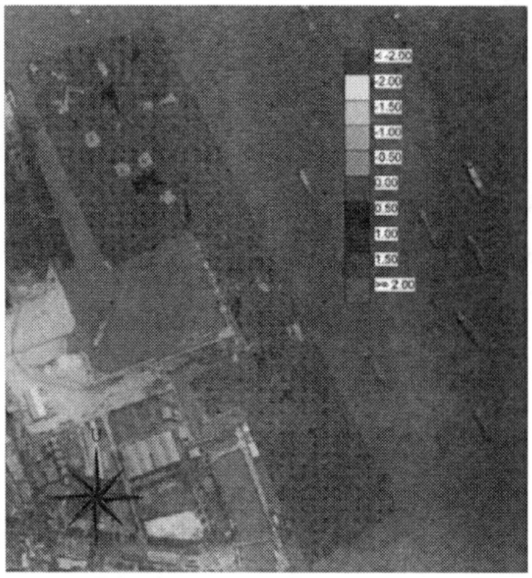

Figure 7. Map of the Depth Difference of SBES and MBES.

3.4 Dredge Volume Difference Analysis

A comparative analysis between the cross-section samples and the volume calculation was carried out with a reference depth of 15 m. The sample surface's difference per cross-section between SBES and MBES can be seen in Figure 8. The white line shows the MBES surface, while the blue line indicates the SBES surface. Furthermore, if seen from several cross-section samples made, both the SBES and MBES data in each section have no significant difference.

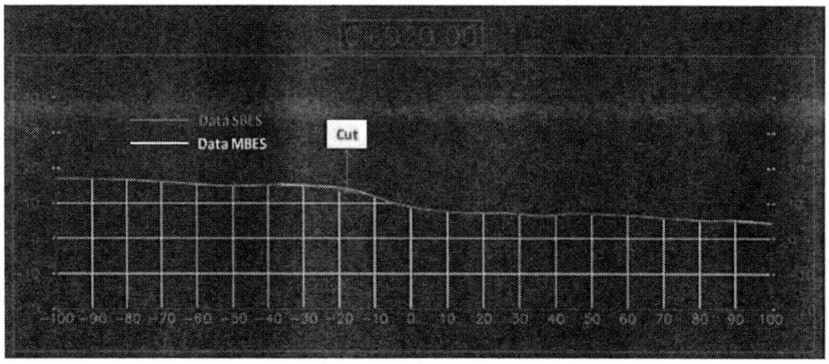

Figure 8. Cross Section Sample.

The volume calculation is carried out to see the total volume between the SBES and MBES data surface in each cross-section. This calculation uses the MBES data as a reference for the existing surface. Based on the analysis of differences in volume with a depth of 15 meters, the study area is 220985.95 m^2 resulting in a cut difference of 28557.5 m^3. Where there is the difference in fill is 4940.5 m^3 for SBES data. If the difference between cut and fill is totaled to see, the actual difference is 23617 m^3. These results, if divided by the area, every 1 m^2 has a difference of 0.11 m^3. This volume shows that the volume difference is not that much in a narrow space, but if the survey area is large enough, it can offer a massive difference in volume to affect the next work process significantly.

4. Conclusion and suggestion

In this study, conclusions and suggestions are as follows:

4.1 Conclusion

In the depth comparison test using statistical tests with the Z test for large samples of SBES and MBES data, the Z value (Z_0) is 0.018 m. Therefore, the single-beam and multibeam echosounder data in the case at Gresik Port are not significantly different. Based on the bathymetric map, the MBES surface is more detailed than the SBES, and the SBES surface is shallower than the MBES. So this shows that the SBES data has a shallower majority depth than the MBES data, which is 21078.81 m^3 MBES as a reference surface. The dredging volume differences with a depth of 15 m as a reference, where the area of the study area is 220985.95 m^2, between SBES and MBES data is 23617 m^3. Moreover, if these results are considered average, then the location of each 1 m^2 has a difference of 0.11 m^3.

4.2 Suggestion

The suggestions are given for further research namely, Comparison of echosounder and singlebeam data This multibeam echosounder is only done for study on the Gresik Port area which is shallow waters so it is desirable for research then the comparison of the two measurement data SBES and MBES can be carried out in water areas deep sea.

5. References

[1] IHO 2008 *IHO Standards for Hydrographic Surveys*, (Bureau: International Hydrographic)
[2] Ridhwan dkk 2016 *Analisa neraca jasa : Studi Kasus Industri Transportasi Maritim*, (Jakarta: Bank Indonesia)
[3] Triatmodjo Bambang 2010 *Perencanaan Pelabuhan* (Yogyakarta: Beta Offset)
[4] Djunarsjah E *Survei Hidrografi* (Bandung: PT. Refika Aditama)
[5] Badan Standarisasi Nasional 2010 *Survei Hidrografi Menggunakan Singlebeam Echosounder* (Jakarta: BSN)
[6] Poerbandono 2005 *Survei Hidrografi* (Bandung: Refika Aditama)
[7] Adil I 1998 *Kontrol Kualitas pada Pemetaan Batimetri* (Jakarta: Forum Ilmiah Tahunan Ikatan Surveyor Indonesia)
[8] Sidik Rahmat 2014 *Analisis Perbandingan Data Batimetri Hasil Pengukuran Multibeam Echosounder dan Singlebeam Echosounder Untuk Wilayah Perairan Dangkal* (Yogyakarta: Universitas Gadjah Mada)
[9] Widjajanti 2011 *Modul Kuliah Statistik dan Teori Kesalahan* (Yogyakarta : Universitas Gadjah Mada)
[10] Purwohardjo 1989 *Pengukuran Topografi* (Bandung: Institut Teknologi Bandung)

IOP Conf. Series: Materials Science and Engineering 1052 (2021) 012016 doi:10.1088/1757-899X/1052/1/012016

Analysis of Stern Shape Effect on Pre-Duct Propeller Performance Based on Numerical Simulation

R B Prihandanu ,I M Ariana, and D W Handani

Department of Marine Engineering, Institut Teknologi Sepuluh Nopember (ITS), Surabaya 60111, Indonesia

Abstract. In process and development of ship building, it always lead to the design that is environmentally friendly and has high efficiency values on the energy and economic side. The factor is the selection and determination of the propulsion system. In propulsion system, propeller and its flow is very influential, such as flow at the stern, the flow that will enter the propeller, and the propeller design used. This can increase the propulsive coefficient of the propulsion system. For single screw ship, the hull efficiency is usually between 1.1 - 1.4, while for twin-screw with conventional hull forms, the hull efficiency is around 0.95 - 1.05. The value of relative rotative efficiency is between 1.0 - 1.07 and the efficiency of the shaft is between 0.96 - 0.995. Meanwhile, the value of open water efficiency and hull efficiency greatly affect to ship performance. From the simulation, with the additional of pre-duct The hull efficiency value has increased by 2.2% - 4.3%, this is due to the increase of wake fraction value higher than the thrust deduction value. The relative rotative efficiency also increases by 0.8% - 1% for each variation of the block coefficient or type of ship. But, for the propulsive coefficient, the resulting gap was decreasing by 2% - 6%. Because of that the distance between the hull and the pre-duct, the design of the propeller, the distance between the pre-duct and the propeller, and the shape of the stern are important references for the pre-duct design.

1. Introduction

In process and development of ship building, it always lead to the design that is environmentally friendly and has high efficiency values on the energy and economic side. The factor is the selection and determination of the propulsion system. In propulsion system, propeller and its flow is very influential, such as flow at the stern, the flow that will enter the propeller, and the propeller design used. So, the addition of a device between the hull and the propeller is important to increase the propulsive coefficient, which will be discussed in this study.

The propulsive coefficient can be obtained by hearing the multiplication of the hull efficiency, the propeller efficiency, and the rotative relative efficiency. Changes in these values will affect the performance of the propulsion system, but each efficiency value has a different vulnerability. For single screw ship, the hull efficiency is usually between 1.1 - 1.4, while for twin-screw with conventional hull forms, the hull efficiency is around 0.95 - 1.05. The propeller efficiency values range from 0.35 - 0.75. The value of relative rotative efficiency is between 1.0 - 1.07 and the efficiency of the shaft is between 0.96 - 0.995[1]. Meanwhile, the value of open water efficiency and hull efficiency greatly affect to ship performance.

Of the three ESD Zones in Figure 1, ESD is needed in zone 1 to improve the performance of the flow and propulsion system at the stern of the ship. This ESD will have several forms, such as Pre-Duct, Pre-Swirl, and a combination of the two. Due to the impact of energy loss from the propeller, hull, and stern form, so that the ESD (Energy Saving Device) placement before the propeller can be used to reduce axial and rotational phenomena that contribute greatly to energy loss in the propulsion system[2].In this

IOP Conf. Series: Materials Science and Engineering 1052 (2021) 012016 doi:10.1088/1757-899X/1052/1/012016

study a circular duct design will be made on several types of vessels or different block coefficients to analysis the hull propeller interaction with circular duct design and without pre-duct for each variation. Base on numerical simulation, it will used computational fluid dynamic method with propeller actuator disk. By comparing the propulsive coefficient and wake flow with and without pre-duct for each variation of ship, the hull-propeller interaction in different coefficient block will be obtain.

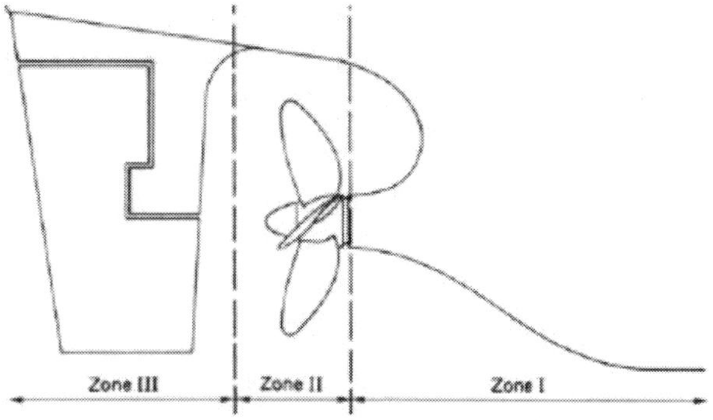

Figure 1. Zone classification of ESD [3].

From the previous research, it was explained that adding the length of the parallel middle body for a ship with the same stern will increase the ship's resistance coefficient. The shape of the stern change from U-type to V-type also affects the drag coefficient and wake flow [4]. So that the value of Cb can be an influence on ship resistance, stern shape, and wake flow. However, the addition of ESD does not have a substantial advantage even though the power reduction can occur up to 7%. This happens because the flow conditioning made by the Pre-Duct can increase the efficiency of the propeller [5][6][7]. However, changing the angle of attack can significantly increase propulsive efficiency [6][8]. Therefore the Pre-Duct and hull designs cannot be implemented carelessly. A match between the two is required for better efficiency in the propulsion system.

2. Hull-Propeller Interaction

2.1 Efficiency of Propeller

The propulsive coefficient is the multiplication of hull efficiency, open water efficiency, and relative rotative efficiency. This value is used to calculate the power requirements of the ship to reach a certain speed.

$$PC = \frac{P_E}{P_D} = \eta \mathrm{H} \times \eta \mathrm{R} \times \eta \mathrm{O} \tag{1}$$

However, this value is influenced by the value of gastric efficiency, propeller efficiency, and rotative relative efficiency. The efficiency value has a different vulnerability. Hull efficiency (ηH) is defined as the ratio of the effective power and thrust at the propeller or can be defined as follows,

$$\eta \mathrm{H} = \frac{EHP}{THP} = \frac{Rt \times V}{T \times Va} = \frac{Rt}{T} \times \frac{V}{Va} = \frac{1-t}{1-w} \tag{2}$$

For single-screw vessels, the hull efficiency (ηH) is usually between 1.1 and 1.4. For twin-screw boats with a conventional hull shape, the hull efficiency (ηH) is approx. 0.95 to 1.05. The higher the block coefficient value, the higher the hull efficiency value [1].

Propeller efficiency (**ηO**) is related to the Open Water Test, which is a propeller that works on a homogeneous wake field without any hull or barrier in front of it. Propeller efficiency depends on the flow, design, rotation speed, and size of the propeller itself. The propeller efficiency values range from

0.35 - 0.75. The value of propeller efficiency can be seen by conducting an open water test and reading the KT KQ J diagram [1]

$$\eta 0 = \frac{TV_a}{2\pi nQ} = \frac{KT_o J}{KQ_o 2\pi} \quad (3)$$

The actual velocity of the water flow to the propeller behind the hull is not constant or always at a right angle to the propeller area, but has a kind of rotational flow. Therefore, compared to the open water test conditions, propeller efficiency is also influenced by the propeller relative rotative efficiency (ηR). It can be defined as efficiency behind hull (ηB).

$$\eta \mathrm{B} = \eta \mathrm{R} \times \eta 0 \quad (4)$$

If resistance and propulsion model tests are performed, then the relative rotative efficiency is determined at model scale from the measurements of thrust and torque with the propeller operating behind the model. Using the non-dimensional thrust coefficient as input data, the values J and torque coefficient are read off from the open water curve of the model propeller used in the propulsion test. The torque coefficient of the propeller working behind the model is derived from

$$KQ_b = \frac{Q_m}{\rho n^2 D^5} \quad (5)$$

$$\eta \mathrm{R} = \frac{KQ_o}{KQ_b} \quad (6)$$

In single-screw vessels the relative rotative efficiency values range from 1.0 to 1.07, in other words, water rotation has a beneficial effect. The relative rotative efficiency of ships with conventional hull shapes and on twin-screw will usually be smaller, approximately 0.98 [1].

From the various efficiencies that affect the propulsive coefficient, it can be concluded that the efficiency value of the hull and propeller has the widest susceptibility, so this value will greatly affect the propeller design. Wake Fraction (w) and Thrust Deduction (t). For single-screw vessels, the wake fraction coefficient is usually between 0.20 to 0.45. The bigger the block coefficient, the bigger the wake fraction coefficient. In a twin-screw vessel with a conventional hull shape, the propeller will be positioned outside the stern centerline of the hull, which will lower the wake fraction coefficient [1]. The wake fraction coefficient will increase the risk of cavitation on the propeller because the water velocity distribution on the propeller is generally not homogeneous. The more homogeneous the wake fraction coefficient is, the higher the advance velocity will be so that a design modification on the stern of the hull is needed to produce an optimal wake field [1].

$$Rt = (1 - t)T \quad (7)$$

From the perspective of the hull studied, the addition of 3 vortex generators on each side made the ship's vibrations reduce by 50% in the longitudinal and vertical directions. The addition of an energy-saving device (ESD) will affect the shape of the wake field thereby increasing the propeller performance [9], [10]. Wake-adaption has an impact on propeller performance, both resulting in greater efficiency, vibration, and minimum noise characteristics. But behind-hull accessories will also contribute to wake flow. So, the hull and propeller compatibility must be analyzed to get the best efficiency.

In general, the thrust deduction coefficient size increases as the wake friction coefficient increases. The shape of the hull may have a significant effect, as in the bulbous bow, under certain conditions (low ship speed), reducing the thrust deduction coefficient. The thrust deduction coefficient for single-screw vessels is usually between 0.12 and 0.30. Ships with a large block coefficient have a large thrust deduction coefficient. For twin-screw vessels and conventional hull forms, the thrust deduction coefficient will be much lower because the suction process from the propeller occurs further from the stern centerline of the hull [1].

$$Va = (1 - t)Vs \tag{8}$$

2.2 *Energy Saving Device*

Energy-saving devices (ESD) are devices used to recover and minimize energy losses and reduce ship resistance or improve the propulsion efficiency of ships[8]. In some existing ESD, the source of energy loss to the propeller and hull is based on axial, rotational, and friction losses. The use of the ESD device near the propeller is mainly aimed at reducing axial and rotational phenomena which do contribute to energy losses. Several ESD models are grouped into certain zones as in Table 2.1.

Table 1. ESD Zone [3].

Energy-Saving Device	Zone(s) of Operation
Wake equalizing duct	I
Asymmetric stern	I
Grothues spoilers	I
Stern tunnels, semi- or partial ducts	I
Mewis duct	I
Reaction fins	I/II
Mitsui integrated ducted propellers	I/II
Hitachi Zosen nozzle	I/II
Increased diameter/low rpm propellers	II
Grim vane wheels	II
Propellers with end plates	II
Propeller boss fins	II/III
Additional thrusting fins	III
Rudder-bulb fins	III

Of the three existing zones, zone 1 is the zone that is most closely related to hull propeller interaction. In concept, the flow in the back of the ship will converge towards the propeller, besides that the propeller also absorbs the incoming flow to the propeller. So that the flow from the hull gathers at the stern and is utilized by the propeller. The faster the flow that enters the propeller, the better the propeller performance. By installing the Pre-Duct as in Figure 2, the flow will be used as an addition to the force of the push because of the foil shape on the Pre-Duct. With the trailing edge or tail foil size of 0.7R propeller for radius ducting, the water flow can be focused on the part of the propeller that is most constrained to produce thrust. As in Figure 2.

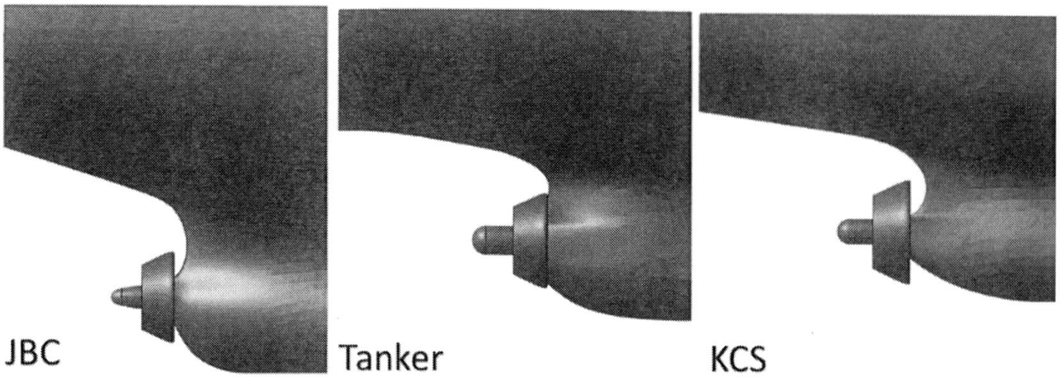

(a)

(b)

Figure 2. Pre-Duct addition refers to the distance of the propeller; (a) position; (b) water flow.

3. Validation

Due to simulation-based model testing, it is necessary to validate the comparison of experimental results with simulation results. The allowable error value cannot be more than 5%-7%. To get this value, the simulation is carried out at the model size so that the parameters being compared can have the same value. The following are the model parameters used on each ship.

Table 2. General Parameter of ships.

Main particulars		JBC	KCS	Tanker
Length between perpendiculars	L_{PP} (m)	280	230	100.9
Length of waterline	L_{WL} (m)	285	232.5	103.5
Maximum beam of waterline	B_{WL} (m)	45	32.2	19.2
Depth	D (m)	25	19	9.3
Draft	T (m)	16.5	10.8	6
Displacement volume	∇ (m^3)	178369.9	52030	9291.97
Wetted surface area w/o rudder	S_W (m^2)	19556.1	9424	2753.83

Block coefficient (CB)	$\nabla/(L_{PP}B_{WL}T)$	0.858	0.6505	0.779
Midship section coefficient (CM)		0.9981	0.9849	0.996
Diameter Propeller	D (m)	8.12	7.9	3.6
Number of Blade		5	5	4

Simulations were performed using the CFD method. The size of domain is adjusted to the distance from the AP of the ship. Forward 2.5 L, backward 4.75 L with a depth of 1.8 L over 1.3 L, and Sides 2 L as in Figure 3. Simulations are carried out with the same domain size and meshing settings for each type of vessel with or without Pre-Duct. This is to get a minimal error when adding Pre-Ducts. The meshing density is carried out on the propeller as shown in Figure 4.

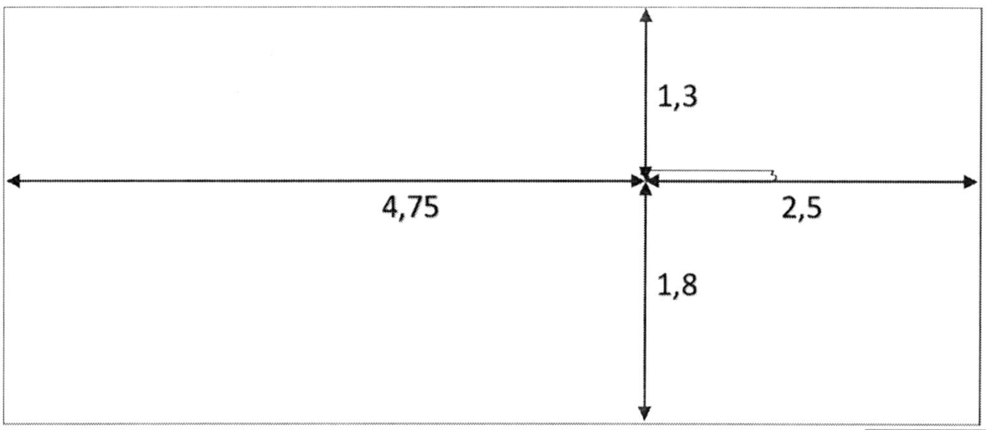

Figure 3. Computation Domain.

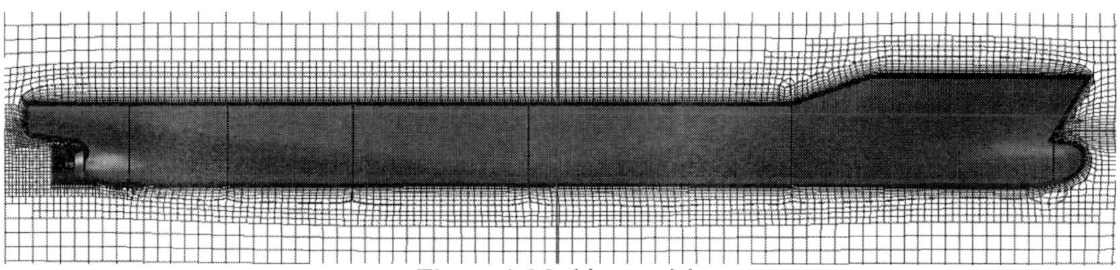

Figure 4. Meshing model.

From these parameters, a simulation is carried out and the following are the results of the simulations that have been carried out for model validation. Validation is done on JBC, Tanker and KCS without Pre-Duct as Table 3-5.

Table 3. Error Simulation Japan Bulk Carrier.

Type of Vessels	Name	n	J	KT	KQ	1-w	1-t
Japan Bulk Carrier (JBC)	EFD	7.8	0.405	0.217	0.0279	0.552	0.812
	CFD	7.52	0.403	0.220	0.0286	0.522	0.831
	Err. %	3.6%	0.4%	-1.5%	-2.5%	5.4%	-2.4%

Table 4. Error Simulation Tanker.

Type of Vessels	Name	RT	T	n	t
Tanker	EFD	237000	305000	3.27	0.22
	CFD	229383	300712	3.39	0.24
	Err. %	3.2%	1.4%	-3.6%	-5.9%

Table 5. Error Simulation KRISO Container Ship.

Type of Vessels	Name	n	KT	KQ
KRISO Container Ship (KCS)	EFD	9.500	0.170	0.0288
	CFD	9.325	0.165	0.0283
	Err. %	1.8%	3.2%	1.7%

4. Comparison and Analysis

From the simulation results that have been carried out on 3 types of ships with different block coefficients, with and without pre-ducts, the following is a comparison of the results.

Table 6. Simulation Model.

Type of Vessels (Cb)	Name	*1-w*	*1-t*	ηH	ηR	ηO	*PC*
Japan Bulk Carrier (JBC) (0.858)	Without Pre-Duct	0.572	0.830	1.452	1.054	0.415	0.635
	With Pre-Duct	0.527	0.781	1.484	1.065	0.386	0.610
	Gap	-7.9%	-5.9%	2.2%	1.0%	-6.9%	-3.9%
Tanker (0.779)	Without Pre-Duct	0.548	0.832	1.519	1.026	0.379	0.591
	With Pre-Duct	0.509	0.803	1.578	1.034	0.355	0.579
	Gap	-7.2%	-3.6%	3.9%	0.8%	-6.4%	-2.0%
KRISO Container Ship (KCS) (0.6505)	Without Pre-Duct	0.793	0.892	1.125	1.016	0.627	0.716
	With Pre-Duct	0.676	0.793	1.174	1.026	0.559	0.674
	Gap	-14.8%	-11.1%	4.3%	1.0%	-10.8%	-6.0%

Based on the results of the simulations obtained as in Table 4. The hull efficiency value has increased by 2.2% - 4.3%, this is due to the increase of wake fraction value higher than the thrust deduction value. The relative rotative efficiency also increases by 0.8% - 1% for each variation of the block coefficient or type of ship. But, for the propulsive coefficient, the resulting gap was decreasing by 2% - 6%. This is due to the reduction in the efficiency of open water which is not proportional to the increase in hull efficiency. Changes in advance velocity also make the advance coefficient shift. So the advance velocity value and the efficiency of open water is reduced. However, the reduction in propeller efficiency is higher than the increase in hull efficiency, so that the propulsive coefficient value decreases.

The suitability of the Pre-Duct shape and placement for each vessel type is different. The Pre-Duct design must be adjusted accordingly to get a better propulsive coefficient value. The pitch-diameter ratio or area ratio can be adjusted to keep open water efficiency in same value for advanced coefficient of propeller with pre-duct

5. Conclusion

CFD method has been feasible to analyse the benefit use of ESD such as the implementation of pre-duct on different type of vessels. It confirm by the validation data from the experiment and the simulation. In general, the addition of the Pre-Duct will increase the hull efficiency by 2.2% - 4.3% and the relative rotative efficiency by 0.8% - 1%. The addition of Pre-Duct has a good impact on the hull efficiency of ships with low coefficient block. but the value of open water efficiency decreases and it is not proportional to the increase in hull efficiency. Therefore the propulsive coefficient value decreases. The distance between the hull and the pre-duct, the design of the propeller, the distance between the pre-duct and the propeller, and the shape of the stern are important references for the pre-duct design.

References

[1] MAN B&W Diesel A/S, "Basic Principles of Ship Propulsion," *Man Diesel Turbo*, pp. 1–42, 2013, doi: 10.1017/CBO9781107415324.004.

[2] A. S. Olsen, "Energy coefficients for a propeller series," *Ocean Eng.*, vol. 31, no. 3–4, pp. 401–416, 2004, doi: 10.1016/j.oceaneng.2003.06.002.

[3] J. S. Carlton, *Theoretical and Analytical Methods Relating to Propeller Action*, no. 1865. 2012.

[4] K. Suastika, F. Nugraha, and I. K. A. P. Utama, "Parallel-middle-body and stern-form relative significance in the wake formation of single-screw large ships," *Int. J. Technol.*, vol. 8, no. 1, pp. 92–101, 2017, doi: 10.14716/ijtech.v8i1.3542.

[5] T. van Terwisga, "On the working principles of Energy Saving Devices," *Third Int. Symp. Mar. Propulsors*, no. May, pp. 510–518, 2013.

[6] F. Çelik, "A numerical study for effectiveness of a wake equalizing duct," *Ocean Eng.*, vol. 34, no. 16, pp. 2138–2145, 2007, doi: 10.1016/j.oceaneng.2007.04.006.

[7] H. Schneekluth, "Wake equalizing duct," *Nav. Archit.*, vol. 102, pp. 147–150, 1986.

[8] A Munazid, I M Ariana, and K A P Utama, "Numerical Study of Pre-Duct on Traditional Fishing Vessels as Energy Saving Device (ESD) Numerical Study of Pre-Duct on Traditional Fishing Vessels as Energy Saving Device (ESD)," in *Maritime Safety International Conference*, 2020, doi: 10.1088/1755-1315/557/1/012049.

[9] L. X. Hou, C. H. Wang, A. K. Hu, and F. L. Han, "Wake-adapted design of fixed guide vane type energy saving device for marine propeller," *Ocean Eng.*, vol. 110, pp. 11–17, 2015, doi: 10.1016/j.oceaneng.2015.09.036.

[10] J. H. Kim, J. E. Choi, B. J. Choi, S. H. Chung, and H. W. Seo, "Development of Energy-Saving devices for a full Slow-Speed ship through improving propulsion performance," *Int. J. Nav. Archit. Ocean Eng.*, vol. 7, no. 2, pp. 390–398, 2015, doi: 10.1515/ijnaoe-2015-0027.

IOP Conf. Series: Materials Science and Engineering 1052 (2021) 012017 doi:10.1088/1757-899X/1052/1/012017

Numerical Prediction of Foils Configuration in A Design of Buoy Glider System for Supporting Tsunami Early Warning

Arifin , H N Wibowo , H Buddin , and W Bambang

Pusat Teknologi Rekayasa Industri Maritim (PTRIM) - BPPT

Abstract. Geographically, Indonesia is a disaster-prone country, with the potential to cause damage to public infrastructure not only cause enormous loss of life. Based on the tsunami risk map released by BNPB in 2018, it needs more attention so that mitigation can be implemented properly. It necessary to revitalize nationally developed by BPPT the buoy-based early warning system for tsunami disasters. Early generation tsunami buoy designs were successfully developed and operated in several waters in Indonesia. In order to improve the performance of the tsunami buoy, a new vehicle system known as the Wave Buoy Glider (WBG) is proposed. It is hoped that this WBG design can cover a wider monitoring area as well as more wave elevation measurement data related to tsunami events. In this research some of the buoy glider-design parameters such as sizes and configurations of foil are studied in order to obtain the optimal pulling force of the wave glider to the surface floater. A numerical calculation using the CFD technique is used to determine the hydrodynamic force. So that from this study, it can be seen that an increasing number used foils, the drag and lift forces also tend to increase. The larger the span foil, the greater the value of the drag and lift forces, especially at the larger AoA angle. The positive values of drag forces at any lift forces confirm that wave glider always move forward with every vertical movement of the floater.

1. Introduction

These Geographically, Indonesia is a disaster-prone country, with the potential to cause damage to public infrastructure and basic connectivity not only cause enormous loss of life. Based on the tsunami risk map released by BNPB in 2018, it needs more attention so that mitigation can be implemented properly. An Early Warning System is required for minimizing of the tsunami risk.

In accordance with the mitigation policy as stated by Indonesia Government, it is urgently to revitalize the buoy-based early warning system developed by BPPT. Early generation tsunami buoy designs were successfully developed and operated in several waters in Indonesia. However, the designed buoy has a weakness in the power supply capacity and the data recording area is not wide enough.

In order to improve the performance of the tsunami buoy system, a new vehicle system known as the Wave Buoy Glider (WBG) is proposed. There is a unique characteristic of this wave glider. It utilize a harvest environmental energy such as wave energy as the propulsion of the vehicle and solar energy as electronic, communication and sensor power supply [1].

IOP Conf. Series: Materials Science and Engineering 1052 (2021) 012017 doi:10.1088/1757-899X/1052/1/012017

This marine vehicle consists of two main parts; a floating part which is located on the surface of the ocean (the float) and the submerged part (the glider) which is gliding approximately eight meters below the surface of the ocean. These parts are connected via (umbilical) a tether cable which is mainly responsible to transfer the forward motion of the submerged glider to the surface floater as shown in Figure 1. Since the surface floater elevates with the waves in the ocean, it pulls the submerged glider up. Lift force generated by the surface floater will be converted into forward thrust because of the aerofoil shaped fins located on the submerged glider.

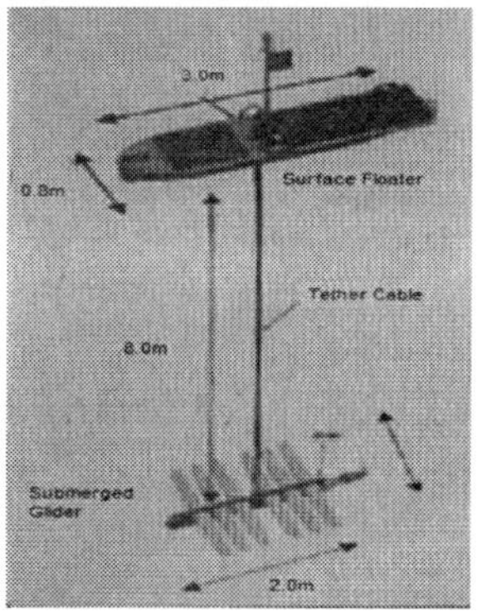

Figure 1. Wave Buoy Glider (WBG).

According to the capability of the designed buoy glider, it is hoped that this buoy glider design can cover a wider monitoring area as well as more wave elevation measurement data related to tsunami events. The WBG can also be designed for another marine applications in order to gather useful data such as ocean's salinity, temperature, climate, fishery, environmental assessment and oil and gas, etc. [2][3]. The designed buoy glider capability improvement can be achieved by involving parameters of hydrodynamic performance in the WBG design such as the drag and lift coefficients [4]. Up to now, significant effort s have been taken on the wave-induced motion analysis of the wave glider. Manley and Willcox [5]

As previously explained, the forward motion of the surface glider is caused by the pulling force of the wave glider which is influenced by the heave motion of the surface floater. In order the surface floater can cause a pulling force on the floater, it is necessary to design the foil configuration on the wave glider correctly. There are several foil configurations with specific aspect ratios that are proposed for the wave glider design.

In this paper, a numerical method will be introduced to describe dynamic behaviour of this glider correlate to the propulsion system of the surface floater. Furthermore, a final CFD model will be introduced and the results will be discussed accordingly.

2. Background

A number studies of the dynamic behaviour of marine vehicles has been done by numerous scientists. For a marine vehicle similar to the WBG vehicle, there are only a few papers describing the dynamic modelling of such systems.

An ASC (Autonomous Surface Craft) ARTEMIS was developed firstly by MIT Sea Grant College Program [6]. One of the main disadvantage of the design is its small size, which limits its applications and durability. In June 2000, the mechanical systems of the ASC were heavily modified in such that some potential improvements in the basic platform design were identified as desirable [7].

Kraus and Bingham [8] proposed an estimation algorithm and an extended Kalman filter, using two body dynamics to describe the potential of the Wave Glider for precise localization. The focus of the mentioned work was on the relative motions of the float and glider. Nicholas D. Kraus [9] published their work considering six-degree-of-freedom in the dynamics equations of motion of the Wave Glider. The work was consistent with the common method of dynamic modeling of marine vehicles which can be found in numerous marine literatures. The focus of the work was to analyze the kinetics of the Wave Glider with respect to three different coordinate systems located on each part independently, then transfer all parameters to the main coordinate frame located on the tethered (umbilical) between the float and the glider and describe the motion with respect to that origin. The main issue with that work was the inconsistent results from simulation results compared to the real time data. Lack of enough hydrodynamic parameters especially for the submerged part and also complicated propulsion system of this vehicle could be mentioned as one of the valid reasons for such inconsistency.

In this paper, two parts of the WBG vehicle are considered separately. The submerged glider which plays the role of the propulsion system of the vehicle is simplified and it is replaced with its effective forces only. The surface floater is considered as an object sailing through the ocean which uses the power supplied by the glider's propulsive force.

3. Related Theory

3.1 Regular wave

Where ζ (x,t) is the wave elevation at time t and position x, ω is the frequency of the wave, ε is the phase difference and k is the wave number which is defined by $2\pi/\lambda$, in which λ is the wavelength. Figure 2 illustrates a regular wave with described formula and its parameters.

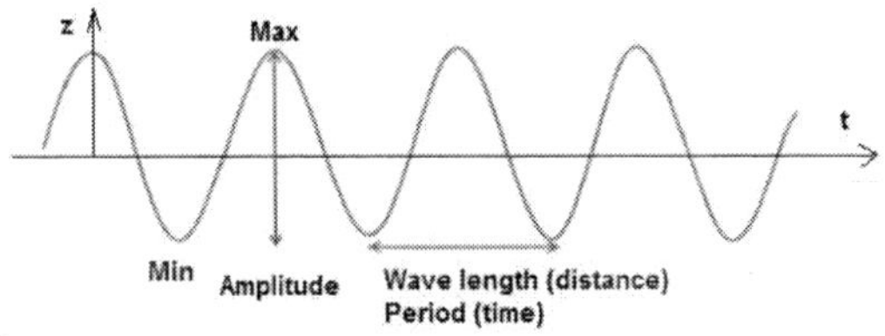

Figure 2. Wave Parameters of Regular Waves.

In figure 2, H is defined as height of a regular wave. In practice, to work with a better accuracy, significant wave height is usually reported as $H1/3$ which is the average of the heights of the largest one third of the waves.

Regular wave theory describes ocean waves as sinusoidal waves. Regular wave mathematic is presented in Equation 1 [10].

$$\varsigma(x,t) = \varsigma_a \sin(\omega t - kx + \varepsilon) \tag{1}$$

For fluid dynamic analysis as well as dynamic modeling of the WBG vehicle, it is necessary to consider the condition of the elevation of the vehicle in the ocean. For that matter, characteristics of sea state 3 is considered.

3.2. Numerical Modelling.

As previously explained, the Submerged Glider is planned to consist of a rigid structure equipped with a number of foil-shaped wings. The submerged glider is mainly responsible for providing thrust for the surface floater and keeping it on a predefined course. Basically, When a foil-shaped structure moves in a fluid, 2 (two) forces will arise which are referred to as drag and lift forces, as illustrated by Figure 3 below.

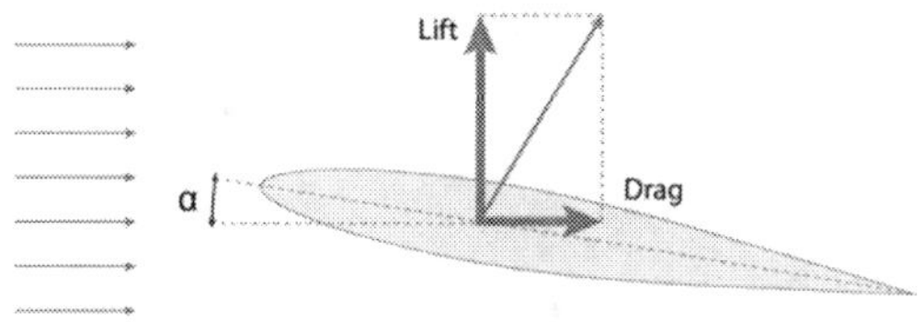

Figure 3. Drag and Lift Force of Glider Foil.

The drag and lift forces acting on the foil can be determined using equations (3) and (4) as follows.

$$F_D = \frac{1}{2} C_D \rho u^2 A \tag{2}$$

$$F_L = \frac{1}{2} C_L \rho u^2 A \tag{3}$$

Where:

F_D : Drag force
F_L : Lift force
C_D : Drag coefficient
C_L : Lift coefficient
ρ : Fluid density
u : Fluid velocity
A : Surface area of foil

The Conservation of Momentum

Imagine a Newton second law, with the famous definition:

$$F = m.a \tag{4}$$

The definition of force itself basicaly is the rate of change of momentum

$$F = \frac{dp}{dt} = m\frac{dv}{dt} \tag{5}$$

The sum of momentum (mass * velocity) of the collided body will always be constant or conserved. This law also states the total forces exerted on the body is the sum of the forces from external as well as from the internal body itself (inertial force, gravity, pressure, viscosity etc.).

In the fluid dynamics equation, the forces are divided by volume and then can be rearranged as the following equation (in the x-direction):

$$\frac{\partial(\rho u)}{\partial t} + \vec{\nabla}.\left(\rho u \vec{V}\right) = -\frac{\partial p}{\partial x} + \frac{\partial \tau_{xx}}{\partial x} + \frac{\partial \tau_{yx}}{\partial y} + \frac{\partial \tau_{zx}}{\partial z} + \rho f_x \tag{6}$$

The above equation is known as the famous Navier-Stokes equation. But, the Navier-Stokes equation sometimes refers to the continuity and energy equation as well.

The Navier-Stokes equations for the incompressible flow around the foil are discretized and solved by a finite volume method over a circular domain large of 50 chords having the NACA airfoil at the center. A structured mesh is built, coming from refined cells near the airfoil wall, where the gradients are expected to be important, to coarse cells at the external boundaries. Uniform velocity inlet and pressure outlet boundary conditions are assumed on the fluid boundaries, while a smooth wall no-slip condition is applied over the solid boundary, just as shown in Figure 4.

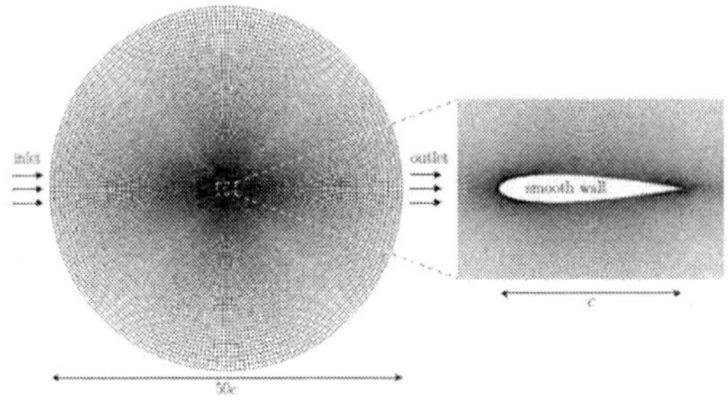

Figure 4. Structured Mesh Built over the Computational Domain Around the Airfoil.

A Reynolds Averaged Navier-Stokes (RANS) approach is chosen for the modeling of the turbulence. The one-equation Spalart-Allmaras model (designed specifically for airfoil simulations) and the two-equation k-ω SST model (providing better results for separating flows) are compared in the validations cases. Two near-wall treatment strategies are tested: the *wall function* approach, where a model is assumed instead of solving the equations down deep into the boundary layers, and the *solve wall* approach, which requires a very high mesh refinement for the first cells near the wall.

The designed submerged Glider as it is shown in Figure 5 represents the WBG's propulsion system which is motivated with propulsion force, via the foils. These foils are designed to move the vehicle by converting the lift force of the glider into thrust.

Considering only the wings of the submerged glider and applying fluid dynamic analysis on them, the thrust forces have been obtained. This work will be done with the help of a Computational Fluid Dynamic (CFD) software package such as ANSYS. In this work, these forces have been considered as the propulsive forces of the vehicle. The boundary conditions in numerical modeling of CFD wave buoy glider are as shown in Figure 6 below.

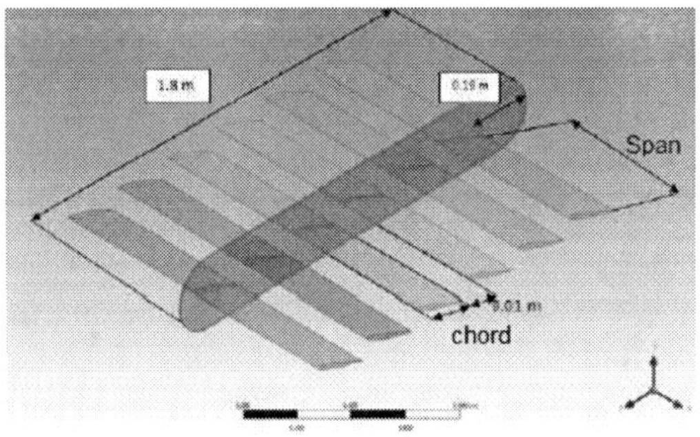

Figure 5. Submerged Glider Propulsion System.

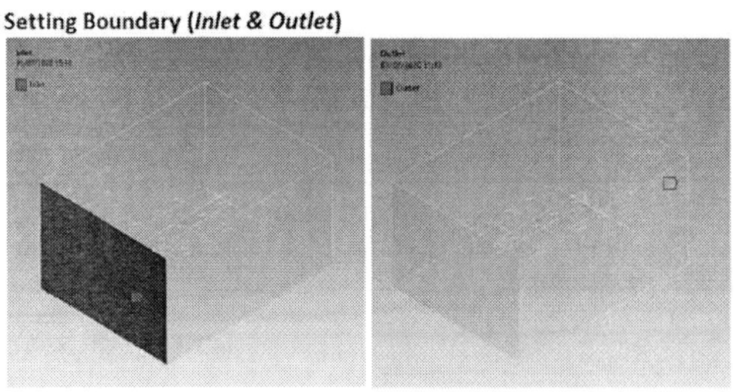

Figure 6. Setting Boundary Condition in CFD.

It is necessary to make the model as simplified as it is possible for the purpose of fluid dynamic analysis. These simplifications resulted in delivering the outputs of the software in shorter time. In addition, on some cases the results are more reliable. In this work, since the model is symmetric and all the foils have equal dimensions, the analysis can be done for one foil only and then repeated for other foils.

In addition, it is necessary to make appropriate assumptions. These assumptions are listed as follows:

- The submerged glider is negative buoyant proportional regarding to its volume. This means that the force caused by weight of the submerged glider has higher value than its buoyancy.
- The foils are the only components which supply the glider with a thrust force.
- The submerged glider always has at higher acceleration than the surface floater. In other words, the tethered umbilical is always in tension.
- The ocean waves assumed to follow the exact ideal pattern (regular wave) which is presented by numerous marine literatures.
- Constant values for the fluid analysis are considered for chosen sea states characteristics based on data obtained from Bales [11].
- Heading angles are assumed to be the same for the float and the glider.

4. Results and Discussion

Based on a numerical calculation has been done, it has been obtained lift and drag forces and then be converted to lift and drag coefficients as function change of foil Angle of Attack as shown in Table 1, 2 and 3 and presented in Figure 7 and 8 below.

Table 1. Result of 2 Wings Configuration.

AoA (deg)	Lift Force (N)	Lift Coeff	Drag Force (N)	Drag Coeff
30	562.6330	0.0206	353.1206	0.0129
20	483.5112	0.0177	219.1964	0.0080
10	240.0983	0.0088	95.8139	0.0035
0	-0.3972	0.0000	57.3500	0.0021
-10	-228.2380	-0.0083	92.9270	0.0034
-20	-513.7380	-0.0188	329.3490	0.0120
-30	-535.5480	-0.0197	468.1607	0.0172

Table 2. Result of 4 Wings Configuration.

AoA (deg)	Lift Force (N)	Lift Coeff	Drag Force (N)	Drag Coeff
30	815.1746	0.0259	527.2376	0.0167
20	645.0772	0.0205	307.9832	0.0098
10	387.3090	0.0123	156.9679	0.0050
0	0.1317	0.0000	86.6205	0.0027
-10	-371.0019	-0.0118	153.4016	0.0049
-20	-784.6629	-0.0249	513.9262	0.0163
-30	-865.8700	-0.0275	771.2266	0.0245

Table 3. Result of 6 Wings Configuration.

AoA (deg)	Lift Force (N)	Lift Coeff	Drag Force (N)	Drag Coeff
30	862.4542	0.0274	570.3803	0.0181
20	671.1469	0.0213	332.5493	0.0105
10	351.9175	0.0112	157.2340	0.0050
0	-0.0653	0.0000	81.4905	0.0026
-10	-350.8834	-0.0111	158.8921	0.0050
-20	-861.3293	-0.0273	572.7470	0.0182
-30	-928.5336	-0.0295	831.8585	0.0264

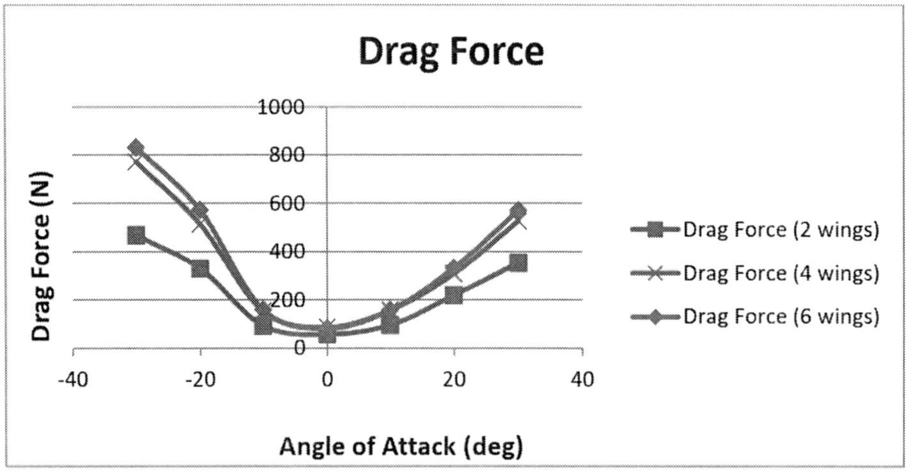

Figure 7. Drag Force Comparison of 2, 4 and 6 Foils Configuration.

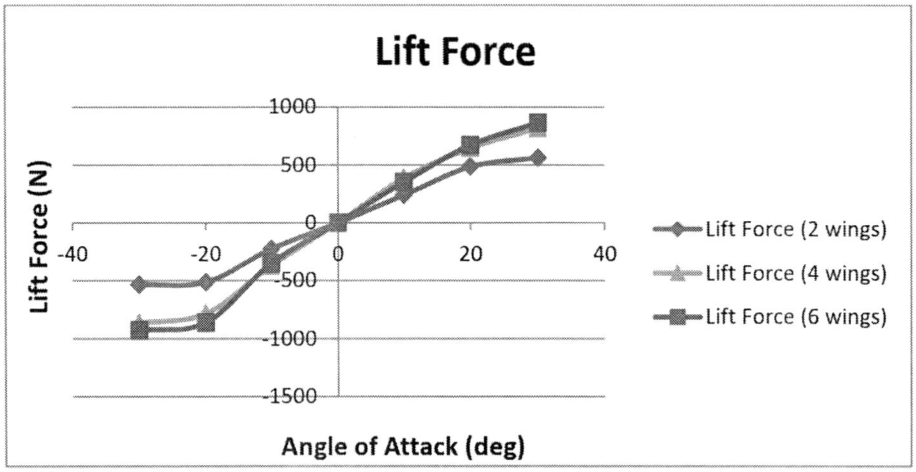

Figure 8. Lift Force Comparison of 2, 4 and 6 Foils Configuration.

Based on these figures above, It can be seen that the greater the Angle of Attack (AoA) of the foil, the greater drag and lift forces. Likewise, in the case of changes in the configuration of 2, 4 and 6 foils, the more the number of foils used, the drag and lift forces also tend to increase.
The numerical calculation results of the lift and drag forces on the wave glider design which are influenced by changes in the span of foils are shown in Tables 4 - 7, and these results are presented in graphical form as shown in Figures 9 and 10 below.

Table 4. Result of 6 Wings Configuration (Span = 550mm).

AoA (deg)	Lift Force (N)	Lift Coeff	Drag Force (N)	Drag Coeff
30	485.1623	0.0154	286.2960	0.0091
20	484.1353	0.0154	190.9680	0.0061
10	336.2130	0.0107	91.3220	0.0029
0	7.8590	0.0002	42.9990	0.0014
-10	-325.9520	-0.0103	89.3592	0.0028
-20	-487.9460	-0.0154	192.0476	0.0061
-30	-501.7479	-0.0159	293.7624	0.0093

Table 5. Result of 6 Wings Configuration (Span = 750mm).

AoA (deg)	Lift Force (N)	Lift Coeff	Drag Force (N)	Drag Coeff
30	677.9137	0.0215	401.0426	0.0127
20	677.3550	0.0215	267.2990	0.0085
10	484.8560	0.0153	127.3302	0.0040
0	6.3130	0.0002	56.9460	0.0018
-10	-472.2380	-0.0150	120.5500	0.0038
-20	-705.1210	-0.0224	270.9725	0.0086
-30	-719.7525	-0.0228	414.3294	0.0131

Table 6. Result of 6 Wings Configuration (Span = 850mm).

AoA (deg)	Lift Force (N)	Lift Coeff	Drag Force (N)	Drag Coeff
30	778.8572	0.0247	458.9685	0.0146
20	789.7500	0.0250	309.5400	0.0098
10	570.5960	0.0181	144.8290	0.0046
0	2.5294	0.0001	63.6520	0.0020
-10	-570.0085	-0.0180	141.1603	0.0045
-20	-814.5370	-0.0258	309.8979	0.0098
-30	-823.1083	-0.0261	476.0105	0.0151

Table 7. Result of 6 Wings Configuration (Span = 1000mm).

AoA (deg)	Lift Force (N)	Lift Coeff	Drag Force (N)	Drag Coeff
30	967.6928	0.0307	574.5480	0.0182
20	973.0994	0.0308	380.7926	0.0121
10	736.3180	0.0234	182.5451	0.0058
0	34.2860	0.0011	74.3293	0.0024
-10	-707.1829	-0.0224	168.7563	0.0054
-20	-1007.2590	-0.0320	383.7394	0.0122
-30	-1024.9670	-0.0325	588.4294	0.0187

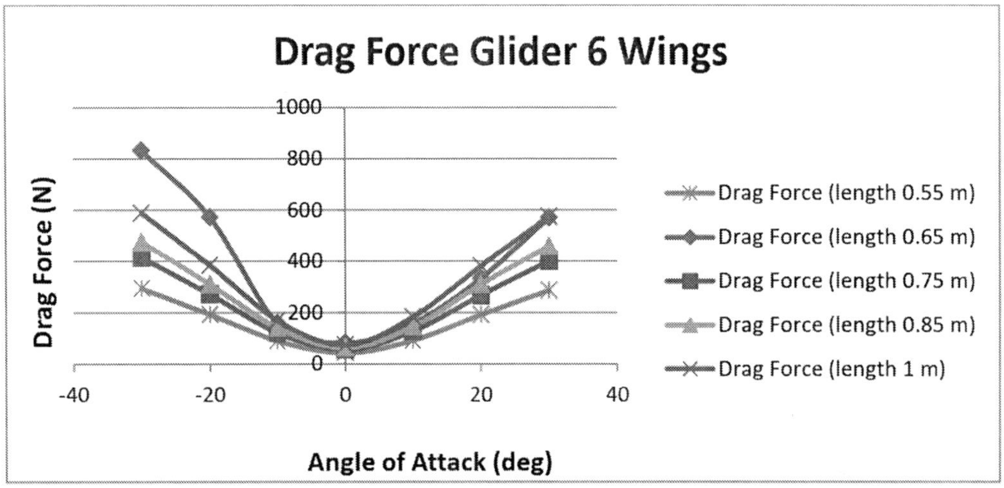

Figure 9. Drag Force Comparison of Different Span Foils Configuration.

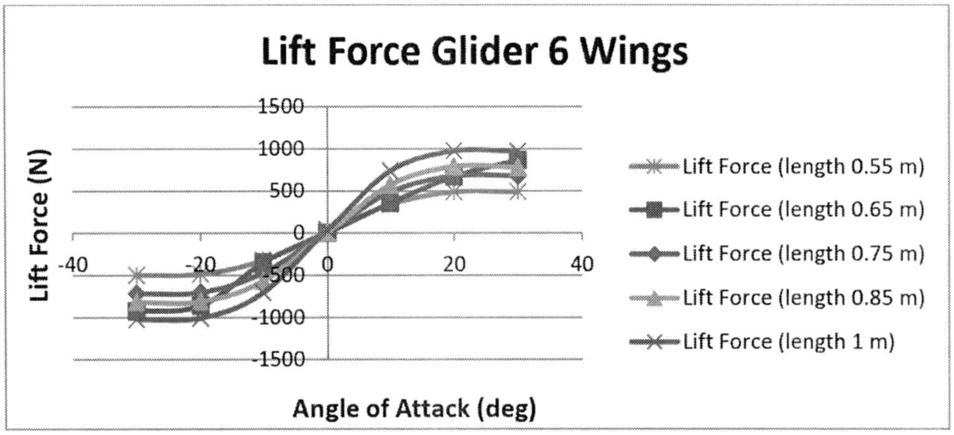

Figure 10. Lift Force Comparison of Different Span Foils Configuration.

Based on these figures above, It can be seen that the greater the Angle of Attack (AoA) of the foil, the greater drag force. However, increasing the lift force effectively occurs at a change in Angle of Attact of foil from 0 to 20 degrees. At a larger angle it does not affect the lift force so much.
The larger the span foil, the greater the value of the drag and lift forces, especially at the larger AoA angle.

5. Conclusion

In the design of the Wave Buoy Glider vehicle to support the provision of early warning tools against tsunami hazards, a numerical simulation of the effect of the foil configuration on the drag and lift forces has been carried out.
Based on the analysis of the results of numerical calculations that have been carried out, the following conclusions can be drawn:

- The greater the Angle of Attack (AoA) of the foil, the greater of drag and lift forces. Likewise, in the case of changes in the configuration of 2, 4 and 6 foils, the more the number of foils used, the drag and lift forces also tend to increase.
- The larger the span foil, the greater the value of the drag and lift forces, especially at the larger AoA angle.
- The positive values of drag forces at any lift forces confirm that wave glider always move forward with every vertical movement of the floater.
- From this research, it was clear that a foil configuration play an important role in the design of Wave Glider Buoy, in order to determine an optimum dimension of the designed foils for WBG.
- It is necessary to do another research by modeling the excitation wave force to get a more realistic pulling force of wave glider to surface floater.

6. Acknowledgements

This research was financially supported by the Agency of Assessment and Application oftechnology, Ministry of Research Technology, Republic Indonesia through Program of Rancang Bangun Buoy dan OBU (Inovasi Sistem dan Teknologi INA-TEWS), PTRIM, TIRBR – BPPT FY2020.

7. References

[1] Autonomous Undersea Vehicle Applications Center, (n.d.) *AUV SystemSpec Sheet – SHARC Wave Glider configuration* . Available: www.auvac.org/configurations/view/247.

[2] Hine R., S. Willcox, G. Hine, and T. Richardson, 2009. The wave glider: A wave-powered autonomous marine vehicle," in *OCEANS 2009, MTS/IEEE Biloxi-Marine Technology for Our Future: Global and Local Challenges*, pp. 1-6.

[3] B. Coxsworth, 2013. *Liquid Robotics announces next generation of Wave Glider sea-going robot* [Online]. Available: http://www.gizmag.com/liquid-robotics-wave-glider-sv3-aquatic/26977/. April.

[4] Nicholas D. Kraus, 2012. Wave Glider Dynamic Modeling, Parameter Identification and Simulation", *Master Science Thesis,* Mechanical Engineering Department.

[5] J. Manley , S. Willcox , The Wave Glider: a persistent platform for ocean science, Oceans (2010) 1–5 .

[6] Manley J.E, 1997, Development of the autonomous surface craft , ACES, *Proc. MTS/IEEE Conf. OCEANS*, vol. 2, Oct., pp. 827_832.

[7] Manley J.E, 2008, Unmanned surface vehicles, 15 years of development, *Proc. IEEE OCEANS*, Sep., pp. 1_4.

[8] Kraus and B. Bingham, 2011. Estimation of Wave Glider dynamics for precise positioning," in *OCEANS*, pp. 1-9.
[9] Nicholas D. Kraus, 2012. Towards a Dynamic Model of an Autonomous Wave Glider.
[10] Perez T., 2005. Ship Motion Control; Course Keeping and Roll Stabilization Using Rudder and Fins," *Advances in Industrial Control. Springer".*
[11] Bales, 1983. Wind and Wave Data for Seakeeping Performance Assessment, *"Prepared for Seakeeping Comittee ITTC*", Athens, Greece, Technical Report.

Impact of seabed slope on steel catenary riser touchdown zone response

A. M. Ogbeifun[1], S. Oterkus[1], J. Race[1], H. Naik[2], D. Moorthy[3], S. Bhowmik[2], J. Ingram[3]

[1]Naval Architecture, Ocean and Marine Engineering, University of Strathclyde, Glasgow, G4 0LZ, UK.
[2]McDermott International, 40 Eastbourne Terrace, Paddington, London, W2 6LG, UK.
[3]McDermott International, 757 N. Eldridge Parkway, Houston, TX 77079, USA.

Abstract. Several factors can affect the response of steel catenary risers (SCR) around its touch down zone (TDZ). These include the stiffness of the soil, the soil suction force on the riser TDZ, the soil degradation with cyclic TDZ loading, etc. Riser strength and fatigue response computation are usually performed considering flat seabed and with the use of rigid or linear (spring) riser soil interaction model. However, bathymetric information obtained for the SCR lay path on the seabed reveal complex seabed profile variation, indicating that the seabed is far from being flat around the SCR TDZ. This paper presents findings from an investigation conducted on the influence of seabed slopes on the strength and fatigue response of SCRs, using a non-linear (NL) riser soil interaction model. The responses of SCRs on positively and negatively sloped seabed (rotated about the static touch down point on flat seabed) are compared with responses of SCRs on flat seabed. From the results, it can be deduced that the SCR dynamic and fatigue responses may be overpredicted or underpredicted in magnitudes dependent on the slope deviation of actual seabed from a flat seabed.

1. Introduction

Steel catenary risers are the most attractive riser solution because of their simplicity and robustness in application [1]. A major challenge with SCR application is the high stress and fatigue damage incurred around its critical sections, which are the hang off (HO) and the TDZ, shown in Figure 1. Prediction of SCR stress and fatigue response around the TDZ was in the early history of SCR design performed using a rigid soil model. This approach was then progressively improved by the development and implementation of linear (spring) soil-riser interaction models. More recently, considerable efforts have been made to improve the SCR TDZ interaction model by the development of the non-linear models [2-4]. The NL riser soil interaction model provides a more realistic approach for evaluating riser response interactions with the seabed.

It is a common practice during the design of SCR to assume a flat seabed [5]. Realistically, seabed can hardly be flat. Bathymetric data reveal complexities in the topography of the seabed and the variation of the seabed profiles along riser azimuth. Sloped seabed, especially in the immediate region surrounding TDZ, may impact SCR configurations calculated based on the flat seabed assumption, leading to SCR TDP mismatch [6]. The mismatch is the distance between the SCR TDP on an assumed flat seabed and the SCR TDP on realistic sloped seabed. This work investigates the influence of seabed slope on the flat seabed static SCR configurations and the impact sloped seabed can have on the strength and fatigue response around the SCR TDZ.

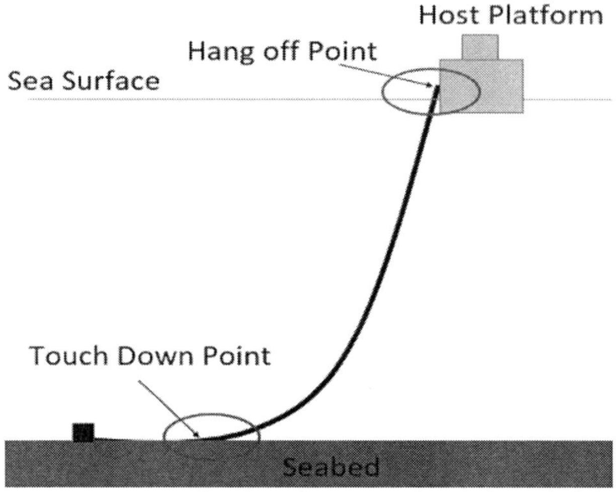

Figure 1. Conventional steel catenary riser schematic [7].

The impact of seabed slope on SCR response is investigated, considering the hysteretic non-linear soil interaction model developed by Randolph and Quiggin [2]. The sloped seabed in this investigation extends linearly before the TDP to the SCR seabed termination. The slopes of the linear seabed profiles considered are 0deg (flat seabed), ± 2deg, ± 4deg, ± 7deg and ± 10deg. Three groups of SCRs with hang-off angles of 8deg, 12deg and 16deg are implemented on theses profiled seabed. The SCRs static, dynamic and fatigue responses are calculated considering nominal configurations of the risers, i.e. no vessel offsets. Findings from this study will provide not only relevant information about SCRs behaviour on the sloping seabed but also possible seabed modifications to improve SCR response.

2. Analysis data

There are three groups of SCRs characterised by their hang-off angles (with the vertical), which are 8deg, 12deg and 16deg, respectively. Each group consists of the SCR resting on nine linear profiled seabed with seabed slopes of 0deg (flat seabed), ± 2deg, ± 4deg, ± 7deg and ± 10deg as presented in Table 1. For each hang off group of SCR configuration, the sloped seabed is made to pass through the static touch down point of the SCR configuration on the flat seabed for that group. The SCRs are made up of 12inch pipe joints, hosted by a floating production system in a water depth of 1500m, and are conveying fluid of density 600kg/m^3. The SCRs wall-thickness (27.5mm) is calculated based on DNV-OS-F201 criteria [8], considering burst and collapse requirements for a design pressure of 10ksi. The hang-off connection stiffness of each SCRs has a linear value of 12kN.m/deg. Note that the HO angles given in Table 1 are the angles the riser makes with the vertical at the riser vessel interface.

Table 1. Riser group for analysis.

Group	Hang off-angle (deg)	Seabed Slope (deg)
1	8	
2	12	0, ±2, ±4, ±7, ±10
3	16	

2.1. Storm and fatigue load data

Presented in Table 2 are the selected storm and fatigue wave loads under which we investigate the responses of the groups of SCRs. Since the study is comparative rather than project design-based, we consider here a total exposure time of the SCRs to the fatigue load to be 30years, neglecting stress concentration factors and other safety factors.

Table 2. Storm and fatigue wave load.

Analysis	Hs(m)	Tp (sec)	gamma
Storm	8.0	13	1.6
Fatigue	1.5	5.5	1.8

2.2. Seabed data

The non-linear hysteretic riser-soil interaction model developed by Randolph and Quiggin (R&Q) is applied for this investigation. Table 3 presents the default model data for the R&Q model. Details of the NL model can be found in [2]. Recall that the seabed profiles investigated in this study are 0deg (flat seabed), ± 2deg, ± 4deg, ± 7deg and ± 10deg.

Table 3. Non-linear soil model data [2].

Model Parameter	Value
Mudline shear strength	5kPa
Shear strength gradient	1.5 kPa/m
Power law parameters (a, b)	(6, 0.25)
Normalised maximum stiffness	200
Suction ratio	0.6
Suction decay parameter	0.4
Penetration parameter	0.2

3. Methodology

In this study, we conducted the analysis calculations using OrcaFlex finite element (FE) software package. The initial flat seabed SCR configurations for the FE modelling is calculated using the catenary equations (1) to (5), with equations (4) and (5) derived from the basic catenary equations (1) to (3). Where s is the catenary length from HO to TDP, x is the horizontal span from HO to TDP, z is the vertical span from HO to TDP, w is the submerged unit weight of pipe section, θ is the HO angle with the horizontal, which can be expressed as $(tan^{-1}(V/H))$, H is the horizontal components of tension, V is the vertical components of tension, and T is the resultant tension force of H and V. These equations are developed based on the assumptions of infinite axial stiffness and negligible bending stiffness for the lines and are suitable for calculating initial line configurations.

$$z = \frac{H}{w}\left[Cosh\left(\frac{wx}{H}\right) - 1\right] \tag{1}$$

$$s = \frac{H}{w}Sinh\left(\frac{wx}{H}\right) \tag{2}$$

$$T = H + wz \tag{3}$$

$$H = \frac{wz}{(Tan\theta)^2}(1 + Sec\theta) \tag{4}$$

$$x = \frac{HSinh^{-1}(tan\theta)}{w} \tag{5}$$

3.1. *Seabed Slope modelling*

The objective of this study is to investigate the impact of the sloped seabed on SCR TDZ responses and compared them with those of the SCR on the flat seabed. Therefore, we first calculate the initial SCR configuration on flat seabed using the catenary equations presented in the previous section, and then rotate the seabed about the static SCR TDP to impose the seabed slope influence on the TDZ section. We apply this technique for the modelling of the three groups of the SCRs of different HO angles resting on the nine seabed slopes (see Table 1). The following conventions were adopted for the sloped seabed modelling:

- The seabed slope is positive when the seabed increases in the positive direction (upward relative to the TDP) from the SCR TDP to the riser seabed termination.
- The seabed slope is negative when the seabed increases in the negative direction (downward relative to the TDP), from the SCR TDP to the seabed termination.

Consider a flat seabed (AD), as shown in Figure 2. We will need to determine the coordinates of point B and C to be able to define the linear sloped seabed profile BC. The coordinates of B and C will be such that the line joining them passes through the static TDP of the SCR on a flat seabed. By performing this rotation, we can impose the effect of the sloping seabed on the SCR configuration. This approach represents the scenario during riser preliminary layout design where the static TDP is known for an assumed flat seabed, which is eventually found not to be flat but sloped, or a scenario where a flat seabed is assumed. Referencing all points from the hang-off point (*O*) on the mean sea level (MSL) in Figure 2, z coordinates of point B and C can be expressed as:

$$Z_B = Z_A - (X_A - X_{TDP})\, tan\, \alpha \tag{6}$$

$$Z_C = Z_D + (X_D - X_{TDP})\, tan\, \alpha \tag{7}$$

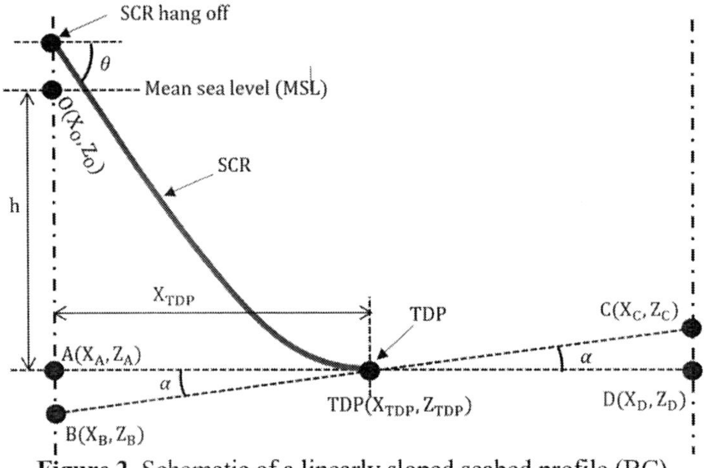

Figure 2. Schematic of a linearly sloped seabed profile (BC).

Where $X_A = X_B$ is the *x* coordinate of point A and B, $Z_A = Z_{TDP} = Z_D$ is the elevation of the seabed referenced from the riser hang-off point, X_{TDP} is the x coordinate of the static TDP of the SCR on a flat seabed, and α is the seabed slope. OrcaFlex FE software possesses the capability to create profiled seabed as well as model the NL riser soil interaction. With the Z coordinates of point B and C calculated from equation (6) and (7), the seabed profile can be defined in OrcaFlex. Examples of SCRs configurations on the flat, positive and negative sloped seabed are shown in Figure 3 (a), (b) and (c). The seabed incidence angle, β is the angle the riser makes with the vertical at the static TDP. Under dynamic conditions, the dynamic incidence angle ($\beta_{dynamic}$) varies about the static incidence angle (β_{static}). Increasing the HO angle with the vertical and the angle of the positively sloped seabed will result in increasing β_{static}, vice versa as observed in Figure 3 (a), (b) and (c), where $\beta_1 > \beta_2 > \beta_3$. β is an essential SCR geometric parameter, influenced by the HO angle, seabed slope and riser motion. It will be shown later how β correlates with the SCR TDZ response under dynamic loading conditions.

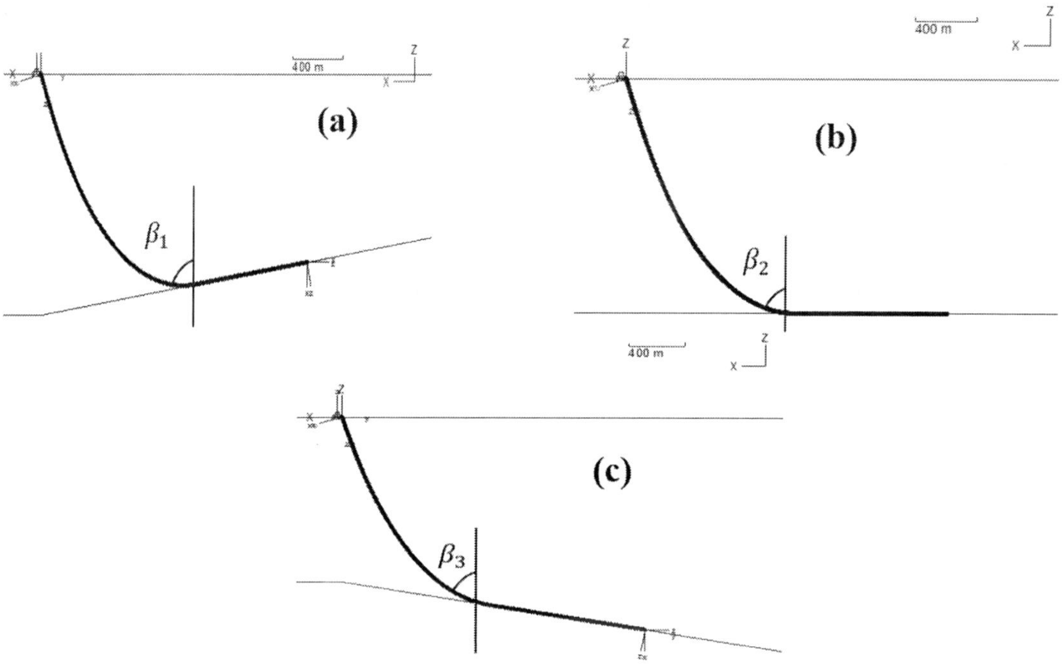

Figure 3. Example: OrcaFlex SCR model with HO angle with vertical = 16deg on (a) flat seabed (0deg), (b) positive slopped seabed (+10deg) and (c) negative sloped seabed (-10deg).

3.2. Numerical solution approach and response calculation

The FE simulations are conducted in the time domain, which is fully non-linear and couples the riser dynamics with the vessel motions. The implicit integration scheme is applied. The modelling, simulation and post-processing are automated using MATLAB programs integrated with the OrcaFlex high-level programming interface (OrcFxAPI [9]). The program calculates the initial SCR configurations using the catenary equations and creates the OrcaFlex SCRs models on flat seabed. The static calculation is then conducted to obtain the SCRs static TDPs coordinate on the flat seabed. The TDP coordinate and the seabed slope are then used to calculate point B and C using equations (6) and (7), such that the straight line from B to C passes through the calculated flat seabed static TDP coordinate (see Figure 2). The linear profile defined by line BC is then used to develop and update the seabed profile, inducing the sloped seabed profile on the configurations of SCR initially resting on the flat seabed. The program repeats this for other SCRs across the three groups, generating the required models. The storm and fatigue irregular wave loads (see Table 2) are modelled using the JONSWAP spectrum and simulated for all SCRs throughout 1200sec. The vessel RAOs and riser pipe content are set as well. The simulation period of 1200sec was considered to be sufficient since the study is a comparative one, same wave loads are imposed on all SCR models across the three groups, and the possibility of resonance in the SCRs are not considered. Numerical results post-processed by the program include the SCR configurations, effective tensions, stress utilisation, and fatigue damage along the SCRs.

The SCRs stress utilisations are calculated using the DNV-OS-F201 combined load resistance factor design criteria. Detailed information about the DNV-OS-F201criteria can be found in [8]. The fatigue damage is computed from the variation of stresses in the risers. The rain flow counting technique [10] is applied to express the varying SCR stress spectrum as a histogram of stress reversal. The Miner's rule, presented in equation (8) is applied to calculate the cumulated fatigue damage, *D*.

$$D = \sum_i \frac{n_i}{N_i} \tag{8}$$

Where n_i is the ith stress range amplitude components of the stress range histogram, N_i is the number of cycles to failure associated with the ith stress range as obtained from the S-N curve [11].

4. Results and discussions

4.1. Static touch down point (TDP) offsets

Figure 4 presents the static TDP arc length locations of the SCR on sloped seabed (s_α) relative to the static TDP arc length for the same SCR on flat seabed (s_{0deg}). The relative static TDP arc lengths ($s_\alpha - s_{0deg}$) is the offsets of the TDP on the sloped seabed from the static TDP on flat seabed, and can be seen as the error incurred when flat seabed are assumed rather than sloped seabed around the SCR TDZ. Positive lengths indicate longer static TDP arc lengths from hang-off (HO) relative to the flat seabed static TDP arc length, and negative lengths indicate shorter TDP arc lengths relative to the flat seabed static TDP arc length.

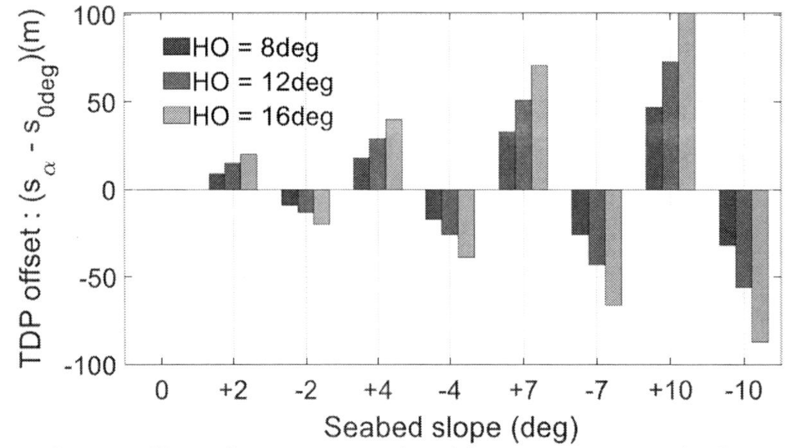

Figure 4. Static TDP offsets of SCRs TDP on sloped from SCRs TDP on the flat seabed (0deg slope).

From Figure 4, it is observed that the SCRs will first touch the seabed (i.e. at shorter arc length from HO) on the negatively sloped seabed. The trend shows a progressive increase in TDP offset as the riser hang-off angle increases, and as the seabed slope increases in both positive and negative sense. TDP offset between 40m and 100m is possible for seabed slope between ±4deg and ±10deg. For deeper water where the SCR TDP is farther from the HO, the TDP offset can be even more significant. The SCR TDP offset errors occur when a flat seabed is assumed during riser design rather than the realistic sloped seabed. Such errors can impact on the design of the SCR whose sections around the TDP are to be cladded with carbon resistive alloys (CRA) to reduce corrosion induced fatigue damage. The cladding of the internal riser section with CRA is expensive and requires accurate estimation of the arc length positions of the sections to be cladded. This underpins the importance of seabed slope consideration during the SCR design.

4.2. Static top tension

The static top tension of SCR is a combination of the weight of its hanging section, including its content ($600kg/m^3$ used in this study), less its displacement (buoyancy). The external and internal pressure effects are accounted for in the case of effective tension calculation. The difference in the lengths of the hanging section (from HO and TDP) caused by the seabed slope can result in a difference in the static top effective tension (T_{eff}). The difference between the static top tension of SCRs on the sloped seabed and those of SCRs on the flat seabed are presented in Figure 5. Table 4 in the appendix presents the numerical values of the top tension. It could be observed from these results that both positively and negatively sloped seabed can cause an increase in the top tension compared with SCRs on the flat seabed. The increase in top tensions is higher for SCRs on seabed slope beyond $\pm$4deg. However, the top tension increase is observed to be smaller in magnitude for positively sloped seabed when compared with SCRs on the negatively sloped seabed. For example, with seabed slope of -10deg, the static top tension of SCRs on the flat seabed can be increased by more than 20kN, when compared with flat seabed SCRs. The bar chart plot trend also indicates that the change in top tension of each sloped seabed SCRs relative to their corresponding flat seabed SCRs increase and decrease respectively for negatively and positively sloped seabed, as the SCR HO angle increases.

Recall that the SCRs on negatively sloped seabed touched down first (i.e. TDP occurs at shorter arc length from HO) before those of positively sloped seabed, with the flat seabed SCRs in between (see Figure 4). This is supposed to mean that the top tension of SCR on the negatively sloped seabed will be smaller than the same SCR on the positively sloped seabed. However, as can be seen in Figure 5, this is not the case. A detailed look at Figure 6 which is the configuration plot for SCR with 16deg HO angle resting on $\pm$ 10deg sloped seabed, reveals the reason why it is so. Although the TDP of the riser on negative seabed touches down first, the riser section beyond the TDP on the seabed still contributes to the SCR hanging weight until the riser seabed frictional force cancels it. On the other hand, the weight beyond the TDP of the SCR resting on the positively sloped seabed is applied on the seabed, not the hanging riser section.

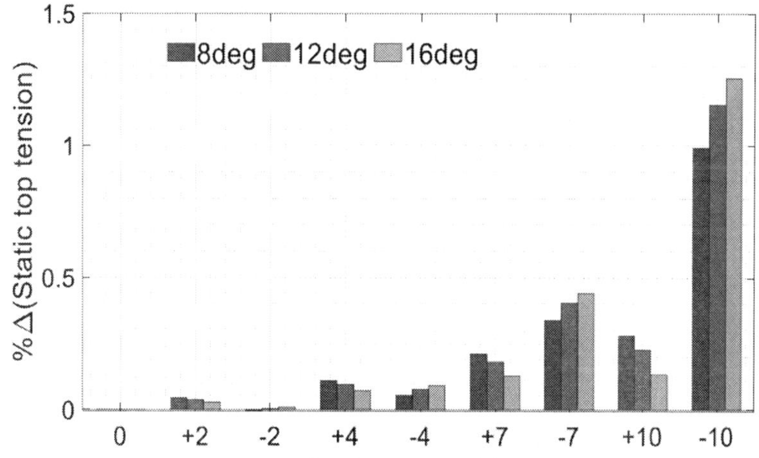

Figure 5. Percentage change in static top tension of SCRs on sloped seabed relative to SCRs on a flat seabed.

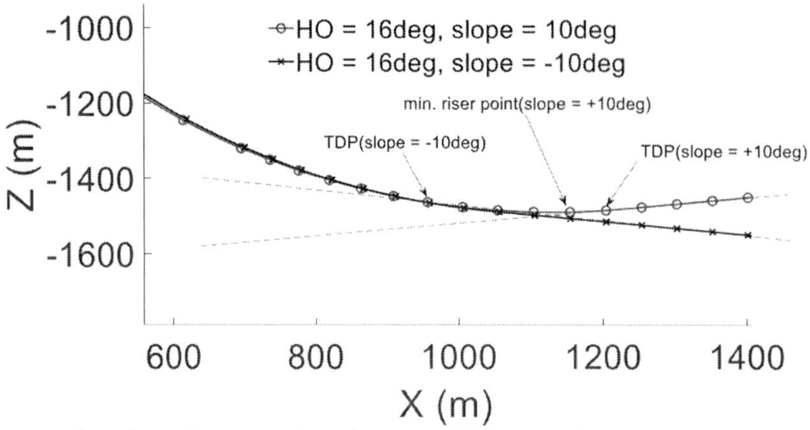

Figure 6. Expanded view of TDZ sections for SCR with HO angle = 16deg, seabed slope = -10deg and +10deg.

4.3. Dynamic analysis – Maximum top effective tension

The range graph maximum effective tension is the vector of nodal maximums over the simulation time. The maximum top tension occurs during the upward motion of the vessel where the SCRs are pulled upward, causing a longer hanging section of the SCRs. It is observed from Table 5 and Figure 7 of the appendix that the negatively sloped seabed SCRs have higher top tension when compared with the flat and positively sloped seabed SCRs. For example, the results presented in Table 5 and Figure 7 indicate more than 5% increase in SCR top tension for SCRs on -10deg sloped seabed compared with SCRs on the flat seabed. Top tensions of positively sloped seabed SCRs have values slightly less than SCRs on the flat seabed. SCRs top tensions are seen to be more sensitive to increasing negatively sloped seabed than positively sloped seabed. The hang-off tensions are also observed to be higher relative to SCRs on the fat seabed for increasing HO angles. These observations can be translated to some sort of positive (for positively sloped seabed) or negative (for negatively sloped seabed) impact on the design of topside supporting structures for the SCRs when a flat seabed is assumed during the SCR design rather than the profiled seabed. As the water depth increases, the top tensions are expected to increase. Similarly, the impact on top tension of flat seabed SCRs by an imposed negatively sloped seabed is expected to also increase.

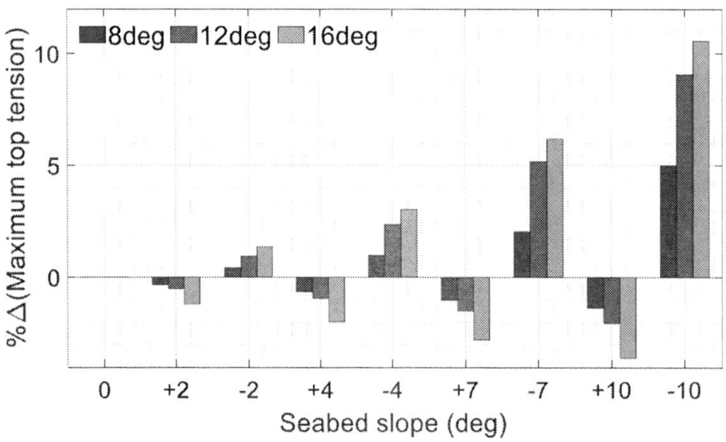

Figure 7. Percentage change in maximum dynamic top tension of sloped seabed SCRs relative to SCRs on a flat seabed.

Note: Negative values mean reduction in maximum top tension in the sloped seabed SCRs relative to the flat seabed SCRs.

4.4. Dynamic analysis – TDZ compression

The negativity of the effective tension value around the SCR TDZ is an indication of riser compression. As observed in Table 6 in the appendix and Figure 8, the SCR TDZ compression is sensitive to seabed slope around the TDZ. The level of compression for flat seabed SCRs are significantly amplified for a small increase of inclination of the negatively sloped seabed. On the other hand, compression observed for flat seabed SCRs is improved (or reduced) significantly when the seabed becomes positively sloped. Note that the abscissa of Figure 8 contains both positive and negative values, where, the positive values indicate an increase in compression (increased negativity of effective tension in the sloped seabed SCRs) relative with the flat seabed SCRs, and the negative values represent the converse. As mentioned earlier, the SCRs resting on negatively and positively sloped seabed have smaller and larger seabed incidence angle (β) respectively, compared with incidence angle on flat seabed SCRs. Under the dynamic condition, this appears to have translated to higher SCR TDZ compressions or higher buckling tendencies for SCRs having smaller β as will be seen later in the most critical compressive point on the riser presented in Figure 11 (b) and (c) of section 4.6. Increased compression for negatively sloped seabed SCRs may be critical for SCRs designs where TDZ buckling is a major challenge. The results show that we under- and over-predict the TDZ buckling when we assume a flat seabed rather than the profiled seabed around the SCR TDZ for negatively and positively sloped seabed, respectively. For example, SCR of HO angle 16deg, encountering -4deg seabed slope around its TDP, can have up to 500% increase in TDZ compression unaccounted for if we assumed a flat seabed profile for the design. On the other hand, with a positively sloped seabed profile, a significant benefit in the form of reduced TDP compression or buckling is observed. In general, it is observed for SCRs resting on the same sloped seabed, that compression increases with decreasing hang off-angle

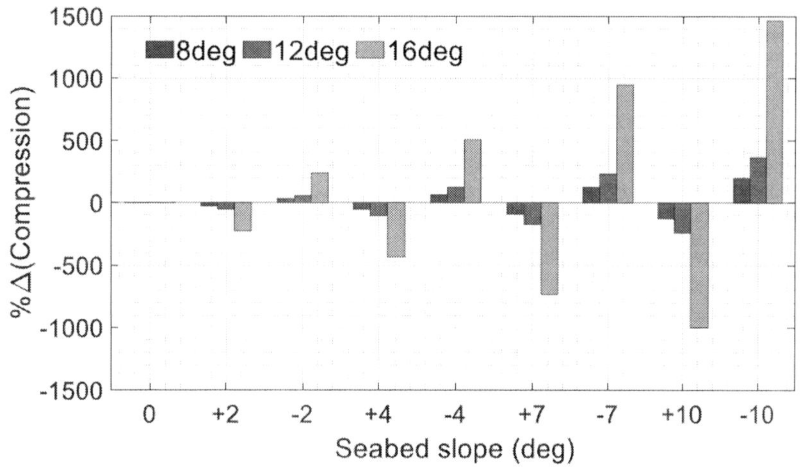

Figure 8. Percentage change in SCR TDZ compression relative to SCRs on a flat seabed.

Note: Negative values mean reduction in compression in the sloped seabed SCRs relative to the flat seabed SCRs.

4.5. Dynamic analysis – Stress utilisation

The range graph maximum of the DNV-OS-F201 dynamic stress utilisation around the SCR TDZ are presented in Figure 9 (a), (b) and (c).

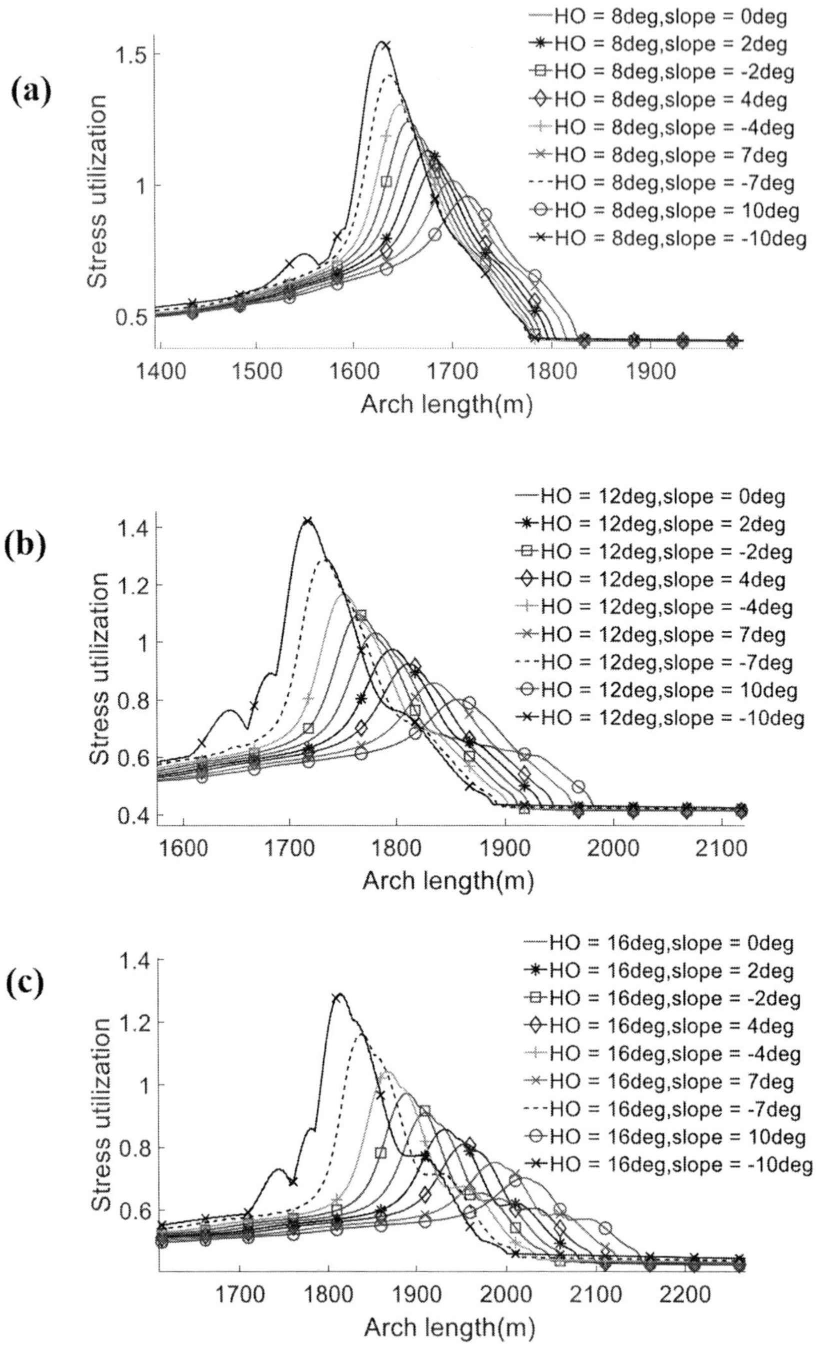

Figure 9. Maximum DNV-OS-F201 stress utilization in TDZ for risers: (a) HO angle = 8deg, (b) HO angle = 12deg, (c) HO angle= 16deg.

The stress utilisation is a combination of stress component contributed to by the axial tension, bending moment and pressure loads. The most dominating component of the stress utilisation around the TDZ is the bending moment component. Since higher buckling tendencies were observed in the TDZ of SCRs resting on the negatively sloped seabed, it is expected that the stress utilisation within this region is higher for the negatively sloped seabed. This is confirmed by the results presented in Figure 9 (a), (b), and (c). It can be observed that SCRs resting on negatively sloped seabed incur higher TDZ stress utilisation compared with same SCRs resting on the flat seabed. The converse is the case for the positively sloped seabed. It is also observed in general (both static and dynamic cases) that the point of highest TDZ stress utilisations shifts towards the HO direction for increasing negatively sloped seabed and towards the seabed anchor direction for the positively sloped seabed. This again is an indication of the challenge that may be encountered during the design specification of clad (CRA) section on the SCR TDZ if the seabed slope around the TDZ is not correctly considered. For SCRs on the flat seabed, the TDZ stress utilisation is observed to be higher for risers with smaller HO angles since this implies smaller β and higher bending stress. However, it is possible to have higher TDZ stress utilisation for SCRs with larger HO angle compared with SCRs with smaller hang off-angle if the seabed slope on which the former risers rest result in smaller β than the later risers. This may imply that β value, which is driven by a combination of HO angle and the seabed slope (α) is the single parameter indicator of the level of bending moments, stress and buckling around the TDZ.

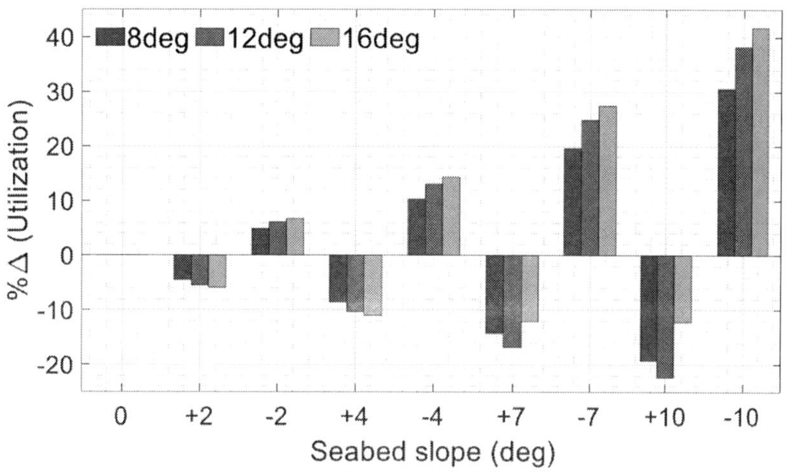

Figure 10. Percentage change in SCRs TDZ stress utilisation relative to SCRs on a flat seabed.

Note that the negative values mean reduction in stress utilisation in the sloped seabed SCRs relative to the flat seabed SCRs. Like the observations made on results for SCRs maximum top tensions and TDZ compressions, the positively sloped seabed provides benefits in the form of the reduction in stress utilisation. On the other hand, assuming a flat seabed in SCR design, when indeed the seabed is negatively sloped around the SCR TDZ, stress utilisation can be underestimated.

4.6. Dynamic analysis - Time history of the most critical point

Compression or buckling in the SCR TDZ depends on the velocity of the TDZ section during riser excitations. The magnitude of the global vertical component of the SCR TDZ velocity (GZ velocity) gives an indication of the level of impact energy and deformation of the SCR when it contacts the seabed at every load cycle. TDZ compression can occur when the difference between the downward velocity of the riser TDZ section are higher compared with the section terminal velocity, and when the TDZ section impact on the seabed at higher acceleration. The time history of the seabed incidence angle (β), the GZ velocity, the effective tensions and stress utilisation at the critical points within the SCRs TDZ sections are presented in Figure 11(a), (b), (c) and (d).

(a)

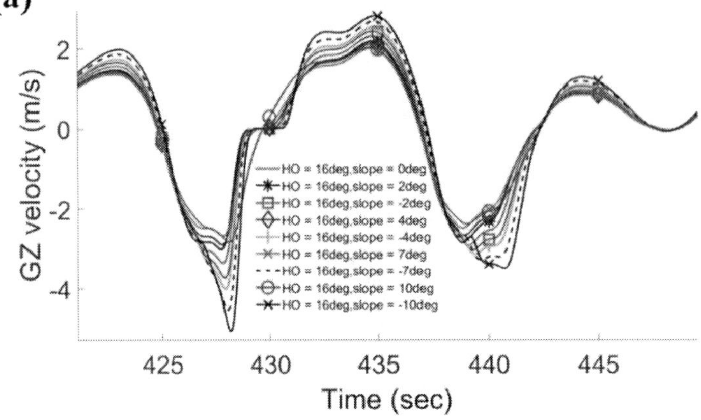
GZ velocity (m/s)
Time (sec)
HO = 16deg,slope = 0deg
HO = 16deg,slope = 2deg
HO = 16deg,slope = -2deg
HO = 16deg,slope = 4deg
HO = 16deg,slope = -4deg
HO = 16deg,slope = 7deg
HO = 16deg,slope = -7deg
HO = 16deg,slope = 10deg
HO = 16deg,slope = -10deg

(b)

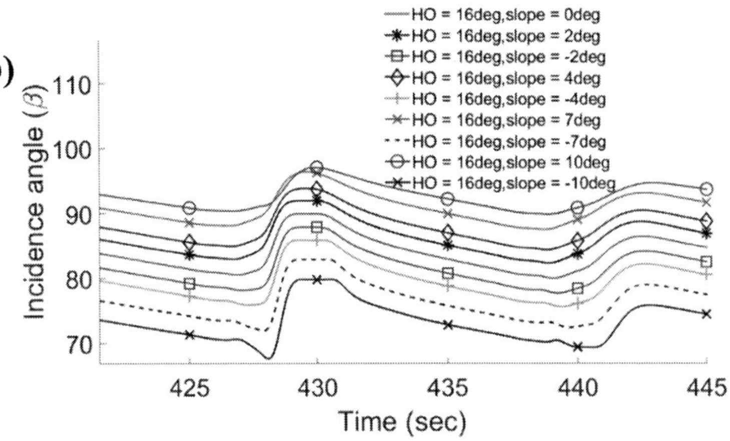
Incidence angle (β)
Time (sec)
HO = 16deg,slope = 0deg
HO = 16deg,slope = 2deg
HO = 16deg,slope = -2deg
HO = 16deg,slope = 4deg
HO = 16deg,slope = -4deg
HO = 16deg,slope = 7deg
HO = 16deg,slope = -7deg
HO = 16deg,slope = 10deg
HO = 16deg,slope = -10deg

(c)

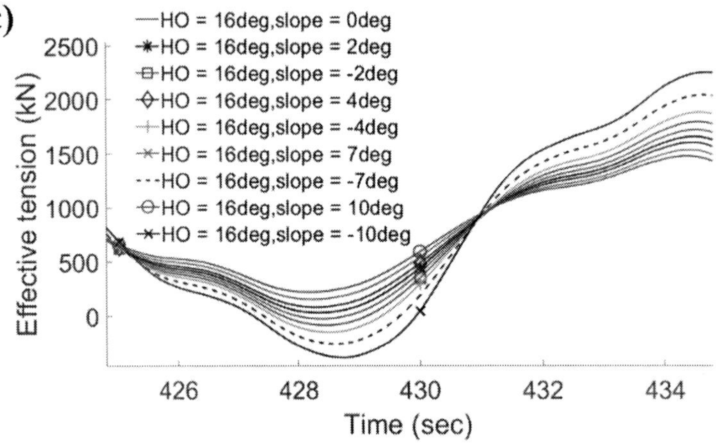
Effective tension (kN)
Time (sec)
HO = 16deg,slope = 0deg
HO = 16deg,slope = 2deg
HO = 16deg,slope = -2deg
HO = 16deg,slope = 4deg
HO = 16deg,slope = -4deg
HO = 16deg,slope = 7deg
HO = 16deg,slope = -7deg
HO = 16deg,slope = 10deg
HO = 16deg,slope = -10deg

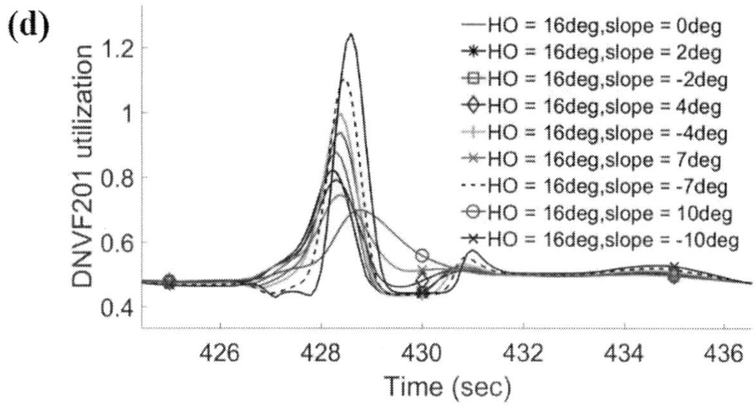

Figure 11. Critical TDZ point time history (a) GZ velocity, (b) seabed incidence angle (β), (c) Effective tension, (d) DNV-OS-F201stress utilisation.

The critical points are the arc lengths along the SCRs measured from the HO, where the highest stress utilisation occur for the SCRs. It is observed that β have the smallest values at the time when the downward GZ velocities are highest. This indicates a correlation between the impact velocity and the resulting SCR curvature characterised by β. On the other hand, value of β are higher for lower values of downward GZ velocities. Hence, higher impact velocities at these points on the seabed resulted in lower values of β, which in turn causes higher dynamic curvature, bending moments, compressions, and stress utilisations. The positively sloped seabed is characterised by smaller β than flat seabed and negatively sloped seabed in that order, and hence reduced compression and stress utilisation in the same order.

4.7. Fatigue analysis – fatigue damage

The range graph fatigue damage per year for the SCRs are presented in Figure 12 (a), (b) and (c). The results revealed that with increasing positively sloped seabed (increasing β), the fatigue damage reduces compared with same SCRs on the flat seabed. The converse is the case for negatively sloped seabed with smaller β. The percentage change in the fatigue damage of the sloped seabed SCRs relative to the flat seabed SCRs are presented in Figure 13. Numerical values of Figure 13 are presented in Table 8 of the appendix. These results indicate imposed benefits on the SCRs fatigue lives by the positively sloped seabed and reduction on the SCRs lives by negatively sloped seabed around the SCR TDZ. The results indicate that we may be under- and over-predicting the TDZ fatigue damage when we assume a flat seabed rather than the profiled seabed around the SCR TDZ for negatively and positively sloped seabed respectively. For example, it could be observed from the results that positively sloped seabed can reduce the fatigues damage response by more than 50%, for SCRs resting on the positively sloped seabed of 10deg. On the other hand, the fatigue damage can be increased by more than 20% for the same SCR configurations, relative to the flat seabed SCRs fatigue damage response. This may indicate that a deliberate imposition of positively sloped seabed through modification of the riser TDZ – seabed contact profile can improve the SCR fatigue response around the TDZ.

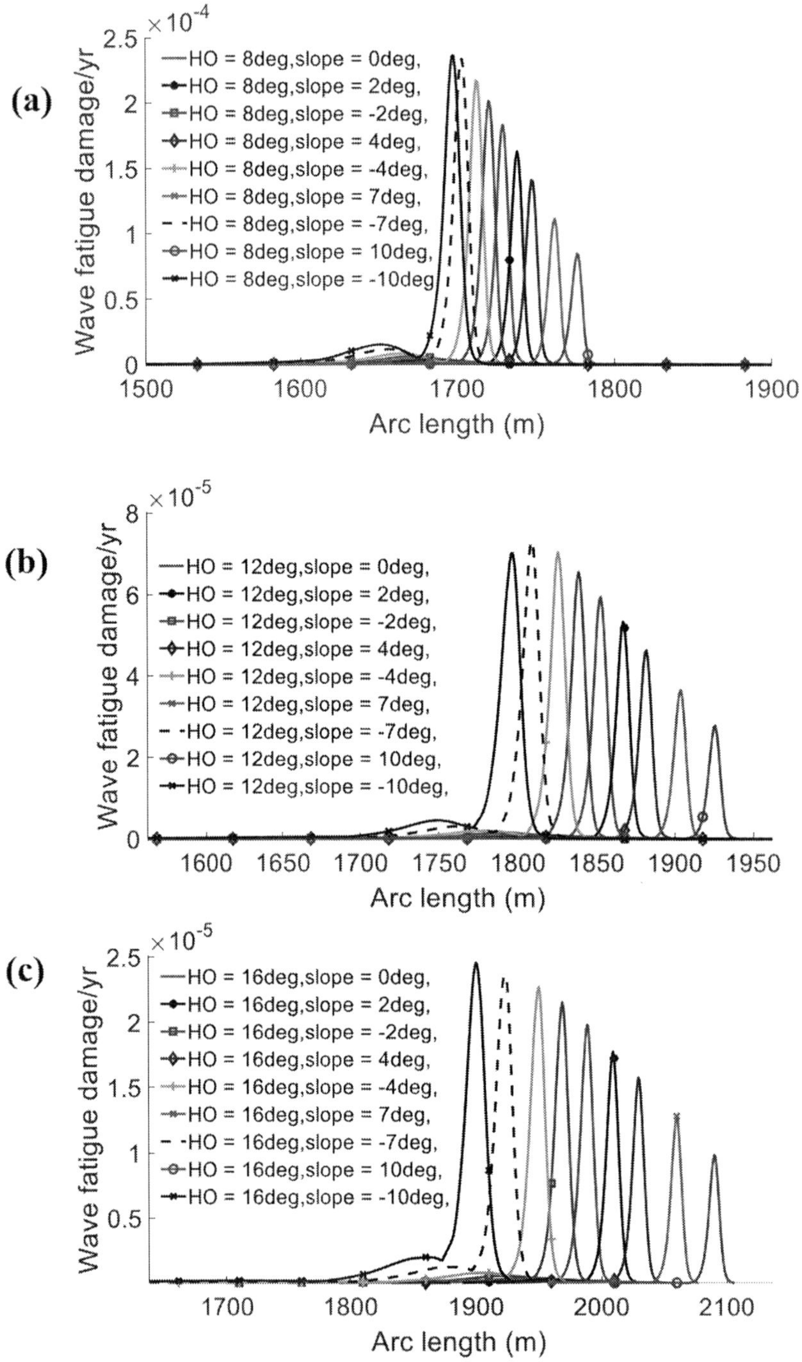

Figure 12. Fatigue damage around riser TDZ for riser groups: (a) HO angle = 8deg, (b) HO angle= 12deg, (c) HO angle= 16deg Configuration.

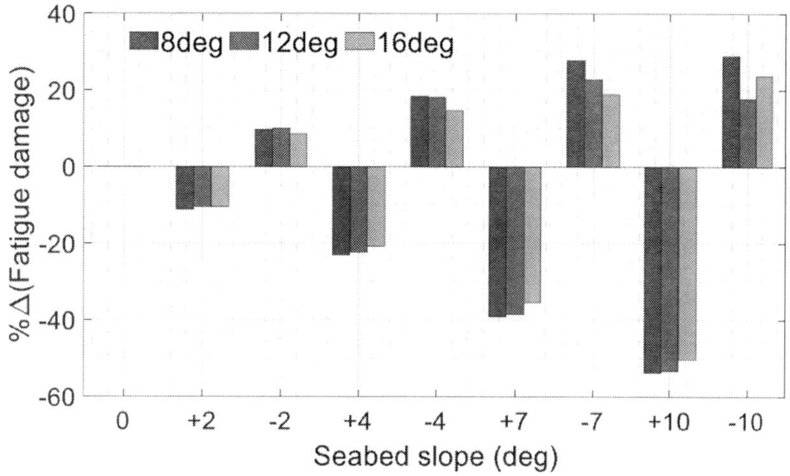

Figure 13. Percentage change in SCRs TDZ stress utilisation relative to SCRs on a flat seabed.

Note: Negative values mean reduction in fatigue damage in the sloped seabed SCRs relative to the flat seabed SCRs.

5. Conclusion

The impact of seabed slope on SCR TDP offset, TDZ compression, stress utilisation and fatigue response have been investigated in this work. The study considers SCR design scenarios were flat seabed are assumed rather than the realistic profiled seabed around the TDZ for SCR design analysis, and then investigate the impact on the SCR TDZ responses for such design assumption. The positively and negatively sloped seabed of different slopes, $\alpha = (0, \pm 2, \pm 4 \pm 7 \pm 10)$deg were imposed on SCR configurations initially modelled on the flat seabed. Three SCRs groups were considered with HO angles of 8deg, 12deg and 16deg. For flat seabed SCR configurations, it is possible to relate the stress and fatigue response at the SCR TDZ to variation in the SCR HO angle, since it is highly likely that smaller HO angle will cause higher TDZ stress utilisation and fatigue damage than higher HO angle. However, when the seabed is sloped, a combination of the HO angle and seabed slope drives the SCR TDZ response. This can be related to a single parameter resulting from such combination, referred to in this work as the SCR incidence angle on the seabed, β. Positively sloped seabed SCRs are characterised by higher β values resulting in lower TDZ stress utilisation and fatigue damage. The converse is the case for the negatively sloped seabed.

Increasing values of α for an actually sloped seabed within the vicinity of the TDP can result in higher TDP offsets compared with an equivalent model where the seabed is assumed to be flat. This can cause errors in the arc length specified for internal cladding of the SCR as well as errors in the arc length at which the maximum compression, stress utilisation and fatigue damage occur along with the SCR. Care should be taking during SCR design in the assumption of the seabed profile around the SCR TDZ.

In ongoing work, the possibility of introducing the structural form to impose a positively sloped seabed-like profile within the SCR TDZ is being investigated. The objective is to access the benefits (on SCR TDZ response) of the imposed positive sloped seabed on the SCR TDP vicinity.

6. Recommendation

A simplified linear seabed profile is assumed in this study, and a one-directional wave load is imposed on the model in the riser azimuth direction (beam sea), where the roll and heave components of the vessel response is expected to be highest. However, the seabed profile can be more complex than the assumed linear and contains local undulations within the SCR TDZ. This is worth investigating, as well as different wave load data and direction of impact on the host platform.

7. Acknowledgement
The authors acknowledge the support of the University of Strathclyde and McDermott International for their support. Results in this work were obtained using the ARCHIE-WeST High-Performance Computer (www.archie-west.ac.uk) based at the University of Strathclyde.

Appendix

Table 4. Change in static top tension of sloped seabed SCRs relative to their respective flat seabed SCRs.

Seabed slope (deg)	Static top tension T_{eff} (kN)			% change in static top T_{eff} relative with flat seabed risers		
	HO = 8deg	HO = 12deg	HO = 16deg	HO = 8deg	HO = 12deg	HO = 16deg
0	2436.01	2646.48	2892.39	0.00	0.00	0.00
2	2437.10	2647.49	2893.21	0.04	0.04	0.03
-2	2435.88	2646.58	2892.70	-0.01	0.00	0.01
4	2438.69	2649.01	2894.48	0.11	0.10	0.07
-4	2437.34	2648.54	2895.06	0.05	0.08	0.09
7	2441.16	2651.28	2896.11	0.21	0.18	0.13
-7	2444.24	2657.19	2905.16	0.34	0.40	0.44
10	2442.84	2652.52	2896.26	0.28	0.23	0.13
-10	2460.15	2677.08	2928.70	0.99	1.16	1.26

Table 5. Change in maximum dynamic top tension of sloped seabed SCRs relative to their respective flat seabed SCRs.

Seabed slope (deg)	Maximum T_{eff} (kN)			% change in max. top T_{eff} relative with flat seabed risers		
	HO = 8deg	HO = 12deg	HO = 16deg	HO = 8deg	HO = 12deg	HO = 16deg
0	3250.95	3543.84	3918.20	0.00	0.00	0.00
2	3239.66	3525.53	3872.10	-0.35	-0.52	-1.18
-2	3264.59	3577.08	3972.48	0.42	0.94	1.39
4	3229.84	3510.62	3840.49	-0.65	-0.94	-1.98
-4	3283.05	3628.04	4037.24	0.99	2.38	3.04
7	3218.00	3490.86	3808.71	-1.01	-1.49	-2.79
-7	3317.74	3727.86	4160.45	2.05	5.19	6.18
10	3206.50	3471.89	3778.03	-1.37	-2.03	-3.58
-10	3413.68	3865.71	4332.41	5.01	9.08	10.57

Table 6. Change in TDZ compression of sloped seabed SCRs relative to their respective flat seabed SCRs.

Seabed slope (deg)	Minimum T_{eff} (kN)			% change in min. T_{eff} relative with flat seabed (compression)		
	HO = 8deg	HO = 12deg	HO = 16deg	HO = 8deg	HO = 12deg	HO = 16deg
0	-105.51	-80.07	-24.60	-28.49	-54.29	-226.08
2	-75.46	-36.60	31.02	31.23	58.91	241.43
-2	-138.46	-127.24	-84.00	-54.94	-104.73	-435.52
4	-47.54	3.79	82.54	66.15	123.89	506.06

-4	-175.31	-179.26	-149.10	-91.74	-174.77	-730.81
7	-8.72	59.86	155.19	126.49	235.10	948.78
-7	-238.97	-268.31	-258.02	-125.67	-239.84	-1002.66
10	27.08	111.96	222.07	200.61	366.98	1462.03
-10	-317.17	-373.90	-384.29	-28.49	-54.29	-226.08

Table 7. Change in stress utilisation of sloped seabed SCRs relative to their respective flat seabed SCRs.

Seabed slope (deg)	Maximum U			% change in U relative with the flat seabed		
	HO = 8deg	HO = 12deg	HO = 16deg	HO = 8deg	HO = 12deg	HO = 16deg
0	1.2	1.0	0.9	0.00	0.00	0.00
2	1.1	1.0	0.9	-4.48	-5.51	-5.90
-2	1.2	1.1	1.0	4.95	6.19	6.73
4	1.1	0.9	0.8	-8.59	-10.38	-11.05
-4	1.3	1.2	1.0	10.32	13.05	14.32
7	1.0	0.9	0.8	-14.24	-16.84	-12.10
-7	1.4	1.3	1.2	19.65	24.80	27.41
10	1.0	0.8	0.7	-19.25	-22.34	-12.15
-10	1.5	1.4	1.3	30.55	38.19	41.71

Table 8. Change in fatigue damage of sloped seabed SCRs relative to their respective flat seabed SCRs.

Seabed slope (deg)	Fatigue damage/year			% change in fatigue damage relative with the flat seabed		
	HO = 8deg	HO = 12deg	HO = 16deg	HO = 8deg	HO = 12deg	HO = 16deg
0	1.8E-04	5.9E-05	2.0E-05	0.00	0.00	0.00
2	1.6E-04	5.3E-05	1.8E-05	-11.07	-10.38	-10.49
-2	2.0E-04	6.6E-05	2.2E-05	9.89	10.18	8.67
4	1.4E-04	4.6E-05	1.6E-05	-22.92	-22.20	-20.63
-4	2.2E-04	7.0E-05	2.3E-05	18.47	18.24	14.71
7	1.1E-04	3.7E-05	1.3E-05	-39.06	-38.51	-35.34
-7	2.3E-04	7.3E-05	2.4E-05	27.85	22.87	18.97
10	8.5E-05	2.8E-05	9.9E-06	-53.64	-53.27	-50.21
-10	2.4E-04	7.0E-05	2.5E-05	28.99	17.82	23.79

References

[1] Larsen C M and Hanson T 1999 Optimisation of catenary risers

[2] Randolph M and Quiggin P 2009 Non-linear hysteretic seabed model for catenary pipeline contact. In: *ASME 2009 28th International Conference on Ocean, Offshore and Arctic Engineering*: American Society of Mechanical Engineers) pp 145-54

[3] Zargar E, Kimiaei M and Randolph M 2019 A new hysteretic seabed model for riser-soil interaction *Marine Structures* 64 360-78

[4] Aubeny C P and Biscontin G 2008 Interaction Model for Steel Compliant Riser on Soft Seabed. In: *Offshore Technology Conference,* (Houston, Texas, USA: Offshore Technology Conference)

[5] Song R and Stanton P 2009 Advances in Deepwater Steel Catenary Riser Technology State-of-the-Art: Part II—Analysis. In: *ASME 2009 28th International Conference on Ocean, Offshore and Arctic Engineering*: American Society of Mechanical Engineers Digital Collection) pp 285-96
[6] Huang C, Hu G and Yin F 2018 Investigation of touchdown point mismatch during installation for catenary risers *Ocean Systems Engineering* 8 313-27
[7] Ogbeifun A M, Oterkus S, Race J, Naik H, Decnop E and Dakshina M The Branched Riser Systems: Concept Development. In: *ASME 2019 38th International Conference on Ocean, Offshore and Arctic Engineering*: American Society of Mechanical Engineers Digital Collection)
[8] Standard D-O F201, 2010 *Dynamic Risers. DNV: Norway*
[9] Heffernan D 2016 An introduction to the Python interface to OrcaFlex. Orcina, Tech. Rep)
[10] Matsuishi M and Endo T 1968 Fatigue of metals subjected to varying stress *Japan Society of Mechanical Engineers, Fukuoka, Japan* 68 37-40
[11] As D N V 2010 DNV RP C203 Fatigue Design of Offshore Steel Structures. Norway)

Characteristic Of Containers In Dengue Haemorrhagic Fever Endemic Area In Coastal Area (Kedung Cowek Village), Bulak District, Surabaya

R Simatupang[1] and W Wicharana[2]
[1]) Parasitology Department, Medical Faculty of Hang Tuah University
[2]) Student of Medical Faculty of Hang Tuah University

Abstract. A container is a water reservoir that can become a breeding place for the *Aedes aegypti* mosquito as a vector for transmitted DHF, especially man-made containers. Kedung Cowek is a coastal area that is located in Surabaya and this area is endemic to DHF. Dengue Haemorrhagic Fever (DHF) is a disease caused by the Dengue virus and transmitted by *Aedes aegypti* mosquitoes. This research aimed to know the characteristics of the containers in Kedung Cowek Village, Bulak District, Surabaya. This is descriptive research that is in collaboration with our lecturer. The data were analyzed in descriptive form and presented in tabular form. The results showed a total number of containers were 199 containers with details of indoor containers as much as 191 (96%) with 104 (92,9%) of them positive for larvae and 8 (7,1%) of all outdoor containers positive for larvae. The indoor containers that had the highest number and positivity of larvae were bathtubs as much as 94 (42.9%) with 51 (49%) of them positive for larvae. From the 8 outdoor containers, the ones with the highest number and positivity of larvae were buckets as much as 6 (75%) with all of which were larva positive. There were 4 (50%) open and 4 (50%) closed outdoor containers with all of them positive for larvae. By finding *Aedes aegypti* larvae, adult *Aedes aegypti* mosquitoes will also be found that have the potential to transmit DHF in coastal areas.

Keywords: containers, dengue haemorrhagic fever, coastal area

1. BACKGROUND

According to the Indonesian Ministry of Health (2003), a container is a water storage place or a vessel that can become a breeding place for *Aedes aegypti* mosquito [5]. Most of the habitat of the *Aedes aegypti* mosquito in urban areas is a place to collect water produced by humans (man-made containers) and which is useful in human daily life. The existence of containers greatly affects the *Aedes aegypti* density, as a vector. The large number of mosquitoes will increase the risk of contracting Dengue Haemorrhagic Fever. The presence or absence of *Aedes aegypti* larvae in a container is influenced by several factors such as the type of container, the location of the container, the color of the container, the condition of the container lid, the presence of larvae eating fish, container volume, container draining and abatization activities [1], [2]. *Aedes aegypti* likes breeding places in the form of water reservoirs filled with clear, dark water and not directly exposed to sunlight. Such water reservoirs are generally found inside (indoor container) and outside (outdoor container) the home [3].

Dengue Haemorrhagic Fever is a disease caused by the *Dengue* virus and transmitted by *Aedes aegypti* mosquitoes [4]. Around the world, dengue fever still causes unrest in the health sector, even in more than one hundred countries including Africa, America and Southeast Asia it is still endemic [5]. DHF cases always exist and can infect anyone regardless of a certain age. Society has a significant influence on the incidence of DHF in terms of behavior. Besides that, the condition of the environment around the community is also influential. The clinical manifestation of DHF is bleeding which can cause shock and lead to death [4].

Until now, drugs and vaccines for DHF are not yet available, so prevention and control is carried out by means of vector control through the 3M-plus program and optimized through a behavior change communication approach based on local culture.

Kedung Cowek Village is divided into 3 Rukun Warga, covering an area of 91,926 Ha with geographical conditions 1 m above sea level, including lowlands and an average temperature of 36 ° C. This area is included in the coastal area of Surabaya, because its boundaries are directly adjacent to the Madura Strait.

Based on the background of this research, the researcher conducted research on the characteristics of containers in dengue haemorrhagic fever endemic area in coastal area (Kedung Cowek Village), Bulak district, Surabaya.

2. LITERATURE REVIEW

2.1 Dengue Haemorrhagic Fever (DHF)

Dengue Haemorrhagic Fever (DHF) is a disease caused by the Dengue virus and transmitted by the *Aedes aegypti* mosquitoes [4]. Dengue virus originates from the genus Flavivirus, family Flaviviridae, and has four serotypes, namely Dengue-1, Dengue-2, Dengue-3 and Dengue-4 [6], [7].

2.2 Vector

Dengue virus transmission can be through primary vectors and secondary vectors. The primary vectors is *Aedes aegypti* mosquitoes [8].

Aedes aegypti's taxonomy:

- Phylum: Arthropoda
- Class: Insecta
- Order: Diptera
- Sub order: Nematocera
- Family: Culicidae
- Sub family: Culicinae
- Genus: Aedes
- Species: *Aedes aegypti* [9]

2.3 Life cycle: Aedes aegypti

Aedes aegypti undergoes complete metamorphosis (holometabola) in its life cycle which has 4 stages, namely egg, larva (larva), pupa, and adult. Adult female mosquitoes generally lay one hundred eggs at a time and lay their eggs on the inner walls of water reservoirs and above the waterline [10], [11]. The eggs hatch within a few days to several months after being laid. The eggs hatch while in the water. When water is added to the container containing the eggs, it triggers the eggs to hatch. Then enter the larva stage which lasts 6 to 8 days. The next stage is pupa. The pupa then becomes an adult mosquito in two to three days [12], [13].

2.4 The habitat and living habits of the Aedes aegypti mosquito

The pre-adult habitat of the *Aedes aegypti* mosquito is clear water in a man-made container. *Aedes aegypti* mosquitoes prefer places that are inside people's homes and often rest and hide in hanging clothes while waiting for the right time to suck the host's blood [14], [15].

Morning to evening is the time for female *Aedes aegypti* mosquitoes to actively suck the host's blood. After sucking blood, mosquitoes will rest in a dark, humid place and a little wind during the process of maturation their eggs [10], [15].

2.5 Risk factors that affect the occurrence of DHF

The risk factors that influence the occurrence of dengue fever are the home environment, biological environment and social environment. The home environment includes house layout, house distance, containers, climate and altitude [6]. One of the most important home environmental factors is the container which is the breeding place of the dengue vector [16].

2.6 Dengue Haemorrhagic Fever prevention

There is no vaccine to prevent DHF. Prevention leads to control of the dengue mosquito vector through 3M-plus program. 3M such as draining and cleaning water reservoirs, closing water reservoirs and recycling items that can collect rainwater. Plus (plus) by replacing water in flower vases, bird drinking containers, and similar places once a week; repairing water channels that its flow is not smooth or damaged; closing holes in trees or pieces of bamboo; sprinkle larvicide in places that are difficult to drain or areas that are difficult to water; keep fish that eat the larvae in ponds or water reservoirs; installing wire netting; keep the house had good lit and ventilated; using a mosquito net and using a medicine that can prevent mosquito bites [8], [17].

2.7 Container

According to the Indonesian Ministry of Health (2003), a container is a water storage place or a vessel that can become a breeding place for *Aedes aegypti* mosquito [5]. The characteristics of the container including the type, material, location, color, cover, and presence of Aedes aegypti larvae in the container determine the presence of the DHF vector [3]. The research conducted by Hasyimi et al. stated that most of the habitat of the *Aedes sp.* mosquito in urban areas is a place to accommodate water produced by humans and which is useful in human daily life such as bathtubs, drums, jars, and bucket [5].

Figure 1. Bathtub

Figure 2. Drums

3. METHODS

3.1 Study area, design and population

The methode of this study is a descriptive study with cross sectional design which aims to determine the characteristics of containers in DHF endemic areas in coastal area (Kedung Cowek Village), Bulak

District, Surabaya. Mixed research method that both qualitative and quantitative supported one another in discussing the final data for our study. The descriptive method is a method used to describe or analyze a research result but is not used to make broader conclusions. Cross sectional study is research to study the dynamics of correlation between risk factors and effects, by means of an observation approach or data collection at once (poin time approach). That is, each research subject is observed only once and measurements are made of the character status or subject variables at the time of examination[18]. The population in this study were all water storage containers and *Aedes aegypti* larvae in each resident's house in Kedung Cowek Village, Bulak District, Surabaya.

3.2 Data collection

The sample was selected randomly (random sampling). Out of 100 houses a container and larvae survey was conducted. All containers containing water are recorded for location, type, color, material, closed or not. In a container containing larvae, the larvae are taken entirely, counted and identified in the Parasitology Laboratory of Medical Faculty of Hang Tuah University.

3.3 Data analysis

The data were analyzed in descriptive form and presented in tabular form. The application used to analyze data in this study was excel.

4. RESULT

The samples in this study were containers in Dengue Haemorrhagic Fever endemic area in coastal area (Kedung Cowek Village), Bulak District, Surabaya including the location of the container, the type of container, the material of the container, the color of the container, the condition of the container lid, and the number of containers that positive Aedes aegypti larvae.

Below, the container data included in the sample will be presented in tabular form.

Table 1. The location of the containers and the number of containers that are positive for larvae.

Location of the container	Number of containers	Percentage (%)	Number of containers with larvae	Percentage (%)
Indoor	191	96	104	92,9
Outdoor	8	4	8	7,1
Total	199	100	112	100

Based on location, containers are divided into indoor and outdoor containers. The container that has the highest number and positivity for larvae was indoor container, as much as 191 (96%) with 104 (92,9%) of them positive for larvae.

4.1 Indoor containers

Table 2. Types of indoor containers and the number of containers with larvae.

Type of containers	Number of containers	Percentage (%)	Number of containers with larvae	Percentage (%)
Bathtub	94	42,9	51	49
Barrel	59	30,9	36	34,6
Dispenser	8	4,2	2	1,9
Bucket	22	11,5	13	12,5
Tandon	8	4,2	2	1,9
Total	191	100	104	100

The types of indoor containers are bathtub, barrel, dispenser, bucket, and tandon. Type that has the highest number and positive for larvae was bathtubs, as much as 94 (42,9%) with 51 (49%) of them positive for larvae.

Table 3. Indoor container material and the number of positive containers larvae.

Container material	Number of containers	Percentage (%)	Number of containers with larvae	Percentage (%)
Cement	64	33,5	31	29,8
Ceramic	37	19,4	22	21,2
Plastic	88	46,1	49	47,1
Clay	2	1	2	1,9
Total	191	100	104	100

The materials of indoor containers are cement, ceramic, plastic, and clay. Material that has the highest number and positive for larvae was plastic, as much as 88 (46.1%) with 49 (47,1%) of them positive for larvae.

Table 4. Indoor container color and number of containers with larvae.

Container color	Number of containers	Percentage (%)	Number of containers with larvae	Percentage (%)
Blue	33	17,3	22	21,2
Gray	78	40,8	40	38,5
White	18	9,4	9	8,7
Black	14	7,3	8	7,7
Green	22	11,5	12	11,5
Red	18	9,4	6	5,8
Pink	4	2,1	3	2,9
Chocolate	2	1	2	1,9
Light Blue	2	1	2	1,9
Total	191	100	104	100

The colors of indoor containers are blue, gray, white, black, green, red, pink, chocolate and light blue. Color that has the highest number and positive for larvae was gray, as much as 78 (40.8%) with 40 (38,5%) of them positive for larvae.

Table 5. The existence of a lid on indoor containers and the number of containers with larvae.

Containers with lid	Number of containers	Percentage (%)	Number of containers with larvae	Percentage (%)
Open	132	69,1	76	73,1

Not Tightly Closed	24	12,6	19	18,3
Tightly Closed	35	18,3	9	8,7
Total	191	100	104	100

The indoor containers divided into open, not tightly closed, and tightly closed containers. Open containers have the highest number and positive for larvae, as much as 132 (69,1%) with 76 (73,1%) of them positive for larvae.

4.2 Outdoor container

Table 6. Types of outdoor containers and the number of containers with larvae.

Types of container	Number of containers	Percentage (%)	Number of containers with larvae	Percentage (%)
Bucket	6	75	6	75
Barrel	2	25	2	25
Total	8	100	8	100

Total number of outdoor containers are 8 containers with all of them positive for larvae. The types of outdoor containers are bucket and barrel. Type that has the highest number and positive for larvae was bucket, as much as 6 (75%) with all of them positive for larvae.

Table 7. Outdoor container materials and the number of containers with larvae.

Container materials	Number of containers	Percentage (%)	Number of containers with larvae	Percentage (%)
Plastic	6	75	6	75
Clay	2	25	2	25
Total	8	100	8	100

The materials of outdoor containers are plastic and clay. Material that has the highest number and positive for larvae was plastic, as much as 6 (75%) with all of them positive for larvae.

Table 8. Color of outdoor containers and number of containers with larvae.

Container colour	Number of containers	Percentage (%)	Number of containers with larvae	Percentage (%)
Red	2	25	2	25
Gray	4	50	4	50
Black	2	25	2	25
Total	8	100	8	100

The colors of outdoor containers are red, gray, and black. Color that has the highest number and positive for larvae was gray, as much as 4 (50%) with all of them positive for larvae.

Table 9. Existence of outdoor container with lids and the number of containers with larvae.

Lid of container	Number of containers	Percentage (%)	Number of containers with larvae	Percentage (%)
Open	4	50	4	50
Not Tightly Closed	4	50	4	50
Total	8	100	8	100

The presence of lids for outdoor containers, as much as 4 open containers (50%) and 4 (50%) non-tightly closed containers. All of outdoor containers were positive for larvae.

5. DISCUSSION

In this study, it was found that most indoor containers were bathtubs. This can be caused by the habit of people having wet bathrooms so that almost every house has a bath. According to research conducted by Aniq and Suharyo, the most common containers used by people in both dengue-endemic and non-endemic areas are bathtubs [19]. The type of indoor container in this study that had the highest number of positivity larvae was the bathtub, which was 51 tubs (49%). This is in line with Budiman's research which shows that there are 15 tubs positive for larvae (61.9%) and is the type of container with the most positive larvae [3]. This can be caused by the need for continuous water and different people's routines regarding draining the bathtub. A bathtub that is rarely drained has the potential to become a breeding place for *Aedes aegypti* mosquitoes. This is reinforced by Muda and Haqi's research which shows that larvae are related to the activity of draining the tub once a week [13]. It is known that the life cycle in the pre-adult stage of *Aedes aegypti* mosquitoes takes place in water. The life cycle of the *Aedes aegypti* mosquito is preceded by the egg stage. The eggs hatch within a few days to several months after being laid. Then enter the larva stage which lasts for 6 to 8 days. Then enter the pupa phase. Pupae in two to three days will become adult mosquitoes. It is hoped that draining the bathtub at least once a week can break the life cycle of the *Aedes aegypti* mosquito [12], [13].

For outdoor containers that had the highest number and positivity larvae are buckets. The bucket in the research area is used to collect water. This is in line with research conducted by Hasyimi et al. that most of the habitat of the *Aedes sp.* mosquito in urban areas is a place to accommodate water produced by humans and which is useful in human daily life such as bathtubs, drums, jars, and bucket [5].

Prevention DHF leads to control of the dengue mosquito vector through 3M-plus program. 3M such as draining and cleaning water reservoirs, closing water reservoirs and recycling items that can collect rainwater. Plus (plus) by replacing water in flower vases, bird drinking containers, and similar places once a week; repairing water channels that its flow is not smooth or damaged; closing holes in trees or pieces of bamboo; sprinkle larvicide in places that are difficult to drain or areas that are difficult to water; keep fish that eat the larvae in ponds or water reservoirs; installing wire netting; keep the house had good lit and ventilated; using a mosquito net and using a medicine that can prevent mosquito bites [8], [17].

The results of this study indicate that many containers still positive for larvae, therefore, it can be used as a reference for evaluating the DHF prevention program by the authorities so that in the future the program implemented is better than before in preventing dengue. Besides that, it can also be used as a source of other similar research.

6. CONCLUSION AND SUGGESTION

6.1 Conclusion

Location and number of containers positive for larvae, containers that had the highest number and positivity for larvae were indoor containers, as much as 191 (96%) with 104 (92,9%) of them positive for larvae. The indoor containers that had the highest number and positivity of larvae were bathtubs as much as 94 (42.9%) with 51 (49%) of them positive for larvae; plastic material as much as 88 (46.1%) with 49 (47,1%) of them positive for larvae; gray color as much as 78 (40.8%) with 40 (38,5%) of them positive for larvae; and as much as 132 (69.1%) open containers with 76 (73,1%) of them positive for larvae. Of the 8 outdoor containers, the ones with the highest number and positivity of larvae were buckets as much as 6 (75%) with all of which were larva positive; plastic material as much as 6 (75%) with all of which were larva positive, and gray color as many as 4 (50%) with all of which were larva positive. There were 4 (50%) open and 4 (50%) closed outdoor containers with all of them positive for larvae.

6.2 Suggestion

Based on the research that has been done, the following suggestions are obtained:

a. Many people still do not know that we actually provide the breeding place for *Aedes aegypti* mosquitoes by seeing the largest number of bathtubs and bathtubs available all the time, so cooperation is needed from the primary health care and the community regarding the implementation of 3M-plus program to control the DHF vector.
b. The next researchers are expected to pay more attention to other environmental factors that affect the presence of *Aedes aegypti* larvae in water reservoirs.

7. REFERENCES

[1] Alifariki L O and Mubarak 2017 Hubungan karakteristik kontainer dengan keberadaan jentik nyamuk aedes aegypti di wilayah kerja puskesmas poasia kota kendari *E-Issn.* **5**(1) pp 388–393.

[2] Ramadhan B I and Achmadi U F 2020 Keberadaan jentik aedes aegypti dan aedes albopictus berdasarkan karakteristik kontainer di sekolah dasar, keluharan duren sawit, jakarta timur, tahun 2018 *Jurnal Nasional Kesehatan Lingkungan Global.* **1**(1) pp 28–35.

[3] Budiman and Hamidah 2017 Aedes agypti di wilayah kerja puskesmas bulili *Jurnal Kesehatan Masyarakat.* **7** pp 107–112.

[4] Sandra T. *et al.* 2019 Faktor-faktor yang berpengaruh terhadap kejadian demam berdarah dengue pada anak usia 6-12 tahun di kecamatan tembalang *Jurnal Epidemiologi Kesehatan Komunitas.* **4**(1) p 1.

[5] Arfan I 2019 Keberadaan jentik aedes sp berdasarkan karakteristik kontainer di daerah endemis dan non endemis demam berdarah dengue. *Jurnal Ilmiah Ilmu Kesehatan: Wawasan Kesehatan.* **5**(2) pp 258–266.

[6] Prasetyani, R. D., 2015. Faktor-faktor yang berhubungan dengan kejadian demam berdarah dengue *Jurnal Major.* **4**(7), pp. 61–66.

[7] Simatupang R 2016 Density figure's status and frequency distribution container in the dengue haemorrhagic in *7th ASEAN Congress of Tropical Medicine and Parasitology Congress* ed Dr. dr. Loeki Enggar Fitri, M.Kes., Sp.ParK (Malang : Faculty of Medicine, Brawijaya University).

[8] Nur Y M, Eliza E, and Haria W E 2020 Faktor-faktor predisposisi yang berhubungan dengan pencegahan DBD di tanjung basung wilayah kerja puskesmas pasar usang *Jurnal Akademika Baiturrahim Jambi.* **9**(1) p 131.

[9] Suyanto S D and Astuti D 2011 Hubungan pengetahuan dan sikap dengan praktek pengendalian nyamuk aedes aegypti di kelurahan sangkrah kecamatan pasar kliwon kota surakarta *Jurnal Kesehatan.* **4** pp. 1–13.

[10] Mawardi M and Busra R 2019 Studi perbandingan jenis sumber air terhadap daya tarik nyamuk

aedes aegypti untuk bertelur *Jurnal Serambi Eng.* **4**(2) pp 593–602.

[11] Febriantoro Y, Alvira L, Hanif A H, Hidayat B A, and Juita N 2012 PAP ' Prevent Aedes Pump sebagai alat untuk memutus siklus hidup nyamuk aedes aegypti dan meningkatkan efisiensi pembersihan air di bak mandi skala rumahan *Jurnal Ilmu Mahasiswa.* **2**(2) pp 71–75.

[12] Centers for Disease Control and Prevention 2012 Life cycle : aedes aegypti *Centers Dis. Control Prev.* **2.**

[13] Muda A S and Haqi D N 2019 Determinan yang berhubungan dengan keberadaan jentik di kelurahan rangkah buntu, Surabaya *Indonesia Jurnal Heal. Promot. Heal. Educ.* **7**(1) pp 22–33.

[14] Dania I A 2016 Gambaran penyakit dan vektor Demam Berdarah Dengue (DBD) *Jurnal Warta.* **48** pp 1–15.

[15] Atikasari E and Sulistyorini L 2019 Pengendalian vektor nyamuk aedes aegypti di rumah sakit kota surabaya *Indonesia Jurnal Public Health.* **13**(1) p 73.

[16] Windyaraini D H, Siregar F T, Vanani A, Marsifah T, and Poerwanto S H 202. Identification of culicidae family diversity as vector control management and mosquito-borne disease prevention in universitas gadjah mada, yogyakarta *Jurnal Kesehatan Lingkungan.* **12**(1) p 1.

[17] Rahmania N A *et al.* 2018 Tindakan Pemberantasan Sarang Nyamuk (PSN) dan 3M-Plus sebagai upaya pengendalian vektor dalam pencegahan penyakit demam berdarah dengue *Jurnal agromedicine.* **5** pp 524–528.

[18] Siyoto and Sodik 2015 *Dasar Metodologi Penelitian* (Yogyakarta: Literasi Media Publishing).

[19] Aniq L and Suharyo 2015 *Hubungan karateristik kontainer dengan keberadaan jentik Aedes agypti di wilayah endemis dan non endemis Demam Berdarah Dengue.* Skripsi. *Universitas Dian Nuswantoro.*

IOP Conf. Series: Materials Science and Engineering 1052 (2021) 012020 doi:10.1088/1757-899X/1052/1/012020

Medical Waste Processing Planning on Hospital Ship 124 M

D Jumineti[1,2], A Santoso[1] and Mardianto[2]
[1] Department of Marine Engineering, Institut Teknologi Sepuluh Nopember (ITS)
Jalan Raya ITS, Surabaya 60111, East Java, Indonesia
[2] PT PAL Indonesia (Persero)
Ujung, Surabaya 60155, East Java, Indonesia

Abstract. Indonesia is an archipelagic country which has 16,056 islands in 2018. One of the efforts to equalize health facilities in archipelagic areas and accelerate the delivery of health facilities when disasters and accidents occur using hospital ships. The operation of hospital ships will produce hospital waste that needs to be treated to reduce the adverse impact on the environment. Some hospital wastes that are toxic in nature are categorized as B3 waste (Hazardous and Toxic Materials). This waste treatment needs to be regulated so that the ship's medical area and the ship's operational environment remain healthy, clean and well-maintained. In this study, it is planned to treat hospital ship waste starting from the grouping of waste in the medical area, calculating the amount of waste generated, diagrams of waste treatment to produce a standard operating procedure that is acceptable and applied according to the analysis on the hospital ship. This research was conducted by making a numerical model based on hospital ship data on the BRS (*Bantu Rumah Sakit*) ship which is currently still under construction at PT PAL Indonesia (Persero). The results of the calculation of medical waste obtained in this study were 318 kg / day with details of non-infectious waste 254.4 kg / day, sharp object waste 3.18 kg / day, chemical and pharmaceutical waste 9.54 kg / day, pathological waste and 47.4% infectious and 3.18 kg / day of broken tube and thermometer waste. Based on the results of this calculation, a suitable waste treatment design and SOP are made that can be applied to the BRS ship.

1. Introduction

Indonesia is a maritime country that has more water areas than land areas. The distribution of Indonesia's population from Sabang to Merauke requires a healthy environment for life and activities. The division of health units also needs to be considered in order to be able to support the provision of health facilities in each region.

One of the efforts to equalize health facilities in archipelagic areas and to provide health assistance facilities when a disaster occurs is by using hospital ships. Hospital ships are health facilities located on the coast or ports that can move from one island to another. An example of using a hospital ship during the March 2020 pandemic is as a quarantine and transport area for covid-19 patients.

The uneven growth of hospitals in Indonesia in each region based on [1] is the reason for the need for ship hospitals in providing health facilities. In addition, the ability of the hospital ship to move is used as a reliable medical vessel when a disaster occurs in a certain area. The main mission of the Navy hospital ships is to provide mobile medical capabilities for deployed military personnel, secondary humanitarian aid and disaster response missions have been at the forefront and centered since the 2004 Asian Tsunami [2].

Health facilities in the medical area of the hospital ship produce medical waste that needs to be treated in order to keep the environment healthy, clean and sterile. The environment is a place for living things in a certain population, a certain time and a certain area. A healthy environment will have an impact on health and good behavior towards life in it. The environment inhabited by living things includes land,

IOP Conf. Series: Materials Science and Engineering 1052 (2021) 012020 doi:10.1088/1757-899X/1052/1/012020

water and air. Apart from environmental health, human health also needs to be maintained by establishing private or government-owned health facilities.

Optimizing hospital ships in extending operational time is by increasing the volume of fuel tanks, reliable medical equipment and creating waste treatment systems to reduce waste generated in medical and domestic processes. The management of hospital ship waste needs to be planned starting from the grouping and calculation of the types of medical waste produced to the standard operating procedures for processing it. According to its shape, hospital waste is divided into solid, liquid and gas forms.

Treatment of hospital ship waste can be processed on board and on land (according to the IPAL qualifications at the Ministry of Health). Hospital ship waste that can be processed on board can use an incinerator to destroy solid waste and STP (sewage treatment plant) to treat liquid waste before it is discharged into the sea. The objectives of this study are to:

1. Classify and quantify medical waste and accommodation generated on BRS vessels.
2. Creating a diagram of hospital waste treatment and accommodation on the BRS ship.
3. Develop applicable standard operational procedures based on the design of the medical waste treatment system and accommodation on the BRS vessel.

2. Hospital

Hospital is a health service institution that provides complete individual health services that provide inpatient, outpatient and emergency services in accordance with Indonesian Ministry of Health Regulation No.4 of 2018 [3]. Public hospitals are divided into 4 types, namely:

a) Hospital type A

The type A hospital is the central hospital which is the highest referral. Type A hospitals have extensive specialist and subspecialty medical services. For example, the type A hospital in East Java is the DR. *Soetomo.*

b) Hospital type B

Hospital type B is a hospital that has extensive specialist medical services and limited subspecialty medicine. Type B hospitals exist in every provincial capital. Teaching hospitals that are not included in type A hospitals are categorized as type B hospitals. For example, type B hospitals in Surabaya are *Husada Utama* Hospital, *Adi Husada Undaan* Hospital, *Bhayangkara* Hospital, Surabaya Islamic Hospital, RKZ Hospital Surabaya, *Mitra Keluarga* Hospital Surabaya, National Hospital, PHC Hospital, Premier Surabaya Hospital, Siloam Surabaya Hospital, Surabaya Surgical Hospital, Undaan Eye Hospital, *Soewandi* Hospital, Haji Hospital Surabaya.

c) Hospital type C

Type C hospital is a hospital that has limited specialist medical services. The type C hospital is located in the district capital. The specialist services available at the Type C hospital are internal medicine services, surgical services, child health services and obstetrics and gynecology services. For example, type C hospitals in Surabaya are *Adi Husada Kapasari* Hospital, *Al-Irsyad* Hospital, *Bhakti Rahayu* Hospital, *Darmo* Hospital, *Darus Syifa* 'Islamic Hospital, *Kenjeran* Family Partner Hospital, Surabaya Lung Hospital etc.

d) Hospital type D

Hospital type D is a hospital that is transitional in nature with only dental and general services. Hospital type D is able to accommodate referrals from *puskesmas* [4].

3. Hospital Ship

Hospital ships are included in the type of mobile hospital. A mobile hospital is a hospital that is ready to use and is temporary for a certain period of time and can be moved from one location to another [5]. Indonesia has 17,504 islands separated by the ocean. To flatten health facilities throughout the islands

IOP Conf. Series: Materials Science and Engineering 1052 (2021) 012020 doi:10.1088/1757-899X/1052/1/012020

in Indonesia, a floating hospital is needed in the form of a hospital ship, which has facilities equivalent to hospitals on land. An example of a hospital ship owned by Indonesia is KRI DR Soeharso. KRI DR Soeharso has a function as a hospital ship and as a warship. KRI DR Soeharso is a hospital ship with the ability as a level II hospital which in its management is included in the Indonesian Eastern Region Fleet in the ranks of the auxiliary elements [6]. The international hospital ships that have ever existed are the USNS Mercy belonging to the United States, Mercy Ship which is managed by the charity Mercy Ship etc. Apart from hospital ships, smaller health facilities that can operate in river water areas are the floating puskesmas as a health facility in Kutai barat. The hospital ship data used in this study were the BRS (*Bantu Rumah Sakit*) ships with the following specifications:

Project Name	: Kapal Bantu Rumah Sakit (BRS) TNI-AL
LOA	: 124 m
LPP	: 107.450 m
Breadth	: 21.80 m
Depth	: 6.700 m
Speed (Max)	: 18 knots
(Cruising)	: 14 knots
Operating Range	: 10 nm
M/E at MCR	: 2 sets x abt. 5400 kW
Patient	: 159 persons
Medical Staff	: 69 persons
Crew	: 280 persons

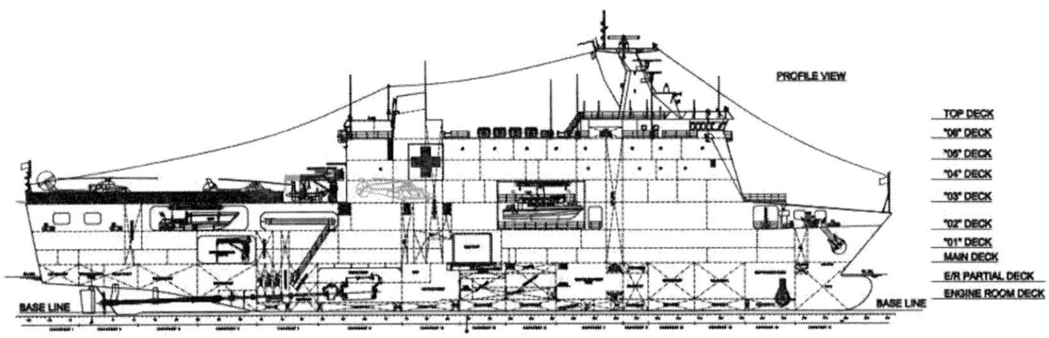

Figure 1. BRS's ship elevation View

4. Waste

Waste is waste generated from community, industrial, institutional or private activities. Hospital waste is all waste generated from hospital activities in the form of solid, liquid and gas [5]. The waste produced by the hospital consists of:

4.1. Hospital Solid Waste

a. Solid Medical Waste

Solid waste consisting of infectious waste, pathological waste, sharp object waste, pharmaceutical waste, cytotoxic waste, chemical waste, radioactive waste, pressure container waste and waste with high heavy metal content.

b. Non Medical Solid Waste

Non-medical solid waste is solid waste generated from activities in hospitals outside of medical originating from kitchens, offices, Tama and yards which are reused if there is technology.

4.2. Liquid Waste

Liquid waste is all wastewater including feces originating from hospital activities which may contain microorganisms, toxic chemicals and radioactive which are harmful to health.

4.3. Gas Waste
Gas waste is all waste in the form of gas that comes from burning activities in hospitals such as incinerators, kitchens, generator equipment, anesthetics and disposal of cytotoxic drugs.

4.4. Infectious Waste
Infectious waste is waste contaminated by pathogenic organisms that do not routinely exist in the environment and these organisms are in sufficient quantity and virulence to transmit disease to susceptible humans.

4.5. Highly Infectious Waste
Highly infectious waste is waste originating from culture and stock of highly infectious materials, autopsies, animal organs and other materials that have been inoculated, infected or in contact with highly infectious materials.

4.6. Cytotoxic Waste
Cytotoxic waste is waste from contaminated materials from the preparation and administration of cytotoxic drugs for cancer chemotherapy which have the ability to kill or inhibit the growth of living cells.

Based on the 6 types of waste produced [5], then the types of waste protection are grouped based on [7]. The classification of waste safeguards is divided into: domestic solid waste, hazardous and toxic waste (B3), liquid waste and gas waste. Hospital waste needs to be treated to eliminate infectious properties and minimized it so as not to damage the environment. In addition to regulations from the Ministry of Health, international regulations also regulate waste minimization on ships, this is stated in MARPOL Annex V regarding Garbage Management in waste minimization which recommends that producers, cargo owners, ports and terminals, ship owners and operators and the government consider waste management related to supplies, provisions, and ship cargo as needed to minimize the emergence of waste in all forms [8].

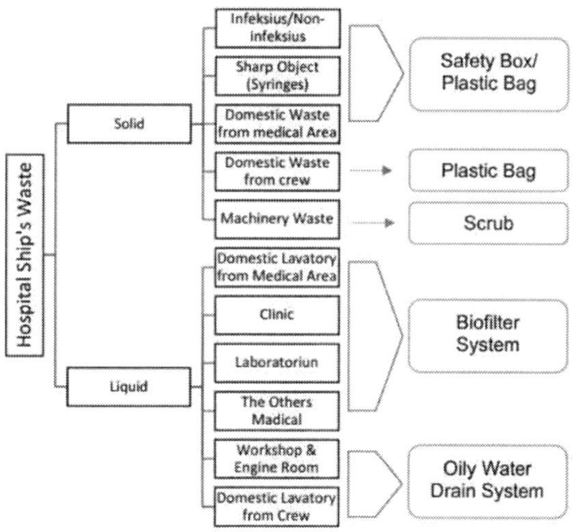

Figure 2. Hospital Ship's Waste

5. Solid Medical Waste

The solid medical waste treatment system can be destroyed by burning it using an incinerator or stored using a safety box for processing on land. The entire solid waste treatment process is burned using an incinerator. The incineration process stages are generally divided into 4:

- Storage process
- Burning process is processed
- Residue combustion handling process
- Process of clearing smoke

The treatment of solid medical waste is basically eliminating infectious properties and minimizing them. The amount of solid waste produced in class C hospitals with beds> 300 is <75 kg / day [9]. To simplify the medical waste treatment process, solid waste needs to be sorted from source using colour coded plastic bags. Plastic bag colour indicator according to[5].

6. Liquid Medical Waste

Liquid medical waste treatment aims to remove contaminants in wastewater so that the processed waste can be reused or not damage the environment when disposed of into the environment[9]. Sources of liquid waste on hospital ships are grouped into:

- Lavatory in the medical area
- Clinic
- Laboratories
- Liquid waste from other medical activities

The output of medical liquid waste that cannot be treated on board is collected in special tanks to be processed onshore. Meanwhile, medical liquid waste that can be processed on ships is processed in the sewage treatment plan (STP) system, where the rest is stored in blackwater, bilge tanks or sludge tank.

7. Calculation of Medical Waste

The calculation of the amount of medical waste uses a reference from a WHO study (1999) which states that the production of daily medical waste generation (TLH) in developing countries is around 1-3 kg /TT.day, while in developed countries (Europe, America) it reaches 5- 8 kg / TT. Days. The composition of this medical waste consists of 80% non-infectious waste, 15% pathological & infectious waste, 1% sharp object waste, 3% chemical and pharmaceutical waste, > 1% tube waste and broken thermometer [10].

7.1. Daily Medical Waste (T_H)

The medical waste generation of the BRS vessel is calculated according to the number of patient beds (TT) provided in 100% average occupancy conditions. The 100% average occupancy is expressed by the notation P = 1

$$T_H = TLH \times TT \times P \tag{1}$$

$$= 2 \times 159 \times 1$$
$$= 318\ kg/day$$

7.2. Non-Infectious Waste (T_{LNI})

Non-infectious waste generation (T_{LNI}) consists of 80% of daily medical waste (T_H), then the value is obtained:

$$T_{LNI} = T_H \times 80\% \tag{2}$$

$$= 318 \times 0{,}8$$
$$= 254{,}4\ kg/day$$

7.3. Pathological and Infectious (T_{PI})

Pathological and infectious waste generation (T_{PI}) consists of 15% of daily medical waste (T_H), then the value is obtained:

$$T_{PI} = T_H \times 15\% \tag{3}$$

$$= 318 \times 0{,}15$$
$$= 47{,}7\ kg/day$$

7.4. Sharps Waste (T_{BJ})

The incidence of sharp waste (T_{BJ}) consists of 1% of daily medical waste (T_H), then the value is obtained:

$$T_{BJ} \quad = T_H \times 1\% \tag{4}$$

$$= 318 \times 0{,}01$$
$$= 3{,}18\ kg/day$$

7.5. Chemical and Pharmaceutical Waste (T_{KF})

Chemical and pharmaceutical waste generation (T_{KF}) consists of 3% of daily medical waste (T_H), then the value is obtained:

$$T_{KF} \quad = T_H \times 3\% \tag{5}$$

$$= 318 \times 0{,}03$$
$$= 9{,}54\ kg/day$$

7.6. Broken Tubes and Thermometers Waste (T_{TTP})

The generated waste of tube waste and broken thermometer (T_{TTP}) consists of 3% of daily medical waste (T_{TTP}), then the value is obtained:

$$T_{TTP} \quad = T_H \times 1\% \tag{6}$$

$$= 318 \times 0{,}01$$
$$= 3{,}18\ kg/day$$

The calculation of the amount of waste still needs to be added with accommodation waste and machinery waste if you want to know the amount of waste on the BRS ship as a whole. The share of medical waste from BRS ships is shown in Figure 3.

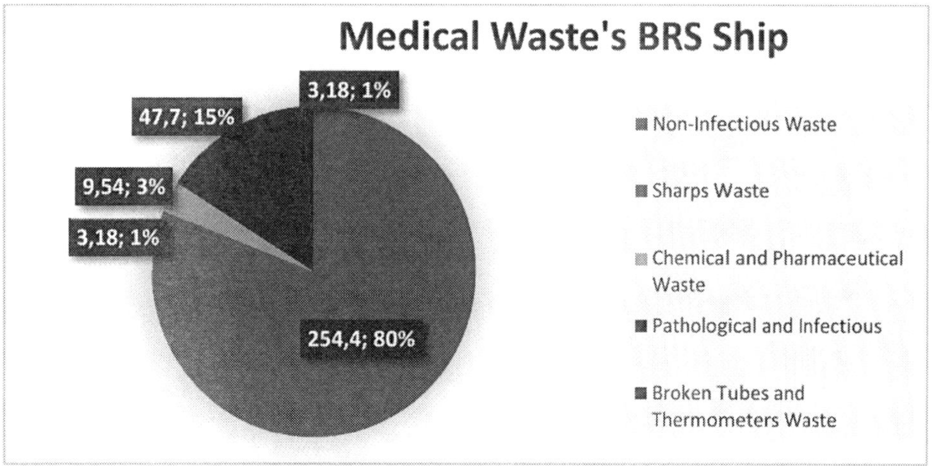

Figure 3. Medical Waste's BRS Ship

8. Designed of Waste Systems

The design of medical waste treatment systems on hospital ships in medical and accommodation areas needs to pay attention to:

- Classification of waste based on the level of hazard and infection.
- For solid wastes which is infectious, it is necessary to take into account the availability of garbage stations for waste that cannot be burned with an incinerator available on board.
- For wastewater, it is necessary to redundant processing equipment such as sewage treatment plants to ensure the system can continue to run during repair conditions.
- A drainage and soil system are provided on each deck to make repairs when a dead end occurs in the pipeline so as not to pollute the accommodation area.
- Reduction of solid waste in garbage stations can be done by burning non-medical waste that does not require a double burner in the incinerator.

9. Conclusion

Based on the results of calculations on the BRS ship with the consideration of patient bed facilities, the calculation results of medical waste in this study were 318 kg / day with details of non-communicable waste 254.4 kg / day, sharp object waste 3.18 kg / day, chemical waste and pharmaceuticals 9.54 kg / day, pathological waste and 47.4% infectious and 3.18 kg / day broken tube and thermometer waste. Based on the results of these calculations, an appropriate waste management design and SOP is made that can be applied to the BRS ship. Waste treatment for solid waste on ships uses a safety box for waste that is not burned in the incinerator and the use of a Sewage treatment plan for treating liquid waste originating from the clinic, operating room, lavatory and funeral room. Standard operational procedures applied to BRS vessels can use solid waste treatment procedures and operational schedules for operation and maintenance of aeration systems at STP for liquid waste. The design of the sewage treatment plant system will become a reference in making a standard operational procedure (SOP). SOP is made by paying attention to the maintenance schedule of the core equipment, namely the Sewage Treatment Plant (STP) according to the maker used and the vacummarator. Making SOP is divided based on usage time in weekly, monthly, annual and every 5 years.

Acknowledgments

In this paper, thank you to PT PAL and Marine Engineering-ITS for allowing us to develop a sewage treatment system on hospital ships. In the future, this research needs to develop a method on how to make a waste treatment system that can be processed 100% onboard efficiently in terms of design, size, operation and availability of materials so that it is possible for Indonesia to equalize Indonesia's health facilities.

10. References

[1] Perhimpunan Rumah Sakit Indonesia, L. Trisnantoro, and E. Listyani, "Jumlah RS di Indonesia-Pertumbuhan RS Publik," no. April, p. 70, 2018.

[2] D. J. Licina, "Hospital ships adrift ? Part 1 : A systematic literature review characterizing US Navy hospital ship humanitarian and disaster response , 2004-2012 . Hospital Ships Adrift ? Part 1 : A Systematic Literature Review Characterizing US Navy Hospital Ship Huma," vol. 28, pp. 2004–2012, 2013.

[3] L. F. Mubin, W. Anggraeni, and R. A. Vinarti, "Prediksi Jumlah Kunjungan Pasien Rawat Jalan Menggunakan Metode Genetic Fuzzy Systems Studi Kasus : Rumah Sakit Usada Sidoarjo," *J. Tek. ITS*, vol. 1, no. 1, pp. 1–6, 2012.

[4] W. Purwonugraho, "Desain Hospital Ship (Kapal Rumah Sakit) Untuk Perairan Indonesia," p. 113, 2015.

[5] Kementerian Kesehatan Republik Indonesia, "Keputusan Menteri Kesehatan No. 1204 Tahun 2004 - Persyaratan Kesehatan Lingkungan Rumah Sakit." p. 64, 2004.

[6] F. S. Rianto and Purwanto, "Implementation of Hospital Aid Vessels of Kri," *J. Prodi Strateg. dan kampanye Militer*, vol. 3, no. 1, pp. 1–23, 2017.

[7] Kemenkes, "Peraturan Menteri Kesehatan Republik Indonesia Nomor 7 Tahun 2019 tentang Kesehatan Lingkungan Rumah Sakit," *Αγαη*, vol. 8, no. 5, p. 55, 2019.
[8] M. Environment *et al.*, "I:\MEPC\71\MEPC 71-17-Add-1.docx," vol. 295, no. July, pp. 1–30, 2017.
[9] R. Rosita, Luwiharsih, R. Gernpari, H. F. Mansyur, Sukirman, and Dkk, "Pedoman Penatalaksanaan Pengelolaan Limbah Padat dan Cair di Rumah Sakit," *Jakarta*, pp. 1–125, 2000.
[10] KLH, "Pedoman Kriteria Teknologi Pengelolaan Limbah Medis Ramah Lingkungan," *Kesehatan*, pp. 1–122, 2014.

Design of Ship Repair Yard Ranking Website Based on Customer Reviews

M S Arif, S R W Pribadi, and A Mahendra

Department of Naval Architecture, Institut Teknologi Sepuluh Nopember, Surabaya, Indonesia

Abstract. Many ship owners currently choose ship repair yards based on conventional ways; there is no specific platform that can provide information objectively ranking the ship repair yards that can be accessed in public. On the other hand, technological developments have shifted in community business activities from an offline way to the online method. The initial process of designing the application is by determining and weighing the criteria for assessing ship repair shipyards using the AHP method and then obtained nine criteria with weights for each standard. After getting the requirements and value, then we make a website-based application design. This application will display shipyard ranking according to the ratings and reviews provided by customers. The calculation of each shipyard's rating and scale will be determined based on the per-criteria value, which is then accumulated based on the weight of each predetermined criterion. After testing the application in a questionnaire on several respondents from shipping companies, this application gets a value of 28.5 out of 30, or 94.9%, which means this application necessary to be applied in the ranking of ship repair yard.

Keywords : Ship Repair, Yard Ranking, Customer Reviews.

1. Introduction

The absence of a ship repair yard ranking system will make it difficult for the ship owner to choose and assess a shipyard. Currently, many owners choose a shipyard based on conventional ways, such as internal selection by each company or using the services of a third party to help evaluate and select a shipyard based on the criteria of each company. Some shipping companies do not perform the shipyard selection process because the company already has a private shipyard or has trusted one shipyard, so it uses a subscription system. From the shipyard selection system above, it appears that there is no particular container that holds all the shipyard assessments objectively that are based on the Owner's point of view.

On the other hand, the rapid development of technology has made the shift in business activities of people around the World from offline way to online, one of which is characterized by the growth of e-commerce, which is very significant from many industry sectors in Indonesia. It will also affect the shipping sector in Indonesia. With the amount of activity in the shipyard in Indonesia, indeed it takes a system/indicator used by the Owner as a reference to choose the right shipyard.

Among the many features of e-commerce, Online Customer review (OCR) can also be referred to as electronic Word of Mouth (eWOM)[1], in the form of reviews from customers given after customers purchase or use products services. This customer review also provides one of the features that have attracted much attention from academics and society as one of the most influential factors in determining consumers' purchasing decisions[2]. User-generated review content is then posted on the online site of a service provider or third-party web website. Also, customer reviews are considered one of the most critical factors in influencing customer trust and interest[3]. Therefore, this system can be applied to the shipping industry also.

2. Literature Study

2.1. Shipyard

The shipyard is a means of constructing or assembling and repairing vessels located by the sea or water. The shipyard is generally located on the seafront and has a functioning platform to launch the ship after completion[4]. The shipyard is also an industry-oriented to produce products in ships, offshore buildings, floating buildings, and others. Most of the shipyard production is based on the specifications and requests of the customer (Owner)[5]. In contrast, the ship is a complex floating structure composed of various constituent components. Vessels can be classified by the primary size, carrying capacity, or the use of the voyage[6].

Shipyards, especially in Indonesia, can be divided into three types[7], namely:

- Building Dock Shipyard, the shipyard that only constructs a new ship.
- Repair Dock Shipyard, which is a shipyard that only performs repair and maintenance of the vessel.
- Building and Repair shipyard, a shipyard that performs the construction of new ships, repairs, and maintenance of vessels.

The shipyard, which is reviewed in this research, is the shipyard that performs the ship repair work, namely Repair Dock Shipyard and Building and Repair shipyard.

2.2. Ship Repair

The ship repair process is a process of repairing or replacing part of the vessel that has not fulfilled the standard of sailing eligibility based on the rules and statutory. reparations on ships generally involve three factors[4], namely:

- Hull and Construction
  Reparation on the ship's body regarding repair and replacement of the plate and profile construction of the vessel body.
- Machinery
  Reparation on machinery parts regarding repair and replacement of machinery parts, piping system, electricity, and ship propulsion system.
- Outfitting
  Reparation in this section concerning repair and replacement of outfitting and hull equipment.

2.3. Customer

The customer is the ultimate destination of the product issued by a company[2]. Referral includes a customer perspective to measure the company's performance in addition to traditional measurements based on the health of the company's financial statements[8]. Company's performance can be reviewed through four perspectives: Financial Perspective, Customer perspective, an internal business perspective, and learning and growth perspective. So, customers are an essential perspective, and the corporation should always strive to satisfy customers[9]. In this study, the customer referred to is the ship owner/shipping company.

The ship owner is the Owner of a merchant ship or commercial vessel involved in the shipping industry[10]. Ship-owners are usually organized in companies (Shipping Line/company), but individuals and investment funds can also own a ship.

2.4. Customer Review and Rating

The Reviews are part of the Electronic Word of Mouth (eWOM), a direct opinion of a person and not an advertisement. Reviews are one of several factors that determine a person's buying decision[1], indicating that people can take several reviews as an indicator of product popularity or value of a product that will affect the willingness to purchase a product. However, not necessarily the more

reviews and ratings mean the product will be purchased by the customer. Many factors are the reason for the decision to purchase a product for the customer.
Ratings are part of a review that uses shapes such as star symbols, thumbs, or numbers rather than text in expressing opinions from customers. Ratings can be interpreted as a user's assessment of a product's preference for their experience, referring to the psychological and emotional state they are living in when interacting with virtual products in mediated environments[11].

2.5. Computer Application
The computer application or software is a collection of electronic data that is stored and regulated by the computer[12]. The electronic data that it contains can be a program or instruction to execute a command. The software serves as a means of interaction between the User and the hardware.

2.6. Website
The Website is an internet facility that connects documents in both local and remote scope. The documents on the Website are called Web pages, while the link in the Website allows users to move from one page to another page, both between the page being stored on the same server or the server around the World[13]. Pages are accessed and read via browsers such as Netscape Safari, Internet Explorer, Mozilla Firefox, Google Chrome, and other browser applications.
Whereas according to the reference, the Website is a collection of related web pages and the files – The files are interconnected. The web consists of the page a page or page and a set of pages named Homepage. The Homepage is in the top position with the linked page underneath. The page under the Homepage is called a child page that contains hyperlinks to other pages within the web.

2.7. Database System
The Database is a logically linked collection of computer data that contains a description of the data and is designed to meet the information needs. Databases are also a large data storage space and can be used by multiple users simultaneously[14]. Also, the Database has many sizes and varying degrees of complexity.
Systems are a collection of elements that interact to achieve a defined goal. The system consists of interconnected parts by accepting inputs and generating outputs in an organized transformation process. [15]

2.8. Analytic Hierarchy Process (AHP)
The Analytic Hierarchy Process (AHP) is a method that can be used in decision making or order priorities from various alternatives to complex problems using multi-level hierarchical structures of objectives, criteria, sub-criteria, and alternative decisions[16]. Essentially, AHP is a method that breaks down a complex and unstructured problem into groups, organizes those groups into a hierarchy, inserts numeric values in place of human perception in relative comparison, and finally, with syntheses, determined which element has the highest priority. AHP provides weights on each factor, variable, and indicator with comparisons between factors, variables, and indicators of each other. The more excellent value of an indicator indicates that the indicator is more important than the other indicators.[17]

3. Methodology
In general, the method of this research can be seen as follows:

- Identification of Problems

The initial stage of the study is identification problems, where researchers begin to conduct observations to identify and determine the background, formulation, purpose, and problem limitations of the current shipyard rating system.

- Literature Review

At this stage, studies are conducted on literature such as journals, books, articles, or previous research-related and may support this research. Literature studies include:
- Shipyard assessment criteria.
- Customer reviews and ratings.
- Website-based computer applications.

▪ Field Study

At this stage, field studies are conducted to obtain supporting data related to the research. Required data includes:
- The condition of the existing ship repair shipyard rating.
- Shipyard assessment criteria.

▪ Data Collection

After the study of literature and field, the following data can be collected:
- Shipyard assessment criteria.
- Points of consideration of each existing criterion.
- Shipyard in East Java.

▪ Data Processing

Once the required data is obtained, the subsequent data processing is done:
- Evaluate the previous criteria so that new criteria are determined based on existing conditions.
- Weight the importance level of the newly defined criteria.

▪ Application Design

After data processing, the application design is done with the following stages:
- Design a customer review-based shipyard rating system.
- Creation of app Mock-Ups.
- Design the ERD, DFD, and Database from the application.
- Design and implementation of the system.

▪ Testing and Analysis

At this stage, testing, verification, and analysis of designed applications are performed. Application testing is carried out with the relevant party, namely the Owner of the ship or Shipping Line.

▪ Conclusions and Recommendation

This stage is the last stage of this research, which is to make conclusions and recommendations. The conclusions obtained from this study should answer the purpose of this study. The recommendation given is used for subsequent research purposes that are related to this research topic.

4. Existing Condition

4.1. Condition of Ship Repair Yard Ranking

Based on the results of the literature study and field study on ship repair shipyard, it is obtained that there is currently no shipyard rating system for the shipyard and new buildings that can represent the assessment from the point of view of the Owner or Owner. World statistics sites such as statista.com to make a rating on the shipyard based on the amount of tax on the company's profits or can be called the Capital Gain Tax (CGT), production of tonnage, and the number of orders year from the shipyard[18]. Due platform to the absence of a platform that facilitates the Owner or the shipping company on the ship Repair shipyard based on customer reviews, then the current selection process is still done in a conventional way by each company.

4.2. Condition of Ship Repair Yard Selection

The shipyard selection process is usually done independently by divisions related to Fleet maintenance such as docking, procurement, or other divisions that in some shipping companies have different division names. Another way to make a shipyard selection is to find a proven independent scheme such as the Product Quality Assessment Manufacturer or MPQA, which is usually provided by the world classification agencies. MPQA can provide an overview of the quality and capabilities of a shipyard. In addition, shipping companies can also use the services of third parties to help evaluate and select a shipyard.

On the other hand, some shipping companies do not perform the shipyard selection process because the shipping company already has a private shipyard so that when it will do the repair work, the company's ship will be directly docking on its private shipyard.
There are also several shipping companies that already trust a particular shipyard, which then performs the "subscription" system at the shipyard. So every time the ship will repair, the shipping company will use the service of his shipyard. Furthermore, for other shipping companies that do not have a private shipyard or a subscription, it is required to perform a shipyard selection each time the company's fleet is about to repair. This election was done in various internal considerations from each company, and there were individual criteria when choosing a shipyard to repair the ship.

5. Application Design

Based on the study and analysis that has been done in the previous chapter, this chapter will be explained from the design and operation of new systems in the form of Website-based applications that can be used to facilitate the process of ship repair shipyard based on customer reviews.

5.1. Ranking System Design

5.1.1. Criteria Identification

The first step in the design of the shipyard review and assessment system is to identify the purpose of the shipyard rating as desired by the shipping company. This identification is used to find out what criteria need to be included in the shipyard rating and review process. Based on the evaluation of the criteria of ship repair shipyard, identification is carried out, as seen in Table 1.

Table 1. Criteria Identification.

No	Criteria	Description
1	Repair Quality	Quality of reparation work in accordance with existing regulations and standards
2	Repair Facilities	Conditions and completeness of supporting facilities of repair work in a shipyard
3	Repair Cost	Fees and payment systems offered by the shipyard
4	Shipyard Experience	A shipyard experience handles similar reparation jobs
5	Shipyard Location	The strategic location of the shipyard
6	Repair Time	Accuracy and availability of docking/repair time
7	Shipyard Labour	Availability and skills of shipyard Labor
8	Shipyard Service	Services and facilities provided by the shipyard to the Owner
9	Shipyard Management	The managerial ability of a shipyard

5.1.2. Criteria Verification

IOP Conf. Series: Materials Science and Engineering 1052 (2021) 012021 doi:10.1088/1757-899X/1052/1/012021

After conducting the identification process of the ship repair shipyard selection, which includes nine criteria, then verification of the criteria of selection of shipbuilding repairs, this verification process aims to determine whether the criteria can be applied, measured, and represents the current shipyard selection conditions by conducting interviews with some expert respondents on this topic so that the points can be considered as points of consideration of each criterion as shown in Table 2 below.

Table 2. Criteria Verifications.

No	Criteria	Points of Consideration
1	Repair Quality	• Quality according to classification regulations • Repair work according to specification and request of Owner • No rejected and re-work on repair work • The vessel can function appropriately after repair
2	Repair Facilities	• Large docking area • Crane capacity is adequate • Adequate workshop facilities • Facilities conform to K3 standard
3	Repair cost	• Inexpensive repair Cost • The cost of repairs is easy to negotiate • The terms system facilitates the Owner • Shipyard includes tax and NPWP • Invoices with exact reparations and details
4	Shipyard Experience	• Experienced in specialized mechanical work • Experienced in similar ship repair work • Have a good track record of shipyard service • Have a good track record at the time of repair • Have a good track record on the quality of repairs • Have a good track record of shipyard facilities • Have a good track record of the location of the shipyard • Have a good track record on the shipyard workforce
5	Shipyard Location	• Dockyard location close to fleet Cruise route • The water depth around the shipyard is sufficient • Easy access to Shipyard
6	Repair Time	• Fast docking/undocking time • Quick Repair Job • Timely repair work • Keep working on holidays.
7	Shipyard Labor	• Experienced labour • Friendly labour • Skilled labour • Enough labour
8	Shipyard Service	• Response to complaints • There is an office for the Owner • There is the consumption for the Owner, • Security and Safety Assurance • There are utilities for ABK • Clean and Neat shipyard Area

No	Criteria	Points of Consideration
9	Shipyard Management	• Clear Shipyard Organizational structure • There is an information system that facilitates transactions • Good communication from the shipyard • The job desk and workflow on the repair project are clear.

5.1.3. Criteria Weighting

The weighting process is carried out to determine the level of exposure of each previously verified ship repair shipyard criteria. Weighting is carried out using Analytic Hierarchy Process or AHP method by submitting expert judgment questionnaires in the form of pairwise comparison to some respondents who are considered experts and experienced in the process of selection of ship repair yard. Based on the results of the AHP questionnaire survey with respondents who have been considered experts from 3 shipping companies, the recapitulation of the criteria weighting questionnaire, as seen in Table 3.

Table 3. AHP Questionnaire Recap.

Criteria	QL	FC	CS	EX	LC	TM	MP	SR	MN
QL		1,44	1,7	1,44	1,81	1	2	2,28	2,15
FC			1,18	1,14	1,31	1,44	1,25	1,58	2,15
CS				1,65	1,1	2	2,62	1,58	1,18
EX					1,58	1,25	1,44	1,58	2
LC						1	1,14	1,25	1,18
TM							1,44	1,25	2,51
MP								1,25	2,08
SR									1,1
MN	Incon: 0,02								

*Table Captions:
QL: Repair Quality
FC: Repair Facilities
CS: Repair Cost
EX: Shipyard Experience
LC: Shipyard Location
TM: Repair Time
SR: Shipyard Service
MN: Shipyard Management
IV: Inconsistency Value

Horizontal value = vertical value
Negative value

After entering and processing the data using decision-making software, it is obtained the weight of the importance level of each criterion, as seen Table 4

Table 4. Weight Criteria Recap.

No	Criteria	Weight	%
1	Repair Quality	0,167	16,7
2	Repair Facilities	0,147	14,7
3	Repair Cost	0,125	12,5
4	Shipyard Experience	0,121	12,1
5	Shipyard Location	0,112	11,2
6	Repair Time	0,104	10,4
7	Shipyard Labor	0,088	8,8
8	Shipyard Service	0,076	7,6
9	Shipyard Management	0,061	6,1
	TOTAL	**1**	**100**

The review and assessment system used for the rating of shipyard reparations will be carried out by providing reviews and values on each criterion. The overall value of a shipyard will be obtained by averaging all values on each criterion. Then the average score of each of these criteria is multiplied by the weight of the per-criteria assessment. The result of the value of each criterion that has been multiplied by each weight is then summed and obtained the final value of a shipyard.

5.2. Application Design

5.2.1. Basic Design Framework

The first step in designing an app is to create a basic framework or design of the application system. This basic framework is needed to make it easier for programmers and users of the application to understand the intent and purpose that the author wants to convey, as seen in Figure 1:

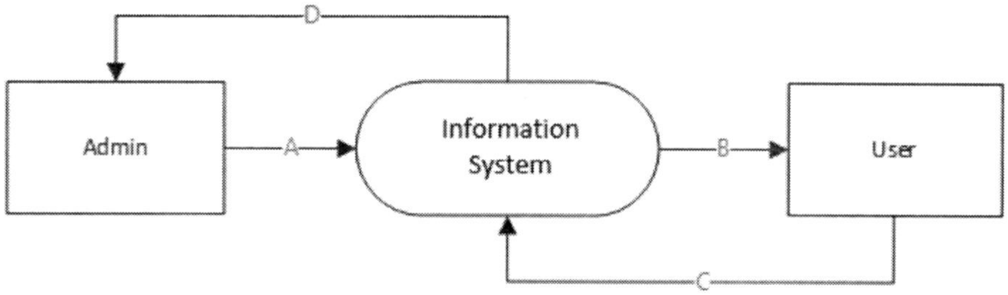

Figure 1. Application Basic Framework.

Description:

- In-process A is done the approval process/rejection of reviews, shipyard accounts, and reports from User by Admin.
- In the process, B obtained output in the form of rating information and shipyard reviews that have been approved by the Admin.
- In process C, user data input is done in the form of reviews, additions to shipyard accounts, and reports.
- In process D, data input by User in process C is sent to the Admin for approval/rejection.

5.2.2. Entity Relationship Diagram (ERD)
Entity Relationship Diagram, or it can be called ERD, is a database designing method database that illustrates relationships between existing entities. Relationships used later can be for the effectiveness of the database system or other system development.

5.2.3. Data Flow Diagram (DFD)
Data Flow diagrams or DFD are the forms of data migrations from one process to another. DFD is also a graphical representation representing a system. DFD serves to describe the components of a system, the flow of data between components, the origin, the purpose, and the storage space of that data. The DFD of the application to be designed can be overhead in the following Figure 2:

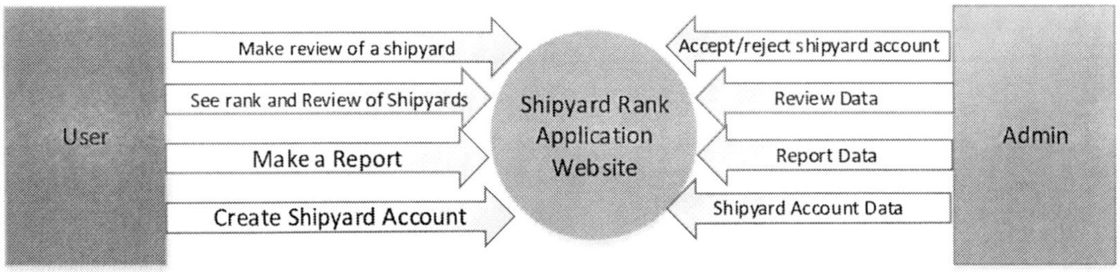

Figure 2. Data Flow Diagram (DFD).

5.2.4. Database Design
Based on the system that has been designed, the database design is done. In the database design database, programmers use phpMyAdmin Application with a hosting server using CPanel application.

5.3. System Design and Implementation
The application was designed named Galangin. Galangin can be described as the first platform in the World that provides information on shipyard-based ship repair and pen ratings from customers, which in the research is still being developed Limited to several shipyards in East Java alone. The presence of Galangin is expected to facilitate the ship owner or shipping company in finding and viewing the reviews and ratings of a shipyard based on the most frequently used criteria to assess a ship repair shipyard. Galangin can be used for free and is accessible at any time using an Internet connection. In addition, Galangin also has a verification system to filter incoming reviews, so there will be no review counterfeiting by irresponsible parties. Therefore, Galangin can be a solution in the selection and removal of shipbuilding repairs.

Galangin application can be accessed by using a Web browser such as Google Chrome or Mozilla Firefox by entering the URL https://galang.in. As for the display of the ship's shipyard rating application based on customer reviews are as follows.

5.3.1. Welcome Page
The Welcome Page is the first page that visitors will find when accessing Galangin's Website. As seen in Figure 3, this page serves as a portal to access the Website, where visitors must Log In for those who already have an account and sign up for those who do not already have an account to register as a user.

5.3.2. Main Page
The main page for the User will display a list and rank from the shipyard, as shown in Figure 4. The main page for the User also has a menu for writing a review called Add a Review. Different from the User, the main page for admins only has a menu to approve the registered shipyard account and accounts, namely the Review control, Shipyard control, and Report control pages.

5.3.3. Shipyard Profile Page

The Shipyard profile page is a page that contains information about a shipyard. Such information includes name, address, logo, contact, description, review per-criterion review, and rating of a shipyard, as shown in Figure 5.

Figure 3. Welcome Page.

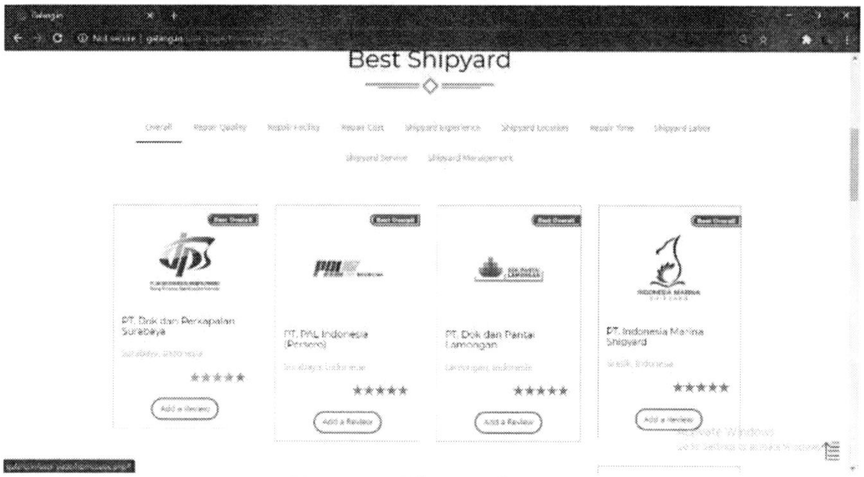

Figure 4. Main Page for User.

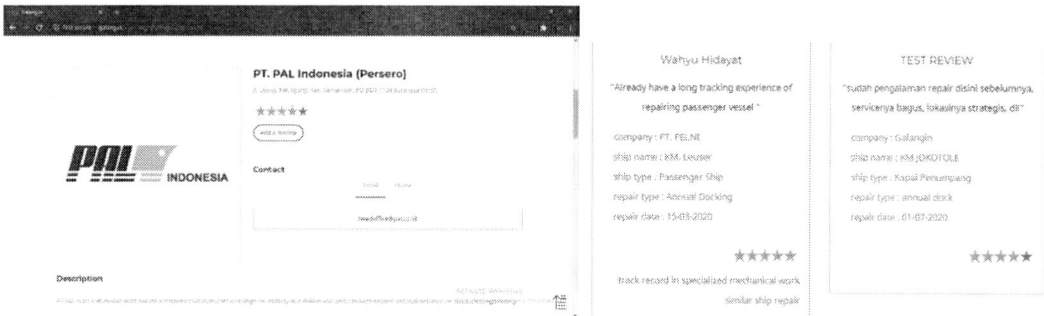

Figure 5. (a) Shipyard profile page and (b) Review display.

5.3.4. "Add a Review" Page

The Add a Review page serves to add or give a review to a shipyard. The display on the Add a Review page can be viewed in Figure 6. The reviewer and the rating are done in each criterion, and the value will be processed based on the weight of each criterion.

Figure 6. Add a Review Page.

Data that must be input by User when creating a review, among others, shipyard name, vessel name, vessel type, repair type, date of repair, and review and the value of each predefined criterion. In addition, the User must also upload a verification file to indicate that the correct repair work is happening and can be held accountable. After uploading the verification file, the User can press the Add a Review button and wait for the reply to be approved by Admin before it can be displayed on the shipyard profile.

5.3.5. "Create Shipyard Account" Page

Besides adding reviews to a shipyard, users can also create a shipyard account. This feature can be run on the Create Shipyard Account menu located on the username handle of each User. When creating a shipyard account, the User must enter data such as username, name, email, description, address, city, country, phone number, and password for his account, as in Figure 7.

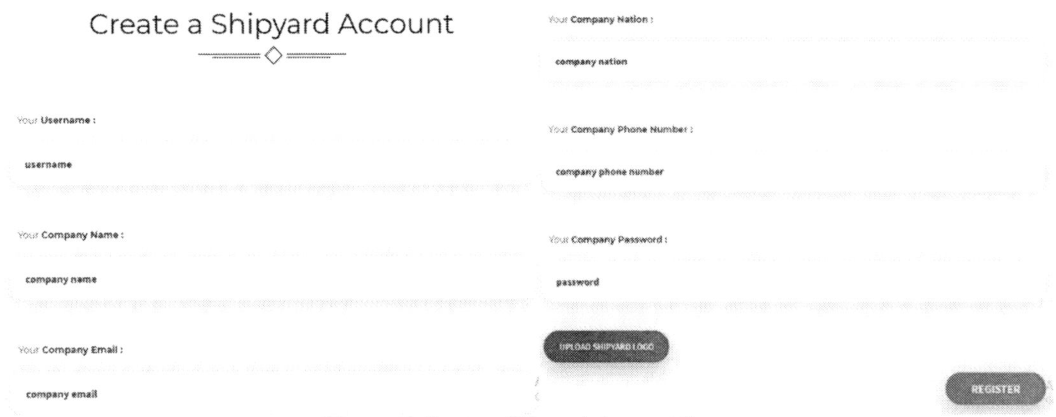

Figure 7. Create a Shipyard Account Page.

6. Analysis and Discussion

6.1. Validation Test

Data in the form of reviews provided by reviewers from cruise companies will be stored on to the server galang.in after the review process is done. Furthermore, the data that can only be accessed by the Admin will be verified the credibility of the review based on the reviewer's profile and the verification document that is the original proof that the company is repairing the ship at the relevant shipyard. Before it is approved by the Admin, the shipyard will also get a notification confirming that the review is indeed from the ship owner that repaired the ship. The shipyard will be given the option to acknowledge the correct or incorrect data of the ship, and then the shipyard recognition input will be sent to the Admin for verification. Reviews that pass the verification process will be submitted to the page and can be viewed in general by all Users on the web "Home" page, while reviews that do not pass the verification process will not be submitted to the User and will remain on the server.

The data of the value or rating per-entry criteria will be calculated automatically by the server based on the weight per-criterion that has been specified by averaging the entire value per-criterion, then the average value per-criterion multiplied by the weight-per-criterion. All average values per criterion that have been multiplied by weights are then summed with the values of other criteria so that the final value is obtained from a shipyard. Review data verified by the Admin will be submitted and accessible online by all Users on the galang.in.

6.2. Application Verification
To analyze the feasibility of the program, the testing and verification of the application with the relevant party is the Owner of the ship. The Owner who helped with this verification process has been competent to assess a shipyard as well as play a role as one of the User on this Website. After testing the app, an interview was then conducted about feedback about the designed app. The interview process is equipped with a questionnaire where this questionnaire contains six questions that aim to know the responses of the parties related to the application of this website program.

Table 5. Verification Questionnaire Recapitulation.

R	Q1	Q2	Q3	Q4	Q5	Q6	TOTAL	Mean	%
1	5	5	4	5	5	5	29	4.83	96.6
2	5	5	4	5	5	4	28	4.66	93.2
						Mean	28.5	4.745	94.9

*note:
R = Respondent
Q = Question

From the results of the questionnaire as seen Table 5, it is obtained that the application gets a value of 28.5 out of 30, or 94.9%, which means the application can be said to be "mandatory" or indispensable to be applied in the rating of the ship repair shipyard.

6.3. Advantages and disadvantages
After testing the application, the advantages and disadvantages of the system that have been made are obtained.

6.3.1. Advantages
The advantages of a customer review-based shipyard repair rating application that have been designed are:

1. The system can be accessed anywhere and anytime online because of the website-based system.
2. The review and rating system is made based on the shipyard repair criteria used by the Owner, so that the system is oriented towards the Owner in the operation of the application.

3. The shipyard value generated in the application is calculated on a per-criterion basis, so that it has different criteria for importance.
4. The ratings, reviews, and values of the shipyard generated in this application can be used as a medium for evaluation and appreciation for the shipyard, so that the shipyard can improve its service or make its value and reviews as a marketing or promotional media for the shipyard.
5. The appearance of the application is inspired by e-commerce applications that are widely used in the world, so that the application is easily understood by many people.
6. The application has a verification system to filter incoming reviews and shipyard accounts, so that irresponsible parties will not falsify reviews and shipyard accounts.

6.3.2. Disadvantages

In addition to the advantages offered, this customer review-based shipyard repair rating application also has the disadvantages of the system that has been designed:

1. Internet access is required to use the application because it is a website-based application.
2. Because it is online, there is a possibility that the application could be attacked by viruses or hackers.
3. The application has not been able to perform verification automatically so it still has to rely on the admin in the verification process.
4. The application has not been able to carry out business activities and is only limited to the review and rating process of the shipyard.
5. Participation from the repair shipyard is still not involved in this research and is only oriented to the point of view of the owner or shipping company.
6. There are still some bugs on the website that was designed, so that the operation of the website is still not optimal

From the above weaknesses, it is hoped that it can become an evaluation in application development at the next stage or research, so that it can become a better application.

7. Conclusion

After the design of computer applications for the rating of ship repair yard based on customer reviews are done, the following conclusions are obtained: There is currently no ship repair yard rating system that can represent the point of view of the Ship Owner. The process of selection of shipyards is still done in a conventional way by each shipping company. There are three ways to process the choice of repair yards: (1) Using private dockyards, (2) Using subscription shipyards, and (3) Selection using internal and external considerations of the company. In addition, the process of evaluating the shipyard is carried out in a conventional way, such as providing complaints or feedback in private for each Owner and the shipyard party.

In designing the ship repair yard rating application, a field survey was conducted to find out what criteria the Owner used in working the selection of the shipyard. Then there are nine criteria and by using the AHP method to weight the criteria, obtained: Repair quality (16.7%), Repair time (14.7%), Repair cost (12.5%), Repair facility (12.1%), Shipyard location (11.2%), Shipyard experience (10.4%), Shipyard workforce (8.8%), Shipyard service (7.6%), and Shipyard management (6.1%). The nine criteria are then used as the basis of a customer review-based shipyard rating system. There are two types of authority in this app, namely Admin and User. Admins have the power to approve or reject a review or shipyard account registered based on the inputted data and its verification files. In contrast, Users have the control to view the list, ratings, reviews, and value of a shipyard. In addition, Users can also create a shipyard account to register new dockyards that are not already on the shipyard list.

The shipyard rating system is implemented in the form of website-based applications. After testing and verifying the application with several respondents from shipping companies by running the application and conducting interviews with respondents using the questionnaire method, it was concluded that from the value range of 0-30, this application gets a value according to the results of the questionnaire of 28.5 out of 30, or 94.9%, which means the application can be said to be "mandatory" or indispensable to be applied in the shipyard rating.

Acknowledgment
We gratefully thank for Department of Naval Architecture for financial support, we also thank the reviewer for the constructive comments and suggestions that substantially improved the paper.

References

[1] E. Ismagilova, Y. K. Dwivedi, E. Slade, and M. D. Williams, *Electronic Word of Mouth (eWOM) in the Marketing Context: A State of the Art Analysis and Future Directions*. Springer International Publishing, 2017.
[2] C. Vasquez, *The Discourse of Online Consumer Reviews*. Bloomsbury Publishing, 2014.
[3] S. M. Mudambi and D. Schuff, "What makes a helpful online review? A study of customer reviews on amazon.com," *MIS Quarterly: Management Information Systems*. 2010, doi: 10.2307/20721420.
[4] *A Guide to Ship Repair Estimates in Man-hours*. 2012.
[5] R. L. Storch, *Ship Production*. Cornell Maritime Press, 1995.
[6] D. J. Eyres, *Ship Construction*. 2007.
[7] G. Bruce, *The Business of Shipbuilding*. 2013.
[8] H. Oh, "Service quality, customer satisfaction, and customer value: A holistic perspective," *Int. J. Hosp. Manag.*, 1999, doi: 10.1016/s0278-4319(98)00047-4.
[9] R. L. Oliver, *Satisfaction: A behavioral perspective on the consumer, Second edition*. 2014.
[10] C. Jahn and T. Bussow, "Best practice ship management," *Ger. Lloyd SE*, 2013.
[11] J. Giese and J. Cote, "Defining consumer satisfaction," *Acad. Mark. Sci. Rev.*, 2000.
[12] T. Ozseven, *Human-computer interaction*. 2019.
[13] N. Brugger, "Website history and the website as an object of study," *New Media Soc.*, 2009, doi: 10.1177/1461444808099574.
[14] H. F. Korth and A. Silberschatz, "Database system concepts," *McGraw-Hill Comput. Sci. Ser. ; McGraw-Hill Ser. Syst.*, 1991, doi: 10.1145/253671.253760.
[15] A. Kruk, T. Westerland, and P. Heller, "Database management systems," *Chem. Eng. (New York)*, 1996, doi: 10.1080/00325481.1984.11698577.
[16] T. L. Saaty, "Decision making with the Analytic Hierarchy Process," *Sci. Iran.*, 2002, doi: 10.1504/ijssci.2008.017590.
[17] "Analytic hierarchy process," in *Multi-Criteria Decision Analysis: Methods and Software*, 2013.
[18] J. Craggs, D. Bloor, B. Tanner, and H. Bullen, "Naval compensated gross tonnage coefficients and shipyard learning," *J. Sh. Prod.*, 2004.

A tabular optimisation technique for steel lazy wave riser

A. M. Ogbeifun[1], S. Oterkus[1], J. Race[1], H. Naik[2], D. Moorthy[3], S. Bhowmik[2], J. Ingram[3]

[1]Naval Architecture, Ocean and Marine Engineering, University of Strathclyde, Glasgow, G4 0LZ, UK.
[2]McDermott International, 40 Eastbourne Terrace, Paddington, London, W2 6LG, UK.
[3]McDermott International, 757 N. Eldridge Parkway, Houston, TX 77079, USA.

Abstract. Steel lazy wave riser (SLWR) is derived from the simple catenary riser (SCR) by the installation of buoyancy modules on its section. Infinite SLWR configurations are possible, and this poses difficulties in determining the best configuration. However, it is possible to capture some suitable configurations which satisfy some given design criteria specific to a project. We referred to these as the optimum configurations for the problem. Several advanced optimization tools and techniques for engineering optimization are available. In this paper, we present a 2D tabular optimization method for SLWR, which is an index-based optimization technique. The approach reduces a multidimensional problem to a 2D type providing a significant reduction in the required computational resources. It combines the design variables in pairs and assigns indices to the resulting design points (configurations) for each combination. The optimum design points are then tracked through index matching using techniques such as data sorting and intersection operations. In the application of the technique to SLWR design, we set the number of design variable for the problem to three. This results in three pair of combinations of the design variables. The design variables are the apparent mass ratio, the sag bend elevation, and the arc height. The output variables of interest to be optimized include the SLWR hanging length, the smeared buoyancy section length, the smeared buoyancy thickness, the riser hang-off tension, the stress utilization and fatigue damage around the bends. Selected optimum SLWR configurations from the optimization process are subjected to an irregular wave simulation to demonstrate the suitability of the approach for such optimisation problems.

1. Introduction

Riser systems are required to transport production fluid between the seabed and the floating production vessel. Risers are the most challenging components in deep-water oil and gas developments because of their extreme dynamic response. The steel catenary riser (SCR) is one of the riser type, which assumes a free hanging catenary configuration between its connecting point to the floating production vessel and the touchdown point (TDP), where it first contacts the seabed. SCRs are attractive riser systems [1] and have been used in many field developments considering their simplicity and relatively lower delivery cost compared to other riser system types [2-4]. However, when subjected to dynamic conditions, the SCR is limited by the high stress and fatigue damage responses experienced around critical sections, which are the hang off (HO) region and the touchdown zone (TDZ) [5], shown in Figure 1. The efforts made to reduce the stress and fatigue responses include but not limited to changing the configuration of SCR by including buoyancy modules on its sections, applying alternative material with better properties compared with steel and developing better riser interaction models to reduce design conservativism [6].

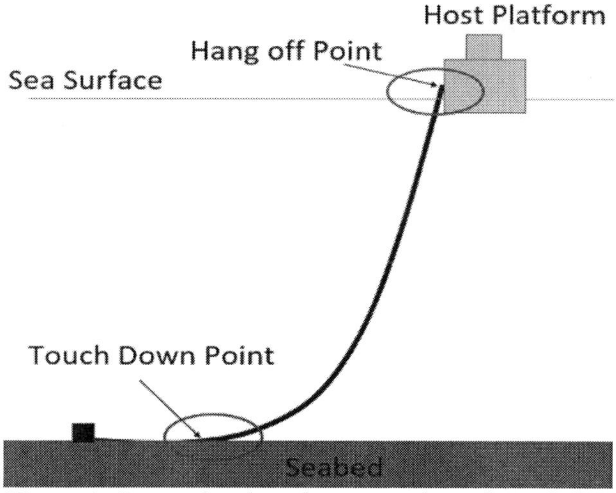

Figure 1. Conventional steel catenary riser schematic [6].

The SLWR solution is derived from SCR configuration by installing buoyancy modules on sections of the riser. The installed buoyancy modules induce compliant bends or 'waves' which help to decouple the vessel motion from the riser TDZ, resulting in the reduction of the stress and fatigue responses. The upward pull of the buoyancy modules on the SLWR section also helps to reduce the riser submerged weight and top tension. The level of benefits that can be derived from SLWR (over SCR) depends on an appropriate combination of its configuration variables. These include the buoyancy module capacity and lengths, the arc height, and the sag bend elevation of the SLWR. Several combinations of the configuration variables are possible, resulting in infinite possible SLWR configurations. However, suitable configurations satisfying a given project design criteria are usually searched for.

This paper presents a simplified 2D tabular optimization approach used to capture a family of suitable SLWR configurations, from the problem's design optimization space. The design space characterized by multi-dimension (3D) is reduced to three 2D spaces defined by a combination of pairs of the three independent configuration variables. These variables are the apparent mass ratio (AMR), the sag elevation (y_{sag}) and the arc height (Δh). Note that for the SLWR example in this paper, a fixed water depth and riser hang-off angle is considered. Unique indices are then assigned to the design points (SLWR configurations) in each of the three 2D design spaces. Output variables (response and geometric variables) are post-processed, arranged in tables, and sorted either in ascending or descending order for minimisation or maximization problems, respectively. The index tables corresponding to the output variables tables are re-arranged accordingly to match the repositioned output variable values in the sorted result tables. The first user-defined number of intersections of the re-arranged index tables point to the family of optimum SLWRs configurations. The detailed procedure for this technique is presented in the methodology section. An example of the SLWR optimization problem is presented in this paper to demonstrate the development, application, and suitability of the tabular optimization technique, which is considered to simplify a supposed complex optimisation process.

2. Methodology

2.1. Lazy wave configuration calculation

A typical SLWR configuration is presented in Figure 2. The highest and the lowest points on the 'wave' bend are referred to as the 'hog' and the 'sag' bend elevations from the seabed, denoted as y_{hog} and y_{sag} respectively. They are related by equation (1), where Δh is the arch height of the SLWR.

$$y_{hog} = y_{sag} + \Delta h \tag{1}$$

Different combinations of y_{sag}, Δh, the height of SLWR (h), the riser hang-off angle with the horizontal (θ) and the buoyancy section capacity results in different SLWR configurations. An example of possible input combinations options for specifying SLWR configurations is given in Table 1 [7]. It should be noted that for a unique SLWR configuration, any of the two input options in Table 1 can be re-expressed in the form of the other through the catenary equations. Option-1 is used in this paper.

Table 1. Design input options for LWR configuration.

Options	Δh	y_{sag}	Section lengths	θ	h
1	✔	✔	-	✔	✔
2	-	-	✔	✔	✔

The SLWR system configuration shown in Figure 2 (a), can be decomposed into five (5) sub catenaries, each with their local coordinate, as shown in Figure 2 (b). The catenary equations used for calculating each sub catenary are presented in equations (2) to (7). The horizontal top tension component (H) is constant everywhere along the SLWR profile, and this ensures continuity in the profile curvature. These equations are used to define the initial configuration of the SLWR for the finite element modelling. The equations are developed based on the assumption that the line has infinite axial stiffness and negligible bending stiffness.

(a)

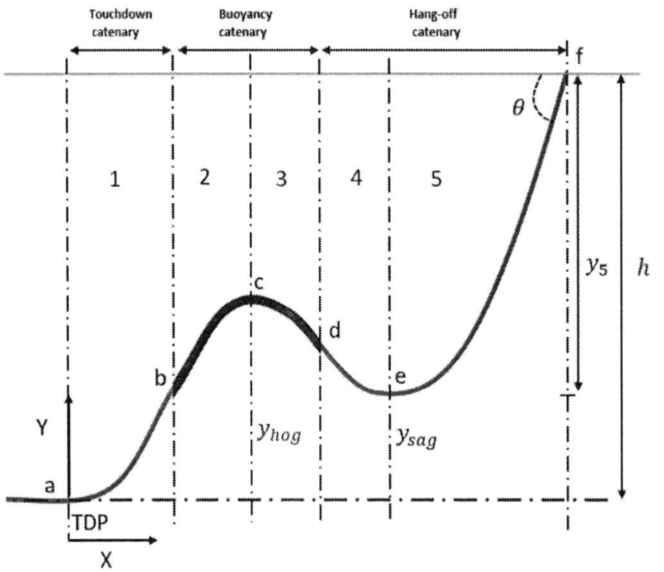

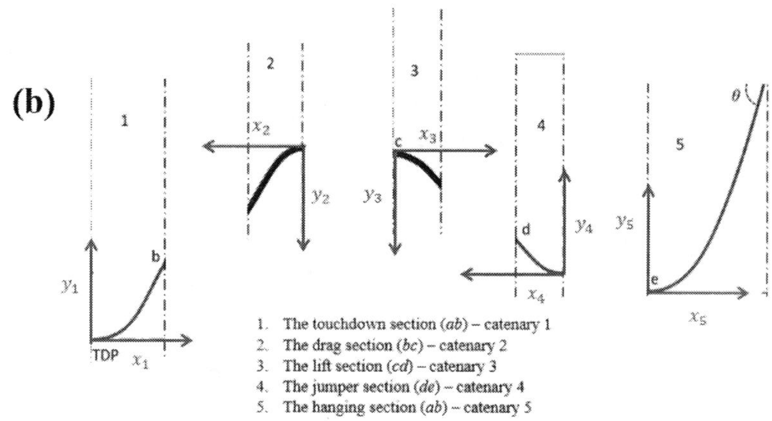

Figure 2. Steel lazy wave riser (SLWR) configuration: (a) Global configuration of SLWR, (b) Sub catenaries of the SLWR global configuration

$$\begin{cases} H = H_5 = \dfrac{w y_5}{(\tan\theta)^2}(1 + \sec\theta) \\ x_5 = \dfrac{H}{w}\cosh^{-1}\left(\dfrac{w y_5}{H} + 1\right) \\ s_5 = \dfrac{H}{w}\sinh\left(\dfrac{w x_5}{H}\right) \end{cases} \tag{2}$$

$$\begin{cases} y_4 = \dfrac{w_b}{w + w_b}(y_{hog} - y_{sag}) \\ x_4 = \dfrac{H}{w}\cosh^{-1}\left(\dfrac{w y_4}{H} + 1\right) \\ s_4 = \dfrac{H}{w}\sinh\left(\dfrac{w x_4}{H}\right) \end{cases} \tag{3}$$

$$\begin{cases} y_3 = \dfrac{w}{w + w_b}(y_{hog} - y_{sag}) \\ x_3 = \dfrac{H}{w}\cosh^{-1}\left(\dfrac{w_b y_3}{H} + 1\right) \\ s_3 = \dfrac{H}{w_b}\sinh\left(\dfrac{w_b x_3}{H}\right) \end{cases} \tag{4}$$

$$\begin{cases} y_2 = \dfrac{w}{w + w_b} y_{hog} \\ x_2 = \dfrac{H}{w_b}\cosh^{-1}\left(\dfrac{w_b y_2}{H} + 1\right) \\ s_2 = \dfrac{H}{w_b}\sinh\left(\dfrac{w_b x_2}{H}\right) \end{cases} \tag{5}$$

$$\begin{cases} y_1 = \dfrac{w_b}{w + w_b} y_{hog} \\ x_1 = \dfrac{H}{w} cosh^{-1}\left(\dfrac{wy_1}{H} + 1\right) \\ s_1 = \dfrac{H}{w} sinh\left(\dfrac{wx_1}{H}\right) \end{cases} \tag{6}$$

$$X_T = \sum_{i=1}^{5} x_i\,;\ h = \sum_{i=1}^{5} y_i\,;\ S_T = \sum_{i=1}^{5} s_i \tag{7}$$

Where s_i, x_i, y_i and H_i are respectively the catenary length, the horizontal span, the vertical span and the horizontal tension components of the ith sub catenary; X_T and S_T are respectively the total horizontal span, and the total riser length from hang-off (HO) to TDP; w and w_b are respectively the submerged unit weight of the bare pipe and the buoyant sections; and θ is the HO angle with the horizontal. The buoyant section capacity is determined by the thickness (or diameter), length, material density and the number of the discrete buoyancy modules installed on the SLWR. For a given buoyancy capacity, buoyancy material density and different values of the buoyancy module thickness (or diameter), there will be different corresponding lengths of the buoyant section. In this work, the discrete buoyancy section is conveniently modelled as an equivalent smeared buoyancy section. This is achieved by deriving an equivalent smeared buoyancy diameter that will provide equal displacements as all the discrete buoyancy modules. The thickness (t_b) or diameter (D_b) of the smeared buoyant section is modelled as the apparent mass ratio (AMR) as will be seen in the next section. Where $2t_b = D_b - D_o$, D_o is the bare pipe section outer diameter.

2.2. *Concept of apparent mass ratio (AMR)*

The apparent mass ratio (AMR) is the ratio of the apparent or submerged weight of the buoyant section, w_b, to the apparent or submerged weight of the bare pipe section, w

$$AMR = \frac{w_b}{w} = \frac{w_p + w_c + w_m - B_b}{w_p + w_c - B_p} \tag{8}$$

Where w_p, w_c, w_m, B_b and B_p are respectively the unit weight of bare pipe, unit weight of riser content, unit weight of buoyancy material, unit buoyancy force provided by the buoyant section and unit buoyancy force provided by the bare pipe section. Equation (8) can be expanded and re-casted as equation (9), which provides a relationship between the AMR, the buoyancy material density (ρ_b) and the smeared buoyant section thickness (t_b). The positive real roots of equation (9) gives the required, t_b for a given AMR and ρ_b. Where D_i and D_o are the inner and outer diameters of the bare pipe section, ρ_c and ρ_w are respectively the content and seawater density.

$$At_b^2 + Bt_b + C = 0 \tag{9}$$

Where:

$$A = 4(\rho_b - \rho_w), B = 4D_o(\rho_b - \rho_w)$$

$$C = (1 - AMR)\,(\rho_s D_o^2 - \rho_s D_i^2 + \rho_c D_i^2 - \rho_w D_o^2)$$

The calculated t_b is used to determine w_b which is substituted into the catenary equations for the SLWR configuration calculation, for a given y_{sag}, Δh, and θ.

2.3. 2D tabular optimisation methodology

For fixed values of the SLWR HO angle with the horizontal, θ and the height of the SLWR, h, let $\mathbf{X}$ be the vector of design input variables i.e. $X_1 = y_{sag}$, $X_2 =$ AMR, and $X_3 = \Delta h$. Let $\mathbf{Y}$ be the design output variables to be optimised i.e. $Y_1 = U_{bend}$, $Y_2 = T_{top}$, $Y_3 = \Delta\sigma_{bend}$, $Y_4 = s_b$, $Y_5 = t_b$ and $Y_6 = S_T$. The optimization problem can be expressed as follows:

$$\text{find } \mathbf{X} = \begin{Bmatrix} AMR \\ y_{sag} \\ \Delta h \end{Bmatrix} \text{ which minimizes } \mathbf{Y} = \begin{Bmatrix} U_{bend} \\ T_{top} \\ \Delta\sigma_{bend} \\ s_b \\ S_T \\ t_b \end{Bmatrix} \tag{10}$$

Subject to the following constraints:

$$AMR < 0, y_{sag} > 0, \Delta h \geq 0$$

Where AMR is the apparent mass ratio of the buoyant sections, y_{sag} is the sag height elevation from the seabed, Δh is the arc height, U_{bend} is the stress utilization in the riser bends, T_{top} is the tension the riser top, $\Delta\sigma_{bend}$ is the stress range in the riser bends (representative of the fatigue damage), s_b is the smeared buoyant section length, S_T is the total hanging length of the SLWR, and t_b is the smeared buoyant section thickness.

From the optimization problem layout, there are three independent design input variables (m =3) and six design output variables (p = 6). The 2D tabular approach presented in this paper requires the joint variation of two (2) variables, while the third is set to an arbitrary practical value. Hence, there will be three combination pairs ($n = \mathrm{C}(m,2) = \mathrm{C}(3,2) = 3$), referred to as the design configurations groups. The number of result tables (denoted as $Y_{p,n}$) will then be $p \times n$. A system of index tables (denoted as $I_{p,n}$) is created of equal dimension with $Y_{p,n}$ to represent the index locations of all results in the tables of $Y_{p,n}$. The layout of the results tables, $Y_{p,n}$ and the index system tables, $I_{p,n}$ are presented in Table 2 and Table 3, respectively. For numerical convenience, each table in $Y_{p,n}$ and $I_{p,n}$ can be converted to columns or vectors, and from this point on can be interchangeably referred to as *result columns* and *index columns* in $Y_{p,n}$ and $I_{p,n}$ respectively.

Table 2. Result tables in each configuration groups.

Configuration group		Output variable tables or columns ($Y_{p,n}$)					
		U_{bend}	T_{top}	$\Delta\sigma_{bend}$	s_b	S_T	t_b
y_{sag}, AMR	$Y_{p,n=1}$	$Y_{1,1}$	$Y_{2,1}$	$Y_{3,1}$	$Y_{4,1}$	$Y_{5,1}$	$Y_{6,1}$
$\Delta h, AMR$	$Y_{p,n=2}$	$Y_{1,2}$	$Y_{2,2}$	$Y_{3,2}$	$Y_{4,2}$	$Y_{5,2}$	$Y_{6,2}$
$y_{sag}, \Delta h$	$Y_{p,n=3}$	$Y_{1,3}$	$Y_{2,3}$	$Y_{3,3}$	$Y_{4,3}$	$Y_{5,3}$	$Y_{6,3}$

Table 3. Index tables system.

Configuration groups		Index system tables or columns ($I_{p,n}$)					
		U_{bend}	T_{top}	$\Delta\sigma_{bend}$	s_b	S_T	t_b
y_{sag}, AMR	$I_{p,1}$	$I_{1,1}$	$I_{2,1}$	$I_{3,1}$	$I_{4,1}$	$I_{5,1}$	$I_{6,1}$
$\Delta h, AMR$	$I_{p,2}$	$I_{1,2}$	$I_{2,2}$	$I_{3,2}$	$I_{4,2}$	$I_{5,2}$	$I_{6,2}$
$y_{sag}, \Delta h$	$I_{p,3}$	$I_{1,3}$	$I_{2,3}$	$I_{3,3}$	$I_{4,3}$	$I_{5,3}$	$I_{6,3}$

Every table (or result column) in $Y_{p,n}$ are sorted in ascending order (for minimisation problem) or descending order for maximisation problems. The corresponding tables (or index columns) in $I_{p,n}$ are likewise re-arranged accordingly to match the repositioned output variable values in the sorted tables in $Y_{p,n}$. Intersection operation is carried out on the re-arranged index columns, and the first user-specified number of index intersections, k, point to a family of optimum SLWRs configurations. The k indices of the optimum SLWR configurations can be generally expressed as equation (11) and depicted in Figure 3, where P is the total number of output variables and N is the total number of pair combinations of the input variables.

$$\cap\left(I_{p,n}\right)_k, 1 \leq p \leq P \text{ and } 1 \leq n \leq N \quad (11)$$

A subset of the intersection in equation (11) can also be searched for. For example, if there is need to know what configurations optimizes U_{bend} (p =1) and s_b (p =4), then equation (12) applies.

$$\cap\left(I_{p,n}\right), p = 1, 4 \text{ and } 1 \leq n \leq 3 \quad (12)$$

The k indices of optimum family members obtained can then be matched to the design and result space to determine the corresponding optimum design variables, which are the AMR, y_{sag} and Δh, and the corresponding values of the design output variable, which are U_{bend}, T_{top}, $\Delta\sigma_{bend}$, s_b, S_T and t_b.

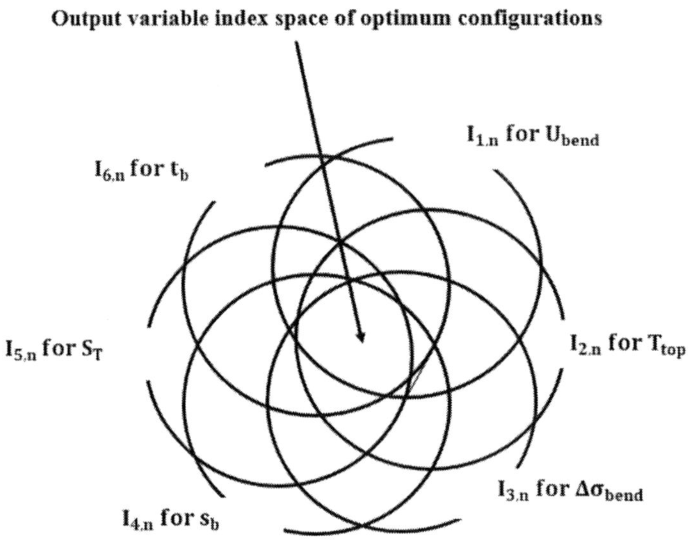

Figure 3. Optimum configuration index space intersections.

3. Analysis data

3.1. Riser model

The SLWR used for this study is made up of a string of 12-inch X70 grade pipes. It is hosted by a generic floating production, storage, and offloading (FPSO) unit at an azimuth of 90deg to the vessel heading. The minimum wall thickness required for burst and collapse pressure resistance was calculated using DNV-OS-F201 criteria [8]. Table 4 presents the design data for the riser.

Table 4. Riser data.

SLWR data	Values
Pipe size	12inch
Internal Design pressure	10ksi
Pipe thickness (fixed)	27.5mm
Hang off angle with the vertical (fixed)	$(90 - \theta) = 12^{\circ}$
Content density	600kg/m3
Hang off stiffness (linear)	12kN.m/deg
Buoyancy material density	500kg/m^3
Water depth	1500m
Buoyant section length	Vary (m) (based on varying AMR)
Smeared buoyancy thickness	Vary (m) (based on varying AMR)
Total riser length	Vary (m)

The design range of values for each of the design input variables consist of 14 entries (design points) presented in Table 5.

Table 5. Design input variable space.

y_{sag}(m)	AMR	Δh(m)	y_{sag}(m)	AMR	Δh(m)
10	-0.10	0	430	-1.75	280
70	-0.25	40	490	-2.00	320
130	-0.50	80	550	-2.25	360
190	-0.75	120	610	-2.50	400
250	-1.00	160	670	-2.75	440
310	-1.25	200	730	-3.00	480
370	-1.50	240	790	-3.25	520

3.2. Environmental, vessel and seabed data

Two irregular wave loads for the combined load and fatigue analysis are presented in Table 6. The optimisation stage is a screening process where hundreds of FE models are simulated to search for a family of optimum configurations. Applying irregular wave loads for the combined load and fatigue analysis during the optimisation process may be prohibitive as huge computation resources will be required to do so. Hence, the equivalent regular wave load data are used during the optimisation stage. Once a family of the optimum SLWR configurations are obtained, the selected optimum configurations will be subjected to the irregular wave analysis for a minimum of 3hrs period.

Table 6. Wave load data.

Analyses	Wave type	Data	Values
Combined	Irregular wave	H_s(m)	8
		T_p(sec)	13
		γ	1.6
	Equivalent regular wave	H_{max}(m)	14.9
		T(sec)	9.5
Fatigue	Irregular wave	H_s(m)	4.5
		T_p(sec)	9.5
		γ	1.8
	Equivalent regular wave	H_{max}(m)	8.4
		T(sec)	7

The equivalent regular wave loads are calculated from the irregular wave loads using equations (13) and (14), where T is the regular wave period, H_{max} is the probable maximum wave height, T_z is the irregular wave zero up crossing period, T_p is the wave peak period and γ is the wave peak shape parameter. Both equations are valid for the JONSWAP spectrum with wave peak shape parameter in the range $1 \leq \gamma < 7$ [9]. All wave loads are considered Beam Sea and current loads are excluded.

$$H_{max} = 1.86H_s, T = T_z \tag{13}$$

$$T_z = T_p(0.6673 + 0.05037\gamma - 0.006230\gamma^2 + 0.0003341\gamma^3) \tag{14}$$

A generic response amplitude operator (RAOs) for the FPSO is implemented. The nonlinear hysteretic riser soil interaction model is used. Details of the default riser soil interaction model data used can be found in [10].

4. Analysis, results, and discussion

4.1. Analysis method

As presented in Table 5, each of the three design variables has 14 entries. There are $n = 3$ combination pairs or configuration groups of the design variables presented in Table 7, resulting in a total number of $2 \times 14 \times 14$ =588 design points. For each of the configuration group, an arbitrary practical value for the third variable is set. As seen in Table 7, $\Delta h = 50m$ for group 1, $y_{sag} = 100m$ for group 2 and $AMR = -1$ for group 3. FE model for each of the 588 design points are built and simulated for both the combined and the fatigue regular wave loads. The numerical simulations of the models were conducted using the OrcaFlex FE software package[11]. Modelling, pre-processing, post-processing, result organisation in tables of $Y_{p,n}$ and the setting up of the corresponding index tables in $I_{p,n}$ are automated using MATLAB/PYTHON programs integrated with OrcaFlex programming interface (OrcFxAPI [12]). The program generates the OrcaFlex models using the equation (2) to equation (7) to define the SLWRs initial configurations.

Table 7. Input variable combination (Configuration groups).

Conf. Group1 $\Delta h = 50$m		Conf. Group2 $y_{sag} = 100$m		Conf. Group3 AMR = -1	
y_{sag}(m)	AMR	Δh(m)	AMR	y_{sag}(m)	Δh(m)
10	-0.10	0	-0.10	10	0
70	-0.25	40	-0.25	70	40
130	-0.50	80	-0.50	130	80
190	-0.75	120	-0.75	190	120
250	-1.00	160	-1.00	250	160
310	-1.25	200	-1.25	310	200
370	-1.50	240	-1.50	370	240
430	-1.75	280	-1.75	430	280
490	-2.00	320	-2.00	490	320
550	-2.25	360	-2.25	550	360
610	-2.50	400	-2.50	610	400
670	-2.75	440	-2.75	670	440
730	-3.00	480	-3.00	730	480
790	-3.25	520	-3.25	790	520

For each model, the output variables: the stress utilization (U_{bend}), the top tension (T_{top}), the stress range ($\Delta\sigma_{bend}$) and the SLWR geometry or dimension (t_b, S_T and s_b) are obtained. Note that stress utilization is calculated based on DNV-OS-F201, considering the combined load resistance factor design criteria for internal and external overpressure conditions [8]. The output ($p = 6$) variables are post-processed, and their extreme or maximum values are obtained and organised in tables in $Y_{p,n}$. The number of result tables or columns in $Y_{p,n}$ will be $n \times p = 18$ (see Table 5). The corresponding system of index tables $I_{p,n}$ will also contain 18 tables or result columns. Figure 6 (a), (b) and (c) present the 2D spatial representation of Table 7, showing the design space index of the three-design variable combination or configuration groups. The design variable values are shaded in 'blue'.

AMR

y_{sag}(m)	-0.1	-0.25	-0.5	-0.75	-1	-1.25	-1.5	-1.75	-2	-2.25	-2.5	-2.75	-3	-3.25
10	1	15	29	43	57	71	85	99	113	127	141	155	169	183
70	2	16	30	44	58	72	86	100	114	128	142	156	170	184
130	3	17	31	45	59	73	87	101	115	129	143	157	171	185
190	4	18	32	46	60	74	88	102	116	130	144	158	172	186
250	5	19	33	47	61	75	89	103	117	131	145	159	173	187
310	6	20	34	48	62	76	90	104	118	132	146	160	174	188
370	7	21	35	49	63	77	91	105	119	133	147	161	175	189
430	8	22	36	50	64	78	92	106	120	134	148	162	176	190
490	9	23	37	51	65	79	93	107	121	135	149	163	177	191
550	10	24	38	52	66	80	94	108	122	136	150	164	178	192
610	11	25	39	53	67	81	95	109	123	137	151	165	179	193
670	12	26	40	54	68	82	96	110	124	138	152	166	180	194
730	13	27	41	55	69	83	97	111	125	139	153	167	181	195
790	14	28	42	56	70	84	98	112	126	140	154	168	182	196

Vectorized indices

Figure 4. Conf. Group 1: Joint variation of y_{sag} and AMR.

AMR

Δh(m)	-0.1	-0.25	-0.5	-0.75	-1	-1.25	-1.5	-1.75	-2	-2.25	-2.5	-2.75	-3	-3.25
0	1	15	29	43	57	71	85	99	113	127	141	155	169	183
40	2	16	30	44	58	72	86	100	114	128	142	156	170	184
80	3	17	31	45	59	73	87	101	115	129	143	157	171	185
120	4	18	32	46	60	74	88	102	116	130	144	158	172	186
160	5	19	33	47	61	75	89	103	117	131	145	159	173	187
200	6	20	34	48	62	76	90	104	118	132	146	160	174	188
240	7	21	35	49	63	77	91	105	119	133	147	161	175	189
280	8	22	36	50	64	78	92	106	120	134	148	162	176	190
320	9	23	37	51	65	79	93	107	121	135	149	163	177	191
360	10	24	38	52	66	80	94	108	122	136	150	164	178	192
400	11	25	39	53	67	81	95	109	123	137	151	165	179	193
440	12	26	40	54	68	82	96	110	124	138	152	166	180	194
480	13	27	41	55	69	83	97	111	125	139	153	167	181	195
520	14	28	42	56	70	84	98	112	126	140	154	168	182	196

Vectorized indices

Figure 5. Conf. Group 2: Joint variation of y_{sag} and Δh.

Δh(m)

y_{sag}(m)	0	40	80	120	160	200	240	280	320	360	400	440	480	520
10	1	15	29	43	57	71	85	99	113	127	141	155	169	183
70	2	16	30	44	58	72	86	100	114	128	142	156	170	184
130	3	17	31	45	59	73	87	101	115	129	143	157	171	185
190	4	18	32	46	60	74	88	102	116	130	144	158	172	186
250	5	19	33	47	61	75	89	103	117	131	145	159	173	187
310	6	20	34	48	62	76	90	104	118	132	146	160	174	188
370	7	21	35	49	63	77	91	105	119	133	147	161	175	189
430	8	22	36	50	64	78	92	106	120	134	148	162	176	190
490	9	23	37	51	65	79	93	107	121	135	149	163	177	191
550	10	24	38	52	66	80	94	108	122	136	150	164	178	192
610	11	25	39	53	67	81	95	109	123	137	151	165	179	193
670	12	26	40	54	68	82	96	110	124	138	152	166	180	194
730	13	27	41	55	69	83	97	111	125	139	153	167	181	195
790	14	28	42	56	70	84	98	112	126	140	154	168	182	196

Vectorized indices

Figure 6. Conf. Group 3: Joint variation of AMR and Δh.

4.2. *Results and discussion*

Recall from Table 2 that each of the ($n = 3$) configuration group contains six result tables or columns, corresponding to each of the ($p = 6$) design output variables. This gives 18 result columns in $Y_{p,n}$. Similarly, there will be 18 associated index tables or columns in $I_{p,n}$ as seen in Table 3. First, the 18 index tables or columns in $I_{p,n}$ are sorted in ascending order, and the associated index tables in $I_{p,n}$ are rearranged accordingly. The intersection of the index columns for each configuration group is then obtained. The second, third and fourth columns of Table 8 present the intersections of tables or index columns in each of the three configuration group. They are respectively $\cap(I_{p,1})$, $\cap(I_{p,2})$ and $\cap(I_{p,3})$. The configuration indices in these columns are pointers to optimum SLWR configurations for the groups. The fifth column presents the combined intersection of the three configuration groups, i.e.:

$$\cap(I_{p,n}) = \cap(I_{p,1}, I_{p,2}, I_{p,3}), \qquad 1 \leq p \leq 6 \tag{15}$$

Table 8. Index numbers intersection (pointers) to optimum configurations.

	Conf. group 1	Conf. group 2	Conf. group 3	Combined
92	[]	[]	[]	[]
93	[]	[]	[]	[]
94	[]	[]	78	[]
95	[]	[]	78	[]
96	[]	[]	[78;92]	[]
97	[]	[]	[78;92]	[]
98	[]	[]	[78;92]	[]
99	[]	[]	[78;91;92]	[]
100	[]	[]	[77;78;91;92]	[]
101	[]	[]	[77;78;91;92]	[]
102	[]	[]	[77;78;91;92]	[]
103	[]	[]	[63;77;78;91;92]	[]
104	[]	[]	[63;77;78;91;92]	[]
105	[]	[]	[63;64;77;78;91;92]	[]
106	[]	86	[63;64;77;78;91;92]	[]
107	[]	[73;86;107]	[63;64;77;78;91;92]	[]
108	[]	[73;86;107]	[63;64;77;78;91;92]	[]
109	[]	[73;86;101;107]	[63;64;77;78;91;92]	[]
110	[]	[73;86;87;101;107]	[63;64;77;78;91;92;106]	[]
111	[]	[73;86;87;101;107]	[63;64;77;78;79;91;92;106]	[]
112	[]	[73;86;87;101;107]	[63;64;77;78;79;91;92;93;106]	[]
113	91	[73;86;87;101;107]	[63;64;77;78;79;91;92;93;105;106]	[]
114	91	[73;86;87;101;107;108]	[63;64;77;78;79;90;91;92;93;105;106]	[]
115	91	[73;86;87;92;101;107;108]	[48;63;64;77;78;79;90;91;92;93;105;106;107]	[]
116	[91;107]	[73;86;87;92;101;107;108]	[48;63;64;77;78;79;90;91;92;93;105;106;107]	107
117	[91;107]	[73;86;87;92;101;107;108]	[48;63;64;77;78;79;90;91;92;93;105;106;107]	107
118	[91;107]	[73;86;87;92;101;107;108]	[48;62;63;64;77;78;79;90;91;92;93;105;106;107]	107
119	[91;94;107]	[73;86;87;92;93;101;107;108]	[48;62;63;64;77;78;79;90;91;92;93;104;105;106;107]	107
120	[91;94;107]	[73;86;87;92;93;101;107;108]	[48;62;63;64;76;77;78;79;90;91;92;93;104;105;106;107]	107
121	[91;94;107]	[73;86;87;92;93;101;107;108;121]	[48;62;63;64;76;77;78;79;90;91;92;93;104;105;106;107]	107
122	[91;94;95;107;108]	[73;86;87;92;93;101;107;108;115;121;122]	[48;62;63;64;76;77;78;79;90;91;92;93;104;105;106;107]	107
123	[91;92;94;95;107;108]	[73;86;87;92;93;101;107;108;115;121;122]	48;62;63;64;76;77;78;79;90;91;92;93;104;105;106;107;108	[92;107;108]
124	[91;92;94;95;107;108]	3;86;87;92;93;101;107;108;115;121;122;12	;62;63;64;76;77;78;79;90;91;92;93;104;105;106;107;108;1	[92;107;108]

From Table 8, the following points should be noted:

- Configuration index 91 appears as the first configuration, which optimises all six output variables in configuration group 1. In Figure 4, Index 91 matches SLWR configuration with AMR = -1.5 and y_{sag} =370m. Index 91 occurs in the first k = 113 row search across the six index columns of configuration group 1 ($Y_{p,1}$). As k increases beyond 115, more configurations index such as 107, 94, 95, 108, etc. is included in the family of optimum SLWR in configuration group 1.
- Configuration index 86 appears as the first configuration, which optimises all six output variables in configuration group 2. In Figure 5, Index 86 matches SLWR configuration with AMR = -1.5 and Δh =40m. Index 86 occurs in the first k = 106 row search across the six index columns of configuration group 2 ($Y_{p,2}$). As k increases beyond 106, more configurations index such as 73, 107, 87,101, etc. is included in the family of optimum SLWR in configuration group 2.

- Configuration index 78 appears as the first configuration, which optimises all six output variables in configuration group 3. In Figure 6, Index 78 matches SLWR configuration with Δh = 200 and y_{sag} =430m. Index 78 occurs in the first k = 94 row search across the six index columns of configuration group 3 ($Y_{p,3}$). As k increases beyond 95, more configurations index such as 92, 91, 77, 63, etc. is included in the family of optimum SLWR in configuration group 3.
- Configuration index 107 appears as the first configuration, which optimises all six output variables in all three configuration groups (Group 1, 2 and 3). Index 107 occurs in the first k = 116 row search across the six index columns vectors of all three configuration groups as indicated by equation (15). Since index 107 is a result of the intersection of the index columns across the three groups, we must therefore match it against the three configuration group design spaces presented respectively in Figure 4, Figure 5 and Figure 6. By doing so, we find out that index 107 matches with the SLWR configurations with y_{sag} = 490m, AMR = -1.75, Δh =320m, 280m. Note that two Δh values occur since Δh appears as the row header in Figure 5 and the column header Figure 6. As k increases beyond 122, more configurations such as 92, 108, etc. are observed to be included in the family of problem's optimum configurations.

4.2.1. Matching family of optimum indices to the design point variables
To demonstrate the suitability of the optimum configurations, we select from the joint intersection indices 92 and 107. These configurations will be checked against the randomly selected configurations 33 and 195. The four configuration indices (92, 107, 33 and 195) are matched against the configuration group design spaces in Figure 4, Figure 5 and Figure 6 to derive Table 9. Table 10 is developed by re-arranging Table 9, such that each row represents a unique SLWR configuration. The randomly selected configurations are highlighted in red.

Table 9. Selected configuration design variable table.

Conf. index	Conf. group 1			Conf. group 2			Conf. group 3		
	y_{sag}(m)	AMR	Δh(m)	y_{sag}(m)	AMR	Δh(m)	y_{sag}(m)	AMR	Δh(m)
92	430	-1.5	-	-	-1.5	280	430	-	240
107	490	-1.75	-	-	-1.75	320	490	-	280
33	250	-0.5	-	-	-0.5	160	250	-	80
195	730	-3.25	-	-	-3.25	480	730	-	520

Table 10. Expanded configuration table.

S/N	Conf. ID	Configurations input variables		
		y_{sag}(m)	AMR	Δh(m)
1	92a	430	-1.5	280
2	92b	430	-1.5	240
3	107a	490	-1.75	320
4	107b	490	-1.75	280
5	33a	250	-0.5	160
6	33b	250	-0.5	80
7	195a	730	-3.25	480
8	195b	730	-3.25	520

4.2.2. Output geometric and configuration analysis for the selected optimum configurations
The static configurations of the SLWRs presented in Table 10 are shown in Figure 7. The selected optimum configurations (92a, 92b, 107a, 107b) are observed to cluster around each other. On the other

hand, the profiles of the randomly selected configurations (conf. 33a, conf.33b, conf. 195a, conf. 195b) are found to deviate widely from those of the optimum solution cluster. Bar plot of the geometric output variable (s_b and S_T) of the configurations are presented in Figure 8. The s_b and S_T values of the randomly selected configurations are observed to be higher than those of the optimum configurations, except s_T of conf 33b and s_b of configuration 195a and 195b. Numerical values are presented in Table 11.

One of the cost drivers of the SLWR is the buoyancy material volume of the buoyant section. The smeared volume, vol_b, can be expressed in terms of t_b, s_b, as shown in equation (16). Recall from equation (16) that t_b and s_b are part of the design output variables to be optimised. Hence, optimising t_b and s_b implies optimization of vol_b.

$$vol_b = \pi t_b s_b (OD + t_b) \quad (16)$$

The bar plot of vol_b is presented in Figure 9. While the smeared buoyancy of configuration 195 requires a higher volume to achieve higher wave bend elevation, configuration 33 require lower buoyancy volume because of its lower wave bend elevation, compared with the selected optimum configurations. However, the problem is set to optimise six output variables (not just s_b and S_T), it will therefore be comprehensive to examine the suitability of the selected optimum configurations in terms of all the six variables.

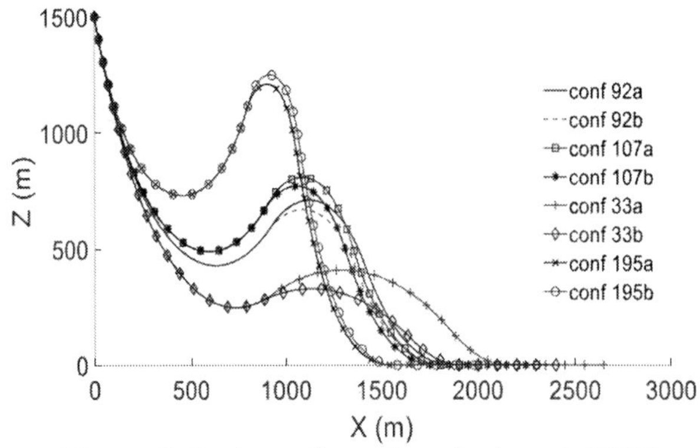

Figure 7. Static configuration of selected SLWRs.

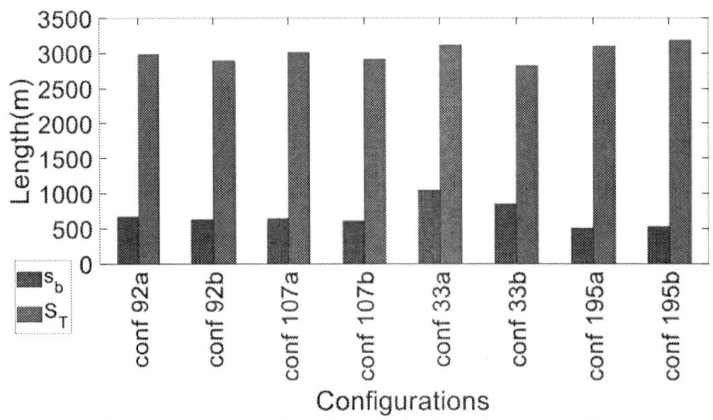

Figure 8. Comparing SLWR hanging length (s_T) and smeared buoyancy section length (s_b).

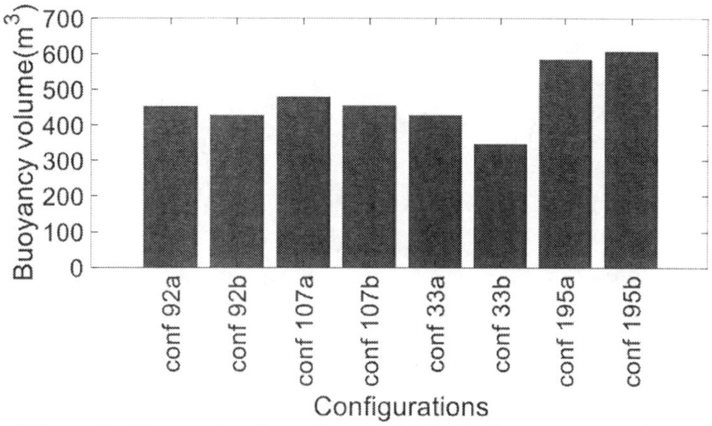

Figure 9. Comparing the volume of SLWRs buoyancy section.

4.2.3. Irregular wave analysis for the selected optimum configuration

The irregular wave loads for combined and fatigue were simulated on the selected SLWR configurations. Figure 10, Figure 11 and Figure 12 respectively present the U_{bend} , the T_{top} and minimum fatigue life of the selected SLWR configurations. Table 11 offers a comparison of maximum values for the design output variables.

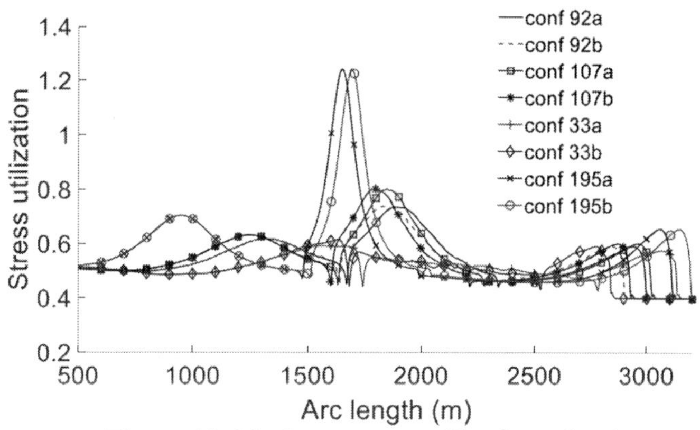

Figure 10. Maximum stress utilization at bends.

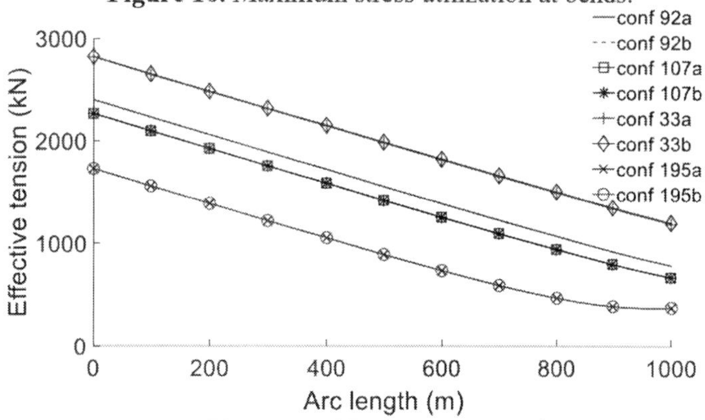

Figure 11. Maximum top tension.

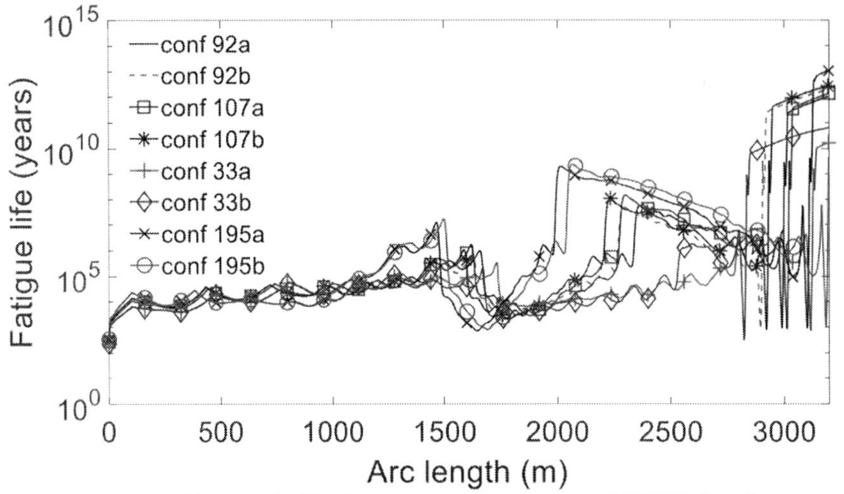

Figure 12. Maximum stress range around SLWR bends.

Table 11. Summary of the configurations output variable.

Conf. ID	Independent Input design variables			Critical values of design output variables						
	y_{sag}(m)	AMR	Δh(m)	U_{bend}	T_{top} (MN)	Min. Life (years)	s_b (m)	S_T (m)	t_b(m)	Vol_b (m³)
92a	430	-1.5	280	0.75	2400.12	254.68	665.87	2986.01	0.34	452.76
92b	430	-1.5	240	0.75	2399.38	256.31	627.47	2890.03	0.34	426.66
107a	490	-1.75	320	0.80	2266.57	277.88	640.42	3008.41	0.36	479.01
107b	490	-1.75	280	0.80	2265.41	277.87	607.04	2916.61	0.36	454.04
33a	250	-0.5	160	0.77	2815.55	226.24	1047.43	3114.76	0.24	427.32
33b	250	-0.5	80	0.77	2820.92	226.13	850.52	2819.41	0.24	346.99
195a	730	-3.25	480	1.24	1728.75	346.59	504.99	3097.08	0.47	583.74
195b	730	-3.25	520	1.24	1729.24	344.97	524.58	3180.36	0.47	606.38

For easy comparison of the performances of the optimum configurations, the values of the output variable for the randomly selected configurations are compared with the mean values of the optimum configuration. The comparison is expressed in percentages presented in Table 12. Positive percentages mean that the value of the randomly selected configuration output variables is higher than the mean of the values of the optimum configurations. The reverse is the case for negative values. The results show that the 2D optimisation technique balances the interest of all six output variables in the selected optimum configurations. For example, while the mean of the fatigue lives of the optimum configurations is lower than those of the randomly selected configuration 33a and 33b by 15%, the top tension of configuration 33a and 33b are higher by 21%. On the other hand, the randomly selected configuration 195a and 195b have stress utilization higher than the mean stress utilization of the optimum configuration by 60%, but with fatigue life higher by about 30% and a reduction in top tension by about 26%. The results indicate the challenges inherent in solving a multi-objective optimisation problem. A unique decision would be possible for a single-objective optimisation problem. However, for a multi-objective problem, with additional considerations from the designer, a given optimum configuration can be selected from the family of optimum solutions. For example, if the design interest is majorly the cost of the SLWR, the total riser length and volume of buoyancy material will be of interest and configuration 195a and 195b will be seen to be more expensive relative

IOP Conf. Series: Materials Science and Engineering 1052 (2021) 012022 doi:10.1088/1757-899X/1052/1/012022

to the current sets optimum configurations. If the designer's major interest is the top tension capacity of the supporting structure for the riser, then configuration 33a and 33b will be seen to provide higher top tension relative to the current sets of optimum configurations. However, for the above two conditions specified, the 2D tabular technique can be reapplied for the subset of considerations and new sets of optimum configuration can be derived for the problem (see the example presented by equation (12)).

Table 12. Percentage change in critical values of the randomly selected configuration relative to the mean of the optimum configuration.

Conf. ID	Independent Input design variables			% change						
	y_{sag}(m)	AMR	Δh(m)	U_{bend}	T_{top}	Min. Life	s_b	S_T (m)	t_b	Vol_b
33a	250	-0.5	160	-0.6	20.7	-15.2	64.9	5.6	-31.4	-5.7
33b	250	-0.5	80	-0.6	20.9	-15.2	33.9	-4.4	-31.4	-23.4
195a	730	-3.25	480	60.0	-25.9	30.0	-20.5	5.0	34.3	28.8
195b	730	-3.25	520	60.0	-25.9	29.4	-17.4	7.8	34.3	33.8

Note: Positive percentages means that the value of the randomly selected configuration output variables is higher than the mean of the values of the optimum configurations. The reverse is the case for negative values.

5. Conclusion

A 2D tabular optimisation technique is presented in this paper. The technique reduces a higher dimensional design space of a problem to sets of two-dimensional design spaces. It then assigns indices (identifiers) to every design point or configuration in the 2D design spaces. The optimum design points are then tracked through index matching, using techniques such as data sorting and intersection operations.

The technique is demonstrated by its application to solve the SLWR riser optimisation problem. For the SLWR example, there are three independent design input variables: y_{sag}, AMR and Δh. The output variables to be optimised are T_{top}, U_{bend}, s_b, S_T, t_b and $\Delta\sigma$ or fatigue damage. Design points or configurations in the 2D design spaces are simulated with the combined and fatigue wave loads. The design output results are post-processed and organized in a system of result tables or columns, $Y_{p,n}$. A corresponding system of index tables or columns, $I_{p,n}$ is set up, which serve as identifiers for results tables in $Y_{p,n}$. The results columns are sorted in ascending order (for minimisation problem), while the system of index tables is re-arranged accordingly to match the new positions for design points in the sorted result columns. The intersection operation is carried out on the index table system, and the first sets of intersected indices indicate the sets of design points or configuration which optimises the design output variables. A few numbers of selected optimum SLWR configurations, along with some randomly selected configurations, were simulated for 3hrs, using the irregular combined and fatigue wave loads. This is conducted to demonstrate the potential of the technique to capture optimum design configuration for the problem. The results showed that the 2D tabular technique could take into consideration a balance of the interest of all output variables, by reporting indices for configurations, which equally optimises the design output. The 2D tabular method is also able to report new sets of configuration candidates if the search criteria or the set of design output variables to be optimised is changed. The tabular approach has robust potentials to cut down computation resources required for a higher dimensional problem. It can also accommodate additional load scenarios and other external design constraints that may be imposed on the optimization process.

6. Recommendation

In the SLWR optimisation example demonstrated in this paper, the tabular optimisation technique reduces the three-dimensional problem to two dimensions. A potential limitation with the technique is with the reduction of higher dimensional space to that of two dimensions, a transformation which needs further investigation. Future work will be to extend this technique to higher dimensional riser optimisation problems. This could significantly improve the efficiency in the riser optimisation computation.

7. Acknowledgement

The authors acknowledge the support of the University of Strathclyde and McDermott International for their support. Results in this work were obtained using the ARCHIE-WeST High-Performance Computer (www.archie-west.ac.uk) based at the University of Strathclyde.

References

[1] Clukey E, Ghosh R, Mokarala P and Dixon M Steel catenary riser (SCR) design issues at touch down area. Int. Society of Offshore and Polar Engineers)
[2] Song R and Stanton P Advances in Deepwater Steel Catenary Riser Technology State-of-the-Art: Part I—Design. American Society of Mechanical Engineers)
[3] Quintin H, Legras J-L, Huang K and Wu M 2007 Steel catenary riser challenges and solutions for deepwater applications. In: *Offshore Tech. Conf.*: (Houston, Texas: Offshore Tech. Conf.)
[4] Quéau L M, Kimiaei M and Randolph M F 2015 Sensitivity studies of SCR fatigue damage in the touchdown zone using an efficient, simplified framework for stress range evaluation *Ocean Engineering* 96 295-311
[5] Gore C T and Mekha B B 2002 Common Sense Requirements (CSRs) for Steel Catenary Risers (SCRs). In: *Offshore Tech. Conf.*, (Houston, Texas: Offshore Tech. Conf.)
[6] Ogbeifun A M, Oterkus S, Race J, Naik H, Decnop E and Dakshina M The Branched Riser Systems: Concept Development. In: *ASME 2019 38th Int.Conf. on Ocean, Offshore and Arctic Engineering*: (American Society of Mechanical Engineers Digital Collection)
[7] Li S and Nguyen C 2010 Dynamic response of deepwater lazy-wave catenary riser *Deep Offshore Tech. Int., Amsterdam, The Netherlands*
[8] Standard D-O F201, 2010 *Dynamic Risers. DNV: Norway*
[9] Veritas D N 2011 Modelling and analysis of marine operations *Offshore Standard*
[10] Randolph M and Quiggin P 2009 Non-linear hysteretic seabed model for catenary pipeline contact. In: *ASME 2009 28th Int. Conf. on Ocean, Offshore and Arctic Eng.*: (American Society of Mechanical Engineers)
[11] Manual O 2012 Online at http://www. orcina. com/SoftwareProducts/OrcaFlex/Documentation *OrcaFlex. pdf*
[12] Heffernan D 2016 An introduction to the Python interface to OrcaFlex. Orcina, Tech. Rep

Experimental Study on Weather Criterion Applied to South Sulawesi Traditional Wooden Boats

M.A.Asis[1*], D.Paroka[2], S.Asri[1], A.H.Muhammad[3], S.Rahman[2]

[1*]Department of Naval Architecture, Hasanuddin University, Gowa, 92171, Indonesia
[2]Department of Ocean Engineering, Hasanuddin University, Gowa, 92171, Indonesia
[3]Department of Marine System Engineering, Hasanuddin University, Gowa, 92171, Indonesia

Abstract. The general stability criteria of International Maritime Organization (IMO) seem to be difficult to apply to the South Sulawesi traditional wooden boats because the down-flooding angle of most traditional wooden boats was smaller than 30 degrees. The weather criterion may be used to assess the stability of traditional wooden boats because limitation of heel angle depending only on the down-flooding angle. However the criterion was developed based on geometric characteristics of ships different from the South Sulawesi traditional wooden boats. In this paper, the hydrodynamics factors correspond to the parameters of weather criterion are determined by model experiments consist of roll decay test and test in regular beam waves. The critical metacentric height for different wave steepness is determined by using the weather criterion with parameters values based on model experiments.The damping factor due to breadth-to-draught ratio was larger than that provide in the weather criterion for corresponding breadth-to-draught ratio of 4.28.The effective wave slope coefficients for wave steepness of 0.004 and 0.006 are smaller than that obtained by the formula of weather criterion. However, a larger value was obtained for wave steepness of 0.008 and 0.01.The bilge keels position with slope of 15 degrees provided the largest damping moment. The South Sulawesi traditional wooden boats comply with the weather criterion when the wave steepness smaller than 0.08. The weather criterion can be applied to the traditional wooden boats if the parameters values of weather criterion formula have been developed.

1. Introduction

Traditional wooden boats safety could not be assessed in the initial design stage, because the ships were built without preliminary design based but solely based on the builder experience. The ship accidents in Indonesia were dominated by the traditional wooden boats compared to the other ships types. Based on the data of National Transportation Safety Committee of Indonesia (KNKT) in the year of 2013, the number of accident due to sunk was 67 casualty [1]. This number of accident is larger than the other types of accidents. The main cause is bad weather and technical factors.The operational limitation of the traditional wooden boats is important in order to know the maximum environmetal condition for the boats safely operated. Paroka, et al [2] applied the second generation intact stability criteria to determine the operational limitation of a traditional wooden boat built in South Sulawesi. The main drawback of those criteria is determination of the hydrodynamics characteristics to calculate the roll angle toward windward due to wave action. The weather criterion of International Maritime Organization (IMO) provided a formula to calculate the roll angle due to wave with parameters values of damping factors as well as effective wave slope coefficient [3]. The values of each parameters in the weather criterion were determined based on ships data with breadth-

IOP Conf. Series: Materials Science and Engineering 1052 (2021) 012023 doi:10.1088/1757-899X/1052/1/012023

to-draught ratio smaller than 3.5, block coefficient smaller than 0.7 with ratio of vertical center of gravity and ship draught between 0.7 and 1.5 as well as the natural roll period up to 30 seconds.

Some research showed that the roll angle due to wave action calculated by using the weather criterion formula tended to be ovesestimate for ships with breadt-to-draught ratio larger than 3.5 and the ratio between the vertical centre of gravity and ship draught larger than 1.5 [4], [5], [6]. In order to applied the weather criterion to a ship with geometric characteristics different with those were used to develop the criterion, IMO suggested to used model experiment [7]. Eventhough, the damping factors correspond to the breadth-to-draught ratio, the block coefficient as well as the bilge keels still use the values given in the weather criterion. The damping factors correspond to the breadth-to-draught ratio of ships with the ratio larger than 3.5 was smaller than that provided in the weather criterion [6]. The damping factor due to bilge keels was also found to be smaller than that given in the weather criterion. Those previous researches showed that an alternative method to determine the values corresponding to the parameters in the weather criterion is necessary in order to apply the criteria to the traditional wooden boats mainly built in South Sulawesi.

Recently the stability of the traditional wooden boats is assessed by means the general intact stability criteria of International Maritime Organization (IMO). Several parameters in this criteria seems to be difficult to apply to the Indonesian traditional wooden boats such as the area under the righting arm curve up to 30 degrees of heel and the area under the righting arm curve between heel angle of 30 degrees and 40 degrees because the down flooding angle is mostly smaller than 30 degrees. If a reliable method to estimate the parameters values corresponding to the formula in the weather criterion is available, this criteria is potentially applied to the Indonesian traditional wooden boats. The weather criterion depend only on the down flooding angle without any limitation of heel angle. The weather criterion was also adopted as the first level of vulnerability criteria of the second generation intact stability criteria of IMO. Here, the ship operational aspects have been introduced. It is recognized that, in order to get a safer ship performance, addressing only design aspect cannot be enough. Operational measures should also be taken into account and operational guidance should be provided to the master. For this reason, both operational limitations and operational guidance have been developed in framework of Second Generation Intact Stability [8].

This paper discusses about application of weather criteria to assess stability of traditional wooden boat with parameters values correspond to the formula to calculate the roll angle due to wave action determined by model experiment. The obtained damping factors correspond to breadth-to-draught ratio as well as correspond to the bilge keels could be used to extend the values in the weather criterion. The obtained critical metacentric height could be used as operational limitation of the traditional wooden boats which is supposed to reduce the capsizing accident due to bad weather. The righting arm characteristics such as the area under the righting arm curve up to a certain angle of heel correponding to the critical metacentric height can be used to develop stability criteria for traditional wooden boats. The results could also be used to provide an operational guidance including the maximum wave steepness for the traditional wooden boats safely operate.

2. Methodology

The ship data used in this paper is a wooden boat built in Bulukumba South Sulawesi with principle dimension and the body plan are shown in Table 1 and Figure 1, respectively.

Table 1. Principle dimension

Parameter	Ship (m)	Model (mm)
Length (L)	22. 8	1140
Breadth (B)	4.8	240
Height (H)	1.6	80
Draught (d)	1.12	56
Vertical center of gravity (KG)	1.58	78.4
Block coefficient (Cb)	0.362	0.362
Displacement (Δ)	40.89	4.52

IOP Conf. Series: Materials Science and Engineering 1052 (2021) 012023 doi:10.1088/1757-899X/1052/1/012023

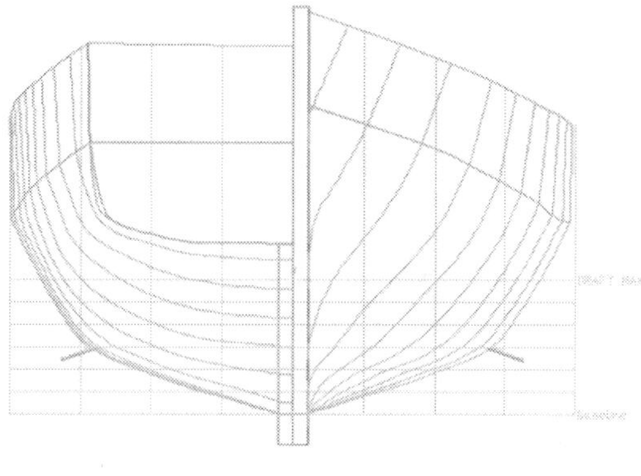

Figure 1. Body Plan

The hydrodynamics factors correspond to the weather criterion was determined by model experiment following the three-step procedure recommended by IMO [9]. The first step is the roll decay test which was performed for ship models without bilge keel and with three different positions of bilge keels consist of the angle of 15 degrees, 30 degrees and 45 degrees, respectively. The bilge keels dimension are determined based on the bilge keels used by the subject ship. The initial roll angle for those tests is 25.0 degrees. The roll motion in time domain is recorded using a dual axis inclinometer and it was stop when the roll amplitude smaller than 0.5 degrees. The extinction coefficients consist of a linear and a quadratic coefficients were statistically determined based on the data of roll decay test. These coefficients were used to determined the Bertin's coefficient by using the following equation [10]:

$$N(\phi_m) = \frac{a}{\phi_m} + b \tag{1}$$

where a is the linear part and b is the quadratic part of extinction coefficients, respectively. ϕ_m is the average of two consecutive roll amplitude of roll decay test.

The second step of three-step procedure is test in regular beam waves. Results of this test were used to estimate the effective wave slope coefficient based on the Bertin's coefficient obtained from the equation (1) by using the equation as follow [9]:

$$r = \frac{{\phi_r}^2 \cdot N(\phi_r) \cdot g \cdot {T_r}^2}{180 \cdot \pi^2 \cdot H_r} \tag{2}$$

where ϕ_r is the roll amplitude obtained in the roll test in regular beam waves with wave height H_r and wave period T_r. $N(\phi_r)$ is the Bertin's coefficient corresponds to the roll amplitude and g is the gravity acceleration. The wave height and wave period were determined based on the wave steepness or ratio between the wave height and the wave length. The roll test in regular beam waves was performed for four different wave steepness of 0.004, 0.006, 0.008 and 0.01 with five different wave frequencies consist of the roll natural frequency, two frequencies smaller and two frequencies larger than the roll natural frequency. The roll angle was measured by using a dual axis inclinometer and the wave elevation was measured by using wave probes located in starboard and portside of the model, respectively as shown in Figure 2. The model is free for sway, heave and roll motions but it was restricted for surge, pitch and yaw motions by a flexible wire rope installed on the steam and the stern with level the same as the vertical centre of gravity of the model.

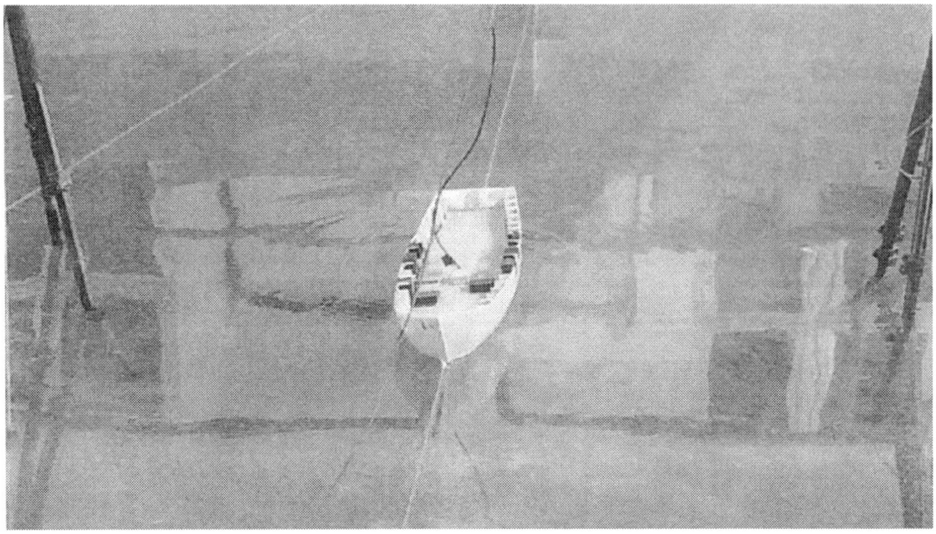

Figure 2. Setting model for roll test in regular beam wave.

The obtained effective wave slope coefficient is used to calculate the roll-back angle in regular wave for the actual wave steepness, following the adjusted value of the weather criteria based on the natural roll period by iteratively solving equation (3) with an initial roll angle of 20.0 degrees.

$$\phi_{1r} = \sqrt{\frac{90\pi rs}{N(\phi_{1r})}} \tag{3}$$

The obtained roll-back angle then used to determine the roll angle toward windward due to wave action by taking into account the actual irregular nature of the sea with correction factor of 0.7. The damping factor corresponds to the breadth-to-draught ratio can be calculated by using the formulae to calculate the roll angle toward windward due to wave action in the weather criterion as follow:

$$X_1 = \frac{\phi_1}{109 \cdot k \cdot X_2 \cdot \sqrt{r \cdot s}} \tag{4}$$

where ϕ_1 is the roll angle to windward due to wave, k is the damping factor due to bilge keels which is taken to be 1.0 if the ship without bilge keels. X_2 indicates the damping factor corresponds to block coefficient.

The damping factor corresponds to the bilge keels is estimated by using the data of model experiment with bilge keels with the following equation:

$$k = \frac{\phi_1}{109 \cdot X_1 \cdot X_2 \cdot \sqrt{r \cdot s}} \tag{5}$$

Here, the damping factor due to breadth-to-draught ratio obtained from the equation (4) is used. The damping factors due to the breadth-to-draught ratio and due to the bilge keels are used to assess the stability of subject ship by using the weather criterion. To obtain the minimum metacentric height comply with the weather criterion, the b/a area ratio is calculated for variation of vertical centre of gravity ranges between 0.8 meters and 2 meters. The minimum righting arm characteristics of the subject ship are then calculated base on the vertical centre of gravity corresponding to the obtained minimum metacentric height.

3. Results and Discussion

The result of roll decay tests of model without bilge keel and those for model with bilge keels are shown in Figure3 and Figure 4, respectively.

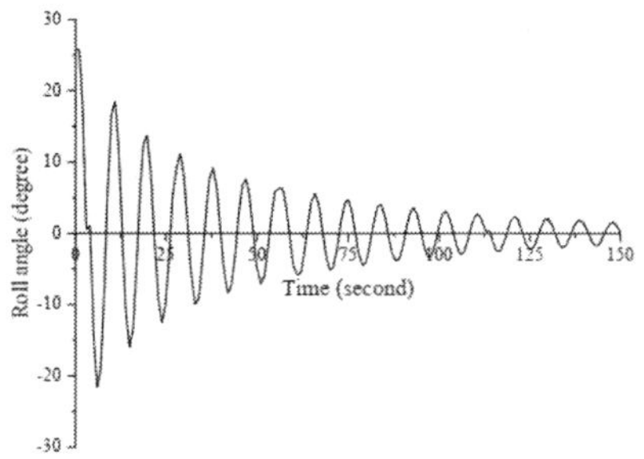

Figure 3. Roll decay test without bilge keels.

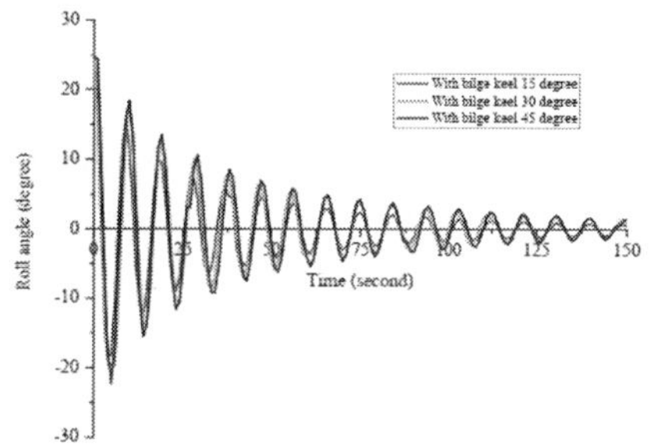

Figure 4. Roll decay test with different position of bilge keels.

The results show that bilge keel can reduce the roll amplitude, where the roll amplitude of model without bilge keels is larger than the roll amplitude of model with bilge keels. The bilge keels position with slope of 15 degrees result in the smallest roll amplitude compared to the position with slope of 30 degrees and 45 degrees, respectively. In order to calculate the Bertin's coefficient, the second order polynomial regression based on results of roll decay tests for ship without bilge keels and with three different positions of bilge keels are developed with the obtained linear and quadratic coefficients shown in Figure 5 and Figure 6, respectively. The presented values were average of five roll decay test for each bilge keels position.

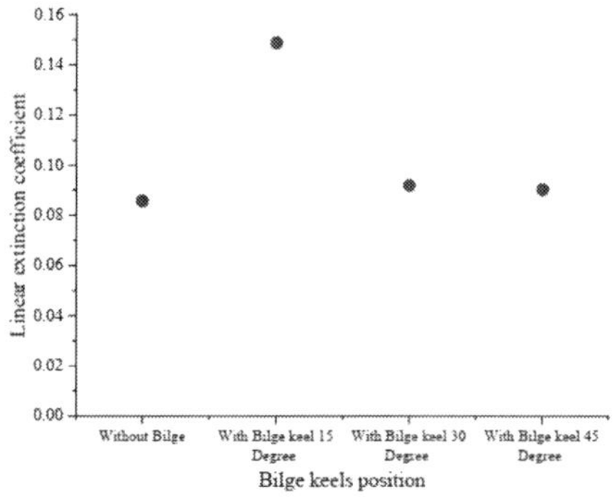

Figure 5. Linier extinction coefficient

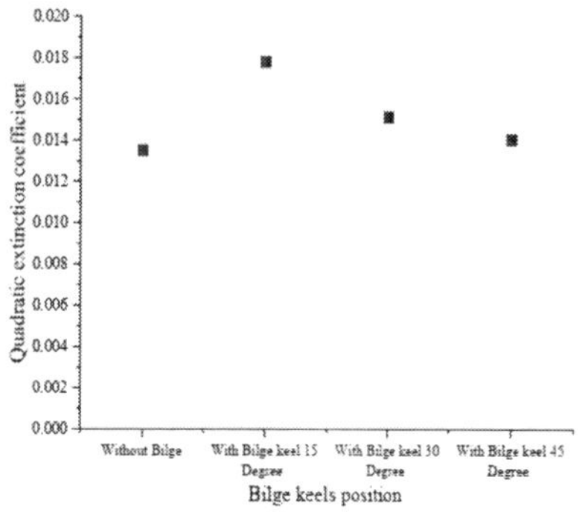

Figure 6. Quadratic extinction coefficient

The extinction coefficients (linear and quadratic coefficients) for ship with bilge keel are larger than that of model without bilge keels. These results indicate that the bilge keels have significant effects on the roll motion of the South Sulawesi traditional wooden boats. The bilge keels position with a slope of 15 degrees induced the largest extinction coefficients compared to the other positions. The extinction coefficients decrease due to increase the slope position of bilge keels. Increasing the linear coefficient due to bilge keels is larger than the increasing the quadratic coefficient for the bilge keels position with slope of 15 degrees. For bilge keels position with slope of 30 degrees and 40 degrees, effect of bilge keels on quadratic coefficient is more significant compated to the linear coefficient. Irkal, et al [11] found that the effect of bilge keels on quadratic damping coefficient was more significant compared to that on linear damping coefficients. The different results mainly for bilge keels position with slope of 15 degrees could be occured due to bilge keels dimension. The present bilge keels area was very small compared to the product of length of water line and breadth. The effect of bilge keels position has been investigated by Gu,et al [12] for FPSO by using CFD and model experiment. They found that the effective position of bilge keels was in the bilge radius of the ship. The damping moment induced by the bilge keels consists of the moment due to the drag force acting on the surface of the bilge keels and the moment due to pressure on the ship hull induced by the bilge keels.The effect of bilge keel on the damping of rolling motion is also influenced by the distance from the bilge keel to the centre of gravity of ship [13].

The damping factor due to the bilge keels depends only on ratio between total area of the bilge keels and product of the length of waterline and the ship breadth in the weather criterion. However, the breadth-to-draught ratio as well as the vertical centre of gravity could have significant effect because the different distance between the bilge keels and centre of gravity of the ship as found by Katayama, et al [13]. The natural roll period of the ship without bilge keels was 4.14 seconds, and that for the ship with the bilge keels position of 15 degree of the slope was 4.18 seconds. The natural roll period was 4.15 seconds when the bilge keels was located with slope of 30 degrees and that of 4.17 seconds for the slope of 45 degrees. The increasing in the roll natural period due to the bilge keels is caused by a reduction in the angular velocity of the roll motion due to the damping provided by the bilge keels as well as increasing the roll moment of inertia due to added inertia induced by the bilge keels. The formula to calculate the natural roll period in the weather criteria did not take into account the influence of bilge keels. The increasing of damping moment induced by the bilge keels can be verified from the roll amplitude obtained in roll test in regular beam wave, as shown in Figure 7 - 10.

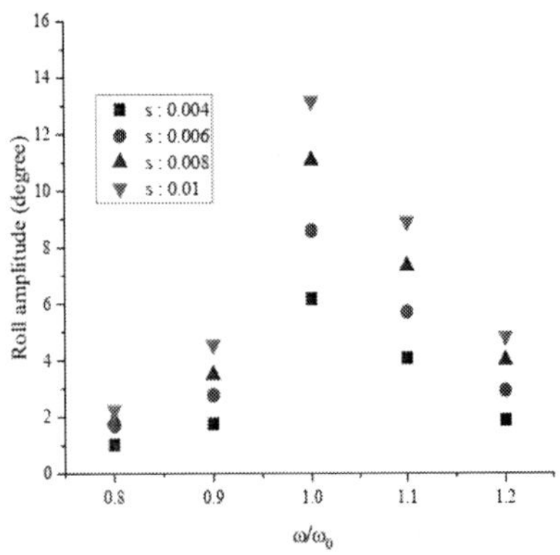

Figure 7. Roll amplitude without bilge keel

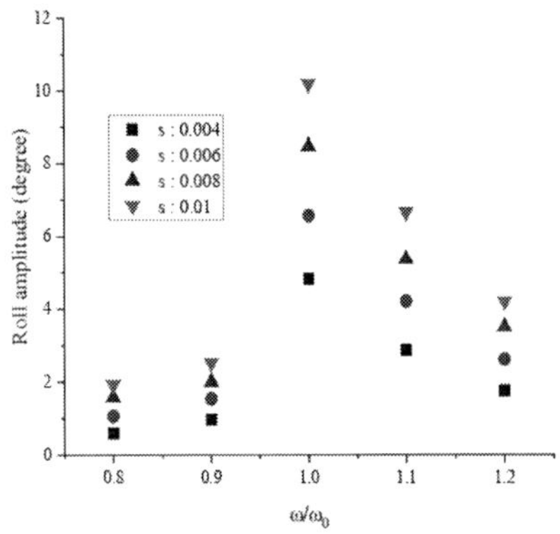

Figure 8. Roll amplitude for bilge keel position with slope of 15 degrees

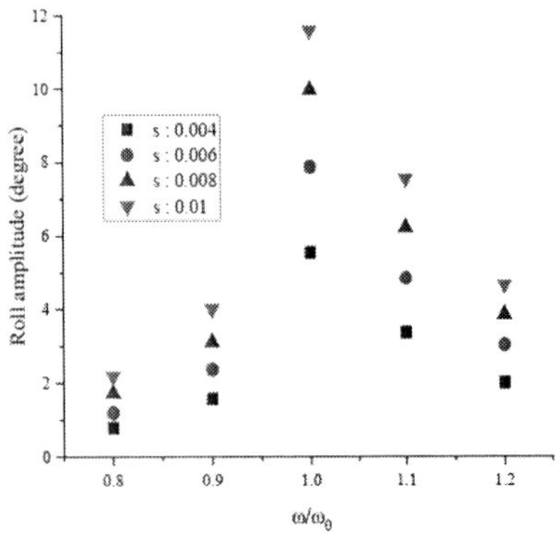

Figure 9. Roll amplitude for bilge keel position with slope of 30 degrees

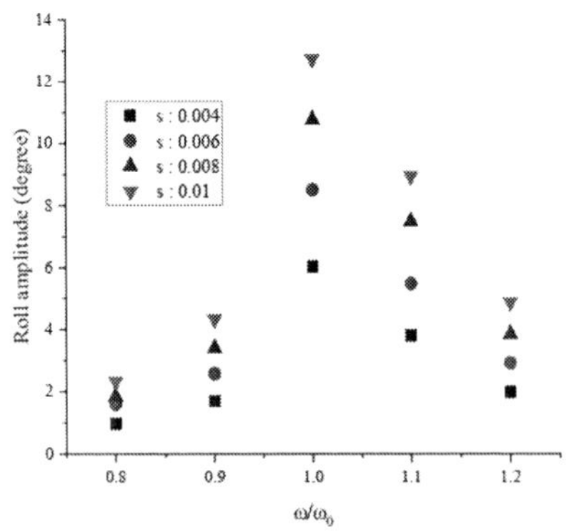

Figure 10. Roll amplitude for bilge keel position with slope of 45 degrees

The roll amplitude for the position of bilge keels with slope of 15 degrees was smaller than that for ship without bilge keel and for ship with the two others positions of bilge keels. The roll amplitude decreases approximately 2.98% for bilge keel sposition with slope of 15 degrees, 1.6% for bilge keels position with slope of 30 degrees and 0.43% for bilge keels position with slope of 45 degrees.

Figure 11 shows the effective wave slope coefficient correspond to the roll natural frequency of the models without and with bilge keels, respectively.

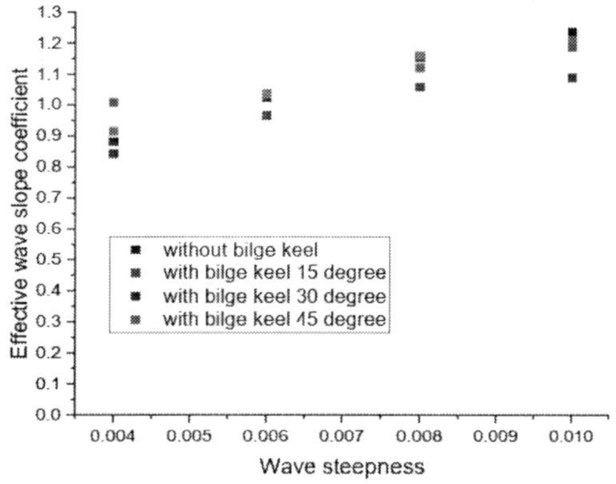

Figure 11. Effective wave slope coefficient

The effective wave slope coefficients for wave steepness of 0.004 and 0.006 are smaller than that obtained by the formula of weather criterion. However, the effective wave slope coefficients for wave steepness of 0.008 and 0.01 are larger than that obtained by the weather criterion formula for both models without and with bilge keels. Some researches showed that the effective wave slope coefficient obtained by the formula of weather criteria is too large if it is applied to ships with geometric characteristics different with those used to develop the criteria [6], [14]. The effective wave slope coefficient in Figure 11 showed that the bilge keels have no significant effect.

The damping factor due to the breadth-to-draught ratio was larger than that provide in the weather criterion for corresponding breadth-to-draught ratio of 4.28 as shown in Figure 12. These results were different with those obtained for differents ships types such as ro-ro ferry [15]. Here, the damping factor due to block coefficient was determined to be same as the value provided in the weather criterion.

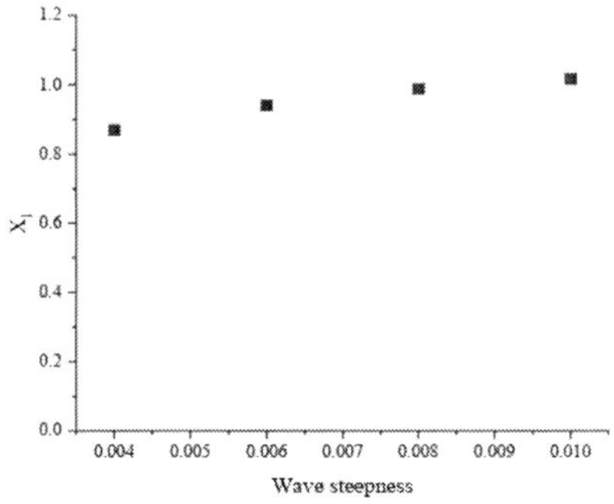

Figure 12. Damping factor due to breadth-to-draught ratio

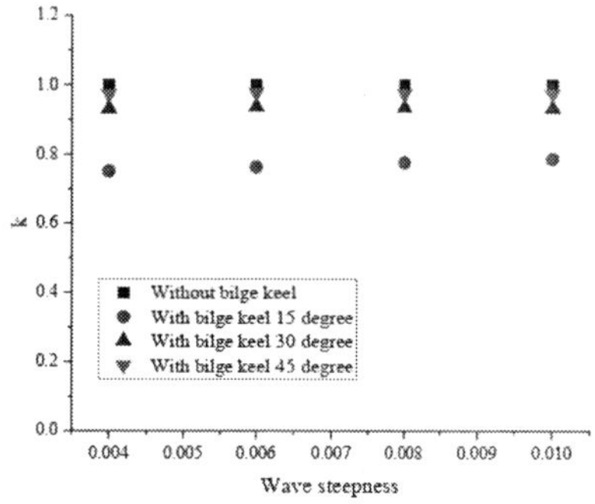

Figure 13. Damping factor due to bilge keels

Figure 13 showed the damping factor due to bilge keels. The damping factor corresponds to the bilge keels was different due to different bilge keels position. The bilge keels position with slope of 15 degrees results in the smallest damping factor. The damping factor for the bilge keels position with slope of 30 degrees and 45 degrees was not significantly different. This results agreed with the extinction coefficients obtained in roll decay test as shown in Figure 5 and Figure 6, respectively. The damping factors correspond to the bilge keels were smaller than that obtained in the weather criterion. The obtained damping factors and the effective wave slope coefficient are used to assess the stability of subject ship using the weather criterion. Here, the angle of down-flooding was determined to be 30 degrees. This angle could be smaller depending on the watertightness of cover construction for opening on weather deck. The wind pressure of 300 Pa corresponding to a wind velocity of 20 m/s was used to calculated the wind moment lever as well as the rol angle toward windward due to wave action. In order to obtained the critical metacentric height, the calculation of area ratio b/a was conducted for vertivcal centre of gravity ranges from 0.8 meters to 2 meters. The corresponding righting arm curve of the variation of vertical centre of gravity is shown in Figure 14.

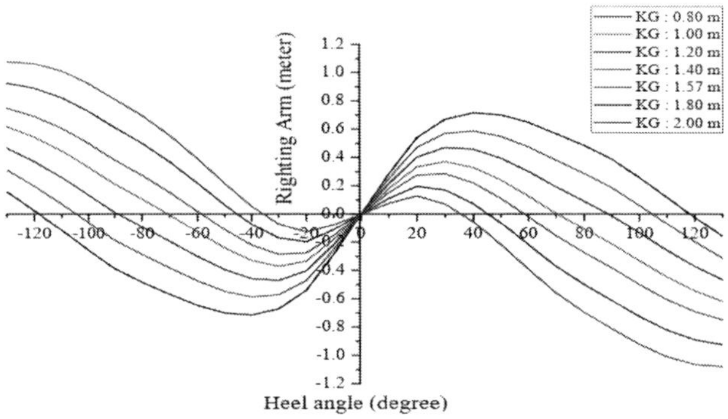

Figure 14. Righting arm for variation of vertical centre of gravity.

The area ratio b/a for wave steepness of 0.06, 0.08 and 0.10 are shown in Figure 15 – 18 for ship without and with bilge keels, respectively.

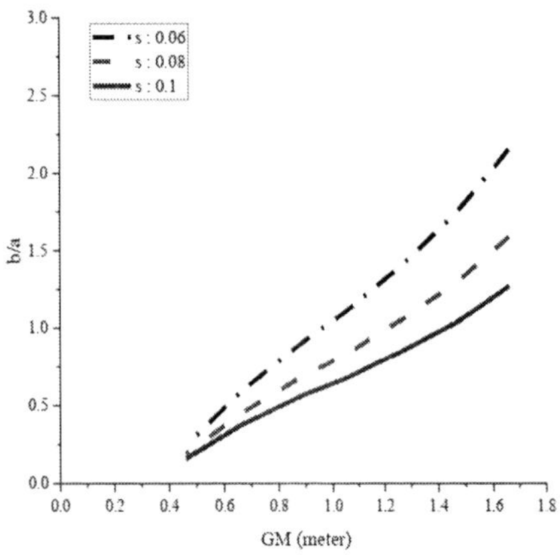

Figure 15. The area ratio b/a for ship without bilge keel.

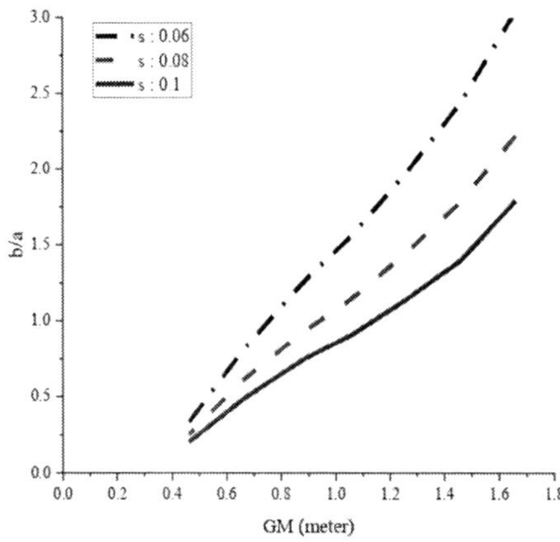

Figure 16. The area ratio b/a for ship with bilge keels position with slope of 15 degrees.

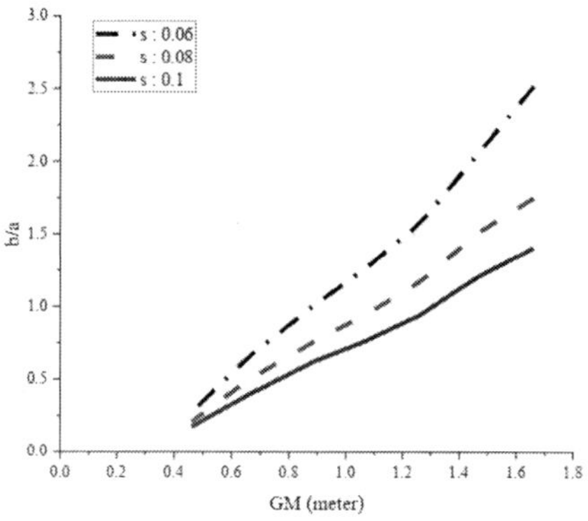

Figure 17. The area ratio b/a for ship with bilge keels position with slope of 30 degrees.

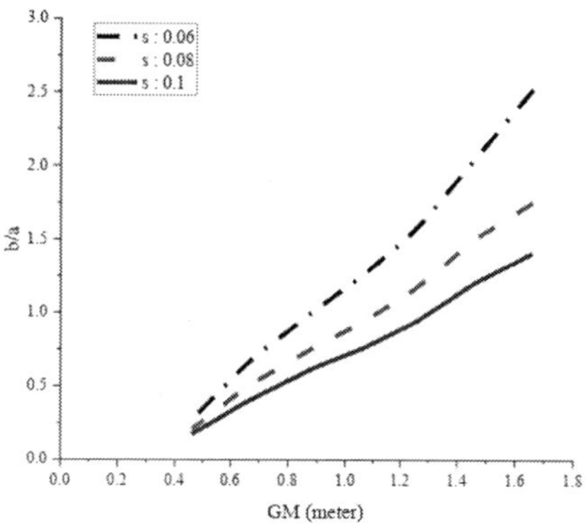

Figure 18. The area ratio b/a for ship with bilge keels position with slope of 45 degrees.

The area ratio b/a decreases when the wave steepness increase. This means that a larger area under the righting arm curve is necessary when the ship operated in the severe weather. The minimum stability requirement was based on the weather criterion with area ratio b/a equal to 1.0. The maximum vertical centre of gravity of ship without bilge keels was 1.51 meters corresponding to the wave steepness of 0.06. A vertical centre of gravity of 1.25 meters was obtained for ship without bilge keels in the wave steepness of 0.08. For wave steepness of 0.10, the maximum vertical centre of gravity for ship without bilge keels was 1.04 meters. The critical metacentric heights (GM) of ship with dan without bilge keels for three different wave steepnes are shown in Figure 19. The bilge keels position with slope of 15 degrees provide the smallest critical metacentric height. The area under the righting arm curve up to heel angle of 30 degrees or the down-flooding angle which one is the smallest corresponding to the critical metacentric height is shown in Table 2.

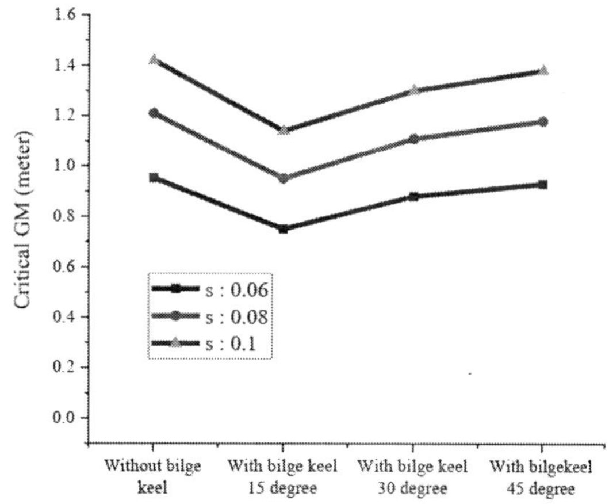

Figure 19. Critical metacentric height.

The required minimum area under the righting arm curve up to the angle of vanishing stability or the down-flooding angle or the angle of 30 degrees which is the smallest can be obtained for the corresponding wave steepness as shown in Table 2. Those area correspond to the critical metacentric height shown in Figure 19. The area under the righting arm curve is larger than that given in the general criteria of IMO [3] for all wave steepness. The critical metacentric height for all wave steepness are also larger than the minimum metacentric height given in the IMO general criteria. This means that the weather criterion should be considered to assess the stability of traditional wooden boats.

Table 2. The area under the righting arm cuver up to heel angle of 30 degrees.

Slope of bilge keels position	Area under the righting arm curve up to 30 degrees (m.deg.)		
	s = 0.06	s = 0.08	s = 0.1
Without bilge keels	6.33	8.30	9.92
15 degrees	4.77	6.31	7.77
30 degrees	5.77	7.54	8.99
45 degrees	6.15	8.07	9.61

4. Conclusion

The stability of a traditional wooden boats built in Tanaberu of South Sulawesi has been assessed by using the weather criterion of IMO with hydrodynamics factors correspond to the formula to calculate the roll angle to windward due to wave determined by model experiment. The bilge keels position with slope of 15 degrees provide the most largest damping coefficient. The maximum wave steepness for the ship safely operated was 0.06 with the critical metacentric height of 0.75 meters corresponding to area under the righting arm curve up to the heel angle of 30 degrees of 4.77 m.deg. The results show that an additional criteria such as weather criterion is necessary in assessing stability of the traditional wooden boats in order to identify the operational limitation. A larger area under the righting arm curve up to heel angle of 30 degrees is need when the ship operated in a higher wave steepness. In order to apply the weather criterion to the traditional wooden boats, the values of parameters to calculate the roll angle to windward due to wave need to be determined for large range of breadth-to-draught ratio, bilge keels geometry as well as block coefficient by using model experiment or computational fluid dynamics.

5. Acknowledgments

The content of this paper is a part of research supported by Directorate General of Higher Eduaction (DIKTI) and Hasanuddin University with grand number 1516/UN4.22/PT.01.03/ 2020. The authors express their sincere gratitude to those institutions for their support.

6. References

[1] Prihartono B 2015 *Development of Toll Laut in the Middle-term National Development Plan 2015 – 2020 and its Implementation in 2015*, National Development Plan Agency of Indonesia, Indonesia (in Indonesian).

[2] Paroka D, Muhammad, AH, Rahman S, Asis MA 2019 *Operational Limitation of Indonesian Traditional Wooden Boat in the Framework of the Second Generation Intact Stability Criteria*, Proc. of Sustainable Island Development Initiative Conf., Surabaya, Indonesia, 2 – 3 September.

[3] IMO 2008 *MSC.267(85) – The International Code on Intact Stability*, London.

[4] Ishida S, Taguchi H, Sawada H 2011 *Evaluation of Weather Criterion by Experiment and its Effect to the Design of a Ropax Ferry, Contemporary Ideas on Ship Stability and Capsizing in Waves* (London: Springer).

[5] Francescutto A 2015*Intact Stability Criteria of Ships – Past, Present and Future*, Proc. of the 12th Int. Conf. on the Stability of Ships and Ocean Vehicles 2015, Glasgow, Scotland, 14 – 19 June,pp. 1199 – 1209.

[6] Paroka D, Asri S, Rosmani, Hamzah 2020 Alternative Assessment of Weather Criterion for Ships with Large Breadth and Draught Ratio by Model Experiment: A Case Study on an Indonesian Ro-Ro Ferry, *Int. J. Mar. Eng.* **162**(1), pp. 55 – 64.

[7] IMO 2013 *SDC 1/INF.6 -Development of Second Generation Intact Stability Criteria: Vulnerability Assessment for Dead-ship Stability Failure Mode*, London.

[8] Petacco N and Gualeni P 2020 IMO Second Generation Intact Stability Criteria General Overview and Focus on Operational Measures, *J. Mar. Sci. Eng.* **8**(7):494, pp. 2 – 20.

[9] IMO 2006 *MSC.1/Circ.1200–Interim Guidelines for Alternative Assessment of the Weather Criterion*, London.

[10] ITTC 2011 *Recommended Procedures: Numerical Estimation of Roll Damping*, Proc. of the 26th ITTC Specialist Committee on Stability in Waves, Rio de Jeneiro, Brazil.

[11] Irkal MAR, Nallayarasu S, Bhatracharyya SK 2019 Numerical Prediction of Roll Damping of Ships with and without Bilge Keel, *Ocean Eng.* **179**,pp. 226-245.

[12] Gu Y, Day S, Boulougouris E 2015 *A Study on the Effect of Bilge Keels on Roll Damping Coefficient*, Proc. of the 12th Int. Conf. on the Stability of Ships and Ocean Vehicles, Glasgow, Scotland, 14 – 19 June, pp. 775 – 783.

[13] Katayama T, Matsuoka M, Ikushima K 2018 *Characteristics of Bilge Keel Roll Damping Component for Shallow Draft*, Proc. of the 13th Int. Conf. on the Stability of Ships and Ocean Vehicles, Kobe, Japan, 16 – 21 September, pp. 350 – 359.

[14] Francescutto A, Serra A, Scarpa S 2001 *A Critical Analysis of Weather Criterion for Intact Stability of Large Passenger Vessels*, Proc. of the 20th Int. Conf. on Ocean, Offshore and Arctic Engineering, Rio de Janeiro, Brazil, 3-8 June, pp. 829-836.

[15] Deakin B 2008 Evaluation of the Roll Prediction Method in the Weather Criterion, *Int. J. Mar. Eng.* **105**(2), pp. 56-67.

IOP Conf. Series: Materials Science and Engineering 1052 (2021) 012024 doi:10.1088/1757-899X/1052/1/012024

Standardization of 30 Gross Tonnage (GT) Danish Seiner

R S Pamungkas[1], H Purwanto[1], Sunardi[2]

[1] Fishing Technology Centre (BBPI), Ministry of Marine Affairs and Fisheries, Tanjung Emas Port, Semarang, Indonesia
[2] Fisheries Department, University of Brawijaya, Ministry of National Education, Malang, Indonesia

Abstract. *Cantrang* – Danish Seiner is growing rapidly in the north coast of Java Seas- Fisheries Management Area 712. In recent year, there have been many issues regarding the prohibition operation of seiner fishing gear impact sustainable fisheries, *Cantrang* owners and fishermen livelihood due to the regulation of fishing gear and the mark down of Gross Tonnage measurement. This research attempt to obtain preliminary data on Danish seiner operating in Indonesia, as set base design for 30 GT Danish seiner standard includes the scope of the principal dimensions of the seiner, main engine powering and the layout of deck for fishing operations. The method uses purposive sampling of 30 GT *Cantrang*, number of population 148 *Cantrang* of 30 GT at 4 regencies; Tegal, Batang, Rembang and Pati. The result of this research are the main measurement value - the ratio of principal dimension, main engine power, and deck area and *Cantrang* Danish seiner operation. The standardized 30 GT *Cantrang* has the following maximum main dimensions: Length (L) = 18.21 meters, Width (B) = 6.58 meters and Deck Height (D) = 2.00 meters with L / B ratio = 3.29-3.33, L / D = 11.2 - 2.74, B / D = 3.96-4.26. The main engine power 1 engine 160-190 HP and for 2 engines 2 x 120 HP. The deck area of 70-80 m^2 significantly impacts *Cantrang* fishing gear operation

1. Introduction

Cantrang is a fishing gear most related to Danish Seiner type of semi-active fishing gear, most of the population is on the northern island of Java. The research data used purposive sampling method of 30 GT *Cantrang* vessel located at 4 regencies, Tegal, Batang, Rembang and Pati, Central Java. The development of *Cantrang* vessels has major impact on capture fisheries resources and generally on fisheries management in Indonesia. The need for a larger vessels size creates potential problems because it can be physically carried out, but on the other hand, there is no permanent law in terms of regulation and legality. This research is a series of studies in creating the Indonesian National Standard (SNI) for the design of *Cantrang* fishing vessels 30 GT which is often a polemic for the management of fishing fleets. This initial research is to collect data on *Cantrang* fishing vessels including the principal dimension of the vessels, the cubic number, the area of the vessels deck, the resistance and propulsion and the maneuverability of the *Cantrang* fishing vessel.

IOP Conf. Series: Materials Science and Engineering 1052 (2021) 012024 doi:10.1088/1757-899X/1052/1/012024

2. Method

The research method in this paper is literature study, using purposive sampling data and analysis of normal distribution and descriptive statistics, the number of sample vessels is 148 *Cantrang* vessels. The 30 GT of *Cantrang* hull modeling was carried out to determine the power requirements of the main engine. Analysis of the vessels motion when operating is carried out by analytical methods. The study was conducted on *Cantrang* vessels with a size of less equal than 30 GT in Pati, Tegal, Batang and Rembang [5].

3. Results and Discussions

The 30 GT of *Cantrang* hull modeling was carried out to predict the power requirements of the main engine. Analysis of the vessels manouver when operating is carried out by analytical methods. The study was conducted on *Cantrang* vessels with a size of less than 30 GT in Pati, Tegal, Batang and Rembang,Central Java

3.1. Principal Dimension

The measurement is carried out by referring to the Minister of Transportation Regulation No. 8 of 2013 concerning Vessels Measurement the main size of the vessels. Statistical data processing with a sample of 148 Danish seiner, obtained the average principal dimension of vessel as follows [5]. :

- Length (L).15.2 meters, standard deviation 0.807 meters
- Breadth (B).5.86 meters, standard deviation 0.346 meters
- Deck Height (D).1.70 meters, standard deviation 0.129 meters
- L/B ratio affects the stability and resistance of the vessel, in case ofthe L/B value is large for the stability and maneuverability of the vessel is insufficient, if the L / B which is greater, the stability of the vessel is adequate, however the value increases the resistance of the vessels. The result of the L / B calculation is 2.6 with standard deviation of 0.241
- L / D affects the longitudinal strength of the vessels, if the L / D value is large, the longitudinal strength of the vessels will be low and vice versa. The result of L / D calculation is 8.72 with standard deviation 0.996
- B / D affects the stability of the vessels, a large ratio value the better the stability of the vessels and vice versa. The calculation result of the B / D ratio is 3.36 with standard deviation of 0.314

By comparison, the ratio of principal sizes from various references [4],[7] for seiners less than 18 meters length as follow : L / B $\leq$ 4.75, L / D$\leq$10.00, B / D$\geq$ 2.10

*3.2. Cubic Number (*CuNO*)*

The value of CuNO [7] is the multiplication of the length of the vessels (L), the width of the vessels (B) and the height of the vessels (D), the product of the multiplication in cubic meters. In this case this value shows the cubic volume of the vessels below the main deck. If the CuNO value is large, the proportion of the building volume above the main deck will decrease with the GT of the same vessels. The calculated value of CuNO is 156.18 m^3 with standard deviation 14.8 m^3

3.3. Deck Area

The deck area is the multiplication of the length of the main deck of the vessels (L), the width of the main deck of the vessels (B) and the coefficient of the waterline (CWL), the product of the product in square meters. In this case, the value of the deck area shows the area of the main deck. If the value of the deck area is large, the proportion of the area required to operate fishing gear and fish handling on board is relatively better. The result of the calculation of the average deck area is 71, 27 m^2 with standard deviation 5,327 m^2

IOP Conf. Series: Materials Science and Engineering 1052 (2021) 012024 doi:10.1088/1757-899X/1052/1/012024

3.4. Resistance Analysis

Resistance analysis is carried out to predict vessels resistance when fishing gear operates, in addition as a method to select the proper vessels main engine. The sample measured and generates 3D modeling with resistance calculation software [2] to simulate existing seiners. In terms of resistance analysis, 1 sample of 30 GT *Cantrang* vessel was taken on Table 1.

Table 1. Measurement Results.

Item	Value	units	Reference/Mean (Max)
Length (L)	14.10	m	15.20 (18.21)
Breadth (B)	5.80	m	5.86 (6.58)
Depth (D)	1.70	m	1.70 (2.00)
Gross Tonnage (GT)	30		30
Main Engine BHP (M/E)	90	HP	160(190)
L/B	2.43		<4.75 (3.33)
L/D	8.29		<10.00 (12.74)
B/D	3.41		>2.10 (4.26)
Deck Area	65.42	m^2	71,27 (82.08)
Cubic Number (CuNO)	139.02	m^3	156.18 (179.72)

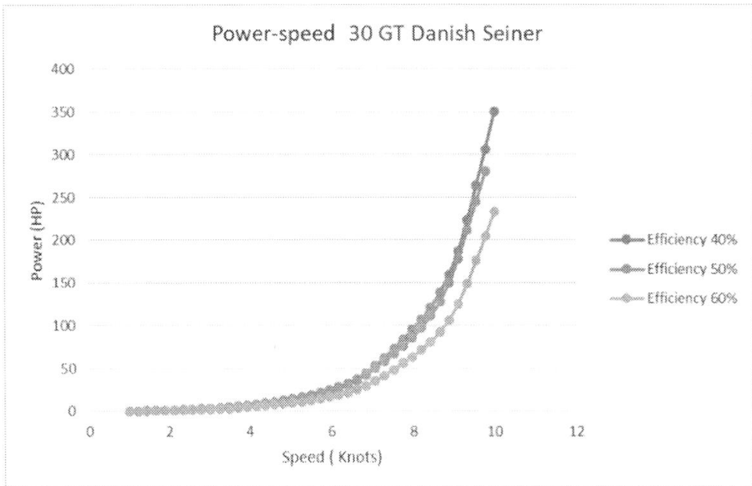

Figure 1. Power-speed Prediction on 30 GT Danish Seiner.

The graph of the relationship between engine power and vessels speed in Figure 1 with different engine efficiency shows the exponential correlation. The optimal vessels speed is in the range of 6-8 knots, but if 8-10 knots are required, the main engine power needs tend to increase. With an engine efficiency of 40%, an effective horsepower of 95 HP is needed for a vessels speed of 8 knots while 350 HP is for a boat speed of 10 knots. With an engine efficiency of 50%, an effective horsepower of 76 HP is needed for a vessels speed of 8 knots while 280 HP for a vessels speed of 10 knots. With an engine efficiency of 60%, an effective horsepower of 63 HP is needed for a vessels speed of 8 knots while 233 HP for a boat speed of 10 knots. *Cantrang* as a semi-active fishing gear that has the potential to produce large amounts of fish, energy consumption per kg is small, even 15% less than the bottom trawl [13] and also produces less environmental impact than trawlers with higher fuel efficiency [9].

Figure 2. 30 GT Danish Seiner.

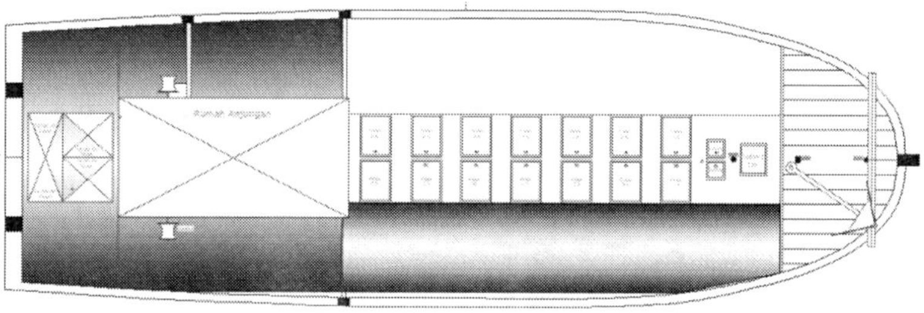

Figure 3. Setting and Hauling Area (blue) 30 GT Danish Seiner.

3.5. *Manoeuvring*

IMO has defined requirements for maneuverability including maneuverability, turnability, and stopping [3]. Some important vessels maneuvering tests are the turning test and zig zag test. This circle motion of the vessels is very important for fishing vessel. The fishing operation begins with knowing a group of target fish, then at a certain distance the net is lowered and the vessels will pull the net around the group of fish. The good circle motion of the vessels will greatly affect the success of fishing operations [8].

From the data based on the classification of the American Bureau of Shipping - ship maneuvering in 2006 which was updated in 2017. The diameter of the *Cantrang* vessels turning cycle under certain conditions:

- IMO standard circle cycle capability to advance is <4.5 * Lwl (66.55 meters)
- IMO standard circle cycle capability for tactical Diameter <5 * Lwl (73.95 meter)
- The minimum standard length [10] of the head rope for purse seiner is 15 * Lwl(221.85 meters length or 70.65 diameter)
- The length of the warp wire is 1000 meters and the length of the head rope is 90 meters, the diameter is 665.60 meters [14]

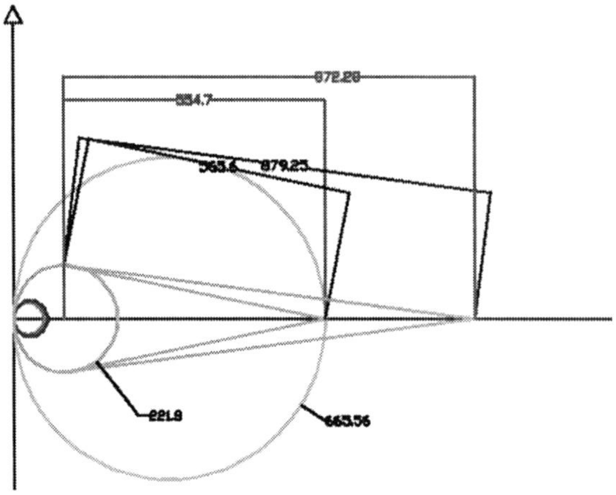

Figure 4. Manoeuvring Analytical Prediction 30 GT Danish Seiner.

4. Conclusions

The principal dimensions measurement value - the ratio of dimensions, main engine power, deck area and the operation of the 30 GT Danish Seiner. The standardized 30 GT Danish Seiner has a maximum value of the principal dimension as follows: Length (L) = 18.21 meters, Breadth (B) = 6.58 meters and Depth (D) = 2.00 meters with L / B ratio = 3.29-3.33, L / D = 11.2 - 12.74, B / D = 3.96-4.26. The main engine power 1 engine 160-190 HP for the use of 2 engines 2 x 120 HP. The deck area of 70-80 m^2 is influenced by the need for warp handling and fishing gear operation

Acknowledgement

Fishing Technology Centre (BBPI), Ministry of Marine Affairs and Fisheries, Tanjung Emas Port, Semarang, Indonesia for all off the supports.

References

[1] Anonymous 2013, Minister of Transportation Regulation No. 8 of 2013
[2] Anoymous 2014, "*Maxsurf Resistance –User Manual*", Bentley System Inc.
[3] Anonymous 2006 update 2017, "Guide for Vessel Manouverability", American Bereau of Shipping,
[4] Anonymous SNI 7797:2013*Alat Penangkapan Ikan – Cantrang,* BSN Arbuckle, James L. 2013 "User Guide" Versi 22.0 IBM SPSS Statistic
[5] BBPI 2017,"Danish seiner Validation Report- *Laporan Hasil Validasi Cantrang"*Fishing Technology Center
[6] *Biro Klasifikasi Indonesia (BKI)* 2004
[7] Fyson, John 1985, "*Design of Small fishing Vessels"* FAO
[8] Liu, J., Hekkenberg, R., Rotteveel, E., & Hopman, H. (2015). Literature review on evaluation and prediction methods of inland vessel manoeuvrability. *Ocean Engineering*, *106*, 458-471.
[9] Noack, T., Frandsen, R. P., Krag, L. A., Mieske, B., & Madsen, N. (2017). Codend selectivity in a commercial Danish anchor seine. Fisheries Research, 186, 283-291.
[10] Prado J 1990, "*Fisherman's Workbook*", FAO
[11] RB. Mulyanto, et.al.2010 "Fishing vessel measurment and calculation - *Perhitungan dan*

IOP Conf. Series: Materials Science and Engineering 1052 (2021) 012024 doi:10.1088/1757-899X/1052/1/012024

Pengukuran Kapal Perikanan", BBPPI, Semarang

[12] Sasmita S 2013 "*Kesesuaian Desain dan Konstruksi Cantrang pada kapal 20 GT untuk peningkatan performa operasional,.* ,Disertasi IPB

[13] Thrane, M. (2004). Environmental impacts from Danish fish products: Hot spots and environmental policies. Institut for Samfundsudvikling og Planlægning, Aalborg Universitet.

[14] Tim BBPI 2017 "*Laporan Uji Operasional Kapal Cantrang >30 GT menggunakan Alat Tangkap Jala Jatuh Berkapal (Cast Net)* , BBPI

Societal risk assessment of terminal and oil refinery unit

AAB Dinariyana[1,2], KB Artana[1,2], DW Handani[1,2] FI Sarasvati[1] and PW Aprilia[1]

[1]Department of Marine Engineering, Institut Teknologi Sepuluh Nopember, Kampus ITS Sukolilo, Surabaya, Indonesia
[2]Center of Excellence for Maritime Safety and Marine Installation, Kampus ITS Sukolilo, Surabaya, Indonesia

Abstract. This paper presents the societal risk assessment of oil refinery in one of the oil companies in East Java. The oil company operates-a terminal for loading and unloading of variety petrochemical products. The products are categorized as flammable which could be hazardous. To prevent unlikely event of hazards such as fire, risk assessment is needed to know the level of risk and propose any mitigation if the risk is unacceptable. The risk assessment is done for all the processes that occurs such as loading and unloading of products and the chemical process in the oil refinery. The system is categorized into 8 nodes based on their function. Firstly, Hazard and Operability (HAZOP) is carried out using BS IEC 61882 standard to identify the deviation of the system, the causes and how to prevent it. The frequency analysis of failure of component is done by using Fault Tree Analysis (FTA) and Event Tree Analysis (ETA) which based on the DNV failure guidance and Oil and Gas Producers (OGP) data as input. In order to know the level of consequence, fire modelling is carried out to simulate how the jet fire and pool fire could impact into the social around the oil refinery in form of heat intensity. The fire modelling for the pool fire and jet fire results the pool radius and flame length respectively, based on the heat intensity of 4 kW/m^2, 12.5 kW / m^2, and 37.5 kW / m^2. Risk is represented using F-N curve with UK-HSE standard. There is unacceptable risk that incurred for jet fire in pipeline and its components from the feedstock tank into pre-cut column, scenario 200 mm hole diameter. Mitigation is done using LOPA to bring the risk to the acceptable region by adding a gas detector which has changed the frequency of system failure from 1.44E-03 to 2.88E-06.

Keywords: Oil refinery, risk assessment, HAZOP, FTA, ETA, fire modelling, LOPA.

1. Introduction

Petroleum or crude oil consists of complex compound of hydrocarbon, found in immense underground reservoirs in liquid, solid, or gaseous form. Petroleum through long chemical engineering process in refinery before it turned into finished product. This paper discusses safety issues in one of the oil companies in East Java which operates large-scale refineries with a capacity of 100 MBCD (Million Barrels per Calendar Day) or 100,000 barrels of condensate per day. The oil products produced in this company include fuels, aromatic products (benzene, toluene, ortho xylene and paraxylene), and LPG (Liquefied Petroleum Gas) and all fluid transfer process is carried out using pipes. Refinery operates 24 hours per day with ± 80 workers per shift and this shift is divided into 2 each day. All processes involving hydrocarbons in the Refinery Unit can cause the occurrence of a potential hazard. Therefore, the risk level in process of fluid transfer which carried out using pipelines needs to be identified and assess the level of risk to avoid safety and damage issues to company and surrounding. One of the risk assessment methods that can be applied to evaluate the safety of personnel and surrounding environment in the refinery area is the Societal Risk Assessment. Societal risk can be described as the relation between frequency an incident occurred, and number of people are exposed to risk in specific population. Social risk assesses the probability of more than one personnel being injured simultaneously in an incident and societal risk is used to calculate the damage caused by major, various fatality risk to maintain level in acceptable region or in ALARP (As Low As Reasonable Practicable)

2. Methodology

Risk assessment in this paper was carried out in four main phases. In the first phase, identifying all the possible hazards from the operation of the system based on BS IEC 61882. In the second phase, by Fault Tree Analysis (FTA) and Event Tree Analysis (ETA) methods, the frequencies of event were calculated and assess the consequences of oil leakage and the radius of heat intensity that possibly appeared. The assessment of the consequences of this study was carried out using fire modelling approach to determine the consequences of an accident. In the third phase, the risk level is represented against the risk acceptance criteria using F-N curve and UK HSE standard. In the fourth phase, if the level of risk of third phase laid in unacceptable region, a mitigation process is necessary. The workflow of methodology is shown in Figure 1.

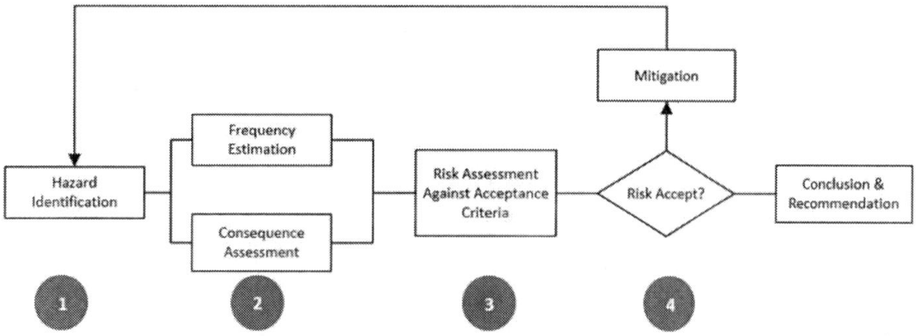

Figure 1. Flowchart of Risk Assessment.

As mention above, the F-N curve with UK HSE standard applied for risk evaluation as a representation of the results of the relationship between F and N. Whereby, F is the value of the frequency obtained from the results of the frequency analysis in events per year, represented in y-axis and N is the number of fatalities of the event or consequence, represented in x-axis. The combination of the F value and the N value is plotted to determine whether the risk is acceptable or not.

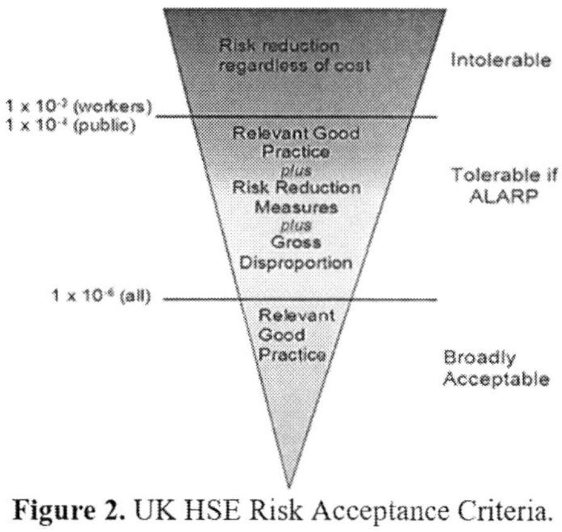

Figure 2. UK HSE Risk Acceptance Criteria.

Risk acceptance of UK HSE standard grouped into 3 regions i.e. broadly acceptable, ALARP, and intolerable/unacceptable area as expressed in Figure 2. The upper bound of risk acceptance has been set at 1.0E-03 per year for non-public or workers. If the risk level exceeds this limit, then the risk level categorized in unacceptable/intolerable area [9].

3. Risk Analysis and Results

3.1 Hazard Identification

Hazard identification is the initial process of societal risk assessment, it used to identify potential hazards that may occur during operation [1]. Hazard and Operability (HAZOP) is a method that commonly used to analyse the specific hazard potential of a system. HAZOP aims to knowing the cause of the potential hazard occurred and generate the type of consequence analysis that must be taken [13]. The processing of crude oil into a finished product is carried out in a large refinery area with a complex system inside. Therefore, the area of refinery needs to be segregated into several parts, this segregation will help create relevant scenario based on the refinery system. The segregated area of refinery unit also can be called as nodes. The node and area segregation are shown in Figure. 3 and defined in Table. 1. On this case there are 8 nodes, the determination of this nodes based on the specific function of each location and the phase of hydrocarbon handled in the system.

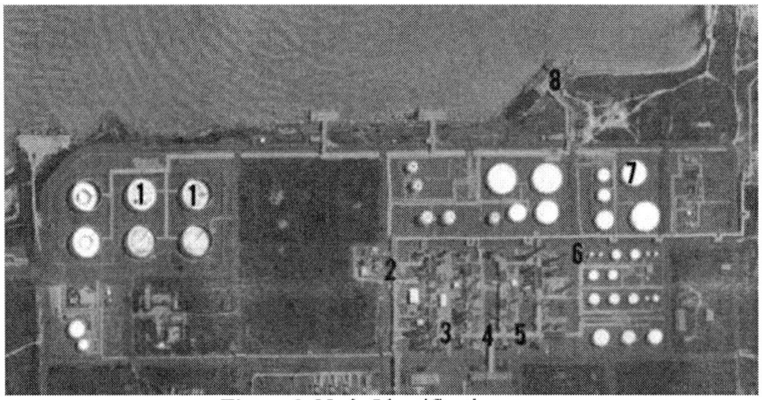

Figure 3. Node Identification.

Table 1. Description of Nodes.

Nodes	Information
Node 1	This node is focused on storing condensate in a feedstock before the transferring system for processing.
Node 2	This node includes the pipeline and its components starting from the feedstock tank to transfer condensate into pre-cut column.
Node 3	This node is focused on the pipeline transfer from the pre-cut column to the condensate splitter.
Node 4	This node is focused on the analysis of components in the pipeline from the condensate splitter to the distillate column.
Node 5	This node is about process of transferring oil from distillate column to diesel stripper
Node 6	Node 6 focused on the transfer pipeline that is transferring diesel oil to diesel storage tank. Diesel oil that has been warmed up many times in the previous stage will be cooled before entering the storage tank.
Node 7	This node is focused on filling and storing diesel oil in diesel storage tank.
Node 8	The transfer process using pipes from the diesel storage tank to the vessel

3.2 Frequency Analysis

In this paper, the Frequency emphasize on two principal frequencies. First, analysis frequency of the oil release and estimated using Fault Tree Analysis (FTA). Second, the Event Frequency and estimated using Event Tree Analysis (ETA).

3.2.1 Fault Tree Analysis

The FTA is used to obtain the initiating event at each node based on the failure frequency of each component at each node. FTA calculating the failure frequency within the case boundaries on Piping and Instrumentation Diagram, such as control valves, safety valves, instrument connections, pumps, feedstock tank, etc [1]. The failure frequency for different hole sizes are calculated based on failure frequency data adopted from DNV Failure Frequency Guidance and International Association of Oil and Gas Producers (IOGP) datasheet. The result of frequency failure for each node with different leakage diameter presented in Table 2.

Table 2. The failure frequency of each node with different leakage diameter.

Node	Top Event	Failure Frequency				
		1-3 mm	3-10 mm	10-50 mm	50-150 mm	>150 mm
1	Oil Release	4.3E-4	1.3E-4	4.2E-5	1.04E-5	1.76E-5
2	Oil Release	6.8E-5	6.7E-6	6.7E-7	1.69E-8	7.9E-9
3	Oil Release	5.6E-3	1.72E-3	5.2E-4	7.5E-5	2.13E-4
4	Oil Release	5.2E-3	1.7E-3	5.3E-4	7.6E-5	5.9E-5
5	Oil Release	5.2E-3	1.7E-3	5.3E-4	7.6E-5	5.9E-5
6	Oil Release	6.1E-3	1.8E-3	6.3E-4	1.1E-4	9.6E-5
7	Oil Release	3.2E-3	2.9E-3	2.8E-3	2.8E-3	2.8E-3
8	Oil Release	4.1E-3	1.2E-3	3.7E-4	7.4E-5	1E-4

3.2.2 Event Tree Analysis

Event Tree Analysis (ETA) is a method used to identify and evaluate possible outcomes if an initiating event occurs due to equipment failure or human error [2]. The purpose of the ETA is to determine whether the initial event will develop into a serious accident or the incident is sufficiently controlled by the safety systems and procedures implemented by the design. The initial event in this case is an oil release. The release of oil can cause a pool fire if there is an ignition from the ignition source. The calculation of oil release is formulated as follows:

$$Q_L = 2{,}1\ x\ 10^{-4} x\ d^2 x \sqrt{\rho_L x P_L} \quad (1)$$

Where:

d = hole diameter (mm)

ρ_L = liquid density (kg/m^3)

P_L = initial pressure of liquid (bar gauge)

The fluid density of the condensate is 828.1 kg / m^3 and the initial pressure is 9.3 bar. The results of the calculation of the oil release are presented in Table 3.

Table 3. The result of oil release calculation.

No.	Hole Diameter (mm)	Density of Condensate (kg/m3)	PL (bar)	Mass of Oil Release at Higher Range of Hole (kg/s)
1	3	828.1	9.3	1.66E-01
2	10	828.1	9.3	1.23E+00
3	50	828.1	9.3	3.07E+01
4	150	828.1	9.3	2.76E+02
5	200	828.1	9.3	4.91E+02

After calculating the ignition probability using equation 1, then determine the frequency for each consequence scenario using the ETA method. The input of ignition probability and probability ignition for immediate and delayed is obtained from OGP (International Association of Oil and Gas Producers). The ignition probability in OGP depends on the type of facility and the amount of gas released per unit time. The result of calculations for each node is shown in Table 4.

Table 4. The result of Event Tree Analysis.

Node	Hole Diameter (mm)	Scenario Frequencies		
		Pool Fire	Evaporation	Nothing Happened
	50	2.87E-06	3.55E-07	3.88E-05
1	150	4.98E-06	6.16E-07	4.80E-06
	>150	1.01E-05	1.25E-06	6.24E-06
	50	4.58E-08	5.66E-09	6.19E-07
2	150	8.10E-09	1.00E-09	7.80E-09
	>150	4.54E-09	5.61E-10	2.80E-09
	50	3.55E-05	4.39E-06	4.80E-04
3	150	3.59E-05	4.44E-06	3.46E-05
	>150	1.22E-04	1.51E-05	7.55E-05
	50	3.62E-05	4.48E-06	4.89E-04
4	150	3.64E-05	4.50E-06	3.51E-05
	>150	3.39E-05	4.19E-06	2.09E-05
	50	3.62E-05	4.48E-06	4.89E-04
5	150	3.64E-05	4.50E-06	3.51E-05
	>150	3.39E-05	4.19E-06	2.09E-05
	50	3.10E-05	3.83E-06	5.95E-04
6	150	4.87E-05	6.02E-06	5.53E-05
	>150	4.93E-05	6.09E-06	4.06E-05
	50	1.38E-04	1.70E-05	2.65E-03
7	150	1.24E-03	1.53E-04	1.53E-04
	>150	1.44E-03	1.78E-04	1.78E-04
	50	1.82E-05	2.25E-06	3.50E-04
8	150	3.28E-05	4.05E-06	3.72E-05
	>150	5.14E-05	6.35E-06	4.23E-05

3.3 Consequence Analysis

Consequence analysis is carried out to determine the effect of the scenario contained in the ETA calculation. In this study, a consequence analysis is implemented to specify the impact of fire on the facilities and personnel in the location of the incident.

3.3.1 Pool Fire

Pool fire occurs when a hydrocarbon liquid that accidentally release accumulates on an area and ignites [2]. The pool will spread until it is constrained by physical impediments such as reach an equilibrium size. When first ignited, the fire spreads quickly over the hydrocarbon pool and continues to consume the fluid at a characteristic burning rate. The flame height will be at variance with the pool diameter and the flame may be inclined by any cross wind. Data on affected areas and the number of personnel on site are required for consequence modelling. Table 4. shown the list of affected areas and the number of personnel.

Table 5. Receiver area and the number of workers.

Receiver	Location	Corresponding Node	Number of Workers
1	Feedstock tank 1	1	2
2	Feedstock tank 2	1	2
3	Storage tank area 1	2	8
4	Central control building	3	25
5	Pre-cut column	3	2
6	Condensate splitter	4	2
7	Power control room	4	2
8	Distillate column	5	2
9	Diesel splitter	6	2
10	Diesel splitter to diesel storage tank	6	4
11	Diesel storage tank	7	2
12	Storage tank area 2	7	6
13	Berth	8	4
14	Marine control building	8	8

The simulation result of pool fire modelling is shown in Figure 4 and the calculation result for pool fire consequnce analysis using fire modelling approach is presented on Table 6.

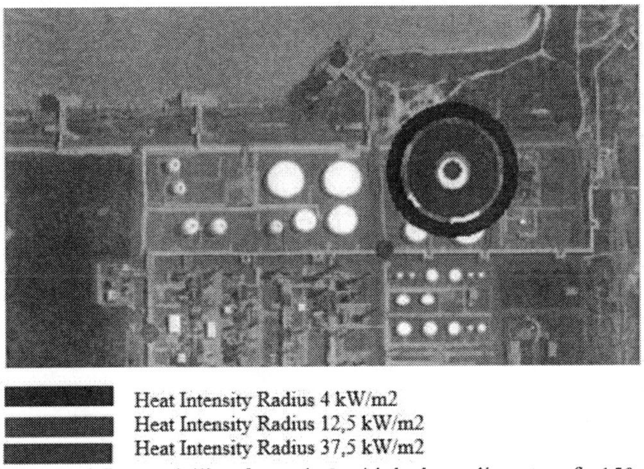

Figure 4. Pool fire modelling for node 2 with leakage diameter of >150 mm.

Table 6. The result of pool fire modelling using fire modelling approach.

No.	Receiver	Leak Bore Diameter (mm)	Correspon-ding Node	Number of People	Pool Diameter (m)	Heat Intensity Radius (m)		
						4 kW/m²	12,5 kW/m²	37,5 kW/m²
1	Feedstock Tank 1	50	1	2	16,87	64.82	42.35	-
		150			52,39	109.49	50.04	-
		200			70,42	132.53	59.46	-
2	Feedstock Tank 2	50	1	2	16,87	64.82	42.35	-
		150			52,39	109.49	50.04	-
		200			70,42	132.53	59.46	-
3	Storage Tank Area 1	50	1	8	52,39	109.49	50.04	-
		150			70,42	132.53	59.46	-
		200			70,42	132.53	59.46	-
4	Central Control Building	50	2	50	5,41	32.11	24.46	13,39
		150			6,60	33.98	25.29	12,14
		200			6,61	33.50	24.81	11,64
5	Precut Column	50	2	2	5,41	32.11	24.46	13,39
		150			6,60	33.98	25.29	12,14
		200			6,61	33.50	24.81	11,64
6	Condensate splitter	50	4	2	5,36	31.98	24.37	13,40
		150			6,56	33.88	25.24	12,26
		200			6,57	33.41	24.76	11,77
7	Power control room	50	4	2	5,36	31.98	24.37	13,40
		150			6,56	33.88	25.24	12,26
		200			6,57	33.41	24.76	11,77
8	Distillate column	50	5	2	5,33	31.88	24.30	13,52
		150			6,48	33.71	25.13	12,20
		200			6,49	33.25	24.66	11.74

3.4 Risk Evaluation

Risk evaluation is concerned to determine the level of risk that produced by refinery operation based on the frequency and consequence analysis. In this paper, risk evaluation using F-N curve and UK HSE standard. The risk evaluation for pool fire with leakage diameter of 200 mm is shown in Figure 5. Based on the result of risk evaluation of pool fire scenario, the risk level of pool fire is categorized in unacceptable area because exceeds the upper bound of risk acceptance (1.0E-03 per year). Therefore, mitigation process is necessary to obtain the risk level in the ALARP area.

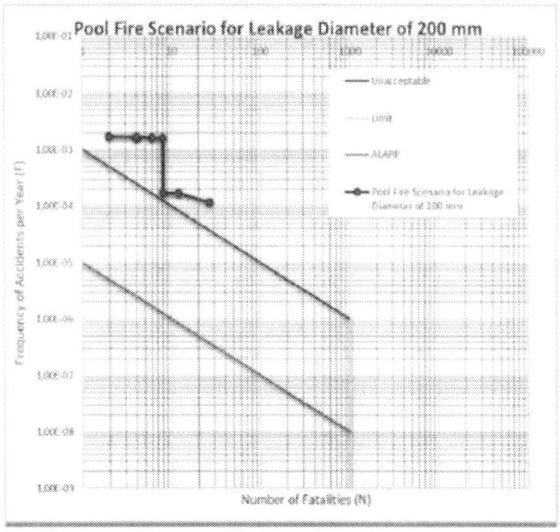

Figure 5. F-N curve for pool fire with leakage diameter of 200 mm.

4. Mitigation

Mitigation is an action that applied when the risk level of risk evaluation categorized as an unacceptable category. Mitigation process commonly suggested in order to reduce risk from hydrocarbon fires at refinery. Mitigation method used in this research is LOPA (Layer of Protection Analysis) method. The main purpose of using LOPA is to identify preventive measures against potential consequences of certain risks by adding an Independent Protection Layer (IPL) as a step to reduce the failure frequency in order to maintain the level risk remain in ALARP area. Mitigation is performed by installing gas detector and CO_2 system as an IPL to the system. The worksheet of LOPA and risk evaluation after applied mitigation process is shown in Figure 6.

Scenario Pool Fire	Scenario Title: Pool Fire on Refinery Unit (200 mm)	Node Number 7	
Date	Description	Probability	Frequenc (Per Yea
Consequence Description	Pipe or equipment leak because some component are failed and lead to fire		
Risk Tolerance Criteria (Frequency)	Action Required		1.00E-04
	Tolerable		1.00E-06
Initiating Event (Frequency)	Oil release from 12 inch pipe in Storage Tank 2 area		2.80E-03
Enabling Event or Condition	N/A		
Conditional Modifiers	Ignition probability	5.77E-01	
	Delayed ignition probability	1.10E-01	
Frequency of Unmitigated Consequence			1.44E-03
Independent Protection Layers (IPL)	Temperature alarm	5.52E-02	
	Pressure alarm	4.22E-02	
	Gas detector	1.00E-01	
	CO_2 system	2.00E-02	
Total PFD for all IPLs		2.00E-03	
Frequency of Mitigated Consequence			2.88E-06
Risk Status		ACCEPTABLE	
Actions Required to Meet Risk Tolerance Criteria	Install gas detector and CO_2 system as IPL to reduce risk		
Notes			
References			

(a)

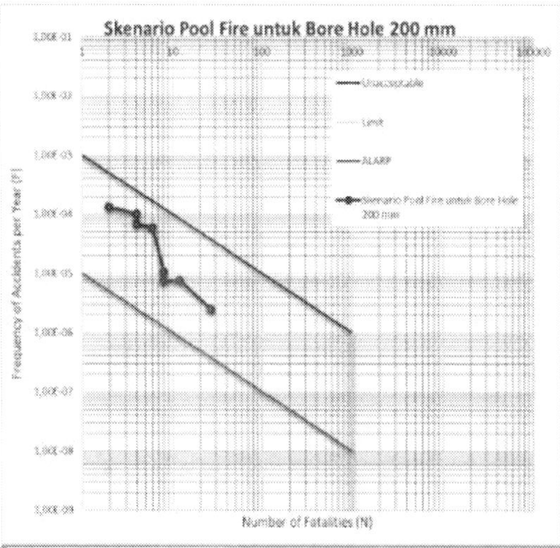

(b)

Figure 6. (a) LOPA Worksheet; (b) Risk evaluation after mitigated.

IOP Conf. Series: Materials Science and Engineering 1052 (2021) 012025 doi:10.1088/1757-899X/1052/1/012025

5. Conclusion

Societal risk assessment has been carried out on refinery operation to quantify the risk level of personnel caused by process hydrocarbon hazard events. The first step of risk assessment is identifying all the possible hazards from the operation of the system based on BS IEC 61882 then the frequencies of event were calculated and assess the consequences of oil leakage and the radius of heat intensity that possibly appeared. The risk level is represented against the risk acceptance criteria using F-N curve and UK HSE standard. Based on result of risk evaluation, the risk level of pool fire consequence is in unacceptable region. Hence, mitigation process was carried out. From mitigation process presented in chapter 4, shows that pool fire scenario is lower than acceptable limit of 1.0E-03. Therefore, we can conclude that the risk level of refinery within ALARP (As Low As Reasonable Practicable) region.

References

[1] Akbar, T. A. (2017). Fire Risk Assessment of Terminal LPG Semarang. Surabaya: Institut Teknologi Sepuluh Nopember.
[2] Alfan, V. N. (2014). Penilaian Risiko Sosial pada Mini CNG Plant Blok Semanggi. Surabaya: Institut Teknologi Sepuluh Nopember.
[3] Artana, K. B., & Dinariyana, A. B. (2013). Teori Keandalan Sistem dan Aplikasinya. Surabaya: Guna Widya.
[4] Armanda, L. D. (2014). Penilaian Risiko Sosial Akibat Kebocoran Pipa pada Jalur Pipa Transfer Bahan Bakar dari Dermaga ke Depo Milik PT. Duta Artha Bahari Nusantara Gresik. Surabaya: Institut Teknologi Sepuluh Nopember.
[5] BS IEC. (2001). Hazard and Operability Studies (HAZOP Studies)-Application Guide BS IEC 61882:2001. London: British Standard.
[6] Buthod, P. (2007). Crude Oil Properties and Condensate Properties and Correlations. In L. W. Lake (Ed.), Petroleum Engineering Handbook (Larry W. Lake ed., pp. 21-1 - 21-20). Tulsa: Society of Petroleum Engineers.
[7] Fuentes-Bargues, J. L., Gonzales-Cruz, M. C., Gonzalez-Gaya, C., & Baixauli-Perez, M. P. (2017). Risk Analysis of a Fuel Storage Terminal using HAZOP and FTA. Spain: International Journal of Environmental Research and Public Health.
[8] Handani, D. W., Dinariyana, A. B., Ariana, I. M., & Fauzi, F. R. (2016). Analisa Risiko Sosial Akibat Jet Fire pada Proses Unloading CNG pada Terminal CNG di Pembangkit Lombok Peaker. Surabaya: Seminar Nasional Kelautan XI.
[9] HSE. (2012). Five Steps to Risk Assessment. London: UK HSE.
[10] Hurley, M. J. (2016). SFPE Handbook of Fire Protection Engineering (5th ed.). London: SFPE.
[11] IMO. (2002). Guidelines for Formal Safety Assessment (FSA) for Use in The IMO Rule-Making Process. London: International Maritime Organization.
[12] International Association of Oil and Gas Producers (OGP). (2010). OGP Process Release Frequencies. London: International Association of Oil and Gas Producers.
[13] Mannan, S. (2005). Lees' Loss Prevention in the Process Industries – Hazard Identification, Assessment, and Control (3rd ed.). Texas: Elsevier Inc.
[14] Massimo, G., Mara, L., & Federica, M. (2013). Risk Analysis in Handling and Storage of Petroleum Products. America: American Journal of Applied Science 10.
[15] Mousselli, A. H. (1981). Offshore Pipeline Design, Analysis, and Methods. Oklahoma: PennWell Books.
[16] Rausand, M., & Hoyland, A. (2004). System Reliability Theory (2nd ed.). Amerika: Wiley-Interscience.
[17] Society of Fire Protection Engineer. (2005). SFPE Engineering Guide to Performance-Based Fire Protection Analysis and Design of Buildings. London: SFPE.
[18] Standards Australia International Limited; Standards Association of Australia. (2004). Risk Management Guidelines Companion to AS/NZS 4360:2004 (3rd ed.). Sydney: Standards Australia International.

Design Propeller Of Fast Missile Boat 60 M By Using Gawn Series

C Kusuma[1,5,*], I M Ariana[1], W H Nugroho[2], E B Djatmiko[3], A A Masroeri[1] and Sutardi[4]
[1]Department of Marine Engineering, FTK, Kampus ITS, Sukolilo, Surabaya
[2]Indonesian Hydrodinamic Laboratory, BPPT, Surabaya
[3]Department of Marine Engineering, FTK, Kampus ITS, Sukolilo, Surabaya
[4]Department of Mechanical Engineering, FTI, Kampus ITS, Sukolilo, Surabaya
[5]Sekolah Tinggi Teknologi Angkatan Laut

Abstract. The 60m Fast Missile Boat (FMB 60m) with a full load displacement of 467T draft 2.6m uses a 3900Kw engine and a maximum speed of 28 knots. In this research, it will be designed using the Gawn series. Choice of the Gawn series is because Gawn has a P / D of 0.4-2.0 and a maximum K_T of 1.1. Powering calculations, performed using the MARIN DESPPC method. Based on the calculation results, the K_{TR} and $J_{Desaign}$ value will be obtained, then all Gawn series propellers that meet the required engine power will be calculated. From the calculation of thrust requirements and the availability of engine power, the Gawn-series propeller is suitable for the propulsion of 60m fast missile boat.

1. Introduction

The Indonesian sea has 2 (two) different characters, the western region which is integrated with the Asian continent has shallow seas (<200m) and the eastern region which is integrated with the Australian continent has a deep-sea (> 200m). The western region with its shallow sea characteristics and thousands of islands is suitable for fast missile boats with a length overall 60m and a displacement of 500GT. This paper focuses on the design of the 60m Fast Missile Boat propeller using the Gawn series propeller application. The choice of the Gawn series propeller as the 60meter fast patrol missile boat propulsor was due to:

a. The Gawn Series was chosen because it is widely used for high-speed ship propellers [1][2][3][4][5][6][7][8]
b. Has a larger pitch and blade area ratio[9][10].

Fast Missile Boat is a multi-role ship, namely as a warship, patrol boat, and search and rescue ship. The most important factor in the design of a fast missile boat is speed. This is what distinguishes fast missile boats from cargo or passenger ships. Speed is a very important factor when functioning as a combat ship. At times as a patrol boat speed is also required in direct pursuit while on patrol.

This paper will discuss the 60m Fast Missile Boat which is applied using a propeller - Gawn series. The design is done using the open water test model diagram [10]. The author designed whether the Gawn Series propeller could be used as propulsion by the initial design required on the 60meter Fast Missile Boat.

2. Propeller Calculation Method

The research object is the 60-meter Fast Missile Boat with the following technical data in table 1 and picture in Figure 1.

Table 1. Specification 60-meter Fast Missile Boat.

Specifications	Value	Unit
Design speed	28	knots
LOA	59.5	meter
Length on waterline	54	meter
LBP	53.71	meter
Length	7.23	meter
Displacement	467	Ton
Block coefficient	0.391	
RPM Engine	600	1/Min
Environmental conditions :		
Temperature	30	m3
Viscosity	1025	kg/
Kinematic viscosity	8.43E-07	m2/s

Figure 1. Fast Missile Boat 60m (https://pal.co.id/publikasi/artikel).

The power requirements (engine) of the fast patrol boat are designed to use 2 (two) main engines with specifications in table 2.

Table 2. Engine Specifications.

Specifications	Value	Unit
Power shaft(Ps)	2x3900	KW
	500-700	Rpm
Gearbox efficiency	97	%
Sea margin	15	%

Sea margin worldwide 15%, the maximum engine power with sea margin is 6630 KW. The efficiency of the gearbox is 97%, then the PE (Engine Power installed). Fast Missile Boat is assumed to be a heavy-duty engine so that the MCR operation is capable of up to 100% Engine load. So that the existing Power Engine is 6431 KW. The calculation of the maximum diameter of the propeller is obtained from the stern design of the ship and the calculation of the empty load so that the maximum diameter of the propeller is 1.8m. The results of testing the MARIN DESPCC software are for the ideal trial condition (no current, wind and, waves). Therefore, it is necessary to add an Allowance Sea Margin to overcome the loss of speed due to waves during the sea trial.

3. Methodology

The calculation begins by entering the ship data including LoA, LwL, B, D, T, J_{desain}. With DESPCC software, the required Thrust value ($T_{requirement}$) will be obtained. By getting the T_R value, the K_T value can be found. Then using the Gawn series will be able to determine the K_{TP} that meets the conditions $K_{TP} > K_{TR}$. After obtaining the K_{TP} value, it can be seen the value of Q, PD, and the efficiency of the propeller.
The research methodology can be seen in the following flow chart Figure 2.

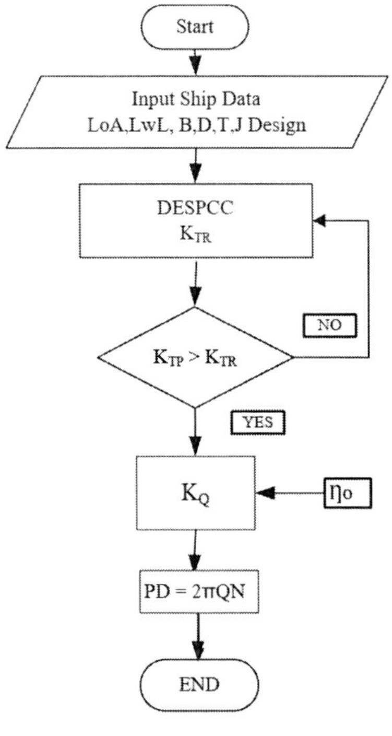

Figure 2. Research Flowchart.

4. Result and Discussion

The 60m Fast Missile Boat resistance is calculated using the MARIN DESPCC software. The calculation is carried out at full load at 2.57m with a speed variation between 20 to 30 knots. The results of resistance testing with the MARIN DESPCC software are in table 3.

Table 3. Results of Resistance Testing With The MARIN DESPCC Software.

VS [knots]	R-TOT [kN]	THRUST [kN]	W [-]	T [-]	ETA-H [-]	ETA-R [-]	ETA-0 [-]	ETA-D [-]
20.00	163.0	186.7	0.053	0.127	0.922	0.968	0.546	0.487
20.50	174.7	200.1	0.053	0.127	0.922	0.968	0.546	0.487
21.00	186.4	213.5	0.053	0.127	0.922	0.968	0.546	0.487
21.50	198.1	226.9	0.053	0.127	0.922	0.968	0.546	0.487
22.00	209.7	240.2	0.053	0.127	0.922	0.968	0.546	0.487
22.50	221.3	253.5	0.053	0.127	0.922	0.968	0.545	0.486
23.00	232.9	266.8	0.053	0.127	0.922	0.968	0.545	0.486
23.50	244.5	280.0	0.053	0.127	0.922	0.968	0.545	0.486

24.00	255.9	293.2	0.053	0.127	0.922	0.968	0.545	0.486
24.50	267.5	306.4	0.053	0.127	0.922	0.968	0.545	0.486
25.00	278.3	318.8	0.052	0.127	0.921	0.968	0.545	0.486
25.50	287.3	329.1	0.052	0.127	0.921	0.968	0.545	0.486
26.00	295.3	338.3	0.052	0.127	0.921	0.968	0.545	0.486
26.50	302.7	346.8	0.052	0.127	0.921	0.968	0.545	0.486
27.00	309.9	355.0	0.052	0.127	0.921	0.968	0.545	0.486
27.50	317.6	363.8	0.052	0.127	0.921	0.968	0.545	0.485
28.00	325.1	372.4	0.052	0.127	0.921	0.968	0.544	0.485
28.50	332.2	380.5	0.052	0.127	0.921	0.968	0.543	0.484
29.00	339.0	388.3	0.052	0.127	0.921	0.968	0.542	0.483
29.50	345.7	396.0	0.052	0.127	0.921	0.968	0.541	0.482
30.00	352.2	403.4	0.052	0.127	0.921	0.968	0.541	0.481

The propeller performance can be expressed in terms of four non-dimensional parameters:

a. Advance ratio, J_S.
b. Thrust coefficient, K_T.
c. Torque coefficient, K_Q.
d. Open water efficiency, η.

$$J_D = \frac{(1-w)V_S}{nD} \quad (1)$$

$$K_T = \frac{T}{\rho n^2 D^4} \quad (2)$$

$$K_Q = \frac{Q}{\rho N^2 D^5} \quad (3)$$

$$\eta = \frac{1}{2\pi} \cdot \frac{K_T J_S}{K_Q} \quad (4)$$

From the calculation results at a speed of 28 knots, the effective w value is 0.052 so that the J_{Design} value is obtained:

$$J_D = \frac{(1-0.052) * 14.4044}{10 * 1.8}$$

$$J_D = 0.758$$

Base on table 2 Thrust propeller (T_P) value required for Full Load load on 2.6m at 28 knot is 372.4 KN, then the appropriate propeller will be selected by getting the required Thrust Coefficient (K_{TR}) value first.

$$\text{Tp} \quad = K_{TR}\, x\, \rho\, x\, n^2 x\, D^4 \quad (5)$$

Tp : 372.4 KN

ρ : 1025Kg/m^3

n : 600Rpm = 10 Rps

D : 1.8m

The K_T value is 0.346 at J = 0.758

The calculation of propeller performance using the open water test experiment method.The gawn propeller series presented by Gawn consists of a series of 37 (3) three-leaf propellers covering a wide range of pitch ratios from 0.4 to 2.0 and blade area ratios from 0.2 to 1.1. When testing the open water test using a propeller model each has a diameter of 503mm (20in.), So that the resulting scale effect due to the small model can be eliminated.[10]. Each type of propeller has a uniform face section pitch, segmental blade section, a constant blade thickness ratio of 0.060, and a boss diameter of 0.2D. The outline blade is elliptical with an inner and outer knot of 0.1R and a blade tip, respectively. The entire series was tested in towing tank No. 2 in A.E.W. Haslar is in the slip range from zero to 100 percent: to achieve this the propeller rotation speed is in the range of 250 to 500rpm. No destructive cavitation characteristics are given for this series. Fast boats that cannot be filled with the B series are recommended to use the Gawn series because they have a higher P / D value, namely 0.6-2. This research only uses the open water test diagram of the experimental results. In table 2 counter curves on pitch ratio inclusive are by experiment. Other counters are derived in Table 4.

Table 4. Gawn Series by experiment [10].

BAR	P/D by experiment
0.2	0.4-1.0
0.35	0.4-1.2
0.5	0.4-2.0
0.65	0.4-2.0
0.8	0.8-1.6
0.95	1.0-1.6
1.1	0.8-1.4

Based on the K_T value of 0.346 at J = 0.758 and by using the open water test diagram [10], several Gawn series blades can be selected in table 5.

Table 5. Available Gawn Series Types.

BAR	P/D	KT	KQ	EFF	PD	PE
				%	KW	KW
0.5	1.6	0.39	0.09	52.3	10947	11275
0.5	1.8	0.469	0.12	47.17	14596	15034
0.5	2	0.53	0.146	43.82	17758	18291
0.65	1.6	0.442	0.103	51.87	12510	12885
0.65	1.8	0.528	0.136	46.96	16507	17002
0.65	2	0.589	0.173	41.09	21042	21674
0.8	1.4	0.377	0.08	56.65	9768	10061
0.8	1.6	0.47	0.114	49.85	13842	14257
0.95	1.4	0.4	0.091	52.81	11121	11454
0.95	1.6	0.521	0.132	47.53	16092	16575
1.1	1.4	0.45	0.106	51.38	12858	13243

From Table 5, it is found that the thrust that meets there are 12 types, and BAR 0.8 with P / D 1.4 is the most efficient with a propeller efficiency value of 56.65% and PD 9768 KW. Assuming a sea margin of 15%[12] and the Fast Missile Boat uses a heavy Improper formatting-duty engine so that it can be loaded with a full load and the efficiency of the gearbox is 0.3 so the PE or engine power needed is 10.061 KW.

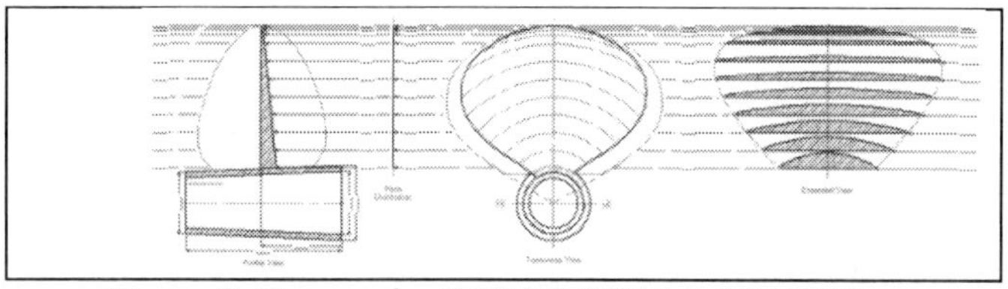

Figure 3. Gawn series BAR 0.8 PDRA 1.4 CAD View.

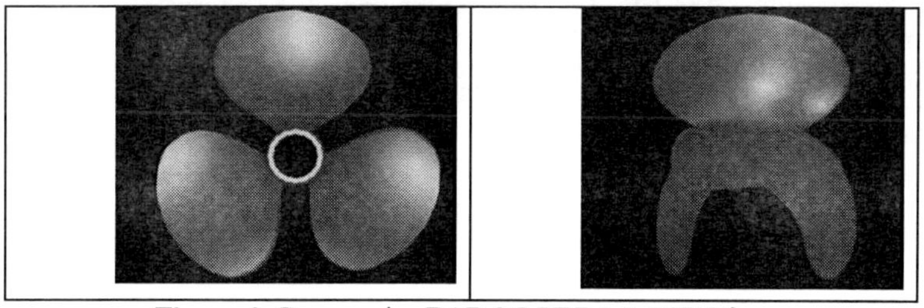

Figure 4. Gawn series BAR 0.8 PDRA 1.4 3D View.

5. Conclusion

From tables 4 and 5 it can be concluded that the Gawn series propeller can be used as a 60m fast patrol boat propulsion using BAR 0.8 on PDRA 1.4 with PE or the required engine power is 10,061 KW. From these data, considering that the efficiency is very low, less than 0.6, it is advisable not to use the Gawn series propeller as a 60m fast missile boat propulsion. However, it should be noted that the propeller efficiency is very low <0.6%, so it is necessary to consider the use of other series that are more efficient than the Gawn series.

6. References

[1] J. Dang, H. J. J. Van Den Boom, and J. T. Ligtelijn, "The wageningen C- and D-series propellers," *FAST 2013 - 12th Int. Conf. Fast Sea Transp.*, 2013.

[2] K. M. Flood, "Propeller Performance Analysis Using Lifting Line Theory," *Mech. Eng.*, no. 1997, p. 60, 2009.

[3] R. Sampson and G. H. G. Mitchell, "Fifty Years of the Kings College Propeller Series at the Emerson Cavitation Tunnel I," no. 2000, 2009.

[4] "1993-1 Hope KJ - Aspects of Modern Propeller Design.pdf.".

[5] D. Radojcic, "Mathematical Model of Segmental Section Propeller Series for Open-Water and Cavitating Conditions Applicable in CAD," *Propellers '88 Symp.*, no. October 1988.

[6] D. Radojcic, A. Simić, and M. Kalajdžić, "Fifty Years of the Gawn-Burrill KCA propeller series," *Trans. R. Inst. Nav. Archit. Part B Int. J. Small Cr. Technol.*, vol. 151, no. 2, pp. 9–17, 2009.

[7] G. H. G. Mitchell, R. Sampson, and M. Atlar, "A Modern Approach to the Representation and use of the KCA Systematic Propeller Series," *Third Int. Symp. Mar. Propulsors smp'13*, no. May, pp. 176–184, 2013.

[8] J. S. Carlton, *Marine Propellers and Propulsion, 2nd edition*, vol. 163. 2010.

[9] J. Holtrop, "Propeller design Maritime Research Institute Netherland, J." Maritime Research Institute Netherland, Netherland.

[10] R. W. L. Gawn, "Effect of pitch and blade width on propeller performance," *RINA Transactions*, vol. 95, no. 1. pp. 157–193, 1953.

[11] Z. E. L. Gem, "Yüksek lisans ödevi."

[12] A. K. Magnussen, "Rational calculation of sea margin," no. June 2017.

Acknowledgments

Authors who want to thank the help or encouragement of colleagues, especially DR.Eng I Made Ariana, DR.Wibowo H Nugroho, Prof. DR. Eko Budi Djatmiko, DR.Eng. A.A.Masroeri, Prof. DR.Sutardi. Acknowledgments that are not numbered immediately after the last numbered section of this paper.

Technical Review of Bollard Pull Calculation to Support Shuttle Tanker from Spread Mooring System FSO

I M Ariana[1,2], F I Prastyasari[1,2,3], and R R Putera[1]

[1] Department of Marine Engineering, Faculty of Marine Technology, Institut TeknologiSepuluh Nopember, Surabaya, Indonesia
[2] Center of Excellences for Maritime Safety and Marine Installation, Institut TeknologiSepuluh Nopember, Surabaya, Indonesia
[3] Department of Urban and Environmental Engineering, Graduate School of Engineering,Kyushu University, Fukuoka, Japan

Abstract. An FSO was designed with Spread Mooring System to substitute the former FSO whose mooring system was designed with Turret Mooring System. Theoretically, the environmental loads affecting the shuttle tanker when the crude oil transfer process from FSO is undergone, it will have a bigger number and require a bigger bollard pull specification of the AHTS vessels to assist the shuttle tanker in place. It is due to the weathervane flexibility that is adjusted to the heading direction of either the FSO or shuttle tanker which found in Turret Mooring System is absent in Spread Mooring System. The calculation of environmental loads in this research is conducted using recommendation from DNV-GL and wind, wave, current, and inertia loads are taken into accounts. The biggest transversal environmental load is recorded by the magnitude of 262.96 MT towards positive vector. Whilst the biggest longitudinal environmental load is recorded by the magnitude of 105.44 MT towards negative vectors. By considering safety factor, age, efficiency, and towing configuration of the AHTS vessels, the recommended specification of the AHTS vessels configuration are the 200 MT AHTS vessel in stern of the shuttle tanker and 220 MT AHTS vessel in port side or starboard side of the shuttle tanker.

1. Introduction

The use of fossil fuels is still very important in our daily life. The statistical data published by PwC Indonesia proofed that the demand of fossil fuels in Indonesia is increasing rapidly every year, yet it is contradicted with the depleting of productivity in existing onshore oil fields [1]. It pushes the government to start exploring oil fields in unconventional and marginal areas, for example the offshore oil fields.

In order to support the day-to-day operations in offshore oil fields whose characteristic is critically isolated, therefore we need highly reliable offshore facilities. One of the offshore facilities that often used is Floating Storage and Offloading (FSO). FSO is a ship shaped floating facility that is moored to keep it static to contain the crude oil obtained from surrounding oil wells via subsea risers [2].

This paper studied about a FSO that will be moored with a Spread Mooring System (SMS) configuration that has been operating since 2018 to substitute the former FSO whose mooring configuration used was the Turret Mooring System. A turret mooring system is an example of a single point mooring that sometimes can be unstable in terms of yaw motions [3]. While the SMS configuration is considered can be operated in all weather conditions and also can have a long lifetime [4]. A replacement of FSO in the middle of production phase has rarely been done and remains novel until now. The complexity of the procedure is the sole reason engineers have chosen other options such as modifying the FSO rather than substituting it. The weather versatility, which is also found in the dynamic positioning system [5], is an issue in the former FSO with Turret Mooring System to adjust its position around the fixed geostatic part of the turret was vanished as the result of changing the FSO with SMS mooring configuration. The static condition of current FSO would lead to higher environmental loads received by the shuttle tanker when undergo crude oil transfer.

The AHTS vessels towing configuration that is not replaced whilst applying the different mooring configuration of FSO will lead to future problems. The lower specification of AHTS vessels being used will be insufficient to assist shuttle tanker when undergoing crude oil transfers from Spread Mooring System FSO and crude oil transfer has to be done in a strict operational window. The company report shows that crude oil transfer was postponed and aborted for multiple times in the last six months, not to mention the near miss accident. Therefore, reanalyzing the environmental loads received by the shuttle tanker when undergoing a crude oil transfer from the FSO is a step that should be taken to assure the safety of crude oil transfer operation and reviewing the AHTS vessels towing configuration.

This objective of this paper is to design a new AHTS vessels configuration to support shuttle tanker from FSO in crude oil transfer that can meet the requirement from calculated bollard pull in accordance with DNV-GL regulations.

2. Methodology

The environmental loads that are taken into consideration on this research are wind, wave, current, and inertia load. The load calculation uses a static hydrodynamic approach based on the environmental load recommendations from DNV-GL.

There are several methods and formulas to calculate wind load and current load, yet the OCIMF (Oil Companies International Marine Forum) formulas is used as recommended by DNV-GL [5]. The formula from the Spanish Standard ROM 0.2-90 is used to calculate the wave load. And last, formula by Chakrabarti is used to calculate inertia load on this research [6].

2.1. Data Collection

The research needs supporting data that can assist the process to complete the research. The obtained data will be processed later to be used as a reference for evaluating the existing AHTS vessels configuration to assist the shuttle tanker from Spread Mooring System FSO. The supporting data includes Metocean Data, AHTS vessels specifications, and shuttle tankers principal dimensions.

The shuttle tanker data being used in this research is obtained by selecting the biggest principal dimension from shuttle tankers that has ever done crude oil transfer from the FSO in the last six months. The selected shuttle tanker is MT Mare Oriens with specification as shown in Table 1.

2.2. Evaluation of Existing Configuration

According to DNV-OS-H202 Section 4: Towing, the value of effective bollard pull of AHTS vessels will decrease every year [5]. The reduction rate of effective bollard pull is 10% + 2% per year from the latest documented bollard pull [7]. Therefore, towing configuration is firstly being evaluated to obtain the actual value of bollard pull on each AHTS vessel.

2.3. Environmental Loads Calculation

Environmental loads are calculated to be the basis of calculating the required bollard pull. The load calculation process uses the calculation steps recommended by DNV -GL [8] [9] and [10] with formulas obtained from OCIMF, the Spanish Standard ROM 0.2-90, and Chakrabarti. It is divided in two conditions, the first one is when shuttle tanker is in connecting process to the FSO, and second is when shuttle tanker is already connected to the FSO.

Table 1. Principal Dimension of MT Mare Oriens

Name	MT Mare Oriens
Type	Oil tanker
Year of build	2008
Class register	RINA
Flag	Italy
IMO	93468773
Call sign	ICDY/247254100
Gross tonnage (GT)	59,611
Deadweight tonnage (DWT)	113,179 ton
Length overall (LOA)	245.55 m
Length between perpendicular (LPP)	235.72 m
Breadth (B)	42.01 m
Height (H)	21.51 m
Freeboard (full)	6.58 m
Freeboard (ballast)	14.68 m
Draught (full)	14.98 m
Draught (ballast)	6.89 m
Speed	16 knots

Transversal wind load:

$$F_{YC} = \frac{C_{YC} \times \gamma_W \times D \times L_{PP} \times V_C^2}{20000} \quad (1)$$

Longitudinal wind load:

$$F_{XW} = C_{XW} \times \rho \times A_L \times \frac{V_W^2}{20000} \quad (2)$$

Transversal wave load:

$$F_{Y\,wave} = C_{fw} \times C_{dw} \times \gamma_{dw} \times H_s^2 \times D'_x \sin\alpha \quad (3)$$

Longitudinal wave load:

$$F_{X\,wave} = C_{fw} \times C_{dw} \times \gamma_{dw} \times H_s^2 \times D'_x \cos\alpha \quad (4)$$

Transversal current load:

$$F_{YC} = \frac{C_{YC} \times \gamma_W \times D \times L_{PP} \times V_C^2}{20000} \quad (5)$$

Longitudinal current load:

$$F_{XC} = \frac{C_{XC} \times \gamma_W \times B \times L_{PP} \times V_C^2}{20000} \quad (6)$$

Transversal inertia load:

$$F_{VR} = \frac{W\left[\pm\frac{4\pi^2\theta_y}{T_R^2 g} + (1 \pm 0.2)\cos\theta\right]}{g \times 1000} \quad (7)$$

Longitudinal inertia load:

$$F_{HR} = \frac{W\left[\pm\frac{4\pi^2\theta_z}{T_R^2 g} + (1 \pm 0.2)\sin\theta\right]}{g \times 1000} \quad (8)$$

2.4. Bollard Pull Calculation

Bollard pull calculation should be performed to know the minimum specification of AHTS vessels to support shuttle tanker from the FSO. Bollard pull calculation is calculated using formula and recommendation from DNV-GL. The bollard pull calculation formula is as follows.

$$\gamma_{TE} = \frac{[80-(18-(0.0417\times LOA\times\sqrt{BP-20}))\times(H_s-1)]}{100} \tag{9}$$

2.5. Bollard Pull Calculation

Data verifying is performed to review whether the existing configuration is sufficient to assist shuttle tanker from the FSO in crude oil transfer. If existing configuration is insufficient, re-establishing a new configuration according to the result of this research should be done.

3. Result and Discussion

3.1. General Information

General information is served to ease the readers depicting the condition in the field. Figure 1 shows the weather direction that used in this research. The angle interval is 45 with 0^0 comes from bow of shuttle tanker. The X-positive axis towards starboard side of shuttle tanker and Y-positive axis towards the stern side of shuttle tanker.

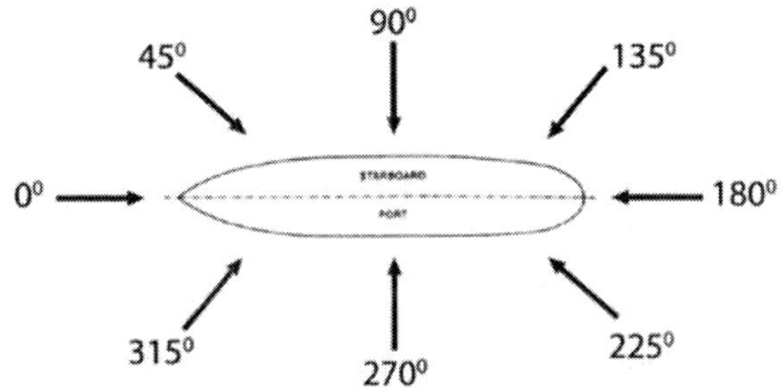

Figure 1. Weather direction

3.2. Actual Bollard Pull Calculation of Existing Configuration

The actual output of bollard pull from AHTS vessels will be reduced in accordance with its age. The reduction factor according to DNV-OS-H202 is 2% per year from the latest documented bollard pull [5]. The equation of actual bollard pull is shown as follows.

$$BP_{riil} = BP_c - (F_R \times BP_c \times n) \tag{10}$$

By using the equation (10), the value of actual bollard pull of each existing AHTS vessel can be acquired. Table 2 shows the actual bollard pull that has been reduced in accordance with its age using recommendation from DNV-GL.

From Table 2, it can be seen that the bollard pull of Catherine Queen is reduced with value of 10.2 MT and SMS Steady is reduced with value of 4.8 MT from what stated in the bollard pull certificates.

Table 2. Existing configuration

Name	Initial bollard pull (T)	Year of built	n	Reduction factor	Existing bollard pull (T)
Catherine Queen	85	2014	6	2%	74.8
SMS Steady	48	2015	5	2%	43.2

3.3. Defining Operations Window

The values of actual bollard pull in Table 2 will later be used as a reference to verify the sufficiency of exisiting configuration based on calculated environmental loads in this research. Due to the operation of crude oil transfer from the FSO lasts less than 72 hours, according to DNV-OS-H101: Marine Operations, General, the marine operation may normally be defined as weather restricted [7]. Weather restricted means that the environmental loads shall be calculated with defined design weather characteristic. The weather characteristic in this paper uses strong breeze characteristic according to *Beaufort Scale*. The design of operation window can be seen on Table 3.

Table 3. Operation window based on the weather condition

Parameter	Weather Directionn							
	0^0	45^0	90^0	135^0	180^0	225^0	270^0	315^0
Hs (m)	3	3	3	3	3	3	3	3
Wind (m/s)	13	13	13	13	13	13	13	13
Current(m/s)	1	1	1	1	1	1	1	1
Water Depth (m)	77							

3.4. Environmental Loads Calculation

This research divided the calculation of environmental loads into two conditions, when shuttle tanker is already connected to the FSO (loaded condition) and when shuttle tanker is in connecting process to the FSO (ballasted condition).

3.4.1. Loaded Condition. Figure 2a. shows the result of transversal environmental loads on loaded condition in all-weather direction from 0^0 to 315^0 with interval 45^0. The peak of wind load is recorded in weather direction comes from 90^0 with value -16.78 MT, wave load in weather direction comes from 270^0 with value 128.62 MT, and current load in weather direction comes from 90^0 and 270^0 with value -117.63 MT and 117.63 MT respectively.

Figure 2b. shows the result of longitudinal environmental loads on loaded condition in all-weather direction from 0^0 to 315^0 with interval 45^0. The peak of wind load is recorded in weather direction comes from 0^0 with value 9.99 MT, wave load in weather direction comes from 225^0 with value -80.10 MT, and current load in weather direction comes from 0^0 and 180^0 with value 17.76 MT and -17.76 MT respectively

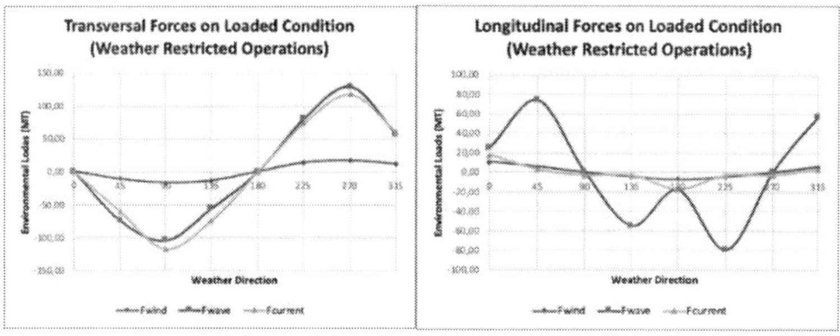

(a) (b)

Figure 2. Force on Loaded Condition

3.4.2. Ballasted Condition. Figure 3a. shows the result of transversal environmental loads on ballasted condition in all-weather direction from 0^0 to 315^0 with interval 45^0. The peak of wind load is recorded in weather direction comes from 270^0 with value 84.23 MT, wave load in weather direction comes from

270^0 with value 26.41 MT, current load in weather direction comes from 90^0 and 270^0 with value -40.79 MT and 40.79 MT respectively, and inertia load 10.59 MT.

Figure 3b. shows the result of longitudinal environmental loads on ballasted condition in all-weather direction from 0^0 to 315^0 with interval 45^0. The peak of wind load is recorded in weather direction comes from 0^0 with value 11.83 MT, wave load in weather direction comes from 225^0 with value -25.84 MT, current load in weather direction comes from 0^0 with value 26.90 MT, and inertia load with value 3.33 MT.

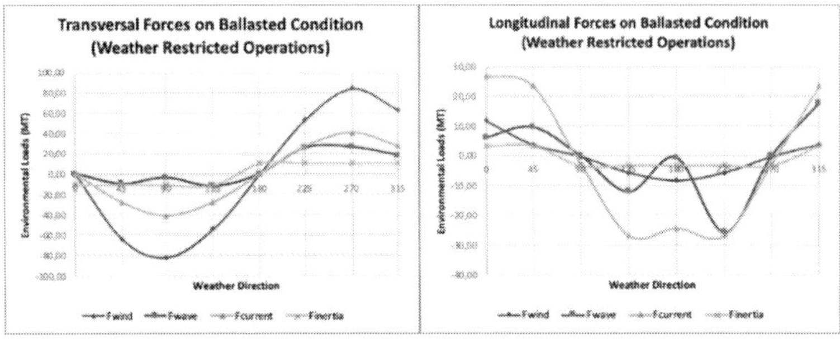

(a) (b)

Figure 3. Force on Ballasted Condition

3.5. Worst Case Scenario

After calculating environmental loads on all weather direction in two shuttle tanker conditions, the next step taken is to choose the weather criteria to be the basis of bollard pull calculation. The choosing of weather criteria options are looking at the worst possible scenario when the biggest wind, wave, and current load occurs in one axis. This step is taken to assure that the recommended operations of crude oil transfer can be overcame when worst weather occurs under the operations window. This research also takes the three biggest resultants of environmental loads into consideration.

Table 4 shows two worst cases of scenarios when the biggest total environmental loads occur under the same axis. Case 1 shows when the biggest total transversal load occurs. The biggest total transversal load occurs when wind, wave, and current load all come from 270°. Case 2 shows the biggest total longitudinal load occurs. The biggest total longitudinal load occurs when current comes from 225° and wind and current both come from 180°.

Table 5 shows three worst case scenarios when the biggest resultant of environmental loads occurs. Case 3 is when wind comes from 180° and both wave and current come from 270°. Case 4 is when all wind, wave, and current all come from 90°. Case 5 is when wind comes from 135° and both wave and current come from 90°.

Table 4. Total Environmental Loads in Worst Case Scenario

Variable		Case 1			Case 2		
		Value		Direction	Value		Direction
Fy	F_{wind}	16.72	MT	270^0	0.00	MT	180^0
	F_{wave}	128.62	MT	270^0	80.10	MT	225^0
	$F_{current}$	117.63	MT	270^0	0.00	MT	180^0
	F_{YT}	262.96	MT		80.10	MT	
Fx	F_{wind}	-0.33	MT	270^0	-7.58	MT	180^0
	F_{wave}	0.00	MT	270^0	-80.10	MT	225^0
	$F_{current}$	-3.30	MT	270^0	-17.76	MT	180^0
	F_{XT}	-3.63	MT		-105.44	MT	

Table 5. Resultants of Environmental Loads in Worst Case Scenario

Variable		Case 3			Case 4			Case 5		
		Value		Direction	Value		Direction	Value		Direction
Fy	F_{wind}	0.00	MT	180^0	-16.78	MT	90^0	-13.61	MT	135^0
	F_{wave}	128.62	MT	270^0	-104.10	MT	90^0	-104.10	MT	90^0
	$F_{current}$	117.63	MT	270^0	-117.63	MT	90^0	-117.63	MT	90^0
	F_{YT}	246.25	MT		-238.51	MT		-235.34	MT	
Fx	F_{wind}	-7.58	MT	180^0	-0.33	MT	90^0	-4.66	MT	135^0
	F_{wave}	0.00	MT	270^0	0.00	MT	90^0	0.00	MT	90^0
	$F_{current}$	-3.3	MT	270^0	-3.30	MT	90^0	-3.30	MT	90^0
	F_{XT}	-10.88	MT		-3.63	MT		-7.96	MT	
Fr		246.49	MT	45^0	238.54	MT	45^0	235.46	MT	45^0

The new configuration is designed under consideration of all five worst case scenarios. Figure 4 shows new recommended configuration with AHTS vessel A is placed in stern of shuttle tanker to counteract transversal and longitudinal load and AHTS vessel B is placed in port side or starboard side of shuttle tanker to counteract merely transversal load. The manoeuvring angle of AHTS vessel A is designed with maximum angle of 600. The designed safety factor to calculate the required bollard pull is determined with value of 1.2.

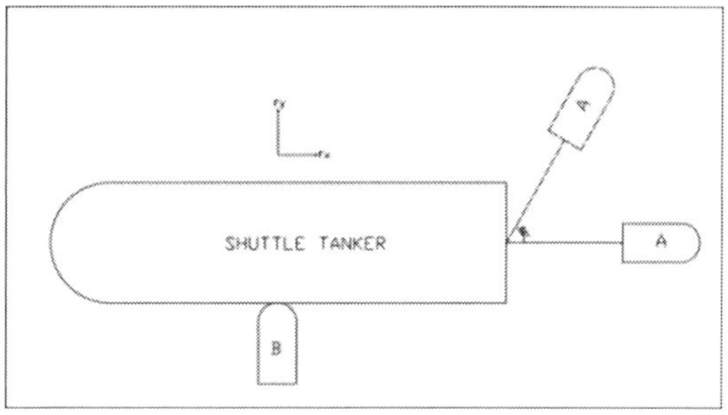

Figure 4. New Configuration

Table 6 shows the specification of AHTS vessels that are available in market in new configuration. BP_{spec} is the bollard pull of AHTS vessel according to what stated in certificate. BP_{riil} is the bollard pull of AHTS vessel after getting reduced with towing efficiency according to DNV-GL recommendation. Fx is the bollard pull output of AHTS vessel in x axis to counteract longitudinal environmental loads. Fy is the bollard pull output of AHTS vessel in y axis to counteract transversal environmental loads.

Data verification is needed to validate whether the existing and new configuration is sufficient to assist shuttle tanker. Existing condition is compared with new condition to verify with requirement under five worst scenarios.

Table 7 shows whether existing condition and new scenario are sufficient to support shuttle tanker under five worst case scenarios. Requirement column indicates the required minimum bollard pull after getting multiplied with safety factor. It can be seen from Table 7 that existing condition with Catherine Queen and SMS Steady is insufficient to support tanker in designed operation window. Contradict to

IOP Conf. Series: Materials Science and Engineering 1052 (2021) 012027 doi:10.1088/1757-899X/1052/1/012027

the existing condition, new condition with 200 MT and 220 MT AHTS vessels is sufficient to support shuttle tanker in all five worst scenarios.

Table 6. Resultants of Environmental Loads in Worst Case Scenario

Variable	Vessel Name		Unit
	AHTS A	AHTS B	
BP_{spec}	200	220	MT
BP_{riil}	164.28	180.71	MT
M/E	4 x 3060	2 x 4000	kW
		2 x 3000	kW
LOA	89.1	85.2	M
B	22	22	M
T	7	7.6	M
Speed	16.4	17	Knots
DWT	3700	4500	MT
Fx	82.14	0.00	MT
Fy	142.27	180.71	MT

Table 7. Verification of New and Existing Condition

Case	Requirement		Existing Condition		New Scenario		Unit
	Fy	Fx	Fy	Fx	Fy	Fx	
Case 1	315.55	4.36	88.69	30.72	322.98	82.14	MT
Case 2	96.12	126.53	35.48	61.44	180.71	164.28	MT
Case 3	246.69		80.18		255.56		MT
Case 4	238.54		80.18		255.56		MT
Case 5	235.46		80.18		255.56		MT

4. Conclusion

Based on the result of environmental loads and bollard pull calculation in this research, the following conclusions are obtained:

1) According to environmental loads calculation in worst case scenario under weather restricted operations, the recorded biggest transversal load is 262.96 MT with all weather directions from 270^0. And for the biggest longitudinal load is -105.44 MT with weather directions from 180^0 and 225^0.
2) The bollard pull calculation indicates that the existing condition with Catherine Queen and SMS Steady is insufficient to support shuttle tanker from the FSO under designed operations window. Thus, a new scenario is established with 200 and 220 MT AHTS vessels.
3) The new recommended towing configuration in this research is 200 MT AHTS vessel in stern of shuttle tanker to counteract transversal and longitudinal load and 220 MT AHTS vessel in port side or starboard side of shuttle tanker to counteract merely transversal load.

5. References

[1] PwC Indonesia. Investment and Taxation Guide 2019 - 10th Edition. 2019.
[2] Shimamura Y. FPSO/FSO: State of the art. Journal of Marine Science and Technology. 2002 Sep 1;7(2):59–70.

[3] Sanchez-Mondragon J, Vázquez-Hernández AO, Cho SK, Sung HG. Yaw motion analysis of a FPSO turret mooring system under wave drift forces. Applied Ocean Research. 2018 May 1;74:170–87.
[4] Mentes A, Helvacioglu IH. Fuzzy decision support system for spread mooring system selection. Expert Systems with Applications. 2012 Feb 15;39(3):3283–97.
[5] Ariana IM, Akbar TA, Prakosa MSB, Prananda J, Tyasayumranani W. Conceptual design of dynamic positioning system for tugboat to improve safer operation in Indonesia, case study: port of Cilacap. IOP Conference Series: Earth and Environmental Science. 2020 Sep 15;557:012067.
[6] Chakrabarti S, editor. Handbook of Offshore Engineering. Vol. 1. Elsevier Science; 2005.
[7] DNV-GL. Sea Transport Operations (VMO Standard - Part 2-2). Sect. Section 4: Towing Jun, 2016.
[8] DNV-GL. Rules for Classification. Ships. Sect. Part 3: Hull Jan, 2017.
[9] DNV-GL. Marine Operations, Design, and Fabrications. DNV-OS-H102 Jan, 2012.
[10] DNV-GL. Marine Operations, General. DNV-OS-H101 Oct, 2011.

Tidal Flow Renewable Energy Potential In The Capalulu Strait

C Kusuma
Sekolah Tinggi Teknologi Angkatan Laut, Surabaya, Indonesia

Abstract. The use of renewable energy that is environmentally friendly and without exhaust emissions will continue to be developed. Such as the utilization of kinetic energy from tidal currents into electrical energy with hydrokinetic turbines. The potential of the tidal current is one of the renewable energy sources given the very large majority of Eastern Indonesia is the ocean and not used optimally. The purpose of writing this paper as a reference counting tidal currents in the islands and strait Indonesia to convert electrical energy this paper calculates the potential of tidal electrical energy in the Capalulu Strait, North Maluku based on tidal stream tables. It is hoped that this paper will serve as an alternative energy source for the development of electrical energy in the province of North Maluku, especially the Sula Islands District (Sulabes, mangoli, and Taliabu) and Taliabu District.

1. Introduction

The eastern part of Indonesia has a very deep sea, but some straits have shallow depths so that it is very potential to be developed as renewable energy by utilizing tidal currents. The availability of electricity will accelerate development in an area. The availability of electricity can use fossil energy (diesel or steam turbines) or renewable energy. One area that needs to be developed is the potential for tidal currents in the Capalulu Strait. The Capalulu Strait is located between 2 districts (Taliabu and Sula Islands) and is located between 3 islands (Taliabu, Mangoli, Sulabes). The availability of electricity in the islands of Sula and Taliabu is not 100%. Even if it is flowing, it is not uncommon for rotating blackouts to occur. The Capalulu Strait is a very strong tidal strait because it is a narrow strait that connects the Pacific Ocean and the Indian Ocean. This strait has the potential to be developed as renewable electricity by utilizing tidal currents.

Currently, on Taliabu Island, only 3 districts have been electrified, namely, Taliabu Barat mines (Holbota, Kawalo, Woyo, Limbo and Lohobubaq), Taliabu Northwest, and Lede District but the conditions have not been so stable due to rotating blackouts [1] [2]. Whereas in 5 Pulau Taliabu Subdistricts namely, North Taliabu, East Taliabu, South Taliabu, Tabona Subdistricts, and South East Taliabu Districts, electricity is not evenly available. [3] [4] [5]The tidal range is the difference in water level at the time maximum tide with minimum low tide, average value ranges from 1-3 m. In several Indonesian seas such as Tanjung Priok(Jakarta), the tidal range is only 1 m, Ambon is about 2 m, Bagan Siapiapi about 4 m, while the highest tide is at the estuary of the Digul River and Muli Strait (southern Papua), which reaches 7-8 m, the strait in the islands of Nusa Tenggara, which reach 2.5 to 3 meters. The straits in East Nusa Tenggara Province are the examples where the flow speed of the marine currents between the islands typically have strong velocities in the range of 2-4 [m/s] or more[6]. The strongest tidal currents recorded in Indonesia are in the Strait Capalulu, between Taliabu Island and Mangole Island, which got the power2up to 5 m[7]. [8]

Figure 1. Capalulu Strait(Google maps).

Tidal currents are currents that are formed due to tidal phenomena. Tides generate ocean currents through the resulting difference in water level, in which case the ocean currents move from a higher water level to a lower one. Based on their direction, tidal currents can be divided into two, namely tidal currents moving towards the land, which occurs during low tide towards the tide and tidal currents that move away from the land, which occurs when the tide goes towards low tide. However, in this paper, the two sides are considered to be the same because it is assumed to use a tidal turbine so that they can adjust to the current characteristics. Tidal current energy has advantages over other types of renewable energy, which is easy to predict. So that it can simplify maintenance, because it can be known with certainty how much electrical power it can generate from time to time. This is very important to know because it is used to find out how much replacement energy must be added to maintain grid stability The amount of potential electric power generated by ocean currents is stated in 4 categories, namely theoretical, technical, practical, and accessible potential. The theoretical potential is a rough estimate of the energy sources of ocean currents in a particular location. The technical potential is an estimate that already involves the efficiency of the tools to be used. The practical potential is an estimate that already involves technical matters that will be faced, such as wave and seabed conditions at the location. The accessible potential is an estimate that already involves accessibility to the location where the ocean current power generation system will be built, such as shipping lanes, etc. in this paper will only discuss the theoretical tidal current potential.

The theoretical potential energy of tidal currents in the Capalulu Strait can be determined by calculating the velocity using the list of tidal currents in the Capalulu Strait issued by the Navy Hydrographic and Oceanographic Service of the Indonesian Navy[9]

2. Theoretical Potential Of Tidal Current

2.1. Potential Tidal in the Capalulu Strait

The Capalulu Strait can be divided into 2 areas, namely on the edge of a depth of 10-20 meters as shown in Figure 2, which has the potential for tidal flow turbines. The center of the strait has a channel depth of 50-77 as a shipping route. The width of the narrow strait connecting the Pacific Ocean and the Indian Ocean in the south makes the tidal currents in the Capalulu Strait very high and even throughout the year. The width of the Capalulu allows the construction of tidal turbines without disturbing shipping.

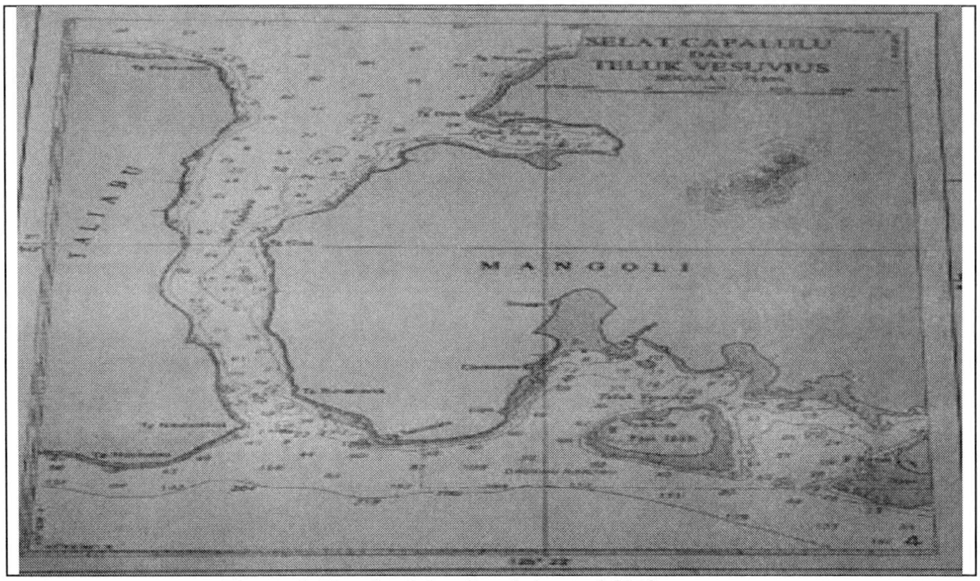

Figure 2. Depth of the Capalulu strait[10].

2.2. Marine Turbine Flow
The estimated global potential for tidal energy is around 500 GW, as shown in Figure 1hTe estimated global potential for tidal energy is around 500 GW, as shown in Figure 3.

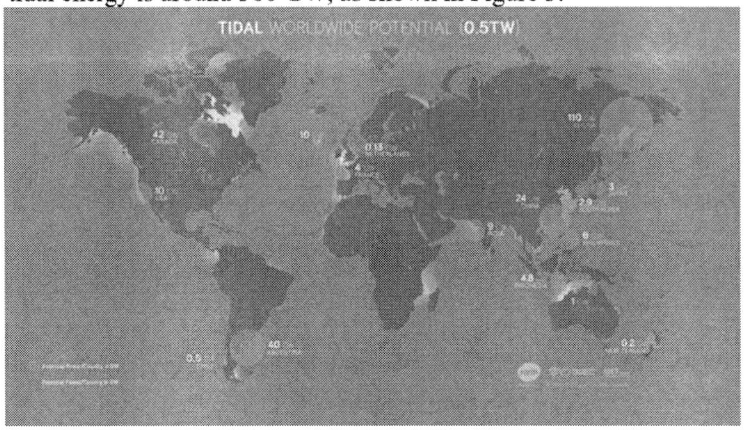

Figure 3. Tidal Worldwide Potential 0.5TW[12].

According to[13] These relatively rapid tidal currents typically have peak velocities at spring tide in the region of 2–3 m=s (4–6 knots) or more. Some locations, such as the Pentland Firth (between Scotland and Orkney) and the Alderney Race (between the Channel Islands and France), or the so-called Big Russell (off Guernsey), have especially intense currents over a large sea area to offer the potential for exploitation on a multi gigawatt scale. Examples of other intense locations include the Severn Estuary (UK's North Devon coast), the straits between Rathlin Island and Northern Ireland, the traits of Messina between Italy and Sicily, and various channels between the Greek islands in the Aegean. Other large marine current resources can be found in regions such as southeast Asia (e.g. the Philippines, Indonesia, Japan), Australia and New Zealand, Canada, and the southeast coast of South Africa. According to Fraenkel [13], the waters in Indonesia have the potential to be developed as a

IOP Conf. Series: Materials Science and Engineering 1052 (2021) 012028 doi:10.1088/1757-899X/1052/1/012028

tidal flow turbine. There are several options for installing turbines in the Capalulu Strait as shown in Figure 4.

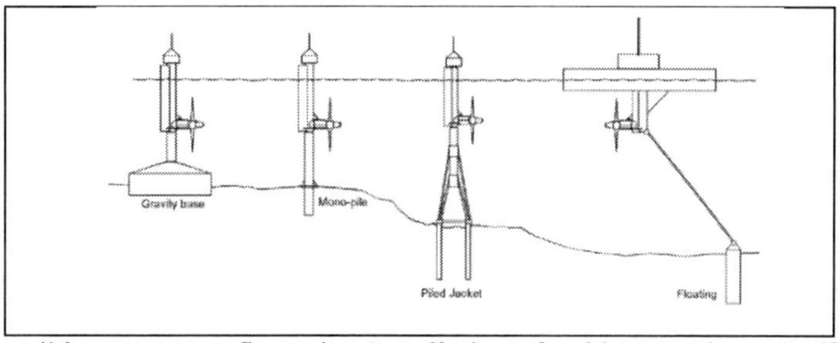

Figure 4. Possible system configuration Installation of turbines on the seabed[12].

In this paper, it is assumed to use a tidal turbine with 3 blades with the assumption that the blade radius is 1 m and the surface area of the blade is $1 m^2$. the electrical system is installed in the turbine as shown in Figure 5.

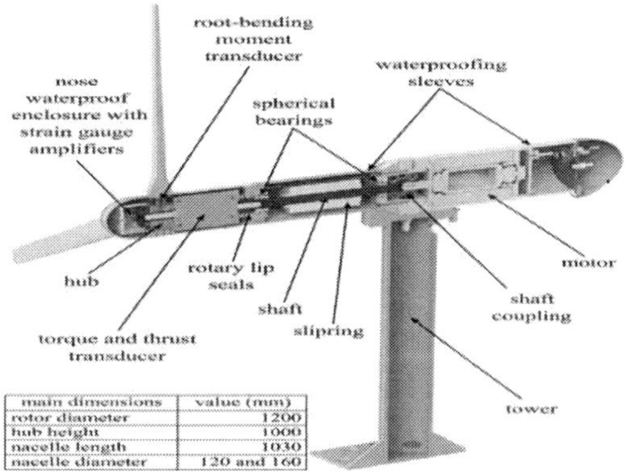

main dimensions	value (mm)
rotor diameter	1200
hub height	1000
nacelle length	1030
nacelle diameter	120 and 160

Figure 5. Small turbines tidal with electricity system[14].

3. Methodology

Assuming the fluid velocity is a discontinuity in the vertical section of the turbine, the average mass of the turbine flow velocity can be estimated : [11]

$$A = \rho * A_r \frac{(V_1 + V_0)}{2} \tag{1}$$

The results of the substance of equation (1) above for the maximum power that can be extracted from the fluid concerning the velocity of the upstream fluid only is like equation (2) [11]

$$P = \text{Cp} * 0.5 * \rho * A_r * V^3 (\text{Watt}) \tag{2}$$

Where:

P = power (watts)
ρ = density of seawater (1025 kg / m^3)
A = cross-sectional area (m^2)
V = flow velocity (m / s)

Cp is the power coefficient which depends on the velocity of the fluid downstream to the upstream which has a theoretical maximum value of 16/27 or 0.59 based on the Bentz limit

4. Result and Discussion

From the tidal stream tables obtained from the Indonesian Navy Hydrographic and Oceanographic Service of the Navy, 2015, [9] the literature shows the potential value of electric current energy in the Capalulu Strait in January 2015 in Figure 6.

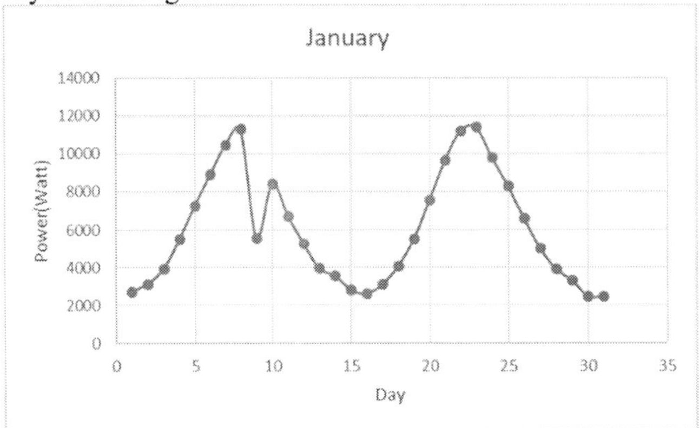

Figure 6. The potential flow of Electrical tidal currents January 2015.

The potential value of electric current energy in the Capalulu Strait in February 2015 in Figure 7.

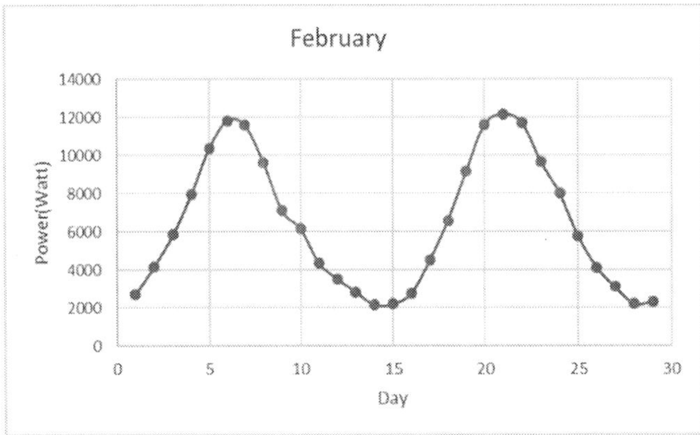

Figure 7. The potential flow of Electrical tidal currents February 2015.

The potential value of electric current energy in the Capalulu Strait in March 2015 in Figure 8.

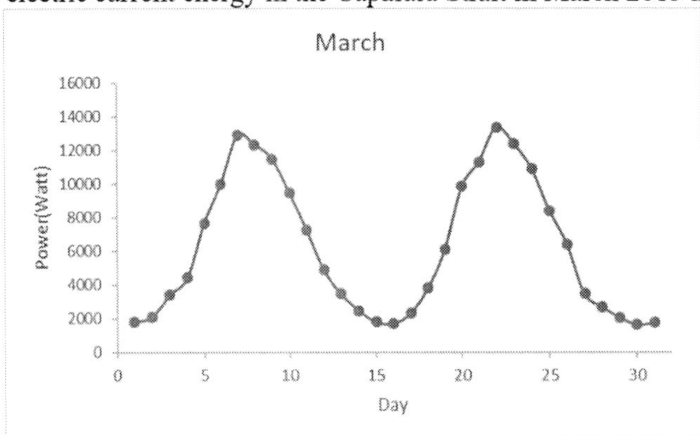

Figure 8. Potential Data of Electrical tidal currents March 2015.

The potential value of electric current energy in the Capalulu Strait in April 2015 in Figure 9.

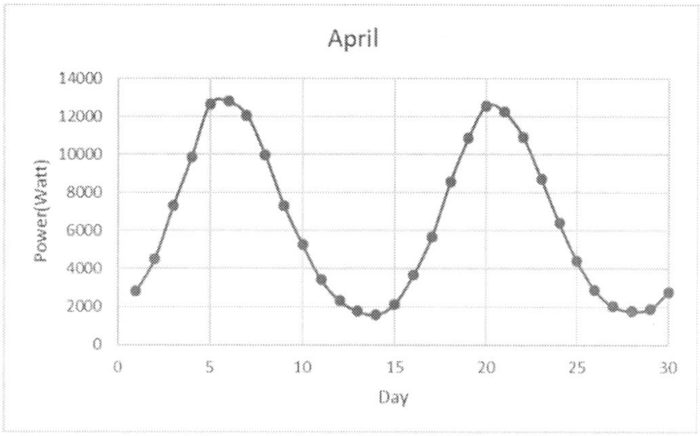

Figure 9. Potential Data of Electrical tidal currents April 2015.

The potential value of electric current energy in the Capalulu Strait in May 2015 in Figure 10.

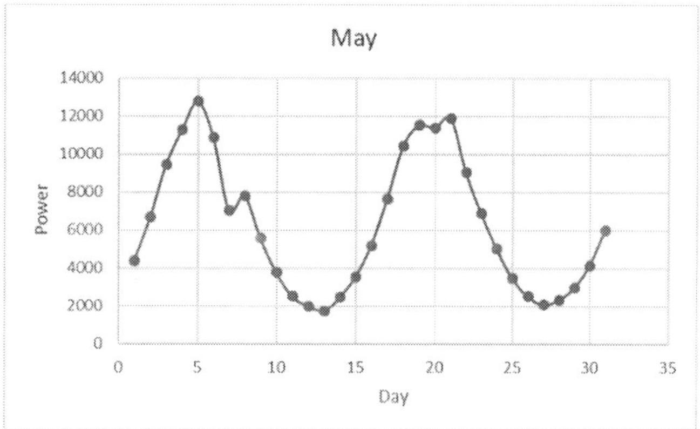

Figure 10. Potential Data of Electrical tidal currents May 2015.

The potential value of electric current energy in the Capalulu Strait in June 2015 in Figure 11.

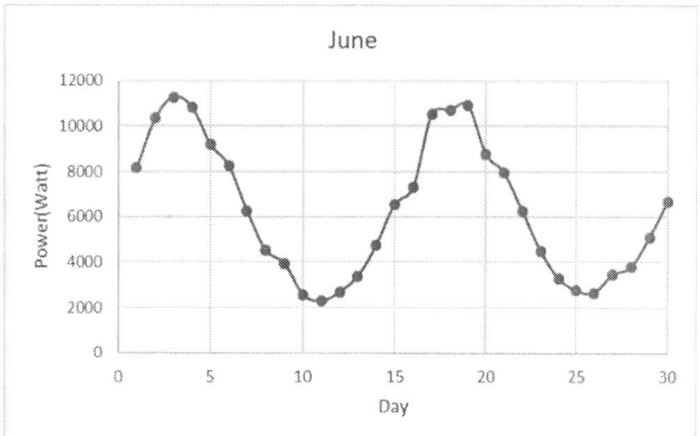

Figure 11. Potential Data of Electrical tidal currents June 2015.

IOP Conf. Series: Materials Science and Engineering 1052 (2021) 012028 doi:10.1088/1757-899X/1052/1/012028

The potential value of electric current energy in the Capalulu Strait in July 2015 in Figure 12.

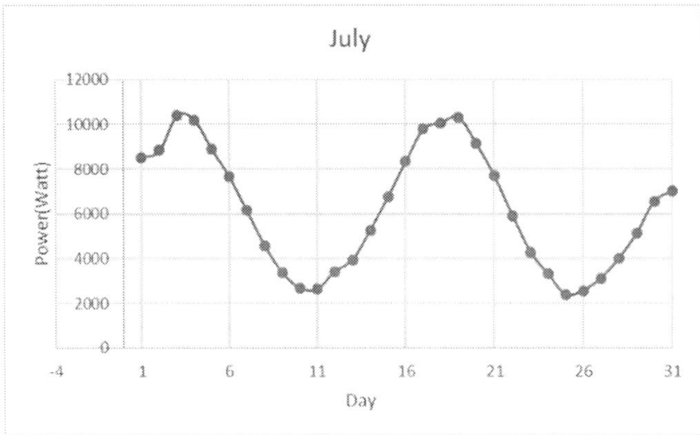

Figure 12. Potential Data of Electrical tidal currents July 2015..

The potential value of electric current energy in the Capalulu Strait in August 2015 in Figure 13.

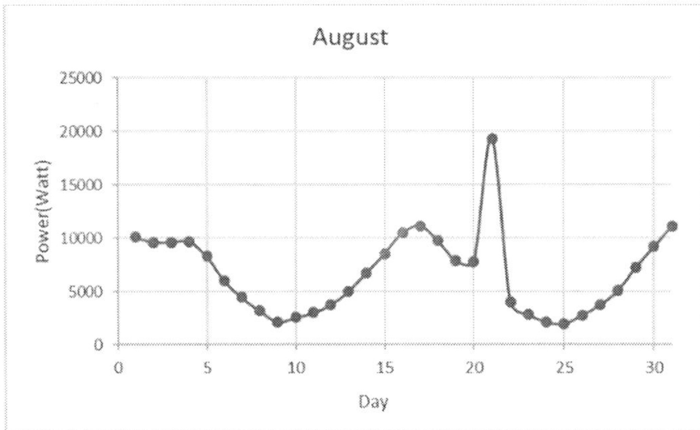

Figure 13. Potential Data of Electrical tidal currents August 2015.

The potential value of electric current energy in the Capalulu Strait in September 2015 in Figure 14.

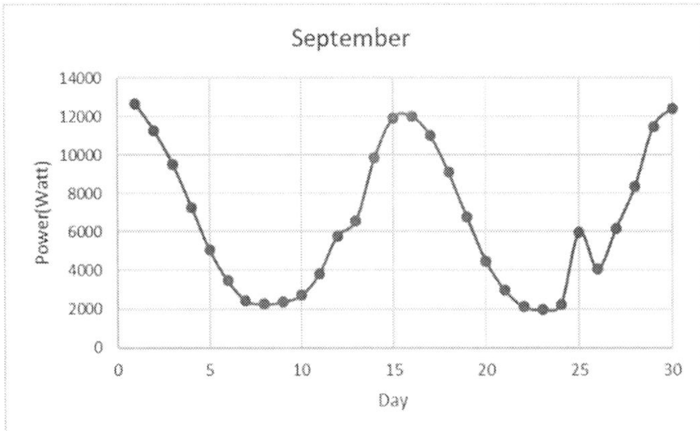

Figure 14. Potential Data of Electrical tidal currents September 2015.

The potential value of electric current energy in the Capalulu Strait in Oktober 2015 in Figure 15.

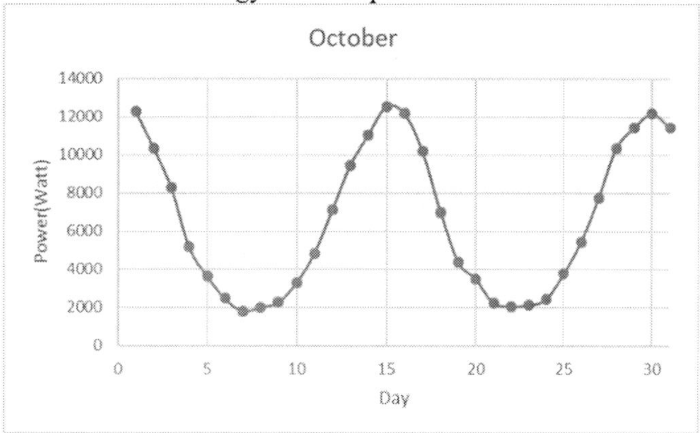

Figure 15. Potential Data of Electrical tidal currents Oktober 2015.

The potential value of electric current energy in the Capalulu Strait in November 2015 in Figure 16.

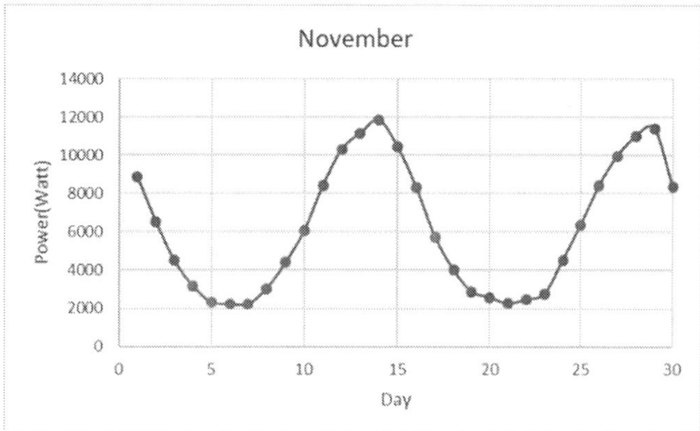

Figure 16. Potential Data of Electrical tidal currents November 2015.

The potential value of electric current energy in the Capalulu Strait in December 2015 in Figure 17.

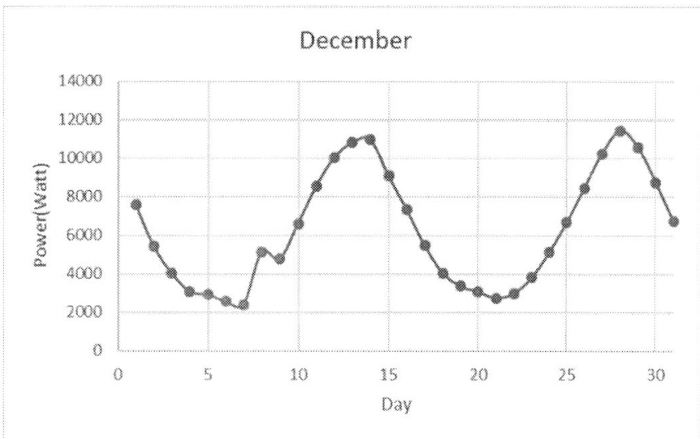

Figure 17. Potential Data of Electrical tidal currents December 2015.

Based on the literature studies that have been obtained, some of the potential of ocean current energy in the Strait of Capalulu can be displayed in the form of a table of average current velocity per month and electricity potential in 2015, between January to December in Table 1 and figure 18.

Table 1. Potential Data of Electrical tidal currents 2015.

Month	Mean Velocity Tidal (Knot)	Power (Watt)
January	3.283	5534.8
February	3.361	6563.3
March	3.241	5323.4
April	3.278	5506.8
May	3.298	5611.9
June	3.343	5840.9
July	3.338	5813.9
August	3.386	6067.6
September	3.352	5889.1
October	3.336	5804.1
November	3.299	5611.8
December	3.341	5829.4
Mean	3.321	5783.1

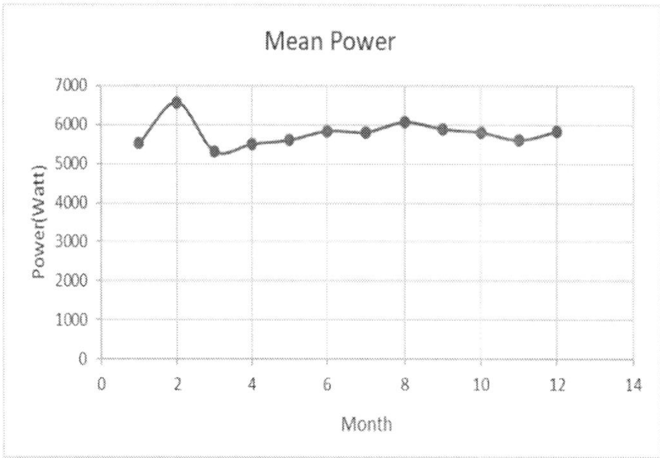

Figure 18. Potential Data of Electrical tidal currents 2015.

Based on the table and graphic images above, the average electrical tidal current velocity in the Capalulu Strait is 3.3 knot or 6.1 m / s with an average potential of 5783 Watt / Hour. The maximum current occurs in February with an average current speed of 3.36 Knot and an electric potential of 6563 Watt / Hour, while the weakest current is in March with an average speed of 3.24 miles/hour and an electric potential of 5323 Watt/hour. From the data and calculations also obtained the potential for evenly distributed flow throughout the year so it is very potential to be developed.

5. Conclusion

Base on the results of the analysis of the discussion above it can be Conclusion:

a. By using 1 tidal turbine with 3 blades, assuming the blades' radius is 1 m and the surface area of all blades is 1m2, an average electric current will be 5783 Watt / Hour. The maximum electric current is obtained in February with an electric potential of 6563 watts, while the weakest electricity potential is in March with an electric potential of 5323 watts.

b. From the data and calculations also obtained the potential for evenly distributed flow throughout the year so it is very potential to be developed.

6. References

1. http://beritamalut.co/2020/08/02/pengusaha-ayam-potong-di-bobong-merugi-karena-listrik-sering-padam/
2. Lentera malut, minggu oktober 18, 2020, https://www.malut.lentera.co.id/2020/09/16/krisis-listrik-pemuda-desa-bobong-minta-pln-maluku-malut-datangkan-mesin-baru-ke-taliabu/
3. Fokus Malut, Minggu, 04 Oktober 2020, https://metro7.co.id/5-kecamatan-di-taliabu-belum-merdeka-listik/.
4. Koran metro, 07 October 2020,10.55 WIB https://metro7.co.id/5-kecamatan-di-taliabu-belum-merdeka-listik/
5. https://www.skrinews1.com/2020/10/skrinews1-5-kecamatan-di-taliabu-belum.html
6. Erwandi, 2011, "The Development Of Indonesian Vertical Axis Marine Current Turbine For The Tidal Power Generation", Indonesia Marine
7. ITB Indonesian Journal of Geospatial Vol. 05, No. 1, 2016, 27-39, Mapping of Sea Current Power Potential in The Strait of Capalulu, The Sula Archipelago, North Maluku
8. Rokhmin Dahuri, 2004 "Pengelolaan Sumber Daya Wilayah Pesisir dan Lautan Secara Terpadu", Pradnya Paramitha, Jakarta
9. Indonesian Navy Hydrographic and Oceanographic Service of the Navy, 2015, List of Tidal Stream Tidal Streams in 2015, Jakarta
10. Indonesian Navy Hydrographic and Oceanographic Service of the Navy, 1993, Map of Indonesia no 10 1993, Jakarta
11. Sandor Bernhard, et.al, (2008), "Flow Investigating in Archard Turbine". The Publishing House of Romanian Academy. Proceeding of Romanian Academy Series A, Volume 9, Number 2/2008, pp 000-000
12. Faisal Wani, Jianning Dong, and Henk Polin, "Tidal Turbine Generators" Submitted: April 23rd 2019Reviewed: November 10th 2019Published: April 1st, 2020, DOI: 10.5772/intechopen.90433
13. P L Fraenkel Marine Current Turbines Limited, 2 Amherst Avenue, Ealing, London W13 8nq, Uk
14. Gregory S. Payne, Tim Stallard b, Rodrigo Martinez,www.elsevier.com/locate/renene,renewable energy 107(2017)312-326 "Design and manufacture of a bed supported tidal turbine model for blade and shaft load measurement in turbulent flow and waves"

Acknowledgments

Authors who want to thank the help or encouragement of colleagues, especially the Sekolah Tinggi Teknologi Angkatan Laut (STTAL). Acknowledgments that are not numbered immediately after the last numbered section of this paper.

Optimum Propeller Interaction Analysis of Harbourtug With Azimuth Thruster

S D Utami[1,2], I M Ariana[1], A Z M Fathallah[1], I P S Asmara[2]
[1]Department of Marine Engineering, Faculty of Marine Technology, Institut Teknologi Sepuluh Nopember, Sukolilo, Surabaya 60111
[2]Study Program of Ship Design and Construction, Department of Shipbuilding, Politeknik Perkapalan Negeri Surabaya, Sukolilo, Surabaya 60111

Abstract. Determine the safety of tug operation at sea, especially in narrow shipping areas such as port areas, it is necessary to pay attention to the maneuverability of ships and the performance of the ship's steering system. In ship maneuvering, a specially designed thruster with 360 degrees side thrust with a small turning angle is needed in order to produce a good ship movement performance. In this study, the authors designed a tugboat propeller ka-series four blades. The propeller can move 360 degrees so that the distribution of power is evenly distributed in all directions. The system is expected to provide extra thrust when needed in the process of maneuvering and docking the ship. Equipped with hydraulics used to quickly lift and lower the thruster and pull it back into the hull when not in use. The analysis results show that each propeller variation such as the difference in thrust value, torque value and resulting efficiency. From the analysis model, the highest thrust value was obtained in the first propeller model with a variation of P/D 1, AE/A0 0.45 valued at 306201.4. The lowest torque value is generated in the third propeller model with a variation of P/D 0.6, AE/A0 0.65 value at -22400.2 Nm. Harbourtug requires a high thrust value, so it is advisable to use the first Propeller model with variations of P/D 1, AE/A0 0.45 The highest efficiency value is obtained in the first propeller model valued at 0.7151.

1. Introduction

Transportation is the activity of moving people or objects from one place to another which takes place in a space. Transportation acts as a support for economic, social, cultural and political activities, so it can strengthen national unity and integrity. The development of transportation is influenced by the nature and level of human life, so it is said that transportation is the cause and effect of the progress of human civilization. Sea transportation mode is more efficient and profitable to carry weight in large quantities for archipelagic areas such as Indonesia [1].

Ships are an important means of sea transportation in the maritime world to promote trade from within and outside the country which has a higher economic value. Increased transportation activities also have an impact on increasing marine incidents and accidents in Indonesia which must be the concern of all parties, not only ship owners but also government, related agencies and the community [2]. Many aspects affect the success of ship operations, namely hull planning, ship engineering system planning, propulsion system planning, calculation of ship maneuverability and other planning [3].

Ships can operate properly if they have a propulsion system design that can overcome total resistance that occurs in order to maintain the service speed as planned [4]. One of the components of the ship propulsion system is a propeller. The propeller transfers powered by changing the rotational force of the propeller into thrust to move the hull with the mass of water by rotating the axial blades on the shaft [5]. The development of the propeller design of course aimed at making the speed of a ship more optimal and effective[6].

Optimization of the propulsion system which is one of the most important parts of a ship because it is closely related to speed. The propulsion system located at the stern of the ship will have a significant impact on the flow from the ship. The irregular distribution of the flow to the propeller can have an influence on the existing propulsion speed [7]. A tugboat is a type of guide ship commonly used to tow and push large ships to enter the port pool area [8].

In the following section, reviews of previous studies of propeller performance scaling methods in general was presented. Next, the specific cases investigated in this study and the methodology used were discussed, followed by the main results. Therefore, in order to assist in proper mooring or berthing operations, a large amount of energy is required that is generated from the propeller thrust.

2. Study area and parameter description

2.1. Characteristics Propeller

An object to be able to move requires a thrust to move the object. Likewise, a ship moving in the waters requires a thrust force that functions to push the ship to be able to move against the forces which in the world of education are called the resistance force of the media through which it passes. Meanwhile, the thrust required to fight this resistance force is known as the propulsive force. In general, the characteristics of the propeller in the open water test as represented in the KT - KQ - J diagram [9]. Each type of ship propeller has different characteristics of performance curves. So that the study of the propeller characteristics of a ship cannot be generalized to the whole shape or type of propeller. The equation model for the performance characteristics of a ship's propeller as follows:

$$K_T = \frac{T}{\rho n^2 D^4} \quad (1)$$

$$K_Q = \frac{Q}{\rho n^2 D^5} \quad (2)$$

$$J = \frac{V_a}{nD} \quad (3)$$

Where KT is the thrust coefficient, KQ is the torque coefficient, J is the advance coefficient, Va is the speed of advance, D is the diameter propeller, N is the rotational speed, T is the thrust propeller, Q is the torque propeller, ρ is the density of the fluid.

2.2. Ducted Propeller

The duct thrust produced arises from the suction force on the curved duct surface. Ducted propellers generally comprise two principal components. The first is an annular duct having an aerofoil cross-section. The second component is a special case of a non-ducted propeller in which the design of the blades has been modified [10]. The propellers for these units can be of either the fixed or controllable pitch type.

Ducted propellers, sometimes referred to as kort nozzles by way of recognition of the Kort Propulsion Company's initial patents and long association with this type of propeller, have found application for many years where high thrust at low or zero ship speed is required as in towing and trawling situations. In these situations, the duct generally contributes a half of the propulsor's total thrust at zero ship speed commonly referred as bollard pull condition [9]. However, this relatively large contribution by the duct falls to more modest proportions with increasing ship speed toward the free running condition, then at very high speeds the propulsor thrust receives a negative contribution from the duct[11].

2.3. *Azimuth Thruster*

Azimuth Thruster is the configuration, which is used in marine vessels to provide necessary thrust in desired direction which give ships better manoeuvrability than fixed propellers and rudder systems. Azimuth thrusters have been in common use for many years and are configured in either non-ducted or ducted propeller arrangements. These units fall into two distinct classes: the first is where a propeller is mounted on a rotatable pod beneath the ship and the second is the Voith Schneider or Kirsten-Boeing propulsion concept, the second where a propeller is mounted on a pod beneath the ships [12].

2.4. *Controllable Pitch Propeller*

The propeller blades are controlled to provide more manoeuvrability to change the blade pitch, the shaft speed can be maintained constant by engaging the generator thereby reducing the variable number of operations. The controllable pitch propeller has found application in most of the propeller types and arrangements so far outlined in this chapter, with the possible exception of the podded propulsors, contrarotating and tandem propellers. The total set of characteristics comprises sets of KT and KQ versus J curves comprising the design pitch angle. When analysing the performance of a controllable pitch propeller at off-design conditions, use should not be made of fixed pitch characteristics beyond say 5 or 10 degrees from design pitch since the effects of section distortion [9]. Different with the single line for fixed pitch propellers, the performance characteristics can be considered as surface forming.

The controllable pitch propeller needs to have greater mechanical complexity. It has several important advantages like maneuver through pitch adjustment and in cases where constant speed engines are used and operational changes can be achieved without accelerating and slowing down the propulsion engine system [13]. In the case of a controllable pitch propeller, there is a limit to the extent of the blade area due to the requirement of passing blade to obtain full reversibility of the blades [14].

3. Methodology

The methodology of this study consisted of several stages as presented on Figure 1.

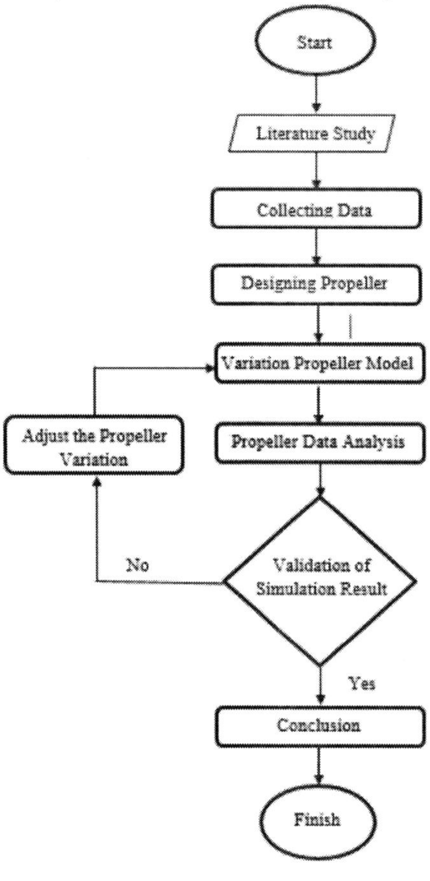

Figure 1. Research flowchart.

3.1 Collecting Data

In this step, the necessary supporting data is collected, including lines plan drawings, general arrangements and other data required according to the analysis carried out, namely maneuver. Apart from that, rules from the class and other supporting data are also needed.

3.2 Resistance Calculation

The calculation steps of the vessel resistance related to the pod body, body-strut interface, and the effect of lift are carried out to stir the propeller flow in succession. This calculation is performed to determine the required power, as well as the dimensions of the thruster to be designed. The Holtrop method is an algorithmic method that is useful for predicting the resistance of tankers, general cargo, fishing vessels, tugs and container ships.

3.3 Propeller Selection

Selection of the specification and type of thruster must consider the calculation of the power of the thruster and also the calculation of all ship resistance using the Holtrop method with the help of maxsurf resistance software. Based on the calculation, the specification can be searched through the thruster catalog according to the required power.

3.4 Designing the Optimal Propeller
In this study using the main propeller data in modeling the Propcad software. After finding the optimal propeller type and design, it is necessary to analyzing data.

3.5 Variation Data Analysis
Analysis in open water condition and analyses the distribution of fluid flow behind the propeller. The variation used are Ae / Ao value, the P / D value and J value.

4. Result

4.1 Model Design
In this study using the main propeller data in modeling the PropCad software. PropCad is software for the geometric modeling of marine propellers, it can generate transverse and profile views, pitch distribution, geometric information, as well as the basic propeller parameters. the results of the visualization of the geometric design as shown in Figure 2.

Figure 2. Propeller Ka-series Geometric.

4.2 Analysis of propeller performance
Propeller model is analyzed for thrust and torque values using the CFD method, giving variations to the predetermined variables. The fixed variable of this study is the propeller, while the variables varied are the Ae / Ao value, the P / D value and J value.

4.2.1 Propeller data analysis 1

The obtained streamline results for P/D 1 and Ae/A0 0.45 were presented on Figure 3, 4, 5, 6, 7 and 8.

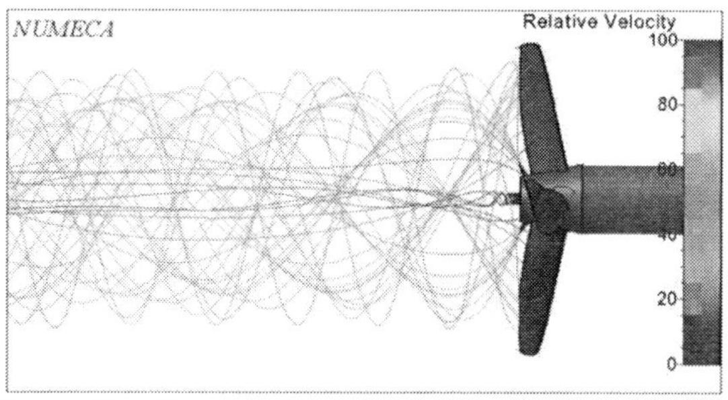

Figure 3. Streamline Result of Ka4-45 with J 0.1, P/D 1 and Ae/A0 0.45.

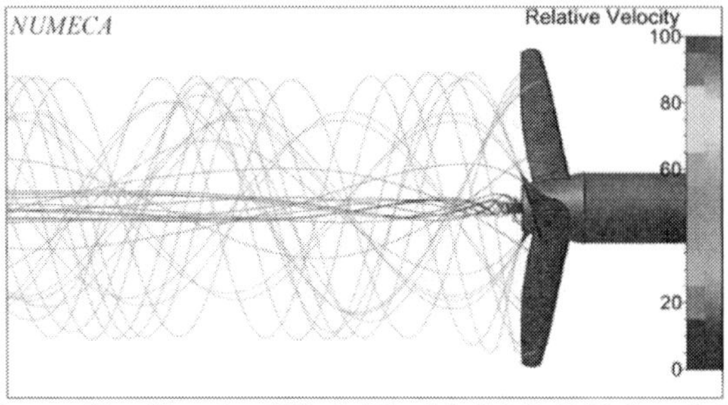

Figure 4. Streamline Result of Ka4-45 with J 0.3, P/D 1 and Ae/A0 0.45.

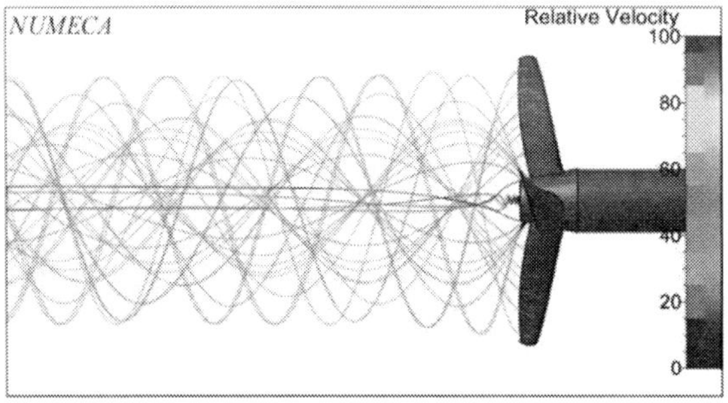

Figure 5. Streamline Result of Ka4-45 with J 0.5, P/D 1 and Ae/A0 0.45.

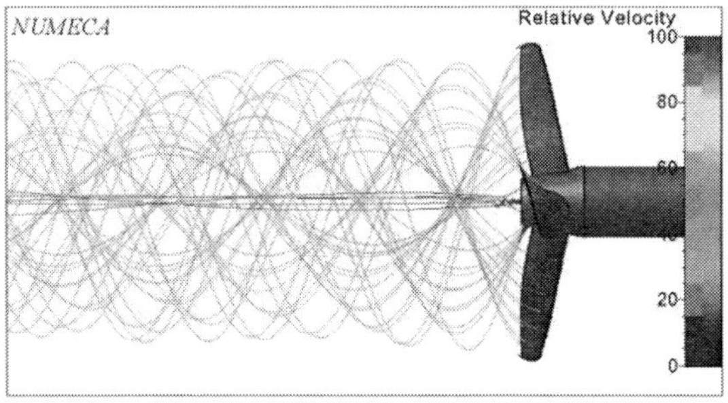

Figure 6. Streamline Result of Ka4-45 with J 0.7, P/D 1 and Ae/A0 0.45.

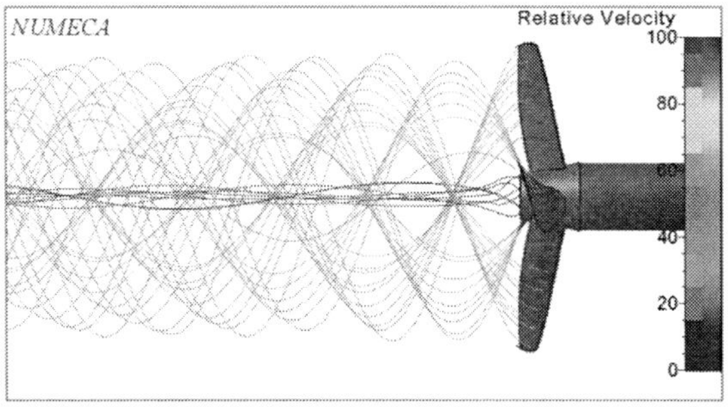

Figure 7. Streamline Result of Ka4-45 with J 0.9, P/D 1 and Ae/A0 0.45.

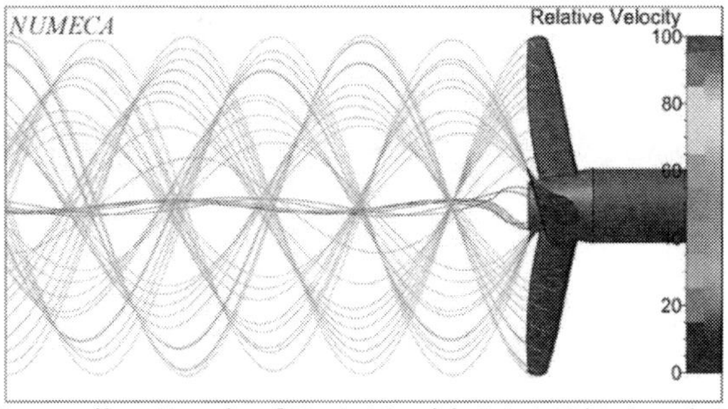

Figure 8. Streamline Result of Ka4-45 with J 1.1, P/D 1 and Ae/A0 0.45.

4.2.2 Propeller data analysis 2

The obtained streamline results for P/D 0.8 and Ae/A0 0.55 were presented on Figure 9, 10, 11, 12, 13 and 14.

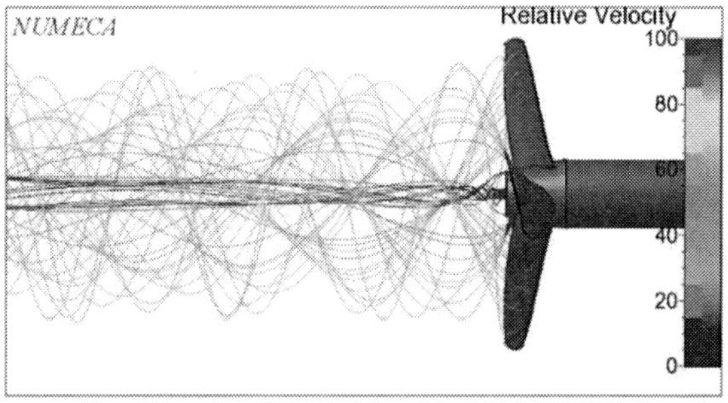

Figure 9. Streamline Result of Ka4-45 with J 0.1, P/D 0.8 and Ae/A0 0.55.

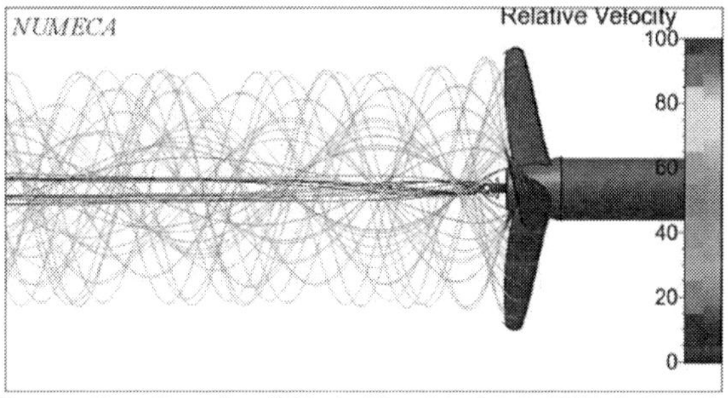

Figure 10. Streamline Result of Ka4-45 with J 0.3, P/D 0.8 and Ae/A0 0.55.

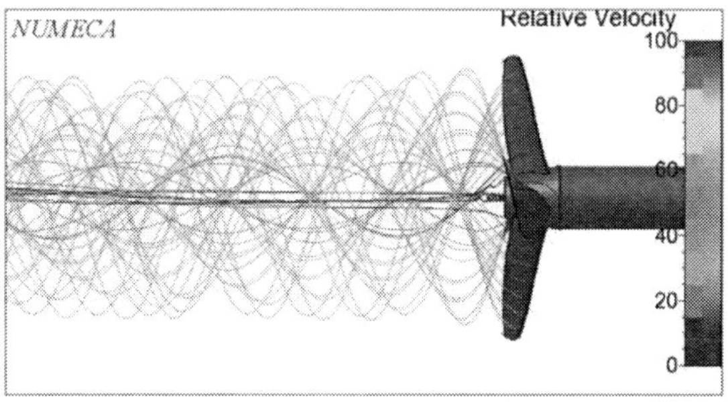

Figure 11. Streamline Result of Ka4-45 with J 0.5, P/D 0.8 and Ae/A0 0.55.

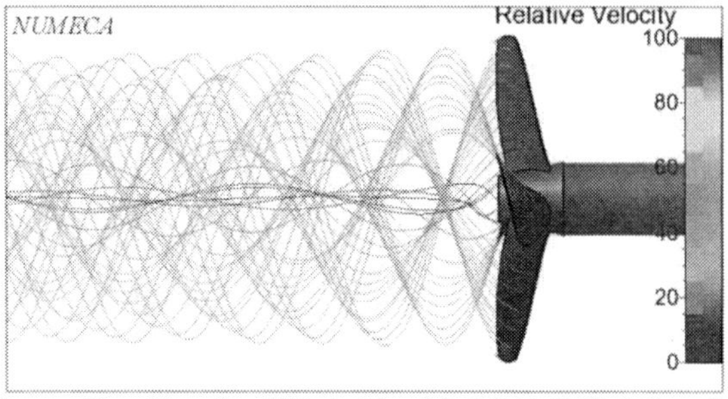

Figure 12. Streamline Result of Ka4-45 with J 0.7, P/D 0.8 and Ae/A0 0.55.

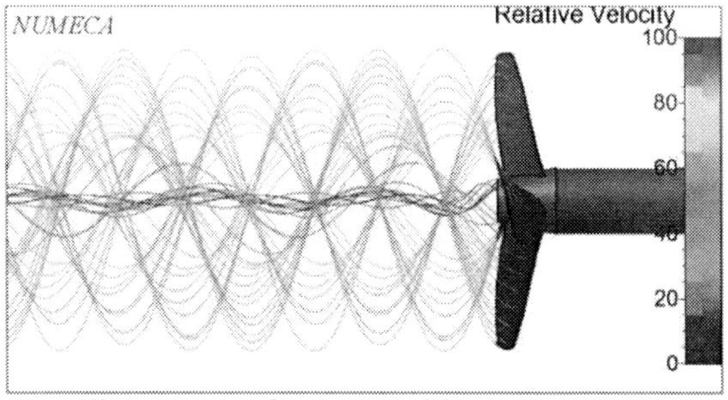

Figure 13. Streamline Result of Ka4-45 with J 0.9, P/D 0.8 and Ae/A0 0.55.

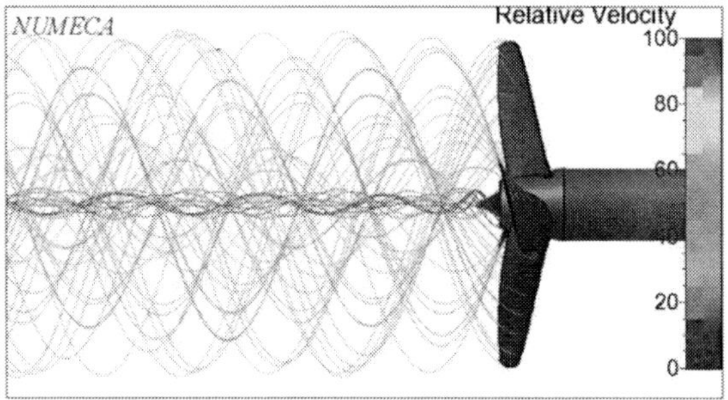

Figure 14. Streamline Result of Ka4-45 with J 1.1, P/D 0.8 and Ae/A0 0.55.

4.2.3 Propeller data analysis 3

The obtained streamline results for P/D 0.6 and Ae/A0 0.65 were presented on Figure 15, 16, 17, 18, 19 and 20.

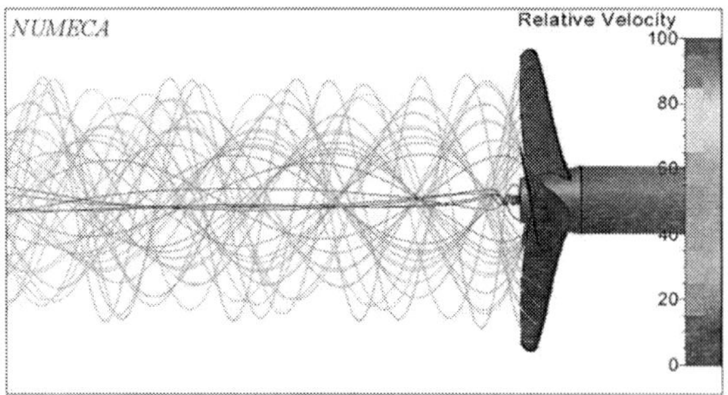

Figure 15. Streamline Result of Ka4-45 with J 0.1, P/D 0.6 and Ae/A0 0.65.

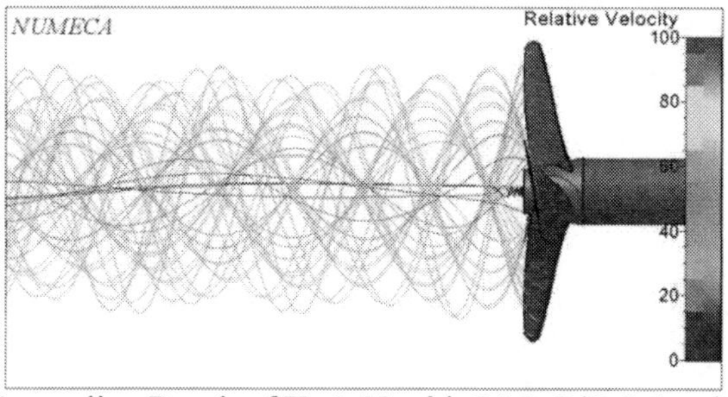

Figure 16. Streamline Result of Ka4-45 with J 0.3, P/D 0.6 and Ae/A0 0.65.

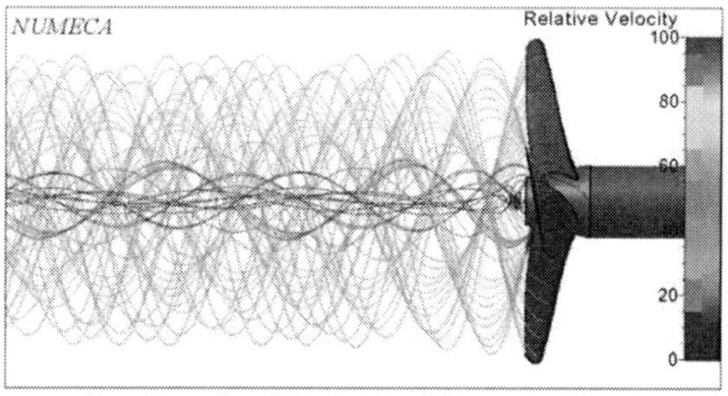

Figure 17. Streamline Result of Ka4-45 with J 0.5, P/D 0.6 and Ae/A0 0.65.

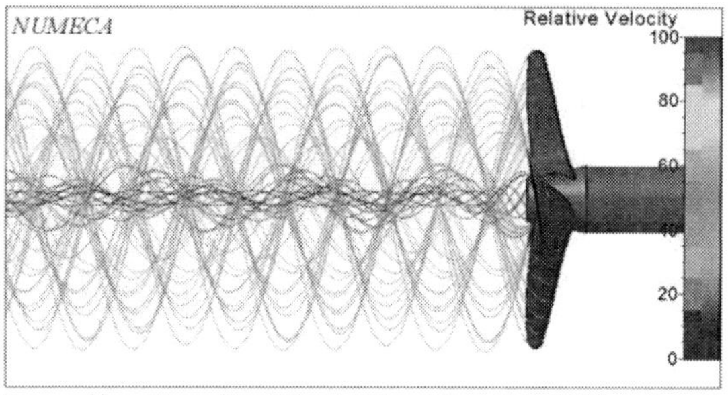

Figure 18. Streamline Result of Ka4-45 with J 0.7, P/D 0.6 and Ae/A0 0.65.

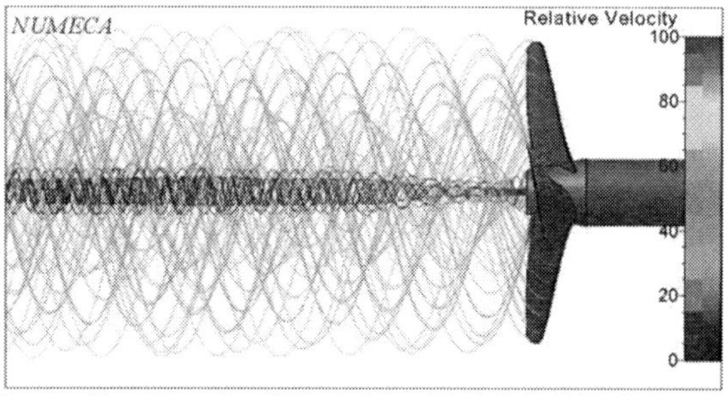

Figure 19. Streamline Result of Ka4-45 with J 0.9, P/D 0.6 and Ae/A0 0.65.

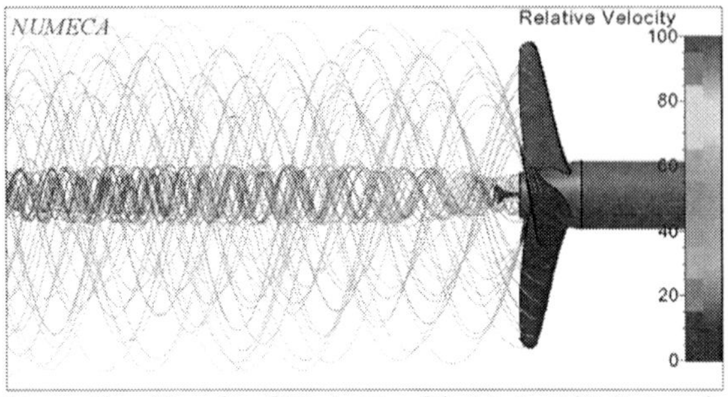

Figure 20. Streamline Result of Ka4-45 with J 1.1, P/D 0.6 and Ae/A0 0.65.

4.2.4 Propeller variation

The result of variation the values efficiency, thrust and torque were presented on Table 1, 2 and 3.

Table 1. Variation on the first propeller model.

J	Va (m/s)	Va (knot)	Vr (m/s)	Thrust (N)	Torque (Nm)	KT	KQ	10 KQ	EFF (%)
0.1	1.81	3.51836	39.82495	306201.4	64523.1	0.4349	0.0633	0.6328	10.94%
0.3	5.43	10.5551	40.15265	263467.2	56662.6	0.3742	0.0556	0.5558	32.16%
0.5	9.05	17.5918	40.80016	211002.5	47502.2	0.2997	0.0466	0.4659	51.21%
0.7	12.67	24.6285	41.7526	153330.5	37608.2	0.2178	0.0369	0.3689	65.80%
0.9	16.29	31.6653	42.98971	89952.3	26102.3	0.1278	0.0256	0.2560	71.51%
1.1	19.91	38.7020	44.48774	13576.4	11747.7	0.0193	0.0115	0.1152	29.31%

Table 2. Variation on the second propeller model.

J	Va (m/s)	Va (knot)	Vr (m/s)	Thrust (N)	Torque (Nm)	KT	KQ	10 KQ	EFF (%)
0.1	1.81	3.5184	39.8250	251446.4	44170.1	0.3571	0.0433	0.4332	13.13%
0.3	5.43	10.5551	40.15265	202428.4	37096.1	0.2875	0.0364	0.3638	37.75%
0.5	9.05	17.5918	40.80016	143619.6	28811.6	0.2040	0.0283	0.2826	57.47%
0.7	12.67	24.6285	41.7526	77935.8	19072.1	0.1107	0.0187	0.1871	65.95%
0.9	16.29	31.6653	42.98971	-2672.1	6847.4	-0.0038	0.0067	0.0672	0.00%
1.1	19.91	38.7020	44.48774	-108377.1	-9594.2	-0.1539	-0.0094	-0.0941	0.00%

Table 3. Variation on the third propeller model.

J	Va (m/s)	Va (knot)	Vr (m/s)	Thrust (N)	Torque (Nm)	KT	KQ	10 KQ	EFF (%)
0.1	1.81	3.51836	39.8250	176819.6	25590.8	0.2511	0.0251	0.25	15.93%
0.3	5.43	10.5551	40.1527	126728.8	20229.9	0.1800	0.0198	0.20	43.33%
0.5	9.05	17.5918	40.8002	64674.8	13342.4	0.0919	0.0131	0.13	55.88%
0.7	12.67	24.6285	41.7526	13997.5	4535.2	0.0199	0.0044	0.04	49.82%
0.9	16.29	31.6653	42.9897	-115873.7	-7077.3	-0.1646	-0.0069	-0.07	0.00%
1.1	19.91	38.7020	44.4877	-245143.8	-22400.2	-0.3482	-0.0220	-0.22	0.00%

4.2.5 Variation comparison graph

The result of comparison the values efficiency, thrust and torque were presented on Figure 21, 22 and 23.

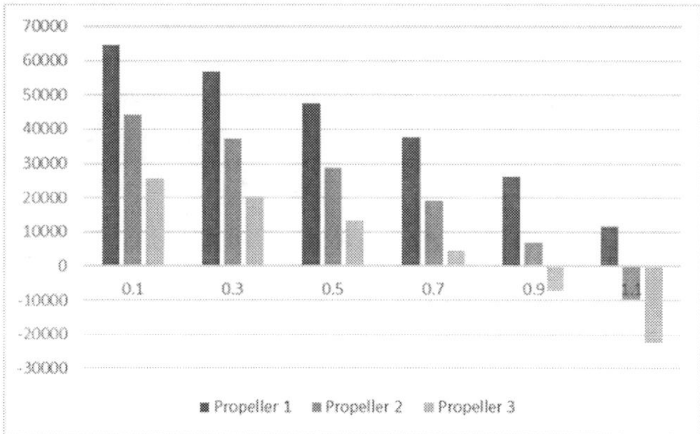

Figure 21. Torque value comparison graph.

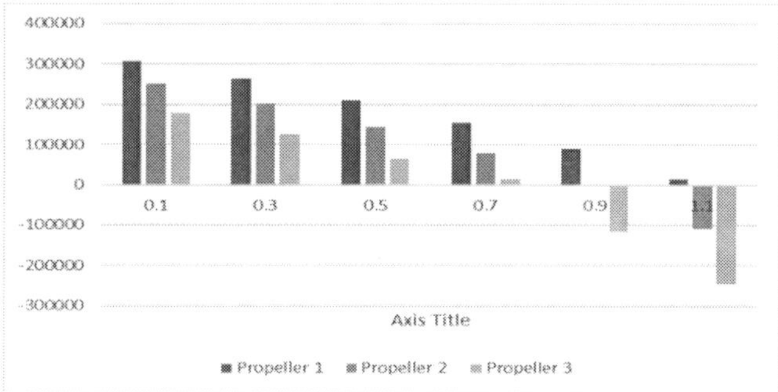

Figure 22. Thrust value comparison graph.

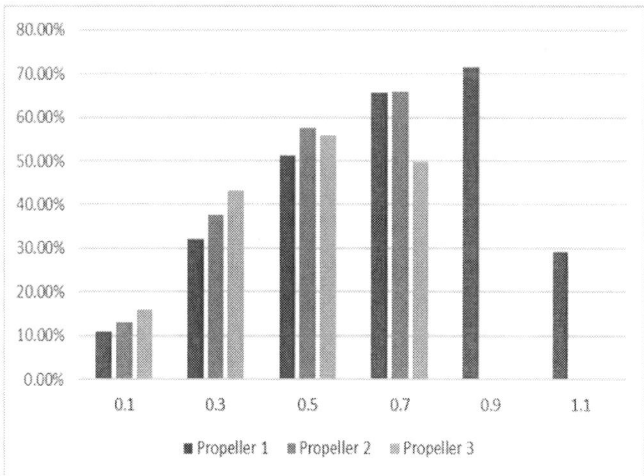

Figure 23. Efficiency value comparison graph.

4.2.6 Propeller visual design
The obtained three variation propeller designs with two views were presented on Figure 24, 25, 26, 27, 28 and 29.

Figure 24. The first propeller visual model perspective view.

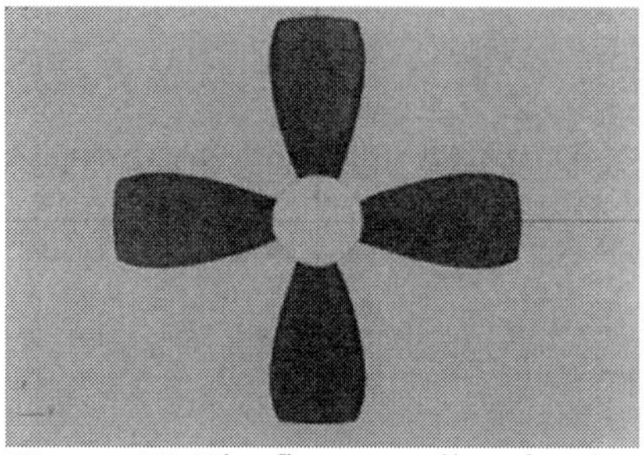

Figure 25. The first propeller visual model front view.

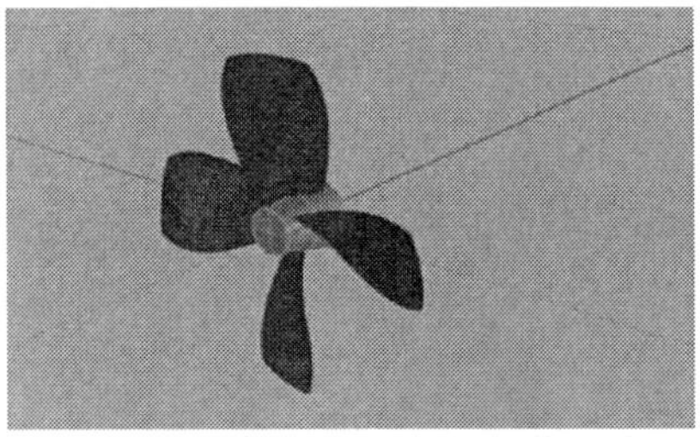

Figure 26. The second propeller visual model perspective view.

Figure 27. The second propeller visual model front view.

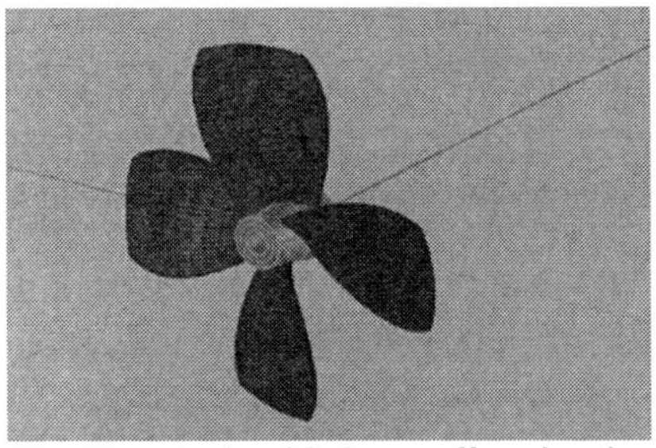

Figure 28. The third propeller visual model perspective view.

Figure 29. The third propeller visual model front view.

4.2.7 Summary report
From the simulation running results on Table 1, Table 2 and Table 3 shows the thrust and torque values at different propeller variations:

1. On the first propeller model has the highest thrust value of 306201.4 compared to other variations of the model.
2. On the third propeller model has the lowest torque value of -22400.2 Nm compared to other variations of the model.
3. On the first propeller model has the highest percentage of efficiency at 71.51% compared to other variations of the model.

5. Conclusion

To minimize the use of engine power on the ship, selecting the propeller from the resulting performance and analyzed the efficiency of the propeller design that has been produced in order to be able to move a tugboat 32.8 meters at a specified speed.

From the 18 variations of the model, the highest thrust value was obtained in the first Propeller model with a variation of P / D 1, AE / A0 0.45 valued at 306201.4. The lowest torque value is generated in the third Propeller model with a variation of P / D 0.6, AE / A0 0.65 value at -22400.2 Nm. Harbourtug requires a high thrust value, so it is advisable to use the first Propeller model with variations of P / D 1, AE / A0 0.45 The highest efficiency value is obtained in the first propeller model valued at 0.7151. Adding variations in propeller data with different blade ratios is suggested for further research.

6. Acknowledgements

The author would like to deliver gratitude to all colleagues from the Department of Marine Engineering, especially the Marine Power Plant (MPP) Laboratory, PT. Pelindo II and the Agency for the Assessment and Application of Technology (BPPT) for the support in carrying out this study.

REFERENCES

[1] Jinca 1953- M Y (Muhammad Y, 2011 *Transportasi laut Indonesia : analisis sistem & studi kasus / M. Yamin Jinca* Surabaya: Brilian Internasional.
[2] Ralahalu K A and Jinca M Y, 2013 The Development Of Indonesia Archipelago Transportation *Int. Ref. J. Eng. Sci.* **2**, 9 p. 12–18.
[3] Jinca M Y, 2012 PELAYARAN KAPAL-KAPAL TRADISIONAL **24** p. 218–231.
[4] ITTC, 2008 Dictionary of Ship Hydrodynamics.
[5] Zhang G Zhao Y Li T and Zhu X, 2014 Propeller Excitation of Longitudinal Vibration Characteristics of Marine Propulsion Shafting System *Shock Vib.* **2014** p. 1–19.
[6] Utami S D Haripamudya H and Kurniawan S B, 2020 Study of passenger vessels design in Berau Regency, East Kalimantan, Indonesia for public transportation *IOP Conf. Ser. Earth Environ. Sci.* **557**, 1.
[7] Nurhadi R Chrismianto D and Rindo G, 2017 Analisa Bentuk Variasi Propulison Module Pada Sistem Propulsi Azipod (Azimuthing Podded Drive) Berbasis Computational Fluid Dynamic (CFD) *J. Tek. Perkapalan; Vol 5, No 1 Januari*.
[8] Trimulyono A, 2015 Analisa Efisiensi Propeller B-Series Dan Kaplan Pada Kapal Tugboat Ari 400 Hp Dengan Variasi Jumlah Daun Dan Sudut Rake Menggunakan Cfd *Kapal* **12**, 2 p. 112–120.
[9] Carlton J S, 2019 *Propeller Performance Characteristics* .
[10] Bhattacharyya A Krasilnikov V and Steen S, 2016 A CFD-based scaling approach for ducted propellers *Ocean Eng.* **123** p. 116–130.
[11] Ariana I M, 1997 *Optimasi penggunaan propeller sebagai penggerak kapal dan pembangkit tenaga listrik pada KLB Maruta Jaya 900 DWT: laporan penelitian* Jurusan Teknik Sistem Perkapalan, Fakultas Teknologi Kelautan, Lembaga Penelitian, Institut Teknologi Sepuluh Nopember.
[12] Manngård M *et al.*, 2019 Estimation of propeller torque in azimuth thrusters *IFAC-PapersOnLine* **52**, 21 p. 140–145.
[13] Bhattacharyya A Krasilnikov V and Steen S, 2016 Scale effects on open water characteristics of a controllable pitch propeller working within different duct designs *Ocean Eng.* **112** p. 226–242.
[14] Anthony I Ekwere W Ogbonnaya E and Ejabefio K, 2013 Design Procedure of 4-Bladed Propeller *West African J. Ind. Acad. Res.* **8**, 1 p. 13–24.

Maneuverability study and economic analysis for tugboat procurement in basin area

I Made Ariana[1,2], A. A. B. Dinariyana D. P.[1,2], Widiastuti Tyasayumranani[3], and Emmy Pratiwi[1,2]

[1] Department of Marine Engineering, Faculty of Marine Technology, Institut Teknologi Sepuluh Nopember, Surabaya, Indonesia
[2] Center of Excellences Maritime Safety and Marine Installation, Institut Teknologi Sepuluh Nopember, Surabaya, Indonesia
[3] Department of Maritime Transportation System, Mokpo National Maritime University, South Korea

Abstract. The growth of the energy industry in Indonesia is increasing rapidly. This study concern about the crude and product oil distribution of one of the biggest oil companies in Indonesia. The distribution of crude and product oil requires large capacity particularly designed bunker transportation, which is an oil tanker. However, the enormous size of the tanker causes the tanker must be assisted by a tugboat to berth at the port of any type of waters. In this study, the simulations were conducted at the Marine Region (MR) III of Tanjung Priok Port, which has a basin-shaped berthing area, to determine the needs of tugboats force based on the size of the tankers. The simulations were carried out in MATLAB software by considering Tankers' principal dimension, environmental data, Tugboat force (Bollard Pull; BP), and Tugboat's heading angle. Then, those forces should be combined with each other into several scenarios and reviewed its economics based on the "buy", "lease", and "combination" scenario, calculated in the Capital Expenditure and Operational Expenditure factors.

1. Introduction

The growth of the energy industry in Indonesia is growing rapidly, along with the increasing number of companies engaged in the energy sector and the growing number of companies that have been established for a long time. The company has a role to explore, produce, develop renewable gas and energy from upstream to downstream. Products from companies in the energy sector will be transported to consumers, hence it needs transportation with large capacity and specially designed bunker, which is Tankers.

The company's Tankers has various size, including VLCC Vessel. Because of its enormous size, the ship must be assisted by a tugboat to berth at the port. Based on Regulation of Minister 57, the year of 2015 article 1 paragraph (2), written that "The uses of Tugboat are to push, pull, or assist a vessel to/from any mooring facilities". Besides that, based on Regulation of Minister 57, the year of 2015 article 38 paragraph (3), written the minimum of Power and Bollard Pull to tow a vessel, according to the Length of the vessel [1]. Those regulations must be followed to enhance the effectiveness and efficiency of any towing process.

Table 1. PM 57 / 2015 Article 38 Paragraph (3)

Length of Vessel	Amount of Tugboat	Power	Bollard Pull
70 – 150 meter	Minimum 1 unit	Minimum 2000 HP	Minimum 24ton *bollard pull*
150 - 250 meter	Minimum 2 units	Minimum 6000 HP	Minimum 65ton *bollard pull*
250 meter or above	Minimum 3 units	Minimum 11000 HP	Minimum 125ton *bollard pull*

According to the passage above, it can be inferred that there is a need to decide the most technically, especially in maneuverability, and economically efficient tugboats, while it still fulfilling the safety factors and government regulations regarding piloting and towing at the company. In this study, a maneuverability simulation for tugboat procurement needs to be carried out. Ship maneuvering should be performed for predicting the navigation safety in restricted areas [2].

The simulations carried out in software by entering the principal dimension of the Tankers, environmental data, the Tugboat force, and the Tugboat's heading angle. The output of those simulations is the minimum Bollard Pull to tow the Tankers, and in the end, it will provide optimal economy in the procurement of tugboats.

2. Methodology

2.1. Company general data

Marine Region (MR) III is located in Tanjung Priok, DKI Jakarta, Indonesia. As a big port in the capital city of Indonesia, MR. III Tanjung Priok is fully equipped with 4 Jetties that facilize Tankers with a minimum of 1,200 DWT and a maximum of 30,000 DWT. Besides that, MR. III Tanjung Priok also has assets such as Lightweight Small Boat, Fire Fighting Equipment, and Oil Leakage Treatment Equipment.

In general, a tugboat has 2 propellers to help the maneuver process in a berthing area. The tugboat will be operated in a towing process. Conventional Propellers have a nozzle to increase the Bollard value and its effectiveness. The nozzle can be mounted on a fixed propeller equipped with a rudder or movable propeller (Kort nozzle) [3], [4].

Tugboats owned or rented by the company uses an Azimuth Thruster type propulsion system that has many advantages over using conventional propulsion as below.

1. Ability to produce optimal thrust in all directions
2. Does not require a rudder
3. Improve ship maneuverability
4. Improve electrical efficiency
5. Provide more space in the engine room
6. Reducing maintenance costs

MATLAB (Matrix Laboratory) is a software used to do numerical computing with computer programming languages and a matrix basis. The software can help to solve mathematical problems by making mathematical simulations and modeling. In this study, it was used to carry out simulations in the selection of alternative scenarios in order to produce recommendations for the most effective and efficient scenario of tugboats and their respective power.

In this process, each scenario will be simulated to get the result of the minimum Bollard Pull to tow the Tankers of each cluster. This simulation needs some data, such as:

1. Tanker Principal Dimension
2. Environmental Data

3. Tugboat Force (Bollard Pull)
4. Tugboat Heading Angle

2.2. *Mathematical Maneuvering Group Model*

Ship maneuvering can be simulated with the hydrodynamic derivatives using a mathematical model such as Maneuvering Modeling Group (MMG) [5]. Maneuvering Modeling Group (MMG) model is one of the most widely used method for ship maneuvering motion simulations because it is possible to take the hull-rudder-propeller interaction into account [5]–[7]. MMG model was firstly developed and proposed by a research group called Maneuvering Modeling Group (MMG) in Japanese Towing Tank Conference (JTTC) [7]. This study refers to the concept for maneuvering simulations presented in MMG model to predict the maneuvering of tugboats.

As shown in figure 1, the simulation model designed based on a few parameters, such as hull, rudder, and propeller as a form of response from environmental factors like wind and waves [8]–[10]. It will use formulas of 3 (three) degrees of freedom such as written below. It will be equipped with an estimation formula to calculate the value of hull forces and hull moments as in the formula below. The β variable is interpreted as a drift angle and the r variable is defined as the turning rate [8], [10].

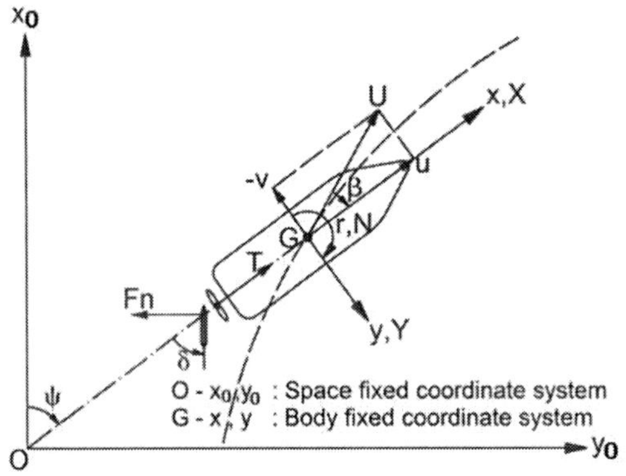

Figure 1. Mathematical Maneuvering Group Model

Surge : $(m + m_x)\dot{u}_G - (m + m_y)v_G r_G$
$= X_H + X_P + X_R + X_W + X_T$

Sway : $(m + m_y)\dot{v}_G + (m + m_x)u_G r_G$
$= Y_H + Y_P + Y_R + Y_W + Y_T$

Yaw : $(I_{zz} + J_{zz})\dot{r} = (N_H + N_P + N_R + N_W + N_T) - x_G (Y_H + Y_P + Y_R + Y_W + Y_T)$

$X_H + m_y v_G r_G$: $0.5\rho L d U^2 \{X'_0 + X'_{\beta\beta}\beta^2 + (X'_{\beta r} - m'_y)\beta r' + (X'_{rr} - x'_G m'_y)r'^2 + X'_{\beta\beta\beta\beta}\beta^4\}$

$Y_H - m_x u_G r_G$: $0.5\rho L d U^2 \{Y'_\beta\beta + (Y'_r - m'_x)r' + Y'_{\beta\beta\beta}\beta^3 + Y'_{\beta\beta r}\beta^2 r' + Y'_{\beta rr}\beta r'^2 + Y'_{rrr}r'^3$

N_H : $0.5\rho L d U^2 \{N'_\beta\beta + N'_r r' + N'_{\beta\beta\beta}\beta^3 + N'_{\beta\beta r}\beta^2 r' + N'_{\beta rr}\beta r'^2 + N'_{rrr}r'^3$

Hull derivates coefficient used to simulate the towing process are equipped with the parameter below [2].

Y'_β : $Y'_{\beta 0}(1 + 0.54\tau'^2)$

$Y'_r - m'_x$: $(Y'_r - m'_x)_0 (1 + 1.82\tau'^2)$

N'_β : $N'_{\beta 0}(1 - 0.85\tau'^2)$

N'_r : $N'_{r0}(1 + 0.33\tau'^2)$

$Y'_{\beta 0}$: $0.5\pi k + 1.4C_b / (L/B)$

$(Y'_r - m'_x)_0$: $0.5C_b / (L/B)$

$N'_{\beta 0}$	: k
N'_{r0}	: $0.54k + k^2$
k	: $2d/L$
$X'_{\beta\beta}$	: $1.15C_b / (L/B) - 0.18$
$X'_{\beta r} - m'_y$	: $-1.91C_b / (L/B) + 0.08$
$X'_{rr} + x'_G m'_y$	: $-0.085C_b / (L/B) + 0.008$
$X'_{\beta\beta\beta\beta}$	: $-6.68C_b / (L/B) + 1.10$
$Y'_{\beta\beta\beta}$	: $0.185L/B + 0.48$
$Y'_{\beta\beta r}$	: $0.97\tau'/C_b - 0.75$
$Y'_{\beta rr}$	: $0.26(1 - C_b)\,L/B + 0.11$
Y'_{rrr}	: $0.069\tau' - 0.051$
$N'_{\beta\beta\beta}$	: $0.69C_b + 0.66$
$N'_{\beta\beta r}$	: $1.55C_b / (L/B) - 0.76$
$N'_{\beta rr}$	: $0.075(1 - C_b)\,L/B - 0.098$
N'_{rrr}	: $0.25C_b / (L/B) - 0.056$

To complete the hull derivates coefficient formula as in the formula below, additional longitudinal mass (mx), transverse additional mass (my), and the mass of additional moment of inertia (Jzz) values are needed, which can be calculated based on Motor Diagram. In this calculation, the values of mx, my, Jzz, and Izz (moment of inertia) can also be calculated through the following equation [8].

mx	: $0.0098\,(\frac{\rho}{2})L^2d$
my	: $0.1370\,(\frac{\rho}{2})\,L^2d$
Jzz	: $0.00634\,(\frac{\rho}{2})\,L^4d$
Izz	: $(\frac{1}{8})\,(\frac{C_b}{L/B})\,(\frac{\rho}{2})\,L^4d \quad = (\frac{1}{16})\,mL^2$

Then, the lateral force and moment that exerted by the propeller can be calculated by some formula or by looking at the KT-Js diagram. In this study, the rudder force can be neglected as the effect of full tugboat assists operation.

2.3. *Economic feasibility*

The economic feasibility calculation will be divided into 2 (two) scenarios, such as the "Buying" Scenario and the "Rent" Scenario. Each scenario will be compared through its Capital Expenditure value and Operational Expenditure value for the next 20 years.

3. Results and discussion

3.1. *Tankers clustering*

To design the cluster of tankers, this study needs the TPPR (Tanker Port Performance Report) data of MR. III Tanjung Priok. According to the TPPR data, can be inferred that the frequency of tankers that berth to MR. III Tanjung Priok in the last 5 (five) years for each mooring facility did not have a significant difference. Each mooring facility has a berthing call on an average of 1 (one) to 3 (three) tankers so that in the cluster design process there is no need to forecast tankers for the next few years. In addition, tankers rely on MR. III Tanjung Priok mooring facilities are mostly vessels owned by companies and subsidiaries.

Table 2. Tankers Cluster Based on 2018's TPPR

Cluster	DWT (ton)	T	LOA (m)	Types of Tanker	Vessel	Arrival
A	1,329 - 6,842	FLUCTUATIVE	65 - 158	Small 1 & 2 + SPOB	27	135
B	7,079 - 23,479		96 - 160	General P.	43	216
C	25,161 – 47,236		133 - 184	Medium R. & Large R.	32	141
TOTAL					102	492

3.2. Real scheme of towing in MR. III Tanjung Priok

Towing activities in MR. III Tanjung Priok is carried out by pulling, pushing, and sliding the tankers until it reaches the mooring facility. MR. III Tanjung Priok has a basin-shaped berthing area, thus tankers must turn off its engine and do a full tugboat assist. Tankers may lean on both its portside and starboard. When a tanker agreed to lean on it portside, it must be rotated in a basin turning area before entering the mooring area.

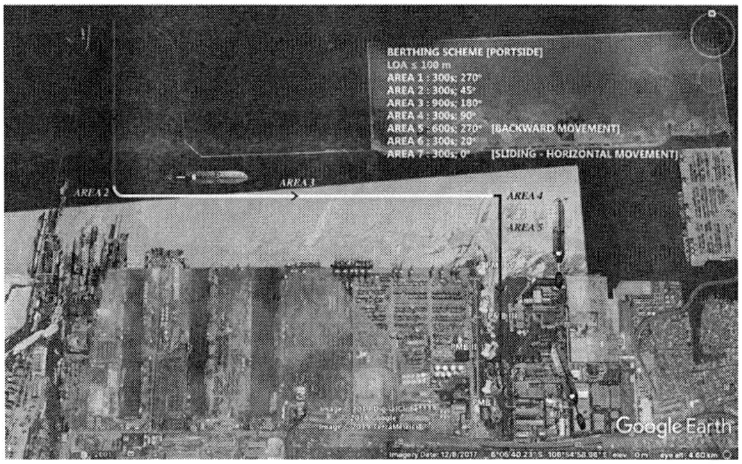

Figure 2. Real Towing Scheme on MR. III Tanjung Priok for Tankers with LOA ≤ 100m

In Figure 2, the scheme shown is the process of berthing a Tanker with LOA ≤ 100m using only one (1) Tugboat. Note that the images of tankers and tugboats shown on the figure are scaled and might not match the actual ship's size and position at sea. By looking at the figure, in area 1, the Tanker is available to still use its own speed with a maximum of 5 knots at a safe speed. In area 2, tugboats are deployed to help maneuver the ship around the turn. Approaching area 4 (basin 4), the ship needs to stop the engine and the tug will pull the tanker so that it will enter backward in area 5. Furthermore, the tug will direct the tanker to lean on areas 6 and 7. Since there's only one (1) Tugboat that assists the ship, the tug needs to create an angle ease the process of leaning the ship to the jetty later on. For the final Tanker leaning process, its tug might pull the ship's aft and rely on the ship's movement influenced by the external factors [11]. Towing is required to keep the ship at a safe speed and not hardly bump the jetty.

While in Figure 3, the scheme shown is the process of berthing a tanker with LOA > 100 m using two (2) Tugboats. As in the previous figure, the images of tankers and tugboats shown on the figure are scaled and might not match the actual ship's size and position at sea. In area 1, the tanker can still use its own speed with a maximum of 5 knots at a safe speed. In area 2, tugboats are deployed to help maneuver the ship around the turn. Approaching area 4 (basin 4), the ship will need to stop the engine and the tug will pull the tanker so that it will enter backward in area 5. Furthermore, the tug will push the tanker to the mooring facility in area 6.

Figure 3. Real Towing Scheme on MR. III Tanjung Priok for Tankers with LOA > 100m

3.3. Tankers data for simulation

To simulate the towing process in MR. III Tanjung Priok, this study needs Tankers to represent each cluster. Tankers shown in Table 3 are the biggest Tankers from each cluster. It can be assumed that if any Tugboat can tow these 3 (three) Tankers, thus those tugboats will surely tow smaller Tankers.

Table 3 Tankers Data for Simulation

Parameter	Cluster A	Cluster B	Cluster C
DWT	6842 ton	23479 ton	47236 ton
LOA	100.5 m	159.98 m	182.5 m
B	18 m	25.6 m	32.2 m
H	9.5 m	16.4 m	19.1 m
F	6.8 m	5.5 m	6.45 m

3.4. Environmental data

The MR. III Tanjung Priok has fairly normal environmental conditions with wind speeds ranging from 4 knots to 15 knots and wave heights of 0.3 – 0.8 meters. The depth of water in the outer basin is approximately 11 meters and in the inner basin is approximately 9 – 10 meters. In addition, tides that occur on the port are in the range of 0.6 to 1.1 meters. The current velocity taken at the mooring facilities areas when at high tide is 0.3 knots from the direction of 45 to 50° and at low tide is 1.1 knots from the direction of 230°.

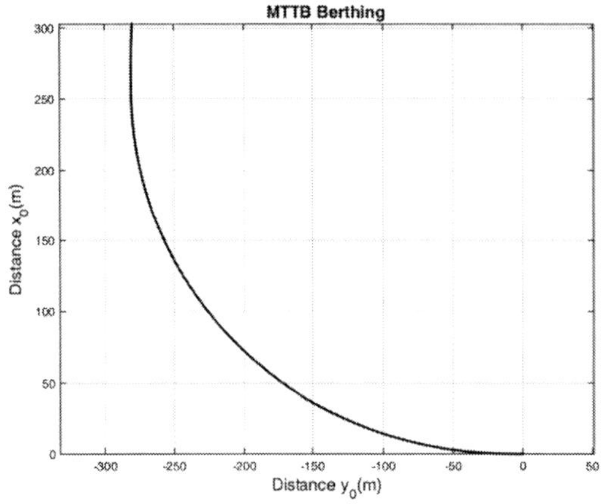

Figure 4. Simulation Result of Cluster A on Area 2 using 1 Tugboat

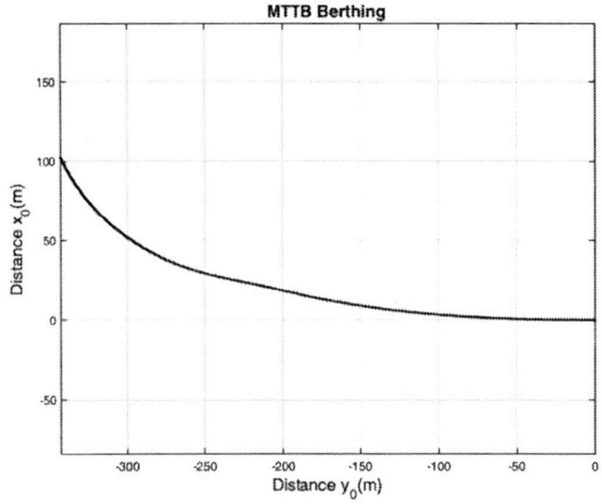

Figure 5. Simulation Result of Cluster B on Area 2 using 1 Tugboat

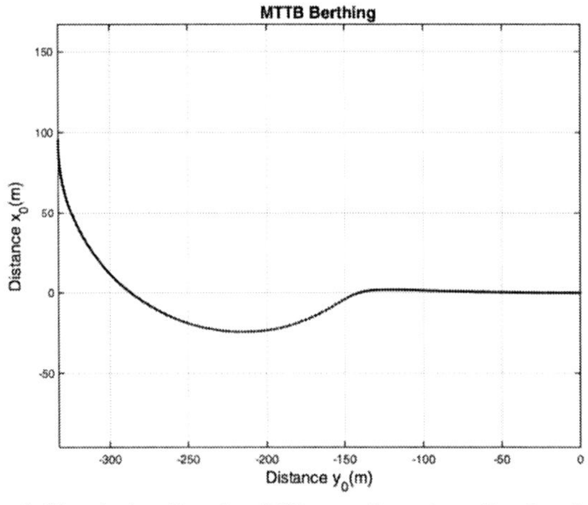

Figure 6. Simulation Result of Cluster C on Area 2 using 1 Tugboat

IOP Conf. Series: Materials Science and Engineering 1052 (2021) 012030 doi:10.1088/1757-899X/1052/1/012030

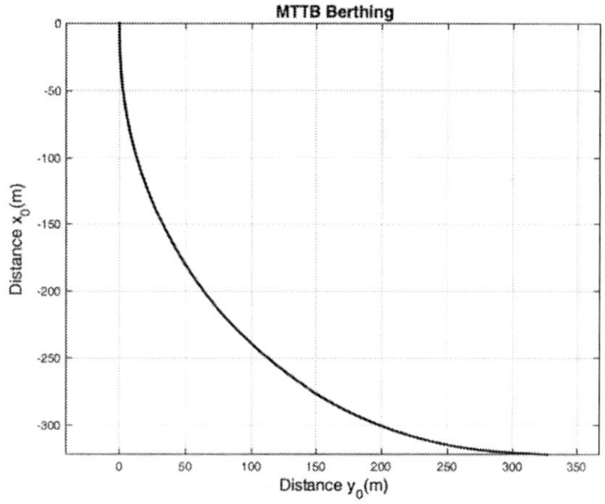

Figure 7. Simulation Result of Cluster A on Area 4 using 1 Tugboat

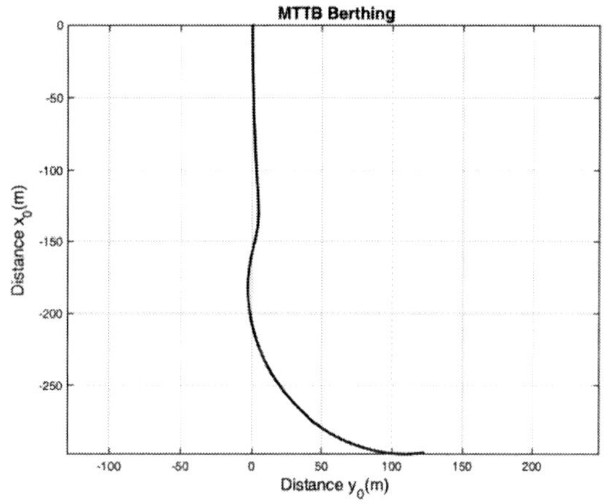

Figure 8. Simulation Result of Cluster B on Area 2 using 1 Tugboat

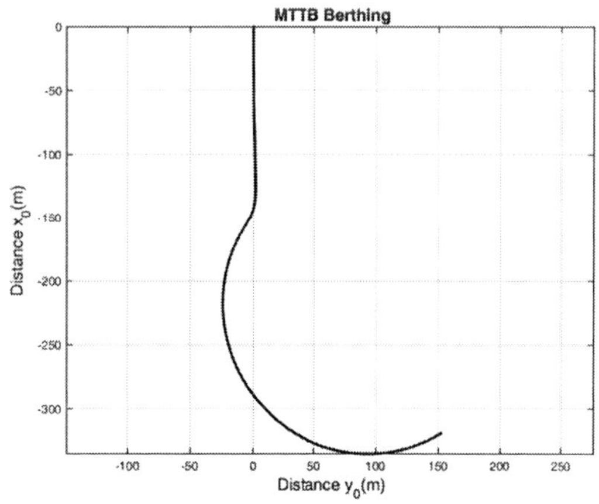

Figure 9. Simulation Result of Cluster C on Area 2 using 1 Tugboat

3.5. Simulation result
The simulation of the towing process in MR. III Tanjung Priok is divided into 2 (two) main areas, such as the Basin Entering Area (Area 2) and Basin Turning Area (Area 4). The output taken from this simulation is the distance of towing represent by the X and Y-axis.

The simulation results as drawn by Figure 4 to Figure 6, are simulated on Area 2 using 1 Tugboat, towed by 135°. In figure 4, Cluster A towed with Tugboat of 22ton Bollard Pull. In figure 5, Cluster B towed with the Tugboat of 32ton Bollard Pull. In figure 6, Cluster C towed with Tugboat of 38ton Bollard Pull.

The simulation results as drawn by Figure 7 to Figure 9, are simulated on Area 4 using 1 Tugboat, towed by 290°. In figure 7, Cluster A towed with Tugboat of 22ton Bollard Pull. In figure 8, Cluster B towed with the Tugboat of 32ton Bollard Pull. In figure 9, Cluster C towed with Tugboat of 38ton Bollard Pull. The movement of the tanker and tugboat is from point 0 on the y0 axis, which Tankers is rotated like in Figure 2 and Figure 3.

From the figures above, it can be inferring that Tankers in each cluster must be towed with a different force, according to its size. The combination that may occur in Economic Feasibility from the simulation above is the buy/rent a combination of 22ton and 32ton, 22ton and 38ton, 32ton and 38ton, or 22ton, 32ton, and 38ton. The simulation can be continued to as much as Tugboat used as possible.

Table 4. Tabulation of Simulation Result

Cluster	1 TUG	2 TUGs	3 TUGs
A	22ton	20ton 15ton	15ton 10ton 10ton
B	32ton	22ton 20ton	20ton 15ton 10ton
C	38ton	32ton 22ton	22ton 20ton 15ton

3.6. Tabulation of simulation result
According to the simulation done by using MATLAB software, this study can infer the minimum value of Bollard Pull to be used by each cluster and succeed to tow Tankers in every turning point. Besides that, the value of minimum Bollard Pulls also exerted by the amount of Tug. The value can be seen in Table 4 about the Tabulation of Simulation Result. Then, this tabulation would be a consideration to determine and inspect the technical analysis in the next step.

3.7. Technical analysis
The tabulation of the minimum Bollard Pull value in Table 4 needs to be evaluated by the parameter of the Tugboat dimension based on Table 5. The tugboat dimension is important to make sure that the Tugboats are on equal height with the Tankers towed. Tugboat with the Bollard Pull of 32ton and 38ton are higher than the Tankers cluster A so that those only can push Tankers cluster B and C.

Table 5. Dimensional Parameter of Tugboat

Bollard Pull	LOA	B	T	Cluster can be Pushed
10ton	15 - 30 m	6 - 7 m	2.5 m	A, B, C
15ton	15 - 30 m	6 - 7 m	3 m	A, B, C
20ton	20 - 30 m	8 - 9 m	3.5 m	A, B, C
22ton	25 - 30 m	8 - 9 m	3.5 m	A, B, C
32ton	26 - 35 m	8 - 9 m	4 m	B, C
38ton	27 - 36 m	8 - 10 m	4.5 m	B, C

Table 6. Summary of technical analysis

TUG	Based on Operational (refer to Table 4)	Based on Dimensional (refer to Table 5)
10ton	Lower force than needed to tow Tankers Cluster C	-
15ton, 20ton, 22ton	May towing the Tankers Cluster C with the addition of bigger forced tugboat	-
32ton, 38ton	-	Higher than Tankers Cluster A

The tabulation of technical analysis could also be seen in Table 6, where each Bollard Pull of Tugboat has its own capability to tow or push Tankers. Then, the result of the Technical Analysis would be a consideration to calculate the economic feasibility.

3.8. Economic Feasibility

In this study, the economic feasibility is calculated based on Capital Expenditure and Operational Expenditure in order to find the cheapest procurement combination of the Tugboat for MR. III Tanjung Priok. The economic feasibility will be divided into 3 (three) scenarios, such as "Buying" Scenario, "Rent" Scenario, and "Buy-Rent" Scenario. Furthermore, each scenario will be calculated for the next 20 years. Table 7 representing the buy scenario costs of each Tugboat, while Table 8 representing the rental costs of each Tugboat. The total costs of the "Buying" scenario are an accumulation of investment costs in its first year with some operational costs, escalated by 2% each year. Operational costs including the prices of BBM, daily operational, periodic survey, port management, insurance, and administration.

Table 7. Buying scenario costs for 20 years of each Tugboat

HP Tug	BP Tug (ton)	Operational + Fuel (per month) x IDR 1,000,000	Buying Price + Additional Cost x IDR 1,000,000	Buying Price + ((Operational + Fuel) 20 Years) x IDR 1,000,000
2 x 600 HP	10	623	11,110	193,756
2 x 650 HP	15	653	12,120	203,762
2 x 800 HP	20	775	15,150	242,528
2 x 1200 HP	22	858	23,230	275,803
2 x 1500 HP	32	892	36,360	299,979
2 x 1600 HP	38	964	38,380	323,333

Table 8. Rental Costs for 20 Years of Each Tugboat

HP Tug	BP Tug (ton)	Rent/month (Without Fuel) x IDR 1,000,000	Rent for 20 Years + Fuel x IDR 1,000,000
2 x 600 HP	10	666	220,358
2 x 650 HP	15	704	231,478
2 x 800 HP	20	818	264,835
2 x 1200 HP	22	960	306,252
2 x 1500 HP	32	1,002	318,418
2 x 1600 HP	38	1,093	344,833

Table 9. Economic Feasibility Result

	TUG	Price of TUG1 x IDR 1000000	Price of TUG2 x IDR 1000000	Price of TUG3 x IDR 1000000	Total Price 20 Year + BBM x IDR 1000000
Buy	22ton-32ton	275,803	299,979		575,782
Buy-Rent	22ton-32ton	**[Buy]** 275,803	**[Rent]** 318,418		**[Buy-Rent]** 594,221
Buy	22ton-38ton	275,803	323,333		599,134
Rent-Buy	22ton-32ton	**[Rent]** 306,252	**[Buy]** 299,979		606,231
Buy-Rent	22ton-38ton	**[Buy]** 275,803	**[Rent]** 344,833		620,636
Rent	22ton-32ton	306,252	318,418		624,670
Rent-Buy	22ton-38ton	**[Rent]** 306,252	**[Buy]** 323,333		629,585
Rent	22ton-38ton	306,252	344,833		651,086
Buy	15ton-22ton-20ton	203,762	275,803	242,528	722,094
Buy	15ton-32ton-20ton	203,762	299,979	242,528	746,269

The total costs of the "Rental" scenario are an accumulation of investment costs each year, escalated by 2% each year, with its fuel costs which also escalated by 2% each year. Investment costs including the fuel, daily operational, periodic survey, port management, insurance, and administration, and the margin for the renter company.

4. Conclusion

1. Towing activities in MR. III Tanjung Priok allows Tankers to lean on both its portside and starboard. When a tanker agreed to lean on it portside, it must be rotated in a basin turning area before entering the mooring area.
2. Based on the simulation, each cluster can be towed by 1 (one) to 3 (three) Tugboats with a different force (ton) and heading angle. For Tankers Cluster A, it can be towed by minimum Tugboats with 15ton-10ton-10ton Bollard Pull, while Tankers Cluster B can be towed by

minimum Tugboats with 20ton-15ton-10ton Bollard Pull. Then, Tankers Cluster C can be towed by minimum Tugboats with 22ton-20ton-15ton Bollard Pull.

3. The economic feasibility was calculated by both of its economical parameter and regulation parameters. Tugboats must still refer to the Government Regulation to maintain safety and security factors. The 3 (three) cheapest combination that the company can get should be the "Buy" Scenario of 22ton-32ton, "Buy-Rent" Scenario of "22ton-32ton" and "Buy" Scenario of "22ton-38ton".

5. Acknowledgments

The authors express gratitude to the Center of Excellences Maritime Safety and Marine Installation (PUI-KEKAL) for funding this paper publication and collaboration during the research.

6. References

[1] Kementerian Perhubungan Republik Indonesia, *PM 57 Tahun 2015*. 2015.
[2] E. Tannuri and G. Martins, "Application of a maneuvering simulation center and pilots expertise to the design of new ports and terminals and infrastructure optimization in Brazil," presented at the PIANC-World Congress Panama City, 2018.
[3] I. Ariana, T. Akbar, M. Prakosa, J. Prananda, and W. Tyasayumranani, "Conceptual design of dynamic positioning system for tugboat to improve safer operation in Indonesia, case study: port of Cilacap," 2020, vol. 557, no. 1, p. 012067.
[4] Zahalka, C. P ., "Bollard Pull," Association of Hanseatic Marine Underwriters, 2007.
[5] O. F. Sukas, O. K. Kinaci, and S. Bal, "System-based prediction of maneuvering performance of twin-propeller and twin-rudder ship using a modular mathematical model," *Applied Ocean Research*, vol. 84, pp. 145–162, 2019.
[6] A. Ogawa and H. Kasai, "On the mathematical model of manoeuvring motion of ships," *International Shipbuilding Progress*, vol. 25, no. 292, pp. 306–319, 1978.
[7] H. Yasukawa and Y. Yoshimura, "Introduction of MMG standard method for ship maneuvering predictions," *Journal of Marine Science and Technology*, vol. 20, no. 1, pp. 37–52, 2015.
[8] I. P. S. ASMARA, "Development of Navigation Safety Analysis for Developing Port Area using Maneuvering Simulation and AIS Data," 2020.
[9] G. DNV, "DNV-RP-H103: Modelling and Analysis of Marine Operations," *DNV GL*, 2014.
[10] T. Fujiwara, M. Ueno, and T. Nimura, "Estimation of wind forces and moments acting on ships," *Journal of the Society of Naval Architects of Japan*, vol. 1998, no. 183, pp. 77–90, 1998.
[11] H. Agerschou, "CHAPTER 2 Facilities requirements," in *Planning and design of ports and marine terminals*, Thomas Telford Publishing, 2004, pp. 5–20.

IOP Conf. Series: Materials Science and Engineering 1052 (2021) 012031 doi:10.1088/1757-899X/1052/1/012031

Review of the risk analysis for the availability of equipment components the construction of ro-ro ferry in Indonesia

Z Ariany[1,3], T Pitana[1], I Vanany[2]

[1]Departement of Marine Engineering, Institut Teknologi Sepuluh Nopember, Surabaya, Indonesia

[2]Department of System Engineering and Industry, Institut Teknologi Sepuluh Nopember, Surabaya, Indonesia

[3]Departement of Industry, Vocational School, Diponegoro University, Indonesia

Abstract. This paper examined the risk assessment adapted to the construction of government-owned Ro-Ro ferries. The data were obtained and developed from domestic shipyards that built these government-owned ships. Empirical studies on ship building have not been widely carried out, especially for domestic Ro-Ro ferries builders, the consequences of which included delays in ship delivery time and occurred higher costs. Some literature stated that shipyard companies agreed on the importance of supply chain risk management, but it has not been applied, so that, the results were still far from expected. Supply chain risk management was a risk assessment by identification, assessment and evaluation, one of the parts raised was the availability of ship materials and components in production activities, which was one of the contributing factors for the high percentage of ship completion delays. The analysis of the potential development in the component industry that was enabled to be developed in the domestic area by considering the readiness of the industry, including the availability of resources (human resources and raw materials), the existence of the industry, government policies, and economies of scale that could be achieved by developing the component industry with technological innovations

Key word: Ship Equipment; Ship Component; Risk Analysis; Risk Assessment; Shipbuilding.

1. Introduction

Roro ferries are a crossing-transportation inter-island that load the passengers and the vehicles. Generally, roro ferries have two ramp doors that are in the forward and stern positions which used for vehicle entry and exit needs. Besides it uses to load the truck, it also loads the passenger's car, the motorcycle along with the passengers.

Ferries are a short distance transportation ship. Ferries have an important role in the loading system transportation on mostly coastal cities; creating direct transit between two aims with a smaller cost than bridge or binoculars. Ferry is a ship that meets the requirements for shipping at sea and it's used to maintain permanent transportation; for example, inter-island. Ro-Ro ferries are mainly used for crossing facilities, including crossing vehicles. [1].

IOP Conf. Series: Materials Science and Engineering 1052 (2021) 012031 doi:10.1088/1757-899X/1052/1/012031

The government through the Ministry of Transportation, Directorate General of Land Transportation, Directorate of TSDP (River, Lake Transportation and Crossing) has the construction and ship rehabilitation program; in this case ferries, pioneer ship, and water bus. [2].

Figure 1. Ferry 500GT. Source PT. DTPS picture.

The collected data of several shipyards have indicators that show a high percentage of ship delivery delays, exceeding the initial contract which cause in delayed benefits for the community and increased costs. The project delays in new shipbuilding will occur, if the material components and sub-components are not handled properly as stated in the paper. [3]. In the paper, it is also found that in several countries such as Brazil, Latvia and other, such as the military shipbuilding project at Brazil shipyard's has repeatedly occurred, due to delays in the completion of construction which increased their costs. [4].

Running a business always comes with a risk, inability to face risks can have serious consequences such as financial losses, lower stock (price) and loss of business reputation or even the worst, bankruptcy. For the very first time in Latvia, the research with the main risk analysis is identified and analyzed in industry shipbuilding and ship maintenance. The aim of that research is to identify, analyze and give the main risk rating in shipbuilding in Latvia; based on the possibility of unwanted results and the possible damage level [5], in Liu and Zhao writing that many uncertainties are found in the shipping industry. Data obtained shows that the time difference between the construction contract and the ship delivery varies from 2 to 5 years. [6].

The delays in construction of the government-owned ships can be anticipated by conducting a risk assessment of the ship components installed during construction, starting from identification, assessment, and evaluation to mitigation that needs to be carried out. In a book published by BPPT, it is stated that domestic ship production policies need to be followed by efforts to develop the local ship component industry, based on the industry potential and the availability of the local raw material. The measurement standardization and ship technical specifications also provide convenience and economic value for the component industry to produce components with the same specifications in large quantities, according to the needs of the number of ships to be built [7], the data and information from the government website [8] furthermore, a risk assessment will be carried out on the components of the equipment components installed in the construction of the government-owned ferry. The aim of the research to be carried out is to regulate and determine the component model of the ferry during construction; so that, it can anticipate delays in ship handover according to the construction contract that has been carried out.

2. Literature Review

The development of marine resources depends on their strong marine engineering industry. The hazardous marine environment poses the highest risk for all maritime works. There is a way in decreasing the related risk with maritime works; by using some analytical methods. One of the published studies discusses the integrated Interpretive Structural Modeling (ISM) with Bayesian Network (BN) approach in risk assessment context. Mitigation and management of maritime risk depend on the experience and knowledge of the domain expert. ISM can be used for combining the knowledge systematically and help to apply the setting; also brief on the complex relationships that exist between system elements; then, determine the relationship of engineering risk factors represented by a causal diagram that builds the BN structure. Then, the subjective assessment is changed to a conditional probability for inclusion in BN. For example, BN application has applied to evaluate the risks of two offshore pipeline projects in Taiwan. The results show that BN can provide risk information explicitly to support better project management. [8]. The writer tries to apply this method to the construction preparation and installation of a government-owned ferry in Indonesia.

Globally, the shipbuilding industry is influenced by demand or need activities and develops along with economic development. In the industrial world, the terms supply chain risk management, shipbuilding and repairing industries are included in the Engineering to Order category, which is a product procurement system by producing product subject to the specification of customer orders. [9].

In a global market overview, the shipbuilding industry is influenced by demand and global economic activity. Basically, whenever the economy is in turmoil, international trade increases then sea transportation will increases also. This could result in a positive trend for many industries, including the shipbuilding industry, which can be seen in the increase number in new ship orders [10], especially Indonesia, which is an archipelagic country. Sea transportation is still considered cheaper rather than by air transportation for connectivity of inter-island, especially the remote and outer islands of Indonesia which has not been reached by air transportation.

2.1. The Risk on Shipbuilding Industry

The shipbuilding industry is influential and related to other industries, including sea transportation and goods supply services, fisheries, and defense. It also affects its supporting industries, which include machinery, electronic components, steel, and other material supplies. The stability of the user industry and its support is direct influential in the shipyard. The multilateral institution, government, and related institution have an important role in supporting the shipbuilding activity, either directly or indirectly affecting the demand for new ship orders. [6].

The shipping industry has unique characteristics compared to other industries. This industry is a strategic industry that has a multiplier effect on other industries such as having a large demand for steel materials and other supporting industries. There are other industries relevant to the shipping industry, as follows: steel, power generator, ports, offshore, fishing, tourism, and security defense. [11].

According to Kurniawati and Triwilaswandio paper, Supply chain performance is largely determined by the internal conditions of the builder shipyard itself, identification of the factors that because problems can help to find the right solution; so that, the shipyard can design the improvement strategies to increase the productivity of new shipbuilding [12], In a large project such as the production of a new ship building, this risk is very possible, so it can be a loss result if this risk occurs and cannot be controlled properly [13] therefore, it is necessary to conduct a risk assessment study on the installed equipment components, as one of the contributors to the highest percentage of ship delivery delays.[14].

3. Proposed Framework & Discussion

In Yuri Kochetkov and B. Aliev's writing, the research methodology used is the statistical data analysis, shipyard industry system analysis and its environment. In the research framework has been found that most of the risk in shipbuilding industry and maintenance in Latvia could be associated with very small, small and medium risks. The final decision in adopting and optimization of the risk in industry company become the management prerogative right in a company.

The management of industrial companies should pay special attention to the marketing department, as their responsibilities include both commercial and financial risks. The commercial risk is very small, but it can have a very large probability of damage. Other things that are very important to pay attention are attention to political risks, especially on an international scale, as well as monitoring changes in the Latvian tax laws. [5].

3.1. SCRM (Supply Chain Risk management)

The SCRM process is described in a similar manner in the literature, which can vary from risk identification and /or measurement, risk assessment and / or risk evaluation to mitigation plans and control and / or monitoring. [10]. Supply chain risk identification is considered a fundamental stage of overall risk management. The determination of risk ranking is based on the terminology of the probability of failure and the impact of failure which is arranged in a risk matrix. Using a risk matrix, hull outfitting, hull construction and machinery outfitting work are categorized as high risk. The high risk category needs a mitigation process to lower the risk rating. [14].

This approach involves a comprehensive and structured determination of the potential supply chain risks associated with a given problem. By grouping the work on the components of the ship's machinery, electricity and other outfitting; so, that mitigation actions can be prepared to keep the construction time and ship handover fulfilling the project construction contract.

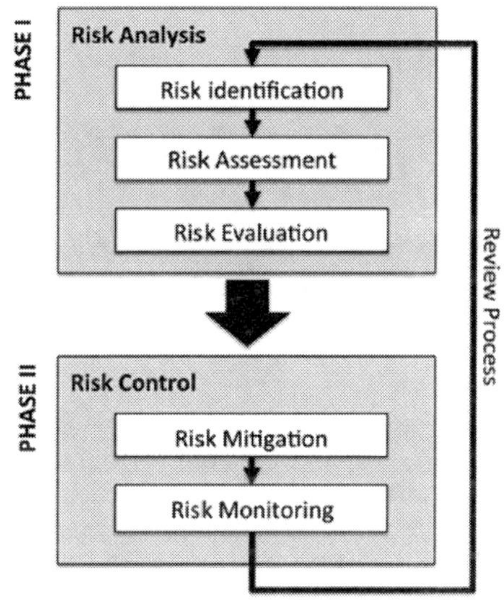

Figure 2. RSCM General Process.
Source. F. de A. L. Ferreira, etl, 2018.

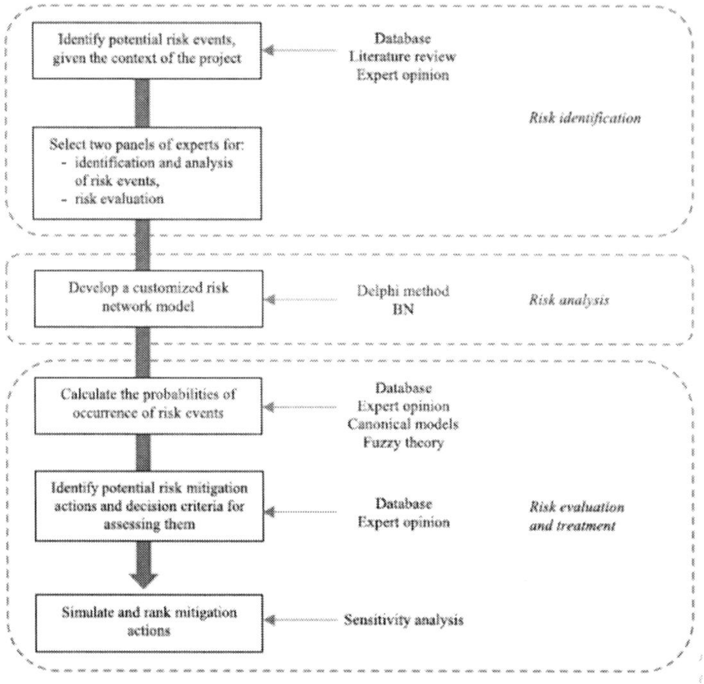

Figure 3. The framework of risk assessment chart.
Source: J. Crispim, J. Fernandes, and N. Rego, 2020.

The framework of risk start from identification of risks, risk analysis and end by mitigation risk and treatment.

3.2. Research Design

This paper follows a systematic approach in writing by Felipe Da 2018, analysis and risk management that is applied when installing ship components in new shipbuilding in the context of the supply chain.

The research design will be carried out on the construction of the Indonesian government-owned Ro-Ro Ferry at a domestic shipyard. The approach recommended by several previous research publications (e.g. [10], [13] [15] [16] [18])

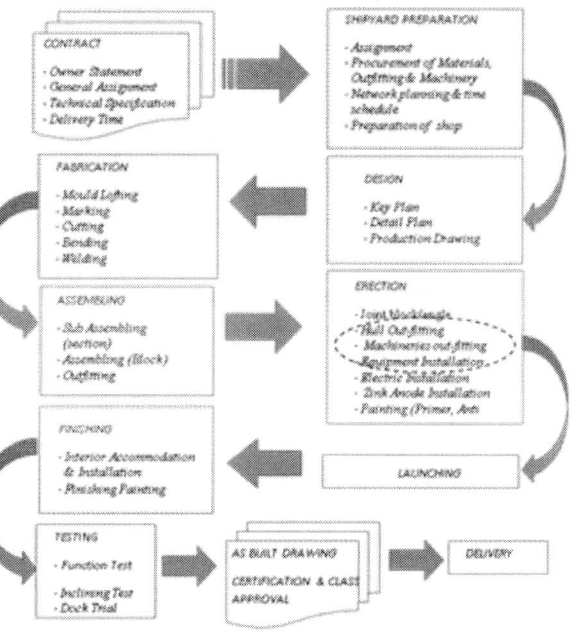

Figure 4. Flow Chart Production.

Figure 4 shows the production flow chart of the Ro-Ro ferry at a domestic shipyard when building a government-owned project, for a ferry weighing 500GT, usually the construction contract to handover is 15 months.

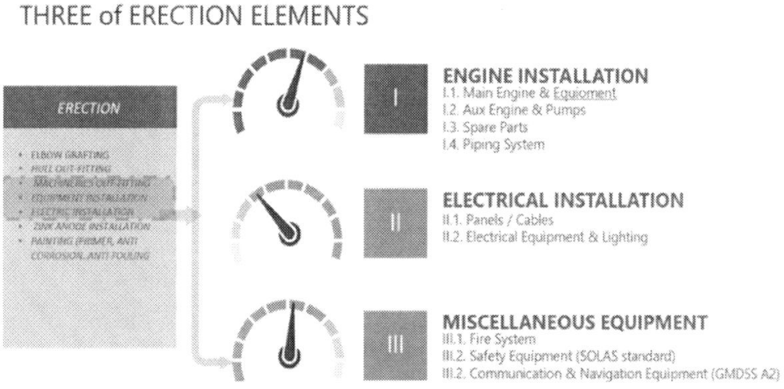

Figure 5. Three Erection Elements.

Figure 5 shows three component elements that considered during risk analysis of tools and ship equipment's installation. Previous data stated that national-owned shipyards and private shipyards have been facing delivery delay based on the delivery time that stated on the letter. This study will describe the crucial factors that caused the delays by risk management analysis during the components and installations of the machinery, electricity and other tools.
Research is being carried out on national and private shipyards that still on-build process of government-owned RoRo ferries and located in Indonesia.

4. Conclusion

The domestic shipyard as one of the components in sea industry should get serious attention; so that become the competitive industry. On a certain aspect and scale, its capabilities are not inferior to world-class capabilities, but further development is needed. Considering that in reality, many needs for domestic vessels but their repairs still depend on foreign shipyards.

The analysis of the potential development in the component industry that was enabled to be developed in the domestic area by considering the readiness of the industry, including the availability of resources (human resources and raw materials), the existence of the industry, government policies, and economies of scale that could be achieved by developing the component industry.

The risk model assessment was categorized to be design activities, procurement of materials activities and production activities. The intended production activity was the process of preparation and installing any sequence of equipment that had to be installed on a ship. On time delivery of the equipment and component of ship building will take biggest portion on ship delivery time.

The determination of risk ranking was based on the terminology of the probability of failure and the impact of failure which was arranged in a risk matrix. Using a risk matrix, hull outfitting, hull construction and machinery outfitting work were categorized as high risk. The high-risk category required a mitigation process to lower the risk rating.

Technological innovations must continue to be developed to support development programs that are on time, reduce costs related to delays in development time and realize the ideals of the State of equality and social justice.

Acknowledgment

This research is a series of dissertation research under the BPPDN scholarship The Education Ministry of Indonesian. Research data is sourced and collaborative from Transportation Ministry of Indonesia, several national and private shipyards in Indonesia. Thank you very much, this will continue for the advancement of the nation.

References

[1] *https://id.wikipedia.org/wiki/Kapal_feri*, 2020. [Online]. Available: https://id.wikipedia.org/wiki/Kapal_feri.
[2] Kementrian Perhubungan, "Perhubungan Darat Dalam Angka 2018," 2019.
[3] M. Basuki, D. Manfaat, S. Nugroho, and A. Dinariyana, "Improvement of the Process of New Business of Ship Building Industry," *J. Econ. Business, Account. | Ventur.*, vol. 15, no. 2, p. 187, 2012.
[4] J. Crispim, J. Fernandes, and N. Rego, "Customized risk assessment in military shipbuilding," *Reliab. Eng. Syst. Saf.*, vol. 197, no. August 2019, p. 106809, 2020.
[5] Yuri Kochetkov and B. Aliev, "Risks in the Shipbuilding and Ship Repair Industry in Latvia," *Manag.* ., vol. 28, no. 1, p. p9–13. 5p, 2016.
[6] J. Liu and J. Zhao, "Bullwhip Effect andthe Value of Information Sharing in Supply Chain -- Illustrated bu a Shipbuilding Supply Chain," *IEEE*, vol. 0, no. 1, 2007.
[7] Buana Ma'ruf, *Strategi Pengembangan Industri kapal Nasional Berbasis Teknologi Produksi dan Pasar Domestik*. 2014.
[8] K. P. Republik Indonesia, "Pengembangan Industri Perkapalan Dapat Prioritas," *https://kemenperin.go.id/artikel/21493/Pengembangan-Industri-Perkapalan-Dapat-Prioritas*, 2020.
[9] F. Adrodegari, A. Bacchetti, R. Pinto, F. Pirola, and M. Zanardini, "Engineer-to-order (ETO) production planning and control: An empirical framework for machinery-building companies," *Prod. Plan. Control*, vol. 26, no. 11, pp. 910–932, 2015.
[10] F. de A. L. Ferreira, L. F. Scavarda, P. S. Ceryno, and A. Leiras, "Supply chain risk analysis: a shipbuilding industry case," *Int. J. Logist. Res. Appl.*, vol. 21, no. 5, pp. 542–556, 2018.
[11] W. Bagenda, E. T. Sule, E. Febrian, and D. Sari, "The influence of unique capability, functional

strategy on company performance of shipbuilding industry in Indonesia," *Int. J. Bus. Glob.*, vol. 20, no. 3, pp. 273–291, 2018.

[12] L. Syaravina and Sunaryo, "Analisa Rantai Pasok Material Pada Kawasan Industri Maritim Terhadap Produktivitas Industri Perkapalan," *Proceeding Semin. Nas. Tah. Tek. Mesin XIV*, no. SNTTM XIV, pp. 7–8, 2015.

[13] C. P. P, M. Basuki, and E. Pranatal, "Analisa Risiko Operasional Proses Bangunan Kapal Baru (Studi Kasus Pembangunan Kapal Lpd 124 M Di Pt . Pal Indonesia (Persero))," *Semin. Nas. Sains dan Teknol. Terap. IV 2018 ITATS*, vol. IV, 2018.

[14] M. Basuki and A. . W. Putra, "Penilaian Risiko Pekerjaan Bangunan Baru Pada Galangan Kapal Klaster Jawa Menggunakan Matrik Risiko," in *Prosiding Seminar Nasional Aplikasi Sains & Teknologi (SNAST) 2014*, 2013, pp. 1–8.

[15] D. Matei, "Analysis of Operational Risks in Shipbuilding Industry," *Risk Contemp. Econ.*, vol. 1, no. January 2012, pp. 121–130, 2012.

[16] T. and M. B. Hidayat, "Improvement Analysis of Component Local Content on Domestic New Building," *J. Wave*, vol. 10, no. Desember 2016, pp. 61–66, 2016.

[17] T. Prasetyo, M. Buana, and A. Sulisetyono, "Analysis of Local Ship Component Industry Development," *J. Wave*, vol. 10 Nomor 2, no. Desember 2016, pp. 39–46, 2016.

[18] Ma'ruf Buana, "Ship ' s Standardization of Domestic Shipping for Competitiveness of the National Shipping Industry," in *Kongres Teknologi Nasional 2016*, 2016, pp. 25–27.

IOP Conf. Series: Materials Science and Engineering 1052 (2021) 012032 doi:10.1088/1757-899X/1052/1/012032

Design of laminated bamboo fishing boat with local cultural heritage using electric motor to support fishing tourism in Pasuruan, East Java

M H N Aliffrananda[1], A R Safaruddin[1], H Supomo[1] and S Regitasyali[1]

[1]Department of Naval Architecture, Institut Teknologi Sepuluh Nopember, Surabaya, 60111, Indonesia

Abstract. Fishing has been one of the main sources of livelihood in Indonesia, to the point that it gives birth to numerous fishing culture preserved by the locals in the country. However, the development of Indonesia's fishing tourism is quite slow, despite its identity as an archipelagic state. In attempt to boost the fishing tourism sector, this study proposes a design of a fishing boat with a touch of Pasuruan cultural heritage, a region in East Java. The design is hoped to attract the tourists' attention by presenting them a local heritage fishing experience. Laminated bamboo is utilised as the boat's material so it can be recycled as furniture when no longer in use. The boat's main dimensions are obtained by means of intuitive design method. Technical calculations including the boat's resistance and propulsion, required power and stability are carried out subsequently. The main dimensions for this boat are: LoA = 8.2 m; B = 1.1 m; H = 0.6 m; and T = 0.42 m. The service speed is set to be 7.5 knots. While the required power to move this boat is around 7.481 kW to power an electrical motor with 25 HP.

1. Introduction

Traditional fishing boats have been part of the Indonesian culture. Not only because of the historical and anthropological value of the nation, these wooden boats are still used up to this day by local fishermen. Despite the technological development in shipbuilding, wood is still mainly used in fishing boats construction. One of the underlying reasons is because wooden ships are relatively cheaper when compared to ships made of steel. It is apparent that wood is one of the highly demanded commodities in Indonesia.

However, a study stated that Indonesia had lost 8.4% of its forest, equivalent to 15 million hectares between 2000 and 2012. Of the 98% forest loss in Indonesia, deforestation occurred in high density forest conservation areas such as in Sumatra, Kalimantan and Sulawesi. The significantly expanding areas reserved for palm oil plantations over the last 20 years is the main cause of forest destruction. Illegal loggings also took part in worsening the deforestation in Indonesia [1]. Wood scarcity began to occur, causing the price of wood to increase. The higher price of wood resulted in an increased production cost for fishing boats and directly threatened the sustainability of the small and medium scale industry (*Industri Kecil Menengah*), such as the local shipbuilding industry. This condition then triggered *multiplier effects* on the smallholder shipbuilding industry chain. A substitute material for wood was highly needed. A substitute material for wood was highly needed.

Eventually, it was later found that bamboo could serve as a substitute for wood as the main material used in fishing boat construction, given its positive characteristics such as [2]: its abundance throughout the archipelago, its relatively easy cultivation and preservation nature, rapid growth (15-30 cm/day) [3], its period for construction $\geq$ 3 years, and its price being relatively cheaper than solid wood with adequate

IOP Conf. Series: Materials Science and Engineering 1052 (2021) 012032 doi:10.1088/1757-899X/1052/1/012032

mechanical properties. Compared to wood which takes decades to harvest and use as the main material for ships, bamboo, on the other hand, only takes about 3-4 years with the same capability for construction.

A survey in East Java [4] found that bamboo is abundant, in terms of species and quantity. There are two types of bamboo that can be used as a ship construction material: Betung Bamboo (Dendrocalamus asper) and Ori Bamboo (Bambusa Arudinacea). Betung bamboo species (Dendrocalamus asper) is spread only in the highlands (350m - 700 m asl) in almost every mountain in Java, while Ori bamboo (Bambusa Arundinacea) is widely distributed in the lowlands (0 – 300 m asl), as seen in Figure 1.

a) b)

Figure 1 The abundance of Ori bamboo in Indonesia is shown through a) and b) where Ori bamboos are found near rice field [4].

Currently, the construction reliability level of laminated bamboo as ship material has been studied, including its mechanical properties of bamboo and its use in accordance with the classification rules of the Indonesian Classification Bureau (BKI) 2013. A flexural study found that laminated bamboo is very suitable to be used as a fishing boat material, construction wise [4]. Bamboos are also cheap, environmentally friendly, strong, renewable, flexible and sustainable.

Despite the proven capabilities of the laminated bamboo's mechanical properties, it is important to produce a prototype of a Jukung fishing boat for the purpose of socialization to the local people. This prototype needs to be developed to build the fishermen's trust in the bamboo laminated construction. With this prototype, it is hoped that the fishermen and the community shipyards will be more than encouraged to opt for laminated bamboo over solid wood.

2. Design Methodology

2.1. Spiral Design

The designing process of a ship is a continuous process, first introduced in 1959, as seen in Figure 2.

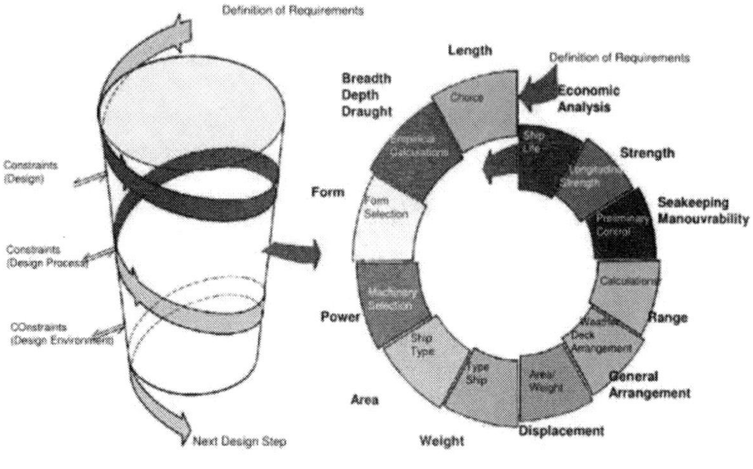

Figure 2 Ship design spiral depicted as a continuous cycle [5]

IOP Conf. Series: Materials Science and Engineering 1052 (2021) 012032 doi:10.1088/1757-899X/1052/1/012032

The ship design will undergo continuous correction until the ship's requirement is completed and is ready to be launched. The stages are divided into *Concept Design, Preliminary Design, Contract Design* to *Detail Design*. The ship design is also divided into several main parts, namely: basic design, hull design, hull outfitting design, machinery outfitting, electrical outfitting and miscellaneous.

2.2. Fishing boat
Based on the Regulation of the Ministry of Marine Affairs and Fisheries of the Republic of Indonesia No. 16 in 2010, fishing boat is a special vessel for fishing activities that has a driving force in the form of a paddle, wind, or combustion engine [6]. Fishing boats are a mode of transportation in the sea which plays a vital role for coastal communities whose livelihoods are fishermen. Fishing boats in general are still being built traditionally by following the previous shipbuilding method which tends to be without proper planning even though with the development of shipping technology, every ship construction including fishing boats can be planned according to the demands and needs of the owners.

By maintaining traditional technology in wooden shipbuilding without any planning, the impact on the use of wood is excessive, namely ± 20 to 50 m^3 for ships with a keel length of 10 to 20 m. This is a waste of wood raw materials. Meanwhile, the production of teak wood, for example, has experienced a drastic decline due to illegal logging and the time it takes for its reproduction is quite long, which is ± 50 years. Hence, the production cost for one unit of a traditional fishing wooden boat is very high.

Traditional shipbuilding techniques in Indonesia are still fixated on previous ship building experiences without the basis for proper and accurate ship planning and design. Such development process has several shortcomings, including low ship speed in relation to the amount of resistance. This is due to the hull shape is less smooth because of the absence of a lines plan and design plan. Apart of that, the use of rudder is not effective because its position does not align with the propeller and if the size of the rudder is not proportional, it can lead to poor ship maneuverability.

2.3. Jukung fishing boat
The characteristic of these boats is its small size, it is not so wide but long, so it is suitable for in local rivers and swamps. Usually, the Jukung fishing boat is made of solid wood and dredged to get the shape of the desired hull. This will be a waste of wood causing the production of boat to decrease as the wood population decreases. Several types of the Jukung fishing boats include:

2.3.1. Jukung Untul and Jukung Polokan. The Jukung Untul type in Figure 3 has a principal size of 3.50 m long, 0.62 m width and 0.35 m depth, distributed around the areas of Malang, Blitar, Tulungagung, Trenggalek, Pacitan, Tuban, Sampang, Pamekasan, Pasuruan, Probolinggo and Situbondo. The Jukung of the Polokan type in Figure 4 generally has a principal dimension of 8 m long, 0.85 m wide and 0.30 m deep, distributed around areas of Trenggalek, Pacitan, Banyuwangi, Jember and Malang.

Figure 3 Jukung Untul

Figure 4 Jukung Polokan

2.3.2. Jukung Kunting and Jukung Selentik. The Jukung Kunting type in Figure 5 generally has a principal size of 3.50 m long, 0.50 m width and 0.30 m depth, distributed in Trenggalek and Pacitan. The Jukung Selentik type in Figure 6 generally has a principal size of 8 m long, 0.85 m wide and 0.30 m deep, found in Tuban (Tambakboyo) and Gresik.

Figure 5 Jukung Kunting

Figure 6 Jukung Selentik

2.3.3. Jukung Banat and Jukung Pandik. The Jukung Banat type in Figure 7 generally has a principal size of 8 m long, 0.8 m wide and 0.40 m deep, found in of Tuban (Tambakboyo) with an almost extinct population condition. The Jukung Pandik type in Figure 8 generally has a principal size of 12 m long, 0.6 m wide and 0.30 m deep, found in Gresik and an almost extinct population condition as well.

Figure 7 Jukung Banat with only one outrigger

Figure 8 Jukung Pandik with no outiggers

2.4. Bamboo lamination

2.4.1. Advantages of laminated bamboo. Laminated bamboo is one of the most promising and renewable technology for producing future construction materials [7]. A study included the manufacture of laminated bamboo with thin blades followed by tensile and bending tests. In 2011, it was found that the parallel tensile strength of the fibers of Giganthloa strictus can reach more than 200 MPa, with MOR flexural strength up to more than 100 MPa, MOE more than 15 GPa. Another study about the strength and possibility of flooring production using a layer of Phyllosachys edulis bamboo slats [8] discussed the advantages of bamboo laminated material when compared to the use of solid wood in terms of durability and sustainability.

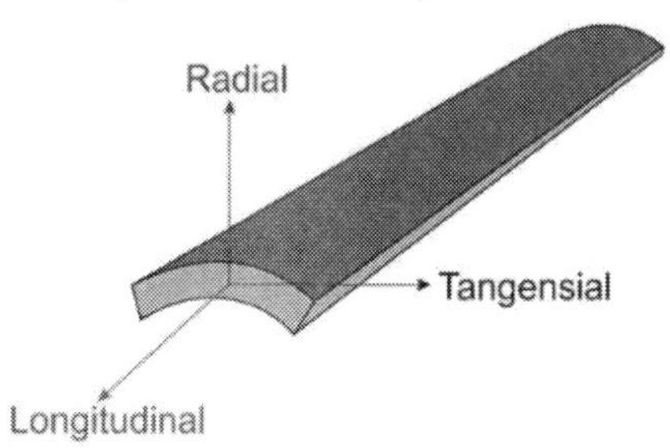

Figure 9 Direction of stress on a blade of bamboo.

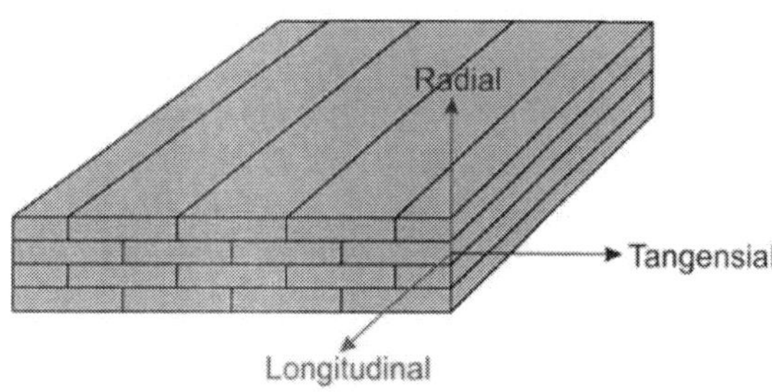

Figure 10 Direction of stress on a laminated bamboo.

Laminated bamboo can also be produced using bamboo fiber or bamboo lumber [9]. This construction also possesses a higher tensile and bending strength when compared to solid teak wood. An initial study in 2015 [4] found that the tensile strength of the laminated blades of Bambusa arundinacea parallel to the fibers reached up to 184 MPa with MOR flexural strength of 84.6 MPa and MOE around 24 GPa. This would be very profitable when used to construct a fishing boat. The direction of tensile and bending stress under study can be described as shown in Figure 9 and Figure 10.

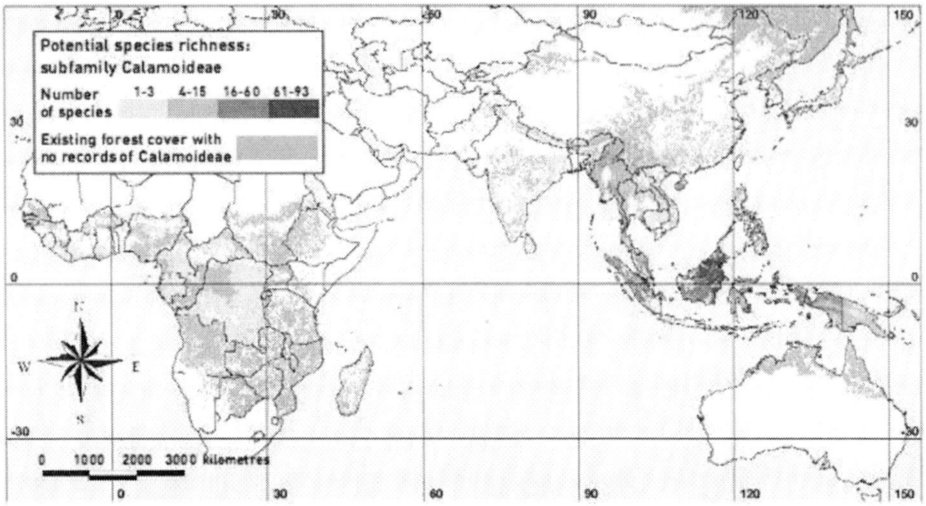

Figure 11 Distribution map of bamboo in the world. [10]

Based on the distribution map, bamboo species and densities can be seen in Figure 11 [10]. Indonesia has a very large density of bamboo species when compared to other countries in the world. This is one of the greatest potentials that bamboo in Indonesia is very abundant in number.

2.4.2. Mechanical properties of bamboo. The mechanical properties of bamboo in general depend on the following factors [11]: The type of bamboo, the age of the bamboo at the time of logging, the moisture (equilibrium moisture content) of the bamboo stems, the part of the bamboo stems used (legs, middle, head), location and distance of used segments (sections of the segment are less resistant to compressive and bending forces). The determination of the mechanical properties of bamboo is based on the prerequisite that the bamboo used in construction is a dry building material with a moisture content of 12%, an equilibrium moisture content at 70% humidity which is considered a reasonable average in tropical climates.In several studies [2], [12] and [13], the mechanical properties of bamboo including tensile strength, compressive strength and shear strength were discussed. In Indonesia, the allowable tensile stress parallel to the direction of the average fiber without mentioning the type of bamboo is 29.4 N/mm^2, while in Indonesia, the required bending stress is 9.80 N / mm ^{2}.

3. Building Process

3.1. Design of Jukung Boat

The Jukung ship design that will be used is the Jukung type in the area around Pasuruan-East Java. This Jukung fish boat is redesigned according to local cultural heritage. The initial design (preliminary design) was made based on the main capacity and size of the Jukung. After the main size and capacity were determined, a basic Jukung design was made from laminated bamboo with an output in the form of Lines Plan, Hydrostatic & Bonjean Curves, General Arrangement, Cross Section, Construction Profile and Scantling, according to BKI regulations with local cultural heritage in mind.

3.2. Mold production

This laminated bamboo Jukung fishing boat mold is made according to the Lines Plan design. With the Mold Lofting method, the shape of the mal is obtained at each position of the Jukung ship. The shape of each cross section is then made using multiplex board. The boards were then arranged lengthwise according to its order with the help of markings and were connected to one another with wood. This mold for Jukung boat with laminated bamboo material can be used repeatedly as long as the main dimension of the boat is identical.

3.3. Fabrication of construction components
The fabrication stage is the manufacture of the components of the Jukung boat construction elements, consisting of keel and frame. This fabrication process was made based on the mold of the Jukung boat with laminated bamboo material.

3.4. Assembly and erection
The assembly and erection process of the Jukung prototype consisted of assembling the keel and the frame, continued by installing a shell made from laminated bamboo. After the Jukung body was formed, the left and right outriggers were installed before applying paint and anti-fouling protection to the hull. The next process is the installation of the boat's electric motor.

3.5. Trial and performance test
After the production has been completed, the next step is to launch the Jukung, followed by trial and performance test to assess the stability of Jukung, maneuvering, speed and resistance of Jukung when operating. The trial and performance data results are presented and socialized to marginalized fishing communities.

4. Results and Discussion
4.1. Design of Jukung Boat
4.1.1. Survey. The survey was carried out in the Sendang Biru area, Malang, a fishery center with a TPI (Fish Auction Place). This survey was carried out to picture an initial view of the design proposed in this study. One of the boats, "Tiga Saudara" shown in Figure 12 had the following main dimensions: 10 m long, 1.1 m wide, 0.75 m high and 30 mm shell thickness.

Figure 12 Tiga Saudara Jukung Boat

4.1.2. Jukung design process. The process of designing a Jukung fishing boat can be carried out based on the offset value from the "Tiga Saudara" Jukung Boat. The final main dimensions have a slight difference than the "Tiga Saudara" Jukung Boat.

Table 1 The Main Dimensions of Jukung Fishing Boat Prototype

Parameter	Dimension	Unit
Length Overall (LOA)	8.2	m
Breadth (B)	1.1	m
Depth (H)	0.6	m
Draught (T)	0.42	m
Service Speed (Vs)	7.5	knot

The first design to be carried out was the Lines Plan design, which was the design of the hull shape of the ship. The Lines Plan design consists of a drawing of the ship's body plan, which is a cross section of the shape of the hull when viewed from the front. This shape is then projected into a sheer plan and half breadth plan so that it becomes a Lines Plan from three views (front, top and side). The design of the ship lines plan can be seen in Figure 13.

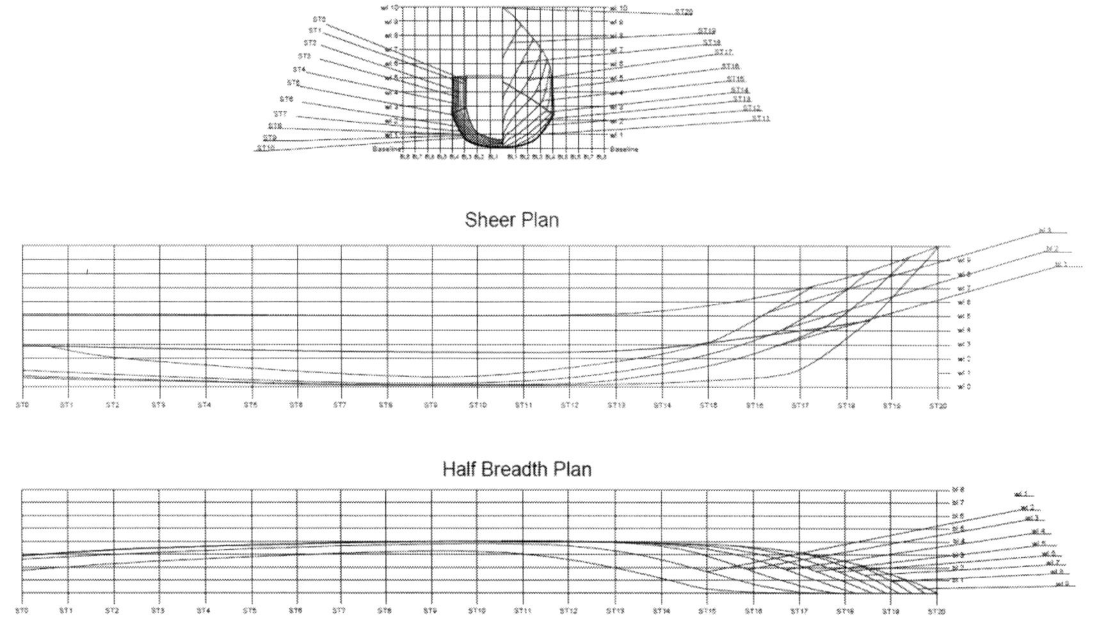

Figure 13 Lines plan of Jukung fishing boat shows that the boat has a slender body, enabling it to sail at a high speed.

The Lines Plan were transformed into a 3D computational model as can be seen in Figure 14. The model is made according to the main dimensions, as can be seen in Table 1. Once the dimensions are set, the curvature of each frame is used as a guideline in creating an accurate replica of the boat by using a 3D design software.

Figure 14 Jukung Fishing Boat 3D Model from perspective view.

Using the 3D model, the hydrostatic analysis of the boat can be obtained directly from the design software. The results are shown in Table 2.

Table 2 Hydrostatic Analysis of Jukung Fishing Boat Prototype

Parameter	Dimension
Displacement	1.864 tonnes
Wetted Surface Area (WSA)	9.4 m^2
Prismatic coefficient (Cp)	0.685
Block coefficient (Cb)	0.562
Maximum sectional area (Cm)	0.834 m^2
Longitudinal Centre of Buoyancy (LCB)	3.165 m
Longitudinal Centre of Floatation (LCF)	3.312 m
Keel to Buoyancy (KB)	0.256 m
Keel to Metacenter (KM)	11.846 m
Immersion (TPc)	0.066 tonne/cm
Momen Trim (MTc)	0.031 tonne.cm
Hull weight, LCG, TCG and KG	1.968 tonnes
LCG	3.769 m
TCG	0 m
KG	0.404 m

When the hydrostatic characteristics of the boat have matched the requirements, the following process include producing a drawing of General Arrangement of the boat as can be seen in Figure 15.

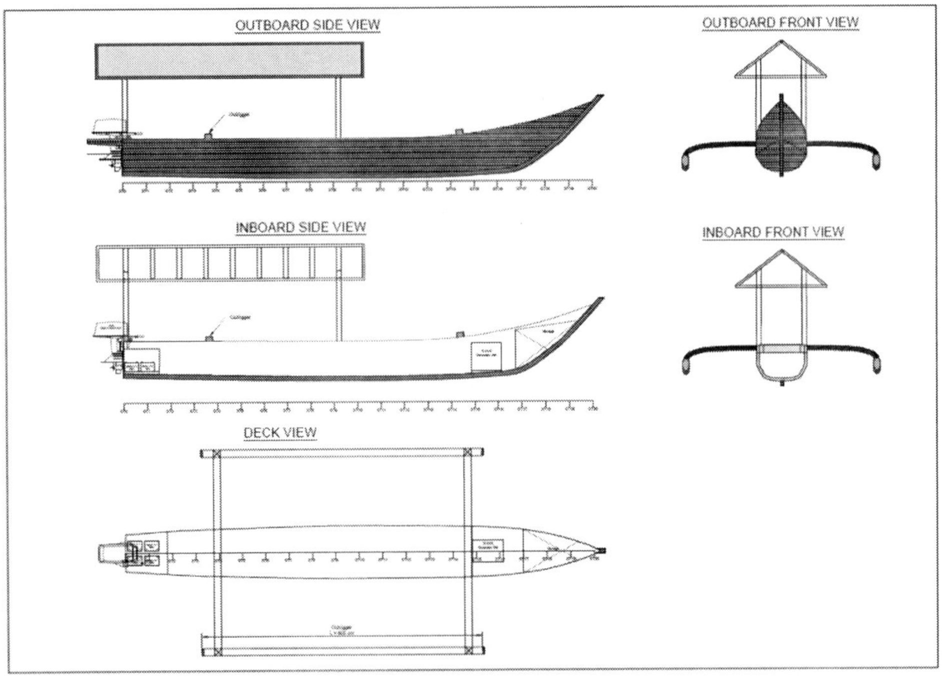

Figure 15 General arrangement of Jukung fishing boat.

The General Arrangement is drawn using the outlines of the boat acquired from the Lines Plan to depict the interior design of the boat. Prior to planning the General Arrangement, a series of calculation is conducted including the calculation of the required power found to be around 7.481 kW. The result was used in determining the capacity of the electric motor, which was found to be 25 HP, supported by 15 batteries of 12 V 200 Ah. The obtained power can move the boat at a speed of 7.5 knots. It is crucial to ensure that the bamboo hull can withhold the stress it experiences during operation. To achieve this, the boat's construction must possess sufficient strength to help support the boat. A series of calculation is done to obtain the adequate construction size to prevent the bamboo hull from experiencing stress beyond its capacity that could lead to deformation during her operation. The calculation was conducted in accordance with [14]. The mechanical properties of the laminated bamboo played an important role in determining the numbers and sizes of the construction elements of the fishing boat. Figure 16 is the layout of the boat's construction.

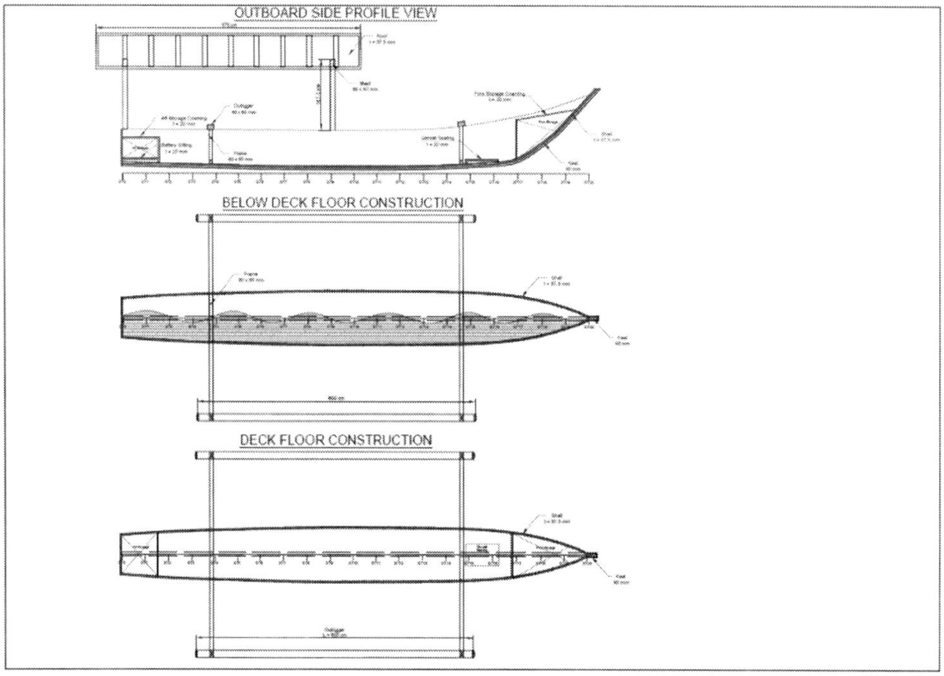

Figure 16 Construction profile of Jukung fishing boat.

This Jukung fishing boat only has two side transverses to support the hull, right below the transverse arms of the outrigger. A detailed drawing of the boat's construction seen from a transverse point of view can be seen in Figure 17 along with its dimensions as listed in Table 3.

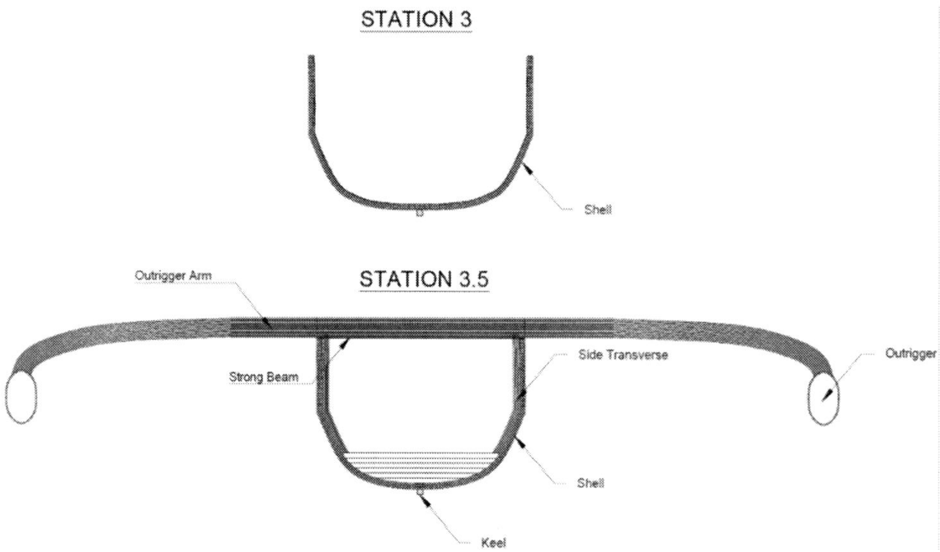

Figure 17 Cross section of Jukung fishing boat.

Table 3 Construction details of Jukung fishing boat.

Construction	Dimension
Strong beam	60 x 60mm
Side Transverse	60 x 60 mm
Outrigger Diameter	130 mm
Outrigger Arm	600 cm
Transverse Outrigger	80 x 60
Shell	37.5 mm
Keel	30 mm

Another crucial point to be considered is the boat's stability. The components of the boat's construction, as well as the hull itself, determine the ability of the boat to turn back to its right position after experiencing an inclination at a certain angle. In this study, the stability analysis is obtained through a simulation using a ship design software.

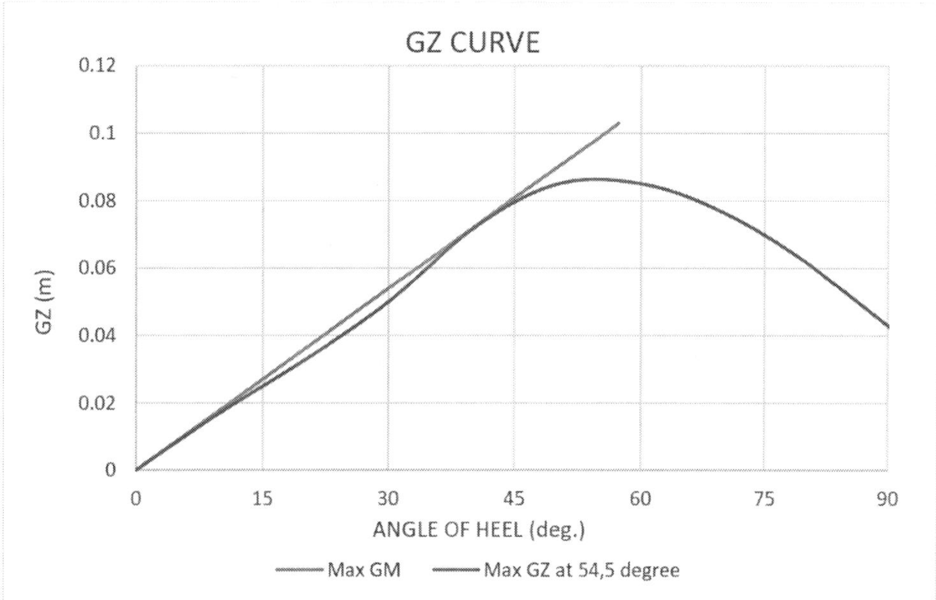

Figure 18 GZ Curve of Jukung Boat without outriggers.

According to Figure 18, the design is shown to have good stability according to its maximum value of GZ, which equals to 0.086 m at 54.5° and its GM_0 value, which equals to 0.103 m. GZ is the length of a righting arm that is needed to return the boat to its equilibrium state when it experiences inclination at a degree. GM is the distance between the center of gravity to an imaginary point of intersection, between the extension of vertical line through the center of buoyancy and a vertical line through the new center of buoyancy when the ship is inclined, also known as the metacentric height.

4.2. Mold production and assembly

The first step to produce a Jukung fishing boat is to make a mold. This mold is made using a multiplex material for its properties that tend to be stronger than ordinary plywood. In addition, multiplex is relatively cheaper and relatively easier to shape than solid wood. The multiplex had the following sizes: 1220 mm wide, 2400 mm long, and 9 mm thick. This multiplex is then cut into shape using a CNC cutting machine as shown in Figure 19, Figure 20, and Figure 21 which is located at the Protomodel Laboratory, Department of Product Design, ITS. The multiplex is cut according to the Lines Plan. There are 20 stations that had to be molded on the multiplex, hence the cutting process chosen to make this Jukung fishing boat is CNC cutting due to its ability to cut in a fast rate while maintaining its cutting accuracy.

Figure 19 Mould production out of multiplex using cutting machine CNC.

Figure 20 Mould production out of multiplex using cutting machine CNC for multiple stations.

Figure 21 Post-cutting of a station of Jukung fishing boat mould.

The subsequent step is to assemble the mold lengthwise of the boat. This series of molds must be placed in the order they are stationed. In addition, the ship molds must be arranged in reverse as in

Figure 22. This was done to simplify the erection process of the Jukung fishing boat. These molds are assembled on straight solid wood to ensure alignment between each mold. The height of each mold must also be adjusted according to the sheer height.

Figure 22 Assembly of Jukung fishing boat mould in reverse order.

4.3. Fabrication of construction components

The bamboos in use need to be fabricated before being assembled and erected into a ship. The bamboos used for ship construction components are Ori bamboo and Betung bamboo which are 1.5 - 2 years old, at this age the bamboo is ripe and ready to be cut down. Logged bamboo is then broken into several strips and cut into 2 meters long, to maintain its alignment. These 2 meters long bamboo strips must be flattened on all four sides (top, bottom, right and left) prior to being laminated. The alignment of the four sides of the bamboo slats is done using a planner machine.

Figure 23 Flattening of bamboo strips using a planner machine.

The bamboo flattening process as shown in Figure 23, is carried out until the entire bamboo surface is flat on all four sides. Apart from being flat on some sides, bamboo must also be clean from the outer skin to maintain the effectiveness of the gluing process.

4.4. Finishing of Jukung Fishing Boat

After undergoing previous steps, the laminated bamboos of the Jukung fishing boat are assembled. The results can be seen in Figure 24 and Figure 25.

a)

b)

Figure 24 Side **a)** and front **b)** view of the Jukung fishing boat after assembly process without the outriggers, as they are still a work in progress at the moment.

a) **b)**

Figure 25 For the finishing touch, the boat's surface is refined by rubbing sandpaper to its construction and applying gel coat to prevent the bamboo from absorbing water as seen in **a)**. Gel coat can also benefit as an exterior shield to keep the bamboo in place even when it experiences crack. It adds aesthetic values to the boat as seen in **b)**.

5. Conclusion

A half complete design and production of a Jukung fishing boat has been conducted as elaborated. Bamboo as the main material of the boat offers more benefits than regular wood. Bamboo is cheaper, as well as more abundant than wood and is found to have similar strength as regular wood. The following were found in this study:

1. The result of this study can be used as a solution to increase the number of Jukung fishing boat owned by marginal fishermen based on a number of reasons:
   a. Laminated bamboo Jukung fishing boat is a sustainable and easily made construction,
   b. It is relatively cheaper than wooden fishing boat for marginalized fishermen,
   c. Bamboo is abundant in Indonesia in terms of species and quantity in Indonesia,
   d. This study can solve the lack of new technology implementation among local fishermen by means of maintaining the local cultural heritage.
2. The main dimensions of the Jukung fishing boat prototype are as listed in Table 1.
3. The results of technical analysis are as follows:
   a. Hydrostatic calculations as listed in Table 2.
   b. Resistance calculations found that the resistance of the Jukung fishing boat is 738.41 N.
   c. Power calculation. With the value of the resistance found, the power required to move the boat can be determined. With a 30% efficiency, the required power is found to be 9.497 kW.
   d. Machinery specification. Using the obtained required power, a 25 HP electrical motor was chosen in order to move the ship with a service speed of 7.5 knots.

Furthermore, the following recommendations can be considered in further research following this study:

1. Perform assembly and erection of all components of the Jukung fishing boat construction made of laminated bamboo,
2. Conduct economic judgment analysis on the construction of the Jukung fishing boat made of laminated bamboo,
3. Hold an educative socialization of the results of innovative laminated bamboo Jukung fishing boat prototype to marginalized fishermen, including the manufacturing methods and sea trials.

References

[1] M. C. Hansen *et al.*, "High-resolution global maps of 21st-century forest cover change," *Science (80-.).*, vol. 342, no. 6160, pp. 850–853, 2013, doi: 10.1126/science.1244693.
[2] Morisco, "Pemberdayaan Bambu untuk Kesejahteraan Rakyat dan Kelestarian Lingkungan," 2006.
[3] S. Cao, J. D. Wang, H. S. Chen, and D. R. Chen, "Progress of marine biofouling and antifouling technologies," *Chinese Sci. Bull.*, vol. 56, no. 7, pp. 598–612, 2011, doi: 10.1007/s11434-010-4158-4.
[4] H. Supomo, D. Manfaat, and A. Zubaydi, "Flexural strength analysis of laminated bamboo slats (Bambusa Arundinacea) for constructing a small fishing boat shells," *Trans. R. Inst. Nav. Archit. Part B Int. J. Small Cr. Technol.*, vol. 157, no. Part B1, pp. 23–31, 2015, doi: 10.3940/nna.ijsct.2015.bl.167.
[5] W. T. Pribadi, S. I. Wahidi, and I. Baihaqi, "Manajemen Produksi Kapal," no. 5 M, 2018.
[6] A. . Fallis, "Peraturan Menteri Kelautan dan Perikanan Republik Indonesia Nomor PER.16/MEN/2010," *J. Chem. Inf. Model.*, vol. 53, no. 9, pp. 1689–1699, 2013.
[7] C. S. Verma and V. M. Chariar, "Development of layered laminate bamboo composite and their mechanical properties," *Compos. Part B Eng.*, vol. 43, no. 3, pp. 1063–1069, 2012, doi: 10.1016/j.compositesb.2011.11.065.
[8] C. H. Lee, M. J. Chung, C. H. Lin, and T. H. Yang, "Effects of layered structure on the physical and mechanical properties of laminated moso bamboo (Phyllosachys edulis) flooring," *Constr. Build. Mater.*, vol. 28, no. 1, pp. 31–35, 2012, doi: 10.1016/j.conbuildmat.2011.08.038.
[9] C. A. Fuentes *et al.*, "Wetting behaviour and surface properties of technical bamboo fibres,"

Colloids Surfaces A Physicochem. Eng. Asp., vol. 380, no. 1–3, pp. 89–99, 2011, doi: 10.1016/j.colsurfa.2011.02.032.

[10] N. Bystriakova, V. Kapos, C. Stapleton, and I. Lysenko, "Bamboo biodiversity: Information for planning conservation and," 2003.

[11] I. Suprijanto, Rusli, and D. Kusmawan, "Standardisasi Bambu Laminasi Sebagai Alternatif Pengganti Kayu Kontruksi," *Pros. PPI Stand. 2009*, no. November, p. hal 1-23, 2009.

[12] K. Janssen, J.J., Boughton, G., Adkoli, N. S., Ranjan, M.P., Sastry, C. B., Ganapathy, P. M., Ramanuja Rao, I., V., Ghavami, K., Ravindran, *Bamboo as an engineering material*, no. 1991. Ottawa, Canada: International Development Research Centre, 1991.

[13] S. Handayani, "Pengujian Sifat Mekanik Bambu (Metode Pengawetan Dengan Boraks)," *Penguji. Sifat Mek. Bambu (Metode Pengawetan Dengan Boraks)*, vol. 9, no. 1, pp. 43–53, 2009, doi: 10.15294/jtsp.v9i1.6921.

[14] Biro Klasifikasi Indonesia, *Volume vii rules for small vessel up to 24 m 2013 edition biro klasifikasi indonesia*, vol. VII. Jakarta: Biro Klasifikasi Indonesia, 2013.

The correlation of tsunami heights and coastal conditions in Palu Bay using the Contingency Coefficient Analysis

I H Hariyanto, D G Pratomo, and M R Darminto

Department of Geomatics Engineering, Institut Teknologi Sepuluh Nopember,Surabaya, 60111, Indonesia

Abstract. The accelerated development of Palu City and its vicinity areas surrounding Palu Bay has changed the land use especially in the coastal area. These changes may reduce the function of the beach as a natural bathing wave. The 2018 Palu Bay tsunami have varying heights. This height variation is probably because of the difference in both geographical and geological conditions in coastal areas which are located surrounding the bay. The research used the contingency coefficient as a method for determining the magnitude of the correlation between tsunami heights and several parameters, such as: land cover, slope, and lithology. The results show the land cover parameter has the most effect the height of the tsunami wave recorded. Based on the significance test using the Chi-Square, the slope of the coastal area is the most representative parameter with a 66% confidence level.

1. Introduction

Population growth in the city of Palu since it was established as a municipality in 1994 increased by 42.55% from 1990 to 2000 and by 18.37% from 2000 - 2010 [1]. This growth requires that development in Palu City and areas along the coast of Palu Bay be accelerated to serve the needs of the community and provide adequate facilities. Based on the earthquake and tsunami catalog, Palu Bay was recorded to have experienced an earthquake and tsunami from 1921 to 2018 with a span of 20 years [2].

Palu Bay itself is bypassed by an active fault called the Palu Koro fault that stretches from the mainland of southern Sulawesi to the northwest to the waters of North Sulawesi [3]. This fracture activity can make earthquake, tsunami, and liquefaction possible later. The earthquake on 28 September 2018 which was centered in Donggala Regency, caused land movements which caused a tsunami due to underwater landslides and liquefaction phenomena in Palu City [4].

The implementation of the post-disaster survey showed the difference in height of the tsunami waves that reached the coast. This difference can be due to changes in land use of coastal areas for development at some point so that it changes the function of the beach as a natural wave absorber [5]. In addition, other geographical and geological parameters at these locations may have a role in changing tsunami wave heights. This study aims to find out how big the correlation of these parameters (land cover, slope, and lithology) to the height of the tsunami wave that has been classified using contingency coefficients as a determinant of the correlation value. The results of this study are expected to be input in evaluating disaster management in terms of natural factors so as to minimize the impact of tsunamis that occur in the future.

2.Data dan methods

2.1. Data

The data used in this study consisted of:

a. Digital Map of Indonesian Coastal Environment (LPI) Number 2015-05
This 1: 50,000 scale map was obtained from the Geospatial Information Agency (BIG). Data taken from the map is bathymetry data for slope processing and land cover data. The bathymetry data itself is the result of compilation of 2014 hydrographic survey data, generalization of a 1: 25,000 scale RBI Map in 2013, and a 1: 200,000 Scale Marine Map in 2013 [5]. Based on technical rules, the vertical datum used in digital LPI maps is the lowest low water level (LLWL) [6].

b. Tsunami Observation Point Survey Data
The 2018 Tsunami Observation Point Survey Data provides authentic field measurements at 24 sample points along the coast of Palu Bay. Administratively, the location is spread in Palu City and Donggala Regency. Wave height data collection is done by using a laser point, GPS Handheld to measure the coordinates of the location of the sample point, and several other supporting equipment [7]. Tsunami wave height measurement methods are measured based on trace water lines that remain on building walls, the presence of dry trees caused by seawater, rubbish or material involved, or coral rocks that have been shed to the mainland.

Table 1. Location.

No	Location	Longitude (°)	Latitude (°)
1	Panggang	119.775	-0.719
2	Lolilondo	119.781	-0.747
3	Lolipesua	119.789	-0.770
4	Lolisaluran	119.811	-0.818
5	Primkopal	119.819	-0.844
6	Tipo	119.829	-0.861
7	Silae	119.835	-0.875
8	Ruko Lere	119.840	-0.881
9	Grandmall Palu	119.843	-0.882
10	Mercure Palu	119.850	-0.884
11	TVRI Palu	119.863	-0.886
12	Kp Nelayan	119.878	-0.864
13	Citraland	119.881	-0.837
14	Tondo	119.880	-0.832
15	Pergudangan	119.882	-0.823
16	Kp Mambaro	119.877	-0.802
17	Poltekes	119.865	-0.790
18	Resort Taipa	119.859	-0.782
19	PLTU Tawaeli	119.855	-0.732
20	Pantoloan	119.852	-0.708
21	Ngada Wani	119.840	-0.695
22	Labuan	119.817	-0.662
23	TPI Lero	119.812	-0.629
24	Pasir Marana	119.789	-0.595

c. Geological Map
Geological Map Review Palu Sheet, Sulawesi scale of 1: 250,000 second print of 1996 from the Center for Geological Research and Development. This map contains some information about lithology or types of rocks making up the coast of Palu Bay and its surroundings, contours, important objects, and other geological information.

2.2. *Equipment*

The equipment that used in this research are:

a. ArcGIS
   This software is used to display and perform spatial data processing.
b. Microsoft Office
   This word and number processing tool is used to do correlation tests and make reports.

2.3. *Methods*

2.3.1. *Identification of Land Cover*

The parameters of land cover are obtained from the LPI Map Number 2015-05. There are several types of land cover that are available. Then identify the observation points for each overlayed land cover type.

2.3.2. *Identification of Slope*

Slope shows the magnitude of the slope angle in percent or degree [8]. In this study, slope is determined from the waters to the point of observation so that the processing uses bathymetry data and height data of several points on the coast of Palu Bay. The entire data is then interpolated into raster using the Topo to Raster method with a cell size of 9×9 m and then reclassify if its result.

Table 2. Classification of Slope. [9]

Class	Slope %
I	0 - 8
II	8 - 25
III	25 - 40
IV	40 - 100

After being classified, it is then identified each observation point to the existing slope.

2.3.3. *Identification of Lithology*

Lithology is a description of rocks in outcrops based on characteristics such as color, mineral composition and grain size [10]. Lithology parameters were obtained from the identification of each sample point against the Geological Map Review of the Palu Sheet, Sulawesi. The process of identifying points of observation of lithology parameters is carried out on ArcGIS by displaying both layers simultaneously.

2.3.4. *Classification of Tsunami's Height*

Because the overall parameter data is nominal, then tsunami height data is also needed for classification to match the type of measurement scale in the parameter data. The recorded tsunami wave heights are then classified according to the following table [11]:

Table 3. Classification of Tsunami's Height.

Class	Magnitude	Energy E_t (ergs) x 10^{22}	Max Height H_t (m)
I	1.0	1	2 - 3
II	1.5	2	3 - 4
III	2.0	4	4 - 6
IV	2.5	8	6 - 8
V	3.0	16	8 - 12

2.3.5. *Correlation Value Determination*

Determination of the magnitude of the correlation of all parameters with the classification of tsunami wave heights uses the contingency coefficient method. The contingency coefficient itself is a statistical test to determine non-parametric correlations using two variables from nominal-scale data [12].

$$C = \sqrt{\frac{X^2}{N+X^2}} \quad (1)$$

Where

$$X^2 = \sum_{i=1}^{b}\sum_{j=1}^{k}\frac{D_{ij}^2}{E_{ij}}, df = (b-1)*(k-1) \quad (2)$$

$$D_{ij} = O_{ij} - E_{ij} \quad (3)$$

Information:
N: total number of observations
C: contingency coefficient
b: number of rows in the contingency table
k: number of columns in the contingency table
i: 1, 2, ..., b
j: 1, 2, ..., k
O_{ij}: observation data the 'i' row and the 'j' column in the contingency table
E_{ij}: expectation frequency values of 'ij' for O_{ij}
X^2: Chi-square calculation results

2.3.6. *Significance Test*

To find out how good the correlation results are, a significance testing method is needed. Testing the results of contingency efficiency can use the chi-square test. Comparison of the chi-square value calculated from the correlation will then be compared with the ideal chi-square value in the coefficient table. Then the conclusion will be obtained how well each calculated correlation result has been calculated.

3. Results and Discussion

3.1. *Land Cover*

Based on the identification of the location of sample points against land cover data through the LPI Map Number 2015-05, the results obtained are as shown in Table 4.

Table 4. Land Cover Result.

Land Cover	Sample Point
Residential Area	11
Bush	7
Farm	1
Field	3
Vacant Land	2

3.2. *Slope*

The results of the bathymetry data obtained from the LPI Map Number 2015-05 will then be processed into a slope or slope with classification and results as shown in Figure 1.

IOP Conf. Series: Materials Science and Engineering 1052 (2021) 012033 doi:10.1088/1757-899X/1052/1/012033

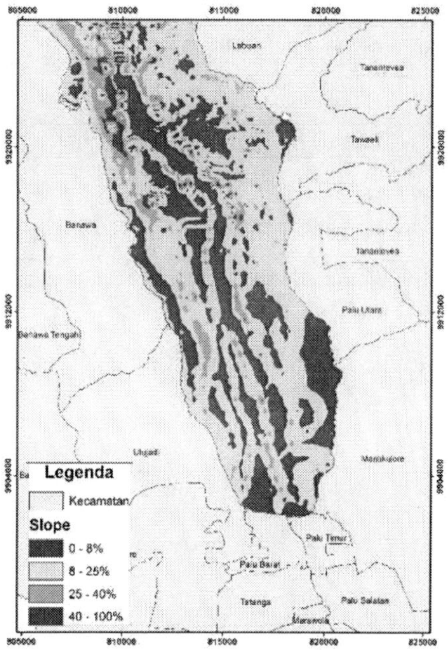

Figure 1. Slope Classification.

Table 5. The Area of Each Class.

Class	Slope %	Area (m^2)
I	0 - 8	58952478.894
II	8 - 25	101433358.648
III	25 - 40	13181428.148
IV	40 - 100	2966008.567

In addition, slope identification results at all sample points are presented in Table 6.

Table 6. Slope Results.

Slope %	Sample Point
0 - 8	1
8 - 25	1
25 - 40	14
40 - 100	8

3.3. Lithology

The results of identification of all sample points for lithology through the Geological Map are shown in the following Table 7.

Table 7. Lithology.

Lithology Type	Sample Point
Alluvium	19
Formation of Tinombo	4
Formation of Pakuli	1

3.4. Tsunami's Height Classification
Tsunami wave heights with vertical datum in the form of mean sea level (MSL) then are classified so that they reach a conclusion as shown in Table 8.

Table 8. Tsunami's Height Result.

Class	Run-up	Sample Point
II	3 - 4	3
III	4 - 6	5
IV	6 - 8	10
V	8 - 12	6

3.5. Correlation
Correlation test obtained from the calculation of contingency coefficients obtained results as in the following Table 9.

Table 9. Correlation Result.

Parameter	Contingency Coefficient
Land Cover	0.561
Slope	0.548
Lithology	0.430

3.6. Significance Test
To determine the suitability of the results of the sample point correlation with the overall population data, a significance test was performed using the Chi-Square analysis method. The results of the analysis of each parameter are shown in the following Table 10.

Table 10. Significance Test Result.

Parameter	Calculated Chi-Square	Chi-Square on the Table	Degree of Freedom %
Land Cover	11.034	10.960	47
Slope	10.286	10.228	66
Lithology	5.432	5.350	50

4. Conclusions
In this study, from the results of the correlation show that the parameters of land cover to be the most influential parameter to the difference in height of the tsunami wave that is equal to 56.12%. While in the significance test, the slope parameter becomes the parameter that has the highest degree of freedom that is equal to 66%. These results indicate that the results of the slope sample correlation data have the closest match to the results of population data processing.

Acknowledgment
This research was made possible thanks to the contribution of the Geospatial Information Agency as the provider of data for the LPI Map Number 2015-05. In the future, this research can be developed in determining the effect of each variable's parameters on tsunami wave heights. Weaknesses in this study can be corrected in subsequent studies in the form of adding measurement parameters, adding sample points, and conducting significance tests at a more varied level of confidence.

References
[1] BPS Kota Palu 2017 Pertumbuhan Penduduk Tahun 2010. Accesed from https://palukota.bps.go.id/statictable/2017/07/05/637/pertumbuhan-penduduk-tahun-2010.html.

[2] PuSGeN 2018 Kajian Gempa Palu Provinsi Sulawesi Tengah 28 September 2018. Bandung: Pusat Penelitian dan Pengembangan Perumahan dan Pemukiman, Badan Penelitian dan Pengembanan, Kementerian Pekerjaan Umum dan Perumahan Rakyat.

[3] Hamilton, W. 1979 Tectonic of Indonesia Region, Geological Survey Professional Paper, UnitedStatesGovernment Printing Office, Washington.

[4] Sukarningsih, N. 2007 Kajian Tingkat Kerentanan Pantai Terhadap Arus dan Gelombang Bedasarkan Citra Satelit dan SIG di Teluk Banten. Bogor: Program Studi Ilmu Kelautan, Institut Pertanian Bogor.

[5] Staff Kantor Presiden. Peta Lingkungan Pantai Indonesia Skala 1:50.000 Wilayah Palu. diakses melalui https://data.go.id/dataset/peta-lingkungan-pantai-indonesia-skala-palu-50000.

[6] Badan Standardisasi Nasional 2001 SNI 19-6726-2002 tentang Peta dasar lingkungan pantai Indonesia skala 1:50 000.

[7] Pribadi, S., dkk. 2018 Merekam Jejak Tsunami Teluk Palu 2018. Laporan Survey Tsunami Badan Meteorologi Klimatologi dan Geofisika.

[8] Sahara, M. 2014 Kajian Kemiringan Lereng dan Curah Hujan Terhadap Tingkat Kerawanan Longsor di Kecamatan Pekuncen Kabupaten Banyumas. Fakultas Keguruan dan Ilmu Pendidikan, Program Studi Geografi. Universitas Muhammadiyah Purwokerto.

[9] Khoiri, M., Jaelani, L. M., dan Widodo, A. 2018 Landslides Hazard Mapping Using Remote Sensing Data in Ponorogo Regency, East Java. Internet Journal of Society for Social Management Systems.

[10] Judefrian. (2013). Litologi Batuan Akuifer. Universitas Sriwijaya Indralaya: Fakultas Pertambangan

[11] Lida, K. 1961 Magnitude, Energy, and Generation Mechanisms of Tsunamis and A Catalogue of Earthquakes Associated with Tsunamis. Tsunamis: Pacific Science Congress Symposium.

[12] Santoso, H. 2009 Analisis Korelasi Berdasarkan Koefisien Kontingensi C Menurut Cramer dan Simulasinya. Semarang: Jurusan Matematika Universitas Negeri Semarang.

CFD analysis of the increase in ship resistance due to biofouling growth represented by roughness length scale

S Regitasyali[1], **M L Hakim**[1] **and I K A P Utama**[1]

[1] Department of Naval Architecture, Institut Teknologi Sepuluh Nopember, Indonesia

Abstract. Fuel consumption is heavily influenced by ship resistance, which mostly consists of friction resistance. During her operation, a ship is susceptible to the attachment of biofouling to its hull. The arbitrary form of biofouling induces an increase in surface roughness, which is responsible for the increase in friction resistance. This research analyses the increase in skin friction resistance due to biofouling, that is represented by a few variations of roughness length scale, in various orders throughout the models. Two-dimensional models of flat plates are utilized to investigate this issue by means of computational fluid dynamics (CFD) analysis. This study uses models with 0.56 m length to be tested in five Reynolds numbers, to be compared to a similar experiment conducted by using wind tunnel. Other models include 30 m and 60 m length to be tested in two Reynolds numbers each. It is found that the increase in skin friction reaches 3.5 – 31.4% for 0.6 m, 68.1 – 206.9% for 30 m with extreme regular roughness and 117.3 – 129.3% for 30 m with extreme irregular roughness, and 59.5 – 179.6% for 60 m with extreme regular roughness and 123.5 – 128.9% for 60 m with extreme irregular roughness.

Keywords: biofouling, CFD, friction resistance, roughness length scale, surface roughness.

1. Introduction

The shipping industry has been the backbone of the world's trade market, as it transports 80 – 90% goods from all around the world. While it is beneficial, it also 89threatens the environment. In 2012, IMO published a research that declared 2.2% of the world's CO_2 emission was a by-product of the world's shipping activities. It is not until April 2018, that IMO adopted the initial strategy on reducing greenhouse gases (GHGs) emission by targeting at least 50% reduction of the 2008 GHGs emission from shipping activities by 2050. It is important to predict the required ship fuel in order to minimize its use. An important component in estimating the use of ship fuel is the ship's resistance. Ships with high resistance values require more fuel use.

Ship resistance is consisted of several components, with friction resistance (R_F) as the biggest component, reaching up to 80-85% of the total resistance (R_T) on low-speed ships and 50% of the total resistance on high-speed ships [1]. This resistance is caused by the friction that occurs between the hull of the ship and the fluid. Ship resistance can increase due to various factors, one of them being the increase in the hull's surface roughness, which in most cases is caused by the attachment and growth of biofouling on the hull.

To analyse the effect of biofouling on the ship's drag, it is important to model the biofouling correctly. However, in numerical simulations, biofouling is almost impossible to model due to the shapes, sizes and the spread of biofouling may vary across the field. Due to this reason, researchers had to model biofouling using other methods. As recommended by ITTC, researchers are encouraged to develop a formula or method, based on experimental data, to predict the effect of coatings and biofouling on ship resistance [2]. A research found that the increase in drag due to biofouling and its distribution reached up to 80% due to the attachment of heavy calcareous fouling and 34% for small calcareous fouling [3]. Numerical simulations have also been conducted to validate these results by means of Computational Fluid Dynamics (CFD) [4]. Another method to investigate the increase in drag due to biofouling involved testing a flat plate in a wind tunnel [5].

In this current study, the effects of biofouling on the increase of friction resistance on a flat plate are analysed by observing the plate's friction resistance coefficient (C_F) using the CFD method. The use of flat plat instead of ship is based on the assumption that the value C_F are similar for both models. Variations in this study include the biofouling's roughness and the plate's size. Each roughness variation has a different effect on the increase of C_F. This difference is expected to provide a realistic insight on the effect of biofouling's roughness in real life situations.

2. Background

2.1. Ship resistance

Ship resistance is consisted of friction resistance (R_F) and residual resistance (R_R). A friction resistance is a force acting tangential to the hull as an accumulation of shear stress induced by the viscosity of the fluid. Meanwhile, one of the components of the residual resistance is induced by pressure: a normal force acting perpendicular to the hull, partly due to viscous pressure resistance (R_{VP}) and wave-making resistance (R_W). If a body is fully submerged in a fluid, the wave-making resistance equals to zero. This is because waves require an interaction between two fluids with different densities to be generated. Meanwhile, the viscosity of the fluid affects the amount of viscous pressure resistance experienced by the ship [6]. The components of the ship resistance can be expressed through (1)**Error! Reference source not found.** and (2).

$$R_T = R_F + R_R + R_A \tag{1}$$

$$R_T = R_F + R_{VP} + R_W + R_A \tag{2}$$

It is common to represent the ship's total resistance in the form of non-dimensional coefficient (C_T) through (3), where ρ is the fluid's density, S is the wetted surface area and V is the body's velocity:

$$C_T = R_T/(0{,}5\rho SV^2) \tag{3}$$

Using (3), each component of the ship's resistance can be rewritten in a non-dimensional form:

$$C_T = C_F + C_{VP} + C_W + C_A \tag{4}$$

Appendages resistance are caused by underwater construction such as propeller, propeller shaft and bilge keels. In this study, the wave-making and appendages resistance are both equal to zero. So, the coefficient of the total resistance becomes:

$$C_D = C_{VP} + C_F \tag{5}$$

$$C_D = (1 + k)C_F \tag{6}$$

Where (1+k) is a form factor used in extrapolation process between a model and a ship. It represents the viscous pressure resistance because the pressure on a body is affected by its own form. Since the models used in this study is a flat plate, the value of form factor is equal to one.

2.2. Boundary layer on a flat plate

The boundary layer is a very thin flow region near the wall of an object where the viscosity and rotational forces cannot be ignored. It is defined as the distance from the wall of the object to a point where the velocity component moving parallel to the wall reaches 99% of the freestream velocity. For flat plates, the higher the freestream speed, the thinner the boundary layer becomes [7]. On flat plates, the boundary layer is divided based on its Reynolds Number (Re), namely the laminar boundary layer and the turbulent boundary layer. Between the laminar and turbulent boundary layers is a transition area, where the laminar flow begins to develop into a turbulent flow, which occurs at Re = 5×10^5.

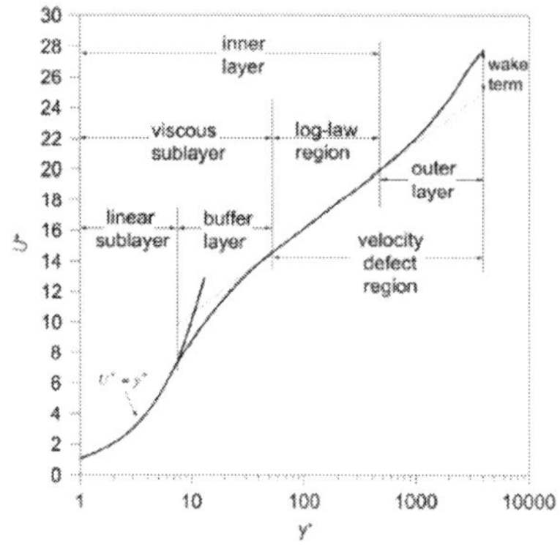

Figure 1 Law of the wall of boundary layer.

Dimensional analysis of a turbulent boundary layer as represented by Figure 1, are as follows:

1. The viscous sublayer is where the viscosity effect of the fluid dominates the flow. This area of fluid is close to the wall and experience direct contact with the surface, thus making the distance, represented by y+, between them equal to zero. Hence, $(y^+ = 0)$. At that exact distance, the fluid is subjected to wall shear stress, rendering it static or in other words, travels with zero velocity $(u^+ = 0)$, so that:

$$u^+ = y^+ \quad (7)$$

2. In the log-law layer, the velocity profile varies in a logarithmic manner. Between the wall and the freestream, both the viscosity and the turbulent effects are equally influential. In this layer, the shear stress (τ) varies along the distance from the wall and is assumed to be constant and equal to the wall shear stress (τ_w). Where, u^+ and y^+ have a logarithmic relationship:

$$u^+ = \frac{1}{\kappa}\ln(y^+) + B = \frac{1}{\kappa}\ln(Ey^+) \quad (8)$$

2.3. Surface roughness

Rough surface increases turbulence resulting in an increase in turbulent stress and wall shear stress. This causes a decrease of fluid velocity in the turbulent boundary layer. The effect of roughness on flow can be seen in the velocity profile [8] as illustrated by Figure 2. The rough surface results in a decrease in the velocity profile of the fluid in the log-law region, indicated by an increase in frictional resistance. This is caused by an increase in the momentum deficit due to roughness and is related directly to the increase in frictional drag on the surface [3]. The derivation of the velocity profile itself is also called ΔU^+ or the roughness function. Hence, the log-law equation becomes:

$$U^+ = \frac{1}{\kappa}\ln(y^+) + B - \Delta U^+ \quad (9)$$

where U^+ = average of scaled viscous velocity, κ = Von Kármán constant (0.41) and B = smooth-wall log-law intercept (5.0).

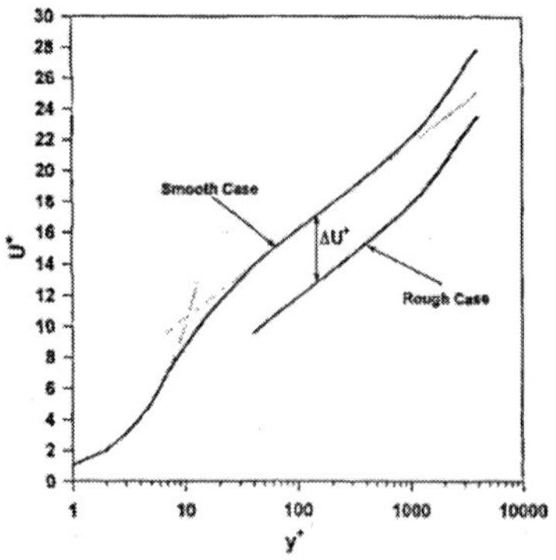

Figure 2 Downward Shift on the Log-law Layer Due to Roughness

3. Methodology

3.1. Mathematical formulation

In this study, the turbulence is modeled using 2-equations, first order RANS Equation. To capture the roughness effect on the viscous sublayer area, the k - ω shear stress transport or k - ω SST turbulence model was used. The turbulence model is a combination of the k - ε turbulence model that models the freestream area, and the k - ω turbulence model, that models the near wall area [9]. In this research, the residual value is set to 10^{-5}.

3.2. Geometry and boundary condition

There are various combinations of boundary conditions that can be applied to solve the case of viscous flow on flat plates with accurate results. However, some combinations require excessive computational effort. A numerical simulation was conducted to validate the frictional resistance experiment due to antifouling paint on flat plates [3]. The domains that were used can be seen in Figure 3 and Figure 4.

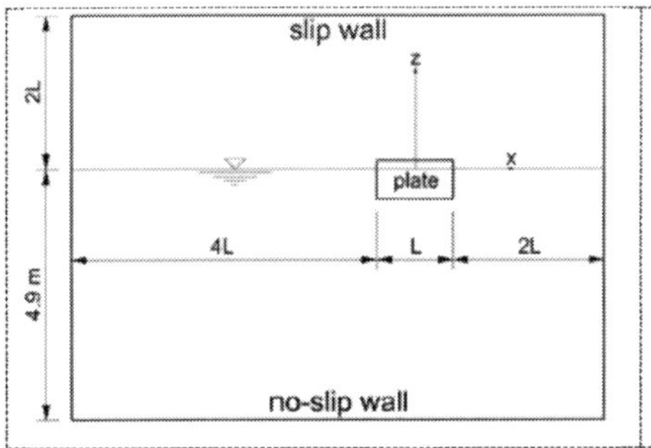

Figure 3. Profile view of validation study domain by Demirel (2014).

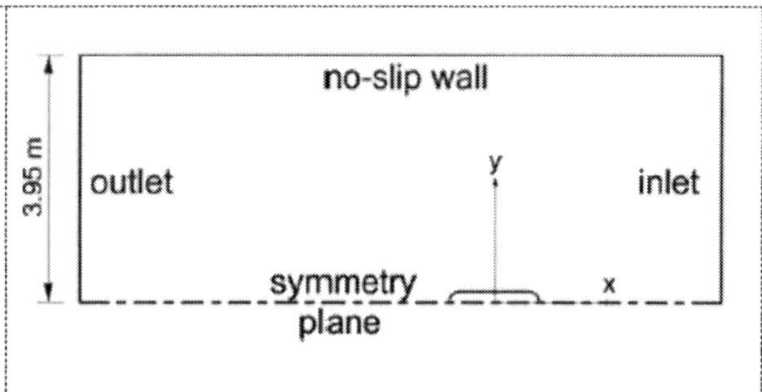

Figure 4. Top view of validation study domain by Demirel (2014).

A study described the techniques needed to validate the William Froude's towing tank experiment [10] using the RANS solver. The area of the edges of the plate only accounted to 0.5% of the total area, allowing the friction resistance of those area to be neglected and allowing the use of a computational 2-dimensional model (Figure 5). The boundary condition was designed as follows: the distance between the leading edge of the plate and the inlet, the distance between the trailing edge of the plate and the outlet, as well as the distance between the upper inlet and the plate reached 2 times the length of the plate (to later be referred to as L). This 2-dimensional model was able to reduce computational effort and time while providing accurate predictions of the drag experienced by the plate.

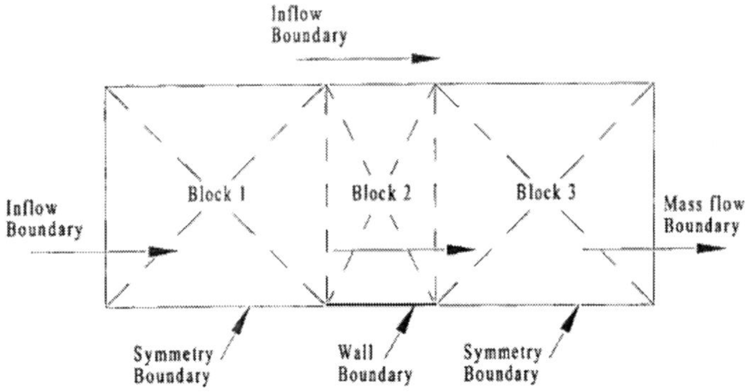

Figure 5 Flat Plate boundary condition to validate Froude's flat plate towing tank experiment.

Recently, a CFD simulation was conducted to investigate the characteristics of drag on a flat plate due to surface roughness caused by biofouling [11]. The study used a similar model of boundary condition as illustrated by Figure 5, with a slight difference: there were no distance between the leading edge of the plate and the trailing edge of the plate to the inlet and outlet of the boundary conditions, respectively as represented by Figure 6.

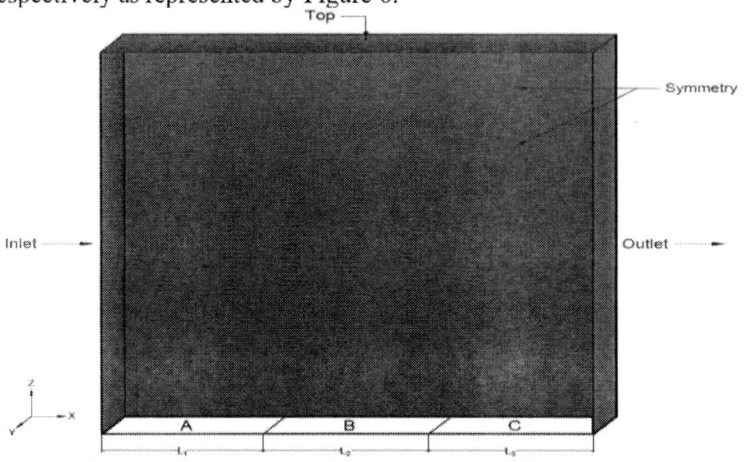

Figure 6 Flat plate boundary condition in a recent CFD study to investigate the characteristics of drag due to surface roughness caused by biofouling.

3.3. *Meshing*

Frictional resistance is dominated by the effect of fluid viscosity in the viscous sublayer. The mesh should be small and dense enough around the viscous sublayer in order to capture the turbulence effect at play. By using y^+, a universal non-dimensional value of distance from the wall to the first grid element (mesh/cell) in a CFD simulation, the value of the first cell height can be obtained by using Flat Plate Boundary Layer equation (10). The appropriate value of y^+ is chosen according to which area of the boundary layer is under observation. For viscous sublayer, the value should lie between $1 \leq y^+ < 5$ according to the law of the wall as illustrated by Figure *1*. In this study, the value of y^+ is equal to 1.

$$y^+ = 0.172 \cdot \left(\frac{\Delta y_P}{L}\right) \cdot Re^{0.9} \tag{10}$$

Since the first cell height value is affected by the Reynolds Number, each simulation in this study had different values as listed in Table 1 and Table 2.

Table 1 First layer height for the CFD validation of wind tunnel experiment.

Length (m)	*No.*	*Velocity* (ms^{-1})	*Re* $(\times 10^5)$	$\Delta y\ (\times 10^5)$
0.56	1	3	1.1	9.6
	2	6	2.2	5.2
	3	9	3.2	3.6
	4	12	4.3	2.8
	5	15	5.4	2.3

Table 2 First layer height for the CFD validation and prediction of characteristics of drag due to surface roughness study.

Length (m)	*No.*	*Velocity* (ms^{-1})	*Re* $(\times 10^8)$	$\Delta y\ (\times 10^6)$
30	1	9.775	2.79	3.8
	2	12	3.4	3.1
Length (m)	No.	Velocity (ms^{-1})	Re $(\times 10^8)$	$\Delta y\ (\times 10^6)$
60	1	9.775	5.6	3.9

	2	12	6.9	3.2

After the first cell heights were determined, the mesh can be generated. In this study, the mesh elements that were formed are hexahedral and quads (Figure 7 - Figure 9). The height of the first cell was set with a growth rate of 1.2 times.

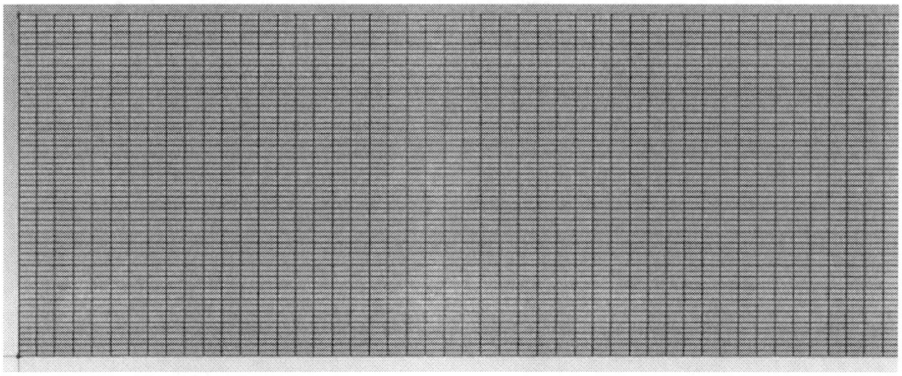

Figure 7 Profile view of meshing at symmetry plane.

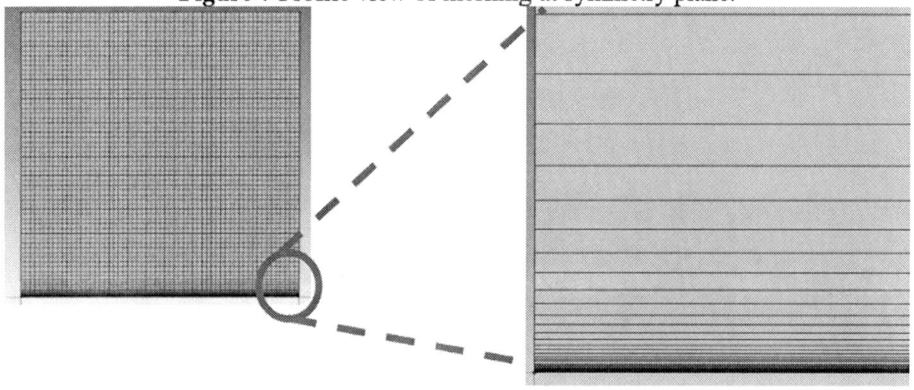

Figure 8 Front view of meshing at the inlet section.

Figure 9 Enlarged view of viscous sublayer at the inlet section.

A good meshing configuration does not compromise the results of the numerical simulation. It is important to ensure the reliability of the mesh by conducting grid independence test. The test is carried out by comparing the numerical results obtained by different number of cell/mesh. The difference should not be significant. In general, the tolerance is set to 2%.

3.4. Surface roughness modeling

Surface roughness is modeled with a roughness length scale based on two parameters, namely the roughness height (k_s) and the roughness constant (C_s). In CFD, this approach can only be used when roughness is constant or remains along the surface of the model. Non-constant roughness can be modeled with user-defined functions.

In this study, simulations were conducted as follows: a validation of a flat plate experiment in a wind tunnel [5], a validation and prediction of the numerical analysis of drag characteristics due to surface roughness on flat plates [11]. In the prediction study, extreme roughness length scales were used to analyze the effect of surface roughness that is in the range of different roughness heights.

3.4.1. Validation of wind tunnel experiment

The roughness height used in the wind tunnel experiment was obtained by scaling the biofoulings that were attached to KM Mentari Pratama's hull at PT PAL during her dry-dock intermediate survey. Biofoulings were found to be the densest around the ship's stern and much lesser around the midship.

This was because biofoulings at the bow had experienced contact with stronger waves than at the stern. Strong waves are capable to wash biofoulings out of the ship's hull. The biofoulings were modeled by sandpapers in the experiment by using a scale factor of 1: 175 [5]. The main dimension of the plate as presented by Table 3, while the roughness length scales can be seen in Table 4.

Table 3 Wind tunnel experiment flat plate size.

No.	Principal Dimension	
1	Length	0.56 m
2	Width	0.75 m
3	Thickness	0.04 m

Table 4 Roughness length scale of biofouling in the first validation study.

Section	ks (mm)	ks (μm)	Sand-paper grit		Notated as
Stern	0.264	264	Medium	P60	Ks 3
Midship	0.111	111	Very Smooth	P150	Ks 1
Bow	0.163	163	Smooth	P100	Ks 2

In the wind tunnel experiment, sandpapers are attached to a flat plate as visualized in Figure 10. As the notation increases, the roughness elements of the sandpaper are rougher and denser to each other. This indicates the increase in the sandpaper's density as the notation increases. However, the sandpaper is generally represented by a "grit", a measurement of the sandpaper's roughness (not to be confused with the notations used in this study). It is inversely related to the actual roughness of the paper. The smaller the grit value, the rougher the sandpaper is.

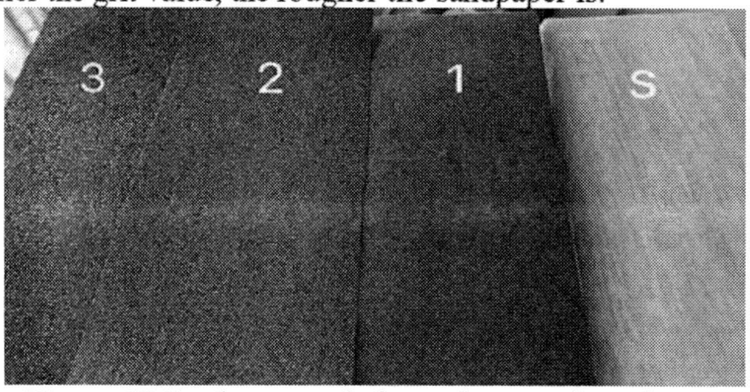

Figure 10 Sand-paper grit for each type of roughness.

Ten variations of surface roughness were tested in this study: smooth, regular roughness and irregular roughness. The variations order that were used in the wind tunnel experiment is as seen in Figure 11.

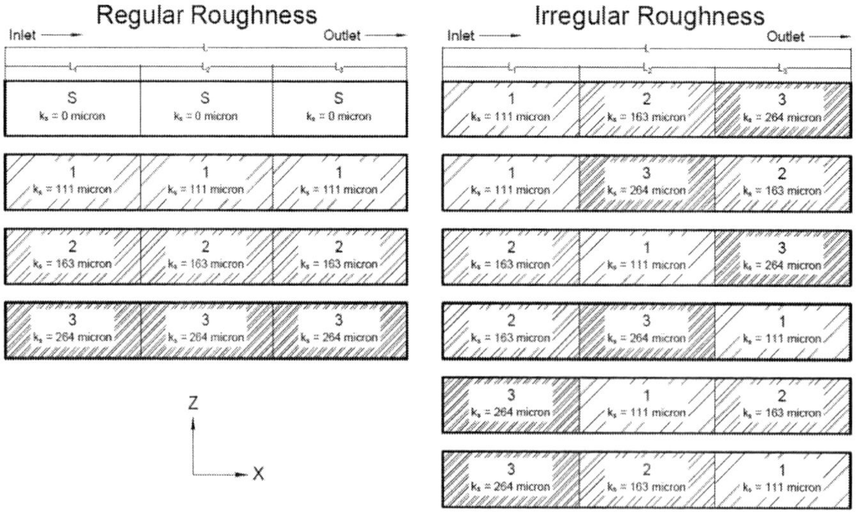

Figure 11 Roughness variation in the first validation study.

The second validation study utilized the following range of roughness shown in Table 5. While the sequence of roughness variations followed the same pattern as demonstrated by Figure 11.

Table 5 Roughness length scale of biofouling in the second validation study.

No.	k_s *(μm)*	*Notated as*
0	0	S
1	81.25	Ks 1 (P)
2	325	Ks 2 (Q)
3	568,75	Ks 3 (R)

Then, a prediction referring to the second study is conducted by using extreme roughness length scales. In this prediction, the model and domain used refer to the model used in the second validation study as listed in Table 6. The models were selected by considering the numerical capacity of the computer device that was used for this study. Each model was tested at two speed variations, namely 19 knots (9.78 m/s) and 23.32 knots (12 m/s).

Table 6 Model sizes used in the drag characteristics validation and prediction study.

No.	**L (m)**	**B (m)**
1	30	3
2	60	6

The extreme values for the prediction of the second study were chosen based on a towing tank experiment in 2007. In the towing tank experiment, it was assumed that the technical roughness value of biofouling could be represented by an equivalent roughness length scale using the roughness heights. The drag characteristics study due to biofouling utilized roughness length scales that represented biofouling in the form of light slime to heavy slime [3]. Meanwhile, the prediction study used extreme roughness length scales (Table 7) that represented biofouling in the form of heavy slime to heavy calcareous fouling. The variation sequence for the prediction study is illustrated by Figure 11.

Table 7 Roughness length scale of biofouling for the prediction of the second study.

No.	*Description of Condition*	*Extreme* k_s *(μm)*	*Notated as*
1	Heavy slime	300	Ks 1
2	Small calcareous fouling or weed	3000	Ks 2
3	Medium to high calcareous fouling	5700	Ks 3

4. Results

4.1. Wind tunnel experiment validation

4.1.1. Smooth plate

The CFD results as listed in Table 8, are compared to the wind tunnel experiment results (WT) [5] and the approximation values by the ITTC formula. As demonstrated in Table 8, the computed skin friction coefficient differed up to 3.89% from the wind tunnel (WT) experiment results. The difference between the CFD simulation and the wind tunnel experiment might be caused by a few unknown variables. One of them being the wind tunnel's unknown turbulence intensity. There were no data available regarding the turbulence intensity of the wind tunnel in which the flat plate experiment was carried out. However, the computed skin friction coefficient was in excellent agreement with the ITTC '57 frictional correlation line, only differing up to 1.71% from the correlation line. Therefore, this CFD simulation of the wind tunnel experiment for the smooth plate condition can be validated and be used in further investigation, namely the regular roughness simulations. The comparison between each approach is shown in Figure 12.

Table 8 Skin friction coefficient for smooth plate comparison between the wind tunnel results and CFD simulation.

No.	V (m/s)	Re	C_F ITTC	C_F CFD	%	C_F WT	C_F CFD	%
1	3	$1{,}1\times10^5$	8,1080	8,0926	0,19	7,7900	8,0926	3,88
2	6	$2{,}2\times10^5$	6,7133	6,6335	1,19	6,4400	6,6335	3,00
3	9	$3{,}2\times10^5$	6,1045	6,0002	1,71	5,8100	6,0002	3,27
4	12	$4{,}3\times10^5$	5,6809	5,6311	0,88	5,4200	5,6311	3,89
5	15	$5{,}4\times10^5$	5,3838	5,3289	1,02	5,1400	5,3289	3,68

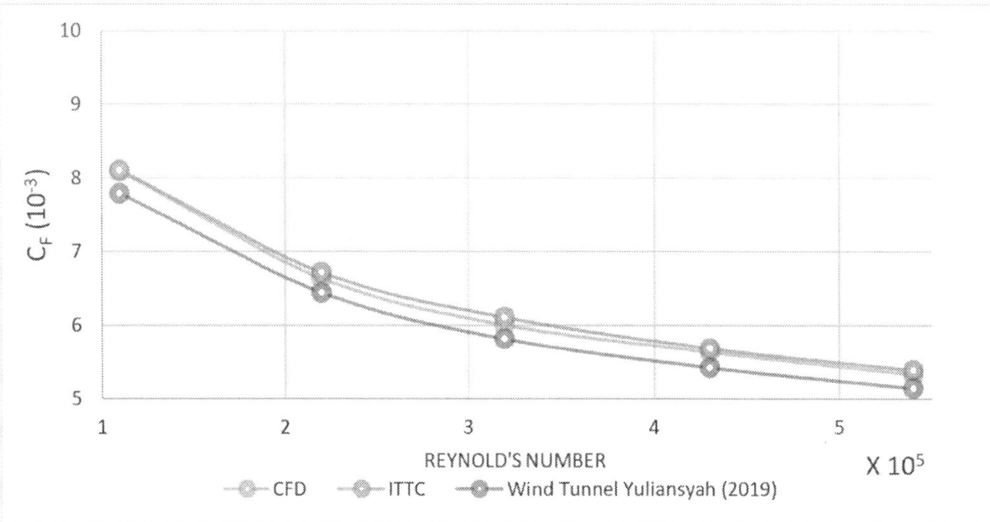

Figure 12 Comparison between skin friction coefficient for smooth plate the wind tunnel results and CFD simulation.

4.1.2. Regular roughness

In addition to the smooth flat plate drag analysis, CFD analysis of flat plate drag due to regular roughness was also carried out. The results can be seen in Table 9.

Table 9 Skin friction coefficient for smooth plate comparison between the wind tunnel results and CFD simulation.

No	V (m/s)	Re	C_F (10^{-3}) Smooth	K1	K2	K3	ΔC_F (%) K1 (%)	K2 (%)	K3 (%)
1	3	$1{,}1\times10^5$	7,79	10,58	14,83	17,06	35,82	90,37	119,00
2	6	$2{,}2\times10^5$	6,44	7,46	8,8	9,94	15,84	36,65	54,35
3	9	$3{,}2\times10^5$	5,81	6,47	7,38	8,1	11,36	27,02	39,42
4	12	$4{,}3\times10^5$	5,42	6,01	6,7	7,34	10,89	23,62	35,42
5	15	$5{,}4\times10^5$	5,14	5,66	6,27	6,82	10,12	21,99	32,69

In the wind tunnel experiment [4], the ΔC_F decreased with the increase of velocity. On the other hand, the ΔC_F increased with the increase of velocity in the CFD analysis. This might be caused by the unknown value of the roughness constant (C_S) value in the wind tunnel experiment [4]. In the CFD simulation, the value was assumed to be equal to be 0.5, which was a default value in the CFD software. Figure 13 is an illustration of the C_F comparison between the wind tunnel experiment [4] and the CFD simulation.

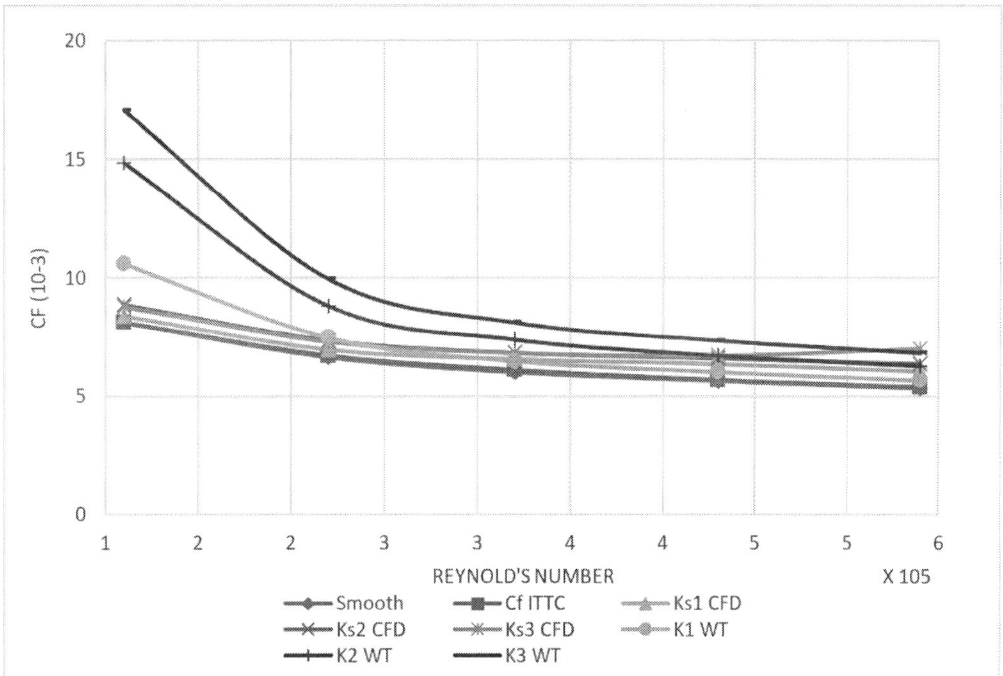

Figure 13 Comparison between skin friction coefficient for smooth plate the wind tunnel results and CFD simulation.

4.2. CFD analysis of drag characteristics due to surface-roughness

4.2.1. Validation study

A validation of a prior CFD analysis of the drag characteristics on a flat plate [11] was also conducted in this study. The previous study analyzed a flat plate with surface conditions as follows: smooth, regular roughness and irregular roughness. The validation was only carried out on the smooth surface conditions and regular roughness. The model used is a flat plate with 30 m in length and a breadth of 3 m, at a Reynolds number of 2.79×10^8. The results of the recent CFD study were found to be in excellent agreement with the prior CFD analysis. The maximum difference between both simulations were below 5% as listed in Table 10. Therefore, this validation study can be verified as valid and may be used in the prediction study.

Table 10 Validation study of CFD analysis [11] on drag characteristics due to surface roughness.

Notation	Ks (micron)	C_F (10^{-3}) [11]	C_F (%)	C_F (10^{-3}) CFD	ΔC_F (%)
S	0	1,8052	-	1,7865	-
Ks1	81,25	2,2499	24,63	2,2536	24,84
Ks2	325	2,9407	62,9	2,9392	62,82
Ks3	568,75	3,2400	79,48	3,2573	80,44

4.2.2. Prediction study

4.2.2.1. Prediction for flat plate of 30 m in length

As shown in Table 11 and Table 12, the increase of C_F due to regular roughness is compared to the value of C_F for smooth surface. These results show a constant change of C_F for each velocity.

Table 11 Prediction study on the friction coefficient due to extreme regular roughness on a 30 m flat plate.

V (m/s)	Re	C_F CFD (10^{-3})	C_F K1 (10^{-3})	C_F K2 (10^{-3})	C_F K3 (10^{-3})
9,775	$2,79 \times 10^8$	1,7865	3,0025	5,2312	5,4131
12	$3,40 \times 10^8$	1,7340	2,9457	5,1176	5,3214

Table 12 Percentage of the increase in friction coefficient due to extreme regular roughness on a 30 m flat plate.

V (m/s)	Re	C_F CFD (10^{-3})	ΔC_F K1 (%)	ΔC_F K2 (%)	ΔC_F K3 (%)
9,775	$2{,}79 \times 10^8$	1,7865	68,07	192,82	203,00
12	$3{,}40 \times 10^8$	1,7340	69,89	195,14	206,89

The constant change in the C_F values are an indication that the surface roughness belongs to the level of the fully rough regime. In the regime, the drag component is entirely caused by the pressure drag experienced by the roughness elements [3]. The changes in the C_F values are illustrated in Figure 14.

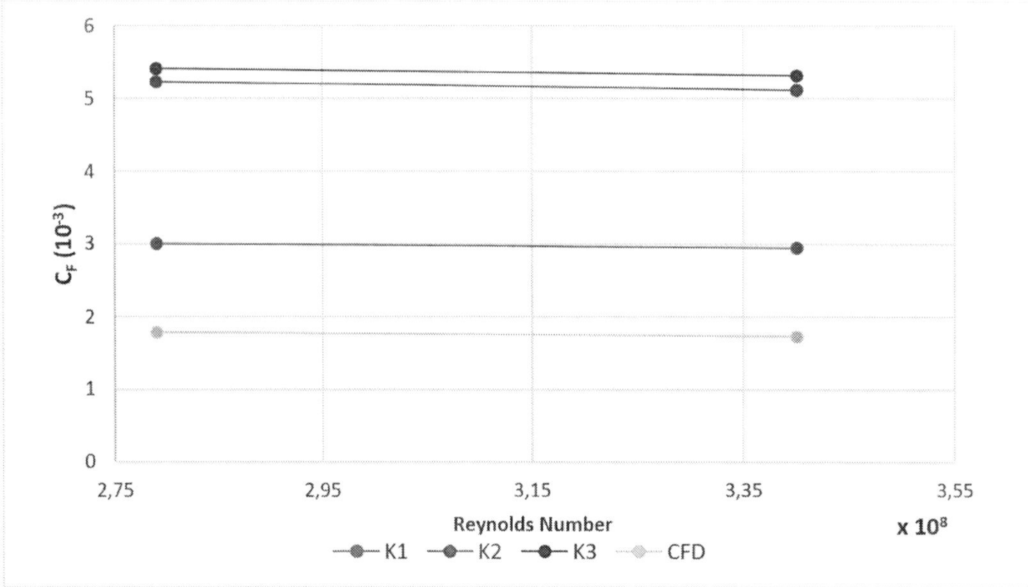

Figure 14 The comparison between the frictional coefficient values due to extreme regular roughness and due to smooth surface of a 30 m flat plate.

Then, CFD predictions were also conducted for irregular extreme roughness which results are shown in Table 13, while the amount of change in the form of a percentage (%) is shown in Table 14.

Table 13 Prediction study on the friction coefficient due to extreme irregular roughness on a 30 m flat plate.

No	V (m/s)	Re	C_F (10^{-3})					
			K123	K132	K213	K231	K312	K321
1	9,775	$2{,}79 \times 10^8$	4,0298	4,0309	4,0759	4,0971	3,8817	4,0971
2	12	$3{,}40 \times 10^8$	3,8076	3,8112	3,8660	3,8690	3,8829	3,8829

Table 14 Percentage of the increase in friction coefficient due to extreme irregular roughness on a 30 m flat plate.

No	V (m/s)	Re	C_F (%)					
			K123	K132	K213	K231	K312	K321
1	9,775	$2{,}79 \times 10^8$	125,57	125,63	128,15	129,34	117,28	129,34
2	12	$3{,}40 \times 10^8$	119,59	119,80	122,96	123,13	123,94	123,94

The extreme irregular roughness variations also belong to the fully rough regime, as can be seen from the constant change in the C_F values. The range of increase in the C_F values caused by irregular roughness was found to be around 117,28 – 129,34% for Reynolds number of $2{,}79 \times 10^8$ and around 119,59 – 123,94%. The constant increase of C_F is more apparent in Figure 15.

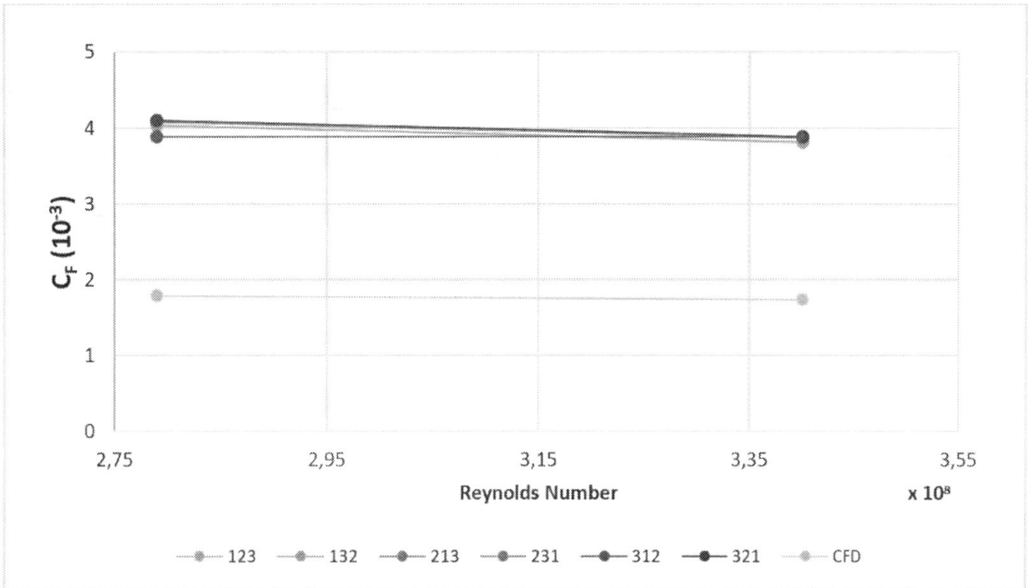

Figure 15 The C_F trend of the 30 m flat plate due to extreme irregular roughness was compared to each other and the smooth plate condition, showing a similar trend of C_F between each extreme irregular roughness.

The prediction results of extreme regular roughness for the flat plate of 30 m in length had greater increase in C_F than the increase in C_F with extreme irregular roughness. In the previous research, it was found that the increase of C_F due to irregular roughness is smaller than the increase of C_F due to regular roughness, which has a similar averaged roughness length scale value. In Figure 14, it can be seen that the curve showing extreme irregular roughness is below the extreme regular roughness Ks 2 (3000 micron) curve and above the Ks 1 (300 micron) extreme regular roughness curve.

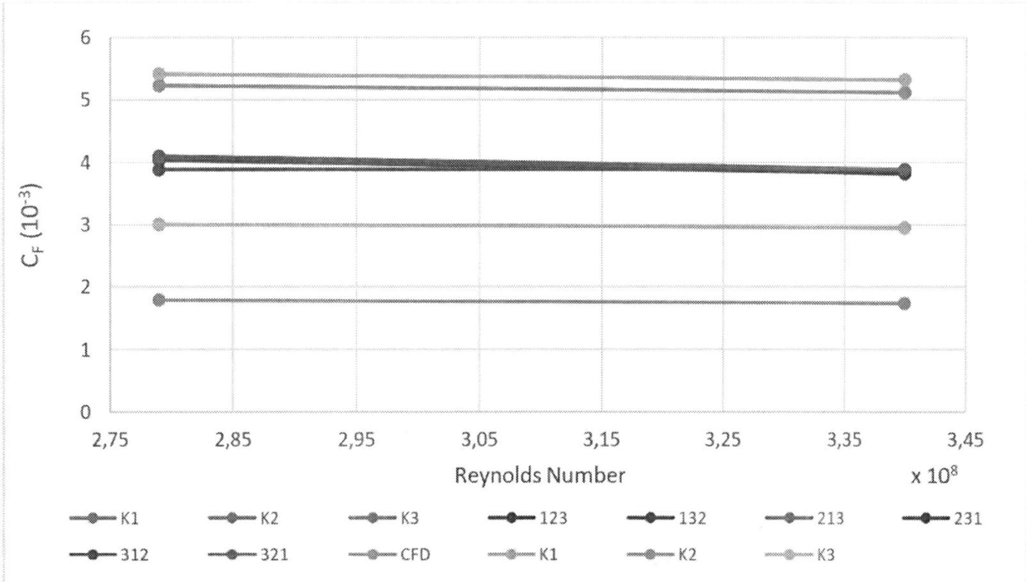

Figure 16 The increase of C_F of the 30 m flat plate due to smooth surface, extreme regular and irregular roughness are compared to each other, showing that the extreme irregular roughness with an averaged roughness length (3000 micron) caused smaller increase in C_F than the extreme regular roughness with the same roughness length (3000 micron).

4.2.2.2. Predictions for flat plate of 60 m in length

Subsequently, extreme roughness CFD simulations were performed for a flat plate of 60 m in length. The increase in C_F is shown in Table 15 along with its change in the form of percentage. The increase of C_F for the 60 m flat plate should be constant like the increase of C_F for the 30 m flat plate. The illustration of this increase in C_F can be seen more clearly in Figure 17.

Table 15 The increase of the friction resistance coefficient due to extreme regular roughness on a 60 m flat plate.

Re	C_F CFD	C_F K1	**C_F K1 (%)**	C_F K2	**C_F K2 (%)**	C_F K3	**C_F K3 (%)**
$5{,}60\times10^8$	1,6960	2,8852	70,12	4,9625	192,61	4,7425	179,63
$6{,}90\times10^8$	1,5601	2,4880	59,47	3,8345	145,78	4,2511	172,50

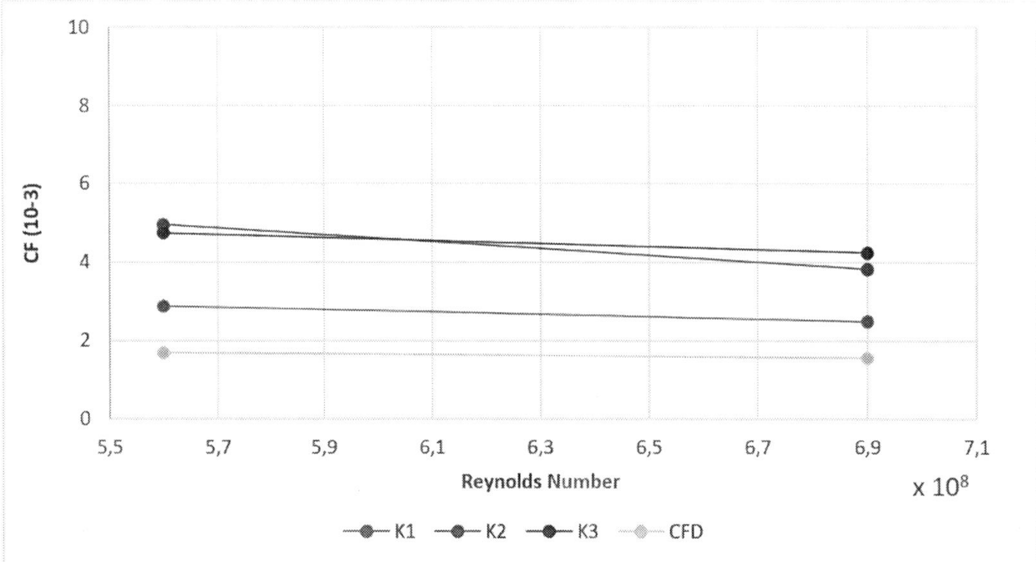

Figure 17 The increase of the friction resistance coefficient of the 60 m flat plate due to extreme regular roughness variation is compared to each other and the smooth plate condition.

Then, CFD simulations of irregular roughness were performed on a flat plate of 60 m in length with extreme irregular roughness variations. The increase of C_F value for each irregular roughness variations on the flat plate of 60 m in length is shown in Table 16. While the ΔC_F in percentage is shown in Table 17.

Table 16 The increase of C_F for flat plate of 60 m in length due to irregular roughness.

Re	C_F					
	123	132	213	231	312	321
$5{,}60\times10^8$	5,2720	5,2403	5,2598	4,6501	4,6390	4,6541
$6{,}90\times10^8$	3,5276	3,5249	3,5541	3,5278	3,5487	3,5392

Table 17 The increase of C_F for flat plate of 60 m in length due to irregular roughness in percentage form.

Re	ΔC_F (%)					
	123	132	213	231	312	321
$5{,}60\times10^8$	5,2720	5,2403	5,2598	4,6501	4,6390	4,6541
$6{,}90\times10^8$	3,5276	3,5249	3,5541	3,5278	3,5487	3,5392

An illustration of the increase of C_F due to extreme irregular roughness for a flat plate with 60 m length can be seen in Figure 18.

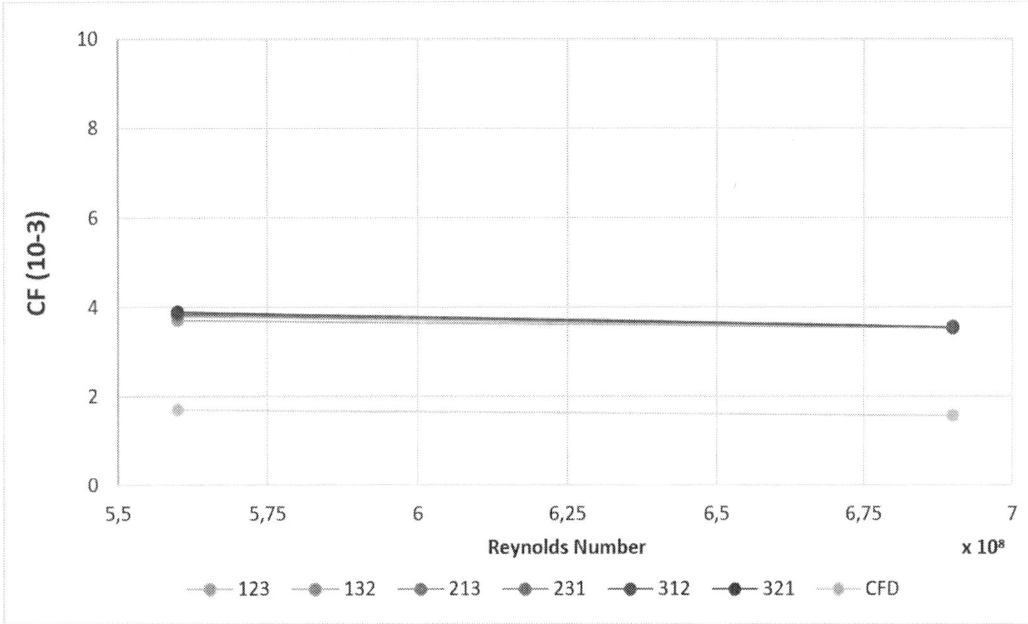

Figure 18 The increase of the friction resistance coefficient of the 60 m flat plate due to extreme irregular roughness variation is compared to each other and the smooth plate condition.

Finally, the prediction results of the CFD simulations for the flat plate with 60 m in length are combined in Figure 19.

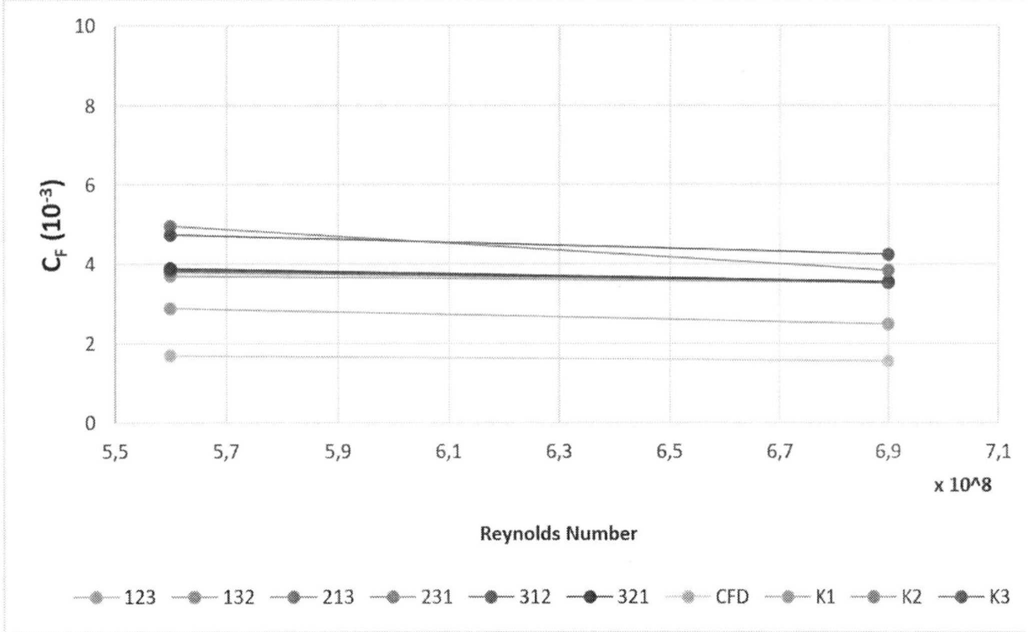

Figure 19 The increase of C_F of the 60 m flat plate due to smooth surface, extreme regular and irregular roughness are compared to each other.

Due to the uneven nature of biofouling spread around a ship, the final assumption of the increase in C_F due to biofouling is based on the values found in the CFD simulations of flat plates with irregular roughness. Thus, it can be said that the rise of C_F on a flat plate due to extreme irregular roughness leads up to 117.3 – 129.34%.

5. Conclusion

Based on the results of this study, the CFD software is found to be sufficient in solving the problems that were raised. The C_F value for the smooth surface of the simulation is close to the C_F ITTC value. The C_F value of the simulation results for regular rough surfaces is also close to the results in the previous CFD study [11]. This study shows that the increase in friction resistance that occurs depends on the roughness conditions, hence:

1. Surface roughness affects the increase of C_F depending on the level of roughness (the smooth regime, transitionally rough regime and fully rough regime) and on the level of the roughness density (C_S),
2. Different variations of regular roughness and irregular roughness have different effects on the increase in C_F. The increase in C_F due to irregular roughness is lower than the increase in C_F due to regular roughness with the same averaged roughness height,
3. The CFD simulation obtained different ΔC_F values from the wind tunnel experiment. This difference may be caused by the unknown value of C_S in the wind tunnel experiment. However, the trend was found to be similar,
4. The CFD simulation for the validation and the prediction of the previous CFD study [11] obtained the appropriate ΔC_F results, as the C_S values that was used in this study was similar with the previous study. The results were also able to obtain the conclusion explained in point number 2.

Acknowledgments

The first author would like to thank Muhammad Luqman Hakim, S.T. as a doctoral student in the Department of Naval Architecture and Shipbuilding Engineering ITS who has helped the writer in understanding and tackling the problems of the CFD simulations in this study.

References

1. SNAME (1988) PNA - Volume II - Resistance, Propulsion and Vibration.
2. Demirel YK, Turan O, Incecik A (2017) Predicting the effect of biofouling on ship resistance using CFD. Appl Ocean Res 62:100–118
3. Schultz MP (2007) Effects of coating roughness and biofouling on ship resistance and powering. Biofouling 23:331–341
4. Demirel YK, Khorasanchi M, Turan O, Incecik A, Schultz MP (2014) A CFD model for the frictional resistance prediction of antifouling coatings. Ocean Eng 89:21–31
5. Yuliansyah MH (2019) Studi eksperimen pengaruh tumbuh biofouling pada lambung kapal terhadap meningkatnya friction resistance. Institut Teknologi Sepuluh Nopember
6. Lackenby H (1962) The Thirty-Fourth Thomas Lowe Gray Lecture: Resistance of Ships, with Special Reference to Skin Friction and Hull Surface Condition. Proc Inst Mech Eng 176:981–1014
7. Cengel, Y.A.; Cimbala JM (2006) Fluid Mechanics: Fundamentals and Applications. The McGraw-Hill Company, New York, USA
8. Schultz MP, Swain GW (2000) The influence of biofilms on skin friction drag. Biofouling 15:129–139
9. Menter FR (1993) Zonal Two Equationk-cl Turbulence Models for Aerodynamic Flows . Florian R . Menter Eloret Institute , Sunnyvale . Mailing Address : 24th Fluid Dynamics Conference ZONAL TWO EQUATION k - o TURBULENCE MODELS FOR AERODYNAMIC FLOWS. 24th Fluid Dyn. Conf. July 6-9, 1993 / Orlando, Florida
10. Date JC, Turnock SR (1999) A study into the techniques needed to accurately predict skin friction using RANS solvers with validation against Froude's historical flat plate experimental data. 62 p
11. Suastika, I.K.; Hakim, M. L.; Nugroho, B.; Nasirudin, A.; Utama, I.K.A.P.; Monty, J. P.; Ganapathisubramani B (2020) Characteristics of drag due to streamwise inhomogeneous roughness Characteristics of drag due to streamwise inhomogeneous roughness.

Design of shell and tube heat exchanger for ballast water treatment applications on ships based on simulation

A Iswantoro, Semin , T Pitana , M B Zaman
Department of Marine Engineering, Marine Power Plant Laboratory, ITS Surabaya, Indonesia

Abstract. Now, environmental issues are the mostly discussed issue and become a top priority in the world. Whether it's environmental damage or environmental pollution. The issue of pollution and damage to the environment and ecology in the region, which have an impact on living organisms, both humans, animals and plants. Environmental damage and pollution can occur on land, water, and air. With various forms of damage and pollution, the methods of handling are different. This research will focus on damage and pollution in waters as well as methods of handling and prevention, especially in marine waters. This research focuses on designing a heat exchanger by varying its types, namely shell and tube heat exchanger. From the modification of model, it will be known which one is more effective in performance. Both temperature exchange, pressure drop, flow rates, dimensions and materials are simulated by software. Where these parameters are influenced by the characteristics of the ship's power of main engine. From 2 model of shell and tube heat exchanger that are one pass-shell and two pass-shell the result for case 1 the effectiveness are 92% & 93%, case 2 are 93% & 94% and case 3 are 94% & 95%.

1. Introduction

Environment issues are mostly discussed issue and a top priority in many countries. The issue of pollution and damage to the environment and ecology in the region, which have an impact on living organisms. Ballast water in the world have become a major cause an ecological imbalance. The International Maritime Organization (IMO) has been actively taking regulated steps to minimize species moved, by adopting the International Convention for Ballast Water Management and Sediments Control in 2004.

1.1 Water Ballast Treatment

In research on ballast water treatment using heat, or thermal treatment, several have been done. As has been done by R. Balaji and O. Yaakob in 2011 with a case study research on the application of Ballast Water Treatment on MT vessels. Bunga Kasturi with engine power of 25090 kW by utilizing heat from the main engine and from a generator with a power of 1020 kW x 3 sets. Resulting in research, the total heat recoverable is 16924.59 kW during sailing. Total heat available 43844.39 kW (unloading). Total heat available 7453.4 kW (loading). [1]

The research was developed in 2014 by R. Balaji and O. Yaakob. One treatment method is heat, which sterilizes ballast water from marine species. Systems that utilize waste heat from ships will provide an economic solution for ballast water treatment. Based on the analysis of waste heat available on existing crude oil tankers, although heat availability is visible, complementary treatment methods are required for high volume treatment. So, combined filtering and heating systems can be combined. Seawater circulating as secondary coolant in the engine is filtered and heated by taking heat from the engine system, rejection steam and exhaust gases. The planned system is a combination of taking heat from multiple sources on board and filtering can be optimized. By considering the components in the system, it can be ascertained that the specific processes associated with each of them (ie heat exchangers and filters) for species mortality / isolation can be carried out. [2]

Then another research, namely the use and utilization of heat from machines to kill microorganisms based on tests and IMO requirements conducted by R. Balaji and O. Yaakob in 2015 resulted in an average heat recovery from exhaust gases, namely between 15% to 33% of the input energy. Tests on species showed > 95% will die (mortality) in the temperature range 55 C to 75 C. [3]

R. Balaji and O. Yaakob then began to conduct research on the design of a heat exchanger to be more optimal in the ballast water heating process in 2017. Optimization of the heating design, using engine exhaust gas as a uid fluid, and ballast water as a uid process, was achieved by using the Lagrangian method, the annual cost is an objective function. Limiting the number of variables, the optimal value is calculated by considering the cost for the utility fluid. In total, four optimal designs and three comparative designs were developed. Heat balance data of the tanker, specific fuel consumption values and fuel costs are considered design variables. Designs are compared based on annual costs, optimal exit temperature and optimal mass from the side of the heat exchanger. The heater design for ballast water treatment uses flue gas for the purpose of retrieving waste heat for ballast water treatment, scope for further improvement includes shell and tube side uids and tubular finned heat exchanger design. The availability and flow limitations of heat depend on the type of vessel, and designs can be worked on for other types of vessels. The increased realization of waste heat will provide a competitive advantage for heat treatment methods, not only in cost but also in increasing the treatment potential of the system. [4]

Research was continued in 2017 by R. Balaji and O. Yaakob regarding the use of heat for ballast water treatment, namely the utilization of waste heat from ship engines as a potential source for heating ballast water. Similar to the schematic arrangement of a ship, a laboratory-scale heat exchanger collects waste heat from jacket water and exhaust gases to test mortality rates for marine species. The result is that mortality is in the range of 80-95% for phytoplankton, zooplankton and bacteria, which can be achieved by maintaining a temperature of 60–65 C for 60 seconds. In addition, the effect of centrifugal pump impellers can also increase the mortality of some species. [5]

After that in 2018 by R. Balaji and O. Yaakob regarding the economic level of heat utilization from the main engine, the utilization of waste heat from ship engines could be a potential source for heating ballast water. Similar to the schematic arrangement of a ship, a laboratory scale heat-engine exchanger that collects waste heat from jacket water and exhaust gases to test species mortality rates. Heat treatment is proven to increase mortality in microorganisms and is more economical because it can save costs. [6]

1.2. Microorganism on Water Ballast

In the 1897, biologists had shown that marine plankton (deep-drifting organisms, most of which are microscopic) could pass through pumps into ships' seawater systems and survive. In 1908 it was reported that in Asia in the North Sea and northern Europe there was an invasion by Chinese mitten crabs believed to be the result of discharging ballast water. It wasn't until the 1970s that scientists began directly sampling the organisms in the ballast water. Many studies have shown that ballast tanks usually contain many species of animals, plants, protozoa, bacteria and viruses, even in quite large numbers. [7]

Small planktonic organisms can be easily pumped in and out of the ballast tank. Plankton can be characterized as holoplankton, meroplankton or tychoplankton. Holoplankton spend their entire life floating in water, and include a variety of bacteria, protozoa, unicellular plants (phytoplankton), and small animals (zooplankton). The latter mainly consist of copepods, mysid shrimp, arrowworms and comb jellies in brine, water fleas and rotifers in freshwater. Meroplankton spend part of their life cycle drifting away in plankton, and include the larvae or eggs of various worms, shellfish, snails, crabs, starfish, fish, and other organisms. Tychoplankton are organisms that usually live on the bottom. Certain other organisms which in the narrow sense are not planktonic can be associated with planktonic hosts, such as certain viruses and nematodes, parasites and flatworms. In addition, some organisms that are non-planktonic in nature can be brought into a ballast tank attached or attached to wood chips or other floating debris, and small fish or shrimp can be carried in through the ballast channel, shown in table 1. [7]

Table 1. Laboratorium test results on water ballast in some location

No.	Location & year	Results
1	Australia 1973	Plankton sampled in 1 ship from Japan included polychaetes, copepods, amphipods, ostracods and chaetognaths (Medcof 1975).
2	Australia 1976-1978	Plankton and fish in 23 woodchip carriers from 13 Japanese ports included 61 species; most common were copepods, molluscs, larvaceans and barnacles. Sediments from 9 woodchip carriers from 7 Japanese ports yielded 32 crustaceans and polychaetes (Williams et al. 1988).
3	Montreal and St. Lawrence River 1980	Plankton samples from 46 ships that had ballasted outside the northwest Atlantic included 132 phytoplankton, 7 protist and 35 invertebrate species (Bio-Environmental Services 1981)
4	North Atlantic 1981	Plankton sampled from a variety of ships and routes included 3 protist, 24 invertebrate and 1 fish species (Carlton *et al.* 1982).
	Australia 1981	Identified 4 fish and reported mysids in ballast water of a domestic bulk carrier (Middleton 1982).
5	Coos Bay, OR 1986-1991	Plankton samples from 159 woodchip carriers from 25 Japanese ports included 402 species in 24 animal, plant and protist phyla, with the most common being copepods, diatoms, polychaetes, barnacles, molluscs and flatworms (Carlton & Geller 1993; Pierce *et al.1997)*
6	Australia 1987-1993	Sediment from ballasted cargo holds in 12 Japanese woodchip carriers arriving in Tasmania in 1987-88 yielded 56 phytoplankton species, including abundant diatoms in 4 ships and dinoflagellates cysts in 7 ships (Hallegraeff et al. 1990). Sediments from 31 out of 83 mainly Japanese woodchip, wheat and ore carriers arriving in Australia in 1987-89 (including the 12 already mentioned) contained dinoflagellate cysts, with toxic species in 4 ships (Hallegraeff & Bolch 1991). 343 ships were sampled by 1990, with sampling continuing through at least 1993 (Hallegraeff & Bolch 1992).
7	Great Lakes and upper St. Lawrence River 1990-1991 Japan 1991	Plankton samples from 86 ships included 110 species of zooplankton in 11 phyla, mainly copepods, cladocerans and rotifers; and 100 species of bacteria, phytoplankton and protists, mainly diatoms and dinoflagellates including 21 bloom-forming, red tide and/or toxic species (Locke et al. 1991, 1993; Subba Rao et al. 1994). Ballast water and sediments sampled in ships at 17 Japanese ports by the Japanese Assoc. for the Prevention of Marine Accidents. Results not published (noted in Kelly 1992).
8	Washington state 1991 Gulf of Mexico Germany 1992-1995	Samples from 6 Japanese woodchip carriers arriving at Tacoma and Port Angeles in 1991 yielded 21 species of phytoplankton and protists from incubated sediments; and at least 8 orders of organisms in ballast water from 3 ships (Kelly 1992, 1993). Ballast water samples in 5 of 19 ships yielded Vibrio cholerae, which genetic analysis found to be identical to the strain responsible for the 1991 South American cholera epidemic and found in oysters in Mobile Bay, Alabama (McCarthy & Khambaty 1994). Plankton sampled in 189 ships, along with organisms in sediment, fouling organisms on tank walls, and larger crabs and fish where possible, included over 350 species, mainly unicellular algae,

IOP Conf. Series: Materials Science and Engineering 1052 (2021) 012035 doi:10.1088/1757-899X/1052/1/012035

		copepods, other crustaceans and molluscs (Gollasch et al., in press).
9	Chesapeake Bay 1993-1994	Plankton net, whole and bottom water samples in 70 ships from foreign ports yielded 275 plant, protist & animal species; and 4 species in sediment from 5 ships (Smith et al. 1996).
10	Hong Kong 1994-1995	Plankton samples from 5 ships from both sides of the North Pacific included 82 species of invertebrates and protists, with copepods being the most common (Chu et al. 1997).
11	Scotland 1994-1995	Plankton sampled from 32 ships and sediment from 24 ships yielded dinoflagellates, diatoms and other organisms. This study is ongoing (Macdonald, in press).
12	Baltimore, MD 1995 New Zealand 1995-1997	Plankton samples from 1 coal carrier from Israel yielded 23 species of dinoflagellates and invertebrates, numerically dominated by copepods, bivalves, polychaetes and gastropods (Wonham et al. 1996). Plankton and bottom water samples from tanks with foreign ballast water in 50 container ships, bulk carriers and break bulk carriers arriving at Lyttelton and Nelson yielded live phytoplankton in 80% of tanks, dominated by diatoms, heterotrophic flagellates and dinoflagellates, and live invertebrates in 83% of tanks with arthropods, molluscs and annelids occurring most frequently (Hay et al. 1997).
13	Valdez, AK 1996 Israel 1996 Various sites	Plankton from 16 domestic and 1 foreign oil tanker included 68 taxa (Ruiz & Hines 1997). Cultured ballast water and sediment samples from 17 ships yielded at least 198 heterotrophs (reported as flagellate, pseudopodial and cilate forms), plus diatoms, cnidarians, turbellarians, nematodes, rotifers, gastrotrichs, polychaetes and copepods (Galil & Hülsmann 1997). Studies are under way or being undertaken in Chesapeake Bay, Long Island Sound, the Port of Morehead City in North Carolina, the Port of Long Beach in California, the Port of Honolulu in Hawaii, the Gulf of St. Lawrence, British Columbia, Sweden and Wales (Gauthier & Steel 1996; Walton & Crowder, 1998; Eldredge 1998; J Carlton, pers. comm.).

1.3. Ballast Water Treatment Method

From year to year, ballast water treatment technology has developed and many kinds. Related companies have even carried out mass production to meet market needs. Figure 1 below are the methods and technology as well as the manufacturers of ballast water treatment equipment.[8]

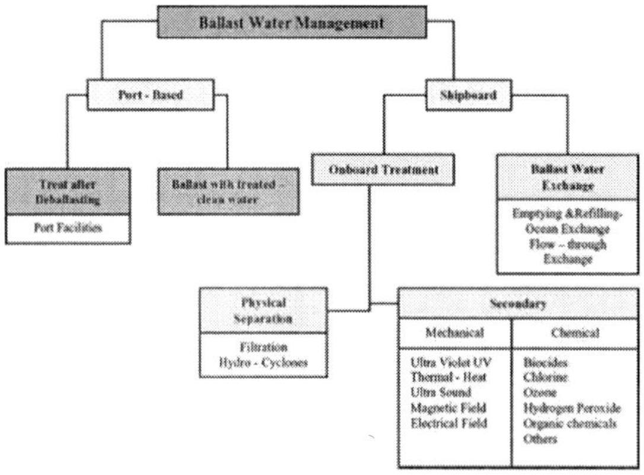

Figure 1. Ballast water treatment method

In this research, ballast water treatment will be carried out using heat, so that it is included in the mechanical method category. Table 2 below are some references obtained from previous research. [9][10]

Table 2. Mechanical methods and its results

Method	Capacity	Microorganism	Effectiveness
UV	NR*	*Gymnodinium* sp.	<6%
		Alexandrium sp	
		Chattonella sp	<40%
UV	2 m^3 h^{-1}	Chlorella	87%
UV	0.2 – 1.6 m^3 h^{-1}	Various	78 – 100%
UV	NR*	Phytoplankton	40 – 99%
		Zooplankton	
		Bacteria	
UV/hydrocyclone	312 – 350 m^3 h^{-1}	Phytoplankton	>85%
		Zooplankton	
		Bacteria	
UV	Regal Princess	Artemia salina	99.50%
	(100 – 3000 m^3 h^{-1})	*Dinoflagellate Prorocentrum*	84.70%
		Tetraselmis sp.	87.60%
UV/Filtration	Vessel Coral Princess (NR*)	Phytoplankton	>70%
		Zooplankton	
		Bacteria	
Heat	Iron Whyalla vessel	Phytoplankton	>98%
	(50,000 of ballast water)	Zooplankton	
		Bacteria	
Heat	Various tests	Phytoplankton	>99%
		Zooplankton	
		Bacteria	
Heat	85 Lmin^{-1}	Zooplankton	90%
		Bacteria	95%
Heat Microwave	1 – 2 L min^{-1}	Microalgae	Complete inactivation
		(Nannochloropsis oculata)	
		Zooplankton *(Artemia)*	
		Oyster larvae	
		(Crassostrea virginica)	
Heat Microwave	1 – 2 L min^{-1}	Artemia cysts	100%
Heat Ultrasound		*Artemia salina*	
		larvae stage	100%
		adults	85%
		cysts	60%
		Dunaliella tertiolecta	40%

IOP Conf. Series: Materials Science and Engineering 1052 (2021) 012035 doi:10.1088/1757-899X/1052/1/012035

2. Methods

2.1 Literature review and collecting data
Problem identification is carried out based on problems that occur in the field, in this case, the problem of pollution due to ballast water. Then from this problem then conduct a literature study and data collection. With the aim of obtaining basic knowledge and data from previous studies that can be used as a reference for further research. At this stage, a study of references contained in papers, journals, proceding, conferences and supporting books is carried out. The collection of various kinds of references serves to strengthen the theoretical basis of thermal ballast water treatment.

2.2 Heat exchanger modeling
The next step after conducting a literature study and gap analysis is to model the heat exchanger with the collected data. Data related to ship engine power and ship speed, this data is used for the input of heat exchanger modeling, so that it will be seen which heat exchanger has the best performance.

2.3 Heat exchanger simulation
After the calculation and modeling process has been carried out, a simulation can be performed. Simulations are carried out based on the data obtained and mathematical calculations.

2.4 Data analysis
The final step is to analyze the result data from heat exchanger modeling. The variable of concern is the performance of the heat exchanger based on thermal efficiency and pressure drop.

3. Results and Discussion
Before doing modeling, there are several things that need to be done first, namely the simulation design. In this case there are 2 calculations, that are for one pass-shell and two pass-shell. Calculation for energy conservation and heat transfer rate equation:

$$q = C_h(t_{h,i} - t_{h,o}) = C_c(t_{c,o} - t_{c,i}) \tag{1}$$

$$q = UA\Delta t_m = \frac{\Delta t_m}{R_o} \tag{2}$$

Where:
Δtm : true mean temperature difference (MTD)
Cc : capacity rate of cold fluid
Ch : capacity rate of hot fluid
tci & tco : cold fluid terminal temperature (in & out)
thi & tho : hot fluid terminal temperature (in & out)
Ro : overall thermal resistance

The overall energy balance for any two-fluid heat exchanger is given by:

$$m_h c_{p,h}(t_{h,i} - t_{h,o}) = m_c c_{p,c}(t_{c,o} - t_{c,i}) \tag{3}$$

Where:
mh & mc : mass flow rate of fluid
cph & cpo : heat capacity of fluid

The expression for maximum possible heat transfer rate *q*max is:

$$q_{max} = C_{min}(t_{h,i} - t_{c,i}) \tag{4}$$

The total of heat transfer rate from hot to cold fluid is:

$$q = \varepsilon C_{min}(t_{h,i} - t_{c,i}) \tag{5}$$

$$\varepsilon = \phi\,(\mathrm{NTU}, C^*, \text{flow arrangement}) \tag{6}$$

$$C^* = \frac{C_{\min}}{C_{\max}} = \frac{(mc_p)_{\min}}{(mc_p)_{\max}} \tag{7}$$

$$\mathrm{NTU} = \frac{UA}{C_{\min}} \tag{8}$$

Where:

ε : heat exchanger effectiveness
C^* : heat capacity ratio
NTU : number of transfer unit

Whereas the following is a formula for finding the Log mean temperature difference:

$$\mathrm{LMTD} = \Delta t_{\mathrm{lm}} = \frac{\Delta t_1 - \Delta t_2}{\ln(\Delta t_1/\Delta t_2)} \tag{9}$$

Based on calculation and simulation can be the following results were obtained.

For case 1: the power of engine 1000 kW for one pass-shell & two pass-shell:

Heat Exchanger Parameters		
Log Mean Temperature Difference	ΔT_{LMTD} =	343.82 C
Mimimum Fluid	C_{min} =	2451 W/K
Cr=Cmin/Cmax	C_r =	0.12
Q max	Q_{max} =	1151811 W
Overall Heat Transfer Coefficient	UA =	6801 W/K
Effectiveness	ε =	0.92
Number of Transfer Units	NTU =	2.78
Required Transfer Coefficient	UA =	3091 W/K
Heat Transfer	Q =	1062651 W
Length	L =	4 m
Number of Tubes	N =	150

(a)

Heat Exchanger Parameters		
Log Mean Temperature Difference	ΔT_{LMTD} =	343.82 C
Mimimum Fluid	C_{min} =	2451 W/K
Cr=Cmin/Cmax	C_r =	0.08
Q max	Q_{max} =	1151811 W
Overall Heat Transfer Coefficient	UA =	6801 W/K
Effectiveness	ε =	0.93
Number of Transfer Units	NTU =	2.78
Required Transfer Coefficient	UA =	3109 W/K
Heat Transfer	Q =	1068824 W
Length	L =	4 m
Number of Tubes	N =	150

(b)

Figure 2. (a) result for one pass-shell, (b) result for two pass-shell

For case 2: the power of engine 2000 kW for one pass-shell & two pass-shell:

Heat Exchanger Parameters		
Log Mean Temperature Difference	ΔT_{LMTD} =	343.82 C
Minimum Fluid	C_{min} =	4679 W/K
Cr=Cmin/Cmax	C_r =	0.12
Q max	Q_{max} =	2198913 W
Overall Heat Transfer Coefficient	UA =	13702 W/K
Effectiveness	ε =	0.93
Number of Transfer Units	NTU =	2.93
Required Transfer Coefficient	UA =	5965 W/K
Heat Transfer	Q =	2050977 W
Length	L =	5 m
Number of Tubes	N =	150

(a)

Heat Exchanger Parameters		
Log Mean Temperature Difference	ΔT_{LMTD} =	343.82 C
Minimum Fluid	C_{min} =	4679 W/K
Cr=Cmin/Cmax	C_r =	0.09
Q max	Q_{max} =	2198913 W
Overall Heat Transfer Coefficient	UA =	13702 W/K
Effectiveness	ε =	0.94
Number of Transfer Units	NTU =	2.93
Required Transfer Coefficient	UA =	5988 W/K
Heat Transfer	Q =	2058774 W
Length	L =	5 m
Number of Tubes	N =	150

(b)

Figure 3. (a) result for one pass-shell, (b) result for two pass-shell

For case 3: the power of engine 3000 kW for one pass-shell & two pass-shell:

Heat Exchanger Parameters		
Log Mean Temperature Difference	ΔT_{LMTD} =	343.82 C
Mimimum Fluid	C_{min} =	7018 W/K
Cr=Cmin/Cmax	C_r =	0.12
Q max	Q_{max} =	3298369 W
Overall Heat Transfer Coefficient	UA =	22073 W/K
Effectiveness	ε =	0.94
Number of Transfer Units	NTU =	3.15
Required Transfer Coefficient	UA =	9054 W/K
Heat Transfer	Q =	3113076 W
Length	L =	6 m
Number of Tubes	N =	150

(a)

Heat Exchanger Parameters		
Log Mean Temperature Difference	ΔT_{LMTD} =	343.82 C
Mimimum Fluid	C_{min} =	7018 W/K
Cr=Cmin/Cmax	C_r =	0.10
Q max	Q_{max} =	3298369 W
Overall Heat Transfer Coefficient	UA =	22073 W/K
Effectiveness	ε =	0.95
Number of Transfer Units	NTU =	3.15
Required Transfer Coefficient	UA =	9081 W/K
Heat Transfer	Q =	3122290 W
Length	L =	6 m
Number of Tubes	N =	150

(b)

Figure 4. (a) result for one pass-shell, (b) result for two pass-shell

4. Conclusion

Based on calculation and simulation, can be take conclusion that:

One pass-shell

With an engine power of around 1000 kW, to get effectiveness 92%, the dimensions of the length of the heat exchanger are 4 meters with 150 tubes. With an engine power of around 2000 kW, to get an effectiveness 93%, the dimensions of the length of the heat exchanger are 5 meters with 150 tubes. With an engine power of around 3000 kW, to get effectiveness 94%, the dimensions of the length of the heat exchanger are 6 meters with 150 tubes.

Two pass-shell

With an engine power of around 1000 kW, to get effectiveness 93%, the dimensions of the length of the heat exchanger are 4 meters with 150 tubes. With an engine power of around 2000 kW, to get an effectiveness 94%, the dimensions of the length of the heat exchanger are 5 meters with 150 tubes. With an engine power of around 3000 kW, to get effectiveness 95%, the dimensions of the length of the heat exchanger are 6 meters with 150 tubes.

5. References

[1]. B Rajoo and B Y Omar. Emerging Ballast Water Treatment Technologies: A Review. Journal of Sustainability Science and Management. 2011. Volume 6, pp. 126-138.

[2]. B Rajoo and B Y Omar. An Analysis of Shipboard Waste Heat Availability For Ballast Water Treatment. Journal of Marine Engineering & Technology. 2014. 11:2, pp. 15-29.

[3]. B Rajoo and B Y Omar. A Ballast Water Treatment System Using Engine Waste Heat: Is It Viable?. Journal of Transport System Engineering. 2015. pp. 31–36.

[4]. B Rajoo and B Y Omar. Validation of Engine Performance for Tests on Ballast Water Heat Treatment Using Engine Waste Heat, International Journal of Marine Engineering Innovation and Research. 2017. Volume 2(1). pp. 16-24.

[5]. B Rajoo and B Y Omar. Laboratory Tests on Heat Treatment of Ballast Water Using Engine Waste Heat. Journal of Environmental Technology. 2017. pp. 1-13.

[6]. B Rajoo and B Y Omar. A Study of Ballast Water Treatment Using Engine Waste Heat. Journal of The Institution of Engineers: Series C. 2018. Springer.

[7]. N C Andrew. Ships' Ballast Water and The Introduction of Exotic Organisms Into The San Francisco Estuary. California Urban Water Agencies. 1998.

[8]. National Research Council. Stemming the Tide: Controlling Introductions of Non-indigenous Species by Ballast Water. National Academy Press. 1996. Washington, DC.

[9]. E Tsolaki and D Evan. Technologies for Ballast Water Treatment: A Review. Journal of Chemical Technology & Biotechnology. 2010. pp. 19-32.

[10]. International Maritime Organization. Ballast Water Management Convention and International Convention for The Control and Management of Ships Ballast Water and Sediments. 2004.

Acknowledgments

The authors thank the marine power plant laboratory staff, Department of Marine Engineering ITS who have helped and supported this research.

Characteristics of Turning Circle and Zig-Zag Manoeuvres of An Indonesian Ferry Ship In Shallow Water

A H Muhammad[1], D Paroka[2], S Rahman[2] M R Firmansyah[3] and T P Putra[4]

[1] Department of Marine Engineering, Faculty of Engineering, Hasanuddin University, Gowa, 50275 Indonesia
[2] Department of Ocean Engineering, Faculty of Engineering, Hasanuddin University, Gowa, 50275 Indonesia
[3] Department of Naval Architecture, Faculty of Engineering, Hasanuddin University, Gowa, 50275 Indonesia
[4] Research and Development Center for Sea, River, Lake and Ferry Transportation, Ministry of Transportation, Jakarta Pusat, 10110 Indonesia

Abstract. Ship manoeuvring performance is very important in navigation safety, especially when ships operate in shallow water. This paper describes the characteristics of turning circle and zig-zag manoeuvres of Indonesian ferry ships in shallow water. A time-domain simulation program was developed for this purpose. Several parameters such as ship speed and water depth levels have been considered in the simulation. The results shows a significant difference in the ship's trajectory and speed when the ship manoeuvring in different water depths. The ship manoeuvre trajectory diameter for shallow water (h/T=1.3) is greater than in deep water (h/T=4.0). Meanwhile, the speed of the ship during turns increases with the decreasing depth of the water (h). Likewise, the 20/20^0 zig-zag manoeuvre of the ship shows that the heading angle in shallow waters has a smaller overshoot angle when compared to the ship manoeuvres in the deep waters.

1. Introduction

Ship Behaviour and performance in shallow water are strongly influenced by hydrodynamic interactions between the hull and the bottom channel. That is, the velocity of flow between the hull and the bottom increases, as the flow pressure decreases. A number of changes experienced by the ship when entering shallow water are expressed by changes in resistance, trim, steering, manoeuvrability and stopping [1]. Many researchers mention that the change in performance is not only influenced by the hydrodynamic interaction between the hull and the bottom channel, but the decrease in water depth also has an impact on the less efficient and not effective of the propeller and rudder action [2] and [3].

Many investigations have been carried out to shows the manoeuvring performance of ships in shallow water through changes in flow velocity or ship speed. Du, et al. [4] has examined ship's speed effect on shallow waters through a system-based method, where a nonlinear transient hydrodynamic and confinement model is implemented to account for the channel bottom effect. The turning manoeuvring model and the confinement model was validated using the experimental data. They concluded that with increasing speed it had increased the hydrodynamic interaction between the hull and the channel bottom. Similar results are also present by Skejic, et al. [5] in his research on water depth's effect on ship manoeuvring through numerical simulations and sea trials. They stated a significant difference in the ship's trajectory and manoeuvring speed when the ship was manoeuvring at different water depths. These were indicated by the increase in tactical diameter and speed of ship manoeuvring along with the decreasing water depth. According to the World Association for Water Transport Infrastructure [6] water depth level is defined as the ratio of water depth including $h/T>3$ (deep water), $1.5 < h/T < 3$ (medium water), $1.2 < h/T < 1.5$ (shallow water); and $h/T < 1.2$ (very shallow water).

Indonesian ferries ship is a type of ship crossing between islands (passengers and vehicles). This ship has unique characteristics compared to commercial ships in general. Some features of Indonesian ferries ship that can affect the manoeuvrability: i) The ship has a low draft (T); ii) The ship operates in confined water (deep and shallow water); iii) The ship has a superstructure with a relatively large wind catch area and; iv) The ship is equipped with a twin propulsion and rudder system (TRTP) with a relatively large ratio of distances between systems. Based on those characteristics, it is crucial view the Indonesian ferries manoeuver efficiency and navigation safety.

The limitation of the speed manoeuvring to the water's depth is an essential consideration in operating ships in confined waters. This paper aims to predict the manoeuvring capabilities (turning circle and zig-zag manoeuver) of Indonesian ferries in deep (*h/T*=4.0) and shallow water (*h/T*=1.3) from the point of view of manoeuver efficiency and navigation safety.

2. Mathematical Model

In the prediction of ship manoeuvring in shallow water conditions through numerical simulations based on the mathematical model, it is important to use hydrodynamic coefficients on which shallow water effects. The models were based on the 3-DOF equations of manoeuvring motion (surge, sway and yaw) in Equation 1, using the coordinate system in Figure 1.

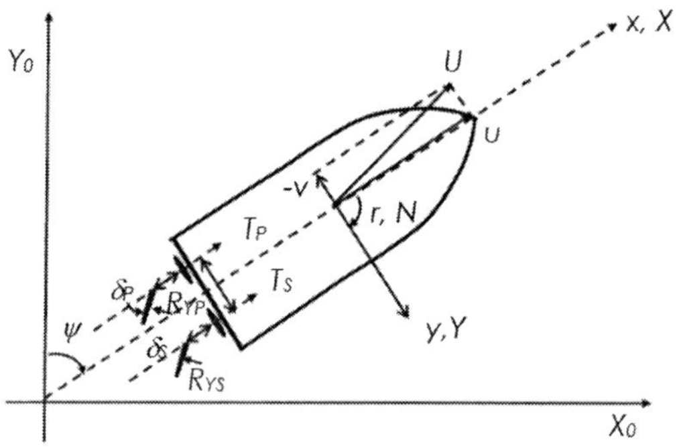

Figure 1. Coordinate system of ship.

$$\begin{aligned}(m+m_x)\dot{u}-(m+m_y)rv &= X_H + X_{P\left\{\begin{smallmatrix}S\\P\end{smallmatrix}\right\}} + X_{R\left\{\begin{smallmatrix}S\\P\end{smallmatrix}\right\}}\\ (m+m_y)\dot{v}+(m+m_x)ru &= Y_H + Y_{P\left\{\begin{smallmatrix}S\\P\end{smallmatrix}\right\}} + Y_{R\left\{\begin{smallmatrix}S\\P\end{smallmatrix}\right\}}\\ (I_{zz}+J_{zz})\dot{r} &= N_H + N_{P\left\{\begin{smallmatrix}S\\P\end{smallmatrix}\right\}} + N_{R\left\{\begin{smallmatrix}S\\P\end{smallmatrix}\right\}}\end{aligned} \tag{1}$$

The notations of m, m_x, and m_y are the mass of ship, and added mass inertia in x-and y-direction respectively; I_{zz} and J_{zz} are moments of inertia and added around z-axis respectively; u, v and r are velocity components at ship's centre of gravity (G). X, Y, and N represent the hydrodynamic forces and moment. The subscript H, P, and R refer to hull, propeller and rudder respectively according to the concept of MMG expression.

Force and moment acting on hull (X_H, Y_H, and N_H) are approximated by the polynomial function of β and r'. The equations are expressed by Kijima, et al. [7] in Equation 2.

$$\begin{aligned}X_H &= \tfrac{1}{2}\rho L d U^2 (X'_{\beta r} r' \sin\beta + X'_{uu}\cos^2\beta)\\ Y_H &= \tfrac{1}{2}\rho L d U^2 (Y'_\beta \beta + Y'_r r' + Y'_{\beta\beta}\beta|\beta| + Y'_{rr} r'|r'| + (Y'_{\beta\beta r}\beta + Y'_{\beta rr} r')\beta r')\\ N_H &= \tfrac{1}{2}\rho L d U^2 N'_\beta \beta + N'_r r' + N'_{\beta\beta}\beta|\beta| + N'_{rr} r'|r'| + (N'_{\beta\beta r}\beta + N'_{\beta rr} r')\beta r')\end{aligned} \tag{2}$$

where: β is the drif angle at midship position by $\tan^{-1}(v/u)$ and r' non-dimensionalized yaw rate by rL/U. X'_{uu}, $X'_{\beta r}$, Y'_{β}, Y'_{r}, $Y'_{\beta\beta}$, Y'_{rr}, $Y'_{\beta\beta r}$, $Y'_{\beta rr}$, N'_{β}, N'_{r}, $N'_{\beta\beta}$, N'_{rr}, $N'_{\beta\beta r}$, and $N'_{\beta rr}$ are called the hydodynamic derivatives.

The mathematical model of forces and moment induced by propeller and rudder (X_P, Y_P, N_P and X_R, Y_R, N_R) can be referred in Khanfir, et al. [8] and Dash, et al, [9].

3. Empirical Equation

3.1. Added Mass and Moment of Inertia

Added masses (m'_x and m'_y) and added moment of inertia (J'_z) are predicted based on the equation developed by Sandakane, et al. [10]. The equation is obtained empirically from the simplified equation of ratio of the added mass and the moment of inertia in shallow water. The equation is represented as Equation 3.

$$\frac{m'_x}{m'_{x0}}, \frac{m'_y}{m'_{y0}}, \frac{J'_{zz}}{J'_{zz0}} = 1 + \alpha\left\{\exp^{(\beta)*d/H)} - 1\right\} \tag{3}$$

Where α and β are coefficient changes according to the ship motion mode and ship form coefficients such as draft-breadth ratio and the block coefficient.

3.2. Hydrodynamic Coefficient

The set of hydrodynamic coefficients is determined based on the equations developed by Kijima, et al. [7]. The equation is obtained semi empirically from a numerical calculation based on lifting surface theory and model tests in full load condition. In shallow water conditions, the coefficient can be obtained by multiplying a correction factor with a coefficient in deep water conditions. The equation is represented as Equation 4.

$$D_{shw} = f(h)D_{dep} \tag{4}$$

Where, D_{shw} is the derivatives in shallow water, D_{dep} is the derivatives in deep water, and $f(h)$ is the correcting factor, $h=d/H$ (d is draught, H is water deep). The correcting factor, $f(h)$ for hydrodynamic coefficient can be predicted by Equation 5 (for Y'_{β}, $Y'_{\beta\beta}$, $Y'_{\beta rr}$, N'_{β}, and N'_{r}) and Equation 6 (for $Y'_{r}-(m'+m'_x)$, Y'_{rr}, $Y'_{\beta\beta r}$, $N'_{\beta\beta}$, N'_{rr}, $N'_{\beta\beta r}$, and $N'_{\beta rr}$).

$$f(h) = \left[1/(1-h)^n\right] - h \tag{5}$$

$$f(h) = 1 + a_1 h + a_2 h^2 + a_3 h^3 \tag{6}$$

Where, n, a_1, a_2, and a_3 are the coefficient of the correcting factor. The resistance and propulsion parameters for simulation are predicted by Holtrop Method [11] and [12]. In shallow water conditions, the resistance coefficient of shallow water $(R'_0)_{shallow}$ is predicted by Equation 7[13].

$$\frac{(R'_0)_{shw}}{(R'_0)_{dep}} = 0.388(T/H)^2 + 1 \tag{7}$$

Where, (R'_0) *deep* is the resistance coefficients in deep water, T is draught and H is water deep.

4. Simulation Program

4.1. Time Domain Simulation

The prediction of ship manoeuvrability can be analysed through the swept paths [14]. The analysis has been carried out using the time domain simulation program of MATLAB-Simulink [15] and [16]. The swept paths of the ship can be obtained by double integrating the acceleration of manoeuvring ship motion mathematical models including hydrodynamic derivatives as the integration process of the 3-DOF equation of motion (surge sway and yaw) in Equation 8, 9 and 10.

$$(m+m_x)\dot{u}-(m+m_y)rv = X_H + X_P\begin{Bmatrix}S\\P\end{Bmatrix} + X_R\begin{Bmatrix}S\\P\end{Bmatrix}$$

$$\dot{u} = X_H + X_P\begin{Bmatrix}S\\P\end{Bmatrix} + X_R\begin{Bmatrix}S\\P\end{Bmatrix} + (m+m_y)rv/(m+m_x) \tag{8}$$

$$u = \int \dot{u}\, dt; x = \int u\, dt$$

$$(m+m_y)\dot{v}+(m+m_x)ru = X_H + X_P\begin{Bmatrix}S\\P\end{Bmatrix} + X_R\begin{Bmatrix}S\\P\end{Bmatrix}$$

$$\dot{v} = X_H + X_P\begin{Bmatrix}S\\P\end{Bmatrix} + X_R\begin{Bmatrix}S\\P\end{Bmatrix} + (m+m_y)ru/(m+m_x) \tag{9}$$

$$v = \int \dot{v}\, dt; y = \int \dot{v}\, dt$$

$$(I_{zz}+J_{zz})\dot{r} = X_H + X_P\begin{Bmatrix}S\\P\end{Bmatrix} + X_R\begin{Bmatrix}S\\P\end{Bmatrix}$$

$$\dot{r} = (X_H + X_P\begin{Bmatrix}S\\P\end{Bmatrix} + X_R\begin{Bmatrix}S\\P\end{Bmatrix})/(I_{zz}+J_{zz}) \tag{10}$$

$$r = \int \dot{r}\, dt;\ \psi = \int r\, dt$$

4.2. Simulation Data

The object of the study is the ferry ship of KMP Bontoharu, owned by PT. (Persero) ASDP Indonesia Ferry. Ship with a capacity of 1050 GT with a length of 54.00 m (*LOA*) is operated at the Bira-Pamatata crossing in South Sulawesi Province. The ship particulars are presented in Table 1.

Table 1. Ship particulars.

Parameter	Value
Length over all (*LOA*), m	54.00
Length between perpendiculars (*Lbp*),m	47.45
Breadth (*B*), m	14
Depth (*D*), m	3.4
Draft (*T*),m	2.45
Ship speed (*V*), m/s^2	6.618
Displacement (*Δ*), Ton	1148
Propeller blade number (*Z*)	2 x 4
Propeller diameter (D_P), m	1.422
Pitch-diameter ratio (*P/D*), m	1.066
Propeller number of revolutions per second(*n*), rps	8.784
Span of rudder, m	1.550
Chord of rudder, m	0.900
Blade area of rudder (A_R), m^2	2 x 1.395
Power of main engine(*BHP*), HP	2 x 1000
Main Engine number of revolutions per minute (*RPM*)	1850

5. Results and Discussion

This study focuses on the manoeuvring prediction of a ferry ship in deep and shallow water conditions. A time domain simulation program by MATLAB Simulink was developed for this purpose. Several parameters such as the initial speed of the ship and water depth level have been considered in the simulation. The information is critical in ship navigation safety, particularly when ship operating in shallow water. Figure 2 shows the simulation results for the 35^0 starboard turn manoeuvring of a KMP Bontoharu in shallow water (*h/T*=1.3), medium shallow water (*h/T*=2.0) and deep water (*h/T*=4.0) at the initial speed of the ship (*U*) 6,618 m/s. It was found that the tactical diameter (D_T) of the ferry ship for shallow water was 10.9 % and 53.1 % larger than that for medium shallow water and deep water respectively. Figure 2 also shows the effect of depths water conditions (h/T=1.3) on the increase in the forward speed of the ship (u) and the turning speed (deg/s) during the ship turning manoeuvring. This trend is similar to the findings of Skejic, et al. [5] Which confirming the characteristics of ship manoeuvring in shallow and deep water.

Figure 3 shows the simulation results of the 20^0/20^0 zig-zag manoeuvring of a KMP Bontoharu at three different water depth conditions (shallow, medium shallow and deep waters). The horizontal and vertical axes express time and heading angle (ψ) respectively. It shows that the ship manoeuvres' heading angle in shallow water has a smaller overshoot angle compared to the ship manoeuvres in medium shallow and deep water. A ship zigzag manoeuvres in shallow water (h/T=1.3), took 65.7 seconds for the 1st overshoot with the heading angles of 3.35^0. These result were faster in 4.0 and 7.9 seconds and heading angles of 6.84^0 and 8.24^0 than in medium shallow and deep water respectively. This result is similar to the findings of Maimun et al. [15], which confirms the characteristics of pusher barge manoeuvring in shallow and deep water. Table 2 and 3 indicates the summary of simulation results of ship turning circle and zigzag manoeuvres in two different water depths.

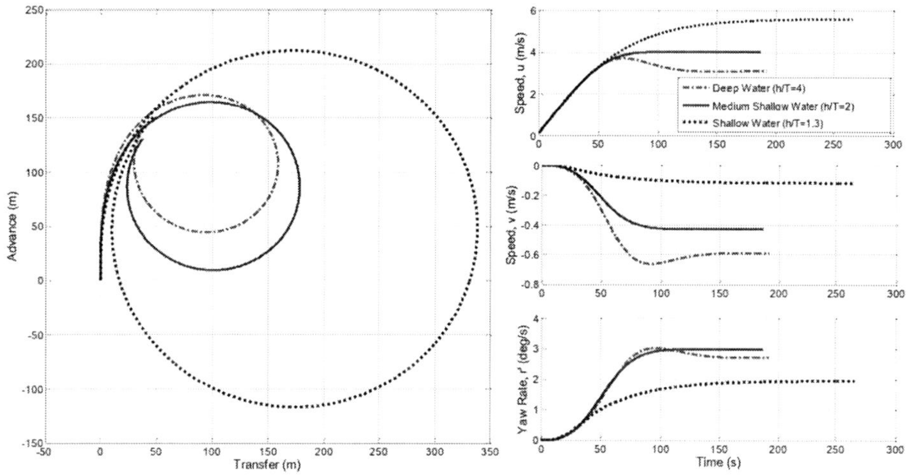

Figure 2. Turning circles in shallow and deep water at initial speed (U) 6.618 m/s.

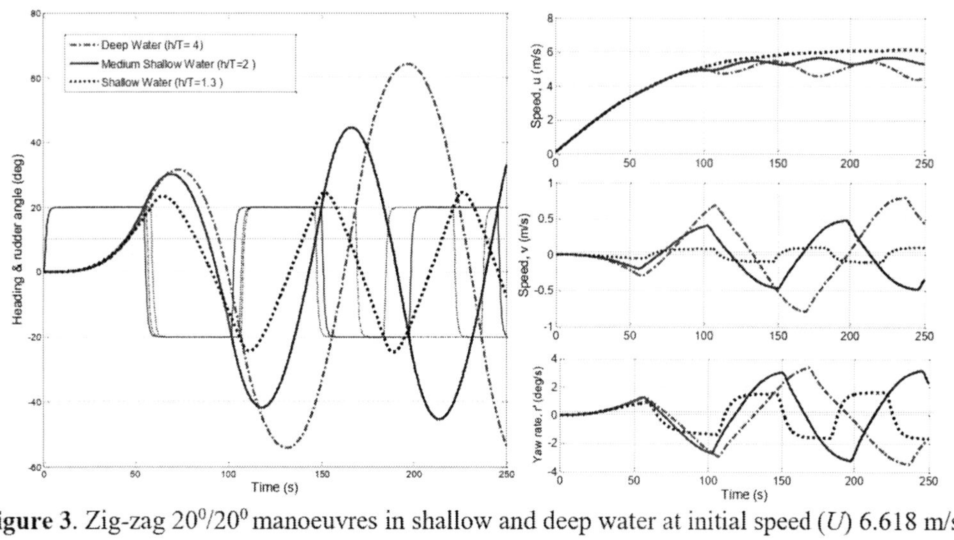

Figure 3. Zig-zag 20^0/20^0 manoeuvres in shallow and deep water at initial speed (U) 6.618 m/s.

Table 2. Turning circle characteristics in shallow and deep water.

No.	Simulation type	δ (deg.)	Vs (m/s)	D_T (m)	A_D (m)	u (m/s)	v (m/s)	r (deg./s)
1	Turning circle (h/T = 4.0)	35	6.618	158.4	171.2	3.10	-0.59	2.72
2	Turning circle (h/T = 2.0)	35	6.618	177.8	164.6	4.00	-0.43	2.98
3	Turning circle (h/T = 1.3)	35	6.618	338.3	212.3	5.88	-0.12	1.95

Table 3. Zig-zag manoeuvres characteristics in shallow and deep water.

No.	Simulation type	δ (deg.)	Vs (m/s)	1st (deg.)	Time (s)	2nd (deg.)	Time (s)	u (m/s)
1	Zig-zag manoeuvre (h/T = 4.0)	20/20	6.618	11.59	73.6	-34.2	131.5	5.41
2	Zig-zag manoeuvre (h/T =2.0)	20/20	6.618	10.19	69.7	-21.8	118.3	5.68
3	Zig-zag manoeuvre (h/T =1.3)	20/20	6.618	3.35	65.7	-4.34	111.6	6.13

6. Conclusions

Prediction of ferry ship Indonesian manoeuvring in shallow water through a time domain simulation program has been analyzed. The numerical simulation results show that the tactical diameter (D_T) of the ferry in shallow waters was 53.1% greater than the tactical diameter in deep water, while the forward speed (U) and turning rate (deg./s) of the ship have increased when the ship is manoeuvring in shallow water. On the other hand, in a zigzag maneuver of 20/20 degrees, the ship heading angle in shallow water has a smaller overshoot angle than ship maneuvering in deep water. So that through reducing the initial speed of the ship when entering shallow water it can improve ship safety,

Acknowledgments

The authors wish to thank Hasanuddin University for funding to this research. This research was funded by Basic Research of UNHAS under grand No. 2649/UN4.1/KEP/2020. The authors would also like to thanks PT. ASDP Indonesia Ferry Brach of Selayar for the sea trial and ship data collection.

References

[1] Duarte H O, Droguett E P, Martins M R, Lutzhoft M, Pereira P S and Lloyd J 2016 *Intl J Maritime Eng*, **158** 177-186.
[2] Varyani K S, Hamoudi B, Hannah S and McGregor R C 1997 *Proc. IFAC Conf. on Manoeuvring and Control of Marine Craft (Croatia)* vol 30 (22) pp 145-150.
[3] Delefortrie G and Vantorre M 2007 *J. Ship Res.* **51(4)** pp287–296.
[4] Du P, Ouahsine A and Sergent P 2018 *J. Comptes Rendus Mécanique* **346(5)** pp390–401.
[5] Skejic R, Kirimoto K, Berg T E and Pedersen E 2012 *Proc. Int. Conf. on Marine Simulation and Ship Manoeuvrability (Croatia)* pp 542 – 550.
[6] PIANC 1992 Capability of Ship Manoeuvring Simulation Models for Approach Channels and Fairways in harbours, Brussels (Belgium), Report of Working Group no. 20 of Permanent Technical Comittee II, Supplement to PIANC *Bulletin no. 77.* 1992
[7] Kijima K, Nakiri Y, Tsusui Y and Matsunaga M 1990 *Proc. MARSIM & ICSM 90 (Japan)* p
[8] Khanfir S, Hasegawa K, Nagarajan V, Shouji K and Lee S K 2011 *J. Mar Sci Technol* **16** pp 472–490. 1982
[9] Dash AK, Nagarajan V and Sha O P 2015 *J. International Shipbuilding Progress* **62** pp 57–111.
[10] Sadakane H, Toda Y and Lee Y S 2001 *J. Japan Institute of Navigation* **105** 11-20.
[11] Holtrop J and Mennen G G J 1982 *J. International Shipbuilding Progress* **29** pp166-170
[12] Holtrop J 1984 *J. International Shipbuilding Progress* **31** pp272-276.
[13] Furukawa Y, Ibaragi H, Nakiri Y and Kijima K 2016 *Proc. 4th MASHCON* (Germany) pp 295-303.

[14] IMO 2002 Standards for Ship Maneuverability *Report of the Maritime Safety Committee on its Seventy-Sixth Session-Annex 6 (Resolution MSC. 137(76))* (London)
[15] Maimun A, Priyanto A, Muhammad A H, Scully C C and Awal Z I 2011 *J. Ocean Engineering* **38(11–12)** pp1277–1356.
[16] Muhammad A H, Hasbullah M, Djabbar M A and Handayani 2015. *International Journal of Technology* **6(3)** pp 452-461.

A Hybrid Methodology for Human Reliability Assessment in Maritime Cargo Accidents

R E Prilana[1], **L P Bowo**[1], **M Furusho**[1]

[1]Graduate School of Maritime Sciences, Kobe University, Kobe, Japan

Abstract. The maritime cargo shipment is always looking for implementing new safety measures to reduce the cargo accident rates. However, it is challenging to find the root cause that led to a specific cargo accident. The accidents result from a complex chain of events, where there is usually more than one factor solely responsible for maritime cargo accidents. The present contribution aims to find the causal factor of the accidents and predict human error probability after the accidents using the hybrid methodology. The hybrid methodology that uses this study to assess human error of maritime cargo accidents is Human Error Assessment and Reduction Technique (HEART) – 4M and TOPSIS with Technique for Risk Perception. The HEART methodology assumes that any predicted reliability of task performance may be modified according to the presence of identified Error Promoting Conditions (EPCs). A technique for Risk Perception decomposes a task into sub-task steps for which quantification data are provided. The corresponding results show that the most common situation that maritime cargo accidents occurred due to the man factor is during the loading and unloading process.

1. Introduction

Nowadays, maritime cargo shipment was developing from time to time. It is also called that the maritime cargo shipment is the backbone of the world economy that more than 90% of world trade is carried by sea [1]. c. Furthermore, total cargo containerization was handled in container ports worldwide, estimated at 793.26 million twenty-foot equivalent units (TEUs), and 64% of container handling occurred in the Asia region [2].

Indonesia is one of the countries in the Asia region that the maritime cargo shipment was increased rapidly. Indonesia is an archipelagic country with an abundance of natural resources, located strategically between the Indian and the Pacific oceans, with approximately two-thirds of its area being the sea [3]. Due to Indonesia's condition, maritime shipping is major transportation in Indonesia [4] and the Indonesian economy's backbone. However, the number of accidents that occurred is directly proportional to the number of maritime cargo shipments. Both the instances above occur during shipping on land or at sea [5].

Safety is an essential subject of maritime transportation. It directly affects human life, the environment, and the transported commodities [6]. Maritime authorities have approved a set of regulatory rules and specifications to improve safety and minimize accidents in the maritime industry [7]. More and more new requirements and amendments to existing regulations for safe maritime transport were introduced by the International Maritime Organizations (IMO) [8]. The human element is one of the essential contributory aspects of the causation and avoidance of accidents, and human error is responsible for more than 80 % of maritime accidents [9], [10].

Not only IMO, but also the International Labour Organization (ILO) and International Association of Classification Societies (IACS) that form maritime regulatory authorities are overly concerned about the possibility of the occurrence of maritime accidents due to human factor issues [11]. Some reliability analysis techniques have been used in every industry to reduce the risk of accidents [12], and one of them is Human Reliability Assessment (HRA). The main reason for HRA to evaluate human reliability and the uncertainty of the data concerning human factors is the complexity of human behavior [13]. HRA is a significant analysis required to analyze the task, identify errors, and reduce error impact if required, making it more than a quantifying index [14].

At this point, this paper's present contribution is to find the causal factors of maritime cargo accidents and to predict the human error probability after the accidents of maritime cargo shipment. The paper integrated the HEART (Human Error Assessment and Reduction Technique) – 4M and TOPSIS with Risk Perception Technique to achieve this purpose.

HEART is a practical tool to calculate Human Error Probability (HEP) [15]. HEART is a simple, flexible, and effective method for determining the human error involved in accidents. Therefore, it has been used in various industries with complex systems, such as nuclear power plants, railway transportation, aviation, off-shore platforms, and the maritime industry [16]. On the HEART – 4M method, the Error Producing Conditions (EPCs) are categorized into four factors: man, machine, media, and management. In this paper, the HEART – 4M method combines with the technique for order preference by similarity to ideal solution (TOPSIS) that development introduced by Bowo [16]. This method evaluates the HEP in maritime cargo accidents at every stage (loading process, sailing process, and unloading process).

After HEP every stage was assessed, the aggregate of HEP of maritime cargo accidents was evaluated using the risk perception technique. Risk Perception is defined as a psychological process that describes the subjective (conscious and unconscious) evaluation (as opposed to objective risk assessment) of the probability to be affected by an imminent undesirable event in a specific situation and an assessment of one's own perceived vulnerability and coping resources [17].

2. Methodology

This paper prompts a hybrid approach to determine HEP's aggregate by integrating HEART – 4M and TOPSIS with Risk Perception Technique. Accordingly, this section provides a brief description of methodologies.

2.1 HEART – 4M

HEART is preliminarily introduced by William [18] to assess human error with defined values. HEART comprises of two fundamental parameters, such as Generic Task (GT) and Error Producing Condition (EPC). The GT parameter consists of nine qualitative descriptions of the appropriate task in the accident process, and every GT provides the value of generic error probability, named nominal human unreliability (NHU). The proposed NHU was based on engineers' or researchers' experience, but William [18] also provided the number of NHU as guidance to the assessor to analyze the task. Table 1 presents the GT and NHU used in this study. The other one is EPC, which indicates the relevant performance shaping factors for humans during a task and can affect HEP's value [16]. There are 38 EPCs defined in the HEART method, directly influencing HEP value for a specific.

Table 1. Generic Tasks (GT).

Generic Tasks (GT)			
Code	**Type of work**	**Condition**	**NHU**
A	Totally unfamiliar	Works performed at speed with no real idea of likely consequences	0.55
B	Restore the system to an original state on a single attempt	Doing it without supervision or procedures	0.26
C	Complex task	Task requires a high level of comprehension and skill	0.16
D	A fairly simple task	Works performed rapidly or given scant attention	0.09
E	The routine, highly practiced, rapid task	Works involving a relatively low level of skill	0.02

F	Restore a system to original	An error occurred even tough following procedures with some checking	0.003
G	Entirely familiar, highly practiced, routine task occurring several times per hour, performed to highest possible standards by a highly motivated, highly trained, and experienced person, totally aware of implications of failure, with time to correct the potential error	However, without the benefit of significant job aids	0.0004
H	Respond correctly to the system command	Even when there is an augmented or automated supervisory system providing an accurate interpretation of system stage	0.00002
M	The miscellaneous task for which no description can be found		0.03

If there is more than one EPC, the assessed proportion of effect is determined. In this context, HEP is calculated with the following equation.

$$HEP_{value} = NHUx\{\Pi_i(EPC_i - 1)APE_i + 1\} \quad (1)$$

In the equation, EPC_i is the ith (i = 1,2,3, …..n; n ≤ 38) error producing condition, and APE_i (0 < i ≤ 1) is assessed of proportion effect, which is weighted the each EPC basic of its importance [15]. As the EPC influence becomes more critical, the value of the APE will be higher.

The HEART – 4M method is the HEART method's developments introduced by Bowo and Furusho [19]. In the HEART – 4M method, there is a categorization of the EPCs into the 4M framework, consisting of man, machine, media, and management. The 38 EPCs established by William describe the working condition, which means the interaction between humans, human-machine interactions, and working environment conditions. William also provides the multiplication for each EPC. Table 2 presents the EPC – 4M categorization and the multiplication number.

Table 2. EPC – 4M categorization.

Man Factors		**X**	EPC 8	Channel overload	6
Physical limitations			EPC 23	Unreliable instruments	1.6
EPC 27	Physical capabilities	1.4	**Media Factors**		**X**
EPC 36	Task Pacing	1.06	EPC 33	Poor environment	1.15
EPC 38	Age	1.02	**Management Factors**		**X**
Psychological limitations			**Communication**		
EPC 21	Dangerous incentives	2	EPC 13	Poor feedback	4
EPC 28	Low meaning	1.4	EPC 14	Delayed/incomplete feedback	3
EPC 29	Emotional stress	1.3	EPC 16	Improvised information	3
EPC 31	Low morale	1.2	EPC 18	Objectives conflict	2.5
EPC 34	Low mental workload	1.1	EPC 19	No diversity information	2.5
Experience			**Coordination**		
EPC 1	Unfamiliarity	17	EPC 2	Time shortage	11
EPC 12	Misperception of risk	4	EPC 6	Model mismatch	8
EPC 22	Lack of experience	1.8	EPC 10	Knowledge transfer	5.5
Skill and Knowledge			EPC 24	Absolute judgments required	1.6
EPC 7	Irreversibility	8	EPC 25	Unclear allocation of function	1.6
EPC 9	Technique unlearning	6	EPC 37	Supernumeraries/ lack of human resources	1.03
EPC 11	Performance ambiguity	5	**Monitoring**		
EPC 15	Operator inexperience	3	EPC 17	Inadequate checking	3
EPC 20	Educational mismatch	2	EPC 26	Progress tracking lack	1.4

Health			**Procedures**		
EPC 30	Ill – health	1.2	EPC 4	Features over-ride allowed	9
EPC 35	Sleep cycles disruption	1.1	EPC 5	Spatial and functional incomaptibilty	8
Machine Factors		**X**	EPC 32	Inconsistency of displays	1.2
EPC 3	Low signal – noise ration	10			

2.2 TOPSIS

TOPSIS is one method of multi-criteria decision-making tool proposed by Hwang and Yoon (1981) to determine the best alternative based on the compromise solution concept [20]. This technique comprises of some steps. The first one is to analyze the decision-making problem and divide it into smaller parts, then a pair-wise comparison matrix is established for each criterion. Table 3 shows that the Saaty's pair-wise comparison scale [21]. The comparison matrix A is a *n x n* real matrix, where *n* denotes the number of evaluated criteria. Each criterion x_{ij} (i = 1, 2, …m, j = 1, 2, …n) inserted in the matrix A represents the ith's relative importance against to the jth. The following formula (2) is used to establish a comparison matrix A.

$$x_{ii} = 1, \quad x_{ij} = 1/x_{ji}, \; x_{ji} \neq 0 \tag{2}$$

Table 3. Saaty's pair-wise comparison scale.

Importance	Definition
1	Equal importance
3	Moderate importance
5	Strong importance
7	Very strong importance
9	Absolute extreme importance
2,4,6,8	Intermediate values

The next step is to calculate the relative weights of criteria. The priority weight (w) can be found with the following formula.

$$w_i = \sqrt{\sum_{i=1}^{m} x_{ij}^2} \tag{3}$$

Calculate the normalized decision matrix (r_{ij}).

$$r_{ij} = \frac{x_{ij}}{w_i} \tag{4}$$

Calculate the weighted normalized decision matrix.

$$P_{ij} = r_{ij} \, X \; x_{ij} \tag{5}$$

The ideal solution is determined.

$$d_{ij}^{+} = \left(P_{ij} - P_{i\,max}\right)^2 \tag{6}$$

The negative solution is also determined.

$$d_{ij}^{-} = \left(P_{ij} - P_{i\,min}\right)^2 \tag{7}$$

The separation from the ideal solution is determined

$$d_i^{+} = \sqrt{\sum_{j=1}^{n} \left(d_{ij}^{+}\right)^2} \tag{8}$$

The separation from the negative ideal solution is determined.

$$d_i^{-} = \sqrt{\sum_{j=1}^{n} \left(d_{ij}^{-}\right)^2} \tag{9}$$

Relative closeness to the ideal solution is calculated.

$$\zeta_i = \frac{d_i^-}{d_i^+ + d_i^-} \tag{10}$$

The normalization value is determined.

$$N = \frac{\zeta_i}{\Sigma} \tag{11}$$

And the final step is is to check the consistency of the decision maker's evaluation. In this step, the consistency of data inserted in matrix A is consistent or not. The consistency index (CI) can be calculated using the following formula.

$$\sum_{j=1}^{n} x_{ij} N = \lambda_{max} N_i \tag{12}$$

$$CI = \frac{\lambda_{max} - n}{n-1} \tag{13}$$

The consistency ratio (CR) should be calculated to check the consistency. The CR value can be found in the following formula.

$$CR = \frac{CI}{RI} \tag{14}$$

In the equation, RI is the random index that Saaty provides the values in table 4.

Table 4. Random index values [22].

n	1	2	3	4	5	6	7	8	9	10
RI	0	0	0.52	0.89	1.11	1.25	1.35	1.40	1.45	1.49

2.3 Risk Perception

Risk perception could be related to attitudes that were found to be correlated with each other and with behavior [23]. In this paper, the aggregate HEP of maritime cargo accidents is determined using HEP every stage on maritime cargo shipment obtained from HEART – 4m and TOPSIS technique. The risk perception technique used is represented by an HRA event tree [24], as shown in figure 1.

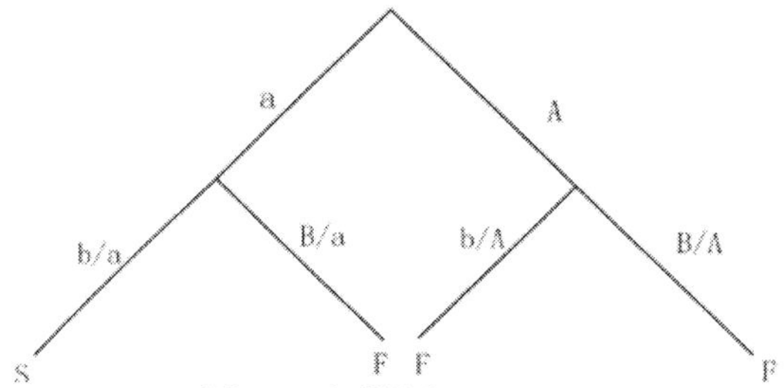

Figure 1. HRA event tree

Using the HRA event tree, a graphical description of the task's procedural steps is illustrated in the logical framework. A binary decision point represents the current action's failure or success [25]. Thus, at every binary branching, the probability of the events must sum to 1.0. Finally, the aggregate HEP was obtained by analyzing the HRA event tree and using the following formula.

$$aggreagate\ HEP = \mathrm{a}(B/\mathrm{a}) + \mathrm{A}(\mathrm{b}/A) + \mathrm{A}(B/A) \tag{15}$$

3. Data

The historical maritime cargo accident data analyzed in this study were obtained from the Indonesia maritime cargo adjusting and surveying company that engages in the marine cargo consultant of insurance. The detail of the marine cargo accident data showed in table 5 below.

Table 5. Maritime Cargo Accidents Data.

IOP Conf. Series: Materials Science and Engineering 1052 (2021) 012037 doi:10.1088/1757-899X/1052/1/012037

No	Vessel's Name	Origin	Destination	Date of Casualty	Cargo	Casualty
1	United Tristan Da Cunha	Kwangyang, South Korea	Jakarta, Indonesia	21 April 2014	Coil Rolled Steel	The truck accident during the unloading process
2	Ambassador Bridge	Bandung, Indonesia	Dhaka, Bangladesh	16 May 2014	Fabric	The truck accident during the loading process
3	Hyundai Highway	Suzhou, China	Batam, Indonesia	28 May 2014	Battery cells	Rough handling during the unloading process
4	Kota Harmuni	Singapore	Bogor, Indonesia	30 May 2014	Haulotte vertical mast	Rough handling during the unloading process
5	Hamburg Bridge	Antwerp, Belgium	Tangerang, Indonesia	20 June 2014	Stray field machine	Rough handling during the sailing process
6	Hyundai	Suzhou, China	Batam, Indonesia	6 June 2014	Battery cells	Rough handling the unloading process
7	KSD 06	Batam, Indonesia	Jakarta, Indonesia	27 December 2015	Steel crap	The vessel capsized

4. Result

4.1 EPC – 4M

There are three shipment processes in maritime cargo shipment: the loading process, sailing process, and unloading process. In this study, the EPC – 4M were determined in every maritime cargo shipment process, and in every process, were found more than one EPC. Thus, 97 EPC – 4M factors were found in Indonesia's seven maritime cargo accidents. Management factors had the most EPCs, 52, consist of communication, coordination, monitoring, and procedures subfactors.

EPC's total in the man factor is 40 consist of physical, psychological, skill and knowledge, and health. Moreover, the other EPCs there are three from machine factor and two EPCs from media factor. The list of EPCs found in the analysis is presented in table 6.

Table 6. EPC – 4M results.

Man Factors	Total	Media Factors	Total
Physical limitations		EPC 33 Poor environment	2
EPC 36 Task Pacing	3	**Total**	**2**
Psychological limitations		**Management Factors**	**Total**
EPC 21 Dangerous incentives	5	Communication	
Experience		EPC 13 Poor feedback	5
EPC 1 Unfamiliarity	3	EPC 14 Delayed/incomplete feedback	7
EPC 12 Misperception of risk	4	EPC 16 Improvised information	5
EPC 22 Lack of experience	4	EPC 19 No diversity information	6
Skill and Knowledge		Coordination	
EPC 7 Irreversibility	4	EPC 2 Time shortage	6
EPC 9 Technique unlearning	4	EPC 10 Knowledge transfer	4
EPC 11 Performance ambiguity	7	EPC 24 Absolute judgments required	3

EPC 15	Operator inexperience	5	EPC 25	Unclear allocation of function	1
Health			Monitoring		
EPC 35	Sleep cycles disruption	1	EPC 17	Inadequate checking	7
Total		**40**	Procedures		
Machine Factors		**Total**	EPC 5	Spatial and functional incompatibility	7
EPC 3	Low signal – noise ration	1	EPC 32	Inconsistency of displays	1
EPC 23	Unreliable instruments	2	**Total**		**52**
Total		**3**			
Sub Total					**97**

4.2 HEP Calculation

In this study, the author calculated the HEP at every process of maritime cargo accidents. However, in this section, we consider only one case in one process. The following calculation description is from case no 1 in the unloading process. There are five EPCs selected, EPC 2, EPC 15, EPC 17, EPC 22, and EPC 35. The data are processed using TOPSIS to get the HEP, and the process as follows:

4.2.1 Pair-wise comparison matrix

The EPCs selected caused the accident constructed in the pair-wise comparison matrix, as shown in table 7. The proportion of every EPC was calculated using formula (2), and the attribute weight (w_i) was calculated using formula (3).

Table 7. Pair-wise comparison matrix and attribute weight.

	EPC 2	EPC 15	EPC 17	EPC 22	EPC 35	w_i
EPC 2	1	5	0.33	0.5	2	5.51
EPC 15	0.2	1	0.17	0.25	0.33	1.11
EPC 17	3	6	1	2	4	8.12
EPC 22	2	4	0.5	1	2	5.02
EPC 35	0.5	3	0.25	0.5	1	3.25

4.2.2 The normalized decision matrix

Table 8 shows the normalized decision matrix was calculated using formula (4).

Table 8. Normalized decision matrix.

	EPC 2	EPC 15	EPC 17	EPC 22	EPC 35
EPC 2	0.18	0.91	0.06	0.09	0.36
EPC 15	0.18	0.90	0.15	0.22	0.30
EPC 17	0.37	0.74	0.12	0.25	0.49
EPC 22	0.40	0.80	0.10	0.20	0.40
EPC 35	0.15	0.92	0.08	0.15	0.31

4.2.3 The weighted normalized decision matrix

After the normalized decision matrix determined, the weighted normalized decision matrix was calculated using formula (5). The maximum weight (max) and minimum weight (min) of every EPC decided from those results, as shown in table 9.

Table 9. Weighted normalized decision matrix and maximum-minimum weight.

	EPC 2	EPC 15	EPC 17	EPC 22	EPC 35	max	min
EPC 2	0.18	4.54	0.02	0.05	0.73	4.54	0.02
EPC 15	0.04	0.90	0.02	0.06	0.10	0.90	0.02
EPC 17	1.11	4.43	0.12	0.49	1.97	4.43	0.49
EPC 22	0.80	3.18	0.05	0.20	0.80	3.18	0.05
EPC 35	0.08	2.77	0.02	0.08	0.31	2.77	0.02

4.2.4 The ideal solutions and its separation

The ideal solution was calculated using formula (6), and its separation was calculated using formula (8). The results show in table 10.

Table 10. Ideal solution matrix and its separation.

	EPC 2	EPC 15	EPC 17	EPC 22	EPC 35
EPC 2	18.97	0	20.40	20.18	14.53
EPC 15	0.74	0	0.76	0.71	0.64
EPC 17	11.05	0	18.56	15.52	6.06
EPC 22	5.70	0	9.82	8.91	5.70
EPC 35	7.25	0	7.56	7.25	6.06
d_i^+	21.86	0	28.56	26.28	16.49

4.2.5 The negative ideal solution matrix and its separation

The negative ideal solution was calculated using formula (7), and its separation was calculated using formula (9). The results show in table 11.

Table 11. Negative ideal solution matrix and its separation.

	EPC 2	EPC 15	EPC 17	EPC 22	EPC 35
EPC 2	0.03	20.40	0	0.0006	0.50
EPC 15	0	0.76	0	0	0.01
EPC 17	0.38	15.52	0.14	0	2.18
EPC 22	0.56	9.82	0	0.02	0.56
EPC 35	0	7.56	0	0	0.08
d_i^-	0.48	27.03	0.07	0.01	1.66

4.2.6 Relative closeness to the ideal solution and normalization

The relative closeness to the ideal solution was calculated using formula (10), and its normalization was calculated using formula (11). The result shows in table 12.

Table 12. Relative closeness to the ideal solution and its normalization.

	EPC 2	EPC 15	EPC 17	EPC 22	EPC 35	Total
ζ_i	0.022	1.00	0.002	0.001	0.09	1.12
N	0.019	0.90	0.002	0.0005	0.08	1.00

4.2.7 Consistency verification

The consistency of data inserted in a pair-wise comparison matrix is checked to verify whether the matrix is consistent or not. This step calculates CI using formula (13) and calculates CR using formula (14). If the CR values are ≤ 0.01, the judgments are considered consistent and reasonable, as shown in table 13.

Table 13. Consistency check.

CI	RI	CR
0.02	1.11	0.022

4.2.8 HEP calculation

After APE's weight was determined using the TOPSIS technique, the HEP can be determined using formula (1), and the result of HEP for case no one at the unloading process shown in table 14.

Table 14. HEP calculation.

EPC 2		EPC 15		EPC 17		EPC 22		EPC 35	
x	APE	x	APE	x	APE	x	APE	x	APE
11	0.019	3	0.90	3	0.002	4	0.0005	1.1	0.08
HEP			0.54						

HEP calculation process, as the way above, was applied to the other cases in all of the maritime cargo shipment process. Table 15 shows HEP's result in seven maritime cargo accidents in all processes (loading process, sailing process, and unloading process).

Table 15. HEP calculation for seven maritime cargo accidents.

Case no	HEP in the loading process	HEP in the sailing process	HEP in the unloading process
1	1	0.42	0.54
2	0.75	0.55	0.82
3	0.13	0.14	0.13
4	0.11	0.17	0.05
5	0.56	0.42	1
6	0.11	0.14	0.16
7	0.79	1	1
Average	**0.49**	**0.41**	**0.53**

4.3 Aggregate HEP calculation

In this step, the aggregate HEP was determined using the risk perception technique, which used the HRA event tree and the formula (15). To clarify this matter, figure 2 shows the example of the HRA event tree of case no 1. A capital letter indicates the shipment process's failures that the value is HEP of each process, as shown in table 15. In contrast, the small letter indicates the successes of the process.

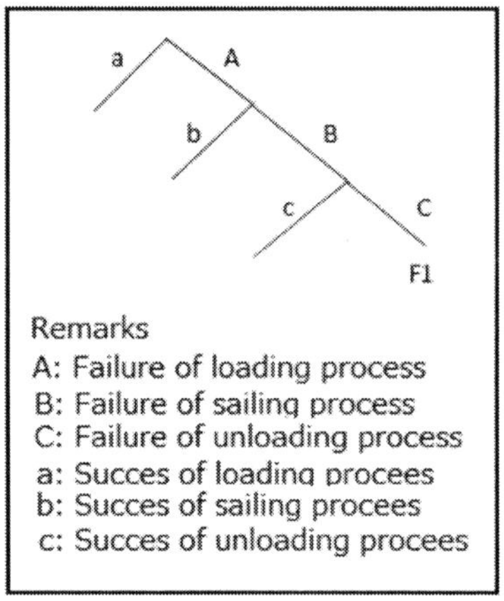

Figure 2. HRA event tree case no 1

In case no 1, the cargo's accidents were shipped from the unloading port to the consignee's place using a truck (unloading process). Using the HRA event tree, the aggregate HEP of case no 1 is F1. The value of F1 was determined by using the formula (15) and was obtained 0.23. The remaining six maritime cargo accidents in Indonesia were analyzed, and the result is shown in table 16.

Table 16. Aggregate HEP of maritime cargo accidents.

Case no	Aggregate Hep	%HEP
1	0.23	22.68
2	0.86	86.28
3	0.002	0.24
4	0.0009	0.09
5	0.56	56
6	0.002	0.25
7	0.79	79
Average	**0.3493**	**34.93**

5. Discussion

The definition of maritime cargo accidents is any accident that occurred within the maritime cargo shipment. Maritime cargo shipment includes all the casualties of any means of transport from the shipper (seller) to the loading port, sailing process, and shipment from the unloading port to the consignee (buyer) does not have to be a ship. The data on maritime cargo accidents are challenging to retrieve because this data is not recorded by the government institution but only the party involved in this accident, like adjusting surveying company and the insurance company. In this study, the human error probability analysis was determined in every process of maritime cargo shipment. The management factor becomes the most common EPC – 4M factor found in these cases. It means that the tasks related to management, such as communication, coordination, monitoring, and procedures, must receive more attention. The management factor could occur because the maritime cargo shipment process needs more intention to coordinate every process.

Based on the HEP calculation result, the HEP value in the unloading process is highest than the other process, followed by the HEP value in the loading process and the sailing process. The handling of the cargo mostly occurred during the loading and unloading process rather than the sailing process. The loading process is the first process of maritime cargo shipment. Thus the operator or other party is more concerned about cargo handling in the loading process. However, it is inversely proportional to the unloading process. Maritime cargo accidents mostly occurred in the unloading process. Therefore the HEP value in the unloading process is higher than the loading process.

In this study, the HEART – 4M combines with the TOPSIS methodology to determine the HEP in every process of maritime cargo shipment. Then the aggregate HEP was assessed using the risk perception technique. The integration of these methods suggests the relation between the EPC – 4M, HEP in every process, and aggregate HEP can involve the accidents.

6. Conclusion

Human reliability analysis is a method that quantifies the probability of human error and provides information for mitigating maritime cargo accidents [19]. In this study, seven maritime cargo accidents in Indonesia were analyzed using the hybrid methodology HEART – 4M and TOPSIS with the risk perception technique. The purpose of this paper is to find the causal factor of maritime cargo accidents in Indonesia. There were 97 EPM – 4M factors found as causal factors, where management factors were the most common factors found.

The other purpose is to predict the HEP of maritime cargo accidents. HEART – 4M and TOPSIS were used to get the HEP value for every maritime cargo shipment process. The average HEP in the loading process was 0.49 (49%), the sailing process was 0.41 (41%), and the unloading process was 0.53 (53%). After the HEP, every process was determined, then the aggregate HEP was determined using the risk perception technique, and the average of aggregate HEP is 0.3493 (34.93%). Finally, the hybrid method proposed in this study provides a practical tool to determine the value of HEP in maritime cargo accidents.

7. References

[1] M. Grote *et al.*, "*Dry bulk cargo shipping — An overlooked threat to the marine environment?*," Mar. Pollut. Bull., **vol. 110**, no. 1, pp. 511–519, 2016.
[2] U. N. C. and T. D. (UNCTAD), *Review of Maritime Transport 2019*, no. October. 2019.
[3] Rumaji and A. Adiliya, "Port Maritime Connectivity in South-East Indonesia: A New Strategic Positioning for Transhipment Port of Tenau Kupang," Asian J. Shipp. Logist., **vol. 35**, no. 4, pp. 172–180, 2019.
[4] N. Tu, D. Adiputranto, X. Fu, and Z. C. Li, "*Shipping network design in a growth market: The case of Indonesia,*" Transp. Res. Part E Logist. Transp. Rev., **vol. 117**, no. January 2017, pp. 108–125, 2018.
[5] R. E. Prilana, L. P. Bowo, and M. Furusho, "*Maritime Cargo Accidents in Indonesia for the Period 2013 - 2018,*" in Asia Navigation Conference 2019, 2019.
[6] E. Akyuz and M. Celik, "*Application of CREAM human reliability model to cargo loading process of LPG tankers,*" J. Loss Prev. Process Ind., **vol. 34,** pp. 39–48, 2015.

[7] E. Akyuz and E. Celik, "*A modified human reliability analysis for cargo operation in single point mooring (SPM) off-shore units*," Appl. Ocean Res., **vol. 58**, pp. 11–20, 2016.

[8] E. Eleftheria, P. Apostolos, and V. Markos, "*Statistical analysis of ship accidents and review of safety level*," Saf. Sci., **vol. 85**, pp. 282–292, 2016.

[9] E. Ashmawy, "*The Maritime Industry and the Human Element Phenomenon*," 13th Annu. Gen. Assem. IAMU Expand. Front. - Challenges Oppor. Marit. Educ. Train., pp. 277–288, 2012.

[10] L. P. Bowo and M. Furusho, "*Human Error Assessment and Reduction Technique for Marine Accident Analysis: The Case of Ship Grounding*," Trans. Navig., **vol. 3**, no. 1, pp. 1–7, 2018.

[11] E. Akyuz, M. Celik, and S. Cebi, "*A phase of comprehensive research to determine marine specific EPC values in human error assessment and reduction technique*," Saf. Sci., **vol. 87**, pp. 63–75, 2016.

[12] J. Tu, W. Lin, and Y. Lin, "*A Bayes-SLIM based methodology for human reliability analysis of lifting operations*," Int. J. Ind. Ergon., **vol. 45**, pp. 48–54, 2015.

[13] M. Konstandinidou, Z. Nivolianitou, C. Kiranoudis, and N. Markatos, "*A fuzzy modeling application of CREAM methodology for human reliability analysis*," Reliab. Eng. Syst. Saf., **vol. 91**, no. 6, pp. 706–716, 2006.

[14] B. Kirwan, R. Kennedy, S. Taylor-Adams, and B. Lambert, "*The validation of three human reliability quantification techniques THERP, HEART and JHEDI: Part II - results of validation exercise*," Appl. Ergon., **vol. 28**, no. 1, pp. 17–25, 1997.

[15] E. Akyuz and M. Celik, "*A hybrid human error probability determination approach: The case of cargo loading operation in oil/chemical tanker ship*," J. Loss Prev. Process Ind., **vol. 43**, pp. 424–431, 2016.

[16] L. P. Bowo, R. E. Prilana, and M. Furusho, "*A Modified HEART – 4M Method with TOPSIS for Analyzing Indonesia Collision Accidents*," **vol. 14**, no. 3, 2020.

[17] M. T. Kinateder, E. D. Kuligowski, P. A. Reneke, and R. D. Peacock, "*Risk perception in fire evacuation behavior revisited: definitions, related concepts, and empirical evidence*," Fire Sci.Rev., **vol. 4**, no. 1, 2015.

[18] J. . Williams, "A *data-based method for assessing and reducing human error to improve operational performance,*" in Conference Record for 1988 IEEE Fourth Conference on Human Factors and Power Plants, 1988, **vol. 72**, no. DOT/FAA/AM-01/3, pp. 436–450.

[19] L. P. Bowo and M. Furusho, "*Usability of Human Error Assessment and Reduction Technique with a 4M framework (HEART–4M) – A Case Study on Ship Grounding Accidents*," J. ETA Marit. Sci., **vol. 2**, no. 1, pp. 41–46, 2019.

[20] C.-L. Hwang and K. Yoon, *Multiple Atribute Decision Making: Methods and Applications*. 1981.

[21] T. L. Saaty, "*Axiomatic Foundation of the Analytic Hierarchy Process*," **vol. 32**, no. 7, pp. 841 855, 1986.

[22] T. L. Saaty, "*How to make a decision: The analytic hierarchy process*," Interfaces (Providence)., **vol. 24**, no. 1, pp. 19–43, 1994.

[23] D. D. Dinh, N. H. Vũ, R. C. McIlroy, K. A. Plant, and N. A. Stanton, "*Effect of attitudes towards traffic safety and risk perceptions on pedestrian behaviours in Vietnam*," IATSS Res., 2020.

[24] D. Swain and H. E. Guttmann, "*Handbook of reliability analysis with emphasis on nuclear plant applications*," no. August, 1983.

[25] F. Vanderhaegen, M. Cassani, and P. C. Cacciabue, *Efficiency of safety barriers facing human errors*, **vol. 11**, no. PART 1. IFAC, 2010.

IOP Conf. Series: Materials Science and Engineering 1052 (2021) 012038 doi:10.1088/1757-899X/1052/1/012038

Analysis of Human Factors in Ship-to-ship Collision Accident: A Semi-quantitative Approach Based on Network Theory

I G M S Adhita[1] and M Furusho[2]

[1,2] Graduate School of Maritime Sciences, Kobe University, 5-1-1 Fukae-minamimachi, Higashinada, Kobe, Japan

Abstract: As long as human beings operate the ship, human factors always have a substantial contribution to ship operation success and failure. Thereby, a study related to human factors, especially about the officer on board, is essential to create better maritime transportation. This research proposed the semi-quantitative analysis to determine how human factors are involved in a ship-to-ship collision using a network theory perspective. The study was conducted by describing the dependency amongst human activities (officer onboard) during ship-ship encounters. The ship-to-ship encounters model has been done in an undirected graph consisting of nodes and arcs. Through the vertex performance rate, system performance could be evaluated. Hence this research aimed to define the weight evolution when the unintended action emerges in a node and distributes its weight to its couplings. In conclusion, the network model for ship-to-ship collision accidents can describe the weight spreading process among functions and how it affects system performance then causes an accident.

1. Introduction

Accidents are always complicated to pin down due to the relation with a lot of uncertainty. Accidents occur randomly and abstractly, as well as the difficulty of gathering post-event information is one of the many uncertainties in conducting investigations. However, the research in this field is essential to enhance the level of safety at work. Enhancing safety will reduce the number of work injuries, increase the company's profitability and credibility, and prevent environmental damage.

In the maritime field, maritime traffic is an indispensable mode of transportation in world trade. Still, it is not as deadly as cars or railroads traffic. This makes it difficult to pay attention to marine traffic when compared to other traffic. However, once an accident occurs, the damage is likely to be enormous not only for human life but also economically and environmentally because the ship size is massive. Hence, the development of the accident model is critical to enhancing ship operations safety.

A systemic model is one of the accident models that have been widely used in many fields of study [1]. A systemic model is a model design to describe system performance on the level of a system as a whole, rather than on the level of specific cause-effect events. Two terms are closely related to this idea, namely the socio-technical system and Safety-II. In a complex environment, the socio-technical term generally describes the interaction between the social and technical systems. Ship as a complex system is composed of people and technical components. In this case, the socio-technical term can be applied in the full sense to analyze ship operation regarding the occurrence of a collision accident. In the maritime industry, a socio-technical system term has been used in analyzing resilience in VTS operation [2] and STAMP based approach for safety management systems [3]. Moreover, research about the possibility of optimizing the organization in ship operation has been done using a socio-technical system approach [4].

Safety-II is a term used to provide a new solution to understand how systems work in everyday operation [5]. Through a systemic model approach, this term has been introduced along with FRAM analysis. This method analysis is focusing on how conditions lead to the emergence of the accident [6]. Given the uncertainty, intractability, and complexity of ship operation, performance adjustments are needed for the system to function [1]. Consequently, failure and success are acknowledge derived from the same resource, namely, everyday performance adjustment.

This article proposed a ship-to-ship collision accident model using an undirected graph model based on qualitative analysis through Safety-II and socio-technical system perspectives. A study based on network theory has been conducted using an undirected graph model to analyze the cascading failure process from a railway accident [7]. Graph analysis has been used in the maritime domain to model a crew and system interaction variability [8]. This graph model presents how the system transmits the information from one function to another function. A function is associated with a vertex, and the edge associates its connection with another function. Each vertex stated has a weight that describes its performance in the system. When the vertex is working with undesirable performance, it can be said that the vertex is overweighted. Since vertices are dependent on the system, this extra weight will spread through the edges and affect the other vertices. This extra weight also affects the vertex's performance to complete its work. Hence, this research aimed to define how the dependency of unexpected system function variation contributes to ship-to-ship collision accidents indicated by a loss of system performance rate.

2. Network Theory

A network, also known as a graph in mathematics, is a set of vertices (V) and edges (E). In a simple manner, it can be written as $G = \{V, E\}$. In graph representation, the vertex is visualized as a dot and the edge as a line that connects two or more vertices. The connection between two vertices can be symmetric, called an undirected graph, or asymmetric, called a directed graph. The directed graph has arcs instead of edges. The edge from two vertices $\{V_1, V_2\}$ in the undirected graph can be written as a pair arcs of $\{V_1, V_2\}$, and $\{V_2, V_1\}$ in the directed graph.

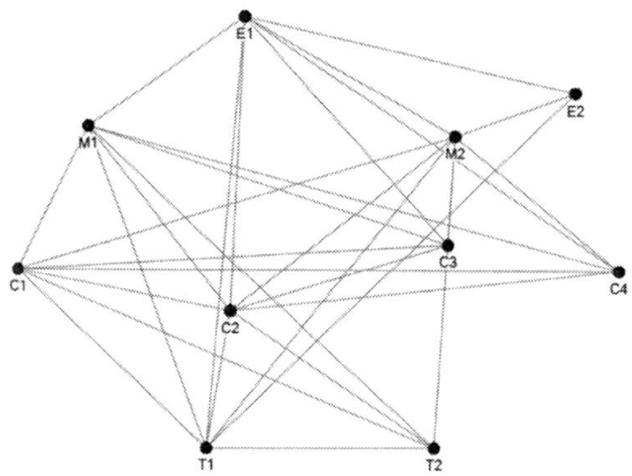

Figure 1. Undirected graph model for activities related to the bridge team during a ship-ship encounter.

3. Data Collection

This study observed the ship-to-ship collision accident that occurred in Indonesia. As an archipelagic country, the maritime industry is a vital sector to support economic development. Hence, it is important to ensure safety during ship operations. The data used in this research was obtained from the National Transportation Safety Committee (NTSC) website. NTSC is an independent Indonesian institution responsible for investigating accidents or incidents in aviation, maritime, railway, and road transport. This analysis using ten ship-to-ship collision accident reports that occurred in the period from 2010 to 2018.

The ship-to-ship collision accident occurrence is sensitive associated with weather conditions, time of the day, size of the ship involved, type of the ship, location, etc. Therefore, it is essential to provide a limitation to select a relevant accident report to avoid bias results. The accidents selected in this study were analyzed ship-to-ship collision involving only merchant ships. Those selected accidents are involving merchant ships equipped with an electronic navigation device (RADAR, ECDIS, AIS, etc.) and relatively in the same size.

4. Ship Collision Accident Analysis

4.1. Qualitative analysis

This qualitative analysis was done based on the Safety-II perspective. In a complex system, uncertainty and complexity are given. Surprisingly, everyday operations were well done very often. But this surprising fact is always ignored. When talking about analysis to enhance safety, we always focus on failure that is practically rare rather than the many works that well-done, and that does not show us any direction to create any better situation.

Things go right in everyday operation because the operator can adjust his performance to the work condition rather than working by imagining how the work should be done. Therefore, the acceptable outcomes and adverse outcomes are acknowledge derived from the same source, namely, everyday performance variability [9]. Thus, a new idea came out to shifting the point of view of safety by looking at the many cases where things go right rather than looking for the few cases where things go wrong to understand how that happens. Later on, this new idea is popularly known as Safety-II [5].

Table 1. Functions for ship-ship encounter situation.

Function	Code
Weather Condition	E1
Time of day	E2
Lookout through an electronic device	M1
Direct lookout	M2
Bridge team communication	C1
Bridge to bridge communication	C2
Bridge to port communication	C3
VTS Communication	C4
Maneuvering	T1
Engine control	T2

To strengthening the concept of Safety-II, a method called Functional Resonance Analysis Method (FRAM) was introduce in 2004 [6] and renewed in 2012 [10]. Along with that, this analysis also applies FRAM properties to describe the system by dividing it into a set of activities called function. A function is defined as something that must be done, but more than just a task. Its present activity with the flexibility to be adjusted by humans as its capacity.

Moreover, one of the socio-technical system property that is emergence phenomenon also applied to this analysis. The emergence phenomenon is a phenomenon where the system has properties that only emerge once its assembled. An accident is acknowledged as a property that emerges from unusual dependencies of system performance rather than a property derived from system components. This phenomenon is widely used in social sciences research, for instance, in the aviation field [11,12]. Unique function's dependency makes the system can produce outcomes beyond their designated outcome. It can be in the form of a wanted outcome or an unwanted outcome. This function's dependency can be used through the resonance phenomenon to trace the problem once the unexpected outcome emerges. Also, this phenomenon can be used to describe how interaction among functions formed.

Based on those principles, there are three steps to conducted qualitative analysis. The first step is to divide ship-ship encounter situations into functions that express the environmental and technological aspects that affect human work (officer onboard). In this analysis, ten functions have been stated: weather condition, time of day, look out through electronic device, direct lookout, bridge team communication, bridge to bridge communication, bridge to port communication, VTS communication, maneuvering, and engine control (Table 1). The second is to determine the upstream and downstream functions. This term is used to describe a temporal relationship between functions. In each accident, the role of upstream and downstream functions may vary regarding their execution time. Finally, the last step is to state two or three functions as a trigger function that describes how the resonance effect occurred between functions and produce an amplifying effect for the system's variability performance. The qualitative analysis of ten ship-to-ship collision accidents used in this research is shown in Table 2.

Table 2. 10 ship-to-ship accidents collected from the NTSC website and qualitative analysis result regarding trigger functions cause initial extra weight for the system.

Accident	Ship involved	Unwanted function variability performance
May 2010	KM. Dian No. 1	M1: too late and slightly imprecise; M2: too late and slightly imprecise.
	MT. Soechi Chemical XIX	M1: slightly late and slightly imprecise; M2: slightly late and slightly imprecise; C1: too late and imprecise.
June 2010	KM. Bosowa VI	M2: too late and slightly imprecise; C2: too late and slightly imprecise.
	KM. Shinpo 18	C1: too late and imprecise; C2: too late and imprecise.
Aug. 2010	KM. Inditama V	M1: slightly late and slightly imprecise; M2: too late and slightly imprecise
	KM. Trisal Pratama	M1: too late and imprecise; M2: too late and imprecise.
Sept. 2012	Bahuga Jaya	M2: too late and imprecise; C2: too late and imprecise.
	Norgas Cathinka	M2: too late and imprecise; C2: too late and imprecise.
May 2013	KM. Lintas Bahari Utama	C1: too late and imprecise; C3: too late and imprecise.
	KM. Lintas Bengkulu	M1: too late and imprecise; M2: too late and imprecise.
June 2015	Leo Perdana	M2: too late and imprecise; T2: too late and imprecise.
	Navigator Aries	M2: too late and imprecise; C1: too late and imprecise.
Des. 2015	Stolt Commitment	M1: too late and imprecise; M2: too late and imprecise; C1: too late and imprecise.
	Thorco Cloud	M1: too late and imprecise; M2: too late and imprecise; C1: too late and imprecise.
Apr. 2017	Baita Jaya Samudra	M1: too late and imprecise; M2: too late and imprecise.
	Elisabet	M1: slightly late and slightly imprecise; M2: too late and imprecise; C2: slightly late and slightly imprecise.
Sept. 2017	Kartika Segara	M1: too late and imprecise; M2: too late and imprecise; C1: slightly late and slightly imprecise.
	JBB De Rong 19	M1: too late and imprecise; M2: too late and imprecise; C1: too late and imprecise.
July. 2018	Bunga Melati 79	M1: too late and imprecise; M2: too late and imprecise.
	Tk. Golden Way 3310	M1: too late and imprecise; M2: too late and imprecise.

4.2. The evolution process of weighted vertex

The system works by delivering the information from one vertex to another vertex inside the graph. While being carried out, each vertex should complete its work to be able to send information properly to its neighborhood. Let give vertex i has L_i that determines its capacity to completing this work and c_i as the tolerance. The relation between c_i and L_i can be written as: $c_i = \alpha_i L_i$, where α_i is a variable coefficient.

Two conditions can occur during the system works. First is when the system can be successful in achieving its purpose. This condition is indicated by each vertex well-performed, and the information can be appropriately delivered from vertex i to vertex j. It can be express as $L_i \leq c_i$. Otherwise, the weight in vertex i exceeds its capacity and causes vertex i to have problems transferring information. It means $Li > c_i$. However, the system has to keep working. Then, extra weight L_i from vertex i at time t must be delivered to vertex j at time $t + 1$. The adjacent vertex could be not only one, but some. By assuming the weight of vertex i evenly distributed that extra weight to its neighbor, the evolution of the weight of vertex j can be calculated as follows [7]:

$$L_j(t+1)\begin{cases} L_j + \sum_{j\in N_j} \frac{L_i(t)}{|N_j|}, & if\ L_i > c_i; \\ L_j, & otherwise. \end{cases} \tag{1}$$

N_j in equation (1) denotes all adjacency vertex j that is connecting with vertex i at time $t + 1$. Changes in weight L_j at time $t + 1$ due to vertex i cannot complete its weight at time t affecting vertex j performance to completing its work. This spreading continuous until all vertices in the network connecting each other. By assuming the system starts with its best condition, where *Performance rate* (0) = 1, the evolution of *Performance rate* in vertex j at time $t + n$ can be defined in a simple manner as:

$$Performance\ rate_j(t+n)\begin{cases} Performance\ rate_j(0)\frac{c_j}{L_j(t+\mathrm{n})}, & if\ L_i > c_i; \\ Performance\ rate_j(0), & otherwise. \end{cases} \tag{2}$$

The global performance of a network (G) can be defined by measuring the average performance rate of each vertex i. It can be written as follows:

$$Performance\ rate(Global) = \frac{\sum Performance\ rate_i}{\sum V_i} \tag{3}$$

4.3. *The Semi-quantitative analysis*

This section intends to describe how to perform a mathematical approach in sub-section 3.2 in the undirected graph model based on the qualitative analysis in sub-section 3.1. The existence of function may vary in each accident, especially for functions related to communication. C3 and C4 may not exist if the accident occurred are not near the port. Then, dependency among vertex was made by considering the upstream and downstream function terms. Also, in natural behavior, some vertices have an exact condition. Vertex E1 only can act as an upstream function. Vertex E2 only can affect by E1 and should always consume as an upstream function by the other vertices. Besides, vertex T2 only can be a downstream function. For vertices M1, M2, C1, C2, C3, C4, and T1 can be both upstream and downstream depends on their activation time.

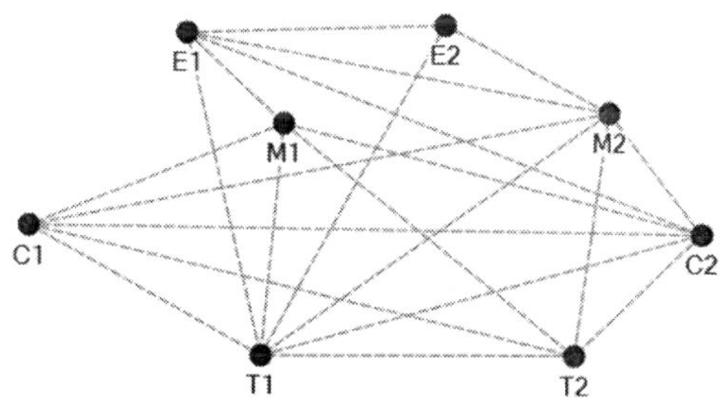

Figure 2. Graph model for functions of Bhaita Jaya Samudra.

For instance, this explanation used an accident case in April 2017, specifically for Bhaita Jaya Samudra. In this accident, vertices C3 and C4 did not exist. As mentions in sub-section 3.1, the interacting trigger functions were identified as M1 and M2. Assume that each function is connected in a given period from t to $t+n$. It started at $t = 0$ where all edges are not yet connected, as shown in Figure 2. At t = 1 the first activated vertex is E1. Follow by E2 at t = 2, M1 and M2 at t =3, C1 at t = 4, and C2 at t = 5. In the end, at t = 6, T1 is activated. Each time vertex i is activated, the dashed line connecting vertex i to the other vertices is changed into a solid line. The illustration of the activated vertex for each period is shown in Figure 3.

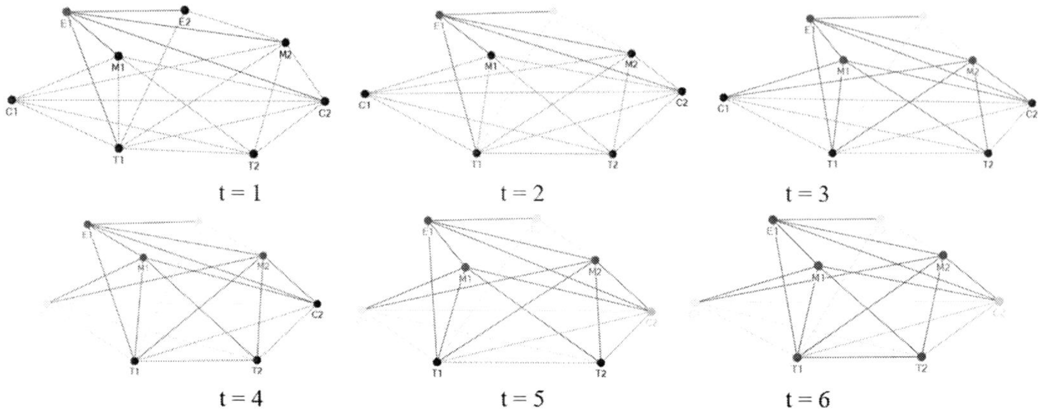

Figure 3. A period of vertex activation in Bhaita Jaya Samudra.

To perform a calculation, first, it necessary to measure weight for each vertex. This analysis does not consider the vertex's hierarchy. Therefore, in normal condition, all vertices are assumed to have $L_i = 1$. L_i says to be changed if the unwanted performance emerges in the vertex. Two phenotypes were used to determine vertex performance, which are time and precision. It is trivial if unwanted variability performance emerges in all functions with a vast number. Thus, this analysis stated five reasonable scores of weight for trigger vertex. Those levels are shown in Table 3.

Table 3. Weight scores for two phenotypes combination of function variability performance.

Phenotype of Function		Weight
Time	Precision	
Too late	Imprecise	5
Too late	Slightly imprecise	4
Slightly late	Slightly imprecise	3
Acceptable	Acceptable	2
On time	Precise	1

The calculation below shows the spreading weight at time t(3) and t(4) to simplify the illustration. T (3) points at vertices M1 and M2 unwell-performed due to their variability performance appearing too late and imprecise. By unique interaction, they start to spread their weights and cause the rest of the functions to work inappropriately.

$t = 3$; change all dash lines connected to M1 and M2 into solid.

$$L_{C1}(3) = 1 + \frac{5}{4} + \frac{5}{4} = 3.5$$
$$L_{C2}(3) = \frac{1}{4} + \frac{5}{4} + \frac{5}{4} = 3$$
$$L_{T1}(3) = \frac{1}{6} + \frac{5}{4} + \frac{5}{4} = 2.8$$
$$L_{T2}(3) = \frac{1}{5} + \frac{5}{4} + \frac{5}{4} = 2.9$$

$t = 4$; change all dash lines connected to C1 into solid.

$$L_{C2}(4) = \frac{1}{4} + \frac{3.5}{3} = 1.42$$
$$L_{T1}(4) = \frac{1}{6} + \frac{3.5}{3} = 1.33$$
$$L_{T2}(4) = \frac{1}{5} + \frac{3.5}{3} = 1.37$$

The weight continues to spread until all dash lines changed into a solid line. Thus, at $t = 6$ when all the dash lines changed into a solid line, each vertex's final weight L achieved is shown in Table 4. The vertex that experiences the weight increment causing its performance rate to decrease. In this case, Bhaita Jaya Samudra's performance overall has decreased to 53%.

Table 4. Weight and performance rate evolution for each vertex in Bhaita Jaya Samudra.

Node i	Variability performance	L_i	c_i	Perf. rate
E1	Precise	1.00	2	100%
E2	Acceptable, Acceptable	2.00	2	100%
M1	Too late, Imprecise	5.00	2	40%
M2	Too late, Imprecise	5.00	2	40%
C1	Slightly late, slightly imprecise	3.50	2	57%
C2	Not at all	4.67	2	43%
T1	Too late, Imprecise	7.00	2	29%
T2	Not at all	14.00	2	14%

5. Results

The result from the analysis of ten ship-to-ship collision accidents in Indonesia found how each ship involved in the accident has decreased its performance. This performance decrement causes the ship to experience collision accidents during the encounter situation. In general, overall ship performance was decreased to 50% - 60%. The interaction between unwanted variability performance of M1 and M2 is the most frequently appear. This interaction also gives the lowest score of performance rate that is 53%.

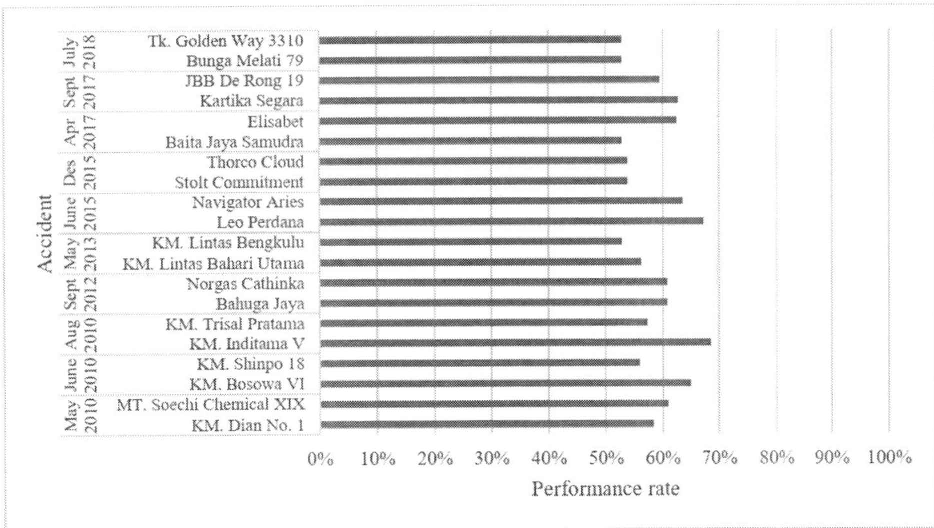

Figure 4. The result of the analysis for 20 ships involved in ship-to-ship collision accidents occurred in Indonesia.

A value of performance rate decrement is strongly affected by the number of available functions in the accident, a kind of variability phenotype combination, and when the unexpected trigger function is active (time emergence)—for example, KM. Inditama V on the accident case occurred in August 2010 and Leo Perdana on the accident case occurred in June 2015. In this accident, the unexpected performance variability appeared on M1 and M2 functions. The time emergence of unexpected trigger function was early, at $t = 3$. Unfortunately, the kind of variability phenotype combination has a relatively low weight, which is 2.5 and 4. In contrast, on the accidents involving Leo Perdana, the time emergence of unexpected trigger function was late, at $t = 4$ and $t = 6$. But the kind of variability phenotype combination has the highest weight from the list, which is 5. These two different conditions cause KM. Inditama V and Leo Perdana experience a performance decrement to be 69% and 67%, respectively.

6. Discussion

This study presents an analysis of 10 ship-to-ship collision accidents involving 20 ships. From the analysis, it can introduce how the dependency among function happened through the weight spreading process. It also shows the contribution of each ship to the occurrence of the accident. In general, each ship's contribution to the occurrence of accidents can be assigned into two groups. First is where both ships experienced a drastic reduction in their performance rate. The second is where only one ship experienced a drastic reduction in its performance rate. The analysis found that the first condition was more common, where out of 10 events, only three events were in the second group.

The variation in function performance indicates that only the interaction between variation with weights 4 and 5 can significantly decrease system performance. If variations with weight 3 appear, then at least three functions appear simultaneously with undesirable performance variations to cause an accident. The transformation of weight in each vertex from the distribution process is also greatly influenced by the vertex's time activation. The earlier vertex to active with unwanted performance variations, the greater the system performance will decrease. It makes sense that if bad weather occurs at night, it would be heavier for other vertices to work since both of these vertices were active in the first place.

This study also emphasizes that emergence properties play an essential role in describing a relation among ship operation functions. Ship accidents cannot be stated as having only one primary cause, but rather a unique interaction of variability performance from multiple functions. The interaction of unexpected variability performance between various functions in the system (ship operations) can produce a new outcome beyond its desired outcome. This unexpected variability performance can state the function has extra weight. Then, this weight is distributed across other functions since the system has to keep working. At one point, the system can no longer organize the extra weight that occurred in each function and cause system performance to drop significantly.

7. Conclusion

In conclusion, this study demonstrates how the unwanted variability of two or three trigger functions resonates and causes the function to be overweight. The undirected graph model presents how extra weight can be distributed from one vertex to another vertex through its connection. Weight evolution can be determined using a simple mathematical approach. The analysis results of 10 ship collision accident reports show that the accidents occurred because each vessel's performance rate fell below 70%.

This approach has the potential to continue to be developed in the future, especially to analyze the other kind of ship accidents, such as ship grounding, ship sinking, or fire accidents. The development also possible by adding a more in-deep qualitative approach to determine the other key functions during the ship-ship encounter situation.

8. References

[1] Hollnagel E 2002 Understanding accidents - From root causes to performance variability *IEEE Conf. Hum. Factors Power Plants* 1–6
[2] Praetorius G, Hollnagel E and Dahlman J 2015 Modelling Vessel Traffic Service to understand resilience in everyday operations *Reliab. Eng. Syst. Saf.* **141** 10–21

[3] Aps R, Fetissov M, Goerlandt F, Helferich J, Kopti M and Kujala P 2015 Towards STAMP Based Dynamic Safety Management of Eco-Socio-Technical Maritime Transport System *Procedia Eng.* **128** 64–73
[4] Bielić T, Mohović R and Ivče R 2011 Sociotechnical model of ship organization effectiveness *Promet - Traffic - Traffico* **23** 49–57
[5] Hollnagel E, Wears R L and Braithwaite J 2015 From Safety-I to Safety-II : A White Paper From Safety-I to Safety-II : A White Paper Professor Erik Hollnagel University of Southern Denmark , Institute for Regional University of Florida Health Science Center Jacksonville , United States of America Prof
[6] Hollnagel E and Goteman Ö 2004 The Functional Resonance Accident Model *Proc. Cogn. Syst. Eng. Process plant* 155–61
[7] Luo Z, Li K, Ma X and Zhou J 2013 A new accident analysis method based on complex network and cascading failure *Discret. Dyn. Nat. Soc.* **2013**
[8] Lee J and Chung H 2018 A new methodology for accident analysis with human and system interaction based on FRAM: Case studies in maritime domain *Saf. Sci.* **109** 57–66
[9] Hollnagel E 2008 Risk + barriers = safety? *Saf. Sci.* **46** 221–9
[10] Hollnagel E 2012 *FRAM: The Functional Resonance Analysis Method* (Denmark: Ashgate)
[11] Passenier D, Mols C, Bím J and Sharpanskykh A 2016 Modeling safety culture as a socially emergent phenomenon: a case study in aircraft maintenance *Comput. Math. Organ. Theory* **22** 487–520
[12] Yang Q, Tian J and Zhao T 2017 Safety is an emergent property: Illustrating functional resonance in Air Traffic Management with formal verification *Saf. Sci.* **93** 162–77

Numerical Analysis of the Resistance Force of Floater of Wave Glider and Flapping Motion of Single Wave Glider NACA-0012 Foil

Erwandi[1] and Widodo[2]

[1] Technology Center of Maritime Industrial Engineering Technology, Agency for the Assessment and Application of Technology.

[2] Indonesian Hydrodynamic Laboratory, Agency for the Assessment and Application of Technology.

Abstract. In 2020, the Agency for the Assessment and Application of Technology – Indonesia (BPPT) is designing wave glider systems for Tsunami Early Warning System. The wave glider systems consist of two components, the floater at the water surface and the wave glider submerged 2 meters or more below the water surface. Wave glider contains several foils that can flap up and down due to the heave and pitch motion of the floater so that the system affords the force to move forward. This paper describes the numerical analysis of the resistance of floater and the selection of the pivot position of the foil where it flaps. The speed of the floater is set from 0.3 m/s up to 1 m/s. NACA-0012 is selected as the first choice of the foil. It flaps in pure pitch motion and coupled pitch and heave motion in the different pivot positions. The flapping frequency is set to 0.1 Hz since the peak period of the wave about 10 seconds. The surge forces due to the flapping motion are then analysed to determine the best position of the pivot. The numerical simulation shows that the pivot position at 2.5% - 5% of chord length from the leading edge of NACA-0012 generates the optimum surge force.

1. Introduction

Indonesia, as an archipelagic country, is part of the well known as "Ring of Fire". The Ring of Fire is the area around the Pacific Ocean where seismic activities are massive. Earthquakes, volcanic eruptions, and tsunamis often hit this area [1][2].

In 2020, the Indonesian Government had been finished mitigating the potential impact of the tsunami. The government will manufacture 13 buoys of Indonesia tsunami early warning systems (INA TEWS) by the end of 2020 and install them in 2021 at several points in southern and eastern Indonesia water.

The conventional buoy type of tsunami early warning system is quite expensive. The system consists of a buoy, an ocean bottom unit (OBU), and a mooring system that connected the buoy and OBU. Buoy, which floats at the surface of the water, contains electronic devices, batteries, antenna, etc. The OBU has the role to detect the sudden pressure increment of the water column due to the under ocean earthquake. In addition, conventional buoy types are also prone to vandalism. Several buoys installed in 2009 had their moorings cut, missing antennas, etc.

In order to reduce the cost and simplified the system, the Agency for the Assessment and Application of Technology (BPPT) proposed to design the wave glider system that can have a function as a tsunami early warning system. According to Liquid Robotics company [3], wave glider is an autonomous unmanned water surface vehicle, that has several capabilities to bear some mission services i.e.: oceanographic and meteorology climate observation, tsunami detection and seismic monitoring, fish and marine mammal monitoring, offshore energy, harbor security, and surveillance for illegal fishing.

It is claimed 24×7 Long-Duration Operations, station keeping or mobile data collection, has the capability for real-time communications and cuts the cost-effective operations. The type of long endurance and low maintenance is a key design goal for the wave glider [4].

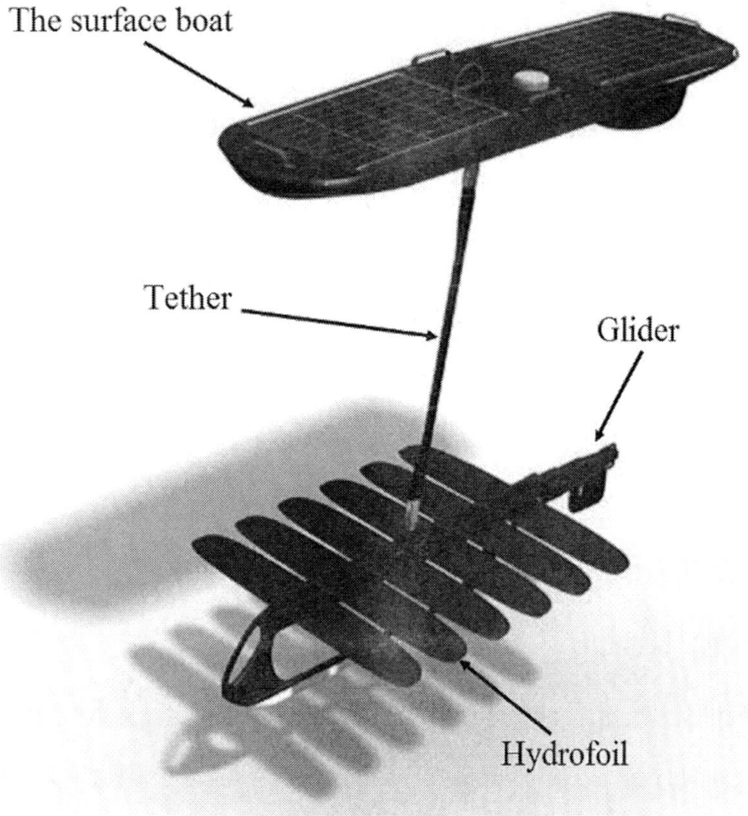

Figure 1. Wave glider system by China National Marine Technology Center.

The wave glider systems consist of two components, the floater at the water surface and the wave glider submerged 2 meters or more below the water surface. The two-body system is coupled with flexible tether or umbilical. Figure 1 shows the wave glider system [5].

Wave glider contains several hydrofoils that can flap up and down due to the heave and pitch motion of the floater so that the system affords the force to move forward.

This paper describes the initial research stage of the design of Indonesian's wave glider. It starts from the numerical analysis of the resistance of floater and the selection of the pivot position of the foil where it flaps.

The speed of the floater is set from 0.3 m/s up to 1 m/s. NACA-0012 is selected as the first choice of the hydrofoil. It flaps in pure pitch motion and coupled pitch and heave motion in the different pivot positions. The flapping frequency is set to 0.1 Hz since the peak period of the wave about 10 seconds. The surge forces due to the flapping motion are then analysed to determine the best position of the pivot.

2. Wave glider operational system

As shown in Figure 2 left [3], when the floater at water surface rides over the wavy water and moves up due to heave motion in the wave crest, the submerged glider will be pulled up thru the tether, it will cause the hydrofoil rotating and flapping downward and induce a thrust to drive the two-body system moving forward. Similarly, when the floater moves down (Figure 2 right), the hydrofoil flapping upward, thus generate a thrust too. The hydrofoil will in a horizontal position, which means there is no thrust when the velocity of the floater is greater than the velocity of the wave glider.

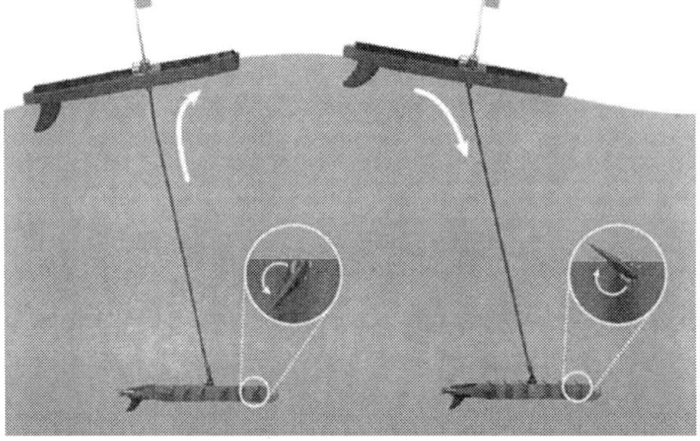

Figure 2. Wave glider operational system

3. Design of the floater

The designed BPPT's floater of the wave glider is shown in Figure 3. The length of waterline (LWL) of floater wave glider is 2.966 m, draft amidship (T) is 0.138 m, beam amidship (B) is 0.675 m, and total displacement 171.3 kg. Figure 3 shows the lines plan of the design of the BPPT's floater.

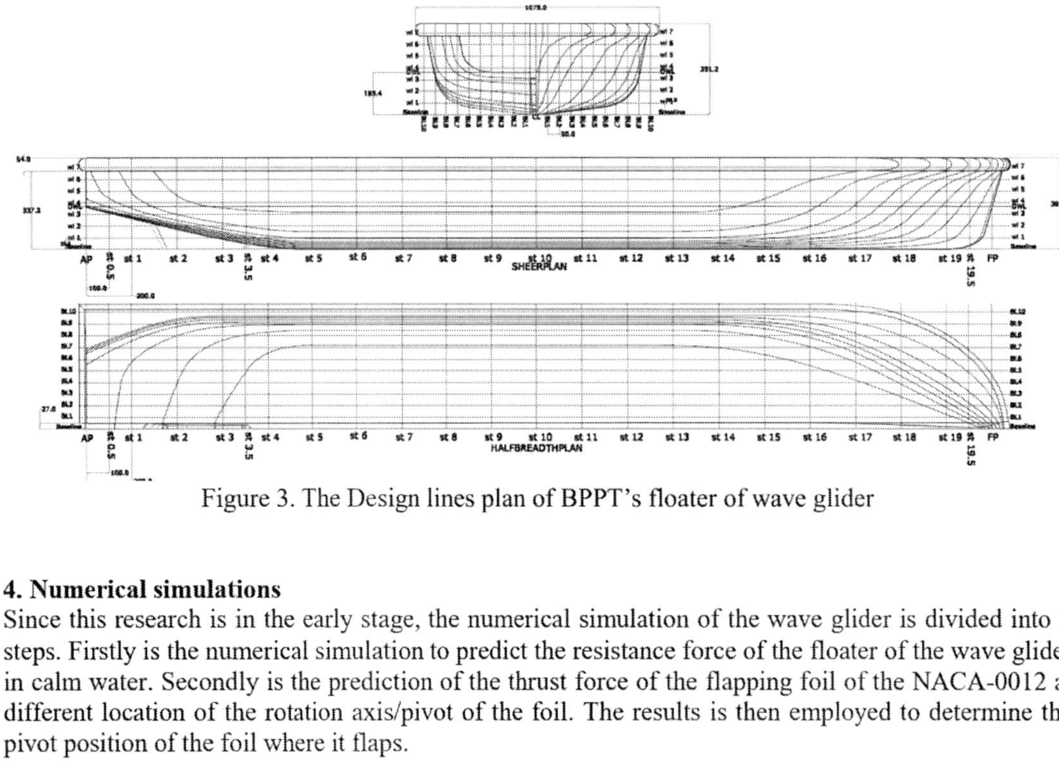

Figure 3. The Design lines plan of BPPT's floater of wave glider

4. Numerical simulations

Since this research is in the early stage, the numerical simulation of the wave glider is divided into 2 steps. Firstly is the numerical simulation to predict the resistance force of the floater of the wave glider in calm water. Secondly is the prediction of the thrust force of the flapping foil of the NACA-0012 at different location of the rotation axis/pivot of the foil. The results is then employed to determine the pivot position of the foil where it flaps.

4.1. Setting-up of the numerical simulation of the resistance force of the floater

The numerical simulation for predicting the resistance force of the floater based on the CFD software Fluent in calm water. The objective of the simulation is to provide data that can be used to design the thrust force of the flapping foil. It is planned by the end of 2020, the physical model will be tested in the towing tank facility of the Indonesian Hydrodynamic Laboratory (IHL) BPPT, therefore the computational domain of the numerical simulation is then adapted to the size of the towing tank. We calls the method is numerical towing tank, with virtual towing tank represents LHI's towing tank. The towing tank belongs to the IHL has the following dimensions: L = 235 meters, B = 11 meters, and water depth H = 5.5 meters.

As shown in Figure 4, the inlet boundary is set 25 m upstream from the point AP of the floater. The pressure outlet is positioned 50 meters. The bottom and side boundaries follow the dimension of the towing tank, 5.5 meters, and 5.5 meters (half of the towing tank width) respectively. The top boundary is set to 3 meter from point AP of the floater.

The flow around the floater is assumed incompressible and turbulent. Eulerian multiphase flow is applied at the Z plane as the water surface, 0.138 meters from baseline at the bottom of the floater. The implicit unsteady Reynolds Average Navier-Stokes (URANS) numerical scheme and the K-ω SST turbulence model are employed to the flow. The structured finite volume method is then used to discretize the computational domain. Figure 5 shows the result of the discretization mesh on the water surface around the floater. The dense mesh in the water surface about the FP of the floater is intended to catch the bow wave and wake when the floater moves forward.

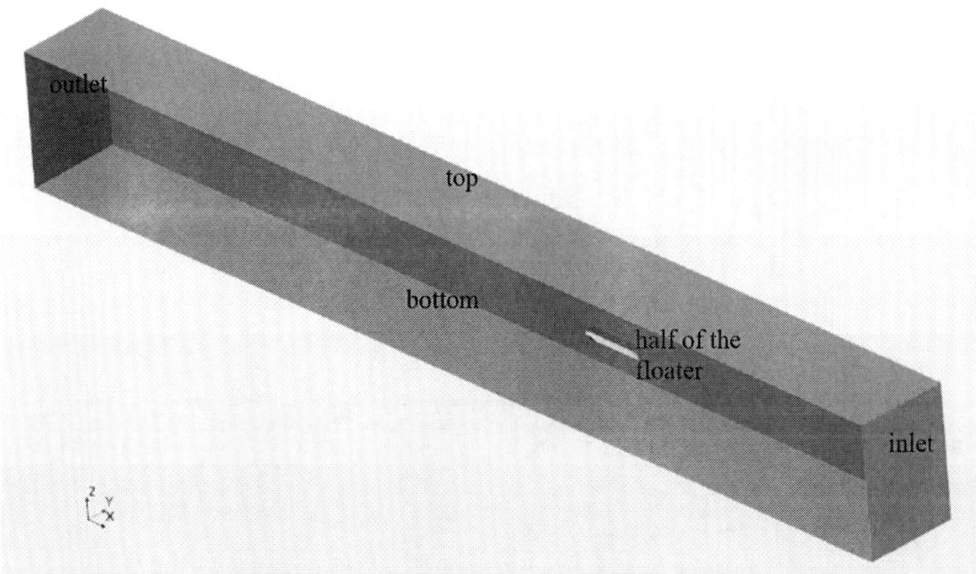

Figure 4. Computational domain of virtual towing tank.

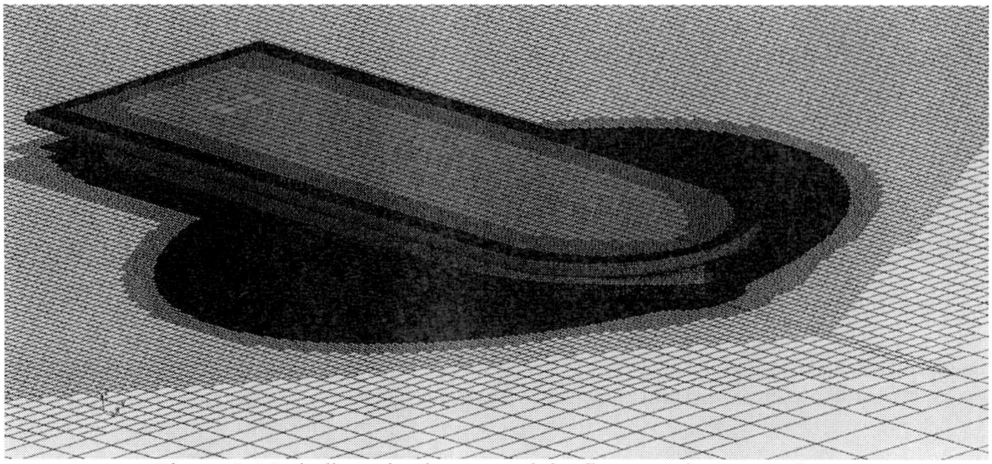

Figure 5. Mesh discretization around the floater and water surface.

4.2. Numerical simulation set-up of the flapping foil of NACA-0012

Flapping foils in nature are very common [7], the tail motion of ocean mammal-like whale or dolphin, or the sinusoidal motion of squid fin are the examples of flapping foil. The flapping foil of the tails and fins of the sea animals are intended to produce thrust or lift force, so they can move forward or maneuver easily. In the air, the birds flap their wings in order to generate lift to fly or to hover while they seek food. Those phenomena attract the curiosity of many scholars to understand the mechanism which can be applied in engineering.

Babu et.al. have applied the flapping foil for the propulsion system of the ship and underwater vehicles [8]. Xu, J., Sun, H., studied experimentally the oscillating foil to harvest energy from marine current [9].

In this research, we conduct the simulation of the passive flapping foil which is affected by the heaving motion of the floater while riding over the wave. The single foil NACA-0012 [5][10] is utilized to flap at a certain position of pivot or rotation axis in the pure pitch motion, pure heave motion, and couple heave and pitch motion. The pivot positions are then varied their distance from the leading edge of NACA-0012 foil. Figure 6 shows the profile of NACA-0012 foil.

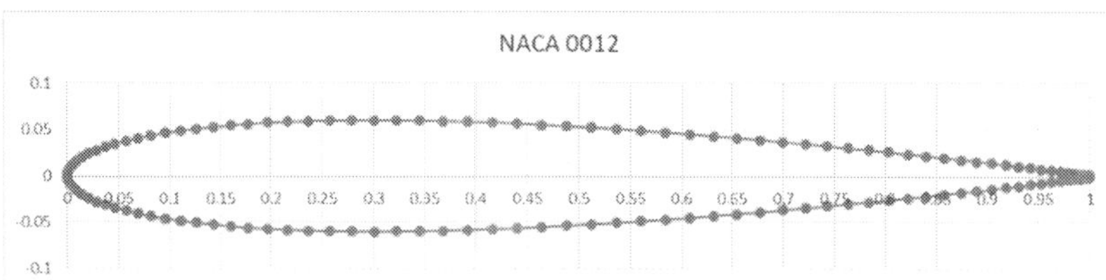

Figure 6. Foil of NACA-0012 profile

A three-dimensional (3D) computational domain is prepared to simulate the motion of NACA-0012 foil as shown in Figure 7. The chimera meshing method is applied for the movement of the foil mesh through computational domain background mesh [11].

Since the problem of the pitch and heave motion of the foil can be simplified in a two-dimensional (2D) computational domain, thus the computational domain of Figure 7 is then converted to a 2D computational domain.

The generated mesh in the 2D computational domain is shown in Figure 8 (a) and (b). Figure 8(a) is a structured background mesh. It surrounds the structured dense mesh around the mesh of the foil, called overlap mesh. Inside the overlap mesh there is denser mesh surround the foil as shown in Figure 8(b), which is the chimera overset mesh. It can move over the background mesh over overlap mesh. The movement of the foil and chimera mesh follows the following equation (1) for pure pitch motion and equation (2) for pure heave motion.

$$\theta(t) = \theta_0 * \sin(2\pi f t) \tag{1}$$

where : $\theta(t)$ is pitch motion (deg).
θ_0 is maximum amplitude of angle of pitch motion (deg).
f is pitch oscillation frequency (Hz).
t is time variable (seconds).

$$h(t) = h_0 * \sin(2\pi f t + \Phi) \tag{2}$$

where : $h(t)$ is heave motion (m).
h_0 is maximum amplitude of heave motion (m).
f is heave oscillation frequency (Hz).
t is time variable (seconds).
Φ is the phase difference between heave and pitch motion.

The frequency oscillation usually is set through kinematic parameter Strouhal number [12]. Figure 9 shows a schematic diagram demonstrating various dimensions and kinematic parameters of the foil motion. The kinematic parameter Strouhal number is expressed in the equation (3).

$$St = \frac{2h_0 f}{U} \tag{3}$$

where : h_0 is maximum amplitude of heave motion (m).
f is heave oscillation frequency (Hz).
U is free stream velocity (m/s).
Φ is the phase difference between heave and pitch motion.

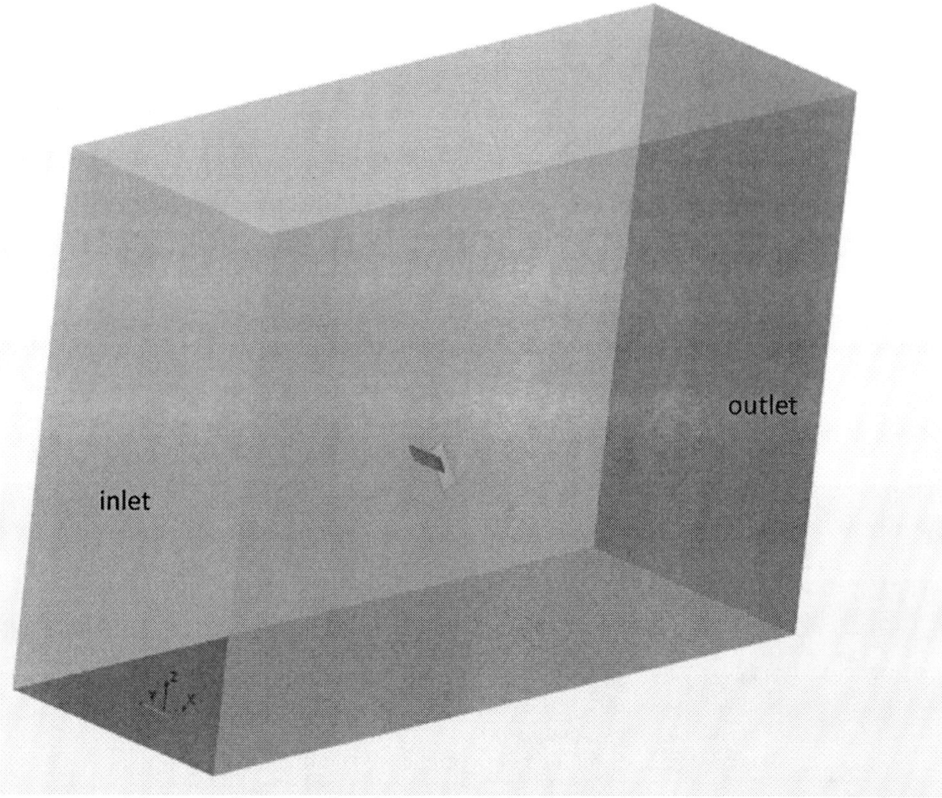

Figure 7. Computational domain of the flapping NACA 0012 foil.

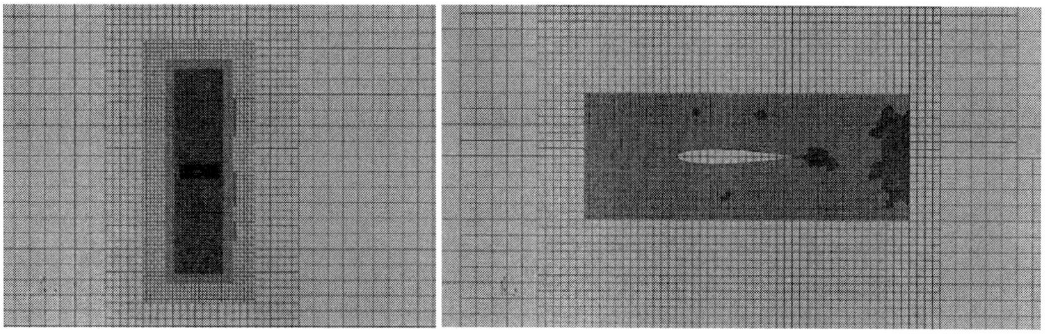

(a) (b)

Figure 8. Generated structure mesh (a) background mesh surrounds overlap mesh, (b) overlap mesh surrounds chimera overset mesh around foil NACA-0012.

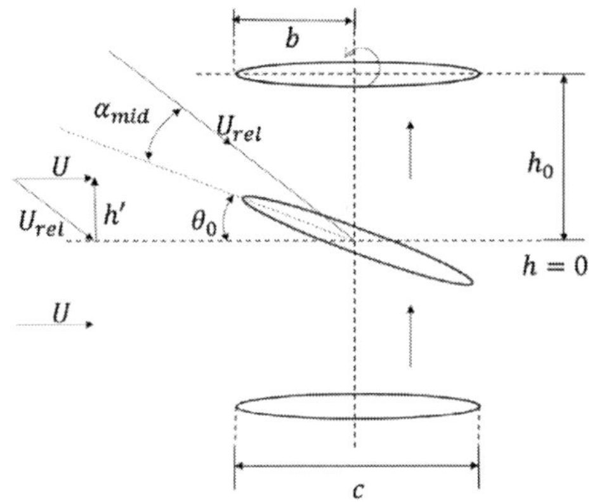

Figure 9. A schematic diagram demonstrating various dimensions and kinematic parameters of the foil motion [12].

5. Results and Discussion

5.1. Results of numerical resistance test

The incremental time of the CFD solver for numerical simulation resistance is set 0.001 seconds (1000 Hz) and generate a snapshot report every 0.01 second (100 Hz). The inner iteration is 20 iterations. It follows the ITTC recommendation for numerical prediction of resistance tests [5].

The visualization of the numerical simulation result of the resistance test in the towing tank at snapshot time 20 seconds is shown in Figure 10. In that figure, the velocity fed to the inlet boundary is 3 knots. It shows the contour of the profile of the wave-making resistance. The prediction of the resistance force of the floater starting from 1.0 knots up to 3.0 knots is shown in Figure 11. It will be compared with the physical model tow resistance test that will conduct at the end of 2020.

The increment of the resistance force is quite rapid starting from a speed of 1.6 knots. It needs a bigger thrust in order for the floater can move more than 1.6 knots. It is mentioned in the specification sheet of the wave glider produced by Liquid Robotics company [3] that the speed of wave glider is typically 1.3 knots.

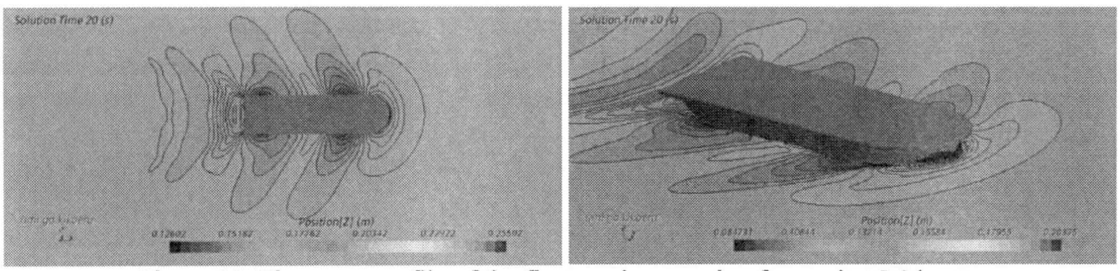

Figure 10. The wave profile of the floater when moving forward at 3.0 knots.

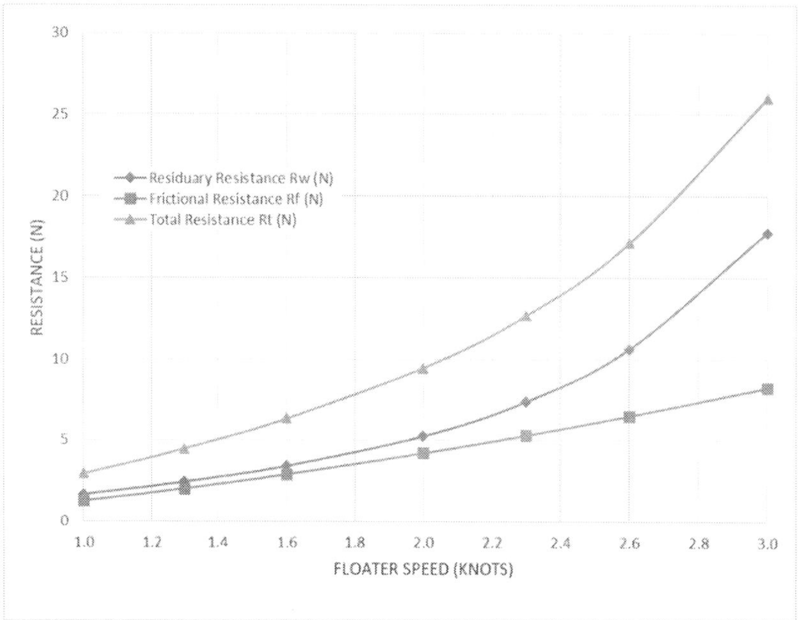

Figure 11. The numerical simulation results of the resistance force of the floater

5.2. Results of numerical flapping NACA-0012 foils

The numerical simulation of the flapping foil of the NACA-0012 is set at different positions of the pivot (rotation axis) of the foil. The pivot coordinate position is determined from the distance of it to the leading edge point of the foil. The distance between leading edge to the pivot point in horizontal axis of the foil is expressed in the non-dimensional length (in %) as shown in equation (4).

$$L(\%) = \frac{X_P - X_{LE}}{C} * 100 \quad (4)$$

Where : L is the distance between pivot point and leading edge point (%).
X_P is the coordinate position of the pivot (m).
X_{LE} is the coordinate position of the leading edge (m).
C is the chord length of the foil (m).

The pivot position is set at -10%, -5%, 0, 2.5%, 5%, 7.5%, 10%, 20%, and 30% from the leading edge for pure pitch. According to Dave et.al. [12], the ideal operating range in terms of thrust and efficiency of the flapping foil has been found to be between St = 0.2 to 0.4, with pitch amplitudes between 40 and 60, however since the peak periods (T_p) of the significant wave height in Indonesian ocean is typically around 9 – 11 seconds, then we take 10 seconds as the peak periods of the wave, so the frequency f is set $1/T_p$ = 0.1 Hz. The velocity in the inlet boundary, pressure outlet boundary, top, and bottom boundaries are all set to 0 since there is no variable that is fed to the boundaries.

The amplitude angle of pitch motion which is fed to numerical simulation is set to 60°. For pure heave, the amplitude heave motion is 0.5 m. Figure 12 shows the velocity magnitude of the fluid around the foil which flapping in pure pitch motion, where the pivot position is 5% of the chord length from the leading edge. The samples of the thrust force produced by the flapping foil is shown in Figure 13 – 15 for the position of the pivot from the leading edge 5%C, 0%C, -5%C respectively. If we see Figure 12, the foil NACA-0012 faces the left or minus direction of the X-axis. It means that the thrust force will give a negative value to propel the foil in the left direction. If we observe the time history of the thrust force in Figure 13 for example, the value of the force is dominantly negative or if we take the average value we will obtain a negative value, it means the flapping foil will thrust the foil forward to the minus direction of X-axis.

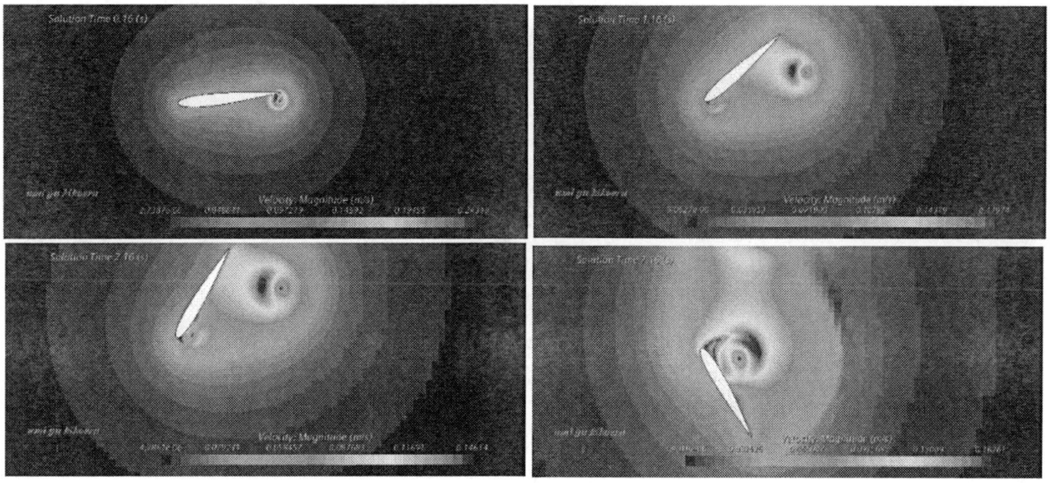

Figure 12. The velocity magnitude distribution of pure pitch motion where pivot position 5%C from leading edge.

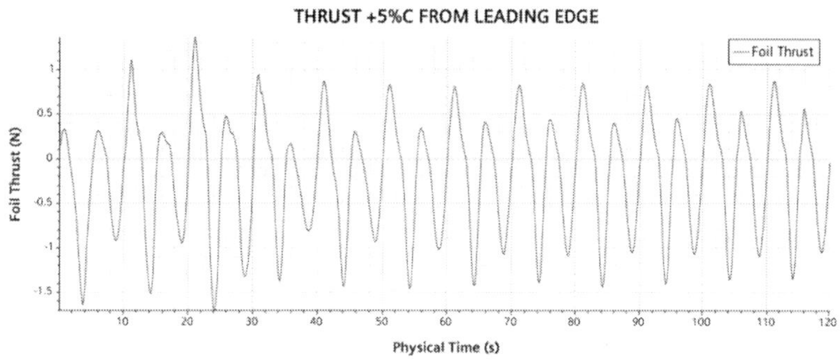

Figure 13. Time history of thrust force of flapping NACA-0012 foil pure pitch motion where pivot position 5%C from leading edge.

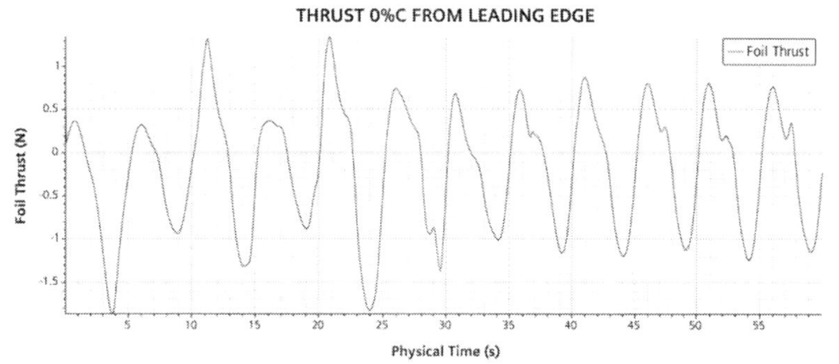

Figure 14. Time history of thrust force of flapping NACA-0012 foil pure pitch motion where pivot position 0%C from leading edge.

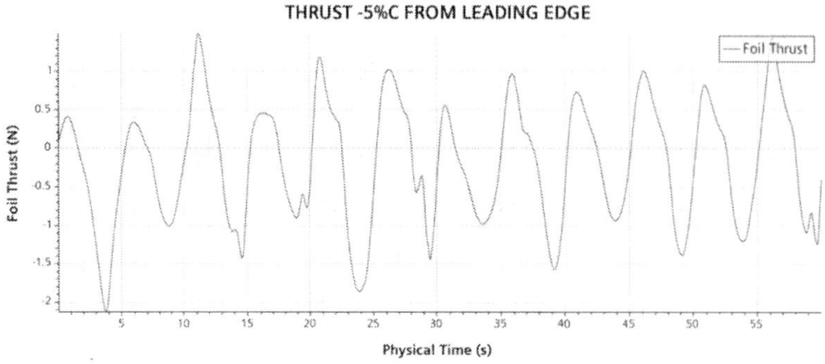

Figure 15. Time history of thrust force of flapping NACA-0012 foil pure pitch motion where pivot position -5%C from leading edge.

If the time histories of the thrust forces at every pivot position are averaged we will know the best pivot position which gives the highest thrust force caused by the pure pitching motion. Figure 16 shows the average thrust force at every pivot position -10%C, -5%C, 0, 2.5%C, 5%C, 7.5%C, 10%C, 20%C, and 30% C from leading edge of the foil. It indicates that the pivot position in the range 2.5%C – 5%C produces the highest thrust force for flapping foil in the pure pitching motion.

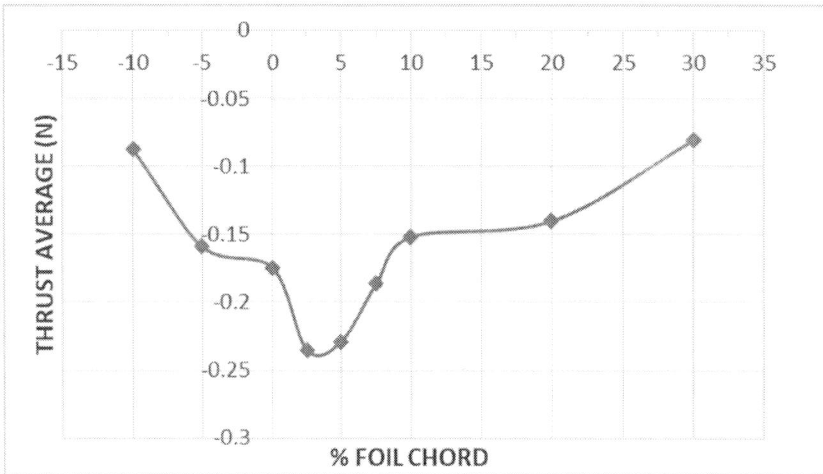

Figure 16. The average thrust forces produced by flapping foil in pure pitch motion at different pivot position.

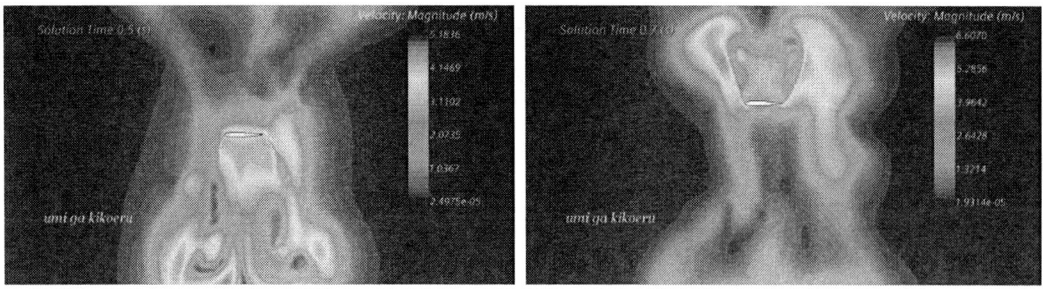

Figure 17. The velocity magnitude in pure heave motion.

The velocity magnitude distribution around NACA-0012 foil which oscillating in pure heave motion is shown in Figure 17. The pivot position is put in the leading edge point of the foil. In heave motion, there is no effect on the difference of the pivot position.

The thrust force generated by oscillating foil in pure heave motion is given in Figure 18. It shows that the thrust force in pure heave motion is a dominantly negative value. It means that the oscillating foil in heave motion produces an optimum thrust force and bring the foil moving forward.

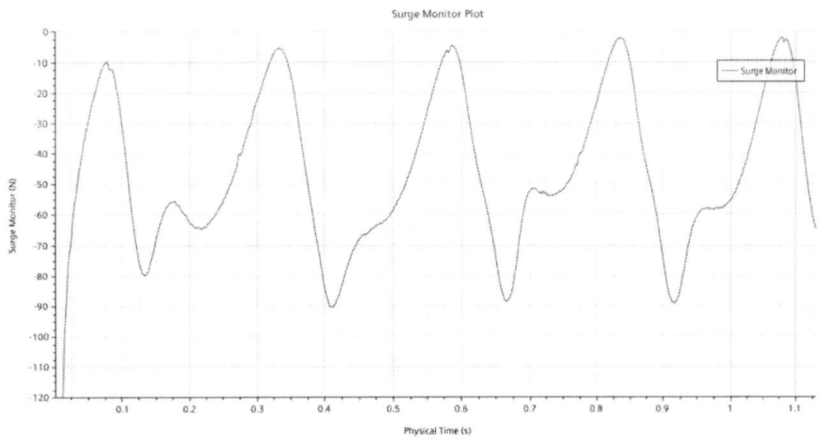

Figure 18. Time history of thrust force of flapping NACA-0012 foil pure heave motion

6. Conclusions

The numerical method for the unsteady RANS simulation for the prediction of the resistance of the floater of the wave glider and the motion of the single NACA-0012 foil is proposed. The method can be harnessed to predict the resistance force of the floater and thrust force generated by a single flapping foil in pure pitch and heave motion. The pivot position of the flapping NACA-0012 foil in pure pitch motion has a big role to produce the thrust forces. However, since the research is in the early-stage more studies are needed for the optimum design for a floater in wavy water and simulation of a group of foils to simulate the thrust force required by the wave glider system.

7. References

[1] National Geographic website 2020- https://www.nationalgeographic.org/encyclopedia/ring-fire
[2] Murjaya J et al 2012 *Tsunami Early Warning Service Guide Book for Indonesian Tsunami Early Warning System* (Jakarta, Indonesian Meteorological, Climatological, and Geophysical Agency, second edition)
[3] Liquid Robotics Inc. 2020 website - http://liquidr.com
[4] Kraus N 2012, *Wave Glider Dynamic Modeling, Parameter Identification and Simulation*, (master thesis in mechanical engineering, University of Hawai'i)
[5] Yang F, Shi W, Zhou X, Guo B and Wang D 2018 Numerical investigation of a wave glider in head seas *J. Ocean Eng.* **164** pp 127–138.
[6] ITTC 2011– *Recommended Procedures and Guidelines* Practical Guidelines for Ship CFD Applications
[7] Wu X, Zhang X, Tian X, Li X and Lu W 2020, A review on fluid dynamics of flapping foils *J. Ocean Eng.* **195** 106712
[8] Babu M, Krishnankutty P and Mallikarjuna J 2014 Experimental Study of Flapping foil Propulsion System for Ships and Underwater vehicles and PIV Study of Caudal Fin

Propulsors, *Conf.: IEEE/OES Autonomous Underwater Vehicles (AUV)*
[9] Xu J and Sun H 2016. Experimental studies of passive oscillating hydrofoil for tidal current energy extracting. *J. Harbin Eng. Univ.* **37** (2), 248–253
[10] Critzos C, Heyson H and Boswinkle R 1955 *Aerodynamics Characteristics of NACA-0012 Airfoil Section at Angle of Attacks from 0 Degrees to 180 Degrees* (Technical Report Archive & Image Library, pp. 79)
[11] Chang X, Ma R, Wang H, Zhao Z and Zhang L 2020 A parallel implicit hole-cutting method based on background mesh for unstructured Chimera grid, *J. Comp. & Fluids* **198**, 104403
[12] Dave M, Spaulding A and Franck J 2019 Variable thrust and high efficiency propulsion with oscillating foils at high Reynolds numbers, *Preprint submitted to Ocean Engineering*

Acknowledgments
The authors acknowledge their appreciation to Mr. Sahlan and Ms. Rina for their encouragement and support in the study. Their thanks also go to the head of the Indonesian Hydrodynamics Laboratory BPPT who help us during the research.

Study of the Interceptors Effect on Ship Manoeuvrability Using the Open Free Running Test Method

M A Hafiz[1], **A Sulisetyono**[1], **A Baharuddin**[2]

[1] Department of Naval Architecture, Institut Teknologi Sepuluh Nopember (ITS), Surabaya, Indonesia.
[2] Laboratorium Hidrodinamika Indonesia – BPPT, Surabaya, Indonesia.

Abstract. Interceptors are plates that are attached to the stern of the ship to produce lift that controls the trim angle of the ship. This paper describes the use of interceptors on the fast boat to enhance its ability to perform the turning manoeuvre. The ship model of High-speed crafts (HSCs) with a scale of 1: 26.9 is tested in an open pool with the open free running test method, namely the movement of the model is controlled from the edge of the pool with a remote control. The ship trajectory data is obtained by using the colour object tracking method, that is, the ship trajectory captured by the camera is processed based on colour filters. The main concern is the varying dimensions of the interceptor including 0% (no interceptor), 50% (half length), and 100% (full length) of the design cord length. The initial trim conditions are examined, such as even keel, trim by bow, and trim by stern with a trim angle of 2 degree. The test results show that the interceptor contributed to improve the manoeuvrability of the boat by decreasing the tactical diameter length respectively at half and full length of the cord by 8.8% and 17.9% in even keel, 6.3% and 15% in bow trim, and 12.9% and 14.8% in stern trim.

1. Introduction

Interceptors are rectangular flat plates attached to the transom of the hull mostly used to control trim angle. Basically, the interceptor works by utilizing a pressure gradient and causing lift at the stern that changes the trim angle [3,9]. Although, the additional effect of the interceptor increases the total resistance at low speed, the rate of increasing resistance of the hull with the interceptor at high speed is less than the hull without the interceptor. Then, the hull moves forward faster and provides about 10–15% fuel efficiency compare to the hull without an interceptor [2].

The interceptor systems were studied with an experimental method on a high-speed hull-scale model with various heights and placement locations of interceptor to investigate effects to hull's resistance. The results show the resistance decrease about 1.50% – 11.30% for Froude number of 0.58 – 1.19, and the trim angle reduced in between 1.60 and 4.70 degrees [1].

Mansoori and Fernandes [5] studied the unwanted movement known as porpoising instability in high speed boat and showed that the interceptor can control trim well so that it can reduce oscillatory pitch motion and avoid the porpoising phenomenon. The interceptor technology further is developed to help maintain the stability of high-speed crafts while manoeuvring in sea way. Since the ship's manoeuvrability can be affected by the magnitude of the ship's trim angle, the interceptor can also be used as a tool in controlling the manoeuvring movements of fast boats.

In this paper, the interceptors are tested experimentally to consider the maneuverability of a high-speed vessel in a model-scale turning circle test. The open free running test is used in combination with the color object tracking method to retrieve track data. This color object tracking method is almost similar to the image processing method, namely image processing based on image retrieval for every second of the test video [11]. In this study, the trajectory of an object was by filtering the color of the video taken with an RGB and HSV based camera [8]. Image or video processing only intends to display a certain color on the desired object within the scope of the camera frame capture, from that color which indicates the location of the object in the frame. The test is carried out in accordance with IMO (International Maritime Organization) regulations concerning Ship Maneuverability Standards [7], namely turning tests that is studied. Although rudder design has an impact on maneuvering performance [10], this is not studied. The variations in the height of the interceptor plate were 100%, 50%, and 0% of the design height of the interceptor based on [6] which included a flat hull trim, bow trim, and aft trim with a 2 degree of trim angle.

2. Experimental setup

The ship model used in this study is the combat ship type with dimensions scale of 1:26.93 from a full-scale ship. The main dimensions of the full-scaled ship and the model are described in Table 1. The ship model is made using material of fiberglass with an axe bow design as shown in Figure 1. The model is equipped with a rudder which functions to control the direction of the ship as presented in Figure 2, with a submerged rudder area of 22.8 cm^2, and 2 blades propeller with a diameter of 4.2 cm is used to push the model ship as shown in Figure 3.

Table 1. Main dimensions of the full-scale ship and the model ship.

	Full-scale Ship	Model Ship
Length o.A. (meter)	18.85	0.7
Length W.L. (meter)	15.89	0.7
Breadth (meter)	4.2	0.2
Depth (meter)	2.8	0.108
Draught (meter)	0.98	0.043
85% Max. Eng. Output (m/s)	15.43	-
Speed Test (m/s)	13.89	3.72

Figure 1. The model ship.

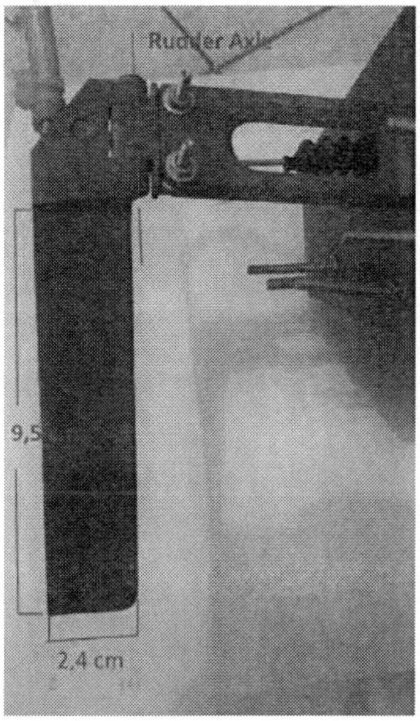

Figure 2. Rudder model.

Figure 3. Propeller model.

The ship model is equipped with several equipment so that it can move under the control of the operator, including the propulsion system, steering system, ship model, control system, interceptor circuits and sensor circuits, as well as cameras, laptops / PCs, and programs for capturing ship track data. Placement of any equipment needed on the hull must consider the location of the ship's centre of gravity and adjust the hull model space. Figures 4 and 5 illustrate the location of the sensor system equipment such as the sensor module and the battery on the superstructure, and the sensor circuit that takes into account preventing the influence of the DC motor heat and preventing contact with outside water. Electrical equipment is located on the inside of the hull model, the motor is positioned right at the centre of gravity of the ship, and other equipment adjusts its position to maintain the ship's centre of gravity, load, heel conditions and even keel conditions, see Figure 6.

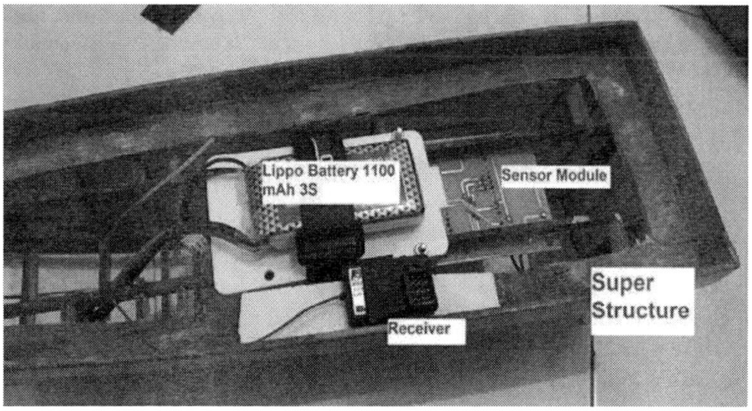

Figure 4. The arrangement of the sensor circuit on the superstructure.

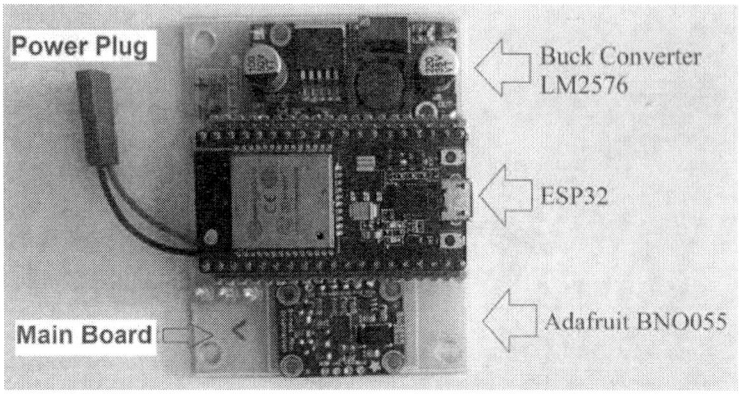

Figure 5. Sensor module circuit.

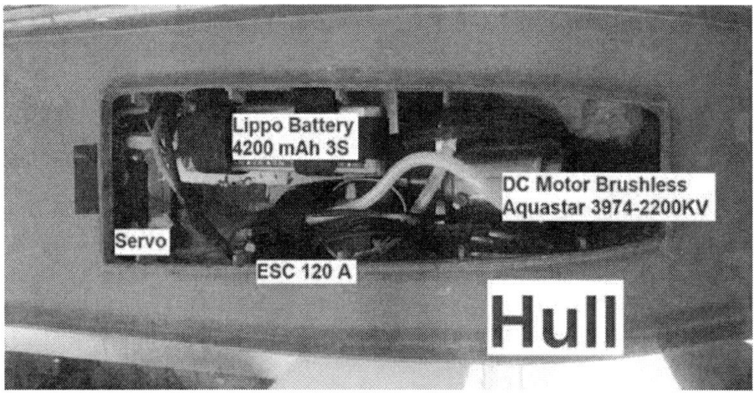

Figure 6. Electrical circuit arrangement.

2.1. Testing pool

The turning circle test for the ship model is carried out in an open pond where at the time of testing the wind speed conditions are calm, which is below 1 knot and the water conditions during the test are calm like glass. Meanwhile, the water temperature during the test is about 28° - 30° Celsius. The test pool area is large enough so that there is not blockage effect from the pool wall on the ship model, see Figure 7. The comparison of the pool depth (h) with the ship draft (T) must be sufficient to avoid the influence of shallow water according to ITTC 2002 rules that the minimum allowable value is h / T = 4. The test pool has dimensions of 4 meters deep, 18 meters wide and 20 meters long, see Figure 7.

2.2. Equipment test

Installation of the interceptor system follows the design referred to in reference [5], the size of the interceptor is designed based on the height of the boundary layer with the formula for boundary layer height (h) as follows $h = 0.382 \times L_{WL} / (Re \times L_{WL} \times 0.2)$. The height of the interceptor plate (d) must not exceed 60% of the thickness of the boundary layer (h) under the transom, for the height of the interceptor plate is 0.1h to 0.6h. Then calculate the interceptor time span (S) by comparing S/d with d/h as shown in Figure 8, so that the span interceptor (S) can be determined.

Based on Figure 8, it can be seen that the height of the interceptor plate used for 100% (full length) is 9.96 mm, for 50% (half length) it is 4.98 mm, and 0% for without using the interceptor, and the length of the interceptor model used is 66.77 mm. The reference in designing the interceptor model is as shown in Figure 9. The interceptor model used in this study is as described in Figure 10 and its size can be seen in Figure 11.

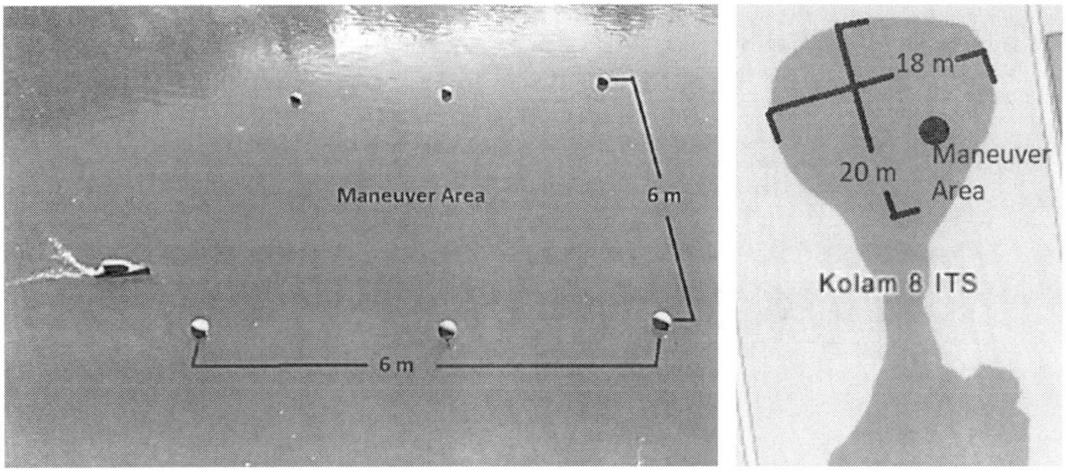

Figure 7. Maneuver area in testing pool.

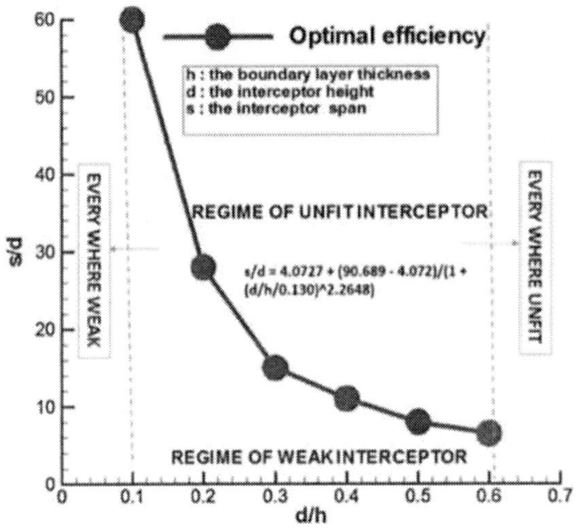

Figure 8. Comparison chart of d/h with S/d [5].

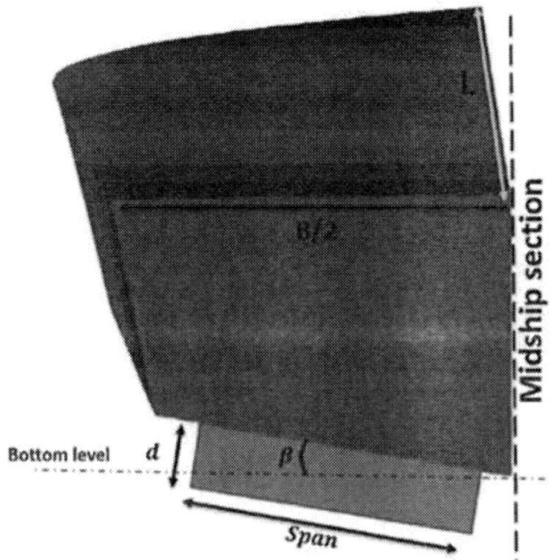

Figure 9. Interceptor design model [5].

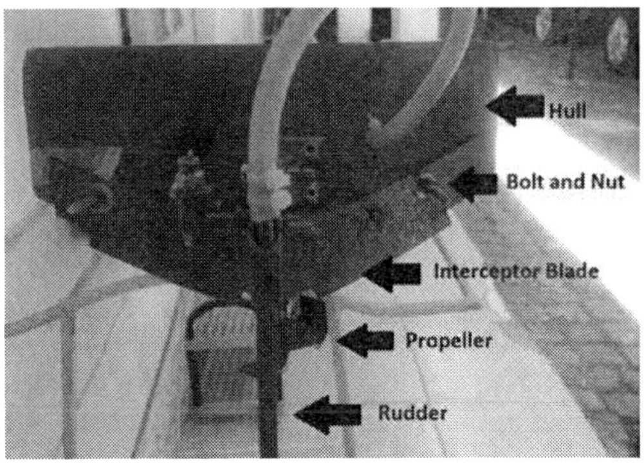

Figure 10. Interceptor model.

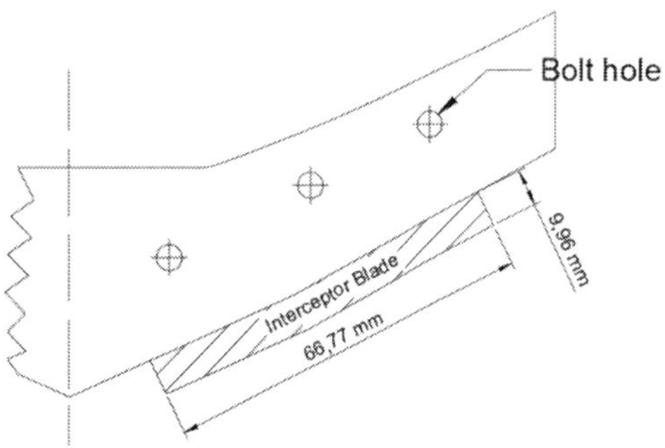

Figure 11. The interceptor model design.

IOP Conf. Series: Materials Science and Engineering 1052 (2021) 012040 doi:10.1088/1757-899X/1052/1/012040

For track tracking, several programs and applications are developed. First, the Mission Logger application, where this application is used as a user interface and records data directly from sensor readings on the ship model. Sensor data is sent over a wireless network connected via a hardware router, and the user interface for this application is shown in Figure 12.

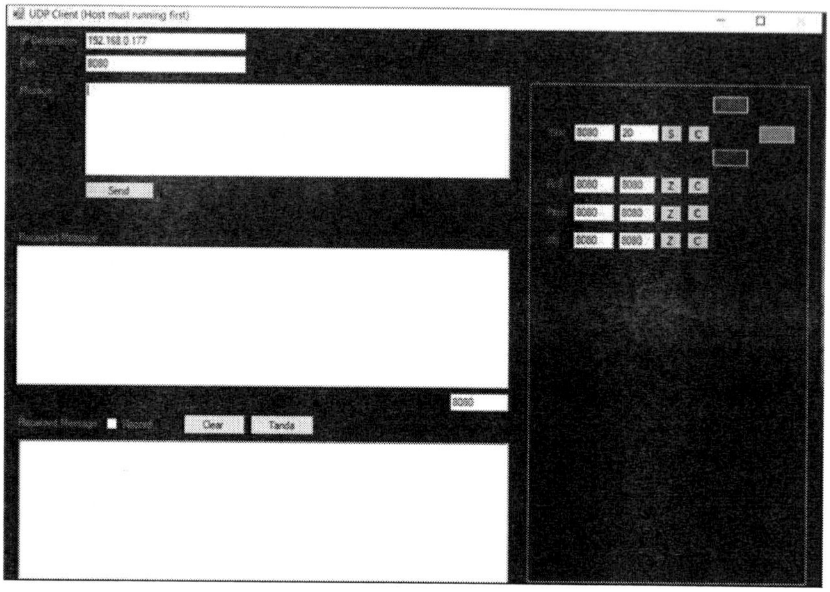

Figure 12. Display user interfaces application "Mission Logger".

Second, the Color Range Detector Program, this program is a program that filters the colors in the image captured by the camera that the "Ship Tracking" program wants to read, this program can set the minimum and maximum range of HSV and RGB values for the colors want to be captured with the program " Ship Tracking ", for an example of color filtering in the image can be seen in Figure 13.

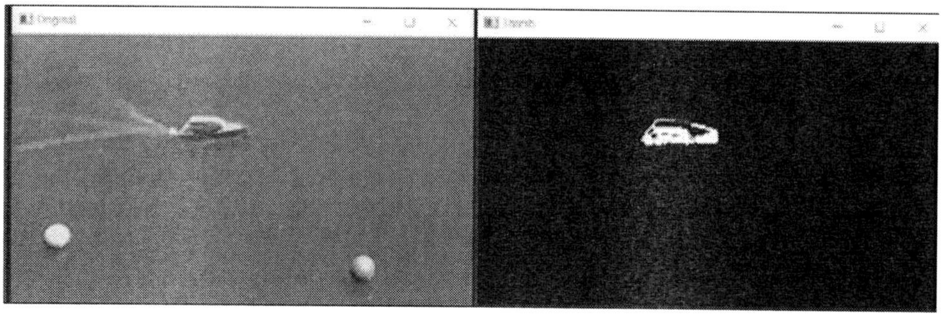

Figure 13. Color filter window program "Range Color Detector".

Third, the Ship Tracking Program, the program functions to read the pattern of ship movement through the pixel coordinates that the ship traversed in the image from the captured camera. This program reads the object's location based on the object's color, where previously the object's color was determined using the "Color Range Detector program", the program will track the object and determine the object's center point as the pixel coordinates that the program reads, then when the object moves, the program will record the object by tracking the coordinates of the pixels that the object traverses.

Fourth, Drafting Application, basically this application aims to plot the actual coordinate points of the ship's movement to form a ship's movement trajectory.

2.3. Testing procedures

The test procedure can be briefly described as follows, first, prepare for the test by measuring wind speed; second, ships must be in a calm water condition and first reset the sensor; third, run the ship model and start recording the model's movements with video and sensors; fourth, the ship enters the buoy area by moving forward in a straight line, and when rotate the rudder marks the start of the rotating motion of the model ship; sixth, the ship rotates in a circle in two rounds and then the ship exits the buoy area, data collection is complete.

2.4. Data Processing

In processing data taken from testing, the steps taken are first, matching camera data with sensor data at the same second of data collection, this is necessary because there are differences in the two data at the same second. This procedure is done by calibrating the time from the sensor data with the time on the test video.

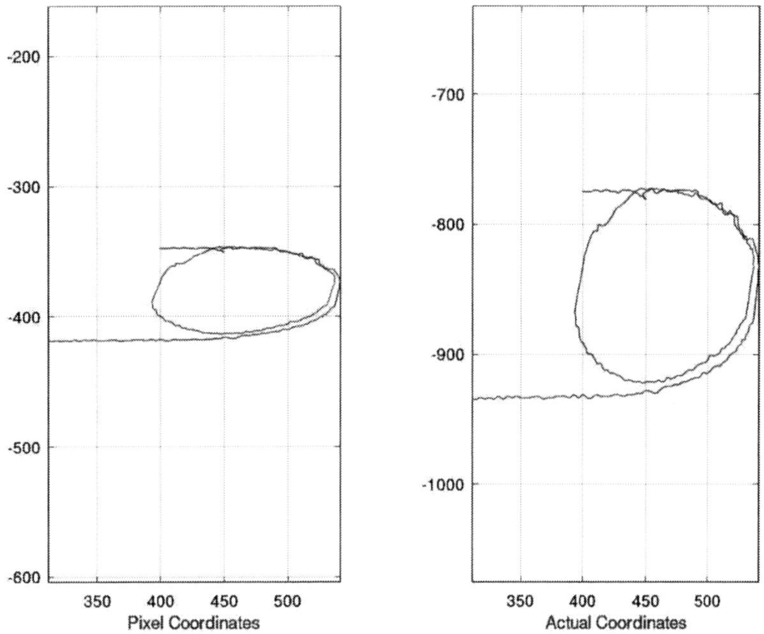

Figure 14. Comparison of the path of the pixel coordinate data with the actual coordinate data.

Second, determining the value of the calibration factor based on the reference buoy placed in the test pool, this is because the position of the camera is located sideways from the ship model object, it is necessary to project the data taken from the test so that the actual coordinate data is obtained. So the path data in the form of pixel coordinates on the video recording is transformed into actual coordinates multiplied by the calibration factor, see Figures 14 and 15. The equation used to determine the value of the calibration factor is shown in equation (1) and (2).

$$\text{Calibration Factor Value at Coordinate X} \qquad (1)$$

$$\frac{\text{Coordinate X}}{(xB - xC)} \text{ x } 6$$

$$\text{Calibration Factor Value at Coordinate Y} \qquad (2)$$

$$\frac{\text{Coordinate y}}{(yB - yC)} \text{ x } 6$$

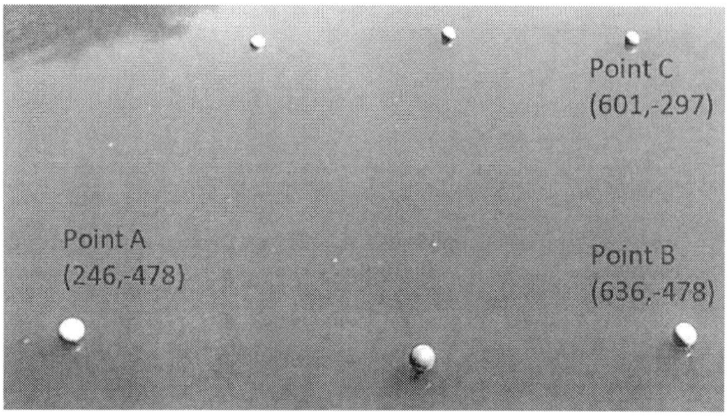

Figure 15. Buoy as a reference for track data calibration.

Third, after obtaining the actual coordinates, filtering or data normalization is carried out in order to obtain a fair turning test trajectory, as shown in Figure 16 for the turning circle results of the model in case of without interceptor device.

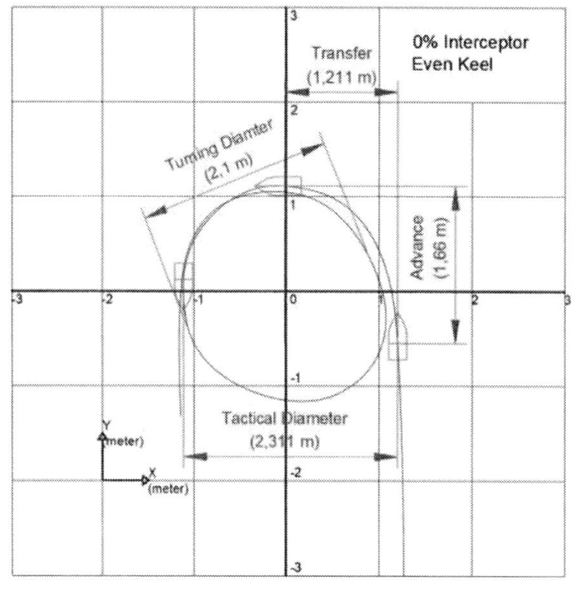

Figure 16. Turning test result of the model without interceptor (0%).

3. Results and discussion

3.1. Verification of test results

Verification of the test results with this technique is carried out by comparing the results with the test results for the type of patrol boat that has a length of 27 meters and 9.5 meters that have been tested at the Indonesian Hydrodynamics Laboratory. The model length of the 27-meter and 9.5-meter Patrol boat models are 3 meters and 1.80 meters respectively, with initial velocities of 22 knots and 10 knots, respectively, and the same rudder angle of 35 degrees. The data being compared are the test results of the ship model without using an interceptor, and the comparison of the results for the three ships is described in Table 2 in term of non-dimensional.

Table 2. Comparison results of the Turning Circle Test for Model Ships 0% Interceptor with the Results of LHI Testing for 27 m Patrol Ships and 9.5 m Patrol Ships.

		Parameter			
		Advance	Transfer	Tactical Diameter	Turning Diameter
Non-Dimensional (x Lpp)	27 m Patrol Boat	3.75	1.29	3.29	3.22
	0% Interceptor Model	2.37	1.71	3.29	3
	9.5 m Patrol Boat	2.39	1.19	3	3.09

From the turning circle test data in Table 2, the tactical diameter of the model without interceptors produces a very close to the test result of a 27 m patrol boat and slightly different from a vessel of 9.5 m by 8.8%. Likewise, the test results in the form of advance length are also close to the results of the patrol boat test of 27 m and 9.5 m, with differences of 36.8% and 6.7%, respectively. The test results of the model ship without interceptors in the form of transfer length have a difference with the results of the 27 m and 9.5 m patrol tests of 24% and 30.5%, respectively. The difference occurred in the test results of a turning diameter between the model ship and the 9.5 m patrol, amounting to 2.9 %, while the difference with 27 m patrol was 6.8%. In general, the model test results are close to both results of the 27 m and 9.5 m patrol boats for the all parameters except a length of transfer. The standards for ship manoeuvrability according IMO Resolution MSC 137(76) 2002 are only parameters of advance and tactical diameter so that the method of model test are satisfied.

3.2. Test results and discussion

Following are the test results with the free running method on a model scale of 26.93, for testing at an initial speed of 27 knots, with the steering angle for turning circle testing is port side 35 degrees. The ship model was tested with interceptor height variations of 9.96 mm (100%), 4.98 mm (50%), and without interceptor (0%), which has the same span length of the interceptor model, namely 66.77 mm. The method of collecting data on this test track is by tracking color objects and wind speed conditions below 1 knot. Table 2 describes the results of the model testing in the early even keel conditions. Tables 3 and 4 describe the results of the tests when the conditions are trim by bow and trim by stern, respectively.

Table 3. Test results for model at even keel condition.

Even Keel Condition		Parameter			
		Advance	Transfer	Tactical Diameter	Turning Diameter
Dimensional (meter)	0% Interceptor Model	1.66	1.2	2.3	2.1
	50% Interceptor Model	1.48	1.03	2.1	2.09
	100% Interceptor Model	1.36	0.75	1.89	1.86
Non-Dimensional (xLpp)	0% Interceptor Model	2.37	1.71	3.29	3
	50% Interceptor Model	2.11	1.47	3	2.99
	100% Interceptor Model	1.94	1.07	2.70	2.66

Table 3 explains that in the initial conditions of the keel event, the interceptors of 100% and 50% provide a reduction in the tactical diameter size of 17.9% and 8.8% respectively on the results of the model test without interceptor.

Table 4. Test results for model at trim by bow.

Bow Trim Condition		**Parameter**			
		Advance	Transfer	Tactical Diameter	Turning Diameter
Dimensional (meter)	0% Interceptor Model	1.57	0.7	1.99	1.88
	50% Interceptor Model	1.39	0.62	1.86	1.78
	100% Interceptor Model	1.28	0.65	1.69	1.64
Non-Dimensional (x Lpp)	0% Interceptor Model	2.24	1	2.84	2.69
	50% Interceptor Model	1.99	0.89	2.66	2.54
	100% Interceptor Model	1.83	0.93	2.41	2.34

At trim by bow conditions as shown in Table 3, the test result in term of tactical diameter for the model with 100% interceptor decrease about 15% from model without interceptor (0%), and the model with 50% interceptor give a tactical diameter 6.3% less than model without interceptor.

Table 5. Test Results for model at trim by stern.

Stern Trim Condition		**Parameter**			
		Advance	Transfer	Tactical Diameter	Turning Diameter
Dimensional (meter)	0% Interceptor Model	1.83	1.17	2.6	2.57
	50% Interceptor Model	1.93	0.95	2.26	2.18
	100% Interceptor Model	1.5	0.95	2.21	2.16
Non-Dimensional (x Lpp)	0% Interceptor Model	2.61	1.67	3.71	3.67
	50% Interceptor Model	2.76	1.36	3.23	3.11
	100% Interceptor Model	2.14	1.36	3.16	3.09

Table 4 show the result test for model at stern trim condition that the model test with 100% and 50% interceptor give about 14.8% and 12.9% less than the model without interceptor results, respectively.

From the results of the test model described in Tables 3, 4, and 5, it is explained that the addition of an interceptor can increase the maneuverability of the ship at high speed, where the interceptor design limit refers to [5], where the d / h value is between 0.1 - 0.6. Besides, the initial condition of the ship trim angle of 2 degrees affects the ability of the ship's turning circle maneuver, namely the trim by bow condition shows the best rotating performance followed by the keel and trim by bow event conditions, respectively. This happens because the location of the LCG point at the time of trim by bow is ahead of the other two trim conditions, so that the yaw force is greater.

However, the addition of interceptors has the most significant effect on the maneuverability of the ship during the initial conditions of the keel event and then the trim by bow and trim by stern conditions are followed, respectively.

4. Conclusions

The accuracy of the fast boat test results using the open free running test method combined with the color object tracking technique in taking the turning circle maneuver data is quite reliable after verification of the similar test results on a patrol boat at the Indonesian Hydrodynamics Laboratory with a difference of 0% in tactical diameter and 6.8% in the turning diameter. In general, the addition of an interceptor in the transom area of the ship improves the ability of the ship's rotary maneuver, where the size of the interceptor of 100% of the design size provides a reduction in tactical diameter length of 14.8% to 17.9% for ships without interceptors (0%) in all initial trim conditions of 2 degrees whereas on ships with an interceptor of 50% of the design size, it provides a reduction in tactical diameter length of 6.3% to 12.9%.

5. References

[1] Avci A G and Barlas B 2019 An experimental investigation of interceptors for a high-speed hull *International Journal of Naval Architecture and Ocean Engineering* 11 256-273

[2] Day A H, Cooper C 2011 An experimental study of interceptors for drag reduction on high-performance sailing yachts *Ocean Engineering* **38** 983-994

[3] De Luca F 2011 Experimental study on interceptor's effectiveness *Dissertation University degli studi Napoli "Federico II"*

[4] Jacobi G, Thill C H, Veer R V, and Huijsmans R H M 2019 Analysis of the Influence of an Interceptor on the Transom Flow of a Fast Ship by Pressure Reconstruction from Stereoscopic Scanning PIV *Ocean Engineering* **181** 281–92

[5] Mansoori M, and Fernandes A C 2016 The interceptor hydrodynamic analysis for controlling the porpoising instability in high speed crafts *Applied Ocean Research* **57** 40–51.

[6] Mansoori M, Fernandes A C, and Ghassemi H 2017. Interceptor design for optimum trim control and minimum resistance of planing boats *Applied Ocean Research* **69** 100–115.

[7] Manoeuvring Committee of 25th ITTC 2008 Standards for Ship Maneuverability *ITTC*

[8] Prabowo, Agung D, and Abdullah D 2018 Deteksi dan Perhitungan Objek Berdasarkan Warna Menggunakan Color Object Tracking *Pseudocode* **5** 85–91

[9] Song K, Guo C, Gong J, Li P, and Wang L 2018 Influence of interceptors, stern flaps, and their combinations on the hydrodynamic performance of a deep-vee ship *Ocean Engineering* **170** 306–320

[10] Sulisetyono A 2014 Development of a Fish Tail Rudder to Improve a Ship's Maneuverability in Seaway *IPTEK, Journal of Proceeding Series* **1**.

[11] Sulisetyono A 2018 The Simple Open Free Running Test for the Evaluation of Turning Ship Ability *6th International Seminar on Ocean and Coastal Engineering, Environmental and Natural Disaster Management* (ISOCEEN 2018) Surabaya Indonesia.

Acknowledgments

The authors would like to thank to Department of Naval Architecture Institut Teknologi Sepuluh Nopember for supporting fee conference.

Problems and technical solutions of capture fisheries in Eastern Indonesian waters

W R Hetharia[1], E R de Fretes[1], F Gaspersz[1], J Tupan[1] and A Feninlambir[1]
[1]Department of Naval Architecture, Faculty of Engineering, Pattimura University, Ambon, Moluccas, Indonesia.

Abstract. Problems concerning capture fisheries in Eastern Indonesian Waters cover the existing of fishes at sea, continuity of catching activities, the use of fishing gears, safety aspects, market, price and recent situation of pandemic Covid-19. This conditions have the impact on decreasing of economy aspect of local fishermen. The purpose of this research is to develop facility of capture fisheries and other related technical aspects to overcome the existing problems and to keep the continuity of fishing activities. The research started by initial field survey by the authors. In addition, related problems from other references are used as input design parameters. Design process end-up with the blue prints and the design specifications of fishing boats and fishing gears. The research was continued with construction of boats prototype, training for the fishermen and sea trials. The results of research were prototypes and design specifications of several types of fishing boats. Fishing boats with hull configurations (monohull, catamaran and trimaran) and fishing gears specification were fixed due to boat operational aspects. In addition, the results of study concerning safety aspect and ship operation in period of Pandemic Covid-19 are presented in this paper. Recommendations of results of this study are presented for future development of fishing boats for local fishermen.
Keywords: Fishing problems, design and specifications

1. Introduction

There are many kind of fishes in Eastern Indonesian waters where those fishes are caught to fulfil international export as well as domestic consumption. The valuable fishes with high price are exported such as Tunas, Skipjack Tuna, Grouper, Indo-Pacific Spanish Mackerel, Red Snappers, any kinds of shrimps. These fishes, together with pelagic and corral fishes are provided also for domestic consumption. Those fishes are caught by local fishermen and high-scale fishing industry companies. The use of fishing gear depends of kind of fish to be caught. Usually the small scale groups of fishermen use small outrigger canoe or small long-boat provided by outboard engine with 2 or 3 crews. Meanwhile, larger vessels such as purse seiners owned by local person operate 15 to 20 crews and skipjack pole and line operate also 18 to 25 crews.

In fact, there are a lot of problems faced by local fishermen in most eastern Indonesian regions. A survey executed by the authors in early 2020 at several spots in Moluccas regions found the important issues concerning the condition of local fishermen. In addition, the effect of Pandemi Covid-19 give additional misery for the local fishermen. Additional survey for secondary data concerning fish potency in East Indonesian waters resumed several special attention of the life of local fishermen [1]. The effect of pandemic Covid-19 recently affect the economic aspect and welfare of local fishermen in many eastern Indonesian coastal regions. Additional data quoted from the following references give an understanding of real problems faced by the local fishermen. Catching process for pelagic fishes at night was presented by Ardi [2] and Suherman [3]. The role of lightning to catch pelagic fishes is presented by Sudirman and Malawa [4] while Usemahu [5] describe the existence of sea and water condition and sea depth as well. The operation of floating fish platform is explained by Kusuma et.al [6] and Purbayanto [7]. Meanwhile, Sudirman and Malawa [4] as well as Baskoro [8] explain the success of the operationg of floating fish platform. The actual problems faced by the fishermen are resumed as follows:

- The use of simple and single fishing gear such are: hand line, trolling line, drifting net, trammel net, circling net, pole and line gives low catching product by the fishermen

- The use of simple kind of boat such as: outrigger canoe and long boat with outboard engine gives the limitation of catching operations
- The existence of fish at sea is limited due to migration period, sea conditions with strong current and waves and full moon, etc.
- Safety and technical aspects of the fishing boats are not provided
- Difficulties of handling of catching due to availability of local market make the fishermen are not willing to operate their vessels.
- Due to Pandemic Covid-19, some purse seiners and skipjack pole and line do not operate.

Those problems need solutions which relate to many aspects. However, this study concentrated on providing the kind boats and their technical aspects to ensure the continuity of the operation.

2. Literature review

This study correlates with design aspects of small fishing boats. Ship design is an iterative process, particularly at early design stages. The output of the first iteration process should be evaluated, analyzed and modified until the design process satisfies the objectives and requirements. An iterative design process is characterized in the design spiral [9] as shown in Figure 1. The designers begin the design proces with the first task then continue the process in sequential steps. When the first iteration (cycle) has finished there will be an evaluation. Then, a second iteration begins and all steps of the cycle are repeated in the same sequence. A number of iterations are needed to reach a satisfactory solution [9], [10]. The design process is divided into several beselines. The first baseline (concept design) generally refers to the activities of a feasibility study. The main objective of this baseline is to clarify the shipowner's requirements (mission) and translates them into naval architectural and engineering characteristics. During the preliminary design phase, ship design parameters are started to develop such as hull dimensions, hull shape, general arrangement, payload, propulsion plant, structure, etc.. This phase refines the major ship characteristics which affects the cost and ship performance. Control factors such as dimensions, power and deadweight would not be expected to change upon completion of this phase. This process is repeated several times. The results obtained at the end of this phase are used in the next design stage. The design process takes several iterations until the best result is achieved. The next step of design process is to develop the results of parametric studies.

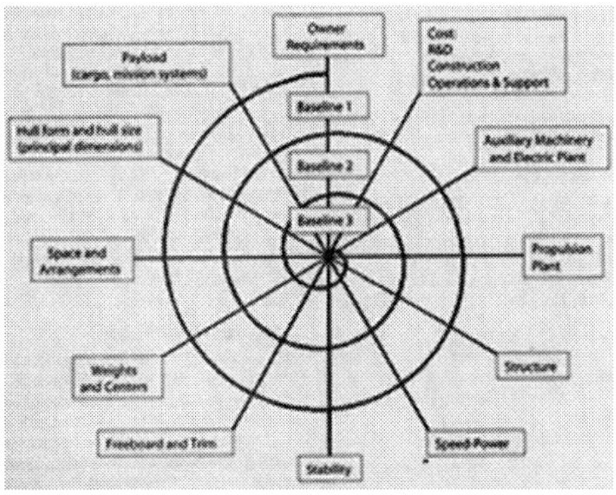

Figure 1. Design spiral

Hydrostatic and stability parameters are considered in this study. The hydrostatics parameter is presented in Figure 1 (a), while stability parameters which covers the vertical centre of gravity (KG), vertical centre of buoyancy (KB), metacentric radius (BM) and initial metacentre (GM) is showed in Figure 2 (b). Basic principle of a ship to be float is weight displacement equals to total weight [11], [12], [13]. This relationship is presented as:

$$\Delta \geq W \tag{1}$$

where: Δ = weight displacement (ton) and W = total weight of ship (ton)

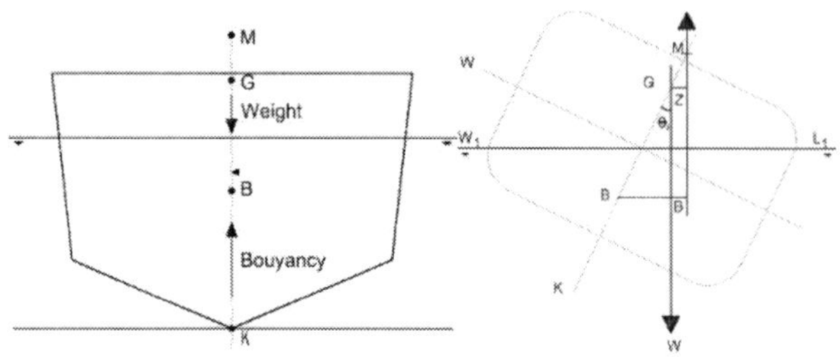

(a) hydrostatics (b) stability

Figure 2. Hydrostatics and Stability Parameters

$$\Delta = \blacktriangledown \times \gamma \tag{2}$$

$$\blacktriangledown = L_{WL} \times B \times T \times C_B \tag{3}$$

where: $\blacktriangledown$ = volume displacement (m^3) and γ = specific weight of sea water (ton/m^3)

L_{WL} = length of waterline (m), B = beam (m) T = draft (m) C_B = block coefficient

Total weight of ship (W) consists of lightweight (LWT) and deadweight (DWT) [14], [10], [15]. Furthermore, LWT consists of hull, engine and supplements, outfitting and fishing gears and DWT consists of cargo (fish and ice), crew and consumption, fuels, fresh water, luggage and fish bait. The critical condition when the boat is filled with water then the cargo is regarding as total incoming water.

$$\text{Cargo} = W_{total} \tag{4}$$

To ensure the boat is in even condition then the position of longitudinal centre of gravity (L_{CG}) should be aligned to longitudinal centre of buoyancy (L_{CB}).

$$L_{CG} = L_{CB} \tag{5}$$

In the case where a boat afloat or sink when the incoming water fill the boat up to the deck (H) then:

$$W_{total} > \Delta = C_B \times L \times B \times H \times \rho \rightarrow \text{sink} \tag{6}$$

$$W_{total} < \Delta = C_B \times L \times B \times H \times \rho \rightarrow \text{float} \tag{7}$$

Some formulas concerning stability parameters of boat are described as follows [11], [12].

$$GM = KB + BM - KG \tag{8}$$

$$BM = I / \blacktriangledown = (L \times B^3/12) / (L \times B \times T \times C_B) = B^2 / (12 \times T \times C_B) \tag{9}$$

where: GM is metacentric height and I is inertia moment to x-axis

$$GZ = GM \times \sin\theta \tag{10}$$

GZ is lever arm for small angle of inclinations ($\theta < 15^O$)

$$\text{Moment statical stability} = W \times GZ = W \times GM \times \sin\theta \tag{11}$$

For the case of trimaran fishing boats, the similar principle of hydrostatics and stability of monohull may be applied with consideration of buoyancy force (ΔB) and moment (ΔMr) from two outriggers.

$$\Delta Mr = \Delta B \times l \tag{12}$$

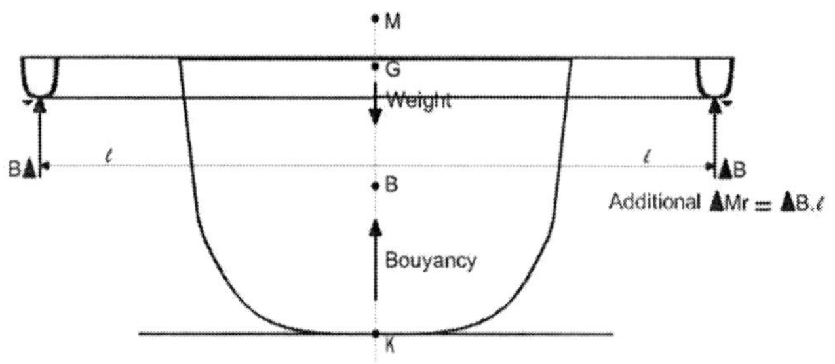

Figure 3. Trimaran configuration

3. Methodology

In this research several fishing boats were designed, constructed and tested at sea. In addition, those boats were developed at difference periods. However, the methodology of the research is similar for those boats which described as follows:

1 *Collecting data base.* The data base of the existing fishing boats owned by local fishermen were collected. The purpose of this task is to get the information concerning their boat as an input for re-designing process.
2 *Re-designing phase.* In this phase, the design process was executed following the standard procedure of ship design. Ship design parameters were determined and evaluated during an iterative process. Some design considerations concerning new concepts were introduced during the process. This concept was introduced in order to gain an added design value.
3 *Developing the ship hull.* The hull form of the fishing boats were developed by using Maxsurf software (Figure 7). In addition, the boat general arrangements were provided for the blue prints. The documents of boat were prepared for the next construction phase. This phase results in design specification of the boats.
4 *Developing boat template.* The output of this task is to find a boat template provided for developing the boat. This is a standard procedure to construct the real boat with the hull material of Fibreglass Reinforced Plastic (FRP).
5 *Construction of real boat.* This task was executed with the aid of boat template. There will be many boats produced by this method.
6 *Sea trial.* This task was executed in order to evaluate results of design in the sea.
7 *Training project.* This task was executed in order to train the local fishermen and boat builders. The implementation of this project is the local boat builders have been developing their products.

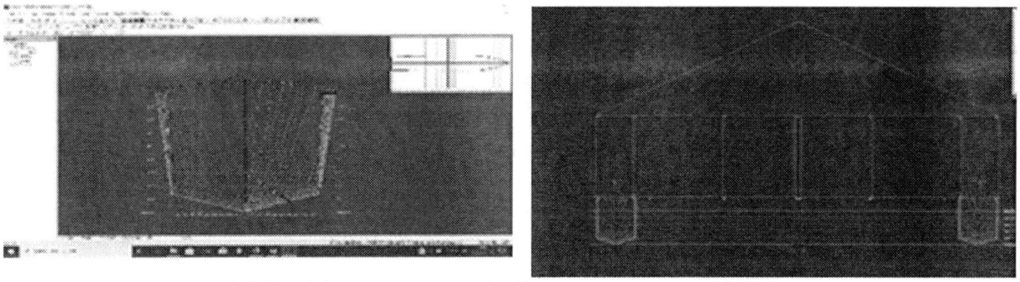

(a) Multi-purpose monohull (b) Trimaran

Figure 4. Lines plan of multi-purpose boat and trimaran fishing platform

4. Results and discussion

4.1. Unsinkable tuna long-boat

The unsinkable tuna long-boat was designed, constructed and operated due to demand of local fishermen to solve their problem of safety [16], [17]. Many accidents occur at sea affected by open sea harsh conditions which end-up with the loss of life and boats at sea. Additional fishing gears such as gill net and bottom vertical line will increase the catching product instead of hand line and trolling line. The unsinkable tuna long-boats were designed and constructed at ship construction workshop, Faculty of Engineering Pattimura University. Sea trial of the boats proved the positive results in safety and increasing the catching product (Figure 5). The training programs were executed for local fishermen and boat builders to product their own boats. The specification of the boats are:

- Length overall L_{OA}: 8.05 m; length of waterline L_{WL}: 7.60 m; beam B: 1.08 m; draft T: 0.39 m
- Three boats the same dimensions and configurations which is 1.5 GT.
- The hull material of Fibre Glass Reinforced Plastic (FRP).
- Capacity: fish payload 0.7 ton, ice 0.35 ton (ratio ice/fish: 0.5)

- Total volume: 4.37 m^3, volume of solid boxes: 0.623 m^3 (14.3 %)
- Prime mover: Outboard engine 25 (speed 12 knots) or 40 HP (speed 15 knots)
- Autonomy: 50 nautical milles with 2 crews
- Fishing gears: hand line, trolling line and additional small gill-net and bottom vertical line
- Fishing grounds: Banda Sea, Ceram Sea and other Moluccas waters
- Floating and stability test
    - At upright condition, full load, the boat still afloat
    - At capsized condition the boat still afloat

(a) (b)
(c)

Figure 5. Unsinkable tuna long-boats

4.2 Multi-purpose fishing boat

The design of this boat is based on principal design of fishing boat [18] as well as principal design of other commercial ships [14], [18]. The main requirements of this kind of boat is designing the ship arrangement to fulfil the operation of multi-fishing gears. This boat was designed and constructed together with group of fishermen partner [19]. This boat were designed and constructed together with a group of local fishermen at the Faculty of Engineering Pattimura University (Figure 6). Multi fishing gears will be applied in this boat. However, this project is hold temporaly di]ue to the financial aspect during the period of Pandemic Covid-19. The specification of the boat are:

- Length overall L_{OA}: 8.50 m; length of waterline L_{WL}: 8.12 m; beam B: 1.20 m; draft T: 0.50 m
- Block coefficient C_B: 0.63; prismatic coefficient C_P: 0,77; midship coefficient C_M: 0.81
- Capacity: fish payload 1.50 ton, ice 0.75 ton (ratio ice/fish: 0.5)
- Gross Tonnage: 2.3 GT
- The hull material of Fibre Glass Reinforced Plastic (FRP).
- Prime mover: Outboard engine 40 HP
- Autonomy: 50 nautical milles with 3 to 4 crews
- Fishing gears: hand line, trolling line, vertical bottom line, gill net, trammel net.
- Fishing grounds: Banda Sea, Ceram Sea and other Moluccas waters

4.3. Mobile fish platform

This platform was designed and constructed in order to be applied as prototype for the local fishermen. Problems with small catching product from the traditional fishing platform (*bagan*) was implemented with introduction of mobile platform. The benenifit of mobile fish platform is application of several fishing gears, cathing operation in many places and any times. Basic design of this fishing platform is based fishing boat design [18] as well as principal design of other commercial ships [10], [14], [18]. Typical hull configuration of fish platform is two-hull vessel (catamaran). The main requirements of this kind of boat is designing the ship arrangement to fulfil the operation of multi-fishing gears. Additional mechanical equipment are provided to this platform to ensure good operation (Figure 7). Other requirements of input design are: netting area should greater the 63 m^2, service speed when operation is about 2.5 knots where normal speed is greater than 6 knots.

Figure 6. Multi-purpose fishing boat

This boat were designed and constructed at the Faculty of Engineering Pattimura University. The specification of the boat are:

- Length overall L_{OA}: 10.10 m; length of waterline L_{WL}: 9.91 m; beam overall B_{OA}: 8.00 m; beam B: 0.90 m; draft T: 0.50 m; deck height H: 1.00 m.
- Block coefficient C_B: 0.63; prismatic coefficient C_P: 0.77; midship coefficient C_M: 0.81;
- Weight displacement 9.40 tons and netting area 67 m^2.
- Capacity: fish payload 2.30 ton, ice 1.15 ton (ratio ice/fish: 0.5)
- The hull material of Fibre Glass Reinforced Plastic (FRP).
- Prime mover: Outboard engine 2x40 HP, electrical power: 10 kVA with winch power 5 kVA
- Autonomy: 25 nautical milles with 5 to 6 crews and fishing ground is near the islands
- Fishing gears: lifting net, gill net, bottom fish trap, vertical bottom line

Figure 7. Mobile fish platform

4.4. Multi-purpose trimaran

This boat was designed to overcome the problem of local fishermen such as the application of single fishing gear with low catching capacity. The application of trimaran-type hull is to allow the boat still operate at "certain wave condition' where compared to its counterpart monohull. Basic design of this multi-purpose trimaran boat is based fishing boat design [18] as well as principal design of other commercial ships [10], [14], [18]. The boat was designed and operated by the authors [20] and [21]. The unique of this boat is the use of multi-fishing gears to keep the continuity of operation by the fishermen. Multi-purpose fishing boat trimaran was constructed together with a company partner which was involved with research fund (Riset Unggulan Kemitraan IV- 2004) (Figure 80. The specifications of the boat are:

- Length overall L_{OA}: 10.84 m; length of waterline L_{WL}: 10.00 m; beam B: 0.90 m; draft T: 0.60 m; deck D: 0.90 m, beam overall B_{OA}: 4.65 m; length outrigger L_{OUT}: 5.64 m; diameter outrigger: d_{OUT}: 0.25 m with Gross tonnage: 4.3 GT
- Capacity: fish payload 2.0 ton, ice 1.0 ton (ratio ice/fish: 0.5)
- The hull material of Fibre Glass Reinforced Plastic (FRP) and wood.
- Prime mover: Outboard engine 40 HP
- Autonomy: 50 nautical milles with 4 to 5 crews
- Fishing gears: hand line, trolling line, vertical bottom line, gill net, trammel net.
- Fishing grounds: Banda Sea, Ceram Sea and other Moluccas waters

Figure 8. Multi-purpose fishing boat trimaran

4.5. *Tuna long-boat trimaran*

This boat was designed, constructed and operated due to the requirements of catching tuna fish at open or harsh sea condition. The ability of this boat to exist at harsh sea condition make it is favourable for the fishermen. The unique of this boat is fast and can resist in wave condition (Figure 9). Previous works by the author to designed and constructed the boat as prototype were applied for many local fishermen to construct their similar boats [22], [23]. Additional fishing gears such as gill net and bottom vertical line will increase the catching product instead of hand line and trolling line. The specification of the fishing boat are:

- Length overall L_{OA}: 9.80 m; length of waterline L_{WL}: 9.30 m; beam B: 0.80 m; draft T: 0.50 m; deck D: 0.85 m, beam overall B_{OA}: 4.06 m; length outrigger L_{OUT}: 4.60 m; diameter outrigger: d_{OUT}: 0.16 m and Gross tonnage: 2.0 GT
- Capacity: fish payload 0.9 ton, ice 0.45 ton (ratio ice/fish: 0.5)
- The hull material of Fibre Glass Reinforced Plastic (FRP) and wood.
- Prime mover: Outboard engine 25 HP with the speed of 14 knots
- Autonomy: 50 nautical milles with 2 to 3 crews
- Fishing gears: hand line, trolling line, gill net and vertical bottom line

Figure 9. Tuna long-boat trimaran

4.6. Solution in pandemic covid-19

Pandemic Covid-19 brings some additional problems for the local fishermen, particularly for a ship with many crew on board such as purse seiner and skipjack pole and line [1]. Discussion with the ship owners and related institution staffs were concluded that some ships did not operate because they can not fulfil the Covid-19 protocol. Meanwhile, some owners operate their ships to meet the Covid-19 protocol by spending additional money. The decisions were made by the authors and shipowners to re-arrange the vessel's general arrangement and additional mechanism. The purse seinere should be provided with the mechanical winch and systems in order to reduce the amount of crew on board. This scenario could increase the amount of catching with additional money for the systems. There should be re-arranged fishing seats on board to keep the distance of crew on board. For example: the process of re-arranging the crew on board sfter discussion with the owner CV Yora as the conterpart of Faculty of Engineering Pattimura University. The discussion end-up with reducing of 12 crews at normal condition (Figure 10.a) to 8 crews (Figure 10. b) in order to keep social distance on board. The result of sea trial showed reducing of catching capacity about 25 % per each trip. However, this drawbacks may be overcome with extending the ship operation period. This study is still contiued in the future for the economic aspects.

Other fishing boat considered in this study was purse seiner. This kind of fishing boat operates a lot of crews on board which are 18 to 22 persons (Figure 11. a). The big amount of crews relates to the need of human power to operate the circling net manually. During the Pandemic Covid-19 most boats stop operating. Some owners operate their boats when all protocol covid-19 was fulfilled. This situation needs additional money for this issue. The solution was made after discussion with head of local fishery board and boat owners. The application of mechanism system could reducing the amount of crew on board (Figure 11.b). Besides, this system could be used for another fishing gears such as bottom long line and gill net. However, it needs further study to confirm re-design results in the future.

The operation of small fishing boats as explained in sub-section 4.1 to 4.5 do not need Protocol Covid-19. This is due to the few crew operate on board and the social distance is fulfilled. Some ship owners follow the Protocol Covid-19 as performed by our counterpart CV Yora such as spraying the disinfectant liquid on board, health test for the crew before boarding and after the trip in order to get the sailing permit and when the boat arrive at other ports the crew stay on board. This procedure spend additional money even so, there was still the profit gained by the company. Future study is required to evaluate this economic aspect of operating the fishing boat.

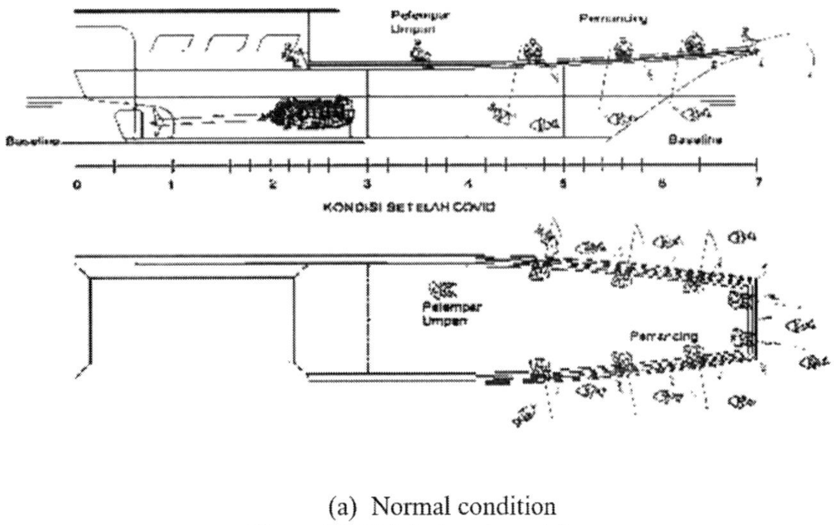

(a) Normal condition
Figure 10. Skipjack pole and line

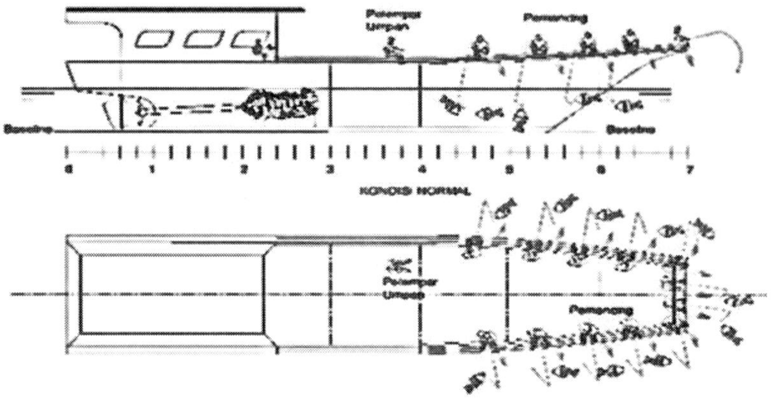

(b) Covid condition
Figure 10. Skipjack pole and line (continued)

(a). The existing purse seiner

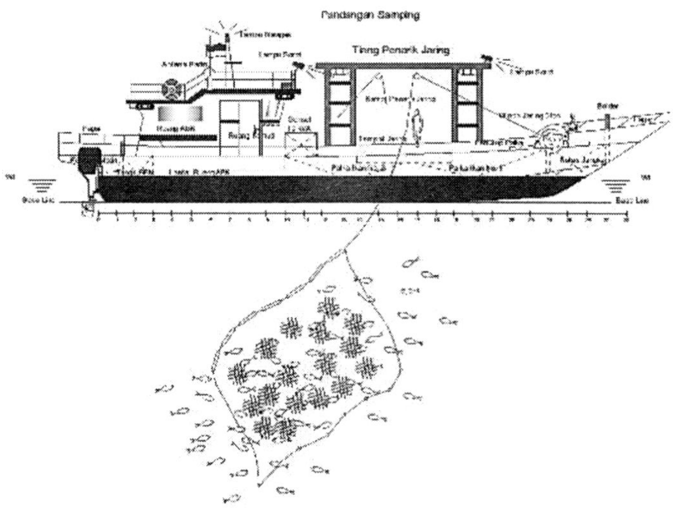

b) Redesign mechanized purse seiner

Figure 11. Purse seiner

4.7. Discussion

Problems concerning the operation of fishing boats in Eastern Indonesian waters were implemented by introducing some new-design boats. Some benefits may be gained from the study includes:

- The application of multi-fishing gears for the new-design boat contribute to keep cathing operation for several kind of fishes.
- Safety aspect of small fishing boats was implemented by introducing solid boxes on board and application trimaran-type of boat instead of monohull.
- The application of mobile fish platform with multi-fishing gears may increase the cathing amout where this boat may operate at any places and any times.
- The record of catching products obtained form the boats operated by the fishermen have proved the succecfull of new design
- Technically those boats and their mechanisms are provided however, other aspects are still become obstacles for the fishermen which are preservation of catching product, harsh sea conditions, the existence of fish at sea, market and price.
- Solution of operation fishing boat during Pandemic Covid-19 was performed at initial stage for two kind ship, which are skipjack pole and line dan purse seiner

5. Conclusion and recommendation

Some conclusions may be drawn from this study concerning the existence of new-small fishing boat for local fishermen.

- Those ships that had been designed, constructed and operated by the local fishermen for some periods which means that the ships are accepted by users.
- The design specification of fishing boats fulfils the requirements for the operation process at sea where the dimensions, configuration and kind of fishing gears may be operated in a good way.
- The application of multi-fishing gears give opportunities for fishermen to continue their operation for other kind of fish
- The design specification provided for any kinds of fishing boat were used by the fishermen when attending training process executed by the Faculty of Engineering Pattimura University.
- It needs some periods to re-arrange purse seiner and skipjack pole and line in order to meet the protocol Covid-19 applied on board.

Meanwhile, the recommendations were made for future study include:

Re-arrange of purse seiner and skipjack pole and line should be considered in a good way to ensure the crew can work safely on board

Main problems which still influence fishermen life are the existence of market, price and product preservation

6. Acknowledgement

Special thanks is dedicated to The Directorate General of Higher Education, Department of National Education, for their financial support to The Maritime Project Faculty of Engineering Pattimura University, 2008

7. References

[1] Laporan Kegiatan PKM, 2020, Fakultas Teknik Unpatti – Survey Potensi Perikanan di Wilayah Maluku.

[2] Ardi, S A H, 2009, Pola Persebaran dan Hasil Tangkapan Bagan di Teluk Pelabuhan Ratu, Skripsi, FMIPA, Universitas Indonesia, Depok.

[3] Suherman, A. dan Aristi D. P., 2005, Penggunaan Lampu Merkury Dalam Perikanan Mini Purse Seine di Jepara, [Laporan Kegiatan Nomor 31/2005] Fakultas Perikanan dan Ilmu Kelautan, Universitas Diponegoro, Semarang.

[4] Sudirman dan Malawa, A., 2004, *Teknik Penangkapan Ikan*, Rineka Cipta, Jakarta.
[5] Usemahu, R. dan Tomasila, L., 2004, Teknik Penangkapan Ikan, Ambon, Departemen Kelautan dan Perikanan
[6] Kusuma, C.P.M; Boesono H, Aristi, D; Purnama, F, 2014, Analisis Hasil Tangkapan Ikan Teri Dengan Alat Tangkap Bagan Perahu Berdasarkan Perbedaan Kedalaman di Perairan Morodemak, Journal of Fisheries Resources Utilization Management and Technology, Volume 3, Nomor 14.
[7] Purbayanto, 2004, Performa Selektivitas Alat Tangkap Bagan Rambo di Perairan Barru Selat Makasar, Penelitian Perikanan Makasar.
[8] Baskoro, M. S, Telussa RF dan Purwangka F, 2006, Efektifitas Bagan Motor di Perairan Waai, Pulau Ambon, Prosiding Seminar Nasional Perikanan Tangkap "Menuju Paradigma Teknologi Perikanan Tangkap yang Bertanggungjawab Dalam Mendukung Revitalisasi Perikanan", Institut Pertanian Bogor.
[9] Gale, P. A., 2003, *The Ship Design Process*, Chapter 5 - Ship Design and Construction, Written by an International Group of Authorities, Thomas Lamb, Editor, SNAME Publication, Vol. 2, Jersey City, New York, USA
[10] Watson, D. G. M., 1998, *Practical Ship Design*, Elsevier Ocean Engineering Book Series, Volume I, ELSEVIER
[11] Barras, B. and Derret, D. R. 2006, *Ship Stability for Masters and Mates*, Butterworth – Heinemann, 6th edition, An Imprint of Elsevier Ltd., L House, Jordan Hill, Oxford OX2 8DP.
[12] Biran, A. B., 2003, *Ship Hysrostatics and Stability*, Technion – Faculty of Mechanical Engineering, 1st publication, Butterworth – Heinemann, An Imprint of Elsevier Ltd., Linacre House, Jordan Hill, Oxford OX2 8DP.
[13] Moore, C. S., 2010, *Intact Stability – The Principles of Naval Architecture Series*, The Society of Naval Architects, SNAME, Jersey City, New Jersey.
[14] Parsons, M. G., 2003, *Parametric Design*, Ship Design and Construction - Chapter 11, Volume. 2., SNAME Publication, Jersey City, NJ, USA.
[15] Lewis, E. V., 1988, *Principles of Naval Architecture, Stability and Strength*, SNAME Publication, Vol. I, Jersey City, New Jersey.
[16] Hetharia, W.R dan De Lima, E. J, 2008, Design and Evaluating of Unsinkable Tuna Long Boats. Proceeding of The 6th Regional Conference on Marine Technology (MARTEC 2008). University of Indonesia, Jakarta 26 – 27 August, 2008.
[17] Hetharia, W. R, 2017, Design and Application of Unsinkable Tuna Longboat for Local Fishermen, OCCASIONAL PAPERS No. 58 (March 2017), Kagoshima University, Japan
[18] Fyson, J., 1985, *Design of Small Fishing Vessels*, FAO of The United Nations', Fishing News Books Ltd., Surrey, England.
[19] Maitimu, NE, Louhenapessy J, Tupan J, 2019, Laporan Pelatihan Pembuatan Prototype Kapal Ikan Long Boat Bagi Kelompok Nelayan Desa Leahari, Fakultas Teknik Unpatti
[20] Hetharia, W. R, 2004, Desain dan Operasional Prototype Kapal Multifungsi Material Fibreglass Reinforced Plastic Untuk Nelayan Pesisir Maluku, Laporan Akhir Riset Unggulan Kemitraan (RUK) IV, Kementerian Riset dan Teknologi Republik Indonesia dan CV. Pipit Perkasa
[21] Hetharia, W. R, de Fretes, E. R dan Camerling, B. J, 2005, Design and Operation of Small Fishing Boat with Bottom Long Line, Prosiding of International Workshop Eco-Friendly Coral Reef Fishery, Unpatti – Tokyo University of Marine Science, Ambon 17 Maret 2005
[22] Hetharia, W.R, de Fretes, E. R dan Matatula, L.A.T., 2003, Desain dan Operasi Prototype Kapal Ikan Tuna Pole and Line Material Fiberglass Reinforced Plastic untuk Nelayan Lokal di Maluku, Prosiding Seminar Nasional Aplikasi Sains dan Teknologi, Direktorat Jenderal Pendidikan Tinggi Departemen Pandidikan Nasional, Jakarta 26 s/d 28 Mei.
[23] Hetharia, W. R, de Fretes, E. R dan Camerling, B. J, 2005, Kajian Tentang Pengembangan Kapal Trimaran di Indonesia, Prosiding Seminar Nasional Teori dan Aplikasi (SENTA) 2007, ISSN 1412-2332 Fakultas Teknologi Kelautan ITS Surabaya, Surabaya 15 Nov 2007.

The development of a risk-based maintenance flowchart to select the correct methodology to develop maintenance strategies of oil and gas equipment

D Priyanta[1], **M B Zaman**[1], **Semin**[1]

[1] Department of Marine Engineering, Faculty of Marine Technology, Institut Teknologi Sepuluh Nopember, Surabaya 60111, INDONESIA

E-mail: priyanta@its.ac.id

Abstract. There are hundreds, even thousands of equipment that are installed on an Oil and Gas Plant. The equipment can be static equipment, rotating machinery, or protective devices. Each type of equipment has a unique failure mechanism and failure mode. Maintenance tasks should be performed to prevent these failures from being happened. The maintenance task should be developed based on the methodology, which will capture the failure mechanism yet to prevent it from being more severe. Before March 2018, the Indonesian government has adopted a time-based maintenance policy for all static equipment operated by oil and gas companies operating in the Republic of Indonesia's territory. Every three years, the static equipment that works must be inspected and recertified. In March 2018, the Minister of Energy and Mineral Resources issued regulation No. 18, which allows the use of risk analysis methods to determine maintenance strategies, known as risk-based maintenance, for all equipment operated by oil and gas companies operating in Indonesia. This paper will develop an initial flowchart to select the correct methodology to develop maintenance based upon the dominant failure mode. A case study to demonstrate the flowchart has also been performed in an Onshore Receiving Facility (ORF).

Keywords. Equipment Criticality Analysis (ECA), Reliability Centered Maintenance (RCM), Risk-Based Inspection (RBI), Risk-Based Maintenance (RBM), Safety Integrity Level (SIL).

1. Introduction

There are hundreds, even thousands of equipment that are installed on an Oil and Gas Plant. Maintenance activities should be performed to prevent failure mechanisms and failure mode being happened. In Oil and Gas practice, there are codes and government regulations as well to have Oil and Gas Companies to perform mandatory maintenance in order the plants are permitted to continue to operate.

Traditional inspection codes for the static equipment are based on prescriptive or time-based requirements. The recommendations require the operator to perform the inspection at a specific time interval regardless of the condition of the equipment. Examples of such offers can be found in American Petroleum Institute (API) inspection codes [1][2][3]. These codes focus only on the probability of failure, and the failure consequence of the equipment is not to be considered. Having these codes implemented, sometimes it is found that the equipment is still fit for service between two inspection intervals.

However, among hundreds or thousands of equipment operated in oil and gas plants, those will not have the same likelihood to fail at the same rate. Also, the consequence caused by every failed equipment does not have the same severity level. There is a concept to represent both the probability of failure and its consequence into a single representation called risk.

API initiated to include the consequence of the failure and the probability of failure of the static equipment in 2000. This concept can be found in API Risk-Based Inspection document guidance [4]. This document has been superseded by the latest document [5] in 2016.

Risk can be defined as a measure of possible loss or injury and is expressed as the combination of the incident probability and its consequences. Risk can be described as the multiplication of the probability of failure (PoF) and the consequence of failure (CoF), as can be seen in Equation 1.

$$Risk = PoF \times CoF \quad (1)$$

In most processing facilities, a large percent of the total risk for a processing unit will be concentrated in a relatively small percentage of the equipment items. These potential high-risk components may require greater attention through a revised inspection plan. The cost of the increased inspection effort may be offset by reducing excessive inspection efforts in the areas identified as having lower risk. Figure 1 shows the typical risk of equipment distribution in a processing facility.

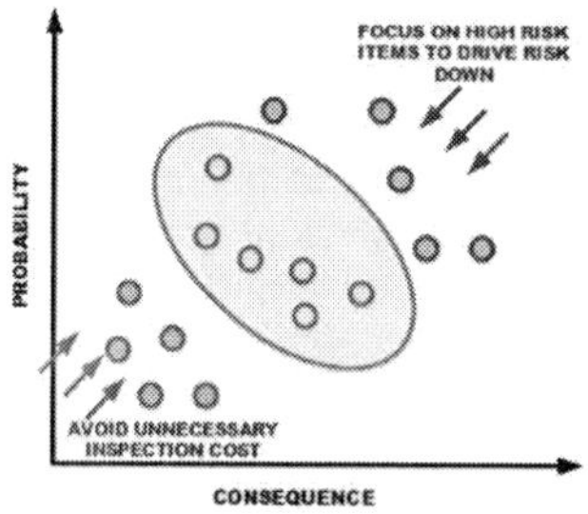

Figure 1. Typical risk of equipment distribution in a processing facility.

Until 2018, the Indonesian government has adopted a time-based maintenance policy for all static equipment operated by the oil and gas companies operating in the Republic of Indonesia's territory. Every three years, static equipment that operates must be inspected and recertified so that it can be operated legally. In March 2018, the Minister of Energy and Mineral Resources issued regulation No. 18, which allows the use of risk analysis methods to determine maintenance strategies, known as risk-based maintenance, for all equipment operated by oil and gas companies operating in Indonesia. For this reason, it is necessary to develop a framework to guide the selection of appropriate analysis methods to determine maintenance strategies for various equipment used in the oil and gas industry. Different methodologies relating to RBM are reviewed, and the implementation sequence is also studied.

2. Risk-based maintenance (RBM)

Maintenance is defined as a combination of all technical, administrative, and managerial actions during the life cycle of an item intended to retain it or restore it to a state in which it can perform the required function [6]. Practically, all companies which operate oil and gas processing plant already have a maintenance strategy to maintain the assets. The maintenance strategies applied generally adopt applicable rules, regulations, and codes relevant to the type of equipment. The maintenance strategies sometimes are also based on manual equipment or recommendations from vendors. The maintenance strategy approach in this way is generally known as the minimum maintenance approach and less attention to the risks that will occur if the equipment fails.

Risk-Based Maintenance (RBM) is a methodology to develop maintenance tasks based on the assets' risk profile. The methods used to create maintenance tasks depend on the equipment failure mode. The failure mode is the effect by which a failure is observed on the failed item [6].

Equipment is a maintenance object. There are hundreds or even thousands of equipment operating in the oil and gas plant. Not all equipment has the same level of risk. High-risk equipment requires more attention compared to low-risk equipment. Therefore, the first step in implementing the RBM is to conduct screening to select medium and high-risk equipment which will be analyzed in detail. Detail analysis aims to develop a maintenance task that is technically feasible and worth doing.

2.1. *Equipment Criticality Analysis (ECA)*

The terminology of equipment criticality analysis (ECA) refers to the initial screening of equipment based on its critical level. ECA is a risk-based screening process to prioritize which equipment should be analyzed further using the appropriate methodology to determine maintenance tasks that are technically feasible and worth doing. Technically feasible is governed by the technical characteristics of the task and of the failure which it is meant to prevent. Whether it is worth doing is governed by how well it deals with the consequences of failure [7].

There are several methods developed to determine the critical level of equipment. The classical reliability textbook discussed this issue under the topic of component importance [8]. The theoretical discussion on how to determine the criticality of the component based on the location of the component in the system as well as the reliability value of the component. The criticality of the component is determined by its component importance. Component importance can be determined by Birnbaum's measure, criticality importance, Vesely – Vussell's measure, or Improvement potential.

The risk matrix is also widely used to determine the criticality level of equipment. The dimensions of a risk matrix depend on the number of probability and consequence ratings. A risk matrix with dimensions of 3 x 3 means that the risk matrix has 3 rating probabilities and 3 rating consequences. In oil and gas industry practice, DNV recommends conducting a screening process for static equipment before analyzing risk-based inspection [9]. This document has been superseded by the latest document [10] in 2017. The screening process is carried out using a 2x2 risk matrix, as shown in Figure 2. The matrix classifies risk into three levels: low risk, medium risk, and high risk.

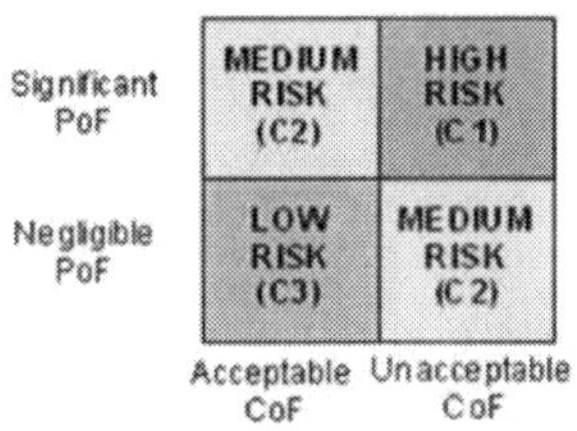

Figure 2. DNV risk screening tool [10].

Norsok recommends a more general criticality analysis process for systems, subsystems, and equipment in the oil and gas processing industry [6]. The critical level is determined based on the level of redundancy and the consequences if the system, subsystem, or equipment fails.

The protocol of the scoring system and risk matrix to determine criticality equipment before RBI analysis can be found in [11]. The protocols were developed based on [10]. ECA categorizes the static mechanical equipment into C1, C2, or C3, which refer to high, medium, or low criticality equipment, respectively. Both probability and consequence are determined by the scoring protocol. The scores will determine the probability level and consequence level. The risk, which represents the criticality of the equipment, is determined by combining the probability level and consequence level.

Efforts to improve the determination of criticality equipment have been carried out. Improvements are made by adding multiple criteria that contribute to the probability of failure and the consequences of failure. Improvements are also made to probability modeling.

A fuzzy inference system (FIS) is the actual mapping process from a given input to an output using fuzzy logic. The FIS application in determining the criticality of equipment can be found in [12]. The application of FIS has been used to rank the criticality of rotating machinery and the instrumentation in the oil and gas processing plant. The equipment criticality can be determined by ranking its functional failure risk (FFR) [12]. The FIS was adopted due to it can minimize the determination of suboptimal priorities from Functional Failure Risk (FFR).

Bayesian networks are a type of probabilistic graphical model that uses Bayesian inference for probability computations. Bayesian networks aim to model conditional dependence, and therefore causation, by representing conditional dependence by edges in a directed graph. Bayesian network applications in determining the criticality of equipment can be found in [13]. The conducted study investigates the safety and operational issues of the constant bottom-hole pressure drilling technique which is used in managed pressure drilling compared to conventional overbalanced drilling. The study first uses bow-tie models to map safety challenges and operating pressure regimes in constant bottom-hole pressure drilling techniques. Due to the difficulties in modeling dependencies and updating the belief on the operational data, the bow-ties are mapped into Bayesian networks. The Bayesian networks are thoroughly analyzed to assess the safety-critical elements of constant bottom-hole pressure drilling techniques and their safe operating pressure regime.

Analytic Hierarchy Process (AHP) is a multicriteria model that provides a methodology for comparing alternatives by structuring criteria into a hierarchy, providing for pair-wise comparisons of criteria at the lowest level of the hierarchy to be entered by the user, and synthesizing the results into a single numerical value. The application of AHP to determine the criticality of equipment can be found in [14]. Four criteria are considered for criticality analysis. They are the effect on the failure of equipment on power generation, environment and safety, frequency of failure, and Maintenance Cost.

The Fine–Kinney method is a comprehensive method for quantitative evaluations to aid in controlling hazards. In this risk assessment method, the risk value is calculated by considering the parameters of the consequence of an accident (C), the exposure or frequency of occurrence of a hazard event that could lead to an accident (E), and the probability of a hazardous event (P). The classical Fine–Kinney method has a limitation in that it assigns equal weight to these three parameters. The improvement of this method incorporating AHP has been developed and it has been implemented to determine the criticality of the equipment. This can be found in [15].

2.2. *Risk-based inspection (RBI)*

Static equipment operated at an oil and gas processing facility has a dominant failure mode of a loss of containment. The typical cause of this failure mode is the metal loss caused by corrosion. The inspection of static equipment is a form of a maintenance task that is commonly carried out to monitor the thickness and corrosion rate that might occur in the static equipment.

In oil and gas industry practice, API released an RBI recommended practice. This recommended practice provides quantitative procedures to establish an inspection using risk-based methods for pressurized fixed equipment, including pressure vessels, piping, tankage, pressure relief devices (PRDs), and heat exchanger tube bundles [5]. API also releases the other document to guide the development of RBI programs on fixed equipment in refining, petrochemical, chemical process plants, and oil and gas production facilities [16]. The intent is for [16] to introduce the principles and present minimum general guidelines for RBI, while document [5] provides quantitative calculation methods to determine an inspection plan.

DNVGL also published RBI recommended practice for oil and gas industry practice [10]. This recommended practice aims to describe a method for establishing and maintaining an RBI plan for offshore pressure systems. It provides guidelines and recommendations which can be used to customize methods & working procedures that support the inspection planning process.

IOP Conf. Series: Materials Science and Engineering 1052 (2021) 012042 doi:10.1088/1757-899X/1052/1/012042

There was a study to implement this RBI methodology [17] with the piping as the object of the RBI study. In refineries and processing plants, the enormous amount of piping is more complicated in distribution than other types of equipment. In general, compared with different types of equipment in these industries, more difficulty in inspection planning is encountered. To lessen the piping risk level, more and more companies have adopted and applied risk-based inspection (RBI) methodology, leading to risk reduction and cost benefits.

A study to incorporate AHP into RBI analysis has been performed by [18]. RBI methodology was proposed to evaluate the maintenance strategy in the industrial process, which was constructed in one of the units of Fujian Oil Refinery ISOMAX unit. All equipment in this unit was evaluated and categorized into five risk zones based on the RBI result, which covered five levels. Also, an application of the analytical hierarchy process (AHP) to select the most practicable maintenance strategy for equipment which was in each risk rating scale, was described.

Another study has also been conducted to couple RBI methodology and Multi-Objective Genetic Algorithm (MOGA) for defining efficient inspection programs in terms of inspection costs and risk level, which also comply with restrictions imposed by international standards and/or local government regulations [19].

2.3. Safety integrity level (SIL)

There are equipment and instrument in oil and gas processing facilities having protective functions or safety functions. The Safety Instrumented Systems (SIS) are the systems responsible for the operating safety and ensuring the emergency stop within limits considered safe whenever the operation exceeds such limits. The main objective is to avoid accidents inside and outside plants, such as fires, explosions, equipment damages, protection of production and property, and, more than that, avoiding life risk or personal health damages and catastrophic impacts to the community. Therefore, the failure mode of this kind of device is the loss of protection.

Safety integrity is defined as 'the likelihood of a safety-related system satisfactorily performing the required safety functions under all the stated conditions, within a stated period of time, and a safety integrity level (SIL) as a discrete level (one of 4) for specifying the safety integrity requirements of safety functions The Safety Integrity Level (SIL) is a statistical representation of the integrity of the SIS when a process demand occurs. Testing the SIS is the only way to ensure the protective device keeps functioning. The frequency of the test depends on the safety integrity level.

There is a standard [20] that provides a basic functional safety standard applicable to all industries. It defines functional safety as: "part of the overall safety relating to the EUC (Equipment Under Control) and the EUC control system which depends on the correct functioning of the Electrical/Electronic/Programmable Electronic (E/E/PE) safety-related systems, other technology safety-related systems, and external risk reduction facilities."

The evaluation of SIS will reveal whether the risk controlled by the protective device in oil and gas processing plants comply with the standard or not. One of the proposed methods to evaluate SIS is using a genetic algorithm as it can be found in [21]. Another method to verify the integrity level of SIS is by using a reliability block diagram and Montecarlo simulation [22]. The verification was done following the requirements of IEC 61508.

A study to simplify and more efficient methodology for the safety assessment of electromechanical SIS in compliance with the Standards IEC 61508 and IEC 61511 can be found in [23]. The proposed technique is based on an alternative implementation of the Reliability Block Diagram (RBD) approach for the performance analysis of the Safety Instrumented System.

2.4. Reliability-centered maintenance (RCM)

For the equipment having failure modes other than loss of containment and loss of protection, there is a methodology to capture different failure modes, analyze them, and recommend maintenance tasks. Such a procedure is reliability-centered maintenance (RCM). Early works of RCM before the formal RCM standard was released can be found in [24].

RCM is a process to ensure that systems continue to do what their user require in their present operating context. It is generally used to achieve improvements in fields such as the establishment of safe minimum levels of maintenance [7]. The technical standard to perform the RCM process can be found in [25][26]. If it is performed correctly, RCM will recommend a maintenance task that is technically feasible and worth doing.

RCM indicates that reliability is the main point of reference for the planning, but the consequences of failures are also assessed. However, uncertainties and risks are to a limited extent addressed by the RCM method. A methodology to incorporate risk as an extension of the RCM analysis was proposed by [27]. The methodology is called reliability and risk centered maintenance (RRCM).

An integration of identification critical components before the implementation of RCM is presented in [28]. A modification of the RCM table to incorporate risk priority numbers before select technically feasible and worth doing maintenance tasks can be found in [29]. The purpose of the modification is to optimize the maintenance task. Another method to optimize RCM planning is to use a genetic algorithm. A multi-objective model is proposed to solve the mathematical problem of optimizing the reliability-centered maintenance planning of an electric power distribution system (EPDS). The main goal is to minimize the preventive maintenance costs while maximizing the index of the reliability of the whole system [30].

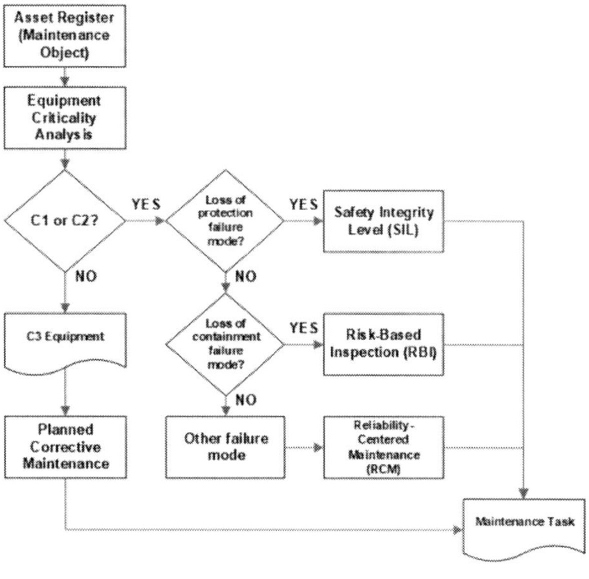

Figure 3. Proposed RBM flowchart to choose maintenance tasks analysis methodology based upon failure mode of the equipment.

3. Proposed RBM flowchart

Maintenance tasks developed from maintenance strategy should be able technically feasible and worth doing to prevent equipment failure being occurred. The maintenance task at least consists of two issues, i.e., what task to do to prevent failure mode being happened and when the task will be executed. Determining what task to do and when to do it requires the understanding of failure modes associated with it and the cause and the mechanism of failure. There are methodologies to deal with these issues.

RBI is a methodology deal with the loss of containment failure mode of static equipment, SIS and SIL methodology deal with the loss of protection of protective device while other methodology to analyze different failure modes are provided by RCM. There are hundreds of even thousands of equipment operate in oil and gas processing plants. Therefore, it is recommended to perform ECA to prioritize the equipment, which can be categorized as having medium and high risk. The results of the ECA can be categorized into high risk, medium risk, or low risk. C1, C2, and C3 refer to the high, medium, and low-risk equipment categories, respectively. Depend on the failure mode of the equipment, equipment with C1 and C2 category will be analyzed further using RBI, SIL, or RCM. The result of the analysis will determine technically feasible and worth doing maintenance tasks of the equipment based upon its failure mode. For equipment with the C3 category, a planned corrective maintenance strategy will be appropriate for it. Figure 3 shows the proposed RBM flowchart to choose maintenance tasks analysis methodology based upon the failure mode of the equipment

4. Case study

A case study regarding the use of the RBM flowchart is presented below in the early implementation of RBM. An onshore receiving facility (ORF) operating in the East Java Province of the Republic of Indonesia has been selected for the case study. The study is still in an early stage of RBM implementation. Therefore, the objective of the case study is to group the equipment based upon its dominant failure mode. Once it is identified, an appropriate methodology to develop a maintenance strategy is recommended.

4.1. System description

Figure 4 shows an overview of the onshore receiving facility being studied. The acid gas from the wellhead is processed in the wellhead platform (WHP). The acid gas is feed to the manifold before it is separated in the liquid-gas separator. The separated fluids are then sent to the floating production storage and offloading (FPSO). The acid gas is processed to become sweet gas in the FPSO and send it back to WHP before it is sent to the ORF.

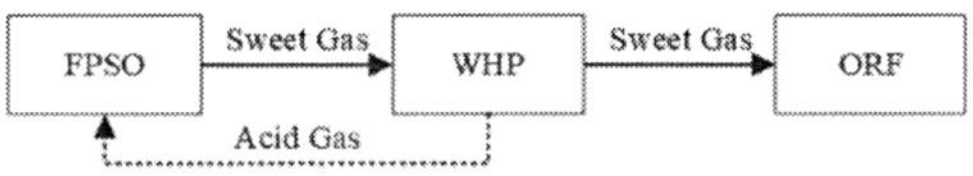

Figure 4. The overview of the onshore receiving facility.

Figure 5 shows the schematic flow diagram of the ORF. The sweet gas from the WHP flows through a subsea pipeline before it is fed into a separator V-001. The separated gas is led to the customer through sales gas metering package M-001. Some of the separated gas is also led to the fuel gas system that is used for the internal utilization of ORF. The separated liquid from V-001 is sent to the liquid burner system and flare system.

Fuel gas scrubber V-002 and fuel gas separator S-001 will process gas from V-001 to become dry gas and send it to fuel gas superheater E-001. This process ensures that there are no liquid contents in the dry gas before it is contained in the instrument gas receiver (V-003). The operation of the ORF is also supported by utility systems such as closed drain and flare system, open-drain system, oily water treatment system, liquid burner system, diesel system, freshwater system, and fire water system.

4.2. The physical asset hierarchy

The first step in conducting this study is to identify the asset hierarchy of the ORF. For the study, the hierarchy of the asset is arranged based upon asset hierarchy as it is recommended in [31]. The source information for the asset hierarchy comes from the company asset register, process flow diagram (PFD) and piping and instrument diagram (P&ID) of the ORF, and other relevant information.

The hierarchy of physical assets can be arranged technically. The technical hierarchy describes the technical structure of the installation by giving functional locations unique identifiers. The technical hierarchy provides an overview of equipment units that belong together technically and shows the physical relationship between main equipment, instruments, valves, etc.

As per standard [31], the physical assets hierarchy is divided into nine levels starting from industry type until part level. The hierarchy starting from the top until the lowest level is as following: industry – business category – installation – plant/unit – section/system – equipment unit – subunit – component / maintainable item – part. An example of the taxonomy of the physical assets can be seen in Table 1.

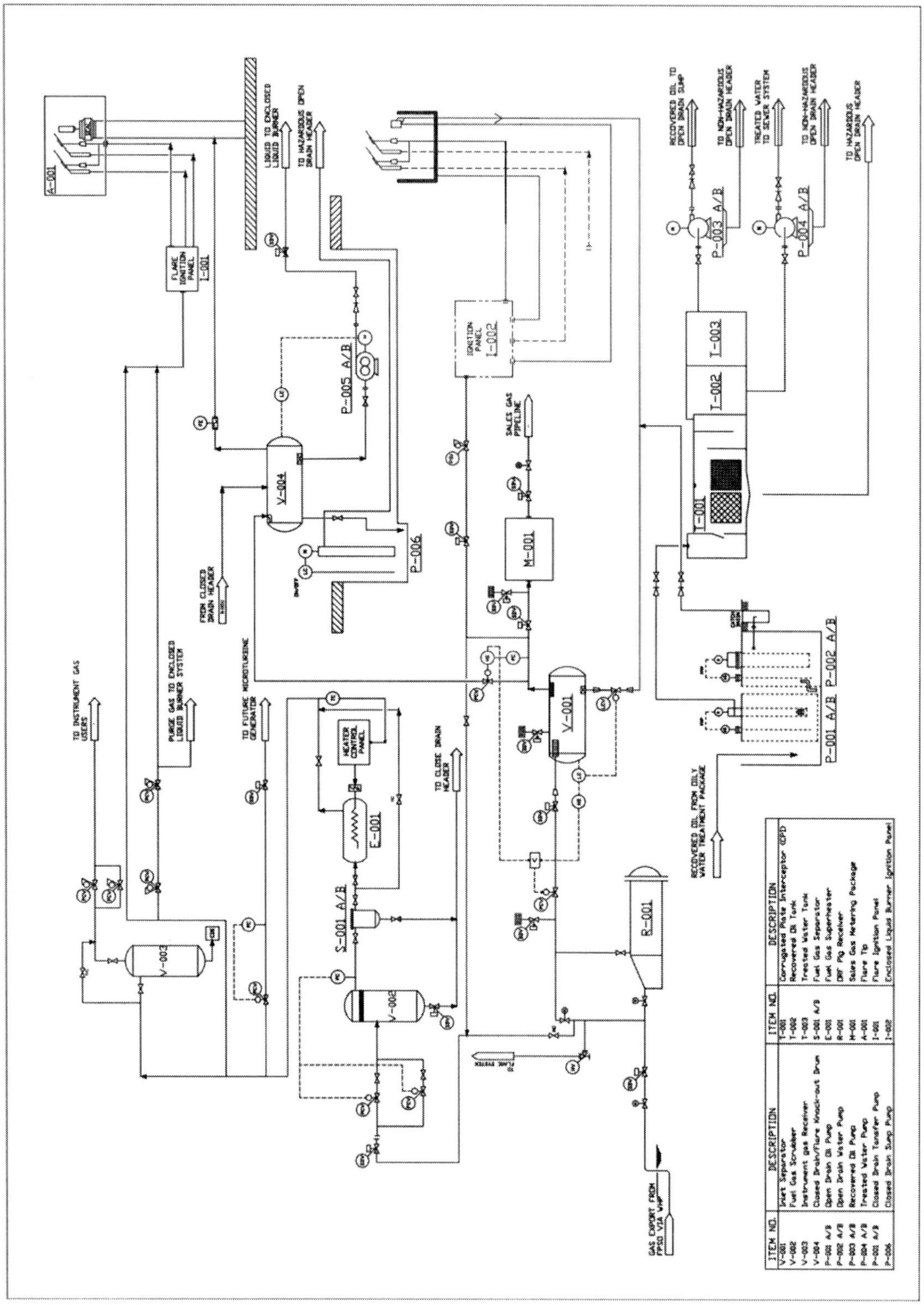

Figure 5. Schematic flow diagram of the ORF.

Table1. The taxonomy of physical assets [31].

Main Category	Taxonomy Level	Taxonomy Hierarchy	Definition
Use / Location data	1	Industry	Type of main industry
	2	Business category	Type of business or processing stream
	3	Installation category	Type of facility
	4	Plant/unit category	Type of plant/unit
	5	Section/ System	Main section/system of the plant
Equipment subdivision	6	Equipment class/unit	Class of similar equipment units. Each equipment class contains comparable equipment units
	7	Subunit	A subsystem necessary for the equipment unit to function
	8	Component/ Maintainable item	The group of parts of the equipment unit that are commonly maintained (repaired/restored) as a whole
	9	Part	A single piece of equipment

For this study, the hierarchy of the physical assets will be broken down until level 8. At this level, the initial maintenance strategy will be determined for the component/maintainable item based on the developed RBM flowchart.

4.3. The study results

The application of the proposed RBM flowchart is initiated by registering all assets based on the hierarchy. The hierarchy of the physical assets will be broken down until level 8. Data in level 1 to level 5 informs the type of industry, business category, installation category, and the plant where the data located. Table 2 shows the list of information at level 5 or the list of systems available in ORF. Level 6 to level 8 describes the equipment information starting from its class/unit until its component. The maintenance strategies might be applied to any equipment at these levels.

Table2. The list of systems in the ORF.

No.	Name of system
1	Close drain system
2	Diesel system
3	Electrical distribution system
4	Flare system
5	Freshwater system
6	Fuel system
7	Instrument gas system
8	Enclosed liquid burner package
9	Custody meter package system
10	Oily water treatment system
11	Open drain system
12	Piping system
13	Gas pig receiver system
14	Fire protection system
15	Gas inlet separator system
16	Telecommunication
17	Workshop

From the process flow diagram (PFD) and piping & instrument diagram (P&ID), it can be identified that the ORF plant consists of 17 systems, 98 equipment class/unit, 137 subunits, and 3490 components / maintainable items. Therefore, there will be massive and complex data in registering the physical hierarchy of the assets. The complexity and integrity of the data will be the issue in handling massive data. The integrity of the data should be stable. A database program will record and manage the hierarchy of the assets as per standard [31]. Figure 6 shows the hierarchy level relationship data in a database program.

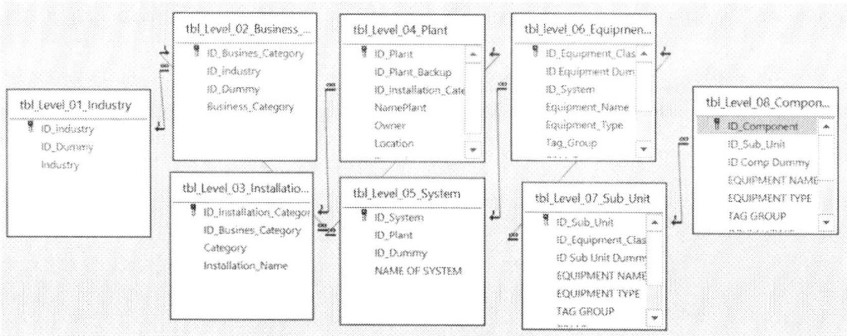

Figure 6. Data relationship.

All data has been processed and stored in the database. Figure 7 shows the ORF's physical asset hierarchy screenshot starting from level 4 to level 8. There are 17 subsystems identified at level 5, as in Table 2. Level 6 groups the equipment in a class or unit. There are 98 equipment classes at this level. In some cases, it is preferable to develop maintenance strategies at this level.

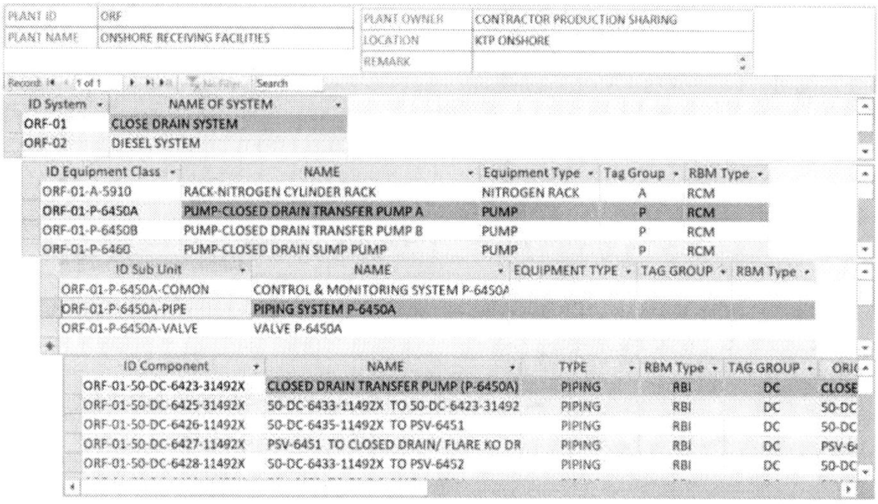

Figure 7. The screenshot of ORF's physical asset hierarchy.

There is a drain transfer pump as an example at level 6. In this case, it is preferable to develop maintenance strategies for this equipment at this level. The pump has dominant failure modes other than loss of containment and loss of protection. Therefore, the maintenance strategies should be developed based on the RCM methodology. Of 98 equipment classes at level 6, 93 of them have been recommended to adopt the RBM methodology to evaluate or improve the maintenance strategies. The use of the proposed flowchart leads to the following recommendations. The RBI, SIL, and RCM methodology will be applied to 13, 29, and 51 equipment, respectively.

Level 7 or subunit is a sublevel necessary for the equipment unit to function. Each equipment unit has different subunits from other equipment units. For example, the pump might consist of power transmission, power unit, control and monitoring system, piping system, and lubrication system. It is rare to develop a maintenance strategy at this level since every subunit might have some components. It is more appropriate to create maintenance strategies at the component level.

Level 8 or component is the group of equipment units that are commonly maintained (repaired/restored). There are 3490 components identified in the ORF. RBM methodology flowchart is applied to all components to obtain an appropriate methodology to develop an effective maintenance strategy. The RBI, SIL, and RCM methodology will be used for 326, 1113, and 2051 components, respectively. Figure 8 shows the distribution of the applied RBM methodology for the assets both at equipment classes (level 6) and at the component level (level 8). Figure 8 summarizes assets grouping based on RBM type both at equipment class (level 6) and component level (level 8).

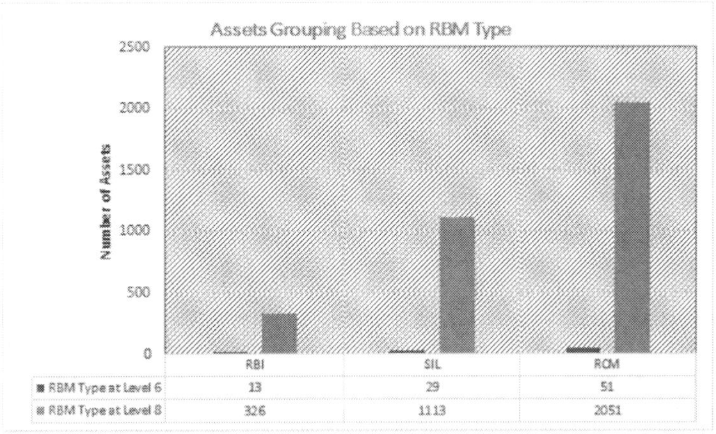

Figure 8. Asset grouping based upon RBM type.

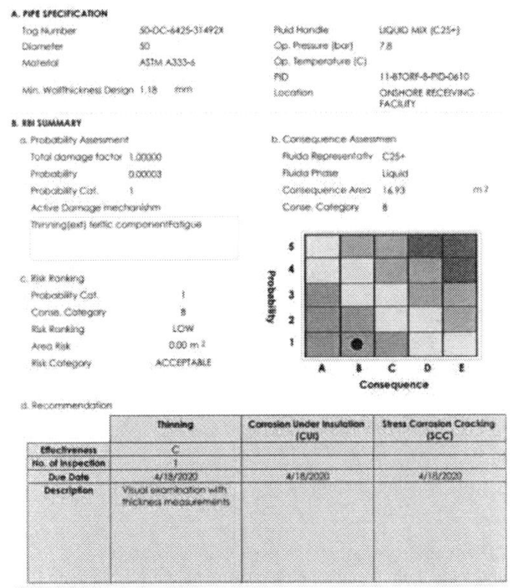

a. RBI analysis result of 50-DC-6425-31492X

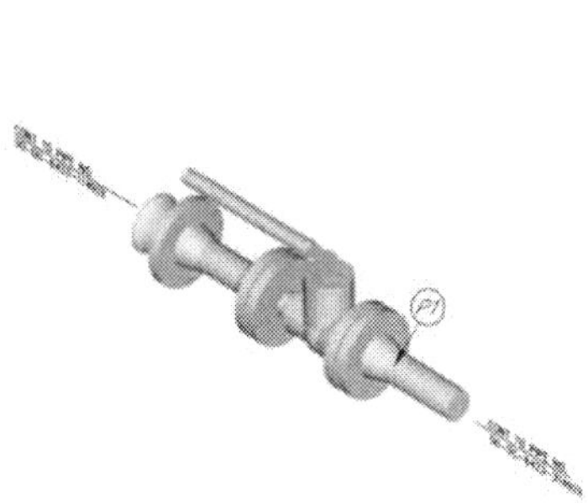

b. Detail inspection plan

Figure 9. RBI analysis result and detail inspection plan.

To demonstrate the application of the proposed RBM flowchart, let's take a drain transfer pump at the equipment class as an example. The pump is recommended to adopt the RCM methodology to develop or to improve its maintenance strategy. In the drain transfer pump, the pump has a piping system subunit (level 7) that consists of many pipings (level 8). The pipings have a dominant failure mode of loss of containment. Therefore, the RBI methodology is recommended to develop the inspection plan. For example, ORF-01-50-DC-6425-31492X is the tag number of piping under the piping system to support the pump. This piping's RBI analysis showed that this piping's risk category is low at the RBI date. An inspection plan has also been developed based on RBI's analysis result. Figure 9 shows the summary of the RBI analysis and detailed inspection plan of the pipe.

5. Discussion and future works

A case study on the application of the proposed RBM flowchart has been completed. The object of the case study is an onshore receiving facility. However, the case study showed that the result of the RBI analysis of the sample component is low risk. The analyst might skip the RBI analysis recommendation for this sample only if the analyst knew that the pipe had been screened before. The method to filter the component and categorize it based on its criticality is known as equipment criticality analysis (ECA).

Further study and the development of ECA methodology will be performed. The ECA study will be focused on the development of ECA protocols for the equipment operated in the oil and gas industries. The method developed must be practical but supported by a strong scientific background. The various ECA methods will be assessed its applicability level in the field, considering that this framework will be implemented for practical industrial purposes.

6. Conclusion

An RBM framework for onshore receiving facility equipment has been developed, and a case study has been performed. The case study proved that the appropriate methodology to develop an effective maintenance strategy is based on the equipment's dominant failure mode. The proposed methods are RBI, SIL, and RCM.

7. Acknowledgment

The authors would like to thank the Department of Marine Engineering – ITS that has supported the initiation of this asset integrity management research for the oil and gas industries, particularly for the facilities operate on offshore platforms.

References

[1] American Petroleum Institute 2018 *API 510 Pressure vessel inspection code: in-service inspection, rating, repair, and alteration* (Washington DC: American Petroleum Institute)

[2] American Petroleum Institute 2016 *API 570 Piping Inspection Code: In-service Inspection, Rating, Repair, and Alteration of Piping Systems* (Washington DC: American Petroleum Institute)

[3] American Petroleum Institute 2014 *API 653 Tank Inspection, Repair, Alteration, and Reconstruction* (Washington DC: American Petroleum Institute)

[4] American Petroleum Institute 2000 *API PUBL 581 Risk-Based Inspection Base Resource Document* (Washington DC: American Petroleum Institute)

[5] American Petroleum Institute 2016 *Risk-based Inspection Methodology* (Washington DC: American Petroleum Institute)

[6] Norsok 2017 *Norok Z-008 Risk-Based Maintenance and Consequence Classification* (Oslo: Norwegian Technology Centre)

[7] Moubray J 1997 *Reliability-Centered Maintenance II* (New York: Industrial Press)

[8] Høyland A. and Rausand M 1994 *System Reliability Theory Models and Statistical Methods* (New York: John Wiley & Sons, Inc) p 183-204

[9] Det Norske Veritas 2002 *DNV-RP-G101 Risk-Based Inspection of Offshore Topsides Static Mechanical Equipment* (Norway: Det Norske Veritas) p 22-24
[10] Det Norske Veritas 2017 *DNV-RP-G101 Risk-Based Inspection of Offshore Topsides Static Mechanical Equipment* (Norway: Det Norske Veritas) p 98-103
[11] Priyanta D 2016 *Asian J of Appl. Sci.* The Development of Equipment Criticality Analysis (ECA) Protocols of Offshore Carbon Steel Static Mechanical Equipment **04** 1258-66
[12] Ratnayake R M C 2014 *J of Loss Prev. in the Process Ind.* Application of a fuzzy inference system for functional failure risk rank estimation: RBM of rotating equipment and instrumentation **29** 216-24
[13] Abimbola M, Khan F, Khakzad N, and Butt S 2015 *Safety Science* Safety and risk analysis of managed pressure drilling operation using Bayesian network **76** 133-44
[14] Jagtap H P and Bewoor A K 2017 *Materialstoday: Proc.* Use of Analytic Hierarchy Process Methodology for Criticality Analysis of Thermal Power Plant Equipments (Hyderabad: Elsevier) **4** p 1927-36
[15] Kokangül A, Polat U, and Dağsuyu C 2017 *Safety Sci.* A new approximation for risk assessment using the AHP and Fine Kinney methodologies **91** 24-32
[16] American Petroleum Institute 2016 *API 580 Risk-Based Inspection* (Washington DC: American Petroleum Institute)
[17] Chang M K, Chang R R, Shu C M, and Lin K.N 2005 *J of Loss Prev. in the Process Ind.* Application of risk-based inspection in refinery and processing piping **18** 397-402
[18] Zhaoyang T, Jianfeng L, Zongzhi W, Jianhu Z, and Weifeng H 2011 *Safety Sci.* An evaluation of maintenance strategy using risk-based inspection **49** 852-860
[19] Moura M d C, Linsa I D, Droguett E L, Soares R F, and Pascual R 2015 *Reliability Engineering & System Safety* A Multi-Objective Genetic Algorithm for determining efficient Risk-Based Inspection programs **133** 253-265
[20] International Electrotechnical Commission 2010 *IEC 61508 Functional safety of electrical / electronic / programmable electronic safety-related systems* (Geneva: International Electrotechnical Commission)
[21] Innal F, Dutuit Y, and Chebila M 2015 *Reliability Engineering & System Safety* Safety and operational integrity evaluation and design optimization of safety instrumented systems **134** 32-50
[22] Kaczor G, Młynarski S, and Szkoda M 2016 *J of Loss Prev. in the Process Ind.* Verification of safety integrity level with the application of Monte Carlo simulation and reliability block diagrams **41** 31-39
[23] Catelani M, Ciani L, and Luongo V 2011 *Microelectronics Reliability* A simplified procedure for the analysis of Safety Instrumented Systems in the process industry application **51** 1503-07
[24] Rausand M 1998 *Reliability Engineering & System Safety* Reliability Centered Maintenance **60** (2) 121-132
[25] Society of Automotive Engineers 2009 *SAE JA1011 Evaluation Criteria for Reliability-Centered Maintenance (RCM) Processes* (Pittsburgh: Society of Automotive Engineers)
[26] Society of Automotive Engineers 2011 *SAE JA1012 A Guide to Reliability-Centered Maintenance (RCM) Standard* (Pittsburgh: Society of Automotive Engineers)
[27] Selvik J T and Aven T 2011 *Reliability Engineering & System Safety* A framework for reliability and risk centered maintenance **96** (2) 324-331
[28] Gupta G and Mishra R P 2018 *Procedia CIRP* Identification of Critical Components Using ANP for Implementation of Reliability Centered Maintenance **69** 905-909
[29] Yssaad B, Khiatb M, and Chaker A 2014 *Int. J of Electrical Power & Energy Systems* Reliability centered maintenance optimization for power distribution systems **55** 108-115
[30] Piasson D, Bíscaro A A P, Leão F B, and Mantovanic J R S 2016 *Electric Power Systems Research* A new approach for reliability-centered maintenance programs in electric power

distribution systems based on a multiobjective genetic algorithm **137** 41-50
[31] International Organization for Standardization 2016 *ISO 14224 Petroleum, petrochemical and natural gas industries -- Collection and exchange of reliability and maintenance data for equipment* (Geneva: International Organization for Standardization)

Energy harvesting capability of single-chamber oscillating water column wave energy device model on controlled wave height and period

Aulia Achmad[1,2] **and F Husain**[1,2]

[1]Ocean Engineering Department, Faculty of Engineering, Hasanuddin University, Makassar, Indonesia
[2]Ocean Technology Laboratory, Ocean Engineering Department, Faculty of Engineering, Hasanuddin University, Makassar, Indonesia

Abstract. The urgency of electricity demand and environmentally friendly power generator made ocean wave energy as one of many renewable energies that could be implemented in Indonesia because many of its coastlines have the power enough to generate electricity locally and regionally. This research is mainly discussing the model's capability to generate electricity based on various wave height and period. This energy harvester model made in the laboratory with 1:16 scale, so the other variable component also following this scale. The turbine used in this experiment using Wells turbine with seven blades following the profile guidelines of NACA 0024. The result of this research shows that the power generated by the turbine correlates to wave height and wave period. The output generated by this energy harvester model could achieve 21.2 kilowatts in one hour on average or 339.4 kilowatts in one hour if upscaled to its real size.

Keywords: Energy Harvester, Renewable Energy, Oscillating Water Column.

1. Introduction

Southeast Asian countries are facing a 'trilemma'—the need to deliver on energy security, economic growth, and development in a sustainable way [1]. As a member of Southeast Asian countries, Indonesia also facing the same problem regarding energy security as mentioned by Lelly et al. [2016], Indonesia's electricity demand grows each year alongside the population growth [2].

Not only to fulfill the electricity demands in the big cities and industrial areas, but Indonesia is also required to satisfy the electricity needs in the 3T (frontier, outermost, and least developed) areas. As one of the largest archipelagic countries, Indonesia has many islands in which most of them are remote but still inhabited by locals. Those inhabited remote islands almost always have the same problems: coastal erosion and limited electricity. OWC devices can be the solution to these problems because this device can also act as a breakwater to protect the area behind it while also harvesting the energy from the waves[10].

The energy contained in the oceans varies greatly, some of which are ocean current energy, wind energy, and tidal and wave energy. Wave energy is adequately used in electricity generation because the power potential contained in wave energy is promising. The remarkable property of water waves, once created by the wind, to propagate energy over very long distances with little loss has naturally given rise to the desire to harness this energy and convert it to a useful form [5].

IOP Conf. Series: Materials Science and Engineering 1052 (2021) 012043 doi:10.1088/1757-899X/1052/1/012043

As briefly mentioned earlier, to convert wave energy into electrical power, one can use an Oscillating Water Column (OWC) system. This system includes three energy converting stages: 1) the OWC inside a chamber forces air alternately into and out of the atmosphere through the duct. 2) a turbine with symmetric blades transforms the bi-directional airflow energy into torque. 3) an electricity generator linked to the turbine transforms the torque into electrical power [3]. This device also can use a water turbine to extract wave energy [9].

The efficiency of oscillating water column (OWC) wave energy devices equipped with Wells turbines is particularly affected by flow oscillations basically for two reasons. First, because of the intrinsically unsteady (reciprocating) flow of air displaced by the oscillating water free surface. Second, because increasing the air flow rate, above a limit depending on, and approximately proportional to, the rotational speed of the turbine, is known to give rise to a rapid drop in the aerodynamic efficiency and in the power output of the turbine. There are plenty of solutions to overcome this obstacle, some of which are determining the turbine's inlet diameter, creating an auxiliary outlet or air valves to control excess air flow, determine the blades' profile, and so on [7]. But these are not the topic of the present study.

The main purpose of this work is to analyze the single-chamber OWC harvesting capability on various wave height and period. Tested in laboratorium conditions, and using in-house built OWC model and 3D printed bespoke Wells turbine model. The tests consist the total of twenty-two variables of wave height and period to observe the energy harvested by the device. The forces acting on the turbine also analyzed using CFD software to calculate the torque produced by the turbine.

2. Experimental study

2.1. OWC chamber model

The OWC model has slanted front and rear side intended to reduce wave reflection and ease the wave force received by the model so it won't break mid testing. The OWC model fixed to a seating mount so it won't move while tested with stronger waves. A schematic of the OWC chamber is shown in Figure 1, where Lf denotes the chamber width, Ds is the draft of the chamber skirt, α is the angle of the chamber's slope, Dw is the still water depth, W is the length of the chamber, and hf is the height of the model. The dimensions of the chamber are summarized in Table 1. The incident wave conditions employed by this experiment are summarized in Table 2. H_m denotes the incident model wave height, H_p is incident full scale wave height, T_m and T_p are for incident wave period at model and full scale respectively.

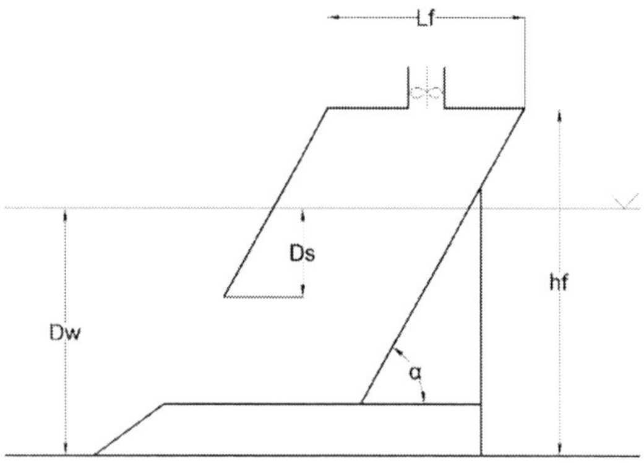

Figure 1. Schematics of the oscillating water column energy device model.

Figure 2. Side profile of the oscillating water column energy device model.

IOP Conf. Series: Materials Science and Engineering 1052 (2021) 012043 doi:10.1088/1757-899X/1052/1/012043

Table 1. OWC device dimensions.

	Lf	Ds	Dw	hf	W	α
Model Scale	0.48 m	0.22 m	0.6 m	0.82 m	0.96 m	60°
Full Scale	7.68 m	3.52 m	9.6 m	13.12 m	15.36 m	60°

Table 2. Summary of incident wave conditions setup

(a) Incident wave height											
Model Scale, H_m	0.15m	0.075m									
Full Scale, H_p	2.4m	1.2m									
(b) Incident wave period											
Model Scale, T_m	1.4s	1.5s	1.6s	1.7s	1.8s	1.9s	2.0s	2.1s	2.2s	2.3s	2.4s
Full Scale, T_p	5.6s	6.0s	6.4s	6.8s	7.2s	7.6s	8.0s	8.4s	8.8s	9.2s	9.6s

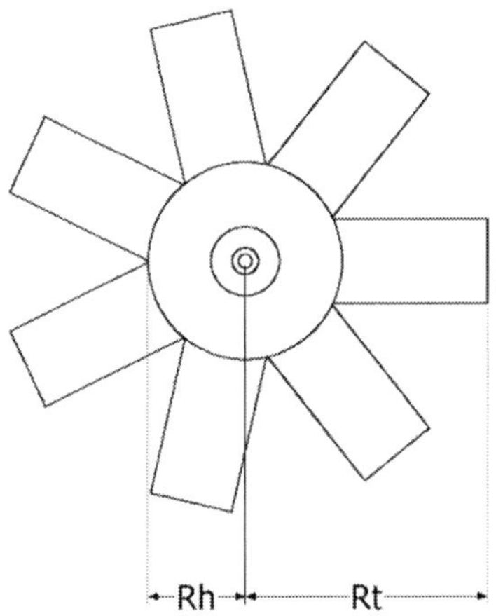

Figure 3. Schematics of Wells turbine rotor

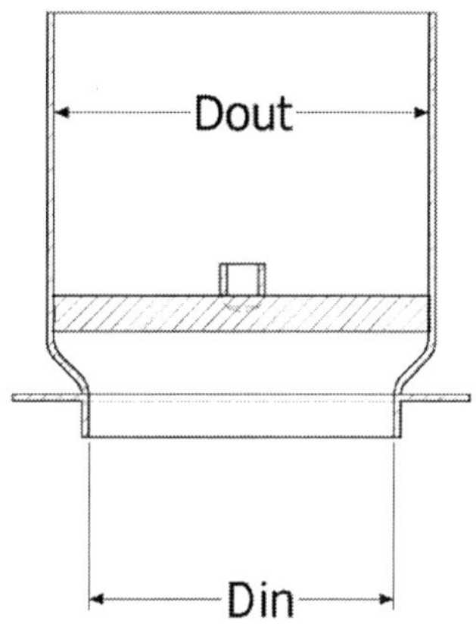

Figure 4. Schematics of the turbine housing

Table 3. Summary of turbine dimensions

(a) Turbine rotor dimensions		
	Rt	Rh
Model Scale	50 mm	20 mm
Full Scale	800 mm	320 mm
(b) Turbine housing dimensions		
	Din	Dout
Model Scale	90 mm	110 mm
Full Scale	1440 mm	1760 mm

2.2. Wells turbine model

The Wells turbine model made using 3D printed ABS plastic filament. The rotor's blade designed following NACA 0024 profile and having seven blades. The study for blade profile and number has been conducted by Bagus et al. [2015], analyzing the optimum blade profile and number to be used in OWC devices. Schematics of turbine rotor and housing shown in Figure 3 and Figure 4 respectively, Rh denotes the hub radius, and Rt is tip radius. In Figure 4, Din denotes the inlet diameter, and Dout is outlet diameter. The dimensions of both schematics are summarized in Table 3.

Wells turbine, which was invented by Prof. Alan Arthur Wells is the most common type of self-rectifying air turbine employed by OWC wave energy devices due to its technical simplicity, reliability, high rotational speed, and design robustness [6]. The Wells turbine is an axial-flow air turbine composed of a rotor with fixed pitch symmetrical airfoil (commonly four digits, double zero, NACA profiles) blades staggered at 0° relative to the plane of rotation around the hub [6].

Due to the use of symmetrical airfoil blades about the chord line, Wells turbine rotates in the same direction by OWC's cyclically reversing outllet flow. Figure 5 illustrates the forces acting on a blade of a Wells turbine and also the basic principle of unidirectional rotation of the Wells turbine.

Where V_A indicates the inlet velocity, V_T is the circumferential velocity at mean radius and W_R is the relative airflow velocity. Angle of incidence, α, is the angle between W_R and blade chord, F_L is the lift force normal to the W_R, and F_D is the drag force parallel to W_R. The forces F_L and F_D are resolved into the axial force, F_A and tangential force, F_T, where F_T direction is independent of the axial flow direction [6].

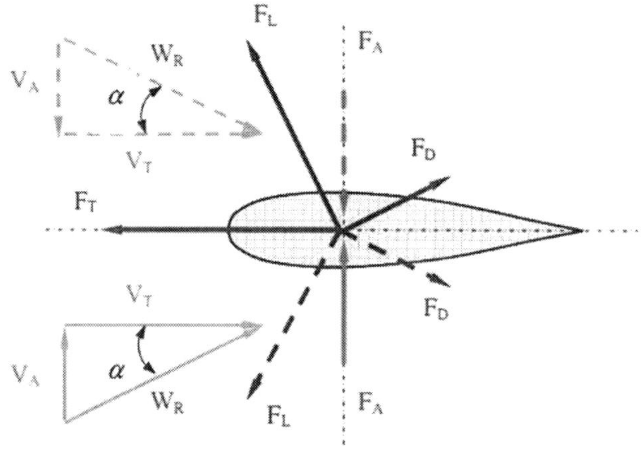

Figure 5. Aerodynamic forces acting on a blade of a Wells turbine rotor [6]

3. Methods

The experiment was conducted in the 24 m long, 1 m wide, and 1.2 m deep wave flume of the Ocean Engineering Department, Engineering Faculty of Hasanuddin University. The OWC model is made of 120 mm thick plywood and polymethyl methacrylate (PMMA) acrylic acting as the front plate. Incident and transmitted waves are measured with five wave probes starting from the front of the wavemaker to the front of the model. The testing variables are as shown in Table 2.

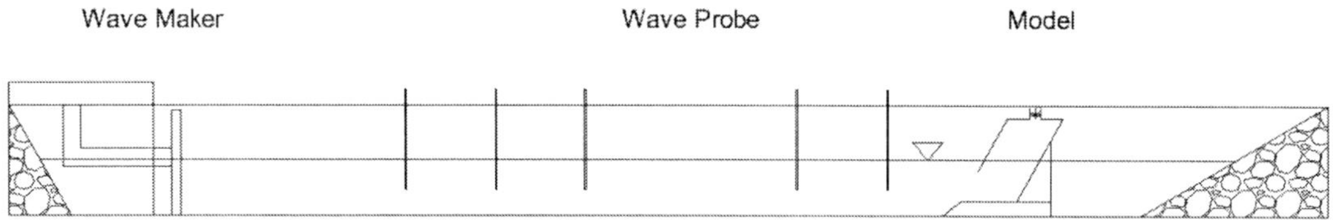

Figure 6. Experiment setup

After wave conditions have been set for the OWC device model inside the flume tank, the experiment could begin. From the wave data, the energy harvest rate could be determined. The wind speed data obtained by using an anemometer near the turbine inlet to determine the torque of the rotor and numerically analyze the force acting on each blades using Computational Fluid Dynamics (CFD) software. The power generated by this model is obtained using an electrical multitester device connected to the turbine's dynamo to obtain voltage, and electric current to be used to calculate the power generation.

4. Experiment data processing

The experiment data summarized in Table 4 and Table 5 are showing the energy harvested by this OWC model, also the lift and torque of the turbine. T denotes wave period, Hi is the incident wave height in running conditions, λ is the wavelength, P_W is the wave power captured by the OWC model, U is wind speed, F_L is lift force of the rotor's blade, τ is the torque of the turbine, and P_E is the electrical power generated by the turbine.

The incident wave in running conditions obtained by measuring the highest and the lowest waveheight in front of the model using wave probe sensor to later calculated using this formula,

$$Hi = \frac{Hmax + Hmin}{2} \tag{1}$$

the wave length obtained from iterating the formula,

$$L = \frac{gT^2}{2\pi} tanh \frac{2\pi d}{L_0} \tag{2}$$

the still water depth is denoted by d, and L_0 can be obtained using the following formula,

$$L_0 = 1{,}56T^2 \tag{3}$$

for wave power can be calculated by using this formula,

$$P_W = \left(\frac{\rho g a^2 . W}{2}\right) . \left(\frac{\lambda}{T}\right) \tag{4}$$

to calculate the torque of a blade, this formula is used,

$$\tau = F_\emptyset . R \tag{5}$$

where Fø is the tangential force, calculated using,

$$F_\emptyset = F_L \sin \alpha - F_L \cos \alpha \tag{6}$$

where,

ρ = Sea water density (kg/m^3)
g = Gravitational acceleration (m/s^2)
a = Amplitude of incident wave (m)
W = Length of the model chamber (m)
λ = Wave length (m)
T = Wave period (s)
F_L = Lift force (N)
R = Blade length (m)
α = Angle of attack

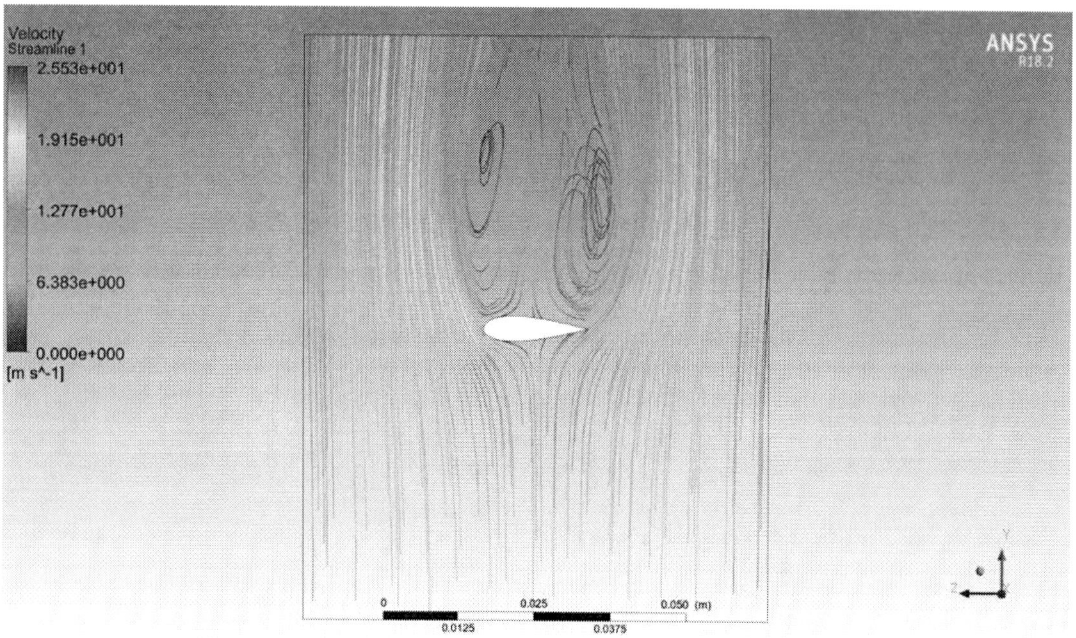

Figure 7. CFD simulation for turbine blade

The lift force for each blade could be acquired using ANSYS Fluent with predetermined wind speed variables from the experiment conducted beforehand. Figure 7 visualise the wind forces acting on one blade. The torque shown below in Table 4 and Table 5 are already in total from all seven blades.

Table 4. Summary of OWC model test result with 0.15 m wave height

T (s)	Hi (m)	λ (m)	P_W (Watt)	U (m/s)	F_L (N)	τ (Nm)	P_E (Watt)	P_E/P_W (%)
1.4	0.16	2.71	59.61	1.82	2×10^{-4}	4.47×10^{-5}	0	0
1.5	0.17	2.99	72.99	3.59	8.65×10^{-4}	1.93×10^{-4}	2×10^{-3}	2×10^{-3}
1.6	0.18	3.27	76.55	5.59	2.3×10^{-3}	5.14×10^{-4}	0.55	0.72
1.7	0.17	3.55	73.46	7.41	4.3×10^{-3}	9.59×10^{-4}	0.68	0.92
1.8	0.14	3.82	53.36	8.16	5.31×10^{-3}	1.19×10^{-3}	0.26	0.49
1.9	0.25	4.09	161.63	10.48	9.19×10^{-3}	2.05×10^{-3}	1.21	0.75
2.0	0.17	4.36	77.33	9.60	7.58×10^{-3}	1.69×10^{-3}	0.45	0.58
2.1	0.28	4.63	210.81	14.92	1.99×10^{-2}	4.45×10^{-3}	5.41	2.57
2.2	0.18	4.89	87.66	10.45	9.13×10^{-3}	2.04×10^{-3}	0.07	0.08
2.3	0.33	5.15	291.77	14.16	1.75×10^{-2}	3.91×10^{-3}	5.89	2.02
2.4	0.20	5.41	108.32	13.11	1.49×10^{-2}	3.32×10^{-3}	0.53	0.49

Table 5. Summary of OWC model test result with 0.075 m wave height

T (s)	Hi (m)	λ (m)	P_W (Watt)	U (m/s)	F_L (N)	τ (Nm)	P_E (Watt)
1.4	0.069	2.71	11.16	1.32	1.05×10^{-4}	2.34×10^{-5}	0
1.5	0.090	2.99	19.70	2.90	5.43×10^{-4}	1.21×10^{-4}	0
1.6	0.081	3.27	16.28	4.07	1.14×10^{-3}	2.55×10^{-4}	0
1.7	0.077	3.55	15.19	4.41	1.36×10^{-3}	3.04×10^{-4}	0
1.8	0.069	3.82	12.31	4.93	1.74×10^{-3}	3.89×10^{-4}	0
1.9	0.098	4.09	24.89	7.46	4.36×10^{-3}	9.73×10^{-4}	0
2.0	0.070	4.36	13.02	5.56	2.27×10^{-3}	5.07×10^{-4}	0
2.1	0.120	4.63	38.75	9.30	7.07×10^{-3}	1.58×10^{-3}	0
2.2	0.071	4.89	13.70	6.36	3.07×10^{-3}	6.84×10^{-4}	0
2.3	0.093	5.15	23.37	8.53	5.85×10^{-3}	1.31×10^{-3}	0
2.4	0.094	5.41	23.99	8.82	6.29×10^{-3}	1.41×10^{-3}	0

5. Results and conclusions

After forty-four different testings, the result of this experiment can be concluded. The wave energy harvest rate by this OWC model is shown in Figure 8. From the graph, we can see there are "spikes" in the fluctuation of energy harvested, and every 0.1-second difference in the period, the energy fluctuates. The highest energy harvest ratio achieved at period 2.1 second with a 2.57 % conversion rate from wave power to electricity.

For harvesting capability on 0.075 m wave height setup, this model with current design iteration and testing variables could not convert any wave power into electricity. Another design approach is necessary for future research, such as reducing the turbine diameter, widens the chamber width, or change the model into multi-chamber type OWC device.

The torque achieved in Figure 9 could translate into how significant the conversion rates are. The highest torque achieved was 4.45×10^{-3} Nm in 2.1 s wave period, the conversion ratio from Figure 8 also achieved peak conversion in 2.1 s wave period. This could be mean that current blades profile is the most effective on converting high wind speed and the power produced doesn't increase linearly in regards to wave periods, hence the need of inlet and turbine diameter adjustments, blades profile redesign, and oscillating chamber reconfiguration for future experiment.

Thus concludes this study, current design iteration still has its flaws and still have many areas that could be improved, but, according to a study conducted by Utami [2010], on Southern of Banten through Southern of East Java coastlines, this model with current design could still convert some energy although very little, theoretically.

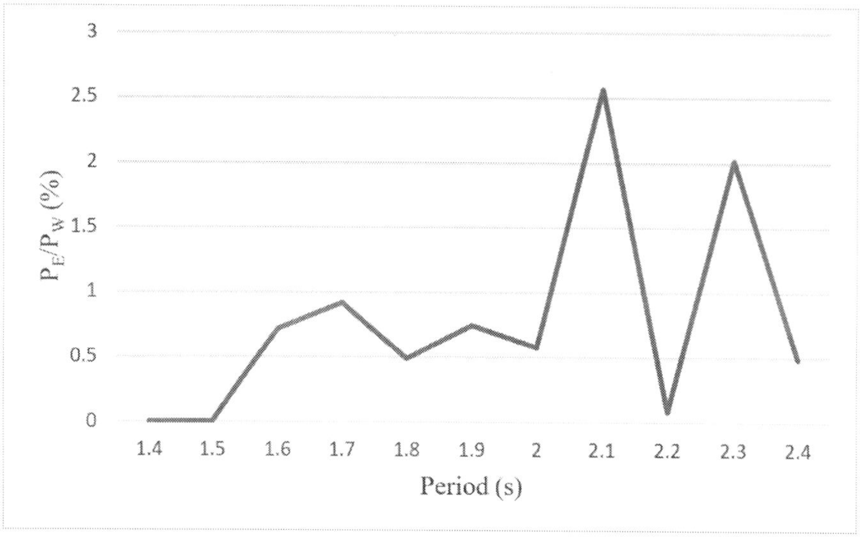

Figure 8. Oscillating water column model harvesting capability on 0.15 m wave height

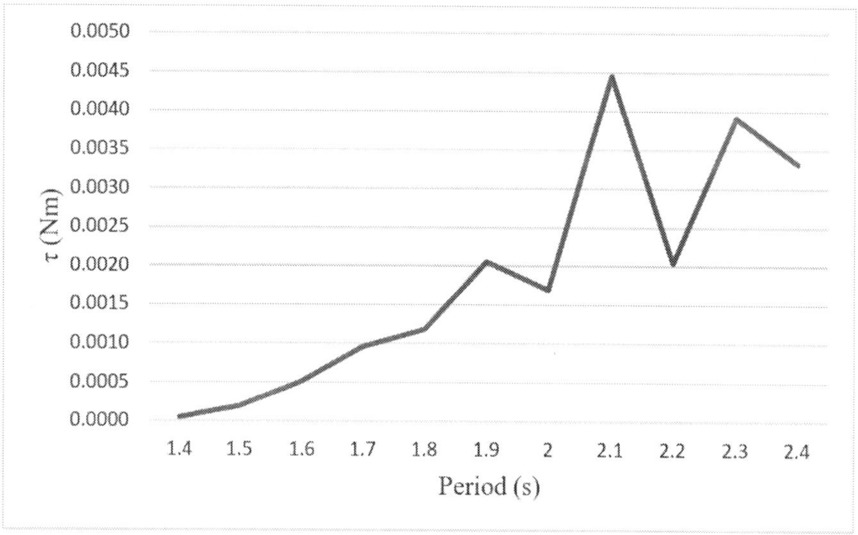

Figure 9. Maximum torque achieved on every wave period (0.15 m wave height)

Acknowledgements
Authors wishing to acknowledge for support from Faculty of Engineering Hasanuddin University and co-author Dr. Eng. Firman Husain, ST. MT. for personally funding this study.

References

[1] Paul A J Bonar, Andrea M Schnabl, Wei-Koon Lee and Thomas A A Adcock 2018 *J. Ocean Engineering and Marine Energy* Vol. 4, pp. 99–109
[2] Lelly Erlita Safitri, Ishak M Jumarang and Apriansyah 2016 *POSITRON* Vol. VI, pp. 08–16
[3] Jiyuan Jin, Beom-Soo Hyun, Keyyong Hong and Zhen Liu 2014 Performance of Oscillating Water Column type Wave Energy Converter in Oblique Waves *J. Korean Society for Marine Environment and Energy* Vol. 17, No. 3 pp. 182–188
[4] Bagus Prasetio, Deddy Chrismianto and Muhammad Iqbal 2015 *J. Teknik Perkapalan* Vol. 3, No. 4 pp. 371–382
[5] Stiassnie Michael, Kadri Usama and Stuhlmeier Raphael 2016 Harnessing Wave Power in Open Seas *J. Ocean Eng. Mar. Energy* Vol. 2, pp. 47–57
[6] Rasool Soltanmohamadi and Esmail Lakzian 2015 Improved design of wells turbine for wave energy conversion using entropy generation *Meccanica* (Springer)
[7] Falcao A F de O and Justino P A P 1999 *Ocean Engineering* Vol. 26, pp. 1275–1295
[8] Siti Rahma Utami 2010 Studi Potensi Pembangkit Listrik Tenaga Gelombang Laut Dengan Menggunakan Sistem Oscillating Water Column di Tiga Puluh Wilayah Perairan Indonesia (Universitas Indonesia)
[9] F Husain, K Inouchi and T Nakamura 2014 Development of a Wave Power Extraction Seawall and Its Effectiveness for Wave Dissipation *Journal of Japan Society of Civil Engineers Ser. B2*, Vol. 70, pp. 1311-1315
[10] Y Torre-Enciso, I Ortubia, L I López de Aguileta and J Marqués 2009 Mutriku Power Plant: from the thinking out to the reality *Proceedings of the 8th European Wave and Tidal Energy Conference* (Uppsala University)

Analysis of the usage of the android-based scantling calculation application for wooden ship construction for survey activity based on BKI vol vii 2013 rules

Sabila Fitri Afsari[1], and Heri Supomo[1]

[1]Naval Architecture Department, Sepuluh Nopember of Institute Technology, Surabaya, Indonesia.

Abstract. In survey activities at wooden shipyards, surveyors often find scantling calculations not comply with the regulations. Because the use of technology has not yet developed, based on traditional technology development so that the weakness occurs in the design and construction. The survey system for scantling calculation is yet manually, surveyors write actual results and carry documents in paper form which is less efficient. Designing media in the form of android application for calculating wooden ship scantling based on BKI Volume VII 2013 rules concerning *'Rules for Small Vessel Up to 24 M'* is needed. The initial step of a design application is creating database structure that normalization BKI table become a system, then, making a mock up for display. After that, create functional as back end system that is processing database and front end that is screen display, and the last step is creating system for reporting result. This application accessed by input the actual data of wood type, mechanical properties and main dimension on each form, then the application will display the results of BKI calculations. The surveyor fills actual data of scantling in column and gives mark if the results comply and will gives a note if not complied then insert a photo to prove. The PELRA ship with length overall (LOA) of 17.65 meters used as the research object. The questionnaire is applied to know response of user about the application. The results, a total percentage is 75.6% to average total score is 37.8 which shows that the application very helpful to assist survey activities. The conclusion, the use of android application for scantling calculation of wooden ship construction is needed because more practical and can accessed anywhere compared to manual survey systems, more efficient in terms of time, and more safety for data storage of report.

1. Introduction

In survey activities at wooden shipyards, surveyors often find scantling calculations not comply with the regulations. One of the reasons during the shipbuilding process, the calculation procedure is carried out only by relying on the ability of the workers where most of the workers do not have an educational background in naval architecture so they do not follow scantling of BKI rules. The design process of the current structure far exceeds the compliance effort. But what happens to the traditional ships technology is the use of technology has not yet developed, based on traditional technology development so that the weakness occurs in the design and construction due to motorization of ship propulsion. Therefore, in order to improve ship design in the future it must be able to know the importance of realizing the best standard of ship design, because ship should be in accordance with the required standards and with the classification standards [1].

Survey system technology for scantling calculation of wooden ship construction is yet limited, surveyor did survey manually it caused the survey and reporting less practical and efficient. In a manual system surveyor writes actual scantling results, in addition, surveyors must carry documents in paper form which is vulnerable and will have an impact on the report if the notes cannot be read. This is can affect flexibility in survey activity in shipyard.

With development of mobile network, people want to achieve more tools software implementation on mobile phones for convenience. Android mobile phone platform is open enough to meet the requirements [2]. Android is a mobile system that is based on modified version if Linux, it was originally developed by a start up of the same name, Android Inc. The main advantage of adopting Android is that offers a unified approach to application development. Developers need only develop for Android, and their applications should be able to run on numerous different devices, as long as the devices are powered using Android [3]. Android platform provides developers huge opportunity to develop applications to cater to the needs of its ever increasing user base. The Android platform includes applications, middleware and the operating system. Developers use the AndroidSDK, which consist of various tools and Application Programming Interfaces (APIs) to develop applications using programming language like Java. Android provides an open marketplace wherein developers can sell and distribute applications instantly [4]. With the routing Android mobile device has access to multiple applications and is available for the latest system updates, greater aesthetic enhancements to the operating system, increased device speed, increased battery life and many more [5].

To assist the surveyor's work, design a media in the form of android-based application is needed. Design android application for calculating wooden ship scantling is based on BKI Volume VII 2013 rules concerning *'Rules for Small Vessel Up to 24 M'*. The rules apply to pleasure craft of length from 6 to 24 m and provided that the pleasure craft classed and approved in accordance therewith are at all times employed exclusively under the conditions for which they have been designed, constructed and approved and that they are in the sense of good seamanship correctly handled and equipped and operated at a speed adopted to the respective seaway conditions. These rules apply to the scantling determination of the structure of sailing craft, motorsailers and motor yachts of normal monohull form, traditionally carvel or clinker built on transverse frames [6].

Only proven boatbuilding wood shall be used for all timber components exposed to water and weather, i. e. timber with good resistance to water and weather, fungal attack and insect infestation, as well as with good mechanical properties that are also suitable for the particular application. Furthermore, it shall have a low swelling and shrinkage properties. The timber divided into durability group from I to V, whereby group I very good, group II good, group III average, group IV moderate, and group V poor [7]. The timber used in boatbuilding shall, if exposed to the weather or used for the primary structural components of a boat, belong to at least durability group III. The aim of this research is to analyze usage of the android-based scantling calculation application for wooden ship construction for survey activity based on BKI. Determination of the component scantlings of structural members shall be carried out in accordance with excerpt tables respecting the scantling numerals in equations (1) or (2):

$$\frac{B}{3} + H \tag{1}$$

$$\mathrm{L}\,(\frac{B}{3} + H) \tag{2}$$

The scantling length is determined as follows:

$$L = \frac{L_H + L_{WL}}{2} \tag{3}$$

Structural members of hulls with larger dimensions or unusual proportions shall have the scantlings determined by individual calculations. The length L_H in [m] of the hull is the horizontal distance between the foremost and the aftermost part of the craft, the length includes structural and integral parts of the craft. The waterline length in [m] is the distance between the foremost and the aftermost intersections of the hull with the flotation plane [6].

Design application is closely related by database; all data must create to be a database and become one system. A database is an ordered collection data, which is normally stored in one or more associated files. The data are structured as tables, where cross references among tables are possible. In this research all database is processing in MySQL My SQL is example relational database system. Such a system includes the programs for managing relational databases. Among the tasks of a relational database system are not only the secure storage of data, but also such jobs as the processing of commands for querying, analyzing, and sorting existing data and for storing new data [8]. The advantage using MySQL is open source support, low overhead, available large table size for database and all feature have good stability [9].

2. Method

The initial step of a design application is creating database structure that normalization table become a system, the data of database is containing scantling calculation from BKI 2013. Then, making a mock up that visualizes the appearance of the application. After that, create functional as back end system that is processing database. The functional is get request data from database and posting data by sending request data to database. The next step is creating front end system, about button, display content from back end system and screen display for user, front end must create synchronic with back end system. The last step is creating system for reporting the result. The figure 1 is all of database that create in MySQL.

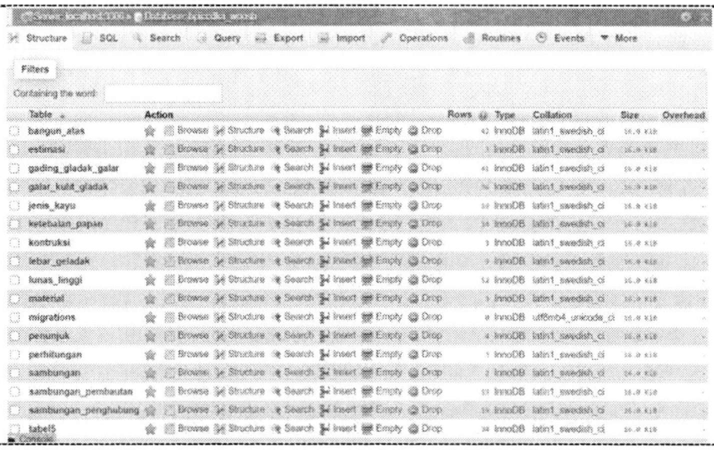

Figure 1. Database in MSQL.

Object of this research is the PELRA ship with length overall (LOA) of 17.65 meters, length water line (LWL) 13.13 meters, breadth (B) 4 meters, height (H) 1.84 meters, draft (T) 1.02 meters and speed (Vs) 7 knot. For validation application, the questionnaire method with rating scale answer is applied to know the user response about the application. The application tested by 5 respondents, that is wooden shipyard workers. Each of respondent fills a questionnaire that containing 10 questions. List question of questionnaire are:

1. Is this android application enough to help surveyors works?
2. Is the use of this application system more practical when compared to the manual system in the shipyard?
3. Are the features in the Android application sufficient to current needs?
4. Is this application needed for survey activities?
5. Does this android application provide significant benefits?
6. Is the time using this android application efficient enough?
7. What is the level of completeness of the features in this application?
8. Is the level of safety in data storage better than the manual system?
9. Is this the application comfortable to use?
10. Is the appearance of this application interesting enough?

Answer from each question have 5 rating scale, that is:

1 (Strongly Disagree)
2 (Disagree)
3 (Neither Agree nor Disagree)
4 (Agree)
5 (Strongly Agree)

IOP Conf. Series: Materials Science and Engineering 1052 (2021) 012044 doi:10.1088/1757-899X/1052/1/012044

Each respondent can provide a total rating of 10 to 50 with the following information:

1. 1-10 : Not helpful at all
2. 11-20 : Not very helpful
3. 21-30 : Moderately helpful
4. 31-40 : Very helpful
5. 41-50 : Extremely helpful

User of this application is surveyor. First step when user finished register then input all the actual data of ship. The actual data are result of material testing that been tested in laboratory, main dimension, and scantling construction and connection data of ship. If user input actual data, the application displays result mechanical properties of material, scantling, and connection from BKI 2013 automatically. After calculating, all of data will recapitulated in history and then history will have recapitulated to the database automatically. To describe how the application can run in figure 2 occur application flow scheme.

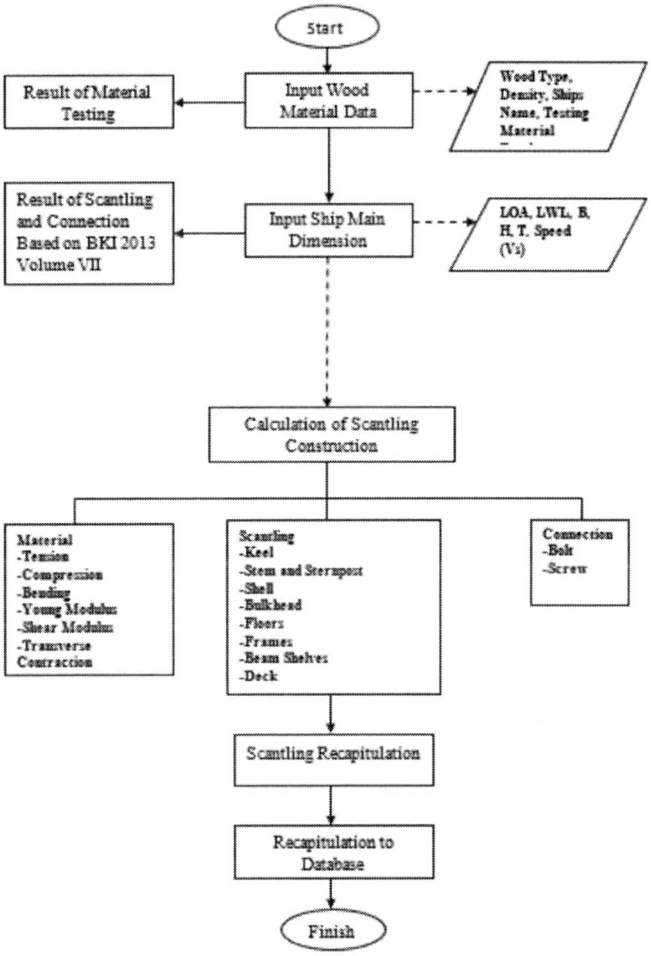

Figure 2. Application Flow Scheme.

3. Discussion and Result

First step, user must register and input personal data for make an account. There is user name, password, user address, phone number and e-mail address. The application will display the main menu see figure

4, then user choose survey menu. Second, user choose material menu and choose type of wood then the application displays standard mechanical properties from BKI that is durability group, density, and testing standard as a tension test, compression test, bending test see figure 6, and user can input the actual data mechanical properties of material. Third, the application automatically changed into scantling menu. User must input the actual main dimension of ship see figure 7, then scantling result from BKI will displayed and user can input all actual scantling data of construction structure member as figure 8 until figure 12. Forth, the connection form displayed automatically then user input the actual connection data see figure 13 and figure 14. The surveyor can give a check mark in check box, attach a picture, information or note in note column. The end of each menu user must push the save button to save all data. The data can have reviewed later in history menu. figure 3 until figure 15 below display how the application work:

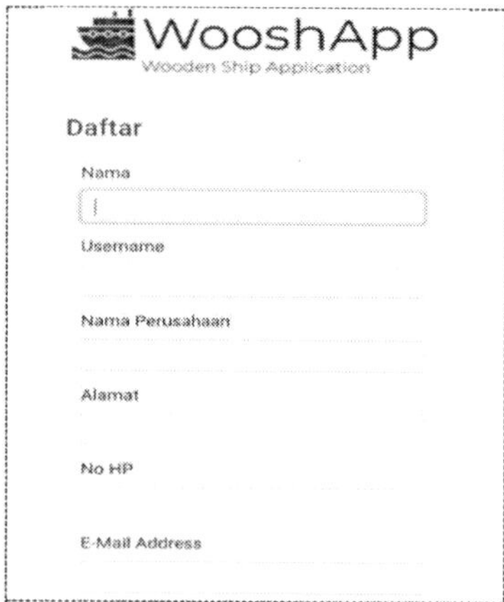

Figure 3. The user must input personal data in register menu. Then user have an account to access the application.

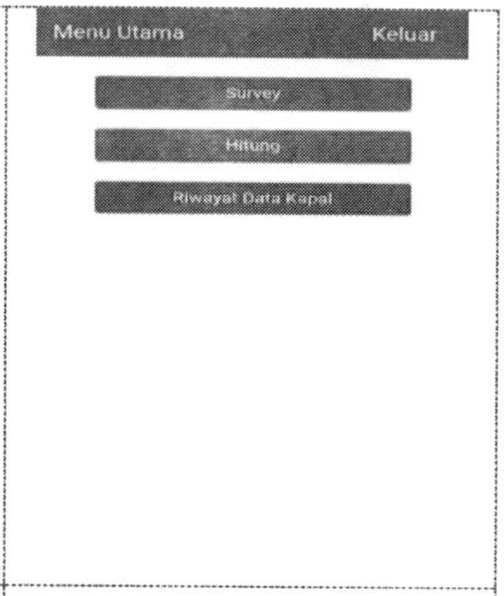

Figure 4. The main menu there are survey, calculate, and history. The history save all personal and calculate data by the ship project.

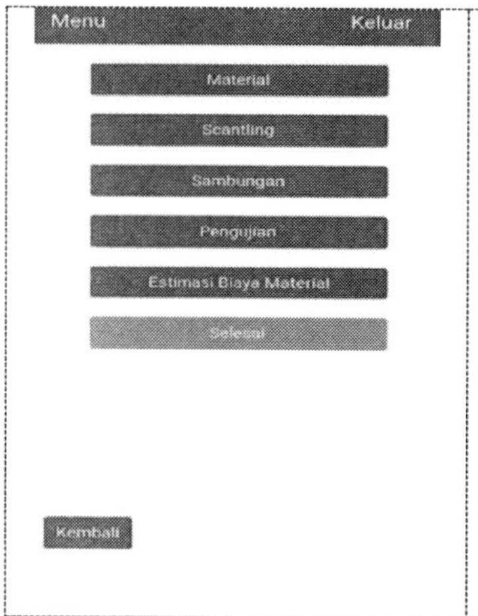

Figure 5. In survey menu, user must choose the material menu first to input the mechanical properties of wood.

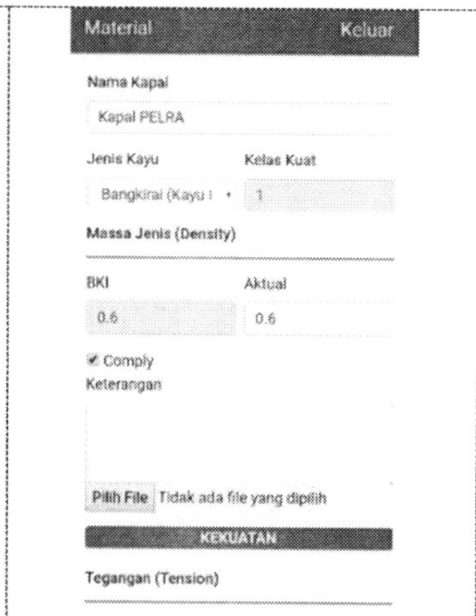

Figure 6. In material menu, user only input the name of ship and wood type. Then mechanical properties of wood will displayed.

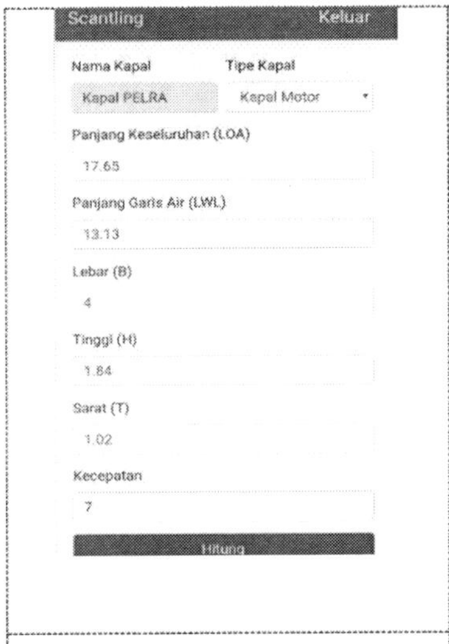

Figure 7. After input mechanical properties to calculate scantling the member of construction, user must input actual data of the main dimension of ship.

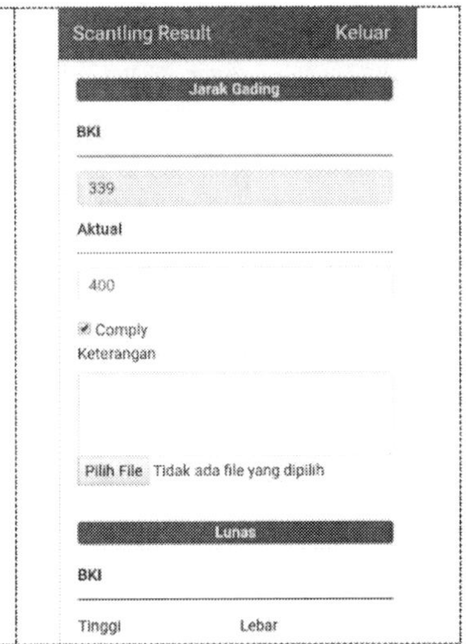

Figure 8. The result of frame space from BKI and actual data. The frame space complying the standard of BKI.

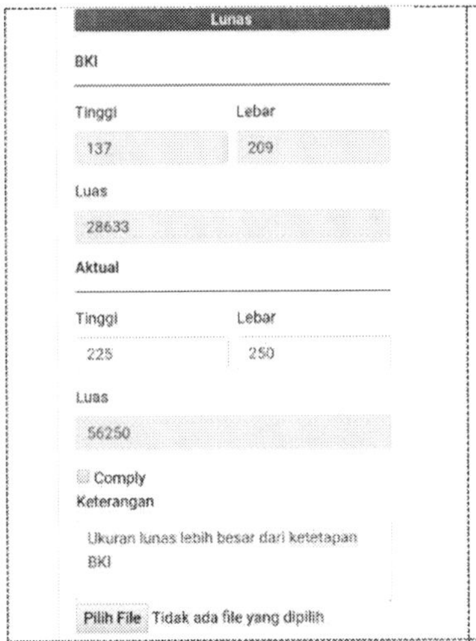

Figure 9. Result of keel calculation from BKI 2013 standard and the actual data from ship.

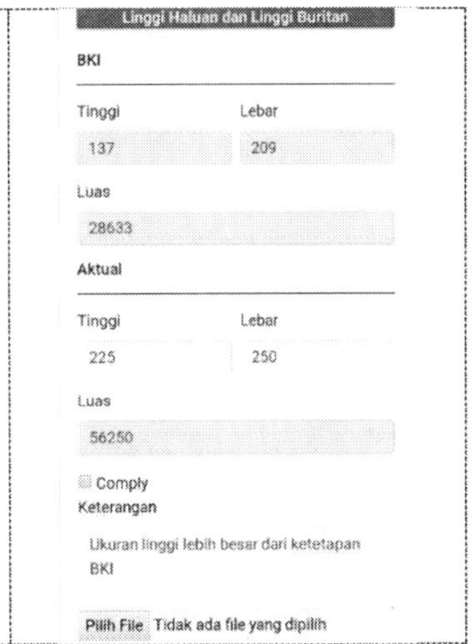

Figure 10. Result from application for stem and sternpost in actual data and BKI 2013 standard.

IOP Conf. Series: Materials Science and Engineering 1052 (2021) 012044 doi:10.1088/1757-899X/1052/1/012044

Figure 11. The result from superstructure side wall that is complying the BKI 2013 standard.

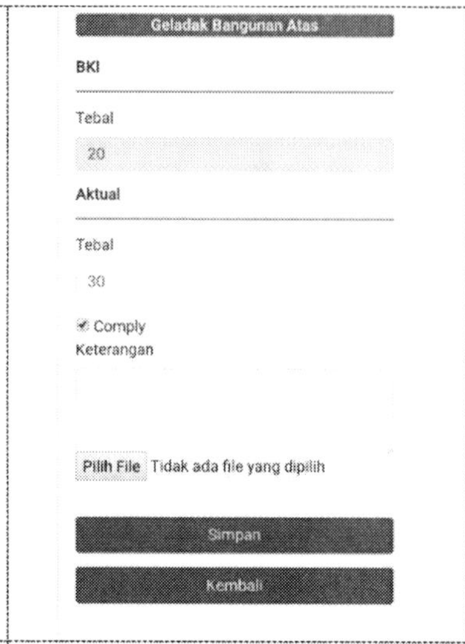

Figure 12. Superstructure deck calculation that the last form from scantling menu, the user must choose save button to saving all calculation data.

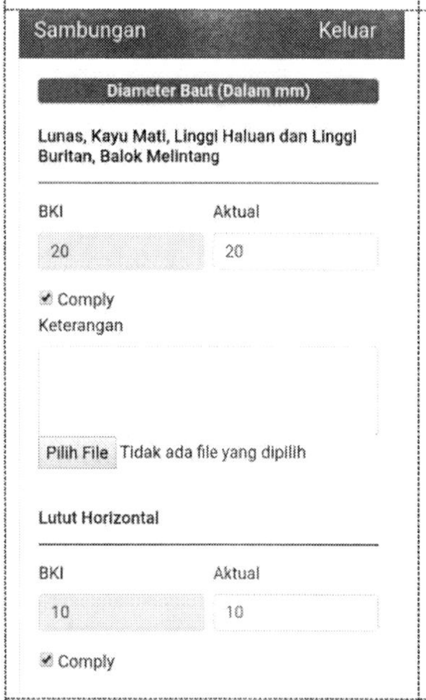

Figure 13. The connection menu form, the bolting of keel, deadwood, transom beam, stem and sternpost.

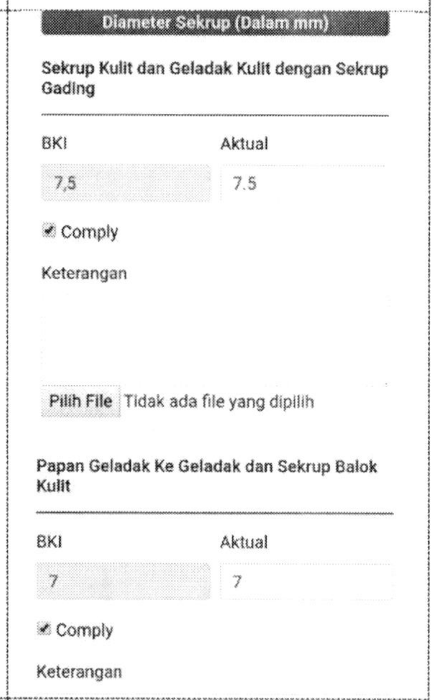

Figure 14. The diameter size of screw for shell with frames and deck planks to deck and shell beams.

IOP Conf. Series: Materials Science and Engineering 1052 (2021) 012044 doi:10.1088/1757-899X/1052/1/012044

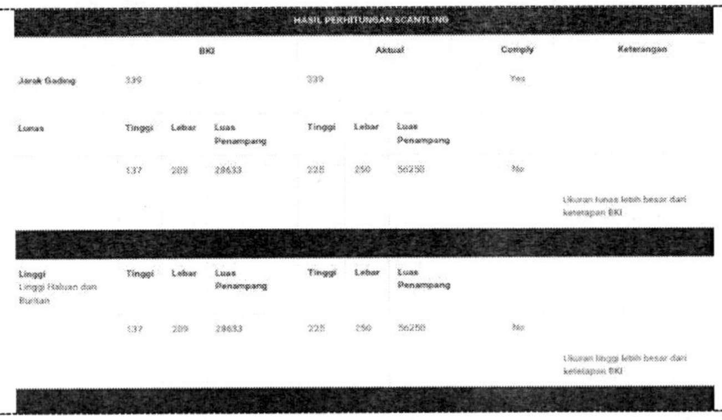

Figure 15. Report result in history menu.

In table 1, all of scantling calculation from BKI 2013 and actual data displayed. Based on the calculation, a few structural member of construction not comply the BKI 2013 volume VII standard. In scantling calculation, beam shelves and thickness of collision bulkhead not comply the standard but in connection, all actual data is complying BKI standard. All result is tabulated in table 1 and table 2 down below:

Table 1. Result of Scantling Calculation.

Structural Member	BKI 2013 (mm)	Actual Data (mm)	Note
Frame Space	339	400	Comply
Keel	Height x Width 137 x 209	Height x Width 225 x 250	Comply
Stem and Sternpost	Height x Width 137 x 209	Height x Width 225 x 250	Comply
Beam Shelves	Height x Width 92 x 71	Height x Width 60 x 140	Not comply
Shell	Thickness x Width 127 x 145	Thickness x Width 140 x 160	Comply
Deck	Thickness x Width 32 x 25	Thickness x Width 40 x 30	Comply
Collision Bulkhead	Thickness 50	Thickness 40	Not Comply
Other Bulkhead	Thickness 30	Thickness 50	Comply
Wrang	Height x Width 141 x 71	Height x Width 150 x 80	Comply
Frame	Height x Width 36 x 28	Height x Width 140 x 60	Comply

Structural Member	BKI 2013 (mm)	Actual Data (mm)	Note
Frame Space	339	400	Comply
Deck Beam	Height x Width 46 x 36	Height x Width 140 x 50	Comply
Superstructure side wall	Thickness 26	Thickness 30	Comply
Deck Superstructure	Thickness 20	Thickness 30	Comply

Table 2. Result of Connection (Bolting and Screw).

Type of Connection	BKI 2013 (diameter in mm)	Actual Data (diameter in mm)	Note
Bolt in keel, stem/sternpost, and transom beam	20	20	Comply
Bolt in horizontal knee	10	10	Comply
Connecting floors with keel and shell and frame; bolt in arms	9	9	Comply
0.8 LWL (in throat)	12	12	Comply
In the end of yacht	12	12	Comply
Screws in shell and deck; shell with frames screws	7.5	7.5	Comply
Screws in shell and deck; shell with frames screws; deck planks to deck and shell beams screws	7	7	Comply
Screwing hanging knees and shelves to frames and deck beams	0	0	Comply

For the validation questionnaire have filled by 5 respondents and the result of questionnaire tabulated in table 3 down below:

Table 3. Score of Questionnaire.

Respondents	Q1	Q2	Q3	Q4	Q5	Q6	Q7	Q8	Q9	Q10	Total Score	Percentage (%)
1	5	4	3	4	4	5	4	4	3	3	39	88
2	4	5	4	5	3	4	3	4	3	4	39	80
3	3	3	3	4	4	4	3	3	4	3	34	76
4	4	5	3	5	4	4	4	4	4	3	40	70
5	5	4	4	5	3	3	3	3	3	4	37	83
Sum	21	21	17	23	18	20	17	18	17	17	189	
				Total							37.8	75.6

4. Conclusion and Recommendation

The highest score is 23 from question number 4 that is the application need for survey activities. The results of questionnaire analysis, a total percentage is 75.6% to average total score is 37.8 which shows that the android application very helpful to assist survey activities. The conclusion, the use of android application for scantling calculation of wooden ship construction is needed because the application more practical and can accessed anywhere compared to manual survey systems, more efficient in terms of time, and more safety for data storage of report. It is recommended to maintain the accuracy of calculation and updated system the application.

Acknowledgement

I would like to express my special thanks of gratitude to my supervisor for his expert advice and I would also like to extend my gratitude to all people for their continuous support and encouragement for my pursuit.

References

[1] Malisan J, Jinca M.Y, Parung H, and Saleng A 2012 *Strength Analysis of Traditional Ships in Efforts to Improve Sea Transportations Safety in Indonesia* International Journal of Engineering & Technology vol 12 no 5 (Pakistan: IJET & IJENS) p 118
[2] Wang H, Zhu X, and Chang J 2011 *Acceleration and Multisensor Pedometer Application Design and Implementation on The Android Platform,* International Conference on Instrumentation, Measurement, Computer, Communication and Control (China: IEEE Computer Society) p 249
[3] Wei M L 2012 *Beginning Android™4 Application Development* (Canada: John Wiley & Sons, Inc.) p 2-3
[4] Himanshu S, Sameer P, Vaibhav D, and Pragya S 2014 *Analysis of Android Vulnerabilities and Moedern Exploitation Techniques,* ICTACT Journal on Communication Technology vol 5 no 1 (India: ICTACT) p 863
[5] Lazarela L and Kore J 2011 *Analysis of the Advantages and Disadvantages of Android and iOS Systems and Converting Applications from Android to iOS Platform and Vice Versa,* American Journal of Software Engineering and Applications vol 6 no 5 (United States of America: Science Publishing Group) p 116-120
[6] Indonesian Classification Bureau (BKI) 2013 *Rules for Small Vessel Up to 24 M* vol 7 (Jakarta: Indonesian Classification Bureau) p 15-97
[7] Indonesian Classification Bureau (BKI) 2014 *Rules for Non-Metallic Material* vol 14 (Jakarta: Indonesian Classification Bureau) p 49-50
[8] Kofler M 2001 *What Is MySQL?* (California: Apress) p 4
[9] Steve S 2002 *MySQL Bible* (New York: Wiley Publishing Inc.) p 12-13

IOP Conf. Series: Materials Science and Engineering 1052 (2021) 012045 doi:10.1088/1757-899X/1052/1/012045

Analysis of Power Requirements and Turning Circle of Amphibi Coach

V D Moganti[1], R E Garetno[1], R Irvana[2] and A Fadillah[2]

[1]Student of Naval Architecture, Darma Persada University, Jl. Taman Malaka Selatan, Jakarta Timur 13450, Jakarta.
[2] Department of Naval Architecture, Darma Persada University, Jl. Taman Malaka Selatan, Jakarta Timur 13450, Jakarta.

Abstract. This research is a continuation of the previous one. It is an analysis to find the power requirements and turning circle of Amphibi Coach. In this analysis, the method used for calculating the propulsion power when in water uses the KR-Barge calculation method. In turning circle analysis using the Hovgaard method, this method is a simple method for estimating turning circle without field testing. The standard used for turning circles refers to the standard Resolution MSC 137 (76) (adopted on 4 December 2002) Standards For Ship Maneuverability. From the calculation of the driving power, the driving power needed when the Amphibi Coach is in the water is 126.99 HP while the Amphibi Coach driving power when on the ground is 420 HP. The final propulsion is determined based on the greatest propulsion, which is 420 HP. The results of the Amphibi Coach turning circle analysis showed that the Amphibi Coach turning circle without Fin Stabilizer was 25.96 m and had a tactical diameter of 51.92 m. For Amphibi Coach turning circle with Fin Stabilizer of 26.34 m and has a tactical diameter of 52.69 m. Amphibi Coach turning circle analysis using the approach formula shows that the results obtained have met the applicable criteria.

1. Introduction

North Sumatra province is one of the top 10 (ten) tourism destinations in Indonesia where there are 339 tourist attractions scattered throughout the region in 33 districts/cities in North Sumatra Province. Currently, only 120 attractions that have been marketed include natural potential, one of which is Lake Toba Area. Lake Toba is a mainstay tourism area, both provincial and national [1].

Amphibi Coach is a bus type vehicle that can be operated on land and water lines designed to support Lake Toba tourism. To support the operation of Amphibi Coach require power adequately. In determining power, it is necessary to determine the resistance elements and the efficiency of the thruster. The resistance elements that work on amphibious vehicles are frictional resistance, shape resistance, and wave-making resistance [2]. The sailing resistance of amphibious vehicles consists of friction drag, form drag and wave drag, friction drag related to water viscosity, water pressure related drag form, and wave drag related to car speed [3].

Increased resistance to shallow water compared to deep water at the same speed is seen as significant. The speed of flow under the hull in shallow waters is higher than in the deep waters because there is an increase in flow under the keel and a decrease in pressure in the region, as a result, the buoyancy decreases and results in sinkage and trim. As the ship sank, the wet surface area increased, as a result, the viscous resistance increased [4].

In general, one of the properties of the ship that needs to be known by the captain or officer of the ship is the ability to move when sailing and when going lean or going off the dock that is the response speed of the steering wheel blade or the control lever of the aircraft engine on the moving platform and also the size or small turning circle of the rotating ship [5]. According to Jamaluddin, the IMO (International Maritime Organization) regulation called "Standard for Ship Maneuverability" should be met at the time of the ship's design and operation [6].

Resistance and maneuvering characteristics are required to predict the speed, acceleration, stop, turn, yaw inspection, and track keeping ability of amphibious ships [7]. In the research journal Renaldi, Amphibi Water School Bus has frictional resistance of 0.176 tons, wave-making resistance is 1.737 tons, and air resistance 0.026 tons so it has a total resistance of 1.939 tons [8]. For the determination of the vehicle's engine while on the ground, the total force of the vehicle is calculated at the desired speed. Traction style is derived from the calculation of total aerodynamic inhibition force, rolling inhibition force, gradient resistance [9].

In the research journal Moganti, Amphibi Coach design has a length size of 13.115 m, width 2.5 m, height 3.8 m equipped with fin stabilizer has met rolling criteria standards according to IMO Level 2 but in this design barrier analysis and turning circle, Amphibi Coach has not been done [10].

From the design analyzed by Moganti, the idea of calculating power requirements and estimates turning circle Amphibi Coach. This research will be conducted to analyze the power requirements and turning circle Amphibi Coach equipped with the addition of fin stabilizer on Amphibi Coach.

2. Methodology

In this study, the object used was the Amphibi Coach of literature [10]. Here is the Amphibi Coach data shown in Table 1.

Table 1. Amphibi coach main size data

LWL	13.115 m
Draft	1 m
Height	3.8 m
Beam Hull	2.5 m
Displacement	27.21 ton
Service Speed	8.2 knot
Fin Stabilizer	Aquarius A25
Fin Length	1.26 m
Fin Wide	0.67 m

Amphibi Coach lines plan is designed using Maxsurf software with a similar shape to the bus in general. Amphibi Coach parts are made almost ship-like to provide good stability. Amphibi Coach is designed using fin stabilizer to provide comfort to its users and equipped with jet ski as supporting recreational rides and at the top of the vehicle is used as a rooftop as a place to relax or sunbathe enjoying the view of Lake Toba. For fin stabilizers use active fin stabilizers that are used only when in water. Fin stabilizer data using fin data from Kongsberg company with type fin designated in Table 1 [11]. Amphibi Coach's design concept and lines plan can be seen in Figure 1.

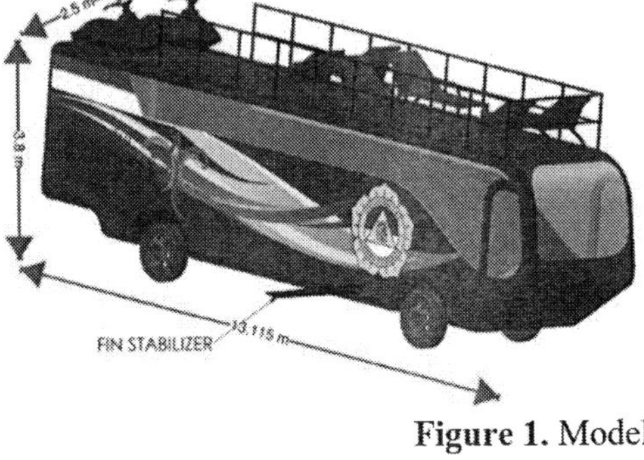

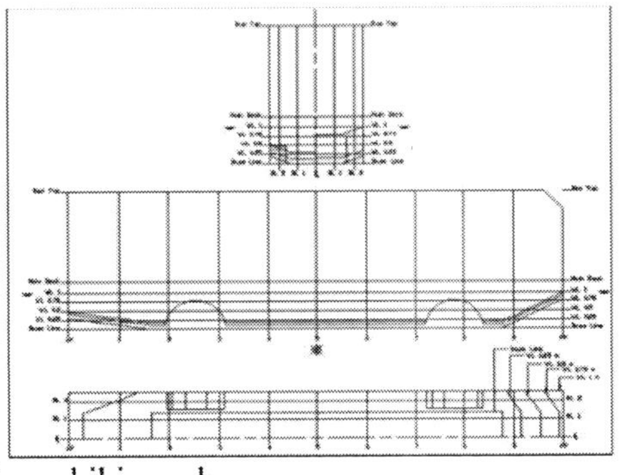

Figure 1. Model amphibi coach

IOP Conf. Series: Materials Science and Engineering 1052 (2021) 012045 doi:10.1088/1757-899X/1052/1/012045

Amphibi Coach drive power analysis is an analysis of the calculation of power while on land and in water. The method of calculating the drive power when in the water is the KR Barge method [12]. This method is used because it sees the shape of the Amphibi Coach that almost resembles a barge.

$$R_t = R_f + R_w + R_a \tag{1}$$

Where

R_t = total resistance (ton).
R_f = frictional resistance (ton).
R_w = wave-making resistance (ton).
R_a = air resistance (ton).

For analysis of power while on land used the calculation formula as follows [13].

$$F_{tot} = F_{Ro} + F_{Ae} = F_{Cl} \tag{2}$$

Where
F_{Ro} = Rolling resistance (N).
F_{Ae} = Aerodynamic Resistance (N).
F_{Cl} = Climbing Resistance (N).

$$T_{aw} = F_{tot} \text{ x rolling radius (m)} \tag{3}$$

Where
T_{aw} = Torque at the wheel (N m).

$$\text{b. p.} = \frac{2\pi TN}{60000} \tag{4}$$

Where
b.p. = brake power (kW).
π = 3.142.
T = engine torque (N m).
N = crankshaft speed (rev/min).

For turning circle calculations Amphibi Coach is done when Amphibi Coach uses fin and without using fin. This calculation uses a formula from Teguh Sastrodiwongso which is the development of the Hovgaard formula[5]. This formula is a simple way of estimating turning circles. Turning circle calculation stipulated by equation (5) :

$$\rho = K_3 \frac{\nabla}{\text{Cn. A} \cos a} \tag{5}$$

K_3 is a coefficient obtained from Table 2.

Table 2. Coefficient K_3

∇/S.L	K_3	∇/S.L	K_3
0.050	1.140	0.10	0.460
0.055	1.285	0.11	0.400
0.060	1.100	0.12	0.370
0.065	0.960	0.13	0.355
0.070	0.845	0.14	0.345
0.080	0.670	0.15	0.340
0.090	0.550		

The formula for calculating the normal style coefficient (Cn) is shown in equations (6) using the Joessel formula.

$$Cn = \frac{0.811 \sin a^{o}}{0.195 + 0.305 \sin a^{o}} \tag{6}$$

Where
∇ = volume displacement (m^3).
A = rudder area (m^2).
S = the area of the middle field extends the ship (m^2).
L = ship waterline length (m).
α = maximum steering leaf angle (35° ~ 37°).

Advance calculation as in equation (7) or distance traveled by ship in its original direction, measured from the point when the rudder was first applied at the turn and Transfer as in equation (8) is the distance of the ship's center of gravity from the original trajectory line when the ship's header is 90^0. It can be calculated using the formula contained in the journal Vessel Maneuverability American Bureau of Shipping (ABS).

$$\frac{Ad}{L} = 0.519 \frac{Td}{L} + 1.33 \tag{7}$$

$$\frac{Tr}{L} = 0.497 \frac{Td}{L} + 0.065 \tag{8}$$

Where
Ad = Advance distance (m).
Td = Tactical Diameter distance (m).
Tr = Transfer distance (m).
L = ship waterline length (m).

3. Result And Discussion

In this study, the power analyzed is the power while in the water at a speed of 8 knots with 60% efficiency and on the 1 m draft with 2 models and power on land with a maximum speed of 100 km/hour. For water power, it is analyzed on 2 models, namely Amphibi Coach without fin stabilizer and Amphibi Coach using Fin Stabilizer.

3.1. Amphibi coach power analysis

Following the data in the methodology, the results of the power analysis when in the water using the Maxsurf Resistance for both models are shown in Table 3.

Table 3. Results of analysis of the power of each model when in water

Model	Speed (knot)	Resist. (kN)	Power (HP)
No Fin	8	11.9	112.653
With Fin	8	13.5	126.992

The results obtained according to Table 3 shows that at the same speed for both models require different power, Amphibi Coach without fin has less power than Amphibi Coach with the fin.

For analysis of onshore power at a constant speed of 100 km/h obtained results as shown in Table 4.

Table 4. Results of calculation of power while on land

Model	Speed (km/jam)	Resist. (N)	Torque (Nm)	Power (HP)
Amphibi Coach	100	5570.25	2935.52	419.06

From the analysis of the above driving power obtained for power in water and on land has a considerable difference in yield. This is due to the difference in the size of resistance while in the water and on land. Therefore, the power determination for Amphibi Coach is taken the most power which is 419.06 HP, and determined to be 420 HP according to the existing catalog [14]. The results of the power obtained do not have a very cyclical difference with the power of Mercedes-Benz Tourismo L bus [15] with a total weight of 24.750 tons which has a drive power of 388.89 HP while amphibi coach weighing 27.21 tons has a power of 420 HP.

From the catalog, the engine size is obtained and is used as a consideration for the shape of the engine room and the transmission line of the Amphibi Coach.

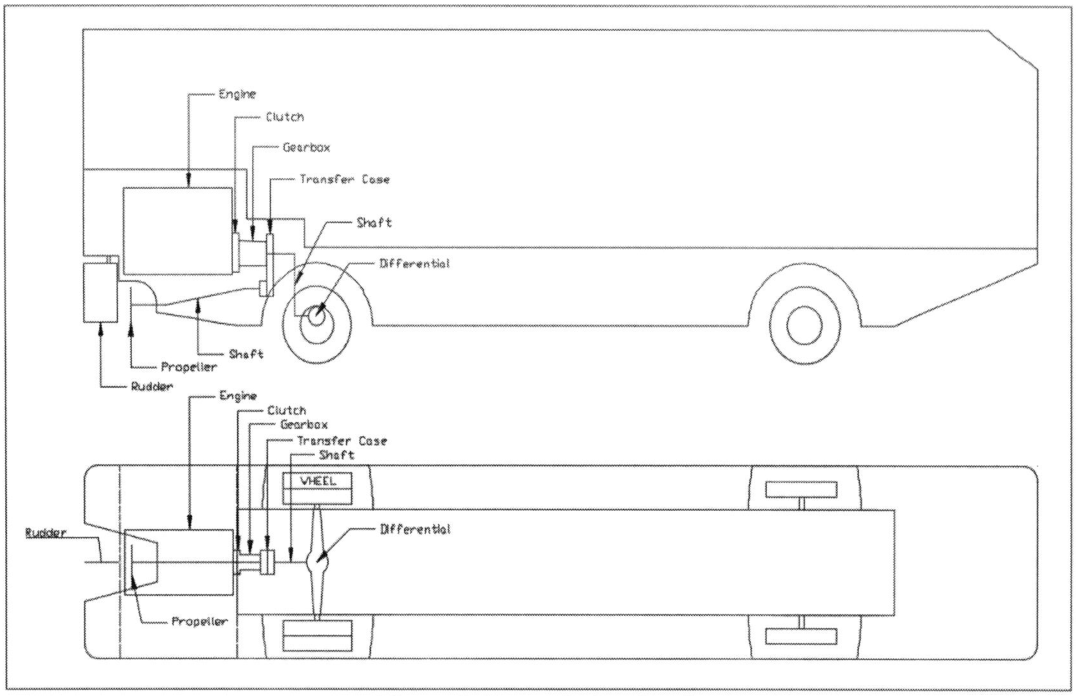

Figure 2. Transmission sketch

For the process of getting into the water, Amphibi Coach uses the ground with a slope of ten degrees.

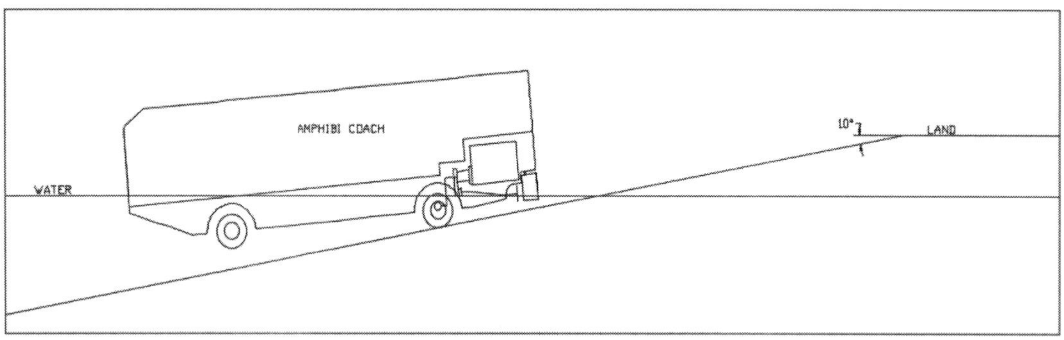

Figure 3. Water entry process

3.2. Turning circle amphibi coach analysis

Amphibi Coach turning circle analysis is carried out on 2 different models namely model 1 without fin stabilizer and model 2 with fin stabilizer. This analysis using a rudder which has an area of 0.534 m^2 with a size of 0.8 m high and 0.47 m wide. For the standard that is the reference calculation turning circle Amphibi Coach refers to resolution standard MSC 137(76) (adopted on 4 December 2002) Standards For Ship Manoeuvrability which in this standard explains that the advance distance of the ship should not be more than 4.5 ship length and tactical diameter distance should not be more than 5 times the length of the ship.

Table 5. Turning circle amphibi coach calculation results

Model	ρ (m)	Td (m)	Ad (m)	Tr (m)	Criteria
No Fin	25.96	51.92	44.90	26.68	PASS
With Fin	26.348	52.69	45.29	27.05	PASS

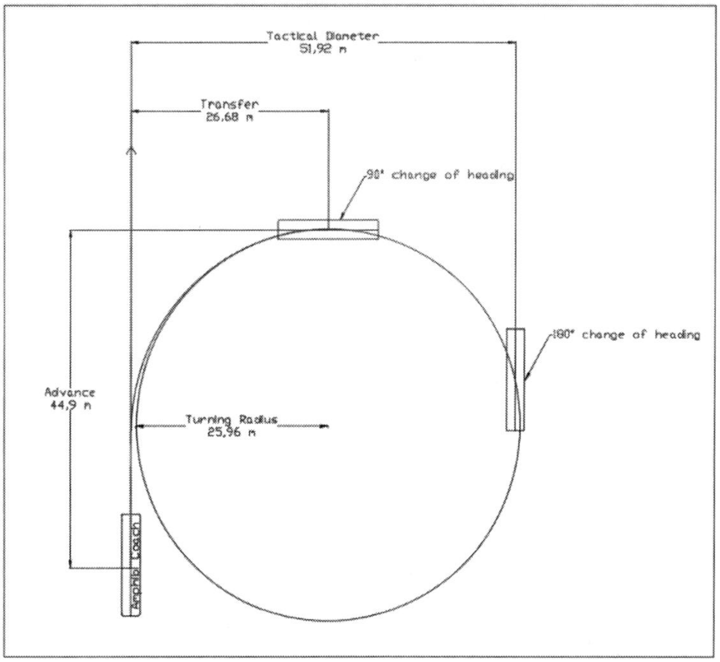

Figure 4. Turning circle amphibi coach without fin stabilizer

From the analysis turning circle, Amphibi Coach obtained the turning circle Amphibi Coach without fin stabilizer of 25.96 m and has a tactical diameter of 51.92 m.

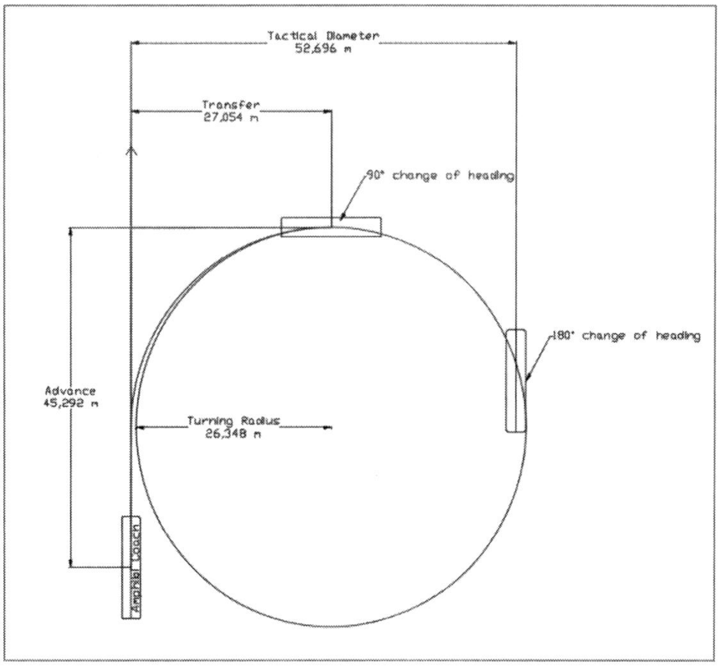

Figure 5. Turning circle amphibi coach with fin stabilizer

Turning circle Amphibi Coach with fin stabilizer of 26.348 m and tactical diameter of 52.696 m. The results obtained show that Amphibi Coach with fin stabilizer has a turning circle larger 0.388 m than Amphibi Coach without fin stabilizer.

4. Conclusion

From the results of the resistance, analysis shows that Amphibi Coach with fin stabilizer requires a power of 126.992 HP while Amphibi Coach without fin stabilizer requires a power of 112.653 HP. The power needed on the ground is 420 HP. The determination of power is taken based on the larger power needs of 420 HP. The analysis of these resistance shows that the effect of using fin stabilizer on Amphibi Coach produces a large resistance that affects the power needs of Amphibi Coach that require the drive power even while from the analysis turning circle Amphibi Coach is obtained turning circle Amphibi Coach without Fin Stabilizer of 25.96 m and has a tactical diameter of 51.92 m. For turning circle Amphibi Coach with fin stabilizer of 26.348 m and tactical diameter 52.696 m. Analysis of turning circle Amphibi Coach using approach formula shows that the results obtained have met the applicable criteria. For further research, It is advisable to check the transmission line of the vehicle along with the engine type, construction and actual model testing.

5. References

[1] K. PUPR 2015 *Profil Kawasan Danau Toba* (Jakarta: Kementrian PUPR)

[2] L. Duan 2015 Study on force characteristics and resistance for water by amphibious vehicle *ICMEIS.* **III** 553-557

[3] Z. Guo 2013 Numerical simulation of an amphibious vehicle sailing resistance *IJCSI.* **10** 33-36

[4] M. Nakisa 2017 Numerical estimation of shallow water effect on multipurpose amphibious vehicle resistance *JNAME.* **14** 1-8

[5] T. Sastrodiwongso 2002 *Kemudi dan Olah Gerak Kapal* (Jakarta: FTK UNSADA)

[6] A. Jamaluddin 2013 Analisa dan evaluasi faktor-faktor yang mempengaruhi karakteristik olah-gerak (maneuver) kapal *Warlit.* **23** 17-26

[7] H. U. A. M. Command 1971 *Engineering Design Handbook Wheeled Amphibians* (Virginia: University Of Virginia Library)

[8] R. R. R. Rinaldi 2018 Desain kapal amfibi water school bus sebagai sarana transportasi pelajar untuk rute pelayaran Kepulauan Seribu - Jakarta Utara *Jurnal Teknik* ITS. **7** G65-69

[9] R. F. Rahman 2018 Perancangan dan analisis karakteristik traksi ada mobil pedesaan serbaguna WAPRODES *Jurnal Teknik ITS.* **VII** 21-26

[10] V. D. Y. Moganti 2020 A study of the effectiveness fin stabilizer on unsada water tour bus to comfort the rolling period to support Toba lake tourism *IOP Conf. Ser.: Earth Environ. Sci.* **557** 012012

[11] Kongsberg 2019 *Marine Products and Systems* (Norway: Kongsberg)

[12] K. Register 2014 *Rules for The Towing Survey of Barges and Tugboats* (Busan: Korean Register)

[13] E. b. D. A. Crolla 2009 *Automotive Engineering Powertrain Chassis System and Vehicle Body* (Burlington: Elsevier's Science & Technology Rights)

[14] MAN 2020 *On-Road Engine and Components* (Vogelweiherstr: MAN Truck & Bus)

[15] Mercedes-Benz 2019 *The Tourismo right-hand drive Technical Information* (Stuttgart: Mercedes-Benz)

Acknowledgments

We would like to thank Darma Persada University who has supported this research and the supervisors who have provided input and suggestions in this research.

IOP Conf. Series: Materials Science and Engineering 1052 (2021) 012046 doi:10.1088/1757-899X/1052/1/012046

Technical and Economical Analysis of Shipyard Re-Layout for Product-Oriented Work Breakdown Structure Implementation

T W Pribadi[1], R N Adam[1] and S I Wahidi[1]

[1] Department of Naval Architecture, Faculty of Marine Technology, Institut Teknologi Sepuluh Nopember, 60111 Surabaya, Indonesia

Abstract. The ship building method used by many shipyards in the world has led to work effectiveness and efficiency. One of them is Product-Oriented Work Breakdown Structure Concept (PWBS). PWBS belongs in group technology. In this method the parts that will be produced for ship building are categorized based on the similarity in form and production process. In Indonesia, not many shipyards have implemented this method. If a shipyard will change its production method, layout adjustments will be required. This research was conducted to analyze the technical and economic changes in the layout of the PT X with study in the case of 2 x 1800 HP tugboat. The technical analysis is done by observing the conditions of the production process and the existential shipyard layout. Furthermore, it also analyzes the investment costs to make changes of the shipyard's layout and analyzes the feasibility value of the investment. Fabrication parts of tugboat are grouped into 4 categories, those are category A: internal parts from plate or part piece of plate, B: parallel parts, C: curved parts (bent plates) and D: shapes (miscellaneous operation). The new layout can reduce the distance of material handling in category A by 2-19%, category B by 29%, category C by 57% and category D by 2-56%. Fabrication costs are estimated to be reduced to IDR 36,152,935 and sub assembly cost reduced to IDR 27,178,200 by applying this group technology concept. Investment costs to make changes the layout is IDR 7,471,954,000. This project is worth to do with 11% IRR value and payback period will occur in the 3rd year.

1. Introduction

Shipbulding method used in many shipyards in the world has led to work effectiveness and efficiency. In the conventional method, the ship is seen as a system. So that the ship is built system by system. This method makes the ship building process slow. This encourages scientists and experts in the field of shipping to create other, better methods. One method is the Product-Oriented Work Breakdown Structure (PWBS). PWBS belongs to the technology group. In this method, the parts to be produced for shipbuilding are grouped based on the similarity in shape and production process.

Product-Oriented Work Breakdown Structure (PWBS) concept has not been widely used by large shipyards in Indonesia to carry out ship building. In PWBS concept the division of labor is oriented towards the products produced by each production workshop, so that work can be carried out based on the division of the ship's zone. In the construction process, the ship is divided into several work packages and classified based on the similarity in design and fabrication processes. Building a ship using this concept will involve an integration of hull construction, outfitting and painting. The purpose of using the PWBS concept is to increase shipyard productivity.

Changes in production technology at the shipyard must be supported by increased facilities and changes in the layout of the shipyard from the existing form. The new layout must be adjusted to the size of the land and the location of the existing production facilities, especially in medium-scale shipyards. Changes in layout as much as possible do not change the position of vital shipyard facilities, such as building birth and dry dock. Layout changes include rearranging the layout of production machines in the workshop, adding an assembly area, setting up work stations, placing warehouse areas and arranging other supporting facilities. In this research, a technical and economic analysis of changes in the layout of the shipyard will be carried out for the application of the PWBS concept in a medium-scale Indonesian shipyard.

2. Literature Review

2.1. Shipbuilding Process

The ship building process is broadly divided into two stages, namely the design stage and the physical construction stage. The design process is the process of translating the owner's requirements into an image that will be used as a reference for ship building. Construction of ship construction includes making parts / components, joining parts / components, building blocks and erection (merging blocks)[1].

In the initial estimation process and design stage, the ship is divided into work packages based on a system or what is called a System-Oriented Work Breakdown Structure (SWBS). However, this division based on SWBS cannot be used to distribute work during the planning, scheduling and physical construction process of ships. Therefore, in these activities, the division of work is carried out based on the product produced or with the Product-Oriented Work Breakdown (PWBS) system. The distribution is adjusted to the design that has been produced from the previous stage [2].

2.2. Shipyard

Shipyard is a place designed to build and / or repair and maintain ships. Generally, shipyards are built on land directly adjacent to the water or sea level. Inside the shipyard there are facilities and facilities used to support ship building activities. Not only that, it also includes facilities for other operational activities. The facilities and facilities owned by the shipyard will affect the performance and capacity of the shipyard in carrying out ship building or ship repair. These facilities include :

- Production workshop
- Building berth
- Launching area
- Warehouse
- Office building

2.3. Shipyard Layout

Careful consideration is required of area availability, topography, hydrography and weather conditions which contribute significantly to the overall characteristics of the layout design. The flow of material is determined from the parallel work stages, which have been carried out and succeeded, so that the end result will be harmonious and balance the production facility. The factors that influence the shape of the shipyard layout are [3] :

- Range of ship types built and the number / year
- Room required
- Available room
- Number of "mechanisation"
- Material handling

Group technology layout or commonly known as process lane construction is an arrangement of work stations in a workshop equipped with production machines, materials and labor to form a construction with the same shape and the same type of machine usage [4]. Figure 1 is tipe of workshop layout which applies group technology concept. Part production processes that are straight with parts requiring a distinguished bending process.

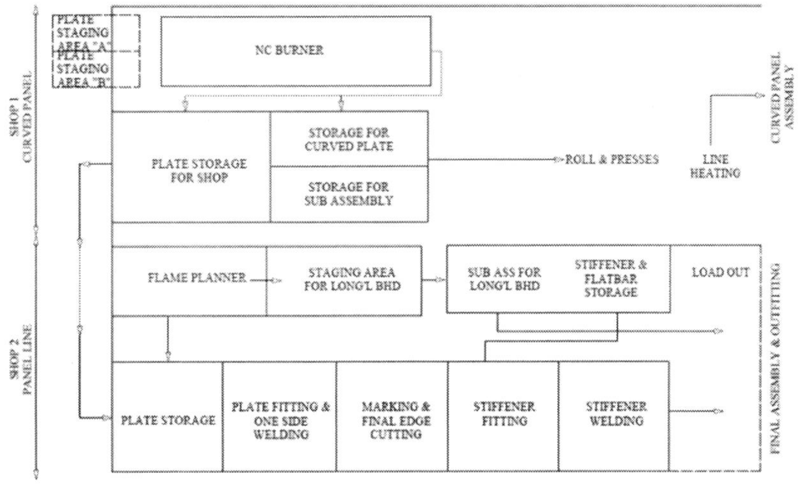

Figure 1. Type of Shipyard Workshop

2.4. Product-Oriented Work Breakdown Structure Concept

The PWBS concept is the result of the development of the technology group on ship building. The ship building process using this concept will divide the ship into work packages. The parameters used to divide the work are shape, size, tolerance, material, type and complexity of the production machine.

PWBS classifies interim products based on the four aspects needed to control the production process. Systems and zones are related to the ship design function, while areas and stages are related to the ship's production function. The four aspects are defined as [2] :

1. The system is a structural or operational characteristic of a product.
2. Zone is a division of production objects based on geographic division.
3. Problem Area is the division of the production process into similar types based on the form, quantity, quality, type of work and others.
4. Stages are the division of the production process based on the sequence of the production process.

2.5. Supervision

Zone-oriented shipbuilding scheduling is necessary to regulate the flow of work over the various process lines so that the manufacture of intermediate (additional) products only when needed. Scheduling must coordinate hull construction, outfitting and painting work so that the completed work package can enter the next stage of work. The goal is to minimize the waiting work packets (buffer) [2].

2.6. Learning Curve Theory

The basic principle of the learning curve is easily explained, such as: if a person performs a task that is repetitive in nature, the time required to complete the next task will decrease until it can no longer be reduced. Initially the learning curve focused on the relationship between cumulative quantities and cumulative average times (the amount of cumulative time divided by the cumulative quantity) [5]. General form of learning curve is

$$Y = aX^b \quad (1)$$

where,
Y = cumulative average times per unit/batch
a = time required for the production of initial quantities
X = cumulative unit of production
b = learning curve index, is log learning curve percentage / log 2
In shipbuilding, in general learning curve is 8-10%, meaning that the time needed to complete each production unit is generally reduced by 8-10% from that required by the previous unit [6].

2.7. Productivity
Productivity is used to measure the performance of operations management. Productivity is a measure of how well a resource is managed and utilized to achieve the desired result. Productivity is generally expressed as the ratio of output to input, or the ratio obtained to resource consumption [7].

3. Research Methodology

3.1. Identification of Problems
The initial stage of working on this research is the identification of the problem which includes activities such as explaining existing problems, determining research boundaries, and describing things that need to be explained in problem solving. In this research, the problem raised is the lack of use of the PWBS concept in the Indonesian shipping industry and planning changes to the shipyard layout that must be done to support the PWBS concept.

3.2. Literature Review
At this stage, relevant literature is collected to support this research. The literature used discusses shipyards, shipyard facilities, layout concepts, learning curves, productivity, and investment. The literature used is in the form of domestic and foreign journals, books and similar research that has been done before.

3.3. Existing Condition Analysis
Analysis of existing conditions is carried out to determine the initial conditions of the shipyard which will be used as a research case study. The analysis is carried out directly by visiting the shipyard which is the object of research and asking for primary data. The data used in this study are the shipyard layout, the layout of the machines in the workshop, a list of shipyard facilities and data on new buildings of ships being built.

3.4. Data Management
The data that has been obtained from the survey results are processed with reference to relevant literature and realistic thinking.

3.5. Technical Planning
After the data is processed, the next stage is technical planning. Technical planning is carried out in the form of shipyard layout planning, planning of new facilities, calculating technical changes that occur. Planning is carried out by looking at the conditions of the shipyard, regulations regarding standard office facilities and the concept of building a ship using PWBS which is the topic of this research.

3.6. Economical Analysis
Economic analysis was conducted to calculate the costs of changing the layout and purchasing new facilities. As well as an analysis of the feasibility of investing on shipyard income based on research case studies.

4. Existing Condition of Production Process and Layout

Observation and survey of existing conditions are needed to obtain real field data regarding the ship production process being carried out and the position of the existing facility layout. This data will be used as a basis for planning a new layout according to the PWBS concept.

4.1. Production Process

The production process in ship building consists of hull construction, outfitting manufacture and installation, to the painting process. The process is carried out in each workshop with the facilities in it. Once a complete or nearly complete construction is formed, the ship will be launched. Then, the finishing, commissioning and sea trials are carried out. At the time of this research, PT X shipyard was building a tugboat 2 x 1800 Hp with the following sizes :

Loa	: 30	m
B	: 11.6	m
H	: 5.1	m
T	: 3.5	m
Main Engine	: 2 x 1800HP	

The tugboat parts were assembled in the south building area and part of the fabrication process was carried out in the south hull workshop. The fabrication process includes marking, cutting, bending and several joining activities. The result of this process is in the form of parts that have a shape according to the previously planned design. Plates and profiles that have gone through the fabrication process will be assembled in the building area. However, there is no clear barrier between the storage area for new material and waste material, so that many piles of new plates are mixed with used plates. In the building area the workers will assemble the fabricated parts into a panel in an empty place and randomly selected so that there is no clear process lane.

4.2. Shipyard Layout

This shipyard has a land area of approximately 57,000 m^2 and a sea area of approximately 70,000 m^2 with a waterline length of 1,200 m. There is an indoor work area of 11,525 m^2 and an outdoor work area of 11,949 m^2.

This shipyard has two large hull workshops in a separate area because apart from serving the construction of new building ships, this workshop is also active in serving ship repair and maintenance requests. The northern hull workshop was used to build ships with large dimensions such as tankers. Meanwhile, the southern hull workshop is used to build smaller ships such as tugboats. Apart from the two hull workshops, this shipyard also has a machine shop and an outfitting workshop. There is also a woodworking shop and a mold loft area which is now no longer used.

Apart from that, this shipyard also has a storage area. Storage facilities are used to store company-owned inventory and store raw materials for the production process. The types of storage facilities at this shipyard are divided into two, namely open storage with an area of 5,250 m^2 and closed storage area of 2,750 m^2. The steel-stock yard at the shipyard is divided into 4 areas. These four areas are quite far apart. For pipe storage an indoor area measuring 29 m x 7.5 m is provided. The paint shop is also an indoor area measuring 29 m x 9 m.

4.3. Shipyard Facility

Ships will not be built with empty-handed. The existing facilities in the shipyard are the foundation for ship building. There are various types of shipyard facilities, such as docking undocking facilities, material handling facilities and workshop facilities.

Docking and undocking facilities are facilities used to raise and lower ships from the water surface to the land. This shipyard has several types of docking undocking facilities including slipway, transverse slipway and floating dock. This shipyard has 4 floating docks which are used in the ship repair process. Meanwhile, for the ship building process, the usual facility used is the slipway.

IOP Conf. Series: Materials Science and Engineering 1052 (2021) 012046 doi:10.1088/1757-899X/1052/1/012046

Material handling facilities owned by this shipyard are various, such as forklifts and cranes. The largest capacity of the means of transport is a floating crane of 75 tons. These cranes are commonly used to transport blocks from the north hull workshop to the transverse slipway in the south and vice versa.

Apart from the above facilities, this shipyard also has various kinds of production machines located in its workshops. These machines include bending machines, cutting machines, drilling machines, lathes and others.

5. Shipyard Layout Design For PWBS Implementation

After analyzing the existing shipyard condition and based on the theory from previous research, the next step is implement the PWBS theory in production process to be the basic of new layout. Design of new workshop layout depends on the production process.

5.1. Production Process

In the ship building process using the PWBS concept, the ship will be divided into several work packages. The ship is divided down to its simplest part. Starting from level 0 which is a complete ship to level 4 which is the raw material. All parts of this tugboat facility are classified into 4 categories where :

- A category is internal parts from plate.
- B category is parallel parts.
- C category is curved parts (bent plates).
- D category is shapes (miscellaneous operation).

Tabel 1. Examples of Fabrication Parts Grouping

Part	Dimension	Category
Keel plate	14	C
Bottom plate	12	C
Tank top plate	12	B
Longitudinal	FB 150x16	D
	L 75x75x6	D
Transverse	FB 150x12	D
	L 100x100x8	D
	6	A
Side girder	10	A
	FB 100x10	D
Centre girder	10	A
	FB 100x8	D
Bracket	9	A
Side shell	10	C
Frame	9	A
Tank plate	9	B
Tank plate stringer	L 100x100x8	D
Bridge deck plate	6	B
Wheel house plate	6	B
Side wall	6	B
Side wall stringer	L 75x75x6	D

The results from analysis in **Tabel 1** of the division of parts category, the percentage of parts in category A was 30%, category B was 24%, category C was 14% and category D was 31%. The output from the fabrication process is in the form of a piece part with edge preparation for the welding process at a later stage. Once formed, the piece parts are joined at the sub-assembly stage to produce a panel. This stage is dominated by fitting and welding activities. Generally, the sub-assembly stage is divided into :

- Sub-assembly for flat panel
- Sub-assembly for curved panel

At the assembly stage, the panels will be connected to form a more complex block unit. In addition, at this stage an outfitting installation is also installed which is included in the on block outfitting. Assembly is carried out from the lowest to the top panel position. Assembly is carried out over the pin jig. Just like the previous stage, the assembly stage is divided into two, namely, assembly for flat blocks and assembly for curved blocks.

The erection stage is the stage of joining blocks resulting from the joining process at the assembly stage. The tugboat erection block process at this shipyard is carried out at the transverse south slipway. Blocks are transported using 2 tower cranes.

5.2. Layout Design

Layout design is carried out by considering the existing layout position. In addition, design is carried out by looking at the needs in ship building using the PWBS concept. Layout planning also takes into account the ease of access to materials, vehicles and employees at work.

Changes to the workshop area were carried out in the hull construction workshop area. These changes are in the form of rearranging the process lane in the workshop. The layout of the workshop is arranged using the group layout technology concept with the type of cell manufacturing production system. In this concept, the production floor will be divided into several process lanes and each process lane contains several work stations with machine and operator facilities in it.

This layout change was made for the two existing hull workshops. The southern hull workshop was used to build ships of smaller dimensions. Meanwhile, the northern hull workshop was used for larger ships. In this design, the south hull workshop will be divided into two areas, namely the fabrication area and the sub assembly area. Several work stations planned for the fabrication stage are pre storage material, marking, cutting, bending, and edge preparation. Figure 2 is comparison between the existing layout with the new layout. The colored lines show the material flow of each category.

In addition, in the sub-assembly area two process lanes are also planned, namely the process lane for the manufacture of the internal structure from the flat panel and the process lane for the manufacture of the internal structure from the curved panel. The panels that have been formed in this workshop will then be brought to the assembly area to be combined into blocks. The assembly area used is the existing assembly area, there is no significant change, only the area of merging the sub-blocks into blocks is arranged in erection order.

5.3. Impact of New Layout Application

Tabel 2. Estimated Time at Fabrication Stage

Learning index	-0,120294234
Number of Fabrication	**Hours**
1	945,2599133
2	869,63912
3	761,9809819
4	644,9407029
5	531,4197923
6	428,3814107

IOP Conf. Series: Materials Science and Engineering 1052 (2021) 012046 doi:10.1088/1757-899X/1052/1/012046

Number of Fabrication	Hours
7	338,9769715
8	263,9572999
9	202,6486922
10	153,6206405

One of the advantages of the technology group concept in the manufacturing industry is that it can shorten the material handling distance. This can happen because the machines are placed close together. In addition, the production process flow is also designed to minimize looping work processes. Based on the layout plan above, the results of the reduction in material handling distance from the existing condition to the new condition are quite significant. For material in category A, the reduction was 2-19%, category B was 29%, category C was 57% and category D was 2-56%.

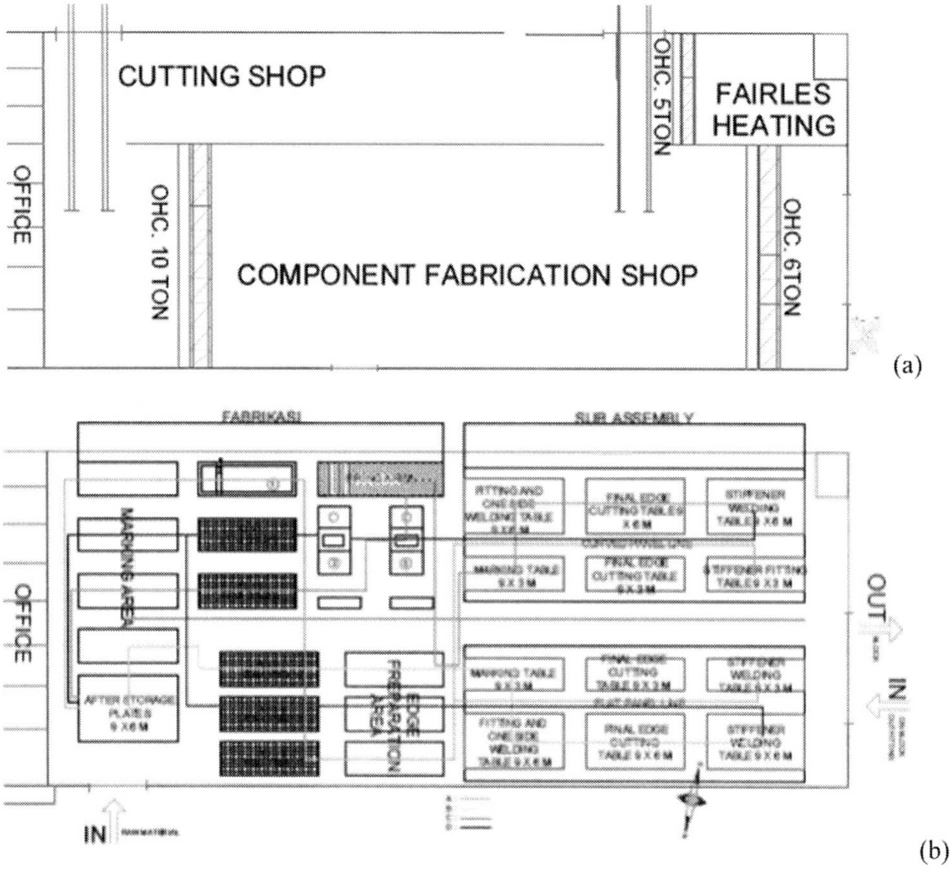

Figure 2. (a) Workshop Layout Existing, (b) New Layout Design

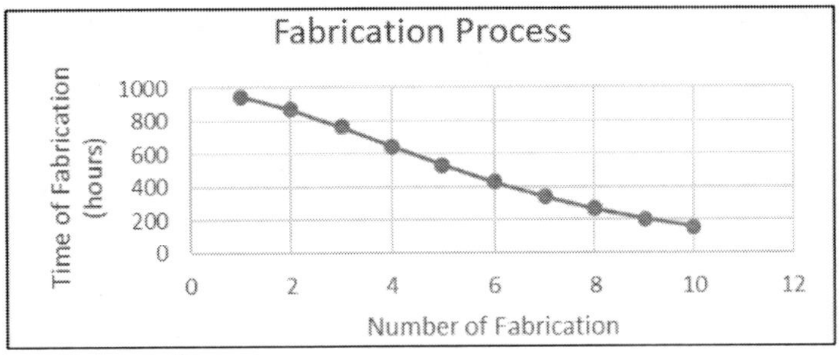

Figure 3. Decreased Time at The Fabrication Stage

Apart from reduced material handling distances, the technology group concept also affects production time. Workers who perform the same type of task over and over will be able to reduce the working time on the next task (learning curve theory). **Tabel 2** and **Tabel 3** is estimated time for ship fabrication and sub assembly tugboat 2x1800 HP. From **Tabel 2** and **Figure 3** also **Tabel 3** and **Figure 4** It is seen that the most significant reduction in time at the fabrication stage occurred in the early days of concept application. However, this production time can go back up if there are workers who are moved or there are worker rotation activities carried out by the shipyard. Not long after that the production time will return to normal again. To maintain the stability of the performance of workers and machines, regular training for workers and maintenance of production machines is still required.

Tabel 3. Estimated Time at Sub Assembly Stage

Learing index	-0,120294234
Number of Sub Assembly	Hours
1	1662,249604
2	1529,269636
3	1339,951656
4	1134,135081
5	934,5073531
6	753,3132639
7	596,0946073
8	464,1717173
9	356,3598792
10	270,143529

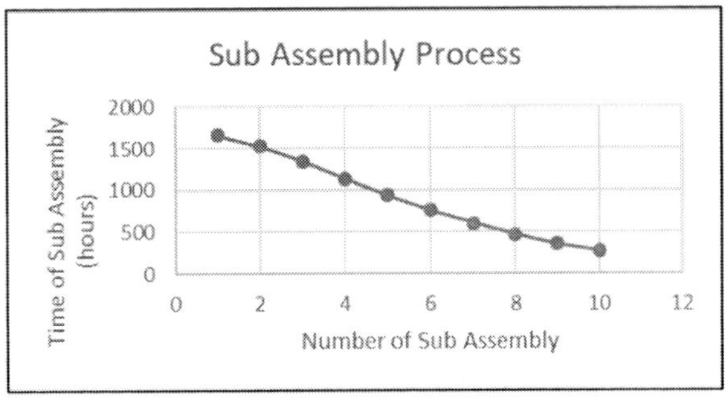

Figure 4. Decreased Time at Sub Assembly Stage

Tabel 4 and **Tabel 5** show that repetitive fabrication of individual components such as brackets can reduce fabrication time for subsequent jobs. This is not limited to the type of ship, because in general the shape of the parts in each ship is the same, sometimes the difference is only in the thickness of the plates used.

Tabel 4. Manhour Value of Bracket and Keel Plate Fabrication

Plate	208 piece bracket (1 plate = 52 bracket)	Keel plate
	Cutting (manhour)	
1	6,87922	7,2
2	6,67284362	6,98400023
3	6,358353432	6,65484527
4	5,982575139	6,26154433
5	-	5,83399982
6	-	5,39227307

Tabel 5. Manhour Value of Bottom Plate Fabrication

Plate	Bottom plate
	Bending (manhour)
1	5,67
2	5,499900181
3	5,240690654
4	4,930966162
5	4,594274856
6	4,246415044
7	3,898396811
8	3,557961865
9	3,230492403
10	2,919613909

5.4. Economical Analysis

Changes in shipyard layout cannot be separated from the calculation of the investment value. It takes a number of costs that must be borne by the shipyard to make changes. In accordance with the results of the technical planning above, the total investment value is IDR 7,471,954,000.

However, changing the layout by applying the technology group concept can also cut ship building times. Costs that can be reduced include the operational costs of material handling facilities and daily costs. The cost reduction in the fabrication stage can reach IDR 36,152,935 and the cost at the sub-assembly stage is reduced to IDR 27,178,200.

So, taking this into account the feasibility value of the investment is calculated. Then this layout change is considered feasible with an IRR value of 11%, it is estimated that the payback period will occur in 3rd year .

6. Conclusion

After analyzing and calculating, the conclusions of this research are as follows:

1. The existing shipyard has two hull workshops which are used for the construction of new building ships. The north hull workshop was used to build ships with large dimensions while the south hull workshop was used to build smaller ships such as tugboats. The fabrication and sub-assembly stages

were carried out in the hull workshop while the assembly and erection stages were carried out in the south building area.

2. The PWBS concept can be applied to the construction of a tugboat by adjusting the layout of the workshop. Adjustments made in the form of production lane design and rearrangement of production machines. It is planned that 4 production lanes in the fabrication area and 2 lanes in the sub assembly area are planned. Each lane contains a work station for each activity. Inside this work station contains rearranged existing machines.
3. By using a workshop layout based on the group technology, costs at the fabrication stage can be reduced to IDR 36,152,935 and sub assembly to IDR 27,178,200. To make changes to the layout, an investment of IDR 7,471,954,000 is required. So that this layout change is considered feasible with an IRR value of 11% and a payback period in 3rd year.

7. Referances

[1] U. Aalto 2016 *Shipyard engineering Lecture 1-1: Shipbuilding process* in *Shipyard Engineering*
[2] R. L. Storch 1995 *Ship Production*. New Jersey
[3] R. A. Shenoi 2007 *Ship Production Technology Southampton*: School of Engineering Sciene pp 1–10
[4] R. Saraswati 2008 *Perancangan Group Technology Layout di PT DPS Surabaya dengan Metode Simulasi dan Taguchi* Jur. Tek. Ind. Univ. W.R. Supratman Surabaya
[5] B. Brookfield 2005 *Management Accounting - Decision Management* no. April, pp 44–46
[6] S. Goldsmith 2016 *Naval Shipbuilding* no. May
[7] T.W. Pribadi 2018 *Manajemen Produksi Kapal*. Surabaya

The effect of low tempering, medium tempering, and high tempering heating temperature variations in the type of medium carbon steel ST 60 on microstructure, hardness, and toughness.

H Yudo[1] , S Jokosisworo[1]
[1] Naval Architecture Department, Diponegoro University, Indonesia.

Abstract. To determine the changes in the microstructure, impact strength, and hardness of the carbon steel medium type ST 60 with variations in temperature: quenching, tempering 200°C, tempering 400°C and tempering 600°C, then a heat treatment was given at 850°C for 30 minutes then quenched with brine solution with a salt content of 10% and continued with tempering for 30 minutes with variations in tempering temperature: 200°C, 400°C and 600°C. The chemical composition of the ST 60 steel specimen obtained that the main constituent elements are iron (Fe) = 98.1%, manganese (Mn) = 0.572%, silisium (Si) = 0.246%, and carbon (C) = 0.362%. Meanwhile, the hardness, impact, and microstructure test results show that the quenching specimen has the hardest and softest material properties. This is indicated by the highest hardness value of 772.67 VHN and the smallest impact value of 1.24 J / mm². The martensite phase structure formed causes the material properties to be the hardest. In the 200°C tempering specimen the hardness began to decrease and the ductility increased as indicated by the hardness value of 625 VHN and the impact value of 2.02 J / mm². In this specimen, a tempered martensite phase was formed, which is softer than martensite. In the 400°C tempering specimen, the hardness value was getting smaller, but the ductility increased as indicated by the hardness value of 448.67 VHN and the impact value of 2.56 J / mm². The bainite phase structure with ductile / malleable properties was formed in this specimen. In the 600°C tempering specimen the properties of the resulting material were the most ductile and softest indicated by the lowest hardness value of 295 VHN and the highest impact value of 3.11 J / mm². The structure of the ferrite and fine pearlite phases with very soft properties was formed in this specimen.
Keywords : Quenching, Tempering, Martensite, Ferrite, Bainite, Pearlite

1. Introduction

The heat treatment process aims to obtain a metal that is hard, soft, ductile, and eliminates residual stress. Heat treatment is often referred to as a way to increase the hardness of the material, in fact it can also be used to change the properties of use or with certain interests for user needs, such as: increasing malleability, restoring elasticity after cold work. Even heat treatment not only changes the properties of the material, but is also able to increase the performance of the material by increasing the strength or certain characteristics of the material that has been heat treated (Beumer, 1985).

Through the tempering process, hardness and ductility can be reduced to meet the requirements of use. If the hardness decreases, the tensile strength decreases while the ductility and toughness of the steel increase. The tempering process consists of reheating the hardened steel to below the critical temperature, followed by cooling. Although this process produces a softer steel, it differs from annealing process because here physical properties can be carefully controlled (Amstead, 1995).

Based on the above background, the writer needs to conduct research studies on the effect of tempering temperature variations on ST 60 steel to determine the effects that occur so that the results can be possible in more optimal use in several applications in the field.

2. Study Methods

The study method used is experimental method. The research material referred to in this includes primary and secondary data as well as theory and references which form the basis of this research. The selected experimental design category is ST 60 steel specimen that has undergone heat treatment with temperature variations, namely tempering.

- *Heating Process*

Heating begins with the preparation of materials and heating kitchen. Heating is done using an electric kitchen. The quenching specimens were heated at 850^{0}C while the tempering process was set at 200^{0}C, 400^{0}C, and 600^{0}C, respectively.

- *Quenching Process*

The quenching process is carried out by cooling all specimens that have been heated at 850^{0}C into a brine solution with a salt content of 10%.

- *Tempering Process*

The tempering process is carried out by reheating the tempered specimens at temperatures of 200^{0}C, 400^{0}C, and 600^{0}C respectively into a heating kitchen then cooled naturally in open air.

- *Testing Specimens*

This specimen test was carried out to determine the characteristics of the material from quenching and tempering at temperature variations of 200^{0}C, 400^{0}C, and 600^{0}C, respectively.

3. Results And Discussion

Chemical Composition Test Results

This test aims to determine the percentage content of the elements in the steel contained in the test object, in particular the carbon content (C). Testing the chemical composition using the WAS Spectrometer by firing argon gas at the test specimen then giving the reading results automatically in the form of the average chemical composition content of the specimen or test object.

Figure 1. Spectrometer Testing Equipment and Test Specimens

From the test results, the chemical composition of the specimen contains the main constituent elements iron (Fe) = 98.1%, manganese (Mn) = 0.572% which is useful for increasing hardness, strength, and can be hardened in steel. Silisium (Si) = 0.246% which is influential in increasing the overall hardening ability, wear resistance, resistance to heat and rust. While other elements obtained are: copper (Cu) = 0.0125%, carbon (C) = 0.362%, aluminum (Al) = 0.0117%, chromium (Cr) = 0.369%, phosphorus (P) = 0.0654%, sulfur (S) = 0.0094%, nickel (Ni) = 0.0135.

Based on the carbon content (C) in the test specimen of 0.369%, the test material is included in the content of medium carbon steel, which is the main requirement for medium carbon steel to contain carbon (C) between 0, 30% - 0.60%.

Testing of Micrographic Structures
The new micro structure will be clearly visible when the surface of the test object is completely flat, smooth and shiny without scratches, and has undergone proper embracing. Observations were carried out under an Olympus Metallurgical Microscope with optimal magnification, while for shooting, the Olympus Photomicrographic System was added.

Microstructure testing is carried out to observe and physically compare the microstructure of each part of the specimen, observe the microstructure of the specimen without, and by experiencing heat treatment followed by a varied cooling process. The main location for taking photos is specimens with quenching treatment and for tempering specimens with temperature variations: 200^{0}C, 400^{0}C and 600^{0}C.

Shooting will be done after the appearance is completely clear or in focus. The number of objects photographed for their micro structure is the same as the number of test objects (tested for hardness) because the hardness test is carried out on objects that have already been photographed for their microstructure. The steps for testing the microstructure are as follows:

1. Inserting the film into the Olympus Photomicrographic System camera that is available.
2. Turn on the appliance by pressing the power button.
3. Prepare test specimens.
4. Place the specimen on the test plane or table.
5. Ensure that the specimen is not skewed.
6. Select the desired magnification.
7. Determine the shooting point so that it can be seen clearly.
8. Carry out the shooting process with the desired magnification

a. *Micro Quenching Photo*

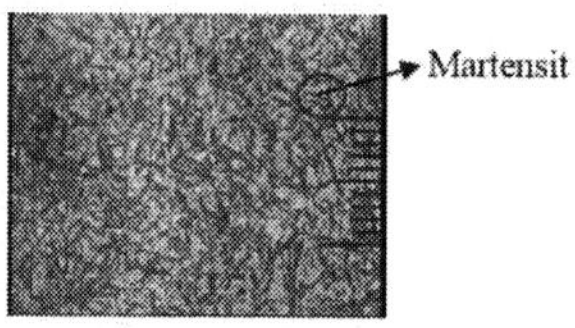

Figure 2. Photo of Micro Structure of Medium Carbon Steel Quenching with Magnification 200x.

This specimen has the highest hardness value, this is due to the presence of a martensite phase (black color with a hardness value> 500 VHN) which is shaped like larger needles that are evenly distributed due to rapid cooling (Quenching) with salt water, causing the properties of the material. hardest and brittle which is one of the main properties of the martensite structure. Although, the hardest and most ductile properties make the specimen tough or malleable or easy to break if the excess energy is applied to it.

b. Photo Micro Tempering 200 °C

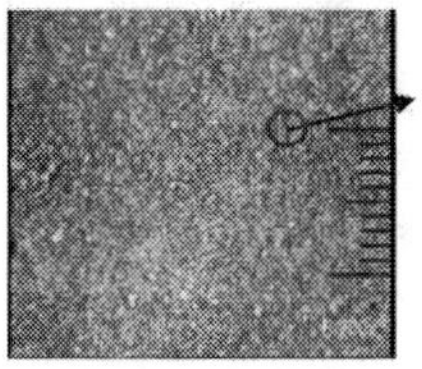

Figure 3. Photo of Medium Carbon Steel Tempering 200°C Micro Structure with 200x Magnification.

At the lowest tempering temperature with a temperature of 200°C, the martensite temper (martemper) phase is obtained which can prove the occurrence of a very fast cooling rate followed by reheating. The structure of tempered martensite (martemper) is formed more tightly and evenly. The fast cooling rate then with the lowest heat input of 200°C produces tempered martensite whose structure is shaped like needles that are evenly distributed (black color and hardness value> 500 VHN) and smaller than the martensite structure, causing a decrease in the hardness value of the test material. The nature of the tempered martensite structure is softer than the quenched martensite structure, because the quenched masrtensite structure has undergone improvement / transformation of the phase structure.

c. Photo Micro Tempering 400 °C

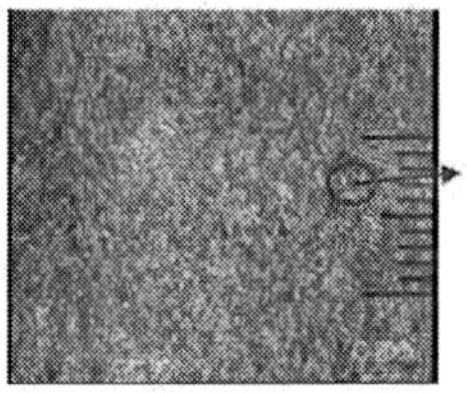

Figure 4. Photo of 400 C Tempering Medium Carbon Steel Micro Structure with 200x Magnification.

There is a predominance of the bainite phase structure (hardness values between 300-400 VHN). The process of forming the bainite structure begins with heating the steel until it reaches the austenite phase with a high temperature of 850°C for 30 minutes which is then followed by rapid immersion / cooling in salt water with a salt content of 10% and then reheated (tempering) at a temperature above the formation Martensite start (Ms) with a temperature of 400°C, held for a while (30 minutes) and continued with normal cooling in the air. So that the austenite phase structure can be transformed into a bainite phase structure. This process causes the hardness value of the test specimen to decrease, but the ductility or toughness increases.

d. Photo Micro Tempering 600 ºC

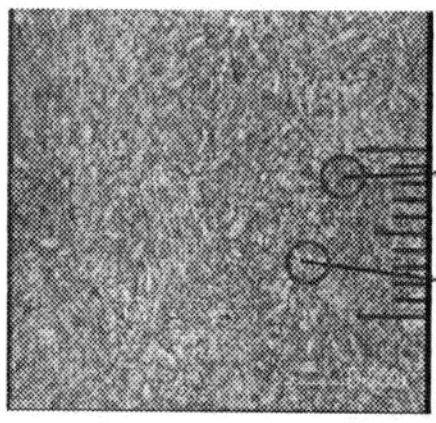

Figure 5. Photo of Micro Structure Medium Carbon Steel Tempering 600°C with 200x Magnification.

It can be seen that the structural forms of the ferrite and fine pearlite phases (hardness between 180-300 VHN) with small and fine grains (fine pearlite). The form of the ferrite structure appears white and soft. While the form of fine pearlite structure is small dark / black granules. The structure of the ferrite and pearlite phases is formed from the transformation of the austenization process (heating the material to a temperature of 850°C) due to slow cooling (re-heating after the quenching treatment until the ferrite and pearlite phases are formed), causing the specimen to be very tough, ductile, and softest which is a common characteristic. both of these phase structures.

a. Impact Test

This test aims to determine the value of resistance, toughness and visibility of the test object against dynamic loads. The impact test is carried out in one stroke for one specimen and 3 specimens for each variation, including specimen A (quenching), specimen B (tempering 200°C), specimen C (tempering 400°C), and specimen D (tempering 600°C).

Things that must be considered when carrying out an impact test are the energy / energy absorbed by the test object or energy to break the test object which can be written in the form of a formula :

$$E = m.g\,(h1\text{-}h2) \tag{1}$$

So,

$$\text{Impact Value} = \frac{\text{energi terserap (j)}}{\text{luas penampang payahan benda uji (mm}^2\text{)}} \tag{2}$$

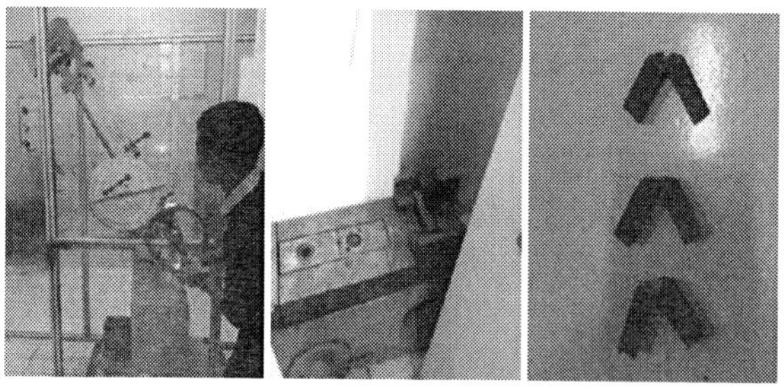

Figure 6. Impact testing process

Table 1. Impact Test Results using the Charpy method

No	Speciment	Wide (mm)	High (mm)	Energy (J)	Impact Value (J/mm²)
A1	*Quenching*	10.32	7.60	88.00	1.12
A2		9.98	7.56	96.00	1.27
A3		10.06	7.10	98.00	1.37
B1	*Quench-Tempering* 200⁰C	10.28	7.40	202.00	2.66
B2		10.10	7.26	134.00	1.83
B3		10.14	7.58	198.00	2.58
C1	*Quench-Tempering* 400⁰C	10.14	7.24	198.00	2.70
C2		10.10	6.80	184.00	2.68
C3		10.12	7.36	172.00	2.31
D1	*Quench-Tempering* 600⁰C	10.30	7.38	226.00	2.97
D2		10.10	7.28	238.00	3.24
D3		10.06	6.98	220.00	3.13

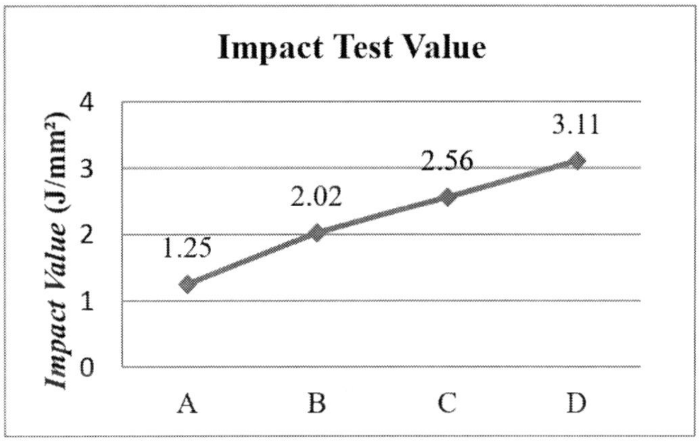

Figure 7. Graph of Mean Value of Impact Test

From the results of the impact test, the highest average toughness value (toughest and most ductile) was the specimen with a tempering treatment of 600⁰C of 3.11 J / mm² and respectively to the lowest position, namely: the specimen of 400⁰C tempering treatment of 2, 56 J / mm², the 200 perlakuanC tempering treatment specimen was 2.02 J / mm² and the lowest (most brittle / soft) specimen was quenched with an impact value of 1.25 J / mm².

Impact Specimen Quenching

It was the most brittle and softest specimen with the lowest mean impact value (1.25 J / mm²). This is due to rapid cooling after heating to a temperature of 850 ⁰C, so that it forms large / coarse metal crystals (martensite), the bonds between the grains are weak and brittle breaks easily. The brittle type fracture can be seen in the following figure with the smallest deformation characteristic (shown by a flat surface fracture).

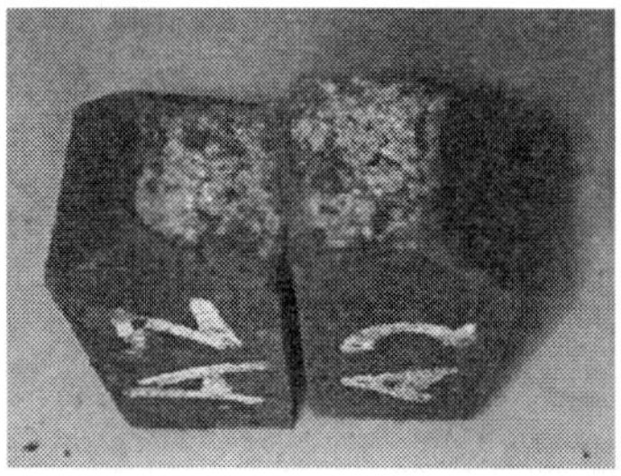

Figure 8. Quenching Specimen Fault

Impact Specimen Tempering 200 ℃
In the 200^{0}C (2.02 J / mm^2) tempering specimen, the tempered martensite phase was stronger, although the hardness decreased because the metal grains were slightly smaller causing the metal bond to be stronger. Brittle-type fractures were seen in this specimen.

Figure 9. Fault of Tempering Specimen 200 ^{0}C

Impact Specimen Tempering 400 C
The impact strength of the 400^{0}C tempering specimen of 2.56 J / mm^2 is greater than that of the 200^{0}C tempering specimen because the bainite phase with a fine carbide dispersion in ferrite makes the metal crystalline grains finer which leads to increased toughness and visible test specimens. The clay type fracture shows high plastic deformation (the fracture surface looks rough and tapered).

Figure 10. Fault of the 400 C Tempering Specimen

Impact Specimen Tempering 600 ℃
The 600^{0}C tempering specimen had the greatest average impact value, at 3.11 J / mm^2. Due to the slowest cooling (with a maximum heat input of 600^{0}C) it produces the finest grains of ferrite and pearlite crystals that make this specimen the toughest and most resilient. The fracture that occurs is a clay fracture, seen from the more sharp and sharp fractures that show higher plastic deformation and greater energy absorption.

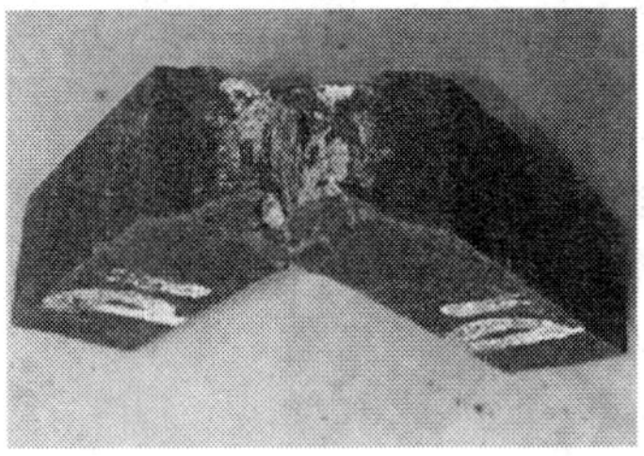

Figure 11. Tempering Specimen Fault 600 ^{0}C

b. Hardness Test

Hardness testing is carried out to determine the level of hardness and resistance of a material to deformation in the local area and material surface. In this case, the hardness test for ST 60 steel uses the Vickers method. The data from the test results are grouped into 4 groups of specimens, namely specimen A (Quenching), specimen B (tempering 200^{0}C), specimen C (tempering 400^{0}C), and specimen D (tempering 600^{0}C) with 3 test points on each specimen.

The hardness testing procedure is by pressing a hardened steel ball or forming a diamond pyramid on its surface. Then the size of the scrap is measured based on the pressure load and the size of the ball or pyramid, so the area of the mark gives a ratio of hardness values. Steel balls are used in the Brinell hardness test and the diamond point in the Vickers pyramid hardness test, which are more suitable for harder metals. In this test, the Vickers method is used with a load of 40 kg. The Vickers hardness test uses a diamond pyramid indenter which is basically a square shape. The angle between the faces of the diamond pyramid facing each other is 136. This value was chosen because it is close to most of the desired ratio values between the indentation diameter and the diameter of the impact ball in the brinell hardness test. The vickers hardness number is defined as the load divided by the surface area of the indentation. Practically. This area is calculated from microscopic measurements of the diagonal length of the trace. VHN can be determined from the following equation :

$$\text{VHN} = \frac{2P\sin(\frac{\theta}{2})}{d^2} = \frac{(1.854)P}{d^2} \tag{3}$$

with :
P = load used (kg)
d = the average diagonal length (mm)
θ = the angle between the facing diamond faces = 136^0

The Vickers hardness test is mostly done in research work because the method gives results in the form of a continuous hardness scale. The shape of the trampling marks that meet the Vickers principle is shown in Figure 12.

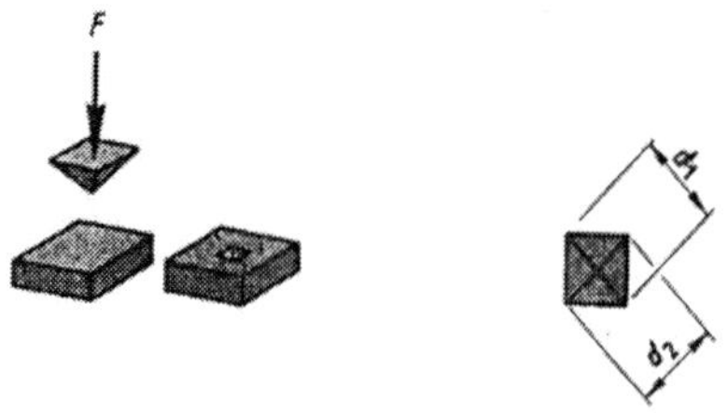

Figure 12. The Principle of the Hardness Testing Process Using the Vickers Method with a load of 1 to 120 kg.

Figure 13. Hardness Testing Process Using the Vickers Method with a Load used of 40 kg

Table 2. Hardness Test Results

No	Specimens	d1 (mm)	d2 (mm)	Average of d (mm)	Hardness Value (VHN)
A1	*Quenching*	0.30	0.31	0.305	797
A2		0.32	0.32	0.320	724
A3		0.31	0.30	0.305	797
B1	*Quench-Tempering* 200°C	0.37	0.35	0.360	572
B2		0.34	0.33	0.335	661
B3		0.34	0.34	0.340	642
C1	*Quench-Tempering* 400°C	0.40	0.40	0.400	464
C2		0.42	0.40	0.410	441
C3		0.42	0.40	0.410	441
D1	*Quench-Tempering* 600°C	0.50	0.50	0.500	297
D2		0.50	0.49	0.495	303
D3		0.51	0.51	0.510	285

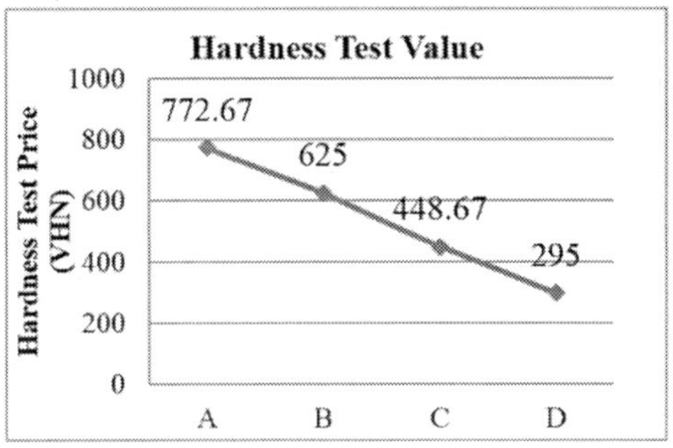

Figure 14. Graph of Average Hardness Test Value

From the results of the hardness test, the highest average hardness price was found in the quenching specimen of 772.67 VHN and respectively to the lowest position, namely: 200°C tempering specimen at 625 VHN, 400°C tempering specimen at 448.67 VHN and the lowest tempering 600°C specimen. amounting to 295 VHN.

Quenching Specimen Hardness
Specimens with treatment, only on quenching without tempering and obtained the highest average hardness price of 772.67 VHN because the quenching treatment has a martensite phase and due to the fastest cooling rate by immersing the test object into a salt solution so that the phase transformation of the austenite runs more very quickly into the martensite phase. And this is evidenced by the highest price of hardness (the smallest diameter of the hole test results).

Tempering Specimen Hardness 200 ℃
The 200 SpesC tempering specimen with an average hardness value of 625 VHN is due to the fastest cooling rate by immersing the test object into a salt solution and then reheating (tempering) with the lowest temperature (200°C) so that the phase transformation goes up to the martensite finish (Mf) until a tempered martensite phase was formed (below 220°C). The nature of this specimen is still classified as hard, indicated by the diameter of the test result hole which is still of little value.

Tempering Specimen Hardness 400 ℃
The average hardness price of 400°C tempering specimen is 448.67 VHN due to the temperature condition of the specimen passing through the martensite start area (Ms) and the temperature getting hotter (400°C) so that the phase transformation from temper martensite enters the bainite phase area (250-550°C) which indicated by a decrease in the hardness value of the test specimen or the softness of the specimen (indicated by the larger value of the hole in the test results).

Tempering Specimen Hardness 600 ℃
The average hardness price of 600°C tempering specimen was 295 VHN. Because at a temperature of 600°C, there is a transformation of austenite to pearlite and ferrite (550-723°C) or from tempered martensite to bainite and then transformed into pearlite. Due to the very high heating temperature (slow cooling) so that the resulting metal grains become fine, small which causes the smallest hardness value (the softest, indicated by the diameter of the test results with the largest value).

Combining Test Results Graphs and Tempering Test Graph Correlation Using Time Transformation Temperature (TTT) Diagrams.

The following is the result of combining the test graph and the correlation of the tempering test graph using the Time Transformation Temperature (TTT) diagram :

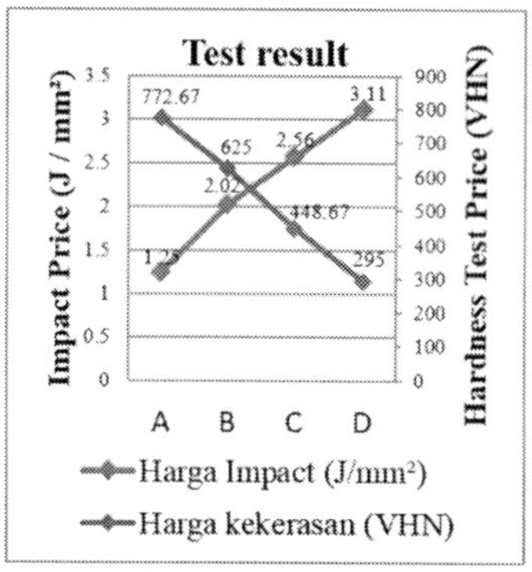

Figure 15. Graph of Hardness and Impact Test Results for St 60 Steel with Tempering Temperature Variations

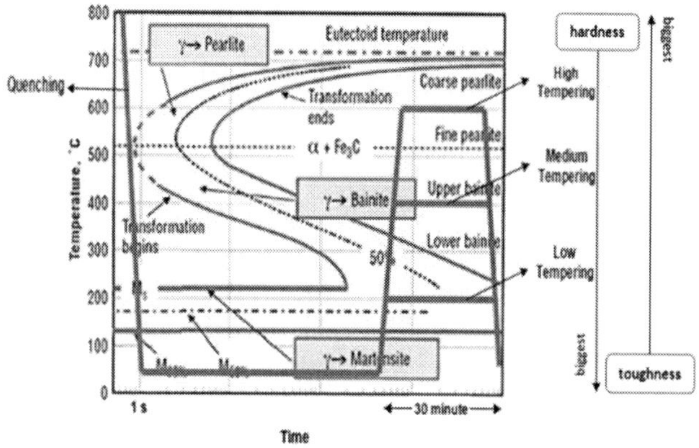

Figure 16. Graph of Tempering Treatment Diagram

The Martensite phase (the hardest character) is formed due to the rapid cooling / immersion process in a brine solution which was previously heated to a temperature of austenite (850°C). Then 200°C tempering by reheating the quenched specimen for 30 minutes to reduce the hardness value by forming a mild temper / martemper martenst phase.Tempering at 400°C enters the formation of the bainite phase. By reheating the quenched test specimen at 400°C for 30 minutes. The formation of the bainite phase gives the material properties to be tougher than the 200°C test specimen, thereby reducing the hardness value and increasing the toughness value of the test specimen. High tempering with the highest temperature, which is 600°C, produces the toughest test specimens with the formation of fine pearlite and ferrite phases, which have soft properties.

4. Conclusions

Based on the research data and the discussion in this study, it can be concluded that the quenching specimen has the hardest and softest material properties, indicated by the highest hardness value of 772.67 VHN and an impact value of 1.24 J / mm². The martensite phase structure formed causes the material properties to be the hardest.

In the 200°C tempering specimen, hardness began to decrease and the toughness / ductility increased, as indicated by the hardness value of 625 VHN and the impact value of 2.02 J / mm². In this specimen a martemper phase (tempered martensite) is formed, which is softer than martensite.

In the 400°C tempering specimen, the hardness value was getting smaller, but the toughness increased. This is indicated by a hardness value of 448.67 VHN and an impact value of 2.56 J / mm². The bainite phase structure with ductile / malleable properties was formed in this specimen.

In the 600°C tempering specimen, the properties of the resulting material were the toughest / ductile and softest, indicated by the lowest hardness value of 295 VHN and the highest impact value of 3.11 J / mm². The structure of the ferrite and fine pearlite phases with very soft properties was formed in this specimen.

5. References

[1] Amstead, B.H., Djaprie, S, 1995, *Teknologi Mekanik*, Edisi ke-7, Jilid I, PT. Erlangga, Jakarta
[2] Amanto Hari dan Daryanto, 1999. *Ilmu Bahan*, Jakarta Smart Grafika Offset. *ASM Metal Handbook Vol. 9, 2004*
[3] *ASTM Standards: A370, The American Society For Testing And Material, 2001, Anual Book Of ASTM Standards.*
[4] Beumer, (1985). *Ilmu Bahan Logam Jilid 1*. Jakarta : Bharata Karya Aksara.
[5] Dieter, G.E., 1987, *Metalurgi Mekanik*, terjemahan. Sriati D., Erlangga, Jakarta.
[6] Djafrie, S. 1985. *Teknologi Mekanik Jilid 1.* Terjemahan dari *Manufacring Process,* Erlangga. Jakarta
[7] Gary, M. 2011. *Heat Treatment.* Makalah Proses Produksi. Universitas Sriwijaya
[8] Groenendijk, G.; Van Der Linde, J.; Sachri, S, 1984, *Pengujian Material*, Cetakan ke-1, CV. Binacipta, Jakarta.
[9] Kurniawan, David. 2009. Pengaruh variasi konsentrasi garam dalam larutan air pencelup pada proses quenching baja ST 60 terhadap kekerasan dalam meningkatkan kualitas baja. Jurusan Teknik Mesin, UM, Malang.
[10] Murtiono, Arief. 2012, *Pengaruh Quenching dan Tempering terhadap kekerasan dan kekuatan tarik serta struktur mikro baja karbon sedang untuk mata pisau permanen sawit*, Departemen Teknik Mesin, Universitas Sumatera Utara, Medan.
[11] Paul, Degarmo, 2003. *Materials and Process in Manufacturing,* Wiley.
[12] Redi, Ridwan, 2017. *Analisa Kekuatan Puntir, Kekuatan Tarik dan Kekerasan Baja ST 60 Sebagai Poros Baling-Baling Kapal (Propeller Shaft) Setelah Proses Tempering)* , Departemen Teknik Perkapalan, Fakultas Teknik, Universitas Diponegoro, Semarang.
[13] Resnick, Halliday (1985) PHYSICS, 3rd Edition. Bandung: Departemen Fisika Institut Teknologi Bandung.
[14] Soedjono. 1987. Pengetahuan Logam I. Departemen Pendidikan dan Kebudayaan. Jakarta.
[15] Supringga, Herdiyan, 2009. *Analisa Kekuatan Puntir, Lentur Putar dan Kekerasan Baja ST 60 untuk Poros Propeller setelah diquenching.* Jurnal Teknik Perkapalan Undip Volume 11 Nomor 2 Fakultas Teknik – Universitas Diponegoro.
[16] Van Vliet, G.L.J, 1984, *Teknologi untuk Bangunan Mesin Bahan-bahan I*, PT. Erlangga, Jakarta.
[17] Van Vlack; Djaprie, S., 1992, Ilmu dan Teknologi Bahan, PT. Erlangga, Jakarta

IOP Conf. Series: Materials Science and Engineering 1052 (2021) 012048 doi:10.1088/1757-899X/1052/1/012048

Fixed Manpower Reduction Analysis to Reduce Indirect Cost of Ship Production

S I Wahidi[1], S R W Pribadi[1], T W Pribadi[1], N L A Wardhana[1] and M S Arif [1].

[1] Department of Naval Architecture, Faculty of Marine Technology, Institut Teknologi Sepuluh Nopember, 60111 Surabaya, Indonesia

Abstract. In the shipping industry, especially shipyards which are used for building and repairing ships, a lot of labor is required. Due to a lot of labor, of course, makes the salary component to be a fairly large cost component in the company. Nowadays, many shipyards have turned to outsourcing-based shipyards, to reduce indirect costs. However, other shipyards have not yet become outsourcing-based shipyards, so the number of workers is still relatively large. The high number of permanent workers in a company indicates the effectiveness and efficiency of the workforce in the company is still lacking, so that the company will experience losses because the workload received by each workforce does not match the salary received by the worker. The research was conducted using the Full-Time Equivalent (FTE) method, which the components required for this method are the effective working days in 1 year and the effective working time in 1 year. After the analysis was carried out, the percentage of reduction in labor remained at 55%, and the percentage reduction in indirect costs is 38%.

1. Introduction

A shipyard is a place that has a function to carry out the process of building a ship (New Building) and repairing a ship and also carrying out ship maintenance. The shipyard can also be used as a shipbuilding process including design, initial frame installation, hull plate installation, equipment installation, checking, feasibility test, to classification by the designated class. The various activities carried out at the shipyard will certainly require sufficient manpower.

Shipyard companies certainly have a lot of activities, both production, and non-production activities. Based on the activities carried out by the shipyard company, the workforce used is of course a different workforce. This is because if a non-production worker is given a portion of the work for production, then the work cannot run properly.

Permanent labor are workers who work in an agency or company permanently based on a decision letter. The existence of permanent workers in a shipyard company will certainly create costs for the company, there are quite some shipyard companies that rely on outsourced labor. The number of permanent workers in the company reached 453 plus 3 Directors in 2017. With that much workforce, of course, it will cause a lot of costs to arise [1].

The determination of the workforce is not only without clear reasons but must be based on the workload received by each worker. Therefore it is necessary to analyze the reduction of permanent manpower so as not to cause cost overruns for the ship production process at the shipyard.

IOP Conf. Series: Materials Science and Engineering 1052 (2021) 012048 doi:10.1088/1757-899X/1052/1/012048

2. Literature Review

2.1. Cost
The cost for a company that produces a product is the price of the production factors that are used to produce its output or its production. The components of the cost, namely:

a. Direct Material Costs

The material costs used are directly used to enter the production process to realize a product result which will be handed over to the owner/customer. Direct material costs for shipyard are divided into 2, namely:

- Staple materials are the raw materials needed to realize production results, including steel plates/profiles, shaft material, wood, paint for rust protection and color paints, main/auxiliary motors, machinery, valves, pipes, navigation equipment, life safety devices. at sea.
- Auxiliary materials are materials needed to process basic materials to achieve a production result, including welding electrodes, oxygen gas, liquid acetylene, carbide, liquid LPG, paint/lime for heaters. [1]

2.2. Direct Labor Costs
Direct labor costs are costs for direct labor that are forged and utilized in handling production process activities which are integrally used to handle all production equipment/facilities so that the production process can be realized. [2]

2.3. Indirect Costs
Indirect costs or overhead costs are indirect material costs and indirect labor and other costs that do not arise and are needed to support the successful completion of the production process.

These indirect material costs include fuel costs for diesel welding motors, electricity costs for driving equipment/production facilities and lighting, equipment costs/occupational safety and health, material costs for smooth work such as lime, paint pen.

Indirect labor costs are labor costs that are not directly utilized for production process activities, but are required to support the success and smoothness of the production process; among others: marketing personnel costs, administrative / personnel costs, calculation/invoice costs, material procurement/storage costs, production design/preparation/supervision personnel costs and others. [1]

2.4. Labor
Labor is any person who is capable of doing work to produce goods or services both to fulfill their own needs and for the community. Some workers are permanent workers and temporary workers. Permanent workers are workers who work in a particular company and are bound by the decisions of that company. Meanwhile, non-permanent workers are workers who only receive income if the employee concerned works, based on the number of working days, the number of work units produced, or the completion of a type of work requested by the employer. [2]

The Labor consist of :

- Direct Labor

  Labor that can convert raw materials directly into finished products and the costs can be charged appropriately to these products.
- Indirect Labor

  Labor that cannot be traced directly to the construction or composition of a finished product. [3]

2.5. Workload
Workload is several work targets or target results that must be achieved in a certain time unit. The workload is the main aspect that forms the basis for calculating employee formation. In this method there are three main stages which will be described below:

- Determine the main output of a function/sub-function and then identify the series of work activities required to produce that output.

- Breaking down a series of activities into more specific task units.
- Calculate the total amount of time required to complete per group of these tasks. [4]

Based on the workload of each workforce, it can be calculated how much labor is needed for an existing position, Department, and Division. The process of calculating this labor requirement can be referred to as resizing, resizing here can increase the effectiveness of the company itself, but it can also harm the company itself. [5]

Several main functions can support new shipbuilding activities [6]:

- Planning
- Engineering
- Purchasing
- Material Control
- Finance
- Production

2.6. Competence

Competence is the ability to carry out or perform a job or task which is based on skills or knowledge and is supported by the work attitude dictated by the job. [7]

2.7. Full Time Equivalent

Full-Time Equivalent is a time-based workload analysis method by measuring the length of time to complete the work then this time is converted into an index of the FTE value. The full-time equivalent (FTE) workload calculation method is a method where the time used to complete various jobs is compared to the available effective work time. FTE aims to simplify work measurement by converting workload hours to the number of people needed to complete a particular job. [8]

To get the FTE value from a work process is as follows:

$$FTE = \frac{(\text{Total Activity Time (a year)} + \text{Allowance})}{\text{Total Available Time (a year)}} \qquad (1)$$

In analysing the workload using the FTE (Full Time Equivalent) method, five steps need to be taken, namely [9]:

1. Determine the work unit and its energy category.
2. Determine the available work time for one year.
3. Developing allowance standards The objective of compiling this data is to determine the allowance for workers, which includes the types of activities and time needed to complete an activity that is not related to the main activity. Activities that are not directly related, for example, are resting, praying, or going to the toilet and several other activities.
4. Establish workload standards, which is the volume of workload felt by workers in completing their work (average time).
5. Calculate the power requirements per unit of work. At this stage, the researcher tries to obtain the number and category of the workforce that work following the workload.

2.8. Allowance

Allowance is the time it takes for workers to carry out activities that can meet their personal needs, relieve fatigue, and other obstacles that cannot be avoided. Allowance is actually needed by workers who as long as measurements are not observed, measured, recorded, or calculated. Therefore, after the measurement and after getting the normal time, Allowance needs to be added to obtain the standard time (standard time) as a basis for determining the workload.

The amount of allowance for each employee varies from one position to another because each position has its own characteristics. Therefore, the amount of Allowance that will be used in calculating the workload must be determined by the company. The allowance that is commonly used is 10% (for manufacturing) up to 20%-25% (for Government Departments/Agencies) of the total working hours per day.

Allowances are given for three things, namely for personal needs, eliminating feelings of fatigue, and obstacles that cannot be avoided. These three factors will be further explained as follows:

1) Allowance for personal needs
   Included in personal needs here are things like drinking moderately to get rid of thirst, going to the toilet, chatting with co-workers just to relieve tension at work. The amount of allowance given to such personal needs varies from one job to another because each job has its own characteristics with different demands. Based on the research, it turns out that the amount of this allowance for male workers is different from that for female workers.
2) Allowance to relieve fatigue
   Fatigue is reflected in, among other things, a decline in production yields both in quantity and quality. If fatigue has come and the worker has to work to produce his normal performance, then the work spent by the worker is greater than normal and this will increase fatigue. If this continues, the limb in question will not be able to perform the work at all even if desired. The things that workers need to get rid of fatigue are stretching the muscles, going outside the room to relieve fatigue and so on.
3) Allowance for inevitable obstacles.
   In carrying out their work, workers will not be free from obstacles that cannot be avoided because they are outside the workers' power to control them. Calculation of allowances for inevitable constraints is carried out using a separate sampling technique because the magnitude of the constraints for such an event varies greatly from one job to another and even from one work station to another. Some examples of unavoidable delays include: receiving instructions from supervisors, making machine adjustments, sharpening cutting tools, and so on. [8]

3. Research Methodology

3.1. Identification of Problems

At this stage before conducting the research, observation activities were carried out to observe the problems faced by PT. X. Problems regarding the number of permanent workers who are too much and the functional ineffectiveness of each existing position for each worker who is in a certain department and section.

3.2. Literature Review

At this stage, a literature study is carried out, which aims to obtain a theoretical basis. The theoretical basis in question is about Human Resources and labor. As well as regarding the indirect costs that have swelled at PT. X. The related literature:

- Human Resource Management
- Cost Analysis: Indirect Costs
- Ship Production: Ship Building Production Process
- Engineering for Ship Production
- Overview of previous studies regarding workload calculations using the FTE (Full Time Equivalent) method

3.3. Existing Condition Analysis

At this stage, an analysis will be carried out on the existing condition of PT. X so that the next process can be carried out, namely calculation analysis.

3.4. Data Management

At this stage data processing will be carried out to facilitate the activities that will be carried out next. Data processing is carried out by compiling a Business Process for new shipbuilding activities, which then the Business Process will be broken down each detail of the activity, to find out what parties can be associated with each of these activities. Then carried out the preparation of systems and procedures for new shipbuilding activities.

3.5. Technical Planning
At the next stage is technical planning, at this stage, the determination of the organizational structure following the company will be carried out, then the weighting of the work is carried out in each position so that the number of workers needed for a position can be determined.

3.6. Economical Analysis
Next is the economic analysis, after reducing the permanent workforce, the salary calculation for each position is carried out.

4. Existing Conditions of Labor and Facilities
Permanent workers are permanent workers who are tied to the organizational structure of the company. The number of permanent workers in PT. X has a different amount for each department and division. This is due to adjustments to a lot of work done by each department and division.

The division of positions below is divided into 4 positions: Senior Manager, Manager, Supervisor, Foreman, and Staff. Each of these positions has its respective responsibilities both on their own work and to their superiors. In the Table 1 is the number of permanent workers before experiencing a reduction:

Table 1. Allocation of Existing Labor.

NO	Dept./Division	Position					
		SM	M	SPV	F	STAFF	
1	Internal Supervisory Unit Department	1					
	- Operational Supervisory Division		1		1	1	
	- Compliance Division						
2	Company Secretary	1					
	- Secretarial, Admin, Public Relations Division		2	2	1	5	
	- Legal Division		1	1	1	1	
3	Finance Department						
	- Budget Division		1	1	1	4	
	- Payment Division		1			3	
	- Accounting Division		1	1		4	
4	Human Resources Department						
	- Personnel Division		1	3	2	8	
	- Empowerment Division		1	1	1	5	
	- Placement Division			1	1	2	
5	Department of Marketing and Business Development	1					
	- Marketing and After Sales Division		1	2	3		
	- Calculation Division		1	2	1	2	
	- Engineering and Information Technology Division		1	1	5	3	
6	Logistics Department	1					
	- Warehouse Division		1	1	2	4	
	- Procurement Division			2	1	3	
	- Facility Division		1	5	13	27	
	Administrations	4	13	18	20	45	
7	Production Department	2					
	- Project Leader		3	5	10		
	- Construction Division		1	4	22	57	
	- Machinery Division		1	3	12	42	
	- Electrical Division		1		7	15	
	- Outfitting Division		1	2	11	17	
	- Dock Master Division		1	4	13	35	
8	Department of Production Planning and Control	1					
	- Production Planning and Supervision Division		1	1	4	7	
	- Quality Control Division		1	2	7	2	
	- SHE Division		1	2	2	7	
	Production	3	12	28	101	209	
	Total	7	25	46	121	254	453

5. Analysis Result and Review
The criteria for reducing the permanent workforce are as follows:

- Workload received by each labor. The greater the workload received by the labor, the more labor required will also increase.
- For work that is technical in the production area, it can be based on the abilities needed to carry out the work, if the skills needed to carry out the work do not require special abilities, then the work can be carried out by outsourced workers.

- For work that is technical in the production area also based on certification, if the work to be carried out requires certification, then the work can be carried out by an outsourced worker. This is because certified outsourcing workers can carry out the work with the same standards.

At the analysis stage, a calculation of the workload that must be received by each position in each department or division is carried out. The calculation is done with a time limit of 1 year so that the workload used is also within 1 year. The calculation method used is to use the FTE (Full Time Equivalent) method. This method uses a time limit for each work done within 1 year. The time in question is the effective working time, according to the Ministry of Manpower (Depnaker) the effective working time is as follows:

1 Year = 52 Weeks
1 Weeks = 5 Days
= 260 working days in 1 year

With working hours adjusted to the working time of PT. X itself is 8 hours of work. Then the workload calculation is carried out in 1 year using the FTE method, here is the formula for the FTE method:

$$FTE = \frac{(\text{Total Activity Time (a year)} + \text{Allowance})}{\text{Total Available Time (a year)}} \quad (2)$$

The FTE value itself has a limit that can be said to be in load, underload, overload, here are the limits for the FTE value:

FTE STANDARD	=	0 - 0,99	*Under load*
		1 - 1,28	*In load*
		> 1,28	*Over load*

What is meant by Allowance here is the time it takes for workers to do their respective needs, such as drinking, defecating, walking to collect materials, worship, and so on. The following is the formula for Allowance:

Allowance = 25% x Effective Working Time x Working Hours of the Day
Then for the total available time is the total working hours in 1 year.
Total Available Time = 8 hours x 260 days

The following in Table *2* are the results of the Marketing Department FTE calculation.

IOP Conf. Series: Materials Science and Engineering 1052 (2021) 012048 doi:10.1088/1757-899X/1052/1/012048

Table 2. Marketing Department FTE Calculation Results.

POSITION	TOTAL ACTIVITY TIME (MINUTE)	FTE	TOTAL MAN POWER
Senior Manager	91,800	0.986	1
Marketing Manager	109,440	1.127	1
SPV (Marketing Division)	89,400	0.966	1
Foreman (Marketing Division)	112,680	1.153	1
Manager (Calculation Division)	91,080	0.980	1
SPV (Calculation Division)	51,240	0.661	1
Foreman (Calculation Division)	31,320	0.501	1
Staff (Calculation Division)	133,440	1.319	2
Manager (Engineering Division)	111,880	1.146	1
SPV (Engineering Division)	118,640	1.201	1
Foreman (Engineering Division) (BASIC DESIGN)	88,700	0.961	1
Foreman (Engineering Division) (HULL DESIGN)	46,820	0.625	1
Foreman (Engineering Division) (HULL OUTFITTING)	104,300	1.086	1
Foreman (Engineering Division) (MACHINERY OUTFITTING)	84,620	0.928	1
Foreman (Engineering Division) (ELECTRICAL OUTFITTING)	82,340	0.910	1
Staff (Engineering Division)	354,920	3.094	3

After reducing the permanent workforce based on the function of the number of projects that have been determined, namely the construction of 2 new ships. However, if at the same time there is an additional project, the company can solve the problem by adding overtime hours and using subcontractor services. Subcontractors required are for the Production Department, especially for the Construction Section, Electrical Parts, Machinery Parts, and Outfitting Parts. The levels of workers who can use the services of subcontractors are welder, machine operator fitter, material handling operator, and helper. Then it can also be done by attracting contract labor. The withdrawal of contract workers is tailored to the needs of the company itself, contract workers can serve as Senior Managers, Managers, Supervisors, Foremen, and Implementers. Attracting contract labor can come from other companies, where the workforce to be drawn has the capabilities needed by the company itself. [10]

Then following Table *3* which is the allocation of permanent workers after analysis using the FTE method:

Table 3. Allocation of Permanent Manpower after Analysis.

NO	Dept. / Division	Position					
		SM	M	SPV	F	Staff	
	Finance and Logistics						
1	Internal Supervisory Unit Department						
	- Operational Supervisory Division		1		1	1	
2	Company Secretary	1					
	- Secretarial, Admin, Public Relations Division		1		1	1	
	- Legal Division		1		1	1	
3	Finance Department						
	- Budget and Payment Division		1	1	1	2	
	- Accounting Division		1	1		2	
4	Human Resources Department						
	- Personnel Division		1	1	1	1	
	- Empowerment Division		1	1	2	2	
5	Logistics Department	1					
	- Warehouse Division		1	1	2	3	
	- Procurement Division			1	1	3	
	Total of Finance and Logistics	**2**	**8**	**6**	**10**	**16**	
	Operational						
6	Marketing Department	1					
	- Marketing Division		1	1	1		
	- Calculation Division		1	1	1	2	
	- Engineering Division		1	1	5	3	
7	Production Department	1					
	- Project Leader			2			
	- Construction Division		1	2	12	4	
	- Machinery Division		1	2	7	23	
	- Electrical Division		1		6	1	
	- Outfitting Division		1	2	8	12	
	- Facility Division		1	2	8	21	
8	SHE Department		1	1	1	6	
9	Quality Control Department		1	2	4	2	
10	PPC Department						
	- Production Planning and Supervision Division		1	1	3	4	
	Total of Operational	**2**	**11**	**17**	**56**	**78**	
	TOTAL	4	19	23	66	94	206

Reduction of the permanent workforce in the positions of Senior Manager, Manager, Supervisor, Foreman, and Executor. The criteria used is the workload received by each workforce. Meanwhile, the reduction of the permanent workforce in the Special Executive position in the Production Department is based on the criteria of workload, skills, and certification. So that if certain jobs do not require special skills, it can be carried out by outsourced workers. Based on the analysis results, Executors who are outsourced workers based on skill criteria are helpers, drilling machine operators, large/small drilling machine operator, saw machine operator, grinding machine operator, cutting machine operator. Executors who are outsourcing workers based on certification criteria are welder operators, fitters, material handling operators and riggers.

Based on the analysis, the results of the percentage reduction in permanent workforce in each position, including Senior Manager from 7 to 4 people (43%), Manager from 25 to 19 people (25%), Supervisor from 46 to 23 people (50%), Foreman from 121 to 66 (45%), and Executor from 254 to 94 (63%). The reduction criteria are used for the positions of Senior Manager, Manager, Supervisor, Foreman, and Staff. Then the results obtained from the total number of permanent workers are 209 people and 124 people as outsourcing workers from Table 4 below:

Table 4. Comparison before and after analysis

NO	Variable	Before	After	Difference	Percentage
1	Total Workforce	456	209	247	55%
2	Cost of Total Workforce	IDR4.239.000.000	IDR2.189.000.000	IDR 2.050.000.000	48%
3	Indirect Cost	IDR1.386.000.000	IDR863.000.000	IDR 523.000.000	38%

After analyzing the reduction of permanent manpower, the total cost of permanent labor as a whole is IDR. 2.189.000.000. in which the percentage reduction is 48% of the total fixed labor cost before analysis is IDR. 4.239.000.000. Then for indirect labor, which affects indirect costs is the position of Senior Manager of all departments, Manager of all Divisions, Supervisor of the Secretarial Division, Administration & Public Relations, Legal Division, Accounting Division, Budget Division, Personnel Division, Training Center, Marketing Division, Calculation Division, Engineering Division, Warehouse Division, Procurement Division, and Sarfas Division, Foreman from Operations Supervisory Division, Secretarial Division, Admin & Public Relations, Legal Division, Personnel Division, and Training Center, Staff from Operations Supervisory Division, Secretarial Division, Admin & Public Relations, Legal Division, Accounting Division, Personnel Division, and Empowerment Division. Based on the analysis, the indirect cost reduction was 38% or IDR. 523.000.000.

6. Acknowledgement
We gratefully thank for Departement of Naval Architecture, Faculty of Marine Technology, Institut Teknologi Sepuluh Nopember Surabaya for financial support, and We also thank the reviewer for the constructive comments and suggestions that substantially improved the paper.

7. Conclusion

After conducting research, the conclusions of this research are as follows:

1. Determining the permanent workforce can be done by using the FTE (Full Time Equivalent) method, this method is done by calculating the workload of each labor based on the completion time of each job in each department and division. After calculating using the FTE method, the result is a reduction in the number of permanent workers by 55% from the previous permanent labor.
2. The workforce reduction process is still based on 3 reduction criteria, namely:
   a. Workload, the smaller the workload, the less labor required.
   b. For work of a technical nature in the production area, consider the skills required to carry out the job.
   c. For work of a technical nature in the field, it must consider labor certification. If the job requires certification, then the workforce can use outsourced labor, because it already has the same standard of certification.

3. After analyzing the reduction of permanent manpower, the total cost of permanent labor as a whole is IDR. 2.189.000.000. in which the percentage reduction is 48% of the total fixed labor cost before analysis is IDR. 4.239.000.000. Based on the analysis, the indirect cost reduction was 38% or IDR. 523.000.000.

8. References

[1] Sasongko, Analisa Biaya, Surabaya: Hang Tuah University Press, 2011.
[2] Priyono, Manajemen Sumber Daya Manusia, Sidoarjo: Zifatama Publisher, 2010.
[3] R. W. M. T. Sri Maryanti, Deskripsi Perencanaan Ketenagakerjaan, 2002.
[4] Hanan, Penentuan Jumlah Tenaga Kerja dengan Metode FTE, Semarang, 2013.
[5] K. P. D. M. Mitchell Lee Marks, Resizing the Organization: Maximizing the Gain While Minimizing the Pain of Layoffs, Divestitures, and Closings., 2001.
[6] T. Lamb, Engineering for Ship Production, Michigan: The Society of Naval Architects and Marine Engineers Ship Production Comittee Education and Training Panel., 1986.
[7] T. W. P. Salsabella S. Sam Aldini, "Perancangan Aplikasi Berbasis Web," Surabaya, 2018.
[8] Hanan, Penentuan Jumlah Tenaga Kerja dengan Metode FTE, Semarang, 2013.
[9] H. P. Herdiana Nur Annisa, "Analisis Beban Pegawai dengan Metode Full Time Equivalent," 2014.
[10] U. Kasim, "Buruh dan Tenaga Kerja," Jakarta, 2013.

Implementation Study of Activity Based Costing System to Define Indirect Costs on Ship Repair Industries

S I Wahidi[1], T W Pribadi[1], S R W Pribadi[1] and S Megawati[1]
[1]Department of Naval Architecture, Faculty of Marine Technology, Institut Teknologi Sepuluh Nopember, Surabaya, Indonesia

Abstract. Good quality, the right turnaround time, and affordable service costs are three important points that must be paid attention to in repair shops offering their services. In determining the selling price of repair services, companies need accurate information about the cost of goods manufactured, which consists of direct costs and indirect costs. In general, the imposition of indirect costs to products uses a traditional cost accounting system so that costs distortions often occur. With the development of existing knowledge, companies must be able to increase their competitiveness and find alternative methods than traditional cost accounting systems. The activity based costing method developed and became a cost accounting method by assigning costs to activities based on the consumption / use of resources. First, activity identification is carried out based on existing business processes, both direct and supporting repair activities. Second, from the results of the identification of existing activities, identification of needs that trigger an activity that can be called an activity driver is carried out. After identifying the activity driver, the type of activity in the ship repair process is obtained which affects the high indirect cost expenditures. By using the activity based costing method, the indirect cost value on ship 1 is IDR 78,994,805, - and on ship 2 of IDR 57,890,169, -. These values are true values and can be traced to each activity through the use of resources. From these results it can be seen that the value of indirect costs when calculated by the activity based costing method produces a smaller value when compared to the existing method. On ship 1, using the activity based costing method, indirect costs decreased by 44% from the calculation of the existing conditions and on ship 2 decreased by 38% from the calculation of the existing conditions.

1. Introduction

Industries world today is experiencing a quite significant development. So this requires a moving company to be able to compete in showing good company credibility. In the shipping industry sector, especially in ship repair services, good quality, correct turnaround time, and service costs are three important points that must be paid attention to by repair shops in offering their services, especially in terms of prices that are able to compete with other repair shipyards. Because if the costs incurred for work are too high it will cause the price of the company to be unable to compete with others. So that there are many things that must be considered in the planning, including the types of repair activities carried out, what business processes occur in them, the use of labor and working hours, the use of shipyard facilities, the use of raw materials and supporting materials must be calculated properly. These are components that trigger costs in the ship repair process and these cost components are divided into direct costs and indirect costs. In general, the imposition of indirect costs to products uses a traditional cost accounting system so that cost distortions often occur [1]. As a result of this inaccurate indirect costing, cost distortion occurs, causing cross-subsidies of the profits earned by the company to cover the indirect costs it incurs.

With the development of existing knowledge, companies must be able to increase their competitiveness and find alternative methods than traditional cost accounting systems. Recently, companies have replaced their traditional cost accounting systems and developed them into activity-based cost management systems, known as Activity Based Costing or ABC [2] which is a method of measuring costs and performance of activities and other objects. So that the cost accounting system with this method is to charge costs to activities based on consumption / use of resources and charge fees to products / services produced based on the use of consumption activities. Therefore, along with the development of existing knowledge, the ship repair industry requires a new method in determining production costs.

Based on the background that has been mentioned, this study aims to, firstly, analyze the methods and systems currently applied by the ship repair industry in determining indirect costs, then identify activities that affect indirect cost expenditures in the ship repair process based on the method. activity based costing, then the last one determines the formulation of the approach in determining indirect costs in the ship repair process based on the activity based costing method.

2. Literature Review

2.1. Ship Repair

In general, ship repair activities in the shipyard based on the Ships Work Breakdown Structure are grouped into several activities consisting of [3]:

1. Drafting Activities
2. Construction work
3. Machining Works and Propulsion Systems
4. Piping Work
5. Electrical work
6. Public Services
7. Cleaning and painting of the stomach
8. Outfitting work / ship equipment

2.2. Cost Concept

In producing goods and services, of course, a company cannot be separated from the production costs it must incur. Some experts argue that this cost concept itself, according to [4] opinion, is defined as cash or cash equivalent value sacrificed to obtain goods or services which are expected to provide current or future benefits to the organization. It is also in line with opinion [5] where cost is defined as an exchange rate, expense, or sacrifice made to ensure the benefit that is reflected as depreciation of cash flows or other assets. The success of a company in planning and controlling costs depends on a thorough understanding of the relationship between costs and existing business activities. Study of the impact of business activities or costs, generally results in the classification of each expense as fixed costs, variable costs and semi-variable costs [6].

2.3. Indirect Costs

Indirect production costs are all manufacturing costs associated with cost objects but cannot be traced to cost objects in an economical manner. In other words, indirect costs are indirect material costs, and indirect labor and other costs that are not visible and are needed to support the successful completion of the production process. Indirect costs are divided into several things including [7]:

1. Indirect material costs are costs incurred for materials that support the success of the production process but do not become an integral part. Indirect material costs in the shipbuilding industry

include fuel costs for diesel welding motors, electricity costs for driving production equipment / facilities and lighting.
2. Indirect labor costs are indirect labor costs used for production process activities, but are needed to support the success and smoothness of the production process, including marketing personnel costs, calculation / invoice costs.
3. Other costs that are included in indirect costs that arise in the completion of the production process and are not classified as indirect material costs or indirect labor costs. In a shipyard company, these costs include maintenance costs, depreciation costs, research and development costs, insurance costs.Activity Based Costing System

2.4. Activity Based Costing System

Activity Based Costing or ABC is defined as a cost calculation system that focuses on activities carried out to produce a product, in other words this system imposes costs on activities based on consumption or use of resources and imposes costs on cost objects [1]. Products or services that are produced are based on the use of activity consumption, so it can be said that this system recognizes a causal relationship between cost drivers and activities. In accounting, cost drivers are factors that give rise to costs and can be used to allocate variable costs and indirect costs to a product [8]. Traditional accounting systems do not have the flexibility to show multiple cost sets and then demonstrate the application of these different costs to the cost groups for a product and service in the shipbuilding industry [9]. By using a variety of indirect cost groups and cost drivers, ABC can provide a more accurate cost description for financing and pricing a company's products and services [9].

This is of course different from the traditional cost calculation system where in this system the costs incurred for the production process are based on products and production volume. So when compared, ABC focuses more on tracking costs based on activities performed and resource use, while traditional cost calculation systems only track raw material costs and labor costs directly to each unit of output and indirect costs are allocated through imposition [10].

3. Research Metodology

3.1 Literature Review and Field Studies

Literature studies and field studies are needed in carrying out research, because the existing studies aim to provide an understanding of the concept of the Activity Based Costing method which will be implemented in determining indirect costs in the ship repair industry as well as knowing problems in the field which are the object of research. The literature study conducted by the author is related to the following topics of discussion:
1. Activity Based Costing and its application in determining the cost of goods manufactured
2. Application of activity based costing in the manufacturing industry
3. The components of the cost of production
4. Definition of indirect costs and their classification

The field studies undertaken to carry out this study consist of:
1. The process carried out in ship repair work
2. The flow of financing for ship repair work
3. Activities carried out during the ship repair process

3.2 Data Collection

This stage is the stage for collecting the data needed in carrying out the research, including:
1. Satisfaction note of ship repair work for 2 tugboats for the period September-December 2019
2. Planning for the number of workers and scheduling ship repair work
3. List of facilities and infrastructure needed for the ship repair process
4. Standard tariff for ship repair work owned by PT. XYZ

3.3 Data Processing

Furthermore, the authors processed the data obtained into several stages, including the following:

1. Identification and classification of ship repair activities
2. Identify the need for the activity or activity driver and classify the activity drivers into cost levels
3. Determine the trigger for the cost of each activity driver
4. Determination of the homogeneous cost groups of the activity drivers
5. Determining the cost rate for each activity driver
6. The calculation of indirect costs for ship repair processes is based on Activity Based Costing

3.4 Comparative Analysis of Existing Conditions with Activity Based Costing Method

From the data processing that has been done, the next step is to analyze the comparison between the existing conditions and the conditions to be developed, namely the ABC method. This stage is needed to determine whether the ABC system to determine the cost of production of ship repair works is better or even the same as the existing conditions. From this comparison, it will be obtained the things that differentiate the existing conditions with the ABC method in determining indirect costs for ship repair work.

4. Existing Condition of Ship Repair Financing

4.1 Indirect Costing Methods

Indirect costs are costs incurred for indirect material needs and indirect labor and other costs that arise and are required to support the success of the production process. At PT. XYZ itself there are different perceptions regarding indirect costs between the calculation department and the company's finance department. The differences in perception are:

1. The Calculation Department considers indirect costs that can be charged to a project as indirect costs related to repair work such as material handling, income tax from a project, the use of electricity and water for a project and the use of supporting materials and labor. project work.
2. The Ministry of Finance considers the indirect costs charged to the project are indirect costs that are not only related to project work but also indirect costs owned by the company, including permanent employee salaries, use of electricity and water for offices, water rental fees , asset depreciation value, corporate tax, building tax.

The difference of opinion between the calculation department and the company's finance department has caused indirect costs not to be included as a component in determining the cost of goods manufactured, so that their distribution is unknown. In practice, companies still experience confusion in positioning indirect costs as a component of the cost of production, because when the indirect costs are included in the calculation, this will make the value of the cost of goods manufactured higher and cause the selling price to be higher as well. With this high selling price, the company does not want to take the risk of being unable to compete with other similar companies' businesses. In addition, in determining the distribution of indirect costs for production purposes, the company still has not found an effective way so that indirect costs that are included as a component of the calculation are only electricity and water used for project purposes. So that until now, the determination of the indirect cost of a repair work has been done by taking a percentage of 20-30% of the existing tariff.

4.2 Method of Pricing for Ship Repair Services

In determining the selling price of the ship repair work owned, PT. XYZ uses the method generally used by other companies, namely still using the conventional method. As has been explained in the previous sub-chapter regarding the calculation of the cost of production for repair work, that the cost of production is the initial capital for production without any profit in it. The calculation of company profit or profit is obtained through the selling price at the standard rate for repair work owned by the company. The formulation of the selling price consists of three components including:

1. Cost of goods manufactured, calculated by the calculation department based on the repair list that has been made.
2. Indirect costs, which include taxes and other costs. In practice, companies take a range between 20% -30% of the fixed cost of production for these indirect costs themselves.
3. Margin, which is defined as the profit or profit earned by the company through ship repair services. The margin percentage is determined by the company directors and the calculation department only follows the provisions in the calculation.

So that from the three components, it can be formulated that the selling price = COGS + Indirect Costs (tax etc.) + Margin.

5. Identification and Classification of Activity

In calculating costs using an activity-based costing system, costs are not charged directly to the product, but use a two-stage process. The first stage is the process of identifying and classifying activities based on process groups and activity levels. Then the cost driver is chosen as the basis for tracking the costs of each activity to the product. In the second stage, the cost of each activity is carried out based on the amount of product consumption of the activities required.

5.1 Identification of Activities and Activity Drivers

Identification is carried out based on activities that occur during the ship repair business process. Activities in the ship repair business process are divided into two activities. First, activities that are directly related to the reparation process are known as direct activities. This type of activity is a type of activity that is easy to identify because it is visible directly on the ground and is listed in the satisfaction note. Second, activities that support the production process are activities classified as indirect activities. These direct support activities include the following:

1. Administrative activities for offering repair services, including searching for orders, preparation and execution of tenders, tender administration, calculation of cost calculations, bidding for repair lists and other activities related to offers of repair services.
2. Material and raw material procurement activities, this activity is related to ordering raw materials or materials, sending materials to shipyards, storing materials to warehouses, distributing materials to each workshop, and controlling the entry and exit of raw materials from the warehouse.
3. Production planning and control activities, this activity comes from the planning and supervision department in planning schedules and manpower requirements for ship repair work, production control, as well as compiling records of work realization and recapitulating the use of man hours.
4. Supervisor class / OS activities, including checking the test results of several ship components such as NDT on the propeller, balancing test on the propeller, and several other tests by the shipyard to the classification bureau. In addition, there is also a class or owner surveyor visiting the shipyard to carry out routine checks on the results of repair work.
5. OHS activities are activities of the K3 department in the supervision of occupational health and safety which are implemented by workers during the ship repair process.
6. Project manager activity, is the activity of the project leader who is in charge of ship repair activities carried out and controls the repair work so that it can be completed on time.
7. The activity of the head of the workshop is the activity of the head of the workshop as the person in charge of the work carried out by the workshop under him.
8. Product quality development activities are training activities for the development of the competence of production workers to maintain and even improve the quality of ship repair work.
9. After-sales administrative activities, activities that are classified into this activity include making realization of cost calculations (rekalbea), clarification and negotiation of recalbeas between companies and ship owners, invoices and payments.
10. Company management and administration activities, this activity is the result of the implementation of internal company support activities carried out by the company management. From the identification results, there are several processes including the executive process, which

is the process of compiling a map of the future business journey, resource management, and the company's administrative support process.

Apart from identifying activities, identification of the needs of each activity is also carried out, such as direct and indirect materials, direct and indirect labor, equipment needs, and use of facilities for the ship repair process. The results of the identification of the activity requirements are called an activity driver or trigger for an activity. Table 1 below is an example of the identification results of activity triggers or activity drivers in the repair list offering activities:

Table 1. Activity Driver Identification on Repair List Offering Activity

Activity:	Repair List Offering
Direct Material	
Indirect Material	Telephone charges, Computer power charges, Internet charges, Software License Cost
Direct Labor	Marketing department executor
Indirect Labor	Marketing department supervisor, Marketing department manager
Machine/Tools/Facility Use	Computer, Marketing Office
Facility Maintenance	Computer Maintenance, Marketing Office Maintenance
Facility Depreciation	Computer Depreciation, Marketing Office Depreciation

5.2 Cost Level Classification

From the results of identification of activities and activity needs, it is obtained the indirect cost components that arise during the ship repair process. The existing activity drivers are grouped simply and then grouped into four levels of activity costs, namely unit, batch, product sustaining, and facility level costs:

1. Unit Level is the level of cost where this cost occurs repeatedly in each unit of product so that this type of cost can be charged to the unit. Types of costs that fall into this level include the use of bending machines, electric drills, office electricity, welding machines, small and large lathes, hand grinders, compressors, cutting machines, pipe bending machines, use of cranes and waterpump floating dock.
2. The Batch Level is the level of costs where this cost occurs in a product group, so that this cost must be distributed according to the number of units in one group. The types of costs included in this level include the salaries of indirect labor for the executive, supervisors and managers from the marketing department, calculations, production plans, project leadership, quality control, K3L, engineering, procurement, warehouse, and finance. In addition, the salaries of indirect workers in the field, such as forklift operators, waterpump operators, crane operators, transfer personnel, dockmasters, workshop checkers, lorry drivers and security personnel. In addition, there are also telephone fees, internet, fuel for lorries and forklifts, water use and use of stationery.
3. Product Sustaining Level, is the level of costs caused by product development activities. In this case, the output produced later is to maintain product quality or improve product quality. Costs that fall into this category are costs for providing production training.
4. Facility Level, is the level of costs at the facility level where these costs arise due to the use of facilities during the production process. Costs that fall into this category include depreciation and maintenance of machines / equipment and offices and production workshops

IOP Conf. Series: Materials Science and Engineering 1052 (2021) 012049 doi:10.1088/1757-899X/1052/1/012049

5.3 Determination of Cost Drivers

After the activity classification results are obtained according to their level, the next step is to determine the cost driver of each activity requirement. In this case, the existing cost drivers are used as the basis for applying activity costs to the product.

5.4 Determination of Homogeneous Cost Groups

The next step is to combine the activities and costs of each activity that have been classified into a homogeneous cost group based on the similarity of activity levels and cost drivers owned by the activity driver. So that from the identification and classification, there are 12 groups of costs including:

1. Manpower of the executive
2. Indirect labor is part of the supervisor
3. Indirect labor is part of the manager
4. Material costs
5. Energy costs
6. Indirect field workers
7. Depreciation of production facilities
8. Depreciation of the production area
9. Maintenance of production facilities
10. Maintenance of the production area
11. Product quality development
12. Internal management and administration support

5.5 Determination of Cost Center Rates

After determining the grouping of activity drivers into several groups, the cost center rate for each activity driver is then calculated. The cost center rate calculation is done by dividing the total cost of each activity driver by the number of each cost trigger from the activity driver. So in this way, the cost center rates are obtained which are shown in Table 2 below:

Table 2. Cost Pool Rates of Each Activity Driver

Group	Cost Type
Labor (Executor)	Marketing
	Calculation
	Production and planning control
	Quality Control
	OHS
	Engineering
	Procurement
	Warehouse
	Finance
Indirect Labor (supervisor)	Marketing
	Calculation
	Project Manager
	Quality Control
	OHS
	Engineering
	Procurement
Indirect Labor (manager)	Marketing
	Calculation
	Production and planning control
	Quality Control
	OHS
	Engineering
	Procurement
	Warehouse
	Finance
Material cost	Telephone charges
	Internet charges
	Software license cost
	Lorry fuels
	Forklift fuels
	Water uses
	Marker uses
Energy cost	Bending machine utilization
	Boring machine utilization
	Office electric utilization
	Welding machine utilization
	Small lathe machine utilization
	Big lathe machine utilization
	Hand grinding machine utilization
	Compressor utilization
	Cutting machine utilization

Group	Cost Type
	Pipe Bending machine utilization
	Crane utilization
	Waterpump floating dock utilization
Indirect field labor	Forklift operators
	Waterpump operators
	Crane operators
	Moving workers
	Dockmaster
	Security personnel
	Lorry driver
	Workshop checker
Depreciation of Production Facilities	Waterpump depreciation
	Lorry depreciation
	Pipe Bending machine depreciation
	Boring machine depreciation
	Welding machine depreciation
	Small lathe machine depreciation
	Big lathe machine depreciation
	Hand grinding machine depreciation
	Compressor depreciation
	Cutting machine depreciation
	Bending machine depreciation
	Floating dock depreciation
	Crane depreciation
	Forklift depreciation
	Computer depreciation
Depreciation of Production Area	Production Office depreciation
	QC office depreciation
	Main office depreciation
	Piping workshop depreciation
	Outfitting workshop depreciation
	Machinery workshop depreciation
	Warehouse depreciation
Maintenance of Production	Computer maintenance
	Waterpump maintenance

Group	Cost Type
Facilities	Lorry maintenance
	Bending machine maintenance
	Boring machine maintenance
	Welding machine maintenance
	Small lathe machine maintenance
	Big lathe machine maintenance
	Hand grinding machine maintenance
	Compressor maintenance
	Cutting machine maintenance
	Pipe Bending machine maintenance
	Floating dock maintenance
	Crane maintenance
	Forklift maintenance
Maintenance of Production Area	Production Office hygiene
	QC office hygiene
	Main office hygiene
	Piping workshop hygiene
	Outfitting workshop hygiene
	Machinery workshop hygiene
	Warehouse hygiene
Product Quality Development	Procurement of production training
Internal Management and Company Administration Support	President director
	Director of operations
	Director of finance and general
	Internal control department unit of compliance supervisory
	Company Secretary department unit of administration, public relations, and quality control

Group	Cost Type	Group	Cost Type
	Company Secretary department unit of law		General and human resource department unit of personnel and monitoring
	Finance department unit of budgeting		General and human resource department unit of empowerment, training, and general
	Finance department unit of financing		
	Finance department unit of accounting		

5.6 Calculation of Activity Costs
The calculation of activity costs is determined by multiplying the calculated cost center rate by the number of cost drivers needed for each activity driver in carrying out an activity. In this calculation, it will be known how the composition of costs between direct costs and indirect costs for each identified ship repair activity.

6 Analysis and Comparison

6.1 Analysis of Indirect Cost Calculation with Activity Based Costing Method
The calculation of the cost of ship repair work using the activity based costing method is different from the calculation of the existing conditions. The calculation of these two methods results in differences in the composition of direct and indirect costs. So that from the calculations carried out, the following results are obtained:

Table 3. Direct Cost and Indirect Cost Composition

	Direct Cost (IDR)	Indirect Cost (IDR)	Total Cost (IDR)
Ship 1	491.868.433	78.994.805	570.863.238
Ship 2	335.244.114	57.890.169	392.638.283

In Table 3 above are the results of calculating direct and indirect costs using the activity based costing method for the same type of work. In performing calculations, the results of identification of activities that have been carried out are used as a reference in determining the needs and the amount of existing costs. So that from the results of these calculations, the cost composition will be obtained. If percentage, the composition of direct costs is 86% and indirect costs are 14%. The value of indirect costs has decreased significantly when compared to the existing conditions.
From these calculations, the results of the composition of direct and indirect costs for each activity are also obtained. So that if you summarize the activities that cause indirect costs during the ship repair process, including the repair list offer, calbea calculations, work schedule planning, material and material ordering, material and material purchase budget planning, material and material delivery to shipyard, material quality checking, inventory control storage of raw materials, and distribution of materials to the workshop, conducting a survey of ship conditions, conducting arrival meetings, handling materials from warehouse to docking area, disposing of waste / waste repairs, conducting tests and clearance, making test and clearance reports, monitoring and recapitulating the use of people's hours , supervision of the implementation of K3L procedures, checking the quality of work results, checking the results of tests and work into classification, production staff allowances (overtime), skills training for production staff, making records of work realization or satisfaction notes, calculating recalculations, clarification and negotiation, invoice, as well as payment.

IOP Conf. Series: Materials Science and Engineering 1052 (2021) 012049 doi:10.1088/1757-899X/1052/1/012049

7 Comparison of Indirect Costs of Existing Conditions with Activity Based Costing Methods

Determination of indirect costs made by PT. XYZ in ship repair work has been using conventional methods, namely by taking a figure of 20% to 30% of the predetermined cost of production. Meanwhile, with the activity based costing method, determining indirect costs is based on activities and the use of resources in carrying out activities. Thus, the two different methods produce different indirect cost values.

Table 4. Cost Distortion Between Existing Condition and ABC on Ship 1

	Ship 1	
	Existing (IDR)	ABC (IDR)
Contract Value	866.441.323	866.441.323
Direct Cost	472.890.279	491.868.433
Indirect Cost	141.867.084	78.994.805
Management and administration	53.125.000	53.125.000
Income tax (PPh 23)	17.328.826	17.328.826
Difference in cost	181.230.134	225.124.259

In Table 4, a comparison of direct costs, indirect costs, management and administration costs, income tax, and the difference between the total costs and the contract value is shown through the existing conditions calculation method and the ABC method on the ship case study 1. With a contract value of IDR 866,441,323, - in the calculation of the existing method, if converted into a percentage, the percentage of direct costs is 55%, indirect costs are 16%, management and administration costs are 6%, income tax is 2% and the difference between total costs and value contract by 21%.

In the calculation using the activity based costing method with the same contract value, the comparison of costs obtained if converted into a percentage is the percentage of direct costs of 57%, indirect costs of 9%, management and administration costs of 6%, income tax of 2%. and the difference between the total cost and the contract value of 26%. When comparing the value of the indirect costs of the two calculation methods, it can be seen that the value of indirect costs on Ship 1 which is calculated using the activity based costing method has decreased by 44% from the calculation of the existing conditions.

Table 5. Cost Distortion Between Existing Condition and ABC on Ship 2

	Ship 2	
	Existing (IDR)	ABC (IDR)
Contract Value	580.991.360	580.991.360
Direct Cost	309.270.183	335.244.114
Indirect Cost	92.781.055	57.890.169
Management and administration	53.125.000	53.125.000
Income tax (PPh 23)	11.619.827	11.619.827
Difference in cost	114.195.296	123.112.250

In Table 5, a comparison of direct costs, indirect costs, management and administration costs, income tax, and the difference between the total costs and the contract value is shown through the existing conditions calculation method and the ABC method on the ship case study 2. With a contract value of IDR 580,991,360, - in the calculation of the existing method, if converted into a percentage, the percentage of direct costs is 53%, indirect costs are 16%, management and administration costs are 9%, income tax is 2% and the difference between total costs and value contract by 20%.

In the calculation using the activity based costing method with the same contract value, the comparison of costs obtained if converted into a percentage is the percentage of direct costs of 58%, indirect costs of 10%, management and administration costs of 9%, income tax of 2%. and the difference between the total cost and the contract value of 21%. When comparing the value of the indirect costs of the two calculation methods, it can be seen that the value of indirect costs on Ship 2 which is calculated by the activity based costing method has decreased by 38% from the calculation of the existing conditions.

In calculating the cost of the two ships using the existing method and the activity based costing method, there are differences in the cost determination mechanism. In the calculation of the existing conditions, the direct cost components are material and labor costs directly involved in ship repair work and the indirect costs include supporting costs for repair activities as well as costs for management and administration. Whereas in the calculation using the activity based costing method, the value of direct and indirect costs is calculated based on the business processes that occur during the repair process, both supporting and direct, so this makes a difference in determining costs that were initially classified as indirect costs. direct cost. As in charging fees for marketing implementers whose value can be charged to the project, this is because the marketing department executor carries out work to find tenders, so that if a repair tender is obtained, the ship repair process can be carried out and vice versa if the tender process is not carried out then the ship repair process is also will not be able to walk.

From the results of calculations according to the activity based costing method, indirect cost components are obtained such as indirect material costs, indirect production and non-production labor costs, maintenance and depreciation costs, management and administration costs and income tax. The management and administrative activities of the company are classified as indirect cost components and are posted as another form of cost. In addition, taxes are classified as indirect costs because taxes are a cost component that must be incurred by the company as a result of ongoing business processes. This is because taxes are a component of production costs that must be paid by companies but are not classified as direct material costs or direct labor, therefore tax is one of the indirect components of production costs. Thus, the only tax calculated in this study is the PPh 23 income tax whose value is calculated at 2% of the contract value. As for the income tax PPh 21 the value is not calculated. This is because the value for PPh 21 is included in the employee salary that has been paid by the company. So that from the salary paid, the company can immediately deduct it to be paid by taxpayers.

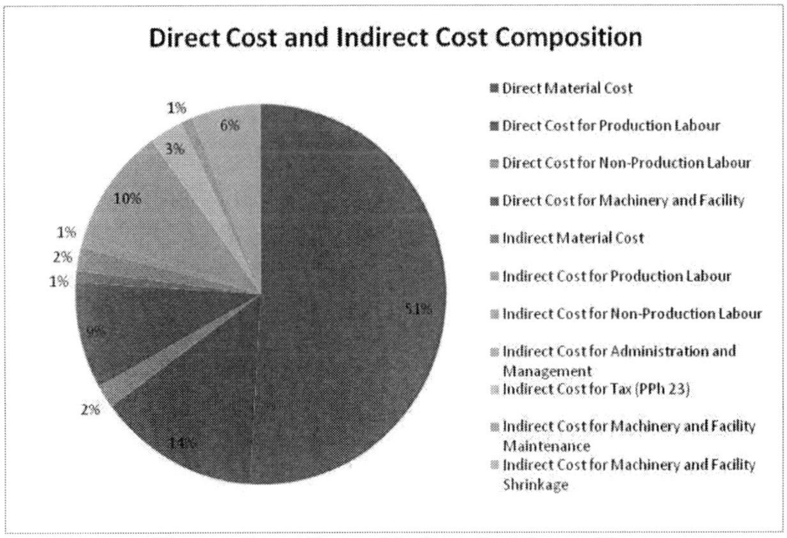

Figure 1. Direct Cost and Indirect Cost Composition

In Figure 1 it can be seen that by using the activity based costing method, the percentage of direct and indirect cost components is obtained. The cost composition in the diagram is the average composition obtained from the 2 ships as the case. The total percentage of direct costs is 76% with cost components in the form of direct material costs of 51%, direct production labor costs of 14%, direct non-production labor costs of 2% and costs of using machines / equipment by 9%. Meanwhile, indirect costs have a total percentage of 24% with cost components in the form of material indirect costs of 1%, indirect production labor costs of 2%, indirect non-production labor costs of 1%, indirect management costs and 10% administration fee, 3% indirect income tax expense, 1% facility maintenance fee, and 6% facility depreciation fee.

8 Conclusions

Conclusion: Based on the results of the research that has been done, it can be concluded that:

1. In determining the indirect costs of ship repair work, PT. XYZ uses a conventional method, namely by taking a value of 20-30% of the cost of production of a predetermined ship repair work. So that from the data that has been obtained and calculated, the results show that the value of the indirect costs for repair work on Ship 1 is IDR 141,867,084, - and on Ship 2 of IDR 92,781,055, -.
2. Through identification that has been carried out based on the activity based costing method, activities in the ship repair process are obtained that affect indirect costs including repair list offers, Kalbea calculations, work schedule planning, material and material ordering, material purchase budget planning, delivery of materials to the shipyard, checking the quality of materials, controlling the inventory for storing raw materials, and distributing materials to the workshop, carrying out a survey on the condition of the ship, conducting the arrival meeting, handling material from the warehouse to the docking area, disposing of waste / waste repairs, conducting tests and clearance, manufacturing test and clearance reports, supervision and recapitulation of the use of human hours, supervision of the implementation of K3L procedures, checking the quality of work results, checking test and job results into classification, skills training for production staff, making records of job realization or satisfaction notes, calculations n recalculation, clarification and negotiation, invoicing, payments, and management and administration activities of the company.
3. The formulation of indirect costs has a total percentage of 24% of production costs with cost components in the form of indirect material costs of 1%, indirect labor costs of production of 2%, indirect non-production labor costs of 1%, costs indirect management and administration costs of 10%, indirect income tax costs of 3%, facility maintenance costs of 1%, and facility depreciation costs of 6%. So that by using the activity based costing method, the value of the indirect cost on ship 1 is IDR 78,994,805, - and on ship 2 of IDR 57,890,169, -.

Acknowledgements

On this occasion, the author would like to express his deepest gratitude to PT. XYZ, a shipyard company operating in the Surabaya area for its assistance in data collection so that this research can run well.

References

[1] M. R. Simbolon, "Sistem Activity Based Costing Sebagai Alternatif Pembebanan Biaya Overhead pada Perusahaan Manufaktur," J. Ekon. dan Bisnis, vol. 2, no. 1, p. 10, 2008.
[2] R. Cooper and R. Kaplan, "Profit Priorities from Activity-based Costing," Harv. Bus. Rev., vol. 69, no. 3, p. 7, 1991.
[3] SPAR Associates, "Generic Ship Repair."
[4] D. R. Hansen and M. M. Mowen, Akuntansi Manajerial, Manajemen Biaya, Akuntansi dan Pengendalian, 8th ed., vol. 1. Jakarta: Salemba Empat, 2013.
[5] W. K. Carter, Akuntansi Biaya (Cost Accounting) Diterjemahkan oleh Krista, 14th ed., vol. 1. Jakarta: Salemba Empat, 2009.
[6] A. Matz and M. Usry, Cost Accounting: Planning and Control. Cincinnati: South Western Publishing, 1995.

[7] B. Sasongko and I. Baroroh, Analisa Biaya Industri Perkapalan. Surabaya: Hang Tuah University Press, 2011.
[8] G. S. Rotikan, "Penerapan Metode Activity Based Costing Dalam Penentuan Harga Pokok Produksi Pada Pt. Tropica Cocoprima," J. Ekon. Manajemen, Bisnis dan Akunt., vol. 1, no. 3, pp. 1019–1029, 2013.
[9] SPAR Associates, "Shipyard Cost Model Using Activity Based Costing Methods," Natl. Shipbuild. Res. Progr., p. 105, 1996.
[10] A. C. Achayit, P. Attarde, and S. Kurkanil, "Activity Based Costing in Construction Projects," Adv. Eng. Res. Stud., p. 6, 2014.

The Application of 2nd Generation Intact Stability Criteria to Ship Operating in Indonesia Waterway: Pureloss Stability

S Anggara[1], R D Maskar[1], M R F Hariadi[1], L M Ichsan[2], M Zaky[1] and A Kurniawan[1]

[1] Research and Development Division, Biro Klasifikasi Indonesia
[2] Statutory Division, Biro Klasifikasi Indonesia

Abstract. The development of 2nd Generation intact stability criteria has entered final chapter since nearly one-decade intensive work. The instrument is intended to measure ship stability in dynamic aspect on wave which has not been considered by its predecessor, IS Code 2008. Prior to enforcement, BKI has taken early preparation to establish inhouse calculation tool independently for assessing ship stability in accordance with the first 2 level of vulnerability criteria, i.e. *Pureloss Stability*. Semi performance based and direct assessment are introduced as the concept behind the calculation, thus the procedure is quite complex and need more time to proceed which leads to utilize automation. Furthermore, a set number of types of ship with various dimension have been tested to the system in respect to Indonesia environmental condition. The results show that typical ships are vulnerable to certain criteria which may beneficial for operator to gain initial precaution.

1. Introduction

First development of stability criteria has been begun in 1939 by Rahola. The background is based on a set of ship accident database which is collected and treated by statistic-regression analysis [2]. Today, this criterion is globally used to determine ship safety stability limit around the world and well known as Intact Stability (IS) Code, adopted by IMO. However, there were remain accident occurred to the ship notwithstanding they have met the criteria. This circumstance lead experts to investigated continuously work on the improvement of safety level. In other side, the technology has evolved during nearly a century, thus ship characteristic built nowadays are significantly different compared to that era. Means that sample behind IS Code may be deviated to condition today [3].

In the beginning of 2002, IMO through sub-committee SLF (Stability and Load line and Fishing vessel safety) initiated first work on the development of 2nd Generation Intact Stability Criteria (2nd GIS). Differently from its predecessor which is based on static condition, 2nd GIS has been considering dynamic effect due to wave. This criterion is dedicated to capture unusual phenomena that not yet covered by IS Code [4]. The safety impact of those phenomena is relative depend on ship characteristic. There are 5 criteria under 2nd GIS, i.e. *pureloss, parametric roll, excessive acceleration, dead ship and surfriding*. The concept of calculation is based on semi performance based where the risk analysis is conducted to serve the result. Each criterion has 3 level, which first level is simple calculation approach and considered as the most conservative. Secondly, less conservative but more proportional approach combining performance and simplification method. The output of first two level is "vulnerability" status of the ship. As for level 3, direct analysis is introduced using numerical or experimental method. If the ship does remain not comply to those 3 tiers, then operational limitation guidance shall be made [5].

Final draft is planned to be finished by this year, and IMO will consider to entry to force this regulation by next years. In this study [6], BKI intend to show the initial works regarding to the readiness prior to its implementation. Calculation program has been established for 5 procedure covering 2 first level criteria. The Program was arranged in spreadsheet that is integrated by macro with Maxsurf Stability Software. Its extensive procedure, especially in level 2, may not allow to use hand calculation because many iterations are required.

This paper is exclusively discussing about *pureloss* stability phenomena which is addressed to ships operated in Indonesia waterway. Therefore, the wave scatter was modified form North Atlantic to Indonesia and then the result is compared for both conditions.

2. Methodology

In accordance with [1], *pureloss* identify the probability of a ship losing stability. The losing righting arm is affected by wave coming from longitudinal direction to the ship. Waves are assumed to have equal length with a ship. Thus, its propagation lead to fluctuation of ship's buoyancy and water plan area (WPA) and bring the righting curve changing over time. When wave crest is amidship, WPA significantly decrease and righting arm become lower, and vice versa. The frequency domain approach is used to captured that every single value. This represents the actual condition of ship operation on sea, that the stability is assumed to be static in the 1st generation criteria.

In order to have real condition, the approach is applied based on wave scatter diagram of intended operation area. 2nd GIS has north Atlantic (NA) condition as default. Thus, if applied to Indonesia (IND) waterway, the scatter shall be modified. The maximum significant wave height, Hs, from scatter diagram is considered as basis to determine Hi in investigation of righting curve. Hi is defined as effective wave height in accordance to wave energy spectrum proposed by Sub-committee. This value may differ depends on ship characteristic including main dimension and hull shaped. Furthermore, The Hi is variated to 10 with same interval and every variation has 10 waves phased with 0.1L increment as if the wave propagates. More variation may be added to obtain reliable result.

Table 1. Effective wave height

Hi NA	0.32	0.64	0.96	1.28	1.60	1.92	2.24	2.56	2.88	3.20
Hi IND	0.21	0.43	0.64	0.85	1.06	1.28	1.49	1.70	1.91	2.13

During draft development, some criteria index was changed several times as procedure changed following updated approach technic. To be noted that this study is developed in accordance with the latest draft of IMO released by sub-committee Ship Design and Construction (SDC) 7. In the draft, level 1 only requires minimum Height of Metacentric (GM) and hull form provision respect to reserve buoyancy. While level 2 use extensive calculation to obtain vulnerability index respect to ship performance and wave condition. The vulnerability index is the function of ship performance (failure condition) and probability occurrence of the wave. That is why level 2 is also known as risk-based analysis method [7]. Both criteria are stated as below. The ship is considered not vulnerable to stability failure if the criteria is accomplished.

Citeria level 1 :

- GM >0.05 m (1)
- Buoyancy reserve in freeboard > WPA x freeboard (2)

Criteria level 2 :

$$max(CR_1; CR_2) \leq 0.06 \quad (3)$$

$$CR_i = C_i x W_i \quad (4)$$

IOP Conf. Series: Materials Science and Engineering 1052 (2021) 012050 doi:10.1088/1757-899X/1052/1/012050

Where W_i is wave probability of occurrences. C_1 and C_2 may be either 1 or 0 depend on fulfilment of required condition. C_1 is 1 if the range angle of stability (RoS) is less than 30°. meanwhile, C_2 is 1 if static heeling angle under action of heeling lever is more than 15°. Otherwise the C_i index are 0. This index then multiplied with W_i. The total summation from each sea state is namely vulnerability index (CR_1 or CR_2).

The are 16 sample to investigate with pureloss stability criteria, consist of several type of ship and variate dimension.

Table 2. Sample ship's main dimension.

No	1	2	3	4	5	6	7	8
Tipe	Container	Container	Container	G. Cargo	Tanker	Tanker	Tanker	Tug Boat
L (m)	105.00	90.80	79.80	69.20	233.00	169.00	87.24	27.13
B (m)	17.00	15.80	21.60	12.50	44.00	32.00	15.40	8.60
H (m)	9.00	7.40	4.40	7.05	21.50	15.10	7.30	4.50
T (m)	5.36	5.95	3.40	5.45	12.70	10.00	5.00	3.50

Table 3. Sample ship's main dimension (cont').

No	9	10	11	12	13	14	15	16
Tipe	Passenger	Fishing V	LCT	SPOB	SPOB	Ferry ro-ro	Ferry ro-ro	Bulk Carrier
L (m)	99.20	46.90	64.00	43.82	103.83	50.31	37.00	130.70
B (m)	19.60	8.70	13.80	8.78	21.95	12.80	15.00	26.40
H (m)	5.60	3.75	4.00	2.39	6.63	3.80	3.00	11.50
T (m)	4.10	3.40	2.50	1.80	4.60	2.90	1.85	5.40

L = length; B = Breadth; H = Depth; T = Draft

3. Discussion

Having investigating level 1 and 2 criteria for corresponding ship, 3 ships show vulnerable to stability failure in accordance with level 1 criteria. Notwithstanding they have adequate stability by fulfilled GM requirement, Regardless the type of ship, some sample does not comply to hull form provision which require freeboard to be fully as reserve buoyancy. The unusual deck design is the main cause, especially for ship with "cut-off" deck which imply to reduce reserve buoyancy such as fishing vessel or ship with watertight coaming that is considered as main deck. This design does not mean deficiency. Therefore level 2 shall be checked as well.

Table 4. Vulnerability result.

No	1	2	3	4	5	6	7	8
Level 1	v	v	nv	nv	nv	nv	nv	nv
Level 2 NA	nv	nv	nv	nv	nv	nv	nv	nv
Level 2 IND	nv	nv	nv	nv	nv	nv	nv	nv

Table 5. Vulnerability result (cont').

No	9	10	11	12	13	14	15	16
Level 1	nv	v	nv	nv	nv	nv	nv	nv
Level 2 NA	nv	nv	nv	nv	nv	nv	nv	nv
Level 2 IND	nv	nv	nv	nv	nv	nv	nv	nv

nv : not vulnerable ; v : vulnerable

The level 1 is not affected by wave, means there is no correlation within updating wave condition. Meanwhile, the level 2 has been done with two wave scatter diagram, NA and IND. For default NA condition, all sample is not vulnerable to this level criteria. Their probability of experiencing stability failure due to pureloss is stated small, where the ships have satisfied the requirement of C_1 and C_2.

The concept of pureloss is investigating righting curve fluctuation by the changing of WPA and displacement. The effective wave leading those circumstance is scaled by the ship characteristic, i.e. main dimension and cross section area. Thus, Hi will be comparable to the ship dimension. As consequence, for the ship with T/H < 0.8 and/or high GM (marked as higher B/H ratio) the changing of curve will not significantly affect the Ros and static heeling angle. In other hand, ship with ordinary hull-shaped design, e.g. there is no "cut off" deck, will also relatively safe to these phenomena [8].

Table 6. Static heel angle Sample 10.

Hs(m)/Tw(s)	2	3	4	5	6
0.5	2.61	1.88	1.15	0.75	0.53
1.5		6.17	3.52	2.26	1.59
2.5		12.13	6.31	3.88	2.64
3.5		26.58	10.76	5.71	3.79
4.5			16.97	10.77	5.03
5.5			27.02	15.15	6.37
6.5				20.89	7.89
7.5				26.27	9.53

Detail view is presented for model sample no.10 which is fishing vessel type. This sample is selected since it is vulnerable respect to level 1. In fact, the ship actually already fails to satisfy to CR2 index. The static heeling angle is already more than 25° for particular Hs and Tw (wave periods). However, need to be remember that the "vulnerability index" is function of C_i and W_i. Coincidentally, the W_i on sea state where the heeling angle fails is zero. Means that, this condition is dangerous for the ship but imposable to occurred. As the result, ship is not vulnerable at all.

Afterward, assessing level 2 criteria with IND wave condition is done. The result shows that C_1 and C_2 requirement are also fulfilled. If compared, NA wave condition capable to reduce up to 80% of ship's RoS and increase up to 70% ship static heel resulted by IND condition.

This may be understood since, basically, average wave height in IND is significantly lower than NA. According to Table 2, average Hi IND is 66% lower than NA. If one see more detail, 52% of IND scatter diagram consist of Hs 1.5 m. Meanwhile, 50% of NA scatter diagram consist of Hs 2.5-4.5 m. The maximum Hs NA and IND is 16.5 m and 5.5 m respectively.

4. Conclusion

Based on investigation result to ship operated in Indonesia, Pureloss is sensitive to certain ship with low reserve buoyancy especially for ship with unusual deck design with "cut-off", such as container and fishing vessel. Some other ship with low GM may also suffer these phenomena. Since the criteria is function of wave probability of occurrence, the ship which does not fulfill the index criteria in certain sea state condition, may not vulnerable if the ship does not experience such condition. It may become operator concern and to be included in ship operation manual.

Indonesia wave height is generally lesser extreme than North Atlantic. This condition lead to give less strict requirement for ship operating locally. It may need to be considered in the future development of 2nd GIS in Indonesia.

5. Acknowledgements

This work is dedicated as part of BKI's contribution to prepare the implementation of this regulation. All activities were supervised by Mr. Fredhi A.P, Head of Research and Development who was directing the methodology and approach.

6. Reference

[1] IMO SDC 7/5 2019 *Finalization of Second Generation Intact Stability Criteria* London UK : 7th Session SDC Meeting
[2] Belenky V, Bessler C C and Spyrou K J 2011 *Development of Second Generation Intact Stability Criteria* US : Naval Surface Warfare Center
[3] Peter W, Blenkey V, Bessler C C et all 2011 *The Second Generation Intact Stability Criteria: An Overview of Development* SNAME pp 121
[4] Tompuri M, Ruponen P, Lindroth D 2016 *Second generation intact stability criteria and operational limitations in initial ship design* Denmark : PRADS
[5] Gualeni P, Petacco N 2020 *IMO Second Generation Intact Stability Criteria: General Overview and Focus on Operational Measures* Italy : Journal of Marine Science and Engineering pp 494
[6] Chung J, Shin D M, Kim W D et all 2020 *Current Status of the 2nd Generation of Intact Stability: Investigation of the Pure Loss of Stability and Parametric Roll Mode* Korea : Journal of Ocean Engineering and Technology pp 34
[7] Turan O, Liu H, Boulougouris E 2015 *Investigation of the 2nd Generation of Intact Stability Criteria in Parametric Rolling and Pure Loss of Stability* UK: STAB
[8] Umeda N, Osugi M, Matsuda A 2017 *Model Experiment on Pure Loss of Stability for a Ship in Astern Waves and Its Relationship with the Second Generation Intact Stability Criteria* Serbia : the 16th International Ship Stability Workshop

IOP Conf. Series: Materials Science and Engineering 1052 (2021) 012051 doi:10.1088/1757-899X/1052/1/012051

Risk Analysis Using the Risk-Based Inspection (RBI) Method for a Pressure Vessel at Offshore Platform

N Cahyono, M B Zaman, N Siswantoro, D Priyanta, T Pitana

Departement of Marine Engineering, Institut Teknologi Sepuluh Nopember, Surabaya, Indonesia

Abstract. Energy sources that can be utilized by humans are oil and natural gas. However, in exploration activities, there are problems regarding the risk level determination of equipment in offshore platforms. The equipment that has an important function for the processing of oil and gas is a pressure vessel. In the contents of the hydrocarbon flow produced by the production well, it consists of various mixtures (gas, crude oil, and water). Therefore, it is necessary to have a separation process between the three fluid phases by using a device, namely a separator (pressure vessel). If the process of separating the three-phase from the hydrocarbon flow fails, it will result in product failure and also threaten the workers and the surrounding environment. So, this study will use the risk-based inspection (RBI) method to determine the level of risk from a pressure vessel. The risk-based inspection (RBI) method was chosen because the oil and gas processing industry have a high level of risk so that safety for business management and environmental conditions is the main consideration. Another advantage of using the risk-based inspection (RBI) method is the reduced shutdown time and the number of equipment inspections.

1. Introduction

Energy is the main requirement used to meet various human needs. Therefore, it requires a large number of energy sources to meet human needs in various fields of life. One of the energy sources that can be used by humans is oil and gas. In this century, oil and gas exploration has been widely available in the high sea, because many oil and gas reserves are found in the depths of the seabed. For this reason, companies engaged in oil and gas mining are competing to build offshore platforms, which are useful for mining oil and gas from the bowels of the earth in the depths of the sea. An offshore platform is a building that operates in the open seas around an oil and gas source to carry out mining activities [1]. Many offshore structures are now widely used [2]: Shallow Water Complex, GBS (Gravity Base Structure), Compliant Towers, FPSO (Floating Production, Storage and Offloading), TLP (Tension Leg Platform), Semi-submersible Platforms, and SPAR.

However, in oil and gas exploration activities, many problems are encountered regarding how to determine the appropriate maintenance strategy for equipment in offshore platforms. So that the authorities in various countries issue regulations regarding the safety of installations and equipment in oil and gas business activities. For this reason, all existing equipment must be guaranteed reliability so as not to cause harm to both workers and the environment. One of the types of equipment that have an important function for the processing of oil and gas is a pressure vessel (separator) that functions to separate the three fluid phases (gas, liquid hydrocarbons/crude oil, and air) [3].

So that this research will explain about the risk-based inspection (RBI) method so that it can predict the criticality level of the pressure vessel. The risk-based inspection (RBI) method was chosen because the oil and gas processing industry have high and dangerous level of risk so that work safety for oil

IOP Conf. Series: Materials Science and Engineering 1052 (2021) 012051 doi:10.1088/1757-899X/1052/1/012051

and gas workers is a major consideration along with environmental conditions. Another advantage of using the risk-based inspection method (RBI) is the reduced plant shutdown time (because the RBI has a 3-5 year period of inspection or maintenance according to the type of equipment), a reduction in the number of inspection equipment (because RBI prioritizes units or equipment that have a risk criticality high/high risk), and reduced inspection and maintenance costs (because the RBI has a 3-5 year inspection or maintenance period according to the type of equipment, only the initial investment is slightly more expensive) [4].

2. Oil and Gas

Since ancient times, the use of petroleum has been widely used by humans around 5000 years ago by the Sumerians, Mesopotamians, and Egyptians [5]. Then in 1859, Colonel Edwin Drake drilled an oil well located in north western Pennsylvania [2]. Colonel Edwin Drake's success led to the start of the search for and mining of oil and natural gas internationally for industrial purposes.

Figure 1. Drake Well [2].

3. Offshore Platform

An offshore platform is a building that operates in the open seas around an oil and gas source to carry out mining activities [1]. According to ABB, there are several types of offshore platforms that are currently widely used. First, there is a shallow water complex, which is a combination of several platforms that have different functions and are connected by a gangway bridge as a connecting bridge between platforms (see Figure 2) [2]. Second, the gravity base structure (GBS) is a platform supported by a large concrete structure embedded in the seafloor (see Figure 3) [2]. Third, compliant towers are platforms that are supported by a relatively flexible foundation structure, so that this platform is suitable for placement in deep waters up to 1000 meters because it can absorb a lot of pressure [2]. Floating production, storage, and offloading or FPSO is a platform that is shaped like a large tanker and is a conversion from the VLCC or ULCC type tankers (see Figure 4) [2]. Tension Leg Platform or TLP is a platform equipped with vertical tendons that are connected to the seabed which is useful as a buffer (see Figure 5) [2]. The semi-submersible platform is a platform similar to the TLP, but without having a tight tie like a tendon (see Figure 5) [2]. Lastly, SPAR is a platform consisting of a single floating cylinder hull that has the function of stabilizing the platform while in the water and allows motion to stabilize the position in case of storms and support the deck platform [2]. However, the cylindrical hull does not extend directly to the seabed but is anchored under the sea by a series of steel cables or the like (see Figure 6).

Figure 2. Shallow Water Complex [2].

Figure 3. Gravity Base Structure [2].

Figure 4. Floating Production, Storage and Offloading [2].

Figure 5. Tension Leg Platform and Semi-submersible Platform [2].

Figure 6. SPAR [2].

IOP Conf. Series: Materials Science and Engineering 1052 (2021) 012051 doi:10.1088/1757-899X/1052/1/012051

4. Pressure Vessel

Equipment that has an important function for oil and gas processing are pressure vessels. Pressure vessels serve to separate the content contained in the hydrocarbon flow produced by production wells that consist of various mixtures, including gas, liquid hydrocarbons (crude oil), and water [3]. Based on its use, there are several types of separators commonly used in the oil and gas processing process, namely: vertical separator (see figure 7 and figure 8), horizontal single tube (see figure 9 and figure 10), horizontal double tube, and spherical [3].

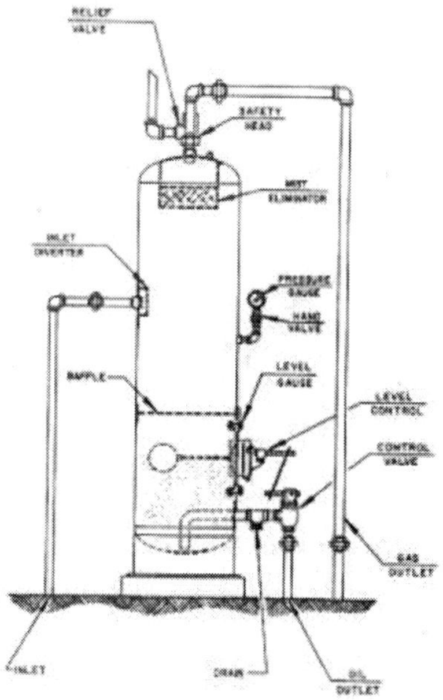

Figure 7. Vertical Low-Pressure Oil and Gas Separator Mechanical Controls [3].

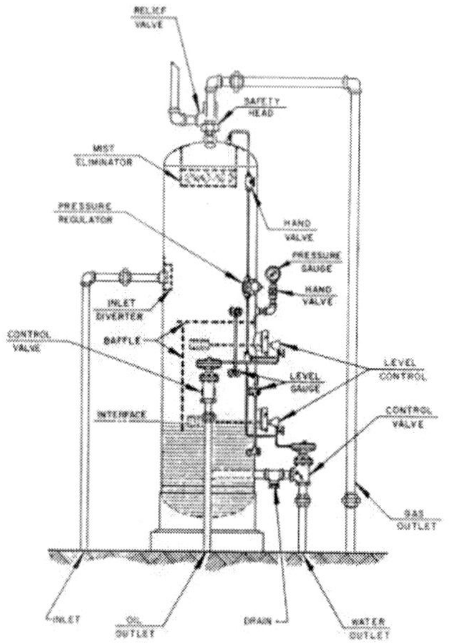

Gambar 8. Vertical HP or LP Oil Gas Water Separation Pneumatic Controls [3].

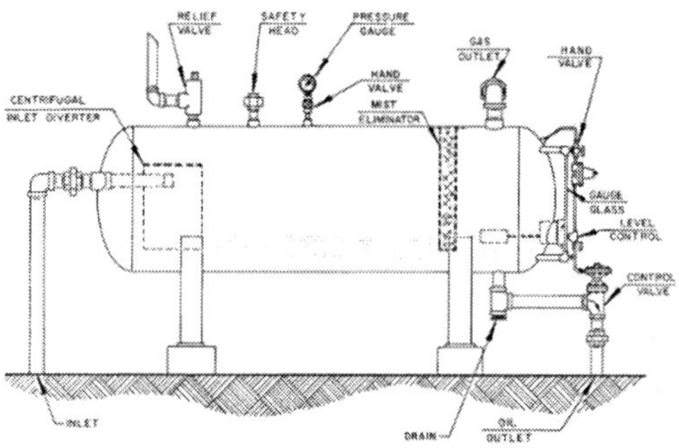

Figure 9. Horizontal HP or LP Oil and Gas Separator Pneumatic Controls [3].

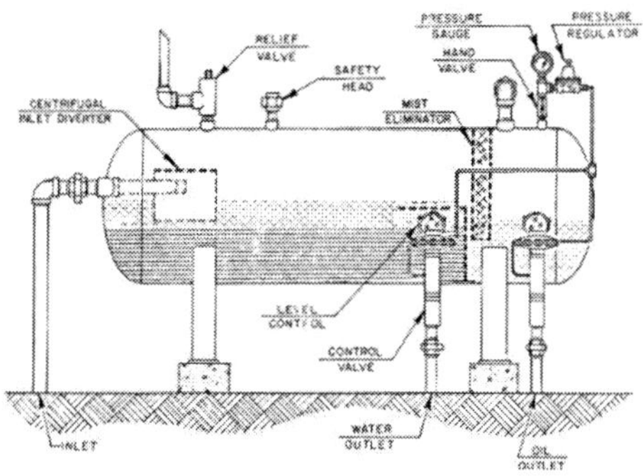

Figure 10. Horizontal HP or LP Oil Gas Water Separator Pneumatic Controls [3].

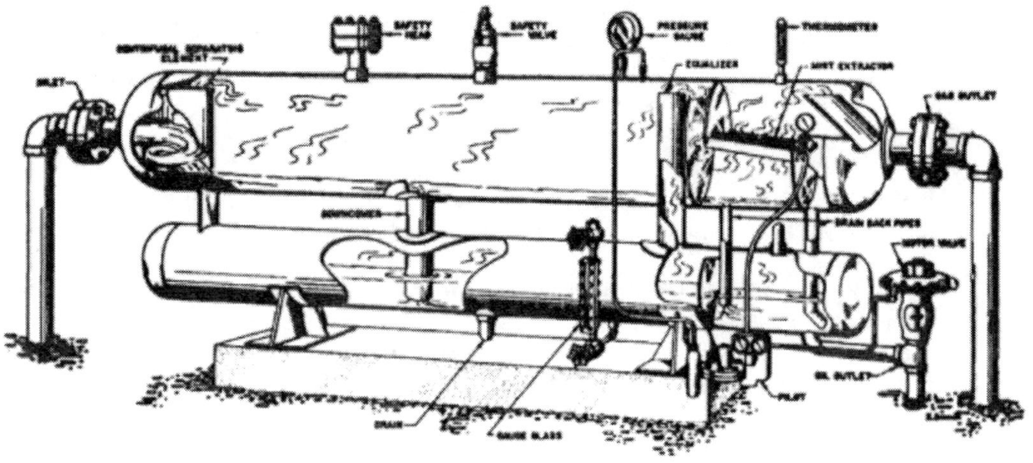

Figure 11. Horizontal HP or LP Oil and Gas Separator Pneumatic Controls [3].

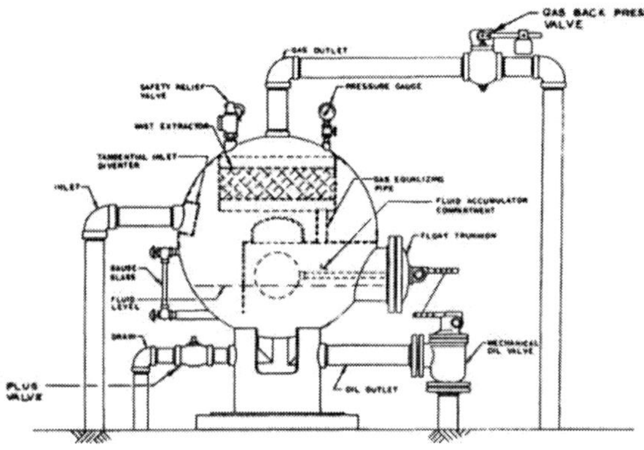

Figure 12. Spherical Separators (Low Pressure Mechanical Controls) [3].

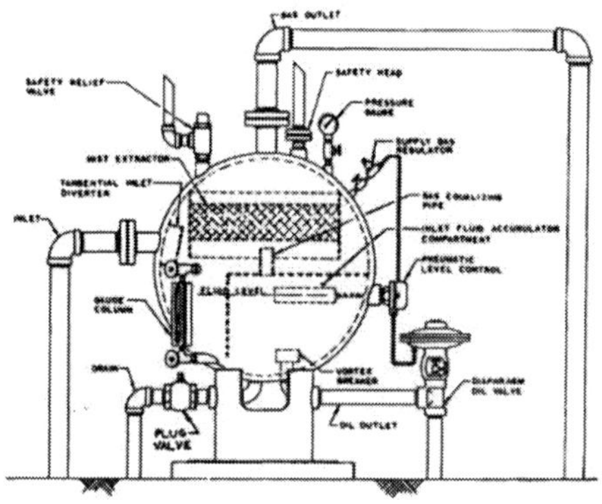

Figure 13. Spherical Separators (High Pressure Pneumatic Controls) [3].

5. Regulation for Processing Oil and Gas

Indonesia is one of the countries that issued regulations regarding safety in oil and gas processing, as follows:

- Regulation No. 1 Year 1970
- Government Regulation No. 11 Year 1979
- Regulation of the Minister of Energy and Mineral Resources of the Republic of Indonesia No. 38 Year 2017

This regulation intends to take preventive measures against system failures that can affect human safety, the environment, and the product itself. The preventive action meant here is to inspect the equipment used to support activities based on the time-based method or the results of risk analysis to ensure success in production and to minimize the occurrence of failures.

IOP Conf. Series: Materials Science and Engineering 1052 (2021) 012051 doi:10.1088/1757-899X/1052/1/012051

6. Risk-Based Inspection (RBI)

Risk-based inspection (RBI) is a risk-based approach to prioritizing and planning inspections, especially in the oil and gas industry because it has high risks [6]. The purpose of the RBI is to identify the damage or defect that could have caused a large-scale accident before it occurred. Also, it can identify incidents that will occur (consequences) when equipment is damaged and how often (probability) these incidents occur [7].

In the RBI, the risk is defined as the product of the probability of an event occurring and the consequences of that event. It can be written systematically as follows [7][12][13]:

Risk = Probability x Consequence (1)

Risk assessment is a process to identify sources of hazards, so that they can estimate risks and evaluate them. This risk assessment process will answer the following questions [8]:

- What caused the problem?
- How often the problem can occur?
- What are the consequences of the problem?

According to API RP 581, risk can be written in the form of a 5 x 5 risk matrix which shows the level of risk into 4 categories, namely:

- Low Risk
- Medium Risk
- Medium High Risk
- High Risk

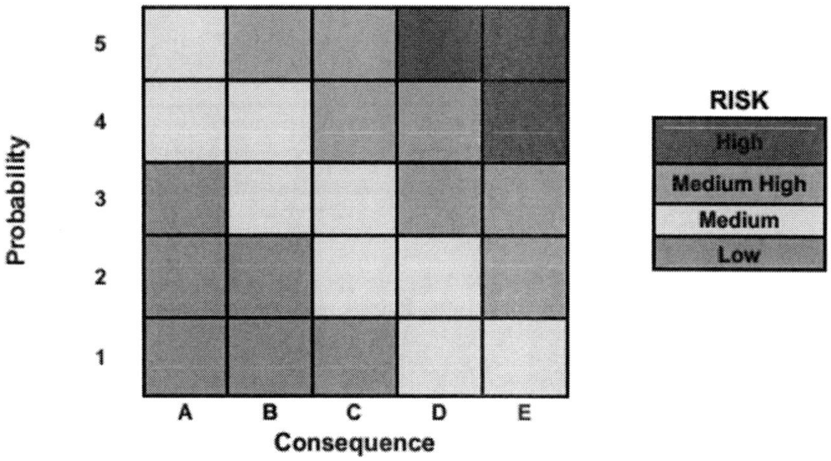

Figure 14. Balance Risk Matrix [9]

7. Risk-Based Inspection (RBI) for Pressure Vessel

A pressure vessel is an equipment that has a temperature and pressure that is different from its environmental conditions [10]. Based on the applicable regulatory requirements, all equipment that uses pressure must be inspected following the inspection code used and is still valid. This inspection activity aims to maintain the appropriateness of the pressure vessel equipment to continue its function. One method that can be used to inspect pressure vessels is risk-based inspection (RBI). In API RP 581, it has been explained that the vessel is divided into 8 component types, namely: KODRUM, COLBTM, FINFAN, FILTER, DRUM, REACTOR, COLTOP, COLMID which must be adjusted to the type of vessel to be analyzed [9]. After that, data related to equipment/components, operating conditions, damage mechanisms, inspection history, and so on will be searched (data and information collection). After the data is obtained, calculations will be carried out related to the probability of failure (PoF) and the consequences of failure (CoF) (risk assessment process). So that the risk level is obtained and can make an inspection plan.

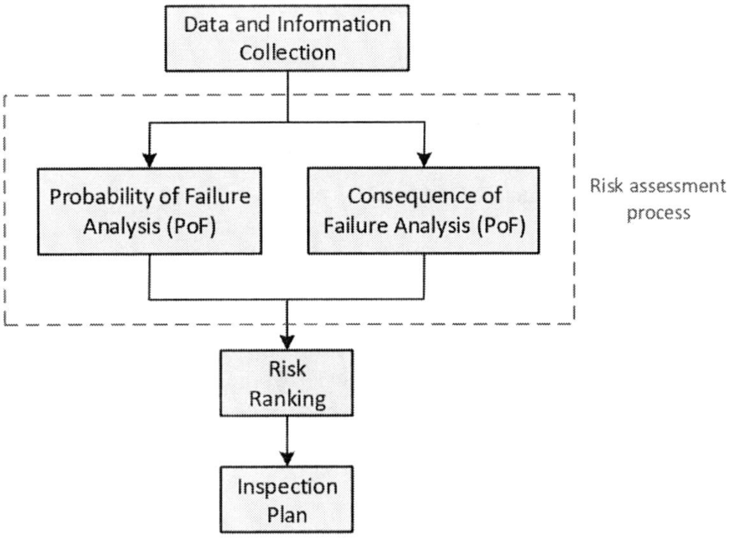

Figure 15. RBI Flowchart [7]

7.1 Data and Information Collection

At this step, data collection related to pressure vessels, such as operating conditions, damage mechanisms, inspection history, chemical compositions of fluid, volume inventory, mass inventory, and so on will be searched related pressure vessel.

7.2 Probability of Failure (P_oF) for Pressure Vessel

For the steps in calculating the probability of failure (PoF) for a pressure vessel are as follows [11]:

- Calculates the generic failure frequencies (gff_n)
- Determine and calculate the damage factors that occur, such as thinning damage factors, cracking damage factors, and so on.
- Calculate the total damage factor ($D_f(t)$)
- Calculate the probability of failure ($P_f(t)$)

7.3 Consequence of Failure (C_oF) for Pressure Vessel

In accordance with the existing problem boundaries above, that the RBI analysis used is level 1 analysis, so the calculation of the consequences of failure (CoF) can be seen as follows [11]:

- Determine representative fluid and fluid properties
- Calculating the release hole size selection, which includes the following steps:
  - Calculate the hole size area (A_n)
  - Calculates generic failure frequencies (gff_n)
- Calculate the release rate
  - Determine the release phase
  - Calculate the liquid release rate (W_n)
- Estimating fluid inventory available for release
  - Calculate the mass of component ($mass_{comp}$)
  - Calculate mass inventory ($mass_{inv}$)
  - Calculating mass available for release ($mass_{avail}$)
- Estimating the release type (t_n)
- Determine the release rate and mass
  - Determine the continuous release rate ($rate_n$)
  - Determine the length of leak duration (ld_n)
  - Determine the instantaneous release mass ($mass_n$)
- Determining the flammable and explosive consequences
  - Calculating the energy efficiency adjustment factor ($eneff_n$)
  - Calculate consequence areas for the equipment damage
  - Calculate consequence areas for personnel injuries
  - Determine factors for blending
  - Calculate blended consequence areas
  - Determine the final weighted consequence of the area
- Determining the toxic consequences
  - Calculate the toxic release rate and release mass
  - Determine toxic leak duration
  - Determine the area of toxic consequence
  - Calculate the weighted toxic consequence for the area
- Determine the non-flammable non-toxic consequences
- Determine component damage and personal injury consequence area
- Financial Consequences
  - Calculate component damage cost
  - Calculate damage costs to surrounding equipment in the affected area
  - Calculate accounts for the time lost in production while the failed equipment is offline
  - Calculate the resulting outage from damage to surrounding equipment
  - Calculate business interruption costs
  - Calculate the potential injury cost
  - Calculate the environmental clean-up cost
  - Calculating the total financial cost and toxic financial cost
- Risk analysis
  - Calculate the area risk
  - Calculate the financial risk

7.4 Risk Raking

For plotting the values of probability and consequence into a graphic, two ways can be used, namely by using a risk matrix. To determine risk by area, the relationship between probability of failure and area-based consequence of failure can be used to form a risk level according to the risk matrix. And to determine risk by financial, it is possible to use the relationship between probability of failure and financial-based consequence of failure, which can be used to form a risk level according to the risk matrix.

Tabel 1. Numerical Values Associated with PoF and Area-Based CoF Categories [7]

Probability Category		**Consequence Category**	
Category	Probability Range	Category	Range (m^2)
1	$P_f(t, I_E) \leq 3.06E-05$	A	$CA \leq 9.29$
2	$3.06E-05 < P_f(t, I_E) \leq 3.06E-04$	B	$9.29 < CA \leq 92.9$
3	$3.06E-04 < P_f(t, I_E) \leq 3.06E-03$	C	$9.29 < CA \leq 929$
4	$3.06E-03 < P_f(t, I_E) \leq 3.06E-02$	D	$929 < CA \leq 9{,}290$
5	$P_f(t, I_E) > 3.06E-02$	E	$CA > 9{,}290$

Tabel 2. Numerical Values Associated with PoF and Financial-Based CoF Categories [7]

Probability Category		**Consequence Category**	
Category	Probability Range	Category	Range (m^2)
1	$P_f(t, I_E) \leq 3.06E-05$	A	$FC \leq 10{,}000$
2	$3.06E-05 < P_f(t, I_E) \leq 3.06E-04$	B	$10{,}000 < FC \leq 100{,}000$
3	$3.06E-04 < P_f(t, I_E) \leq 3.06E-03$	C	$100{,}000 < FC \leq 1{,}000{,}000$
4	$3.06E-03 < P_f(t, I_E) \leq 3.06E-02$	D	$1{,}000{,}000 < FC \leq 10{,}000{,}000$
5	$P_f(t, I_E) > 3.06E-02$	E	$FC > 10{,}000{,}000$

7.5 Inspection Plan

The inspection plan is based on the risk level of a pressure vessel at the time the RBI analysis is carried out. If this inspection is carried out, it will not directly impact on reducing the risk of a pressure vessel. However, at least carrying out an inspection is expected to get actual information from a pressure vessel. The accuracy of this information is used to update the condition of a pressure vessel to reduce uncertainty in the analysis of the probability of failure.

The inspection plan is designed to detect and measure specific types of corrosion that may occur (such as local corrosion and general corrosion), cracking, or other types of damage. Each type of damage has its method of detection and measurement. So that the inspection plan is effective if the inspection method and the area of the inspected can represent the types of damage that may occur.

Tabel 3. Inspection Effectiveness Category [7]

Inspection Effectiveness Category	Inspection Effectiveness Description	Description
A	Highly Effective	The inspection methods will correctly identify the true damage state in nearly every case (or 80-100% confidence).
B	Usually Effective	The inspection methods will correctly identify the true damage state most of the time (or 60-80% confidence).
C	Fairly Effective	The inspection methods will correctly identify the true damage state about half of the time (or 40-60% confidence).
D	Poorly Effective	The inspection methods will provide little information to correctly identify the true damage state (or 20-40% confidence).
E	Ineffective	The inspection method will provide no or almost no information that will correctly identify the true damage state and are considered ineffective for detecting the specific damage mechanism (less than 20% confidence).

8. Conclusion

From this explanation, it can be concluded that a pressure vessel is an equipment that has an important function in terms of processing oil and natural gas. The function of the pressure vessel is to separate the flow of hydrocarbons into 3 types of fluid (gas, oil, and water) for further processing. However, several factors cause failure in this activity. So that several authorities in various countries issued regulations regarding the safety inspection of installations and equipment in the oil and gas industry. Several authorities in various countries have also issued regulations regarding the safety inspection of installations and equipment in the oil and gas industry. For this reason, all existing equipment must be guaranteed reliability so as not to cause harm to workers and the environment. The risk-based inspection (RBI) method was chosen because the oil and gas processing industry have a high and dangerous level of risk. RBI is a method that is considered complex because it can determine and predict the current and future risk levels for pressure vessels.

References

[1] Putra, F. M. A., 2016. *ANALISA TEKNIS DAN EKONOMIS PEMBANGUNAN INDUSTRI MANUFAKTUR BANGUNAN LEPAS PANTAI DI JAWA TIMUR.* Surabaya: Repository ITS.

[2] ABB, 2013. *Oil and gas production handbook an introduction to oil and gas production, transport, refining and petrochemical insdustry.* Olso: ABB Oil and Gas.

[3] Sivalls, C. R., 2009. Oil and Gas Separation Desain Manual. Odessa: Sivalls, INC..

[4] Seo, J. K. et al., 2015. A risk-based inspection planning method for corroded subsea pipelines. *ELSEVIER,* 109(Ocean Engineering), pp. 539-552.

[5] Hassan, D. A., 2013. Review of The Global Oil And Gas Industry: A Concise Journey From Ancient Time to Modern World. *Petroleum Technology Development Journal,* Volume 3, pp. 123-141.

[6] Zhaoyang, T. et al., 2011. An evaluation of maintenance strategy using risk based inspection. *ELSEVIER,* 49(Safety Science), pp. 852-860.

[7] API, 2016. *API RECOMMENDED PRACTICE 580-RISK BASED INSPECTION.* 3th ed. Washington DC: API Publishing Services.

[8] Saifulloh, M., 2018. *ANALISIS PENJADWALAN PROGRAM INSPEKSI PRESSURE RELIEF DEVICE (PRD) DENGAN METODE RISK BASED INSPECTION API 581 PADA CENTRAL PROCESSING PLANT SISTEM PRODUCTION GAS SEPARATOR.* Surabaya: s.n.

[9] API, 2016. *API RECOMMENDED PRACTICE 581-RISK-BASED INSPECTION METHODOLOGY.* 3th ed. Washington DC: API Publishing Services.

[10] Al Qathafi, A. Q. & Sulistijono, 2015. Studi Aplikasi Metode Risk Based Inspection (RBI) Semi-Kuantitatif API 581 pada Production Separator. *JURNAL TEKNIK ITS,* Volume 4, pp. 89-94.

[11] Leta, J. & Mekker, J., n.d. *Example Problems Manual For API RP 581.* s.l.:The Equity Engineering Group, Inc..

[12] Siswantoro, N., Priyanta, D., Zaman, M.B., Semin, 2020. *Failure Mode and Effect Criticality Analysis (FMECA) Fuzzy to Evaluate Critical Level on Main Engine Supporting System. IOP Conference Series: Earth and Environmental Science,* 557, pp. 1-13.

[13] Siswantoro, N., Semin, Zaman, M.B., 2020. *Criticality assessment for marine diesel engine using failure mode and effect criticality analysis (Fmeca) approach: Case study on lubricating oil system. International Review of Mechanical Engineering,* 14, pp. 258-263.

IOP Conf. Series: Materials Science and Engineering 1052 (2021) 012052 doi:10.1088/1757-899X/1052/1/012052

The development of internet-of-things (IoT)-based ship monitoring system in madura strait

A Maulidi[1], T Irmiyana[1]

[1] Lecturer at Teknik Bangunan Kapal, Politeknik Negeri Madura.
Jl. Raya Taddan KM. 4 Camplong Sampang, Jawa Timur, Indonesia.

Abstract. The 4.0 revolution industry has come up with the internet and technology so-called Internet of Things (IoT), used in the navigation and monitoring systems for ships safety. This monitoring system receives data from the AIS Transceiver, the equipment installed, and integrated with the ship's network infrastructure. The equipment transmits signatures tracking the ships sailing and going ashore as the AIS Receiverreceives them. With the increasing number of ships transporting fuels and natural gas from mine-producing areas in the Madura Strait, the development of an IoT-based monitoring system is prominent. The development will ease to monitor movements of ships that transport natural gas. It may also become the basis for further developing ships monitoring systems to improve Madurese fishers' safety operating in the Madura Strait. In the initial system design, the system can transmit and receive static data, such as ships'ID and names. Then it can transmit and receive dynamic data, such as the ships' geographical coordinate, to identify marine traffic around the areas. Later, this system could be developed so that it not only collects data but also proceeds the data to analyze the ships condition.

1. Introduction

As one of the islands in Indonesia, Madura is one of that has a fairly dense population. Most of the Madurese work as fishermen, especially on the southern coast of Madura. Sailing activities for fishing in Sampang is wide enough that requires a ship with sufficient navigation tools with a high level of accuracy. Even though the profession as a fisherman is a hereditary job, fishing boats often get lost in the middle of the vast expanse of sea and don't know which direction to go.

The navigation system is the key of ship safety in shipping. Marine navigation is many used in marine vessels as safety tool, cicerone and communication tool when there is in the middle of the sea, many ships use marine navigation tools such as compasses, maps, radar and GPS as tool to help them when they sail in the oceans. The regulations for ship traffic monitoring equipment are contained in the International Maritime Organization (IMO) Convention on the Safety of Life at Sea (SOLAS) Chapter V, which organize the safety of shipping for all ships [1].

Currently, sailing activities at sea are not only used by traditional ships, but also large ships of bulk carrier (MIGAS). This is caused by enhancement in shipping activities by bulk carrier from exploration in the Madura Strait [2].

In previous research, a prototype Automatic Identification System (AIS) based on Mini Computer (Mini PC) has function as a transmitter of static data and dynamic data of fishing boats via radio waves [3]. Another research that has been done is the development of receiver AIS prototype based mini PC that can receive static and ship dynamic data [4]. However, this research has not reached the level of integrating the static and dynamic data into an information that becomes the basis for a safety monitoring system [5][6][7].

The increasing of shipping activities by bulk carrier ships from result of exploration in Madura Strait, It is rational to begin developing monitoring system IoT-based in Madura Strait [2]. The development of ship monitoring system of IoT-based, will make it easier to monitor the movement of bulk carrier ships in their activities. Another thing that is expected is increasing in fishing communities safety in the Madura Strait.

2. Methods

In the implementation of this research, methodology was arranged as in the picture below:

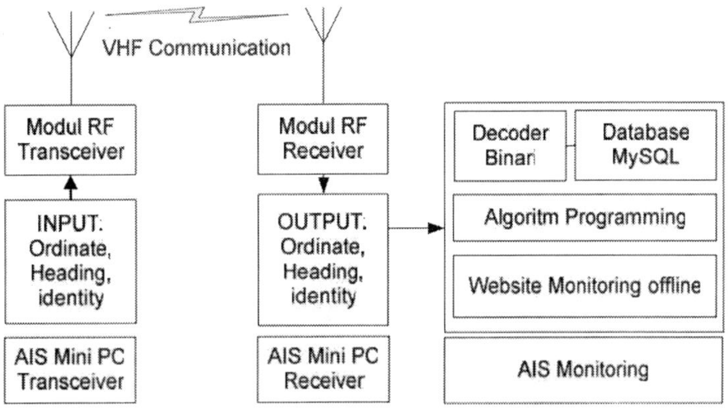

Figure 1. Research Methodology.

The explanation of the research stages is as follows:

1. AIS Transceiver Module: Ordinate data, degrees of direction, and ship identity are generated by the GPS component contained in the AIS Transceiver Module, then with the program algorithm that has been entered on the Arduino mini PC reading the ordinate logger data into binary data that can be transmitted through the module by the VHF Transmitter module is a VHF signal [3].
2. AIS Receiver Module: Ordinate data, degrees of direction, and ship identity are received in the form of logger data, then with the program algorithm that has been previously entered, the Arduino mini PC reads the ordinate logger data into binary data [4].
3. Database creation is basically the combination between AIS data and ships database. In the database creation, it takes time because they have to move the data from the shipping database and convert the data and then inserted into database editor [8].
4. Website Monitoring: PC server, binary data will be stored in the MySQL database as ship sweet data [9]. This data will be obtained using PHP Programming to be displayed as an offline website [10].

IOP Conf. Series: Materials Science and Engineering 1052 (2021) 012052 doi:10.1088/1757-899X/1052/1/012052

3. Result and Discussion

3.1 AIS Data

The results of the execution of the reading algorithm for position (longitude & latitude), and angle are then displayed on the LCD's display which functions as an interfaces between the device and humans that shows the algorithm processing data on the mini PC. Table 1 below is the AIS Transmitter and Receive test data carried out at 3 locations.

Table 1. AIS Transmitter and Receive Test Data.

AIS	Data	Location 1	Location 2	Location 3
Transmitter	Longitude	11316.3770 E	11316.31800 E	11315.8390 E
	Latitude	712.9183 S	712.9339 S	712.7095 S
	Angel	215^0	222^0	141^0
Receiver	Longitude	11316.3770E	11316.31800E	11315.8390E
	Latitude	712.9183S	712.9339S	712.7095S
	Angel	215^0	222^0	141^0

3.2 Database

Figure 2 is display query shipping database on localhost with mysql command that can be opened through browser, but it is running offline. The display can be seen on the number of data bytes in size ship country's. Data can be added to the query function on localhost when adding the ship data.

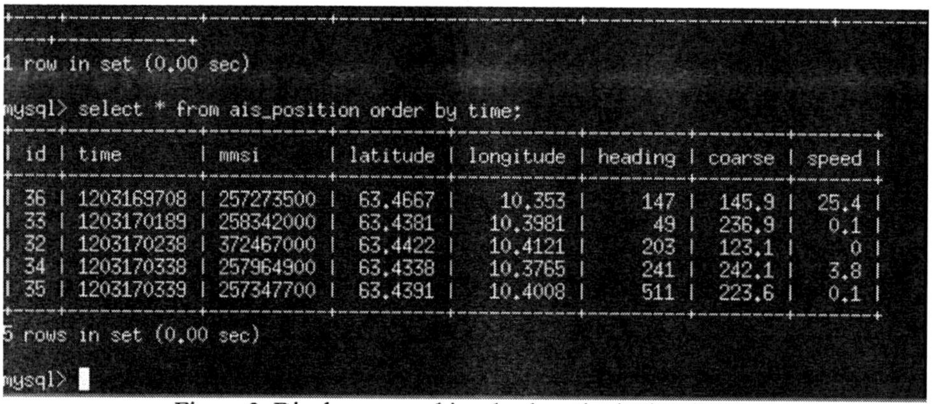

```
1 row in set (0.00 sec)

mysql> select * from ais_position order by time;
+----+------------+-----------+----------+-----------+---------+--------+-------+
| id | time       | mmsi      | latitude | longitude | heading | coarse | speed |
+----+------------+-----------+----------+-----------+---------+--------+-------+
| 36 | 1203169708 | 257273500 | 63.4667  |    10.353 |     147 |  145.9 |  25.4 |
| 33 | 1203170189 | 258342000 | 63.4381  |   10.3981 |      49 |  236.9 |   0.1 |
| 32 | 1203170238 | 372467000 | 63.4422  |   10.4121 |     203 |  123.1 |     0 |
| 34 | 1203170338 | 257964900 | 63.4338  |   10.3765 |     241 |  242.1 |   3.8 |
| 35 | 1203170339 | 257347700 | 63.4391  |   10.4008 |     511 |  223.6 |   0.1 |
+----+------------+-----------+----------+-----------+---------+--------+-------+
5 rows in set (0.00 sec)

mysql>
```

Figure 2. Display query ships database in the localhost.

3.3 Data Record from AIS Decoder

Interpreter/AIS Decoder is used to convert AIS data to be output CSV data every hour. Further results of this AIS data reading device are injected MySQL database. Figure. 3 shows the results of AIS Decoder.

IOP Conf. Series: Materials Science and Engineering 1052 (2021) 012052 doi:10.1088/1757-899X/1052/1/012052

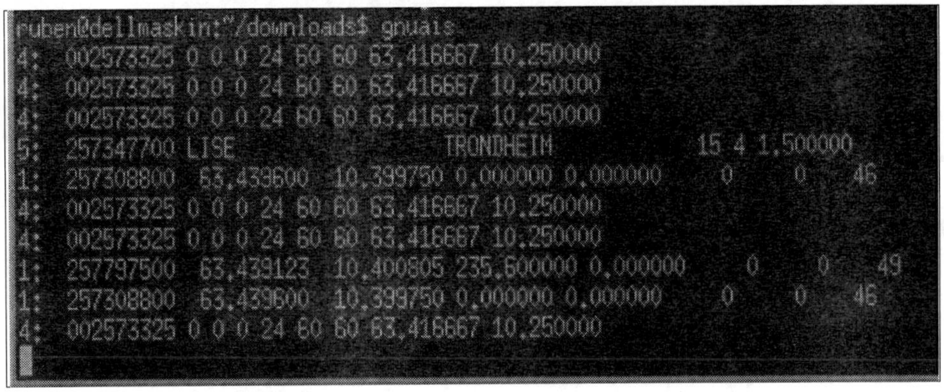

Figure 3. Data converted by interpreter / decoder AIS.

3.4 Display In Web Browser

Figure. 4 shows the different types of vessels that could be seen from the color. The changes and the identity of each ship could be seen by clicking each ship. By using web browser, it could be shown the ship condition, that visited the harbor. The whole ship is summarized in Figure 5 and 6, which describes the vessel data such as vessel name, MMSI, vessel type, engine power, and Speed. On viewing the page, it can also show either the ships safe category or inspection without the need to click around the ship, in other words, it is possible to monitor existing vessels in Madura strait through that page.

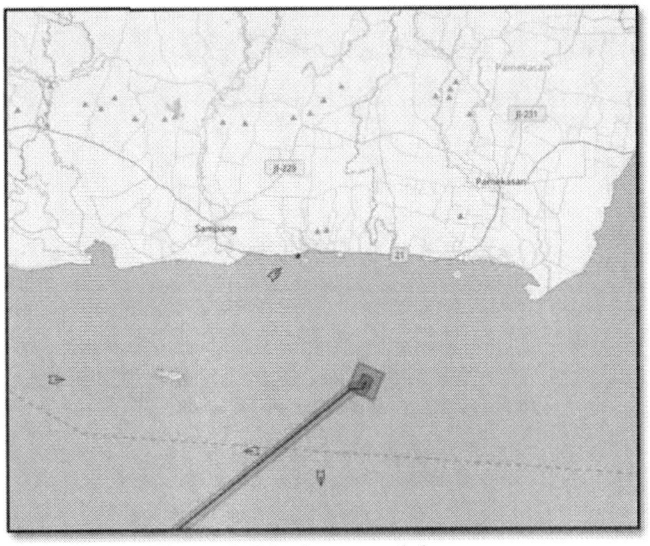
Figure 4. Display on web browser.

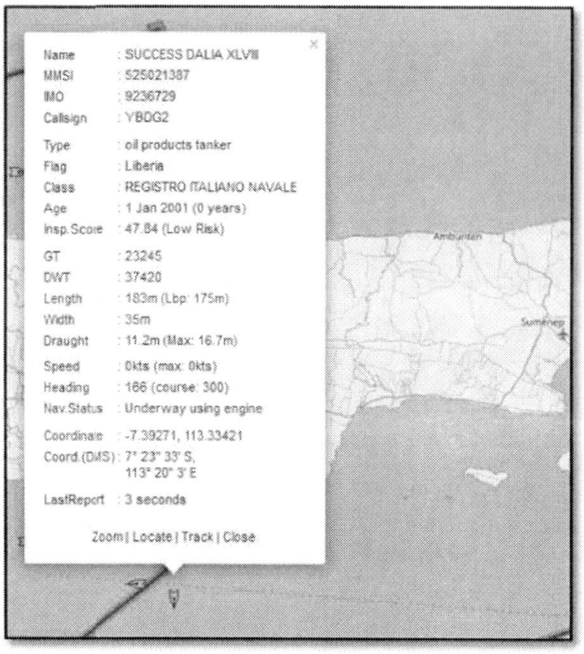

Figure 5. Display ships identity.

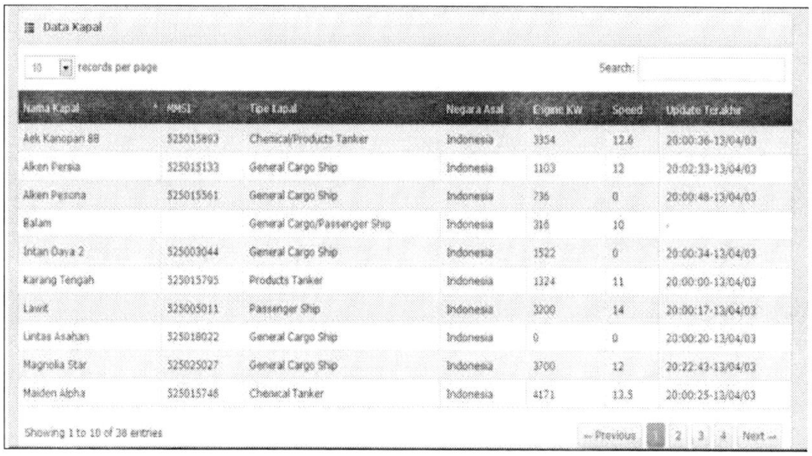

Data Kapal

10 records per page Search:

Nama Kapal	MMSI	Tipe Kapal	Negara Asal	Engine KW	Speed	Update Terakhir
Aek Kanopan 88	525015893	Chemical/Products Tanker	Indonesia	3354	12.6	20:00:36-13/04/03
Alken Persia	525015133	General Cargo Ship	Indonesia	1103	12	20:02:33-13/04/03
Alken Pesona	525015561	General Cargo Ship	Indonesia	736	0	20:00:48-13/04/03
Balam		General Cargo/Passenger Ship	Indonesia	316	10	-
Intan Daya 2	525003044	General Cargo Ship	Indonesia	1522	0	20:00:34-13/04/03
Karang Tengah	525015795	Products Tanker	Indonesia	1324	11	20:00:00-13/04/03
Lawit	525005011	Passenger Ship	Indonesia	3200	14	20:00:17-13/04/03
Lintas Asahan	525018022	General Cargo Ship	Indonesia	0	0	20:00:20-13/04/03
Magnolia Star	525025027	General Cargo Ship	Indonesia	3700	12	20:22:43-13/04/03
Maiden Alpha	525015748	Chemical Tanker	Indonesia	4171	13.5	20:00:25-13/04/03

Showing 1 to 10 of 38 entries ← Previous 1 2 3 4 Next →

Figure 6. Display all ships in the web browser.

4. Conclusion

With the increasing number of ships transporting fuels and natural gas from mine-producing areas in the Madura Strait increasing risk of marine accident, the development of an IoT-based monitoring system is prominent. Research result exemplify mini PC technology can be used as a transmitter and receiver for AIS data that transmits and receives ship static and dynamic data, it is shown by the browser offline. The development will ease to monitor movement of ships that transport natural gas. It may also become the basis for further developing ships monitoring systems to improve Madurese fishers' safety operating in the Madura Strait. In the future, it can be used for many online marine monitoring system.

5. References

[1] SOLAS, 2000, *Safety Of Life At Sea. Carriage Requirements for Shipborne Navigational Systems and Equipment*. s.l. : SOLAS, 2000. Vol. Chapter V.

[2] Artana, Ketut Buda, et al., 2013, *Penlilaian risiko pipa gas bawah laut*. Surabaya : Guna Wijaya.

[3] Maulidi, Akh., Prasetyo, Taufan and Irmiyana, Triyanti., 2019. *Disain Sistem Navigasi Automatic Identification System (Ais) Transceiver Berbasis Mini Computer Pada Kapal Nelayan Tradisional Di Madura*. 01, Bengkalis : INOVTEK, 2019, INOVTEK POLBENG, Vol. IX, pp. 12-17. 2580-2798 .

[4] Maulidi, Akh., Prasetyo, Taufan and LAR., Anauta. 2019, *Sistem Penerima (Receiver) Automatic Identification System (AIS) Berbasis Mini Computer Pada Kapal Nelayan Tradisional Di Madura*. INOVTEK POLBENG, pp. 344-349.

[5] AIS, 2001. *The Complete Guide of Automatic Identification Sistem (AIS)*. s.l. : AIS.

[6] Pitana, Trika, E, Kobayashi and N, Wakabayashi. 2010, *Estimation Of Exhaust Emission Of Marine Traffic Using Automatic Identification System Data (Case Study : Madura Strait Area, Indonesia)*. OCEANS 2010 LEEE, pp. CFP100CF – CDR 978-1-424.

[7] Kobayashi, E., et al., 2010. *Installation of an Asian AIS data receiving system network.* Korea : s.n., Proc of Japan Institute of Navigation.

[8] Maulidi, Akh, et al., 2013. *Integrasi AIS dan Shipping Database Sebagai Dasar Pengembangan Metode Keselamatan Kelautan,* FTK ITS SENTA 2013. pp. D45-D49. Surabaya.

[9] Dinariyana, AA B, et al. 2018, *Pengembangan Prototype Software Real Time Monitoring Berbasis Data Automatic Identification System (AIS).*, Surabaya : IPTEK Journal of Proceedings Series, 2018, Vol. 10.

[10] Maulidi, Akh., Pitana, Trika and Buda, Ketut. 2017, *Pengembangan Sistem Monitoring Automatic Identification System (AIS) Berbasis Website Secara Real Time.* INOVTEK BOLBENG, pp. 153-166.

6. Acknowledgments

We would like to thank the Director of Politeknik Negeri Madura (POLTERA), the Chairperson of the P3M POLTERA, Colleagues of Poltera, All Committee 5th International Conference on Marine Technology (SENTA 2020) which has been approve and publish the results of this research. And my family who always support all activities.

Economic analysis on the LNG Distribution to power plants in Bali and Lombok by utilizing mini-LNG carriers

Emmy Pratiwi[1,2], Dhimas Widhi Handani[1,2], Gede Bagus Dwi Suasti Antara[2], A A B Dinariyana[1,2], and Hayy Nur Abdillah[1]

[1] Department of Marine Engineering, Faculty of Marine Technology, Institut Teknologi Sepuluh Nopember, Surabaya, Indonesia
[2] Center of Excellences Maritime Safety and Marine Installation, Institut Teknologi Sepuluh Nopember, Surabaya, Indonesia

emmy.p@its.ac.id

Abstract. In recent years, small-scale LNG has contributed significantly in supplying the demand from industry users of natural gas as clean fuel. Small-scall LNG has the advantages of providing the greater operational flexibility and lower initial investment cost than conventional LNG or pipeline supply. Focusing on the case in Indonesia, many designed gas-fuel power plants are located in remote area or in the limited depth of water level especially in eastern part of Indonesia, such as Lombok and Bali. Therefore, small-scall LNG is purposed to overcome the natural gas demand in such conditions. Two scenarios were developed in this study in utilizing the conventional LNG carriers to deliver the natural gas from origin located in LNG terminal Benoa, Bali, to several power plants in Bali and Lombok. Both scenarios are new ship LNG building and the conversion of Landing Craft Tank (LCT) into non-conventional LNG carrier. Economic analysis on the LNG Distribution to power plants in Bali and Lombok utilizing mini LNG carriers is the main objective of this study. Capital Expenditure (CAPEX) including mini-LNG carrier, investment in jetty, storage facility, regasification unit and the system distribution to the power plant were considered. While the Operational Expenditure (OPEX) consist of operational and maintenance cost in receiving terminal and LCT. The result of this papers can be concluded that based on the economic perspective, utilizing the mini LNG carriers are feasible and requires relatively small cost to supply the LNG demand in Lombok and Bali.

1. Introduction

Liquid Natural Gas (LNG) has become increased global demand for electricity generation since natural gas is the efficient and the cleanest energy source compared to fossil fuel [1]. The growth of investment in gas infrastructure is an critical issue and become increases in order to meet the gas demand, particularly in countries with potential reserves and those currently with a high dependence on coal [1]. Gas reserves in Asia Pacific, which account for 9% of the global total, are more distributed across countries, including China, Indonesia, Malaysia and Australia [1]. According to BP Statistical Review of World Energy, by the end of 2018, the total proven gas reserve in Indonesia has approximately 2.3 trillion cubic meters [2]. The government of Indonesia committed to increase the consumption of domestic energy sources, including natural gas. Through Government Regulation Number 79 of 2014 concerning National Energy Policy, domestic consumption of natural gas is targeted to increase from the current 19% to 24% in 2050. On the contrary, the use of fossil fuel is reduced from the current 42% to 20% in 2050 [3]. This research supports energy management particularly in Indonesia that is directed towards equitable energy by increasing equal access to energy at affordable prices and a more efficient management of energy supply.

Design of the LNG supply chain is the most critical system to support the natural demand. The LNG supply chain consists of exploration, extraction, liquefaction, transportation, storage and regasification an then delivered to gas costumers as end users [4]. However, as the archipelago country, the locations of gas costumers particularly power plants in Indonesia are scattered at several point. It brings the great challenge to distribute the LNG in such location of users. Small-scale LNG has been much applied to design the LNG supply power plants which are located in remote area or in the limited depth of water level due to its operational flexibility and lower initial investment cost [5], [6]. Small scale LNG generally refers to LNG associated facilities (receiving terminals, storage units, vessel, etc.) with similar characteristics but of a lower magnitude than conventional LNG infrastructure [6]. The vessel used in small scale LNG has smaller dimensions to meet the water depth requirement so that the operational will be easier, more flexible and reduce the investment cost.

This study aims to provide the economic analysis on the LNG distribution to power plants utilizing mini-LNG carriers to support equitable energy, affordable prices and a more efficient management of energy supply. Not only utilizing the conventional LNG carriers only, this study also performs the economic analysis regarding the conversion of LCT vessel into non-conventional LNG carrier. LNG are stored in the ISO tank then transported to the power plant using LCT. Finally, this work provides an overview of the implementation of small scall LNG distribution to small scale power plant in Lombok and Bali island, Indonesia.

2. Design of LNG distribution in Bali and Lombok, Indonesia

Small scale LNG supply chain is adopted in this study to design the LNG distribution. In general, the concept of small-scale LNG is not much different from the medium-large LNG supply chain. In a small-scale LNG, LNG-related facilities have a lower magnitude than conventional LNG infrastructure [6], [7]. LNG can be shipped to smaller satellite terminals using mini LNG carriers with smaller capacities.

Small-scale LNG (SSLNG) carriers is the most common vessel used in shipping the LNG for sea based transports [5], [7]. However, the characteristic of waterway near the location of power plant in Indonesia has limited depth of water level or shallow waterway. Based on the current condition, this study purposes to utilize the converted Landing Craft Tank (LCT) as the non-conventional LNG carriers and also compare the economic aspect between the converted LCT and new building of ships. The conversion of LCT vessel into non-conventional LNG carriers has been proved that it could be implemented in remote area and give better efficiency both in technical and economic aspects [8]. The advantage of the LCT vessel is that it does not require a special port in the loading process because LCT vessel has a ramp door that can be used in loading and unloading the passengers, heavy equipment, cargo, construction materials and transportation equipment. LCT vessel also have a relatively low draft that allows them to operate in shallow waterway [8]. LCT vessel is converted and modified in order to meet requirement particularly ship's construction, lashing system, seating system and stability. ISO Tank is used to store the LNG then to be shipped by using LCT to the user. ISO tank is a tank constructed based on International Organization for Standardization (ISO) standards so that it can be transported using various modes of transportation. This tank has a protective and insulating construction on the surface of the tank.

As shown in Figure 1, LNG is filled to the ISO tank in LNG Terminal (origin) then LNG ISO tank is stored in the LCT vessel. LCT vessel carrying ISO LNG tanks sailing to the MPP plant destination. At the receiving terminal, the ISO LNG tank is unloaded from the ship and then transported by truck to stacking yard. Before entering the regasification unit, LNG is pumped from the ISO LNG tank to the buffer tank. This buffer tank serves to prevent interruption of the LNG supply to the regasification unit due to ISO LNG replacement activities.

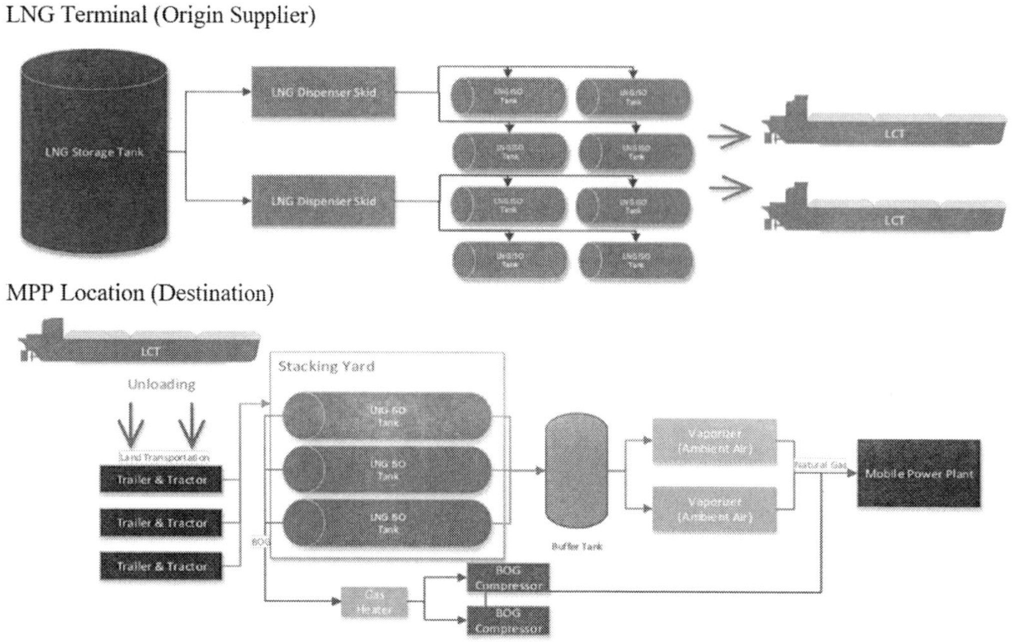

Figure 1. Distribution process diagram with ISO LNG Tank

The design route of mini LNC supply chain actually has been conducted by authors in previous research. This study is a continuation and part of LNG supply chain research of the author's road map. Based on the statement of Directorate General of Oil and Gas of Indonesia at the Green-Eco Energy conference in Bali, distribution in Bali is used as an example of LNG distribution which adapts the conditions in Indonesia which are spread over several islands and separated by waterway. The utilization of Mini Scale LNG carrier is one of the factors that causes LNG distribution in Bali to be carried out in more efficient cost. Therefore, supporting the Bali Green Eco Energy, power plants in Bali and island near to Bali, Lombok island, will be defined as the end users of natural gas in small-scale LNG supply chain. Floating storage and regasification unit (FSRU) in Benoa Terminal, Bali, was selected as a potential LNG supply terminal. Lombok Peaker, MPP Lombok, Gilimanuk and Pemaron power plants located in Bali and Lombok Island are used here as the demand in this case study.

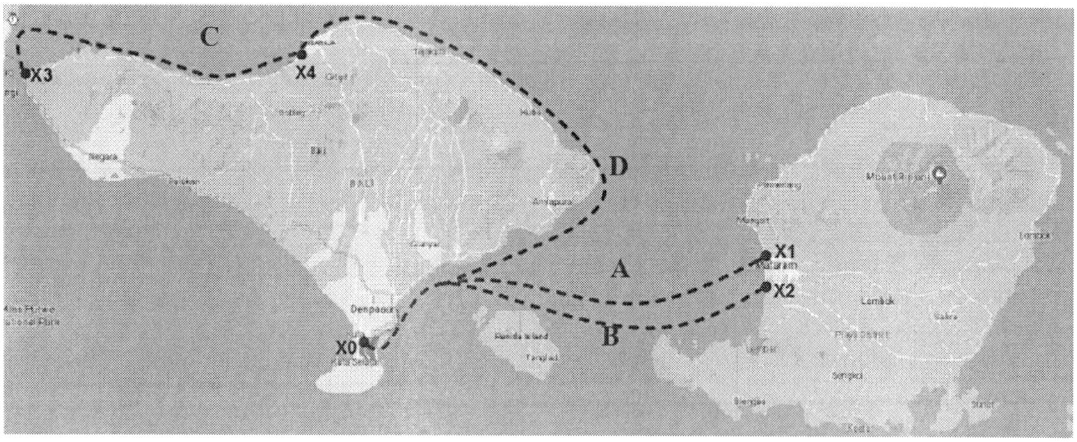

Figure 2. Design of LNG distribution route in Bali and Lombok Island, Indonesia

Mathematical model has been performed previously to solve capacity-based transportation cases. The location of users, Lombok Peaker, MPP Lombok, Gilimanuk and Pemaron power plants in Bali and Lombok Island are denoted by X_1, X_2, X_3 and X_4 as shown in Figure 2. Figure 2 also ilustrates the distribution of LNG in four power plant. Table 1 shows the capacity of power plant and LNG demand. Those 4 powerplant will be supplied by LCT non-conventional as mini LNG carrier with specification as in Table 1.

Table 1. Specification of LCT non-conventional as mini LNG carrier

BHP	1760 kW	LWL	4666 m
SFOC	190 g/kWh	LPP	45,71 m
Vs	10 knots	Depth (moulded)	3.5 m
Gross Tonnage	1200 GT	Draft (moulded)	2.5 m
LOA	50.50 m	Number of ISO Tank	32

Table 2. Capacity of power plants in Bali and Lombok Island, Indonesia

Symbol	Location	Capacity (MW)	MMSCFD	LNG Demands (m^3/day)	ISO Tank Demand (unit/day)
X0	LNG Terminal Benoa				
X1	PLTMG Lombok Peaker	136	9,07	417,07	21
X2	MPP Lombok	50	10,00	460,00	24
X3	Gilimanuk	130	8,70	398,70	20
X4	Pemaron	80	5,30	245,30	13

Considering the capacity of power plants, roundtrip and sailing frequency, the number of mini LNG used to meet the gas demand can be summarized in Table 3.

Table 3. Summary of Number of mini LNG carrier and total trip per year

Route	Route	Number of Mini LNG Carrier	Total Trip/Year	Information
A	X0-X1-X0	1	180	Continuous
		1	180	Intermittent
B	X0-X2-X0	1	120	Continuous
		1	120	Intermittent
C	X0-X3-X0	2	53	Continuous
D	X0-X4-X0	1	80	Continuous

3. Economic Analysis

Economic analysis needs to be performed to evaluate whether the cost occurred by distribution activities of the LNG ships is feasible. The revenue derived from the gas selling with varied price margin is included in the economic analysis. Two parameters, capital expenditure (CAPEX) and operational expenditure (OPEX), are considered in the economic analysis. CAPEX is defined as the total capital for the initial investment. CAPEX consists of two major which are the investment or capital costs for transportation and investment costs for terminal construction. OPEX is all costs occurred to support the operational distribution of LNG, including the operational costs of the receiving terminal and the transportation costs for transporting LNG from the refinery to the receiving terminal, as well as maintenance costs. Related work about economic analysis in small scale LNG distribution concluded that cost of LNG transportation depends on the amount of cargo demand and shipping distance. Several parameter or criteria to analyze the feasibility in the economic aspects are used, Payback Period (PDB), Net present Value (NPV) and Return of Investment (ROI).

The payback period (PDB) is a capital budgeting procedure used to determine the profitability of a project. A discounted payback period gives the number of years it takes to break even from undertaking the initial expenditure, by discounting future cash flows and recognizing the time value of money. Formula for calculating PDB is described as follows:

$$PDB = \frac{Total\ Cost\ of\ Investement}{Project\ Cash\ Flow} \tag{1}$$

The Net Present Value (NPV) is the difference between the present value of cash inflows and the present value of cash outflows. NPV is used in capital budgeting to analyze the profitability of a projected investment or project. Formula for calculating NPV is described as follows:

$$NPV = \sum_{t=0}^{n} \frac{(C)t}{(1+i)t} - \sum_{t=0}^{n} \frac{(Co)t}{(1+i)^{t}} \tag{2}$$

Where:

NPV = Net Present Value
(C)t = Cash Inflow on year t
(Co)t = Cash Outflow on year t
n = Project Duration
i = Discount Rate
t = Number time of Periods

The relationship between NPV and discount rate can be presented as the Internal Rate of Return (IRR). IRR is used to find out at the interest rate (discount rate) at what rate an investment can provide benefits. The IRR also shows how much the discount rate is when the net present value (NPV) is zero. If the interest rate is more than the IRR, it is better not to continue investing. The greater the IRR value than the Weighted Average Cost of Capital (WACC) value, it means that the investment in the project is more feasible.

The Return on investment (ROI) is a performance measure used to evaluate the efficiency of an investment or to compare the efficiency of a number of different investments. ROI measures the amount of return on an investment relative to the investment's cost. To calculate ROI, the benefit (or return) of an investment is divided by the cost of the investment, and the result is expressed as a percentage or a ratio. Formula for calculating ROI is described as follows:

$$ROI = \left(\frac{\sum_{1=0}^{n} CashFlow}{n} \right) \times \frac{1}{Investment} \tag{3}$$

4. Results and discussion

4.1. Capital Expenditure (CAPEX)

CAPEX is calculated for all initial cost of investment from the transportation using mini LNG carrier to gas consumed by power plant, including jetty construction, regasification unit and transportation in receiving terminal. In this study, two scenarios are developed for the mini LNG construction, which is new building ships of mini LNG and conversion from LCT vessel to non-conventional LNG carrier.

Two result will provide different insights which one is feasible and more efficient. In the receiving terminal, the process of loading and unloading LNG ISO tank from mini LNG Carrier to the stockpile is utilized the container crane and stacker as presented in Figure 3. The process applies for all power plants. The scenario of loading and unloading LNG ISO tank is important to be designed because the equipment involved in the process should be considered in the component of CAPEX.

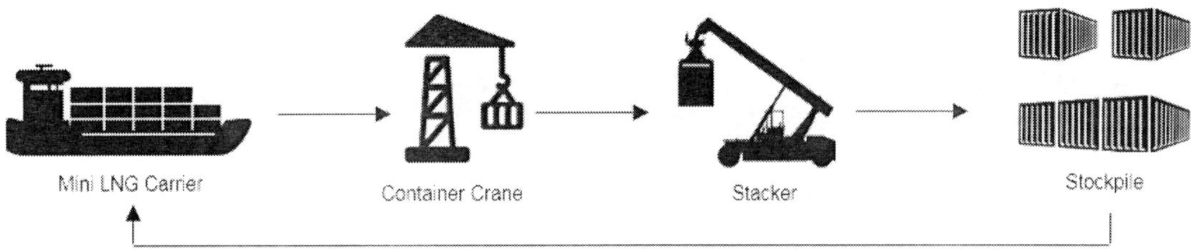

Figure 3. Loading-unloading LNG ISO Tank

The detail component considered in the CAPEX calculation is shown in Table 4. Pricing on the detail calculation for CAPEX supported by qualified vendor which has long experience in related business.

Table 4. Component of CAPEX

SYSTEM	Equipment
Mini LNG Carrier (New Building or Conversion)	
Jetty and Terminal	Jetty Crane Truck & Trailer (40") Jetty Equipment (bollard, wire, etc) Procurement and Civil Works (hardening, cut & fill)
LNG Storage and Stockpile Yard	Land Rent Stacker LNG Metering Unit
Regasification Unit	LNG buffer Tank LNG Vaporizer (Ambient Air Type)
Gas Distribution to Power Plant	Compressor Gas Heater Gas Pipeline (Carbon Steel) Piping System (Pigging, valve, etc) Gas Metering Unit LNG pipeline (stainless steel) Crane and Stacker
Building	Building (control room, monitor room, office)
Fire Fighting Equipment	Hydrant and Portable Fire Fighting Equipment Alarm and Security Systems
Monitoring and Control System	PCS, SCADA (hardware dan software)
Decommissioning	Building (control room, monitor room, office)
Safety Stock (Diesel Oil)	Storage Tank Pump Piping System
Project cost	

Engineering and cost estimate study to calculate the price of building the new mini LNG carrier or conversion from LCT vessel is the most important part because the cost of investment for mini LNG carrier contribute a huge percentage in the components of CAPEX. An approach model of price estimation and construction scheduling of LCT Non-Conventional LNG ISO Tank carriers by making a cost breakdown are as follows:

1. Engineering & Design
   a. Engineering Design
   b. Class Approval Fee (Design & Construction)
2. Procurement Materials & Materials
3. Ship Construction & Installation
4. ISO Tank & Installation of supporting equipment

Detailed cost estimates are mare which include each activity in the Work Breakdown Structures (WBS) structure and the Cost Breakdown Structures (CBS) structure are calculated by estimating the cost of each activity or the material price at that time. According to the calculation of the estimated cost of building a ship, the cost of the LCT vessel for new building and conversion schemes, including materials, construction services and procurement of 20 ft LNG ISO Tank, is US$ 4,199,543 and US$ 2,705,188. The prices are included 10% VAT.

In addition, since the generator in power plant can be operated both natural gas and oil fuel mode, diesel fuel is used as the safety stock in the power plant to mitigate if there is a delay in LNG distribution to the power plant. The storage for diesel fuel and the piping system has been considered in the calculation of CAPEX.

Table 5. Summary of CAPEX for both scenario

CAPEX	Unit	Price	
		Scenario 1	Scenario 2
LNG Distribution to Lombok Peaker	set	$ 18.290.869	$ 18.744.569
LNG Distribution to MPP Lombok	set	$ 18.888.542	$ 18.378.292
LNG Distribution to Pemaron	set	$ 12.575.145	$ 12.098.825
LNG Distribution to Gilimanuk	set	$ 18.412.798	$ 17.902.548
TOTAL		$ 68.167.354	$ 76.969.689

Table 5 presents the summary of CAPEX of LNG distribution to four power plants for both scenario, first scenario which is utilizing new building of LNG ships and the second scenario, the conversion of mini LNG carrier from LCT vessel. Based on Table 5, it can be concluded that there is no significantly difference price between two scenarios because cost of construction services and procurement of 20 ft LNG ISO Tank has major contribution in total cost. However difference of CAPEX value will be further analyze how much CAPEX affect to the NPV and IRR in economic analysis.

4.2. Operational Expenditure (OPEX)

OPEX consist of two major components, operational and maintenance cost of LNG receiving terminal and Mini LNG Carrier. The detail prices for all component in the operational distribution LNG are presented in Table 6 and Table 7 respectively. The operational costs of the receiving terminal consist of maintenance cost, operational cost and labor cost. The components of operational cost are included logistic cost, lant rent and power requirement at all receiving terminal. All the prices are determined based on the literature study. Operational and maintenance cost of mini-LNG carrier are the summed cost for all required number of mini LNG carrier in a year.

Table 6. Operational and Maintenance Cost of LNG Receiving Terminal

O&M Cost of Receiving LNG Terminal	Maintenance Cost	Operational Cost	Labour Cost	TOTAL
	(US$/Year)	(US$/Year)	(US$/Year)	(US$/Year)
PLTMG Lombok Peaker	300.000,00	3.501.103	514.759	4.315.862
MPP Lombok	300.000,00	1.552.759	514.759	2.367.517
Pemaron	300.000,00	2.182.414	514.759	2.997.172
Gilimanuk	300.000,00	3.365.172	514.759	4.179.931
		Total		13.860.483

The operational and maintenance costs of this mini LNG carriers are all costs occured during the operation of the LCT ship per year, including ship maintenance costs. The operational costs in LNG distribution consist of several things as listed in Table 7 and the total cost are included for all 7 mini LNG carriers.

Table 7. Operational and Maintenance Cost of Mini LNG Carrier

O&M Cost of Mini LNG Carrier	Unit	Value
Fuel Oil	US$/year	5.804.848
Lubricating Oil	US$/year	115.019
Fresh Water	US$/year	16.605
Consumable cost	US$/year	204.828
Port Service Fee	US$/year	65.604
Maintenance Cost	US$/year	207.586
Annual crew cost	US$/year	660.414
Insurance and Administration Costs	US$/year	338.738
Total	US$/year	**7.413.641**
Total + margin 30%	**US$/year**	5.804.848

Based on the sum of operational and maintenance cost of LNG terminal and mini-LNG carrier, Table 8 shows a summary of the OPEX costs for each distribution route. There is no difference value of OPEX for both scenario because the operational and maintenance cost for either new built LCT and LCT conversion is same.

Table 8. The summary of OPEX

OPEX		US$/year
Benoa-Lombok Peaker-Benoa	$	6.609.552
Benoa-MPP Lombok-Benoa	$	4.391.043
Benoa-Pemaron-Benoa	$	4.700.767
Benoa-Gilimanuk-Benoa	$	7.796.854
Total	**$**	**32.450.782**

4.3. PDB, NPV and IRR Calculation

In order to carry out economic feasibility studies of LNG distribution in this study, the economic models are conducted by given varies margins/expected profits earned from the sale of LNG to end customer. The scenario of economic calculation is performed by doing a variation of the expected sale price margin namely US$ 2 - US$ 4 with an increase of 0.5 USD.

Table 9. Revenue and Variation of margin rate.

	Margin	Total
	$ 2,00	$ 24.138.667
Revenue (USD/mmbtu)	$ 2,50	$ 30.173.333
	$ 3,00	$ 36.208.000
	$ 3,50	$ 42.242.667
	$ 4,00	$ 48.277.333

Finally, the NPV and IRR are calculated to ensure whether the LNG investment from the financial aspect is feasible. Figure 4 and Figure 5 display the payback period on the LNG investment estimate for both scenario with differing revenue margins. The graph of payback period shows that the larger LNG sale margin, the faster the payback period can be achieved. The payback period graph in Figure 4 shows that with a project duration of 10 years, investment capital can be fulfilled in 4 years at the earliest if the margin is USD 4. However, payback period calculation with margin USD 2 and USD 2.5 per MMBtu cannot be achieved in 10 years of project duration. For the scenario 2 as shown in Figure 5 presents that the payback period has sooner payback period to recover the cost of an investment. Both scenarios show that the margin USD 2 are not feasible because of longer paybacks but margin USD 2.5 still can provide benefits for the second scenario.

Table 10. IRR for both scenarios with different margin

Margin	IRR (Scenario 1)	IRR (Scenario 2)
$ 2,00	-13,7%	-13,3%
$ 2,50	-1,3%	3,0%
$ 3,00	8,5%	14,0%
$ 3,50	16,6%	23,4%
$ 4,00	21,6%	29,5%

To get the profit, the NPV value must be positive in the 10th year, where the IRR value must be greater than the discount rate value. The project will be worth investing if the IRR is above 10% according the government's discount rate. Table 10 represents the summary of IRR for both scenarios for different margin variation. It shows that both scenarios will give benefits if the margin rate is set above USD 3.

According to some references, the margin rate of LNG distribution using small scall LNG is higher than large-scale LNG vessels due to the transportation cost. In the SSLNG vessels, the LNG price dropped into the range of USD 4 [5]. Therefore, focus on minimizing the cost for LNG vessels and design ISO tank can be alternative to reduce the transportation cost in SSLNG distributions to make the projects become feasible.

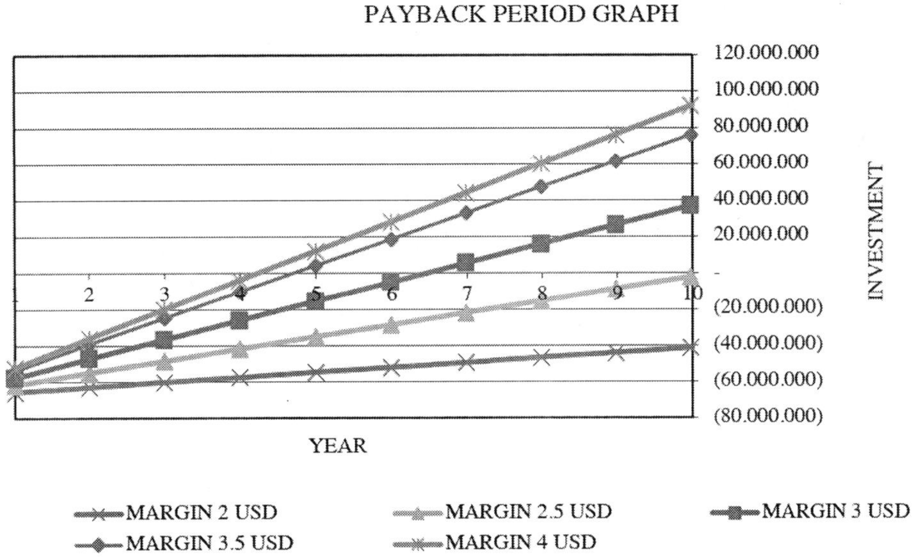

Figure 4. Payback period for economic analysis of scenario 1 (new building)

IRR can be used as a basis for determining the discount rate at which the present value of all future cash flows is equal to the initial investment. The higher an investment's IRR, the more desirable it is to carry on with the investment. Based on the result of NPV and discount rate, IRR for both scenarios with different margin are presented in Table 10. It shows that the scenario 2 has better profitability of investments due to lower investment of cost with the same margin.

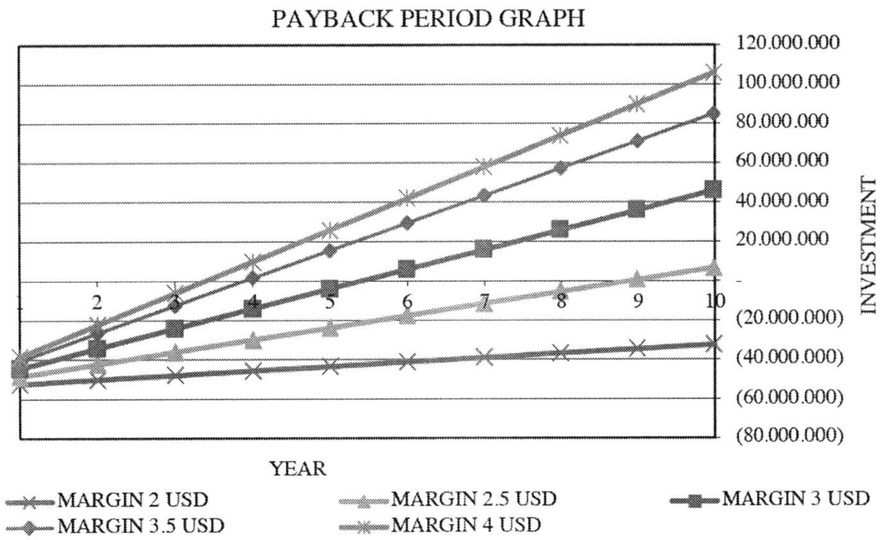

Figure 5. Payback period for economic analysis of scenario 2 (LCT vessel conversion)

Based on the IRR calculation, if the government set the discount rate is 10%, both the scenario 1 and 2 are still feasible if the margin rate is above 3 USD MMBTU.

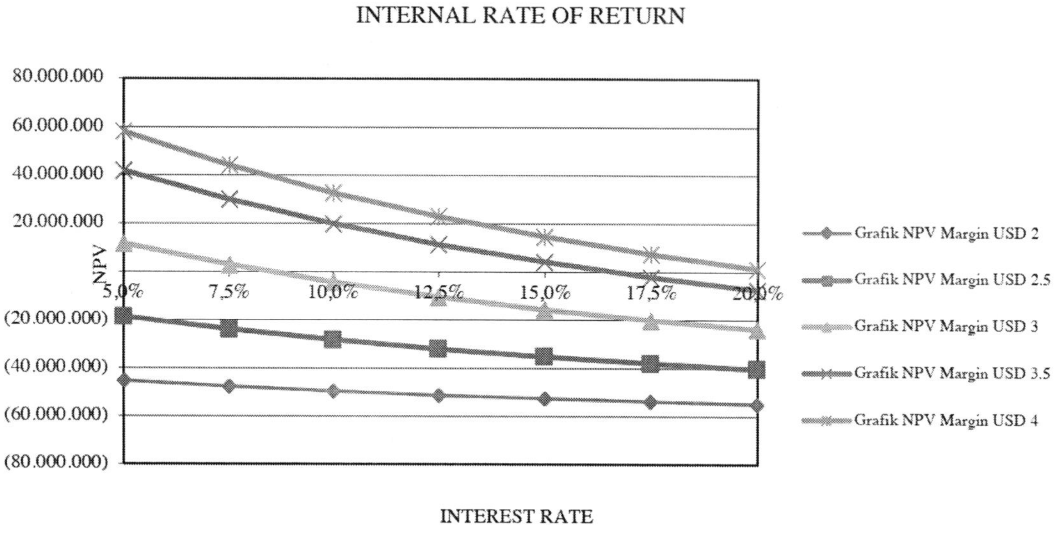

Figure 6. Discount rate and net present value for economic analysis of scenario 1

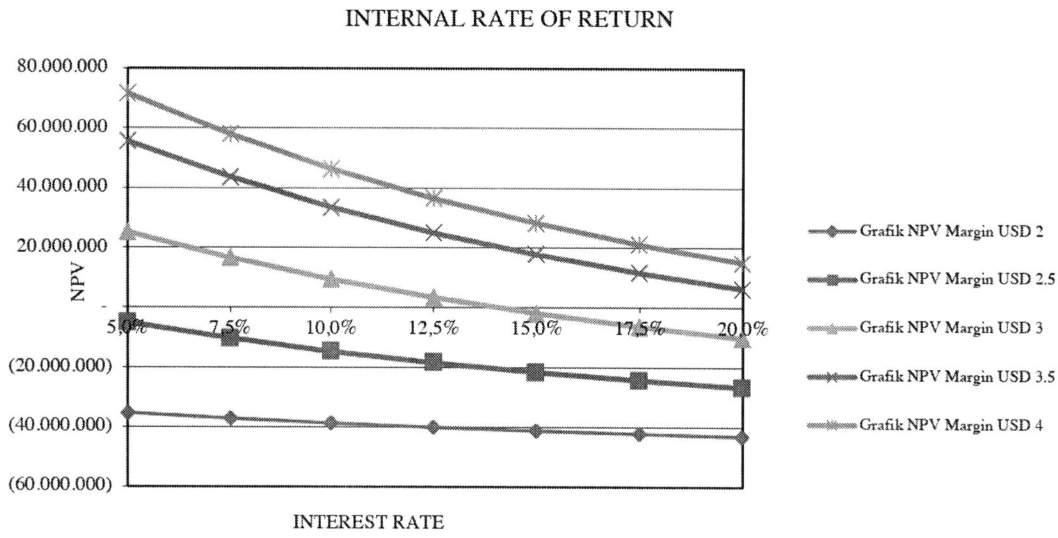

Figure 7. Discount rate and net present value for economic analysis of scenario 2

5. Conclusion

This study presents the implementation of small-scale LNG supply chain by using conventional LCT to supply gas-fuel power plants which are located in remote area or in the limited depth of water level especially in east part of Indonesia, particularly in Lombok and Bali. The utilization of ISO tank to store the LNG then transport them using conventional LCT proves that it provides easier operational. The economic analysis was performed to convince whether the LNG distribution for power plants in Bali and Lombok will be worth investing. Finally, the result of the study can be concluded by utilizing small scall LNG with ISO tank, both scenarios is feasible for investment if the margin rate is above 3 USD/ MMBTU.

6. Acknowledgement

The authors are grateful to the Ministry of Research and Technology/National Research and Innovation Agency (Kemenristek/ BRIN/ LPDP) of Republic Indonesia in publication support by 189/E1/PRN/2020, the Center of Excellence for Maritime Safety and Marine Installation, Institut Teknologi Sepuluh Nopember, Surabaya and our research partner Agency for the Assessment and Application of Technology (BPPT).

References

[1] International Gas Union (IGU), "2020 edition of the Global Gas Report," 2020. [Online]. Available: https://www.igu.org/app/uploads-wp/2020/08/GGR_2020.pdf.
[2] BP, "BP Statistical Review of World Energy 2019 68th edition," 2019. [Online]. Available: https://www.bp.com/content/dam/bp/business-sites/en/global/corporate/pdfs/energy-economics/statistical-review/bp-stats-review-2019-full-report.pdf.
[3] The Ministry of Energy and Mineral Resources Indonesia, "Neraca Gas Bumi Indonesia 2019-2027," Jakarta, 2019.
[4] R. Grønhaug and M. Christiansen, "Supply Chain Optimization for the Liquefied Natural Gas Business," in Innovations in Distribution Logistics, Berlin, Heidelberg, 2009, pp. 195–218.
[5] M. A. Budiyanto, A. Riadi, I. G. N. S. Buana, and G. Kurnia, "Study on the LNG distribution to mobile power plants utilizing small-scale LNG carriers," Heliyon, vol. 6, no. 7, p. e04538, Jul. 2020, doi: 10.1016/j.heliyon.2020.e04538.
[6] APEC Energy Working Group, "Small-scale LNG in Asia Pacific," 2019.
[7] A. Bittante, F. Pettersson, and H. Saxén, "Optimization of a small-scale LNG supply chain," Energy, vol. 148, pp. 79–89, Apr. 2018, doi: 10.1016/j.energy.2018.01.120.
[8] H. N. Abdillah, K. B. Artana, I. M. Ariana, and A. A. Masroeri, "Technical and Economic Analysis of LCT (Landing Craft Tank) Vessel Conversion into Non- Conventional LNG (Liquid Natural Gas) Carriers," p. 8.

IOP Conf. Series: Materials Science and Engineering 1052 (2021) 012054 doi:10.1088/1757-899X/1052/1/012054

IUU fishing and transhipment identification with the miss of AIS data using Neural Networks

A A Masroeri [1,3], **Aulia Siti Aisjah** [2,4], **Muhammad Mukhlis Jamali** [2,5]

[1] Department of Marine Engineering, Faculty of Marine Technology, Institut Teknologi Sepuluh Nopember (ITS), 60111, Keputih, Surabaya, Indonesia
[2] Department of Engineering Physics, Faculty of Industrial Technology and System Engineering, Institut Teknologi Sepuluh Nopember (ITS), 60111, Keputih, Surabaya, Indonesia

[3] masroeri@its.ac.id, [4] auliasa@ep.its.ac.id, [5] m.mukhlisjamali@gmail.com

Abstract. Exploitation of maritime natural resources in Indonesia is still widespread. Efforts to monitor illegal fishing and transshipment practices are still less than optimal due to the limited ability of monitoring instruments. The loss of automatic identification system (AIS) data has an impact on weakness in the ship's motion monitoring system. The weakness of the system in the previous research, without regard to data losses so that in real identification of illegal fishing and transshipment, it becomes less accurate and valid. Losses data it means as missing of some the data in along ship trajectory. This research designs system integration with predictor to identify the occurrence of illegal fishing and transshipment in the presence of missing AIS data. Predictor are designed using recurrent neural networks (RNN) and system integration is designed using artificial neural networks (ANN). Predictor and system integration are simulated, tested and validated using data of real ship that committed illegal fishing and transshipment. Data achieved from the marinetraffic.com and NASDEC-ITS data centers. The validation results show results from the predictor can be used as input for system integration and system integration, and it has a high accuracy.

1. Introduction

Indonesia has geographic potential in a marine area of 77% of the total area of Indonesia. This geographical potential is what makes Indonesia a maritime country. Unfortunately, in the Indonesian seas there are still many exploitations of biological resources. Illegal, unreported, and unregulated (IUU) fishing and transshipment practices are both threats and challenges that arise amidst Indonesia's fishery potential. In the current monitoring of illegal fishing and transshipment activities in Indonesia is carried out through Vessel Traffic System (VTS) at the Head Office of the Ministry of Marine Affairs in Jakarta or from the nearest port. The monitoring is based on automatic identification System (AIS) data information and a long-range radar camera. The capable of monitoring up to 40-60 kilometers, as well as monitoring cameras (CCTV) as far as 4 kilometers [1]. Indonesia's marine area is very large, but the number of monitoring fleets has not met the adequacy standard, for this it needs to be supported by technology that is able to identify shipping as IUU fishing and Transshipment categories.

The existence of AIS technology since the 1980s, as a communication system that functions for anti-collision, vessel traffic services and search and rescue. AIS can also be used to shipping monitor, and very helpful in monitoring shipping activities at sea [2]. The International Maritime Organization (IMO) made a regulation requiring every ship to install AIS equipment in 2000. AIS as an electronic information system is required to be installed on ships with a capacity above 300 GT [2]. AIS is very helpful in navigation throughout the shipping. AIS also can be used to show events around the location of a ship / several ships are located [3]. Several studies in the utilization of AIS data, linking it to avoid collisions, maintain the course, etc. Several methods are offered in designing an automatic anti-collision system based on AIS data. AIS is able to provide information on the density of sea transportation, which allows collisions or navigation diversions along the busy shipping. This shows that AIS is able to support technology in e-navigation and various uses for security and safety in marine transportation [4].

Research on AIS technology with system integration to identify vessels indicated to be carrying out IUU fishing and Transshipment has been previously designed. However, in the previous research there were still weaknesses due to of AIS data available [5]. In the real conditions of illegal fishing and Transshipment occurred, there are loss of AIS data which resulted in inaccuracy and invalidity of the results of the identification of vessel movements categorized as illegal fishing and transshipment. In real conditions, sometimes we find that AIS data is missing for a long time. This becomes a serious problem if the loss of data indicates an IUU incident.

Therefore, this research was conducted to design a system integration with the addition of predictor to identify IUU fishing and transshipment with the loss of AIS data using neural networks. The use of neural networks in the design of system integration for identification. In this study neural networks are able to produce more general identification results than other expert system methods such as fuzzy logic. In the previous study, a fuzzy logic system is only able to identify based on predefined rules. In many cases fuzzy logic is unable to identify something that is outside the rules that have been set in the fuzzy logic. In contrast to neural network, which is able to produce general identification [3]. Neural networks are always able to learn new things as long as there is new data [6].

In this study, a predictor and identification system is proposed to be integrated into one system. The system has two functions, namely (i) as a trajectories predictor, and (ii) identifier a shipping related to IUU. The two sub systems, namely the sub system predictor and the sub system identifier. The predictor system works when there is missing data, and the identifier system works to identify one of the following forms of events, that are: normal shipping, shipping that is categorized as an IUU fishing, and categorized as an IUU transportation.

2. Method

2.1. Data gathering

The AIS data used are dynamic ship data and partly static data, as well as the validation data for IUU fishing and transshipment criteria obtained from NASDEC-ITS and the marinetraffic.com website. The data used for predict the longitude and latitude position, heading, and ship speed. Categorized of ship speed when casting, towing, houling, and ship position of longitude and latitude are used for identification od IUU fishing. If there are two adjacent ship data, namely the latitude, longitude, different headings and speed differences, which are used to identify IUU transshipment [5].

The block diagram of the design system drawing is shown in Figure 3. Sub system 2 of ANN selection used k-fold cross-validation technique [4]. This technique was used in the both of ANN selection and ANN decision. There are of 4515 observations the dataset, which were divided into five sets. Four sets of 80% data or 3612 observations are used to train the neural network. The remaining sets of 20% of observations or 903 observations are stored to test model performance.

In ANN decision Transshipment used of k-fold cross-validation with 4500 observations the dataset. Data divided into five sets. Four sets of 80% of observations or 3600 observations are used to train the neural network, and the remaining sets of 20% of observations or 900 observations are stored to test model performance. In ANN decision fishing design used of 4745 observations dataset, which are divided into five sets. Four sets of 80% observations or 3796 observations are used to train the neural network, and the remaining sets of 20% of observations or 949 observation are stored to test model performance. Several other studies used a percentage amount of data for training approximate 70% and the rest of 30% for validating. The definite provisions of this comparison have not been disclosed by many previous researchers [7].

2.2. Design predictor when the losses of longitude and latitude data

Some of data are missing in time series and statistically modeled, for example Autoregressive (AR). The AR models is widely used in the real world [8]. The existence of missing data will have an impact on the predictor accuracy. As shown in Figure 3 below, there are 3 sub-systems, namely (i) predictor unit, (ii) selection unit, and (iii) identification unit. Predictors function to predict data: longitude and latitude position, direction and speed of ships, due to missing trajectory data. The selection unit is used to select data in categorized as a single data for one ship or multi-ship data. Single data will be sent to the IUU fishing identification unit, while multi-ship data will be sent to the IUU transhipment unit. The identification unit for both IUU fishing and transhipment uses recurrent neural networks (RNN).

Several algorithms were developed to get the best predictor parameters. The best parameter chosen, is the one that will result in high accuracy. The algorithm model with the highest accuracy, in follows:

- Number of LSTM layers: 1
- Number of neuron units in LSTM layers: 5
- Learning rate: 0.005

Flow chart to predict the ship position data when missing data occurs, is shown in Figure 1 below. The target parameter set is Mean Square Error (MSE), and use the Adaptive Moment Estimation (Adam) optimizer. The predictor model is trained with 100 epochs. The performance of the RNN model for predict the missing AIS latitude and longitude data is 0.055% of MAPE and time step is 2.

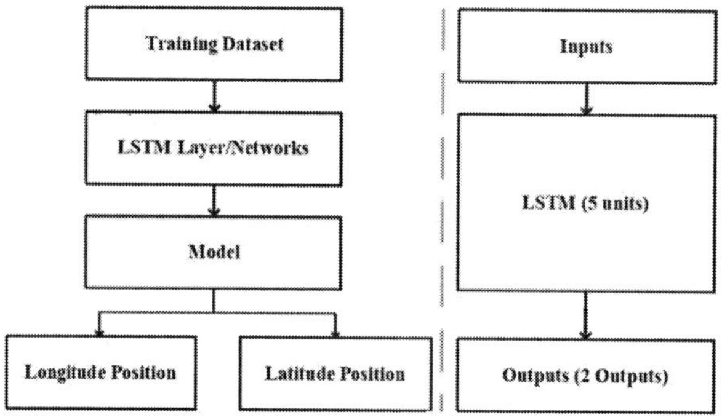

Figure 1. RNN model for predicting the latitude and longitude when missing of AIS data

2.3. Test and validation of heading predictors

The test and validation of predictor of illegal fishing and transshipment vessels is divided into two. In testing, the number of real data used of 10 AIS data as perpetrating illegal fishing and 10 AIS data as perpetrating illegal transshipment. RNN predictor give the results of high accuracy or MAPE value $\leq$ 10%, while for the speed predictor it can be said to have been successful [7].

Test of predictor of AIS heading data have been carried out, the optimal parameters are obtained which can be used to predict the missing AIS heading data. These parameters are:

- Number of LSTM layers: 1
- Number of neuron units in LSTM layers: 10
- Learning rate value: 0.1

Apart from using the above parameters, the network is trained and tested using batch size 2, in addition the input and output layers are dense layers (fully connected). The loss function used is Mean Squared Error, the optimizer used is Adaptive Moment Estimation (Adam), and the model is trained in 100 epochs. The flow chart of From the results of this training and testing, the best model is obtained for the prediction of missing AIS heading data as shown in Figure 3. The performance of the RNN model is MAPE of 8.01% and time step is 2.

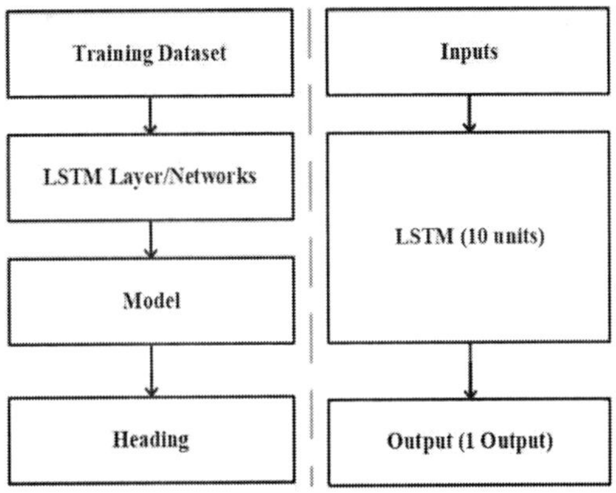

Figure 2. RNN model for predicting the missing heading data of AIS

2.4. Design a system integration for identification (ANN selection and ANN decision)
The input and output design of the system integration for identification designed using artificial neural networks with input from the predictor output (RNN Predictor and Speed Predictor) is shown in Figure 1.

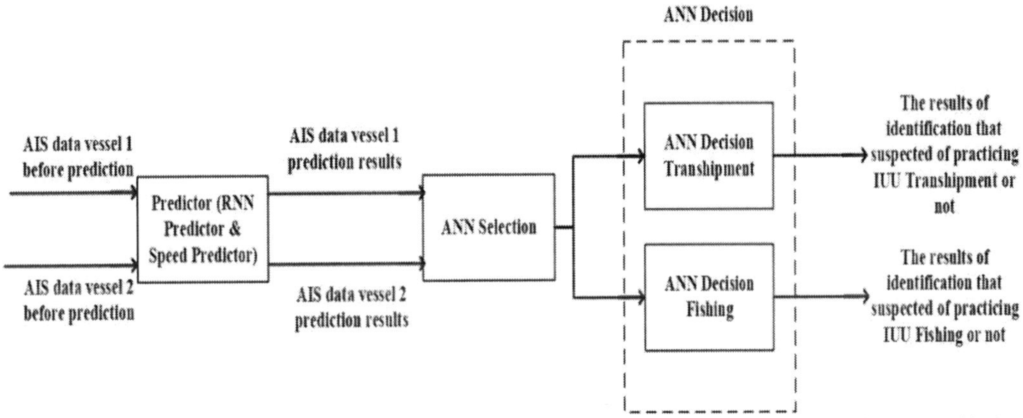

Figure 3. System integration block diagram for identification with input from predictor prediction results.

System integration for identification designed using artificial neural networks is divided into two subsystems, the first is ANN selection and the second is ANN decision. The input from the ANN selection is the output of the predictors, namely AIS data from two ships, in the form of a variable distance between two ships and the difference in the headings of two ships. ANN selection is used to sort vessels suspected of carrying out IUU transhipment or IUU Fishing. ANN selection output is used as input for ANN decisions which include IUU transhipment or IUU Fishing.

ANN decision is divided into two subsystems, namely ANN decision of illegal transhipment and ANN of decision illegal fishing. The input variables used in ANN decision transhipment are the difference in headings, the distance between 2 ships, and the difference in speed between the two vessels. Meanwhile, the input variables in ANN decision fishing are the speed of the ship at the time of casting, the speed of the ship during towing, the speed of the ship during hauling, the rate of change in the position of the latitude and longitude of the ship. The output of both is the result of the decision making it is stated that they are suspected of committing IUU transhipment or IUU fishing or not carrying out the two illegal practices.

2.5. Test results of prediction AIS data

In the process of testing of prediction on illegal fishing and transhipment when the missing AIS speed data, in various of AN architectures. In some test data, both real data and generated data, it shows the characteristics of the predictors, namely the amount of error at ship speed, longitude and latitude position, as well as heading error and average speed as shown in Table 1 below. In this method, predictor works in constant time intervals.

Table 1. The characteristic of predictor

No	Parameter system	Value
1	RMSE speed	0.83 knot
2	MAPE longitude (%)	0.00015
3	MAPE latitude (%)	0.00296
4	MAPE heading (%)	1.26
5	RMSE speed (knot)	1.08

2.6. Test and validation of integration system

Integration system consist of predictor, selection and decision units. In the selection and decision units, validation of both used of k-fold cross validation. The optimal parameters in the selection and decision units are in Table 2 below, and architecture of selection unit in figure 4 (a) and (b).

Table 2. The characteristic of selection and decision units

Parameter Design	Selection Unit	Decision Unit of IUU Transhipment	Decision Unit of IUU Fishing
Number of hidden layers	1	1	1
Number of neurons in hidden layer	10	10	5
Activation function	• rectified linear unit (ReLu) in input and hidden layer, • sigmoid in the output layer	• rectified linear unit (ReLu) in the input layer and hidden layer, • sigmoid in the output layer	• rectified linear unit (ReLu) in the input layer and hidden layer, • sigmoid in the output layer
Optimizer	adaptive moment estimation (Adam)	adaptive moment estimation (Adam)	adaptive moment estimation (Adam)
Learning rate	0.01	0.005	0.001
Number of epochs	100	100	100

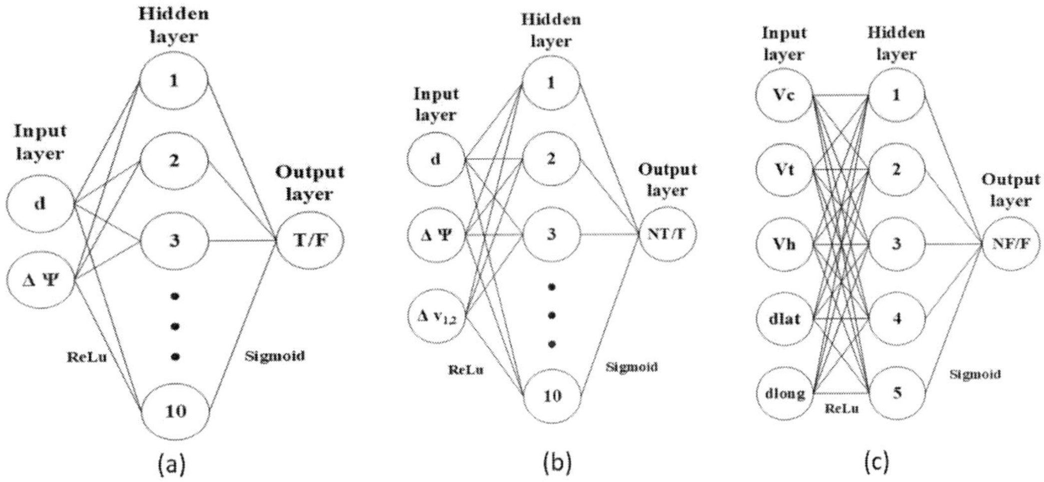

Figure 4. ANN of (a) selection unit, (b) Identification unit of IUU transhipment and (c) Identification of IUU fishing

The system integration that has been designed is tested with a various of ship motion patterns that have been generated. The generation of data on the position of two vessels to test the movement patterns of illegal transhipment. In variety data of heading, speed and position of ships are input of system. Testing is done to see whether the system designed is in accordance with the desired target. In testing the movement of illegal fishing vessels is carried out by varying of pattern in casting, towing, hauling, and rate of change of latitude and longitude. The result of system will identify of trajectory pattern in illegal fishing vessels or not.

Validation is done to test whether the system that has been designed can be said to be valid or appropriate. Validation for the occurrence of illegal transhipment is carried out using system input in the form of ship speed data at the port and changes in ship position in previous studies. The ship's AIS real data is obtained from the marinetraffic.com website, namely Nordic Bahari ship data. This approach is carried out with the assumption that a port is a ship stationary in the middle of the sea waiting for another ship to carry out a transhipment . The validation data is a ship berthed at the port of Sorong, Papua on April 16, 2016 at 11.00 West Indonesia Time (WIT) until April 16, 2016 at 11.15 WIT. The second validation data is AIS data on the Eagle Seville and Lurongyuanyu 105 ships in Batam Waters on 02 May 2016 at 11.30 WIB until 02 May 2016 at 11.45 WIB.

Validation for the occurrence of illegal fishing is carried out using system input in the form of ship speed data in the Singapore Strait, fishing vessel 1 on August 21, 2017 at 02.40 WIB until August 21, 2017 at 05.15 WIB. The second validation data is AIS data from fishing vessel 2 on 21 September 2017 at 06.43 WIB until 21 September 2017 at 10.45 WIB. The ship's AIS real data was obtained from NASDEC-ITS, as many as hundreds of data, and already sorted when there was indication of illegal fishing. This approach is taken because the ship has a motion pattern that is exactly the same as the *cantrang*, which is to make a circular motion. The validation data is real AIS data from vessels caught in illegal Transhipment and illegal fishing.

3. Results and discussion

Several tests were carried out for AIS data on fishing vessels sailing in the Singapore Strait area. The ship's starting position is at a1 and the end point - stops at a20. The ship makes the movement through the closed trajectory, and returns through the starting position on a1 until it stops at a20. It is assumed that data is missing for areas a4 to a9, as well as at a12 to a1. The series data, are as follows (Latitude - L, Longitude - B, Heading - Ψ, Speed - U) in unit of (°, °, °, knot) that are: a1 = (103.78357; 1.18565; 266;

0.6), a2 = (103.78050; 1.18559; 267; 0.6), a3 = (103.77229; 1.18592; 275; 0.6); a4 = (103.76632; 1.18667; 277; 0.6), a9 = (103.74566; 1.18925; 141; 0.6), a10 = (103.75596; 1.18288; 110; 0.6); a11 = (103.75857; 1.18214; 91; 0.6), a12 = (103.76138; 1.18225; 83; 0.6); a17 = (103.78396; 1.18568; 70; 0.6), a18 = (103.78951; 1.18568; 78; 0.6), a19 = (103.79233; 1.18834; 83; 0.9), a20 = (103.79813; 1.18921; 82; 0). The illustration of the presence of missing data is shown in Figure 5 (a) below. The presence of missing data in Figure 5, assuming it is replaced by 4 data. The predictive trajectory pattern is shown in Figure 5.b.

The results of the ship trajectory prediction are shown by the position, heading and speed of the ship as follows, a5 = (103.7646; 1.1868; 277.53; 0.6), a6 = (103.7655; 1.1869; 277.72; 0.6), a7 = (103.7664; 1.1868; 278.25; 0.6), a8 = (103.7669; 1.1868; 279.07; 0.6), a13 = (103.7603; 1.18568; 77.86; 0.6), a14 = (103.7611; 1.1823; 73.40; 0.6), a15 = (103.7618; 1.1824; 74.70; 0.6), a16 = (103.7629; 1.1824; 75.9; 0.6). Figure 5 (c) shows the real trajectory pattern based on AIS data.

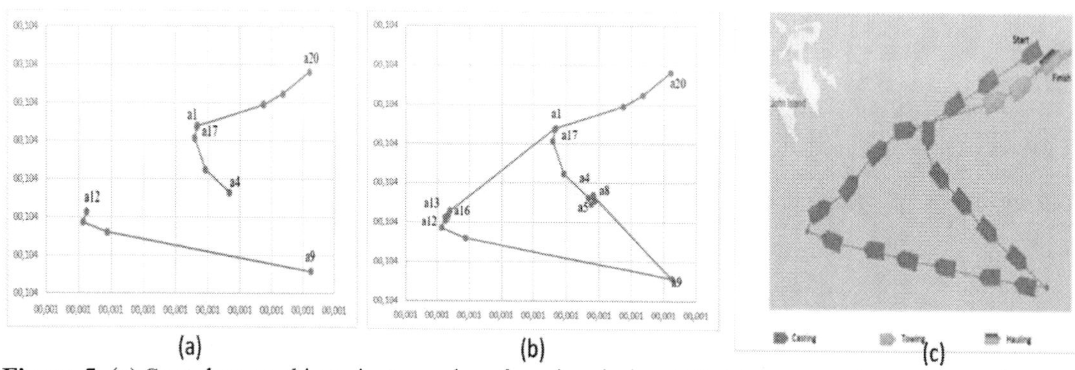

Figure 5. (a) Spot data on ship trajectory when there is missing AIS data on vessel movements from points a4 to a9 and a12 - a1 (position in the Singapore Strait), (b) Ship trajectory results are predictors for ranges a4 - a9, (c) Fishing vessel in real trajectory according to AIS data.

Several tests were carried out on: 30 illegal fishing data and 30 illegal transhipment which be generation patterns. The accuracy was very high, that is 90.3% of IUU transhipment and 89.9% of illegal fishing identification.

4. Conclusions

This study presents the design of system integration with predictor has successfully identified the motion patterns of vessels that do not perform and perform IUU transhipment and fishing, even with the loss of AIS data. The system integration for identification is divided into 2 subsystems, namely, ANN selection for sorting two vessels, and ANN decision as the final identification whether the vessels practice IUU fishing or transhipment . The predictors for the prediction of missing AIS data are divided into 2 subsystems, namely, the RNN predictor for predict the missing AIS longitude, latitude, and heading data of illegal fishing and transhipment vessels, and speed predictors for predict the missing AIS speed data of illegal fishing and transhipment vessels.

This study also demonstrated the ability of Recurrent Neural Networks (RNN) to effectively predict the missing of AIS data and the ability of Artificial Neural Networks (ANN) to effectively identify the motion patterns of vessels. The pattern of in a normal trajectory or in do not perform and in IUU transhipment and fishing.

Acknowledgement

Thank you to Ministry of Education and Culture for funding of this research, and also to Directorate of Research and Community Services ITS.

References

[1] J. Oh, S. Park, and O. Kwon, "ScienceDirect," *UMK Procedia*, vol. 5, pp. 21–31, 2016.

[2] M. Fiorini, A. Capata, and D. D. Bloisi, "ScienceDirect AIS Data Visualization for Maritime Spatial Planning (MSP) *," *UMK Procedia*, vol. 5, pp. 45–60, 2016.

[3] A. Hahn, A. Bolles, M. Fränzle, S. Fröschle, and J. H. Park, "ScienceDirect," *UMK Procedia*, vol. 5, pp. 1–20, 2016.

[4] M. Lensu and F. Goerlandt, "Big maritime data for the Baltic Sea with a focus on the winter navigation system," *Mar. Policy*, vol. 104, no. July 2018, pp. 53–65, 2019.

[5] I. Fransisco, A. S. Aisjah, and A. Masroeri, "Decision Support System Based on Automatic Identification System (AIS) for Illegal Unregulated Unreported (IUU) Fishing Using Fuzzy Logic," in *SENTA - 2017*.

[6] S. Carvalhosa, A. P. Aguiar, and A. Pascoal, "Cooperative Motion Control of Multiple Autonomous Marine Vehicles :," in *IFAC Proceedings Volumes*, 2009, vol. 43, no. 16, pp. 395–400.

[7] B. Erie *et al.*, "A neural network approach for chatter prediction in turning A neural network approach for chatter prediction in turning," *Procedia Manuf.*, vol. 34, pp. 885–892, 2019.

[8] O. Anava, "Online Time Series Prediction with Missing Data," in *Proceedings of the 32nd International Conference on Machine Learning*, 2015, pp. 1–9.

Study on the LNG Distribution to Bali – Nusa Tenggara Power Plants Utilizing Mini LNG Carriers

H N Abdillah[1,3], K B Artana[1,2,3], A A B Dinariyana[1,2], D W Handani[1,2] and P W Aprilia[1]
[1]Department of Marine Engineering, Faculty of Marine Technology, Institut Teknologi Sepuluh Nopember, Surabaya, East Java 60111, Indonesia
[2]Center of Excellence for Maritime Safety and Marine Installation, Institut Teknologi Sepuluh Nopember, Surabaya, East Java 60111, Indonesia
[3]Corresponding author: hayyabdillah19042@mhs.its.ac.id

Abstract. LNG has proven to be a green enegy source and reliable for environmentally, it's combustion residue produces much cleaner pollutants than other fossil fuels. The Ministry of MEMR shows that the increasing demand for national electricity reaches 7.5% each year. Mini LNG carrier is an alternative solution for distribution liquid natural gas to archipelago areas with an operational cost effectiveness challenge. Power plants located in Bali and Nusa Tenggara Island are used as a case study. The aim of this study is to calculate LNG distribution using mini LNG carriers, estimate the operational cost in each power plants and risk assessment. The capacity of vessel and power plants set as the constraint, LCT non-conventional as mini LNG carrier carry out 32 ISO Tank 20ft. Hence its is need 26 mini LNG carrier to supply power plant in Bali and Nusa Tenggara as scenario 1, while 7 mini LNG carrier have to provided to supply power plant in Bali and Lombok as scenario 2. Scenario 1 is inefficient because it requires high investment and operational aspects. The final risk assessment based on hazard analysis due to hydrocarbon handling an F-N curve was found to be in ALARP compared to acceptance criteria.

Keywords. *Distribution, mini LNG carrier, operational cost, risk assessment*

1. Introduction

Global demand for LNG, increasing more than half by 2040, based on IEA [1]. Asia-Pasific region has a potential high demand for liquid natural gas, where this LNG supplied from the Middle East, Indonesia, Australia, and Malaysia [2]. Liquid natural gas is an alternative fuel that produces carbon dioxide and air pollutants than coal when burned to produce electricity. LNG reserves in Indonesia are abundant shown in Figure 1, it can be used for the next 230 years. LNG has proven to be a green enegy source and reliable for environmentally, it's combustion residue produces much cleaner pollutants than other fossil fuels. In the decade between 2008 and 2018, Indonesian power generation rose by 79%, from 149 TWh to 267 TWh. A coal-fired in power plant takes more time to start the engine and stop the process than a gas-fired in the power plant. This flexibility process makes liquid natural gas become an alternative fuel to renewable energy sources like solar and wave power, which are only available when the sun shines and the moving waves [3]. LNG carried by vessel is received by a principal liquid natural gas terminal facilities. Nowadays, terminal facilities are devided into two type facilities, onshore and offshore, offshore facilities set as floating storage & regasification unit called as FSRU [4]. In recent years, many developing industries have rised their investment in LNG infrastructure facilities, however the development of LNG supply chain in many countries has not exhaustive [5].

Chracteristic gas business in Indonesia in the last since 1963 used for industrial plant and gas price set as low price, while in 2002 use of liquid natural gas developed into export materials, power generation, petrochemical industry, fertilizer plants and industrial feedstock. The conduct of natural gas business in the future, gas priority for domestic market LNG. The ministry establish a master plant to integrate the national gas distribution plant in Indonesia. Strategy to preparing gas utilization based on among others are prepare for infrastructure upgrades and technology for natural gas process; natural gas price based on regional authority and utilization; the amount of gas provided by supply and demand of consumers throughout Indonesia; clearness of natural gas Clearness of natural gas utilization policy is needed to coordinate between producers, operational agencies and users. However, the participation of local government is very important in gas utilization; research and development on the academic side for more efficient and effective used of fossil energy [6].

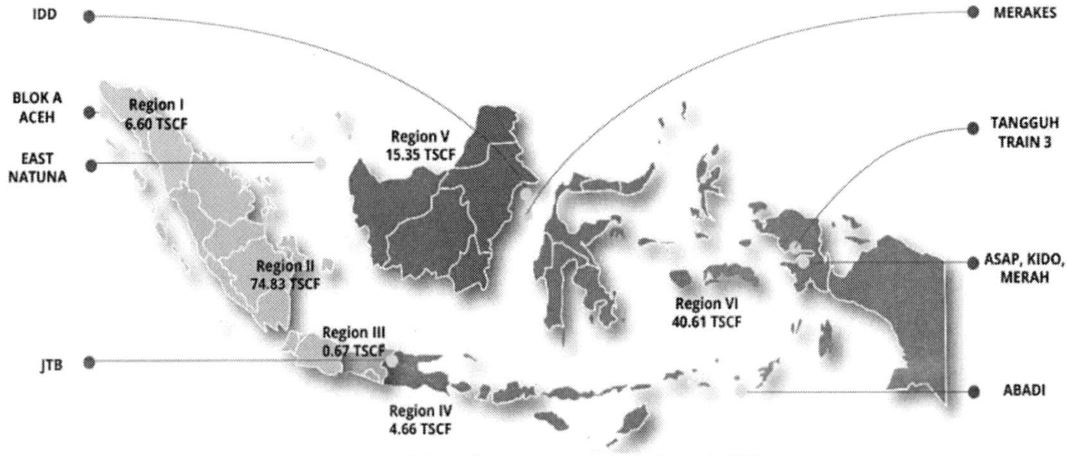

Figure 1. Natural gas reverses in Indonesia [7]

In Indonesia, many power plants that used diesel fuel as the main fuel for power plants are scattered. This trend is causes limited access to accommodate cheaper energy due to geographical aspects. Futhermore, Government subsidy on diesel fuel for power plant is too exaggerated so it needs to be considered in the future. Oil and gas industry in Indonesia still to be an important sector, where set as the main national energy, but also contributes to national incomes, including 12% of the 2014 state budget. These aspects also upgrades the national economic growth. As one of LNG exporting country, the natural gas industry in Indonesia has been experiencing significant transformation. Indonesia is known as one of the natural gas exporting countries which currently occupies the 4th position in the world, however the increasing of economic growth affects the gas consumption in the country. Indoensia should allocate a larger natural gas production drilling to occupy domestic requirement of natural gas. In the future, Indonesia is expected to switch from a gas exporter to a gas importer in order to fulfill the domestic requirement. Moreover, domestic gas utilization in Indonesia is still struggling because the sparse gas infrastructure, fragmented gas market, untransparent domestic pricing regime and uncertain law of oil and gas.

Mathematical model used to solve capacity-based transportation cases. Futhermore, the development of this model is the capacitated distribution problem, with calculate each operational cost. Even though the many research of capacitated distribution problem, not many have to applied it to the case of mini LNG carriers. Mini LNG carrier is an alternative solution for distribution natural gas to archipelago areas with an operational cost effectiveness challenge. In this study there is two alternative scenario, first scenario located in Bali-Nusa Tenggara, while second scenario located in Bali-Lombok. Lombok

Peaker, MPP Lombok, Gilimanuk and Pemaron power plants located in Bali and Lombok Island are used here as a case study (first scenario). There are Lombok Peaker, Sumbawa, Kupang Peaker, MPP Flores, Maumere, Waingapu, MPP Lombok, and Bima power plants set as second scenario, futhermore Benoa set as origin.

Some other researcher described about hazard identification and risk analysis of LNG facilities. Hazard identification arranged by LNG specific scenarios based on past events and studies [8]. Qualitative and qualitative approach used to determine risk assessment and safety aspect of mini LNG carrier. Qualitative approach identify the hydrocarbon hazard and handling system, hence quantitative and qualitative combination approach used to analyze the non-hydrocarbon hazard such as operational system and loading-unloading system. Hazard register devided into fire dan explosion hazard caused by hydrocarbon leakage. The aim of this study is to optimize LNG distribution using mini LNG carriers and calculate the operational cost in each power plants. The capacity of cargo vessel and power plants set as the constraint, LCT non-conventional as mini LNG carrier carry out 32 ISO Tank 20ft. The results of this study will describe the number of mini LNG carriers needed, operational cost and risk assessment for power plants in two scenario, Bali and Nusa Tenggara, Indonesia.

2. Case Study: LNG distribution in Bali and Nusa Tenggara, Indonesia

Power Plant in Bali and Nusa Tenggara will be supplied by LCT non-conventional as mini LNG carrier. Specification of mini LNG carriers, LOA: 50,5m; B: 11,4m; H: 3,5m; T:2,5m design by BPPT. Recently, most mini LNG carrier networks are constructed using smaller vessels, compared to the ones used in conventional networks. Transport distance could be very short mini LNG carrier networks [4]. In archipelagos like the Indonesia region, many islands do not have adequate water depth nor piersor berths capable of handling large LNG tankers. Thus, mini LNG carrier may be used. LNG terminal project is being developed in Indonesia to supply the power plants [9]. This study focus on Bali and Nusa Tenggara Island, the data in Table 1 are shown the capacity of power plant and LNG demand. Sandards conversion used to calculated the LNG capacity in m^3 and the number of ISO tank demand, while each plant consumed 0,2 MMSCFD (Million Standard Cubic Feet per Day) of natural gas per day per 1 MW power rate; 1 MMSCFD is equal to 46 m^3 LNG; 1 MMSCFD equal to 1000 MMBTU; 1 ISO Tank is equivalent to 21 m^3 of LNG. A preliminary study to develop LCT as mini LNG carrier was done in 2018 [10].

Table 1. Capacity of power plants in Bali and Nusa Tenggara, Indonesia [9]

Power Plant	Capacity (MW)	MMSCFD	LNG Demands (m^3/day)	ISO Tank Demand (unit/day)
PLTMG Lombok Peaker	136	9,07	417,07	21
MPP Lombok	50	10,00	460,00	24
PLTG Gilimanuk	130	8,70	398,70	20
PLTG Pemaron	80	5,30	245,30	13
PLTMG Sumbawa	50	10,00	460,00	24
PTLMG Bima	50	10,00	460,00	24
MPP Flores	20	4,00	184,00	10
PLTMG Maumere	40	8,00	368,00	19
PLTMG Waingapu	10	2,00	92,00	5
PLTMG Kupang Peaker	40	2,67	122,67	7

The draft of the mini LNG carrier are made as low as possile because it serves the coastal area, which has a restricted depth. Table 2 shows the matrix of distance at scenario 1 and Table 3 represent the matrix of distance at scenario 2. The speed of mini LNG vessel is 5 knots. LNG distribution will be supplied by Benoa terminal as origin. The capacity of power plant in the range of 10-136 MW, which supplied by LNG in range of 92-460 m^3/day. Type of power plants could be devided into base load and peaker, hence base load power plants will be operated on 100% load and peaker power plants will be operated on 30% load. LNG distribution is challenging, nowadays the are only few LNG suppliers and supporting facilities have not been prepared, while cost for transformation reached 70% of the delivered natural gas price [11].

Table 2. Matrix of distance at the Nusa Tenggara power plants

Item	X0	X1	X2	X3	X4	X5	X6	X7	X8
X0	-	65	62	614	372	552	374	191	291
X1	-	-	3	566	352	448	306	109	202
X2	-	-	-	563	355	451	302	111	205
X3	-	-	-	-	356	539	244	588	405
X4	-	-	-	-	-	182	117	205	93
X5	-	-	-	-	-	-	296	387	272
X6	-	-	-	-	-	-	-	350	163
X7	-	-	-	-	-	-	-	-	127
X8	-	-	-	-	-	-	-	-	-

Notes: X0: Benoa-Origin X2: MPP Lombok X4: MPP Flores X6: Waingapu X8: Bima
X1: Lombok Peaker X3: Kupang Peaker X5: Maumere X7: Sumbawa

Table 3. Matrix of distance at the Bali and Lombok power plants

Item	X0	X1	X2	X9	X10
X0	-	65	62	137	92
X1	-	-	3	135	89
X2	-	-	-	138	92
X9	-	-	-	-	46
X10	-	-	-	-	-

Notes: X0: Benoa-Origin X1: Lombok Peaker X2: MPP Lombok X9: Gilimanuk X10: Pemaron

Figure 2 and Figure 3 ilustrates the distribution of LNG in scenario 1 and scenario 2. LNG distributed from Benoa Bali as origin to power plant in Lombok Peaker, MPP Lombok, Gilimanuk and Pemaron. Distribution using mini LNG carrier is considered effective and efficient to supply power plants with low gas requirements [12]. Determine on routing, sailing frequency and operational cost are importand issues when planning the shipping services [13]. A simple techno-economic approach is provided in this study to determine distribution model for determine sailing frequency, the number of the mini LNG carrier and operating costs include waiting time and shipping time cost. Thomson H., et al [14] considered the parameter in their results clearly show that the cost saving in small LNG distribution strengthens the cost competitiveness of LNG as a fuel.

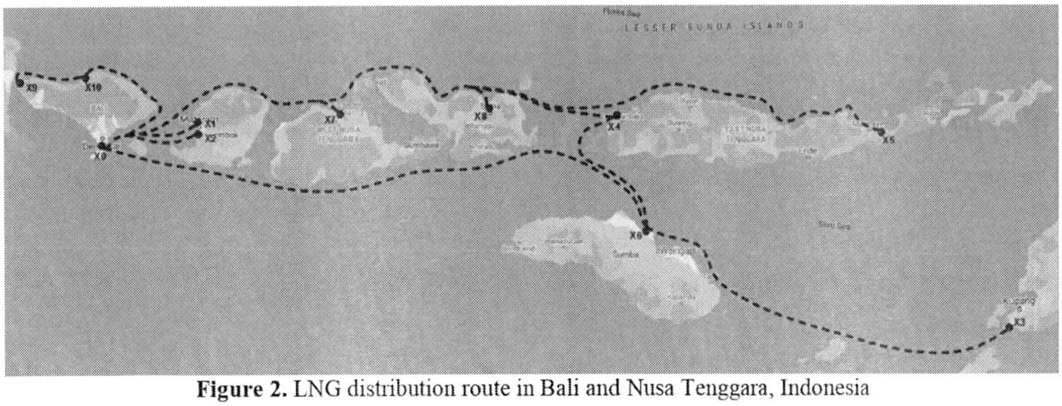

Figure 2. LNG distribution route in Bali and Nusa Tenggara, Indonesia

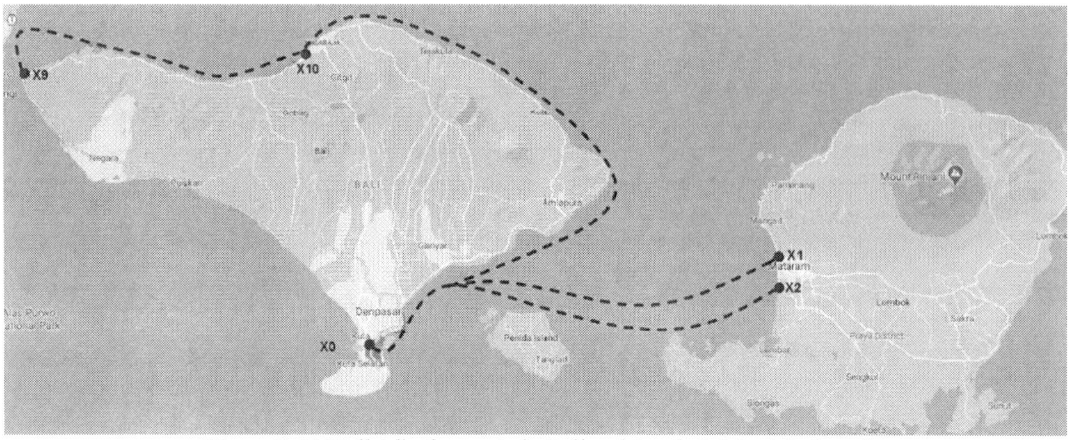

Figure 3. LNG distribution route in Bali and Lombok Island, Indonesia

Frequency analysis of leakage occurrences in components is calculated based on the existing components in the P&ID system . Calculation of component leakage is carried out based on DNV Failure Frequencies Guidelines Process Equipment Leakage Data for Use in QRA. The formula in the quation (1) for calculating component leakage based on these guidelines.

$$F(d) = C\ (1 + aD^{n})\ d^{m} + Frup \quad (1)$$

Where
F : Frequency of leaks (per valve year)
C, m : Constants representing hole size distribution
a, n : Constants representing equipment size dependency
Frup : Additional rupture frequency (per valve year)
D : Valve diameter (mm)
d : Hole diameter (mm)

The component leakage scenario is defined into three conditions, total leaks, full leaks and zero pressure. Frequency analysis of pipe leakage is also carried out which serves as the main medium for fluid transfer. Frequency results for each existing equipment refer to DNV. The calculation is used with the following formula in the equation (2)

$$F(d) = C\ (1 + aD^{n})\ d^{m} + Frup \quad (2)$$

Where
F : Frequency of leaks (per meter year)
C, m : Constants representing hole size distribution
a, n : Constants representing equipment size dependency
Frup : Additional rupture frequency (per meter year)
D : Pipe diameter (mm)
d : Hole diameter (mm)

Frequency analysis is also carried out in the event of a pump leak which is calculated based on the components in the P&ID system. The calculation is used with the following formula in the equation (3)

$$F(d) = Cd^{m} + Frup \quad (3)$$

Where
F : Frequency of leaks (per valve year)

C, m : Constants representing hole size distribution
Frup : Additional rupture frequency (per valve year)
D : Pipe diameter (mm)
d : Hole diameter (mm)

The safeguard functions component as an indicator of flow, pressure and temperature. The formula in the quation (4) for calculating component leakage based on these guidelines.

$$F(d) = Cd^m + \text{Frup} \quad (4)$$

Where
F : Frequency of leaks (per valve year)
C, m : Constants representing hole size distribution
Frup : Additional rupture frequency (per valve year)
D : Pipe diameter (mm)
d : Hole diameter (mm)

After calculating the frequency of leakage and failure of components in the system, then the identification of failures in the system is carried out using Fault Tree Analysis (FTA). Even Tree Analysis (ETA) in this study is to calculate modeling of events that cause fires and explosions. Fires and explosions occured because of ignition or explosion. Various of worker category have been clustered into three as shown in Table 4.

Table 4. Various of worker category

Personnel Group	Worker Category	POB
LNG Carrier	**Normal peration**	**22**
	Master	1
	Chief Officer	1
	Second Officer	1
	Third Officer	1
	Radio Operator	1
Deck	Boatwain	1
	Quartermaster	3
	Sailor	2
	Cook	1
	Mess Boy	1
	Chief Engineer	1
	Second Engineer	1
	Third Engineer	1
Engine	Fourth Engineer	1
	Engine Foremen	1
	Oiler	3
	Wiper	1
Jetty	**Normal Operation**	**8**
HSE	HSSE	1
Mooring	Mooring Gang	2
	Mooring Guard	2
Operator	Operator Crane	1
	Operator Mobile Crane	2
Office	**Normal Operation**	**18**
Principal	Manager	1
Administration	Administration Staff	2
HSE	HSSE	2
Technician	Technican	5
Misc.	Misc. Worker	8

3. Result and Discussion

3.1. LNG distribution using mini LNG carrier

The loading and unloading process of LCT ships carrying LNG includes moving the ISO tank using a crane at the port, then LNG ISO tank carried by stacker to the stockpile to be arranged in the stockpile area. The same process applies to empty ISO tanks that will be carried by the ship to the port of origin (Benoa). Figure 4 shows the loading and unloading process, estimated the capabilities of the components were set to determine total loading unloading process such as lift velocity of the crane and stacker, times to lashing on the ship and stockpile, estimated distances. The total time for loading and unloading is 20 hours, this process is supported by the terminal facility (1 crane and 2 stacker). For loading unloading process, it is assumed that lift velocity by crane 25 m/min.; lift velocity by stacker 24 m/min.; velocity of stacker 350 m/min.; lashing process 5 min./unit.

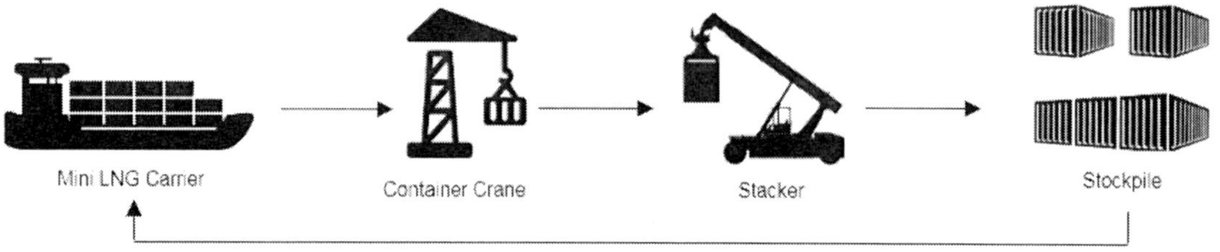

Figure 4. Loading-unloading LNG ISO Tank

The number of mini LNG carrier to supply each route in Bali and Lombok Island find out by calculating the round trip of vessel. Table 5 represent the total of round trip which consist of loading unloading, time at sea, and slack time. The result shows ISO Tank deman and number of mini LNG carriers. Hence its is need seven mini LNG carrier to supply power plant in Bali and Lombok, five numbers of vessels were used continuously, however two others vessels were used intermittently as shown in Table 6.

Table 5. Calculating of round trip estimation scenario 1 and scenario 1

Route	Sea Distance (Nm)	Loading-Unloading (h)	Time at Sea (h)	Round Trip (h)	Slack Time (h)	Total (h)	ISO Tank Demand (unit/round trip)	Number of Mini LNG Carrier
		a	b	c=a+b	d	e=c+d		
Scenario 1								
X0-X1-X0	130	20	26,0	46,0	4,6	50,6	44	1
X0-X2-X0	124	20	24,8	44,8	4,5	49,3	47	1
X0-X3-X0	1228	20	245,6	265,6	26,6	292,2	75	2
X0-X4-X6-X0	873	20	174,6	194,6	19,5	214,1	123	4
X0-X5-X0	1104	20	220,8	240,8	24,1	264,9	204	6
X0-X7-X0	382	20	76,4	96,4	9,6	106,1	102	3
X0-X8-X0	582	20	116,4	136,4	13,6	150,1	144	4
X0-X9-X0	274	20	39,1	59,1	5,9	65,1	54	2
X0-X10-X0	184	20	26,3	46,3	4,6	50,9	26	1
Scenario 2								
X0-X1-X0	130	20	26,0	46,0	4,6	50,6	44	1+1
X0-X2-X0	124	20	24,8	44,8	4,5	49,3	47	1+1
X0-X9-X0	274	20	39,1	59,1	5,9	65,1	54	2
X0-X10-X0	184	20	26,3	46,3	4,6	50,9	26	1

Scenario in Lombok Peaker and MPP Lombok power plants used vessels that operate continously and intermittently moda for each vessels. This alternative is used to minimize opeartional cost. Futhermore scenario in Gilimanuk and Pemaron power plants, each served by 1 vessel and 2 vessels that operate continously. Safety stock is provided by diesel oil in case of delays in LNG distribution to power plan, diesel oil used for safety stock because the type of power plant could accommodate natural gas and diesel oil as a fuel. Hence its is need 26 mini LNG carrier to supply power plant in Bali and Nusa Tenggara as scenario 1, while 7 mini LNG carrier have to provided to supply power plant in Bali and Lombok as scenario 2.

Table 6. Number of mini LNG carrier and total trip per year

Route	Number of Mini LNG Carrier	Total Trip/Year	Information
Scenario 1			Continous
X0-X1-X0	1	180	Continous
X0-X2-X0	1	180	Continous
X0-X3-X0	2	30	Continous
X0-X4-X6-X0	4	45	Continous
X0-X5-X0	6	33	Continous
X0-X7-X0	3	90	Continous
X0-X8-X0	4	60	Continous
X0-X9-X0	2	53	Continous
X0-X10-X0	1	80	Continous
Scenario 2			
X0-X1-X0	1	180	Continous
	1	180	Intermittent
X0-X2-X0	1	120	Continous
	1	120	Intermittent
X0-X9-X0	2	53	Continous
X0-X10-X0	1	80	Continous

3.2. Operational Cost

The calculation of operational costs in this study includes fuel oil (A), lubricanting (B), fresh water (C), port cost (D), ship maintenance (E), consumable (F), administration costs (G) and crew salaries (H). The cost assumptions used in this study to determine the operational costs per year of a mini LNG carrier for LNG distribution in Bali and Nusa Tenggara power plants. The assumptions used are based on latest cost reference. Port service rates are provided by Pelindo III Celukan Bawang, port charges consist of berthing service, mooring and jetty services [15], while HFO rates and lubricant rates based on previous literature studies. Maintenance of the vessels calculated by annual survey cost, futhermore each parameter to calculate the operational cost correlated with yoyal trip per year and sailing type, its continous or intermittent. Operational cost depand on total trip/year as shown in Table 7 and Table 8.

Table 7. Total operational cost scenario 1 [x10^3]

Item	X1	X2	X3	X4-X6	X5	X7	X8	X9	X10	Total/Year[$]
A	1.307	1.273	2.515	5.529	7.526	4.108	5.167	2.240	877	30.543
B	26	25	45	77	121	38	72	44	17	466
C	4	4	1	4	4	6	5	5	3	38
D	40	40	80	161	243	121	161	80	40	967
E	16	16	5	24	18	24	21	21	11	157
F	32	32	63	126	189	95	126	63	32	757
G	100	100	201	401	602	301	401	201	100	2.407
H	51	51	103	206	309	154	206	103	51	1.235
Total										**36.570**
Grand Total (+margin 30%)										**47.541**

By comparison, results of scenario 1 and scenario 2 to supply LNG typically employs different scale vessels, its indicated by total number of the vessels. Higher capacity of mini LNG vessel to be occupy

Bali-Nusatenggara power plants, however scenario 2 that supply Bali-Lombok could be selected. This cluster has power plant that are located close to each other and the capacity. LNG supply in Indonesia can run well if all of stackholder arrange and concern the weaknesses and strengths in this bussiness process, include political will of the Goverment. In Indonesia, LNG price is currently approved by Ministry of Energy and Mineral Resources, so for the contractor, oil and gas company refer to regulation [16]. In another paper, more detailed economic analysis is carried out related to the capex and opex of the lng distribution business scenario.

Table 8. Total operational cost scenario 2

Item	X1	X1-a	X2	X2-b	X3	X4	Total/Year [$]
A	1.306.958	384.826	1.272.863	565.717	2.240.499	876.717	6.647.580
B	25.896	25.221	44..394	17.371	7.625	11.209	131.717
C	4.074	4.074	5.432	2.716	1.200	1.811	19.307
D	40.205	40.205	80.409	40.205	17.757	26.803	245.584
E	16.096	16.096	21.462	10.731	4.740	7.154	76.279
F	29.104	29.104	29.104	29.104	29.104	29.104	203.731
G	92.593	92.593	92.593	92.593	92.593	92.593	648.148
H	47.492	47.492	47.492	47.492	47.492	47.492	332.446
Total							**8.304.792**
Grand Total (+margin 30%)							**10.796.229**

3.3 Risk assessment

The risk assessment is caaried out by calculating the frequency and consequence of potential hazard during operation and loading unloading of LNG ISO tank into the stockpile which the layout as shown in Figure 6 (a).

3.3.1 Scenario and frequency analysis

The potential hazard for hydrocarbon leaks in the LNG system on the mini LNG carrier is caused by several factors. Hydrocarbon leakaged will escalate if there is heat source or spark that could create a fire or explosion hazard. The potential hazard occured when to port facilities and the vessel will be carried out by considering the follows loading-unloading from mini LNG carrier, operational system, port activity and facilities aspects. Frequency analysis is carried out to determine probability of component failure, any system failure that occurs can be dangerous. The methods used to analyze the frequency are Fault Tree Analysis (FTA) and Event Tree Analysis (ETA). The consequences model based on three category LNG carrier, jetty and office. Distribution impact of workers in the terminal will be seen at the result of consequence modeling. Specific location of the workers distribution called by receiver, where each receiver have a number of hazard event scenario. Table 9 describe the summary of fault tree analysis which devided by four hole scenario.

Table 9. Summary of Fault Tree Analysis

No	Hole Scenario			
	3 mm	10 mm	50 mm	150 mm
1	1.242E-03	1.241E-03	1.237E-03	1.24E-03
2	1.41E-03	1.29E-03	1.23E-03	1.21E-03
3	1.34E-03	1.22E-03	1.59E-03	1.14E-03
4	4.01E-04	3.91E-04	3.81E-04	3.79E-04

Event Tree Analysis (ETA) in this study is to calculate modeling of events that cause fires and explosions. ETA modeling based on the gas release event, then find out the frequency value of Jet fire,

Flash fire, VCE and gas dispersion events. Figure 5 illustrated the example of fault tree analysis and event tree analysis for hole scenario 50 mm Summary with same steps, Table 10 to Table 13 are recapitulation of the ETA analysis for each hazard event in jet fire, VCE, flash fire, and gas dispersion. The largest frequency value is gas dispersion event for all leak hole scenarios, while the smallest frequency value is flash fire event for all leak hole scenarios.

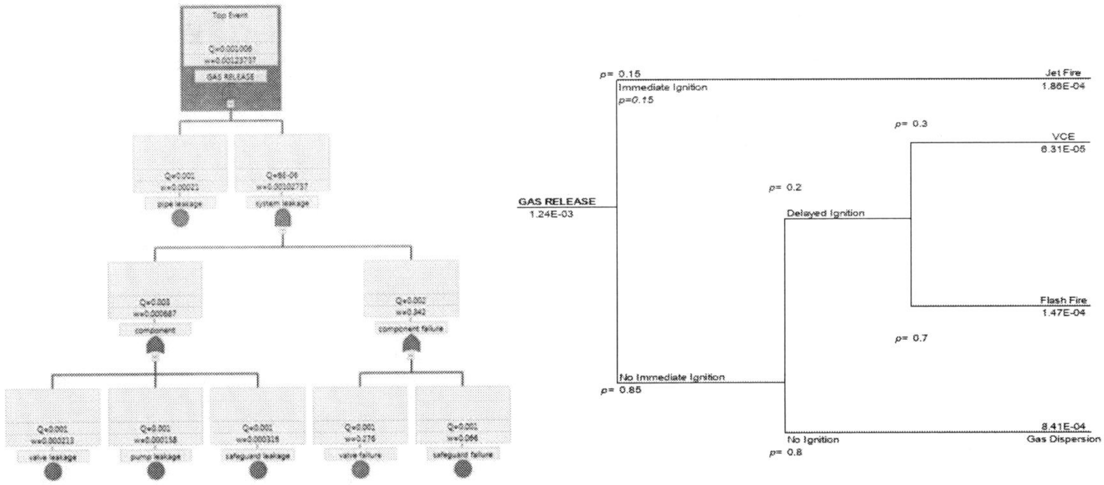

(a) (b)

Figure 5. (a) FTA for hole scenario 50 mm (b) ETA for hole scenario 50 mm

Table 10. Summary of Event Tree Analysis

Potential Hazards	Hole Scenario			
	3 mm	10 mm	50 mm	150 mm
Jet Fire	1.86E-04	1.86E-04	1.86E-04	1.86E-04
VCE	6.34E-05	6.33E-05	6.31E-05	6.31E-05
Flash Fire	1.48E-04	1.48E-04	1.47E-04	1.47E-04
Gas Dispersion	8.45E-04	8.44E-04	8.41E-04	8.41E-04

3.3.2 Consequences analysis

The consequences analysis from potential hazards includes of fatality, partial and total asset losses, disruption of LNG supply, disruption of electricity supply from power plants. Quantitative risk analysis by analyzing the frequency and consequences of potential events that. The next step is risk mapping to determine risk level from the system, evaluation and mitigation are carried out if a scenario is considered unacceptable. The upper fatality risk for workers in the operation of the facilities has been set at 1.0E-03 per year, hence the lower tier set at 1.0E-05 per year and the risk exposure is considered acceptable region. Area between the upper and lower tiers is the intermediate tier. Risk exposures in the intermediate tier should be reduced to "As Low As Reasonably Practicable" (ALARP).

The total individual risk for each worker category is compared against the risk acceptance criteria, in order to assess whether the risks are in ALARP area, or any risk mitigation to reduction measure is required. The accumulation of fatalities in each scenario will be used as input for risk representation. Figure 5 (a) ilustrated the modeling of radiation coverage due to hydrocarbon leaks located near the mini LNG carrier where the loading-unloading process occurs. Figure 5 (b) overview the F-N Curve plot for all fire and explosion scenario. Risk assessment based on hazard analysis due to hydrocarbon handling an F-N curve was found to be in as low as reasonably practicable (ALARP) compared to acceptance criteria.

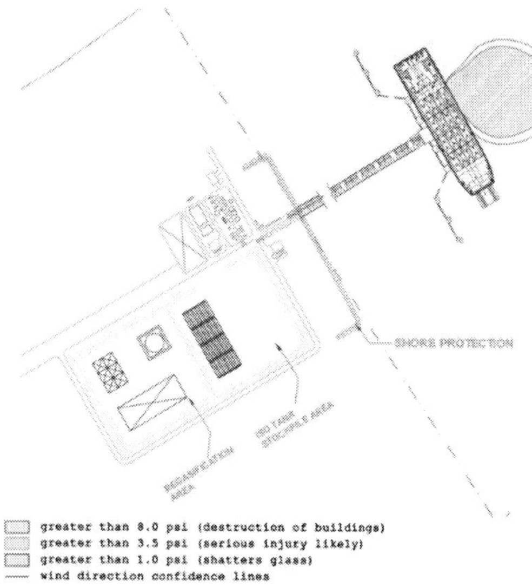

(a)

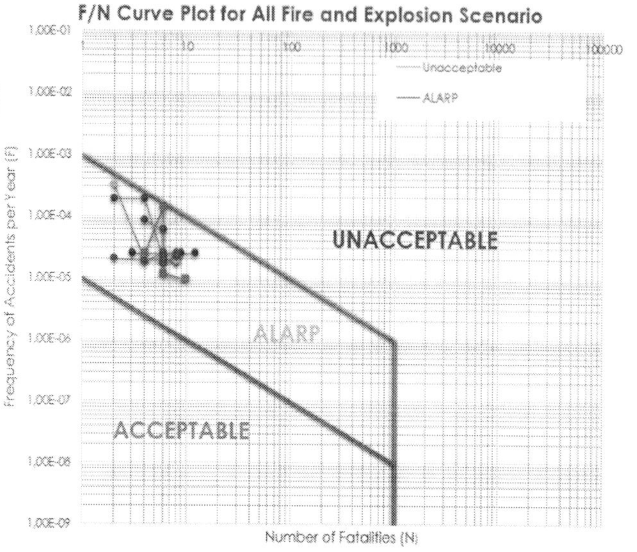

(b)

Figure 6. (a) Modeling of radiation coverage due to hydrocarbon leaks (b) F-N Curve plot for all fire and explosion scenario

4. Conclusion

This study discuss the LNG distribution and cost of shipping in four power plant (Lombok Peaker, MPP Lombok, Gilimanuk and Pemaron). The techno-economic model approached in this paper estimated the operational cost for LNG distribution in Bali and Lombok, which range from $ 19.307 - $ 6.647.580 depending on the sailing frequency for transport distance range from 62–137 Nm. As discussed in the result and discussion parts, an LNG distribution operating cost typically consists of fuel oil, lubricating, consumable, port services, maintenance, crew cost, insurance and administration. In the earlier paper, the authors informed the object of this study is LCT non-conventional as mini LNG carrier, the main dimention LOA: 50,5m; B: 11,4m; H: 3,5m; T:2,5m. Hence its is need 26 mini LNG carrier to supply power plant in Bali and Nusa Tenggara as scenario 1, while 7 mini LNG carrier have to provided to supply power plant in Bali and Lombok as scenario 2. Based on the the number of vessels from scenario 1 and scenario 2, scenario 1 is inefficient because it requires high investment and operational aspects, so the next study will focus on LNG distribution in Bali and Lombok. Scenario 2 supplied by five numbers of vessels were used continuously, however two others vessels were used intermittently. This alternative is used to minimize opeartional cost. The final risk assessment based on hazard analysis due to hydrocarbon handling an F-N curve was found to be in as low as reasonably practicable (ALARP) compared to acceptance criteria. Future work from this research can include the study of more advanced LNG terminal selection and the development of techno-economic models by greedy algorithm to determine optimal LNG distribution with more various destination and constrain.

Acknowledgement

The authors are grateful to the Ministry of Research and Technology/National Research and Innovation Agency (Kemenristek/BRIN/LPDP) of Republic Indonesia in publication support by 189/E1/PRN/2020, the Center of Excellence for Maritime Safety and Marine Installation, Institut Teknologi Sepuluh

Nopember, Surabaya and our research partner Agency for the Assessment and Application of Technology (BPPT).

References

[1] International Energy Agency (IEA), "*Southest Asia Energy Outlook,*" World Energy Outlook Special Report, 2018.

[2] R. Raj, S. Ghandeharium, A. Kumar, J. Geng and M. Linwei, "*A techno-economic study of shipping LNG to the Asia-Pacific from Western Canada by LNG carrier,*" *Journal of Natural Gas Science and Engineering,* vol. 34, pp. 979-992, 2016.

[3] B. Y. Yoo, "*Economic assessment of liquefied natural gas (LNG) as a marine fuel for CO2 cariers compared to marine gas oil (MGO),*" *Journal of Energy,* vol. 121, pp. 772-780, 2017.

[4] T. Uemura and K. Ishigami, "*LNG Supply Chain Infrastructure Configuration - Formulating Policy Options for Promoting Natural Gas Utilization in the East Asia Summit Region Volume II: Supply Side Analysis,*" ERIA , Jakarta, 2018.

[5] Y. Ma and Y. Li, "*Analysis of the supply-demand status of China's natural gas to 2020,*" *Petrol. Sci.* , vol. 7(1), pp. 132-135, 2010.

[6] E. Lisowski and W. Czycki, "*Transport and Storage of LNG in Container Tank,*" *Journal of KONES Powertain and Transport,* vol. 3, pp. 18-27, 2011.

[7] The Ministry of Energy and Mineral Resources, "*Peta Jalan Kebijakan Gas Bumi Nasional,*" Secretariat General, Jakarta, 2018.

[8] S. Lee, "*Quantitative risk assessment of fire & explosion for regasification process of an LNG-FSRU,*" *Journal of Ocean Engineering,* vol. 197, 2020.

[9] The Ministry of Energy and Mineral Resources, "*Rencana Usaha Penyediaan Tenaga Listril (RUPTL) PT Perusahaan Listrik Negara (Persero) Th. 2018 s.d. 2027,*" The Ministry of Energy and Mineral Resources, Jakarta, 2018.

[10] H. N. Abdillah, K. B. Artana, I. M. Ariana and A. A. Masroeri, "*Technical and Economical Analysis of LCT (Landing Craft Tank) Vessel Conversion into Non-Conventional LNG (Liquid Natural Gas) Carriers,*" in *MASTIC,* Bali, 2018.

[11] A. McFarlan, "*Techno-economic assessment of pathways for liquefied natural gas (LNG) to replace diesel in Canadian remote northern communities,*" *Journal of Sustainable Energy Technologies and Assessment,* vol. 42, p. 100821, 2020.

[12] M. A. Budiyanto, A. Riadi , I. G. N. S. Buana and G. Kurnia, "*Study on the LNG distribution to mobile power plants utilizing small-scale LNG carriers,*" *Journal of Heliyon,* vol. 6, p. e04538, 2020.

[13] C. Hsu and Y. Hseieh, "*Routing, ship size, and sailing frequency decision-making for maritime hub-and-spoke network,*" *Journal of Mathematical and Computer Modelling,* vol. 45, pp. 899-916, 2007.

[14] H. Thomson, J. Cobette and J. Winebrake, "*Natural gas as a marine fuel,*" *Journal of Energy Policy,* vol. 87, pp. 153-67, 2015.

[15] PT PELINDO III , "*Tarif Tanda Masuk Pelabuhan di Lingkungan PT Pelabuhan Indonesia III (Persero),*" PT PELINDO III, Surabaya, 2015.

[16] The Ministri of Energy and Mineral Resources, "*PM 8 Th 2020 - Procedures for determining prices of natural gas in the industrial sector,*" ESDM, Jakarta, 2020.

A simulation study of the fleet sizing problem in pilotage and tug services: Case study of Tanjung Perak port, Surabaya

F Hadi[1], **H Supomo**[2] **and T Achmadi**[1]

[1] Department of Marine Transportation Engineeing, Faculty of Marine Technology, Institut Teknologi Sepuluh Nopember, Surabaya

[2] Department of Naval Architecture, Faculty of Marine Technology, Institut Teknologi Sepuluh Nopember, Surabaya

Abstract. Pilotage and tug services are one of the essential services in port operation especially for ports which has channel like Tanjung Perak. There are important interactions between decision on sizing tugboats and decision on utilizing that fleet. This paper combines simulation and assignment model for supporting strategic fleet sizing problem at ports where a Monte Carlo simulation is built to capture the behavior of the real system, whereas the assignment model is built to allocate tugboats near optimal. This approach is quite flexible in the sense that the simulation part can deal with the stochastic aspects with a quite precisions and the assignment model can easily change the tugboats configurations. The performance of the model developed is evaluated by comparing the real data and the simulation results with high precision. The methodology developed in this paper provided valuable decision support on strategic fleet sizing problems in pilotage and tug services at the port.

Key words: fleet sizing problem, assignment problem, pilotage and tug services, port performance

1. Introduction

Pilotage and tug services are one of the essential services in port operation not only to assist ship in the berthing process, but also to escort ships passing through the channel (for ports with channel). Therefore, providing efficient pilotage and tug services will largely determine the overall port performance, especially for ports with heavy traffic. The provision of pilotage and tug services can be viewed from two perspectives, first as a special service for shipping operators and secondly as a public service because of the potential of accidents [1].

To provide an efficient pilotage and tug services, port authority or port operator requires sufficient tugboats not only in term of quantity, but also in term of power. If the number of tugboats is not enough, congestion will arise at the port which result in losses suffered by shipping operators and loss of potential revenue for the port. Conversely, if the number of tugboats is excessive, the utility of tugboat will be low which will result in higher cost due to high fixed costs of tugboats. In addition, when the number of tugboats is insufficient, the port authority or port operator can charter a tugboat from another party to make up for the shortage of tugboats. However, this will have an impact on high cost due to higher charter rate (especially in the short-term charter). Therefore, deciding on the number of tugboats that must be owned by port authority or port operator is an important part of strategic (long-term) fleet size planning [2].

In general, pilotage and tug services required by ships calling at the port can be categorized into 3 types; the first is to assist the ship when entering the port area, the second is to assist the ship when leaving the port area, and the third is to assist the ship when shifting from one location to another in the port area. The number of tugboats required to assist one ship will depend on the size of the ship, measured in length overall (LOA). If the ship is relatively small, then the number of tugboats required is only one unit, whereas for the ship relatively large in size, the number of tugboats required is more than one unit. The problem of fleet sizing planning in the case of tugboats configurations is quite complex since it involves high uncertainty parameters such as inter-arrivals time of the demand for services, the size of the ships and the type of service required. In addition, the potential for a simultaneous arrival of demand for services, makes fleet sizing problem of the tugboats highly stochastic. For this reason, the combination of simulation and assignment model was chosen as the methodology in solving this problem.

The objective of this paper is to present a decision support methodology for strategic planning in fleet sizing problem in the case of tugboat to support port operations. The methodology includes a Monte Cargo simulation model built to describe the uncertainty of the real system behavior represented by probability distributions, whereas the assignment model built to allocate tugboats. There is a strong iterdependencies between strategic, tactical and operational planning levels, therefore, the decision must consider those aspects simultaneously. The methodology in this paper can provide support for port authorities or port operators in decision problems related to tugboats configurations.

2. Literatur Review

Fleet sizing is a complex problem which requires various decision, both strategic and operational point of view. Strategic decisions deal with long-term planning about maximize projected profit such as fleet fleet size, type and compositioan, whereas operational decision deal with routing strategy [2]. In addition, the real-life fleet sizing problem is very depend of the many factors which are difficult to predict and makes this problem highly stochastic. The important interactions between decision fleet sizing and fleet utilisation was studied by GJ Beaujon. The author formulated a model to optimize both side of fleet size and utilisation simultaneously under dynamic and uncertain conditions [3]. Furthermore, the study of interaction between the stochastic behavior of the seaborne trade and the composition of the fleet was studied by Pantuso [4]

The shortcomings of solving fleet sizing problem by using deterministic model was studied by Ali Moradi et al with case study in surface mining operation. They found that at least there are two improtant shortcomings of deterministic model, namely: disregarding the effects of downstream process on the operation and ignoring the fleet management system effects. They developed an integrated simulation-optimization framework to address the fleet sizing problem and target the two shortcomings [5].

The methodology for strategic planning regarding the fleet sizing problem was studied by Fagerholt et al. The authors develop a methodology in decision support for strategic planning in tramp and industrial shipping by combining a Monte Carlo simulation and optimization. The methodoloy was tested on a real case for a major Norwegian shipping company [6].

Hamidreza Eskandari and Ehsan Mahmoodi have developed the methodology for fleet sizing problem in offshore industry by using simulation-based multi-objective optimization which comparing routing based on fixed schedule and routing based on platform demands. A discrete-event simulation combined with simulation-based optimization model is used to evaluate the near-optimal fleet size and composition by minimizing expected total cost and minimum platform service level as a constraint [7].

Another model for fleet sizing problem was presented by K-H Chang et al whose developed a multi-objective vehichel fleet sizing problem by integrates simulation-optimization and data envelopment analysis technique. The model was applied in designing an effective automated materials handling systems which involved uncertainty in the production process [8].

The different method for solving fleet sizing problem was presented by Halvorsen et al where their developed a voyage-based solution with the case study of Statoil – the leading operator on the Norwegian continental shelf. Their solution approach for the problem considered that all candidate voyages the the supply vessels may sail are generated a priori and the composition of the fleet and determining the voyage to sail area decided by solving a voyage-based model [9].

3. Problem definition

This paper focuses on the problem of evaluation the adequacy level of the number of tugboats to support port operations with the case study of the Tanjung Perak port area. The main reason for choosing this port is because this port is the second largest port in Indonesia after port of Tanjung Priok in Jakarta. As a large port, traffic at this port is quite heavy and if not managed properly, the potential for congestion will be very high. The second reason is that it has a channel where ships of a certain size when entering or leaving the port must be assisted by pilotage and tug services. The last one is because there are many terminals alongside the channel which make safety major issues. The whole area of the port of Tanjung Perak is depicted in Figure 1 [10].

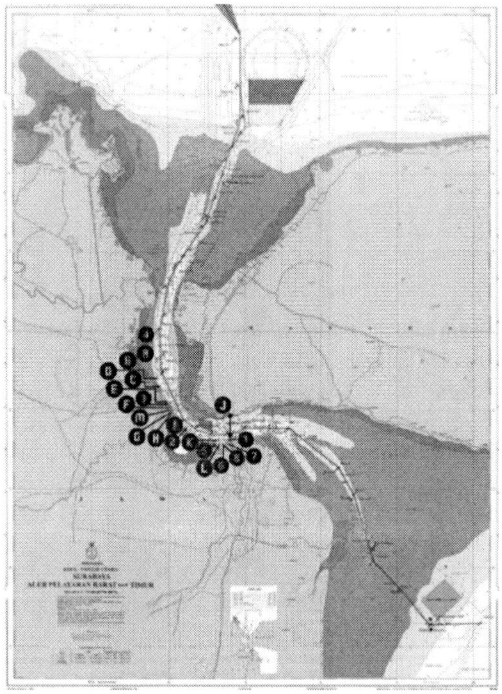

LEGEND:
1. Tanjung Perak Port
2. Teluk Lamong Terminal
3. Gresik Port
4. Manyar Terminal
5. Private Terminal
6. Private Terminal of PT. Pelindo Marine Service
7. Private Terminal of PT. Dock dan Perkapalan Surabaya
8. Private Terminal of PT. PAL Indonesia

A. Private Terminal of PT. Siam Maspion
B. Private Terminal of PT. Karya Indah Alam Sejahtera
C. Private Terminal of PT. Petro Kimia Gresik
D. Private Terminal of PT. Smelting CO Jakarta
E. Private Terminal of Pertamina Asphalt Gresik
F. Private Terminal of PLTU Gresik
G. Private Terminal of Semen Indonesia
H. Private Terminal of PT. Indonesia Marina Shipyard
I. Private Terminal of PT. Wilmar Nabati Indonesia
J. Private Terminal of PT. BEN SANTOSA
K. Private Terminal of PT. Pertamina MOR V
L. Private Terminal of PT. ISM Bogasari
M. Private Terminal of PT. Delta Artha Bahari Nusantara

Figure 1. The area of port of Tanjung Perak.

The provision of pilotage and tug services follows Transportation Ministerial Regulation number 57 of 2015 concerning pilotage and tug services in port area. The requirement for pilotage and tug services are regulated in Article 38 Paragraph 3 of this regulation, stating that ships with a length of 70 m – 150 m must be assisted using minimum 1 unit tugboat with a minimum power of 2.000 HP, whereas as ship with length of 150 m – 250 m assisted by minimum 2 units of tugboats with a minimum power of 6.000 HP, and ship with a length of more than 250 m assisted by minimum 3 tugboats with a minimum of 11.000 HP. The details of this regulation can be seen in Table 1 [11]:

Table 1. Ministerial Regulation No. 57/2015.

No.	Ship's Length Overall	Tugboat		
		No. of Tugboat	Minimum Total Power	Minimum Bollard Pull
1.	70 m – 150 m	1 unit	2.000 HP	24 Ton
2.	150 m – 250 m	2 units	6.000 HP	65 Ton
3.	250 m or more	3 units	11.000 HP	125 Ton

The port authority or port operator is responsible for ensuring the smoothness and safe process of the ships whether entering, leaving or shifting movements. The request for the pilotage and tug services are independent of each other, meaning that requests can either come sequentially or simultaneously (paralel). Meanwhile, the duration of service depends on the type of service and the location of the ship. The farther the distance, the longer it will take for tugboat to be ready as it has to return to base after completing its task. In heavy traffic conditions, the possibility of not having tugboats availabel at the base is quite high, on the other hand, many tugboats are under utilized when the traffic is not so heavy. With these highly uncertain conditions, port authority or port operator must be able to make the right decisions regarding the planning of tugboats configurations should be owned.

Currently, the port authority of port Tanjung Perak area operates 16 tugboats with a composition of 13 owned tugs (2 of which are very old and will no longer operate) and 3 charter tugs which are about to expire, therefore, the total number of tugboats that can be fully operational in only 11 units. To maintain confidentiality, the data of 11 tugboats operated are presented with code as shown in **Table 2**. It is necessary to evaluate the operational performance of the pilotage and tug services with the remaining 11 tugboats and if necessary how many additional tugs are needed so that performance can run optimally.

Table 2. Tugboats configuration.

(a) Existing tugboats available

No.	Tugboat	Power (HP)
1.	TB-01	2 x 2.400
2.	TB-02	2 x 1.800
3.	TB-03	2 x 1.000
4.	TB-04	2 x 1.000
5.	TB-05	2 x 1.000
6.	TB-06	2 x 1.500
7.	TB-07	2 x 1.800
8.	TB-08	2 x 1.800
9.	TB-09	2 x 1.800
10.	TB-10	2 x 1.300
11.	TB-11	2 x 1.500
12.	TB-12*	2 x 1.220
13.	TB-13*	2 x 1.220
14.	TB-14*	2 x 1.100
15.	TB-15**	2 x 750
16.	TB-16**	2 x 600

* *chartered tugboats which are about to expire*
** *owned tugboats which are very old*

(b) Existing tugboats fully operational

No.	Tugboat	Power (HP)
1.	TB-01	2 x 2.400
2.	TB-02	2 x 1.800
3.	TB-03	2 x 1.000
4.	TB-04	2 x 1.000
5.	TB-05	2 x 1.000
6.	TB-06	2 x 1.500
7.	TB-07	2 x 1.800
8.	TB-08	2 x 1.800
9.	TB-09	2 x 1.800
10.	TB-10	2 x 1.300
11.	TB-11	2 x 1.500

4. Proposed method

We propose a framework, as shown in Figure 2 to evaluate fleet sizing problem which consists of two 2 phases. In the phase I, a Monte Carlo simulation is applied to capture the behavior of the real system, whereas in the second phase, if focuses on allocation of tugboats to serve the demands generated in phase 1. More details about phase I and II the framework are as follows:

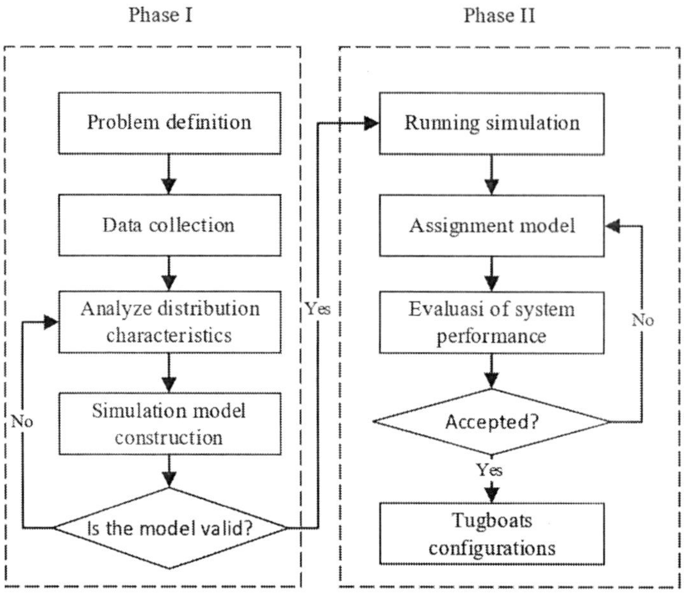

Figure 2. Proposed method.

The phase I of the framework starts with the problem definition which include clear description of how real system works (business process). Next, historical data is collected which are then used to identify the behavior of the real system. Such behavior can obtained by analysing the probability distribution characteristics of several parameters including inter-arrivals time between requests, ship size and duration of services. Result of this analysis will be used to construct a simulation model by applying a Monte Carlo method. In the last stage of phase I, the model need to be verified and validated to ensure the it sufficiently represent the real system.

In the phase II, the assignment model is developed based on the business process and the regulation as shown in Table 1. Before running the assignment model, the scenarios of tugboat configurations are generated, as follows:

- Scenario 1 : the number of tugboats deployed is 11 units, same as the current condition where 3 chartered tugs are not extended and 2 very old tugs are no longer operated.
- Scenario 2 : the number of tugboats deployed is 14 units, 11 of which are currently owned and 3 chartered tugs are extended with the power of 2 x 1.220 HP each.
- Scenario 3 : the number of tugboats deployed is 16 units, 11 of which are currently owned, 3 chartered tugs are extended and replacement of 2 old tugs with the power of 2 x 1.500 HP each.

Next, the system performance is evaluated based on the result of assignment model for each scenario. The evaluation of system performance is done by not only calculate the tugboats utilization but also the number of service requests that must wait.

5. Results and discusssion

The proposed method presented in Section 4 has been tested on the real problem of pilotage and tug services faced by port authority in port of Tanjung Perak area. The characteristics of historical data and general assumpitons are describe in Section 5.1. This is followed by verification and validation model based on characteristics derived from historical data in Section 5.2. In the following sections 5.3 we describe an analysis of results.

5.1. Characteristics of historical data

In the period of January – September 2018, the number of pilotage and tug services reached 15.579 movements. The data shows that pilotage and tug services in port of Tanjung Perak area are dominated by small ships group with a proportion of about 78%. It is followed by medium and large groups having a proportion of 21% and 1% respectively. The detail picture is shown in **Table 3**.

Table 3. Pilotage and Tug Services in Port of Tanjung Perak.

Ship Categories	Service (Movement) Types			
	Entering	Leaving	Shifting	**TOTAL**
Small (70 – 150 m)	3.711	4.179	4.271	**12.161**
Medium (150 – 250 m)	1.030	1.190	1.064	**3.284**
Large (250 m up)	49	49	36	**134**
TOTAL	**4.790**	**5.418**	**5.371**	**15.579**

One of the key factors in developing a model as close to the real conditions as possible is the proces of identifying uncertainty parameters. In this paper, we identified three parameters that greatly affect the system performace and are stochastic in nature, namely the inter-arrival time between requests, ship size and service duration. The results of the identification of these parameters are as follows:

5.1.1. Inter-arrivals time between service requests

The inter-arrival time between service requests is not only one of the factors that determines the number of tugboats needed, but aslo a factor that can describe the density of a port. The inter-arrival time profile based on the historical data is depicted in Table 4.

Table 4. Profile of inter-arrival time.

Statistical properties	Distribution: Exponential	Formula
Mean (hours) : 0.3808 Minimum (hours) : 0.0000 Maximum (hours) : 10.8333 Standard deviation : 0.4198	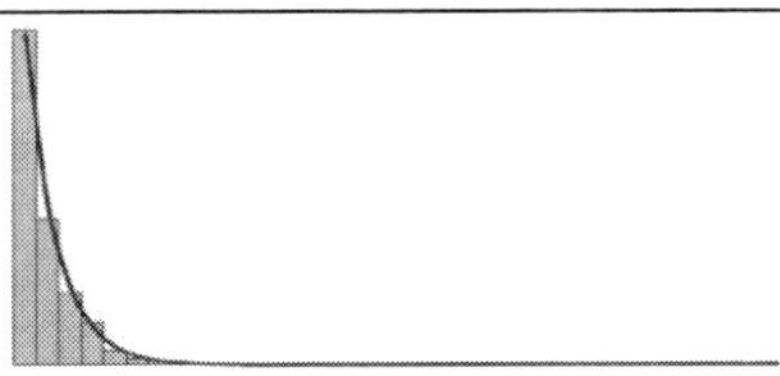	$f(x;\lambda) = \begin{cases} \lambda e^{-\lambda x}, & x \geq 0 \\ 0, & x < 0 \end{cases}$

It can be seen from **Table 4** that minimum inter-arrival time is 0 which mean that some request may come simultaneously. It is recorded that there are 2.074 requests or more than 13% of 15.579 come at the same time.

5.1.2. Ship size

As previously mentioned, ships under 70 m of LOA are excluded in this study, as they do not require pilotage and tug services. The ship size profile derived from historical data can be seen in Table 5.

Table 5. Profile of ship size.

Statistical properties	Distribution: Gamma	Formula
Mean (m) : 124.58 Minimum (m) : 70.00 Maximum (m) : 268.80 Standard deviation : 40.85	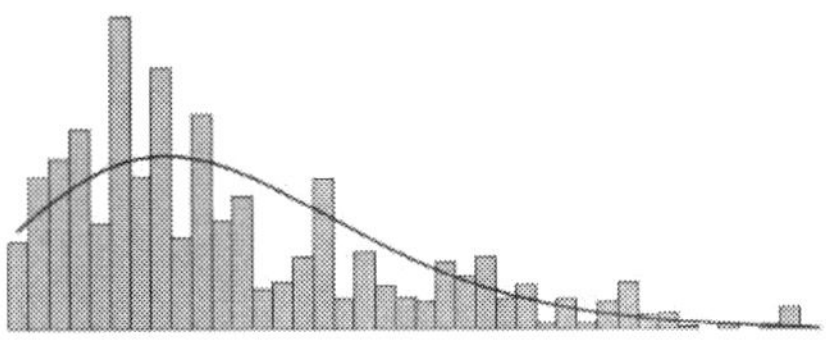	$f(x) = \dfrac{x^{\alpha-1}e^{-x/\beta}}{\beta^{\alpha}\Gamma(\alpha)}$

IOP Conf. Series: Materials Science and Engineering 1052 (2021) 012056 doi:10.1088/1757-899X/1052/1/012056

5.1.3. Service time duration

As the duration of service time are categorized into three groups according to the type of services, namely entering, leaving and shifting movements, the profile of service time duration is also categorized into three groups. The profile of each groups are as follows:

Table 6. Profile of service time duration.

Statistical properties	Distribution: Lognormal	Formula
Entering movements Mean (hours) : 0.689 Minimum (hours) : 0.150 Maximum (hours) : 4.500 Standard deviation : 0.194	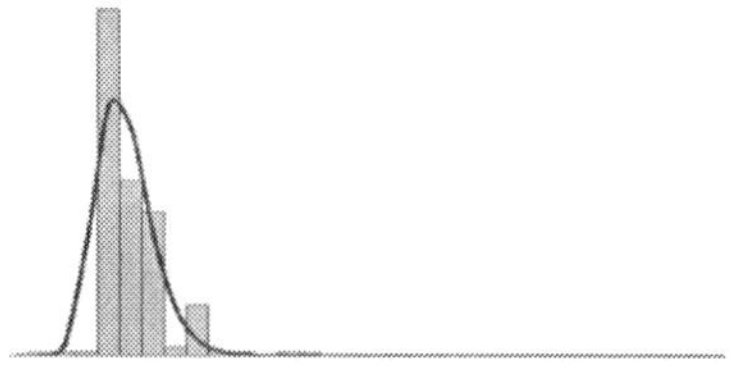	$f(x) = \Phi\left(\frac{(\ln x) - \mu}{\alpha}\right)$
Leaving movements Mean (hours) : 0.683 Minimum (hours) : 0.133 Maximum (hours) : 5.417 Standard deviation : 0.195	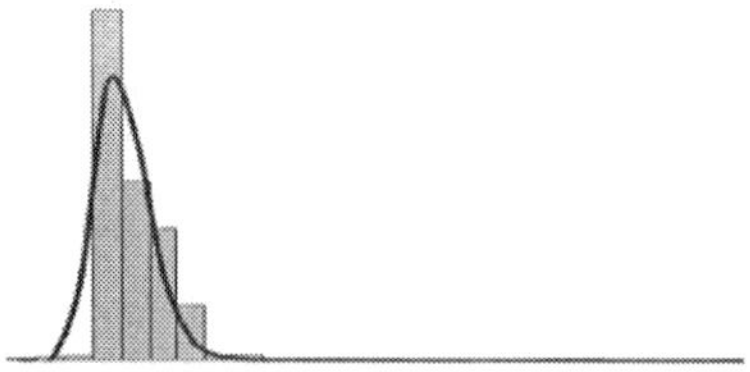	$f(x) = \Phi\left(\frac{(\ln x) - \mu}{\alpha}\right)$
Shifting movements Mean (hours) : 0.686 Minimum (hours) : 0.117 Maximum (hours) : 4.000 Standard deviation : 0.184	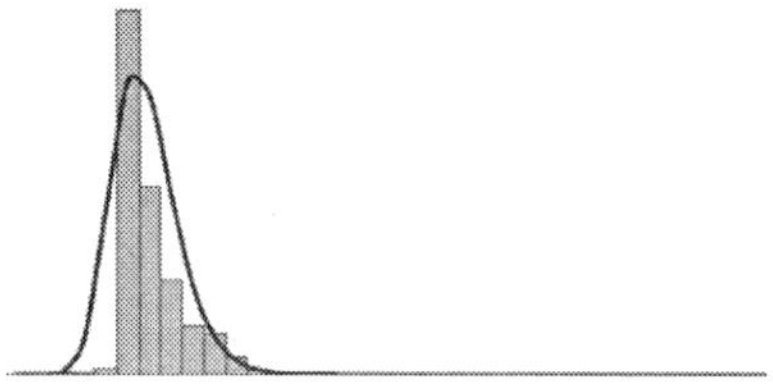	$f(x) = \Phi\left(\frac{(\ln x) - \mu}{\alpha}\right)$

5.2. Model verification and validation

Separation between model verification and validation is very difficult to do because these two processes are closely related and often use the same technique for both. A model can be said verified if the behavior generated by the model is similar to the actual behavior. To verify the model, we use certain condition where the result can be analyzed analytically and compare them with those produced by the model. On the other hand, validation is the process of comparing the representation of a model to the real system. If the comparison is true (with tolerable error), then it is valid.

For validation, the model is run using the aforementioned distribution of 8 months length data and then compare the results with the real data. The results as can be seen in Table 7 shows that the difference between real data and simulation result is only 0.07% which is substantially low. Therefore, we can proceed the model with the characteristics as previously mentioned to predict the performance of the system by applying some scenarios.

Table 7. Comparison between data and simulation result.

(a) Historical data.

Ships Category	Service (movement) types			
	Entering	Leaving	Shifting	TOTAL
Small	3.711	4.179	4.271	12.161
Medium	1.030	1.190	1.064	3.284
Large	49	49	36	134
TOTAL	4.790	5.418	5.371	15.579

(b) Simulation result.

Service (movement) types			
Entering	Leaving	Shifting	TOTAL
3.601	4.261	4.301	12.163
1.059	1.199	1.031	3.289
60	43	35	138
4.720	5.503	5.367	15.590

5.3. Analysis of results

By implementing the probability distribution as presented in previous stage, demand estimation for pilotage and tug services for the next few years can be done by assuming a grwoth of 3%-4% per year. In this paper, we estimate the demand for 2021 or three years from the base year. By replicating the simulation process 30 times, the estimation of demand for pilotage and tug services are as follows:

Table 8. Projected demand in 2021.

# Requests	Mean	Minimum	Maximum	Std-Dev
Total	27.091	26.872	27.262	108
Entering	8.326	8.227	8.481	83
Leaving	9.416	9.246	9.532	84
Shifting	9.350	9.221	9.508	95

The assignment model is then used to allocate tugboats in each scenario to evaluate the performance of the system based on the demand projected. The evaluation of system performance is carried out by taking into account two aspects which include the number of requests that must be queued and the utilization of tugboats. In this paper, the system perfomance is said to be good if there is no queuing requests and the utilization of tugboats is high. The replication process of this assigment model is 30 times, same as the replication of simulation for demand estimation. The results from the methodology provided important decision support for strategic fleet sizing planning as shown in Table 9 - Table 11.

Table 9. Results of scenario 1.

(a) Number of queue.

Mean	Min	Max	StD
401	157	472	84.75

(b) Tugboats utilization.

	TB-1	TB-2	TB-3	TB-4	TB-5	TB-6	TB-7	TB-8	TB-9	TB-10	TB-11
Mean	25%	25%	26%	41%	43%	42%	19%	19%	18%	18%	17%
Min	21%	21%	17%	33%	37%	37%	12%	9%	12%	13%	10%
Max	30%	29%	31%	46%	49%	51%	32%	35%	34%	33%	31%

Table 10. Results of scenario 2.

(a) Number of queue.

Mean	Min	Max	StD
9	4	15	3.83

(b) Tugboats utilization.

	TB1	TB2	TB3	TB4	TB5	TB6	TB7	TB8	TB9	TB10	TB11	TB12	TB13	TB14
Mean	18%	17%	19%	30%	29%	32%	30%	33%	32%	17%	16%	1%	8%	13%
Min	12%	3%	12%	25%	14%	25%	18%	26%	18%	10%	13%	1%	6%	6%
Max	21%	22%	24%	37%	46%	46%	38%	36%	39%	32%	21%	1%	14%	22%

Table 11. Results of scenario 3

(a) Number of queue.

Mean	Min	Max	StD
1	0	2	0.67

(b) Tugboats utilization.

	TB1	TB2	TB3	TB4	TB5	TB6	TB7	TB8	TB9	TB10	TB11	TB12	TB13	TB14	TB15	TB16
Mean	16%	15%	14%	26%	24%	25%	26%	24%	28%	28%	26%	1%	6%	6%	11%	11%
Min	14%	3%	3%	24%	13%	13%	14%	14%	26%	26%	17%	1%	6%	6%	11%	11%
Max	18%	18%	18%	28%	28%	28%	29%	30%	29%	30%	32%	1%	7%	7%	12%	12%

The results provided important decision support for strategic fleet sizing planning. It is showed that increasing the number of tugboats would reduce the number of queues almost exponentially, but on the other hand it also reduce tugboats utilization linearly. This indicates that the trade-off between the impact of additional tugboats on the number of queues and tugboats utilization must be carefully considered in the decision process.

In addition, the results also indicate that the possibility of increasing tugboats utilization without increasing the number of queues is difficult to reached. This is because the number of service request that come simultaneously is quite large, between 13% - 15%. Another factor that makes this objective difficult to reach is because the average inter-arrival time between requests is smaller than the average of the service time, which means that there are new requests before the tugboat has finised carring out the task it is running.

The results is also showed that in the case of port of Tanjung Perat area, scenario 2 gave the best result although there are still some request to wait. This decision is based on the results which show that even though there are still any queues, but the number of queues is tolerable, where the average number of queues is 9 requests which is equivalent to 0.034% of the total requests.

6. Conclusion

In this paper we have presented a methodology to support the decision of tugboats configuration for pilotage and tug services in port operations. Our proposed methodology combines a Monte Carlo simulation and fleet assignment model, where a Monte Carlo simulation is built to generate request types, ships size and duration of services which are then applied in assignment model. There are two major advantages of the method we proposed:

- The model of Monte Carlo simulation makes it possible to deal with the stochastic aspects in the pilotage and tug services in a port with a quite precision
- The assignment model is flexible in the sense that it can easily change the tugboats configurations to provide decision support for strategic fleet sizing planning.

One main disadvantage of this methodology is that the process requires a lot of computer time due to many runs of the model that must be made to provide trustworthy results. The requirement for a relatively large amounts of high quality data to generate probability distributions for the parameters of the real system is another drawback of this methodology. However, we believe that the processing time is not a big deal when compared to the important important strategic decisions that may have an impact several year ahead in time.

We have tested the methodology on a actual pilotage and tug services at port of Tanjung Perak area and proved to provide valuable decision support on important strategic fleet sizing problems faced by port authority or port operator. The issue of cost has not been considered in this methdology, this is an interesting topic for future research. Another interesting area for future work is how to make processing time faster.

7. Acknowledgments

We are grateful to Mr. Yoyon Adriono, research asistant in the marine transportation laboratory of the department of marine transportation engineering for his assistance in developing the macro spreadsheet on this model. We also thank PT. Pelabuhan Indonesia III (Persero) which has provided information and data related to the problem in this paper.

References

[1] Ö. Uğurlu, M. Kaptan, S. Kum, and S. Yildiz, "Pilotage services in Turkey; key issues and ideal pilotage," *J. Mar. Eng. Technol.*, 2017, doi: 10.1080/20464177.2016.1262596.

[2] A. Shyshou, I. Gribkovskaia, and J. Barceló, "A simulation study of the fleet sizing problem arising in offshore anchor handling operations," *Eur. J. Oper. Res.*, vol. 203, no. 1, pp. 230–240, 2010, doi: 10.1016/j.ejor.2009.07.012.

[3] G. J. Beaujon and M. A. Turnquist, "Model for fleet sizing and vehicle allocation," *Transp. Sci.*, vol. 25, no. 1, pp. 19–45, Feb. 1991, doi: 10.1287/trsc.25.1.19.

IOP Conf. Series: Materials Science and Engineering 1052 (2021) 012056 doi:10.1088/1757-899X/1052/1/012056

[4] G. Pantuso, K. Fagerholt, and L. M. Hvattum, "A survey on maritime fleet size and mix problems," *Eur. J. Oper. Res.*, vol. 235, no. 2, pp. 341–349, 2014, doi: 10.1016/j.ejor.2013.04.058.

[5] A. Moradi Afrapoli, M. Tabesh, and H. Askari-Nasab, "A stochastic hybrid simulation-optimization approach towards haul fleet sizing in surface mines," *Min. Technol. Trans. Inst. Min. Metall.*, vol. 128, no. 1, pp. 9–20, 2019, doi: 10.1080/25726668.2018.1473314.

[6] K. Fagerholt, M. Christiansen, L. Magnus Hvattum, T. A. V. Johnsen, and T. J. Vabø, "A decision support methodology for strategic planning in maritime transportation," *Omega*, vol. 38, no. 6, pp. 465–474, 2010, doi: 10.1016/j.omega.2009.12.003.

[7] H. Eskandari and E. Mahmoodi, "A simulation-based multi-Objective optimization study of the fleet sizing problem in the offshore industry," *Marit. Econ. Logist.*, vol. 18, no. 4, pp. 436–457, 2016, doi: 10.1057/mel.2015.21.

[8] K. H. Chang, A. L. Chang, and C. Y. Kuo, "A simulation-based framework for multi-objective vehicle fleet sizing of automated material handling systems: An empirical study," *J. Simul.*, vol. 8, no. 4, pp. 271–280, 2014, doi: 10.1057/jos.2014.6.

[9] E. E. Halvorsen-Weare, K. Fagerholt, L. M. Nonås, and B. E. Asbjornslett, "Optimal fleet composition and periodic routing of offshore supply vessels," *Eur. J. Oper. Res.*, vol. 223, no. 2, pp. 508–517, 2012, doi: 10.1016/j.ejor.2012.06.017.

[10] "Alur pelayaran barat dan timur Surabaya." Pusat Hidragrafi dan Oseanografi - TNI AL, Jakarta, 2016.

[11] "Peraturan Meteri Perhubungan Republik Indonesia No:57/2015." Kementerian Perhubungan Republik Indonesia, Jakarta, 2015.

Modification and Analysis of Twin Screw Ship Stern for Optimizing Propulsive Coefficient Values in Propulsive System

J Rulianto[1,*], **M Indiaryanto**[2], **I M Ariana**[1]
[1]Department of Marine Engineering, Faculty of Marine Technology, Institut Teknologi Sepuluh Nopember (ITS), Surabaya, East Java 60111, Indonesia
[2]Agency for the Assessment and Application of Technology - Hydrodynamics Laboratory Indonesia (BPPT-LHI), Institut Teknologi Sepuluh Nopember (ITS), Surabaya, East Java 60111, Indonesia
*Corresponding author : jangkarulianto95@gmail.com

Abstract. Twin-screw ship has the advantage of being able to reduce the diameter of propeller by using the same rotation and engine power as a single-screw ship. With the result that twin-screw ship can be operated in areas with lower water depth. However, the type of twin-screw ship has a deficiency in the value of hull efficiency (η_{hull}) and the relative rotative efficiency (η_{RR}). This occurs because of the propeller position which is far from the field of the flow of the ship's hull speed (V). So that the value of the water flow to the propeller (advance velocity/ V_a) becomes low. Various studies have been conducted to increase wake fraction (w) and reduce the thrust deduction factor (t) values some of which are by modifying the shape of the ship's stern parts bossing to increase the water speed to the propeller (advance velocity/ V_a) which will have a large effect on the wake fraction (w) and thrust deduction factor (t). Among them is by modifying the shape of the ship's stern bossing so that it can increase the speed of water to the propeller (advance velocity/ V_a) which will greatly affect the value of the wake fraction (w) and thrust deduction factor (t) to increase the propulsive coefficient (PC).In this study, it is known that the QPC value for the ship model tested in the Hydrodynamics Laboratory Indonesia (LHI) is 0.54 while for the four models tested using the CFD method is model one 0.63, model two 0.55, model three 0.63, and model four amount 0.51. Based on the above test it can be concluded that the third hull model has the highest QPC. be in ALARP compared to acceptance criteria.

Keywords. *Build faction, hull efficiency, relative rotative efficiency, thrust deduction factor, twin-screw*

1. Introduction

The value of water flow to the propeller (V_a), also has an impact on the value of the propeller work efficiency behind the ship and the propeller work efficiency during an open water test, known as relative rotative efficiency (η_{RR}). Relative rotative efficiency is a comparison of the propeller work efficiency at the rear of the ship (η_B) compared to the efficiency during the open water test propeller (η_O) [1,2]. This relationship is indicated by the equation $\eta_{RR} = \eta_B/\eta_O$. By knowing the value of the hull efficiency (η_{Hull}), it can provide information that the hull we designed is effective and efficient so that power reduction due to hull and propeller design errors can be minimized [3].

To conduct experiments to determine the effectiveness and characteristics of the hull, this research will discuss the effect of single screw, twin screw hull design, and modification of twin-screw ship bossing on the value of advance velocity (V_a) which will later be used as a determination of value. from the wake fraction and thrust deduction factors to obtain the value of the propulsive coefficient of each model. This research was conducted with a simulation method using *Maxsurf*, *Rhinos*, and *Numeca* software. Variations of changes in ship speed will be carried out to obtain the results of the effectiveness of the change in the bossing of the twin-screw hull. This research is expected to provide new insights, especially in determining the bossing design of the twin-screw hull.

2. Ship Performance Theory and Modification

2.1 Efficiency in Ship Drive System

The ship propulsion system has several definitions of the power transmitted, starting from the power released by the propulsion motor to the power provided by the ship's locomotive to the surrounding fluid. The ratio of these powers is often expressed in terms of efficiency, although in some cases it is not a direct power conversion value.

2.1.1 Hull Efficiency (η_{Hull})

Hull efficiency (η_{Hull}), is the ratio between effective power (P_E) and thrust (P_T). Hull efficiency is a measure of the suitability of the hull design (stern) to the propulsor arrangement so that this efficiency is not a true form of power conversion. Then the value of the efficiency of this hull can be more than one, in general, the figure is taken around 1.05 [4].

$$\eta_{Hull} = \frac{(1-t)}{(1-w)} \tag{1}$$

Where

η_{Hull} : Hull efficiency

t : Thrust deduction factor

w : Wake fraction

t, and w are propulsion parameters, where t can be obtained by the following equation :

$$\eta_{Hull} = 1 - \frac{R}{T} \tag{2}$$

$T_{Standard}$ $= 0{,}5 \times C_P - 0{,}12$; for ships with a single propeller.

$= 0{,}5 \times C_P - 0{,}19$; for ships with a double propeller.

With C_P (prismatic coefficient) :

$$C_p = \frac{V}{L \times B \times T \times C_m} = \frac{V}{L \times A_m} \tag{3}$$

2.1.2 Propeller Efficiency (η_{Prop})

Propeller Efficiency (η_{Prop}) is the ratio between thrust (P_T) and power delivered (P_D). This efficiency can be expressed as power conversion, and the difference in value that occurs is that the measurement of the propeller torque is carried out. in open water conditions (Q_O) or behind the ship conditions (Q_B) [5]. The following equation shows the two conditions of propeller efficiency, as follows :

Propeller Efficiency (Open Water) : $$\eta_O = \frac{T \times V_a}{2\pi Q_O\, n} \tag{4}$$

Propeller Efficiency (Behind The Ship) : $$\eta_B = \frac{P_T}{P_D} = \frac{T \times V_a}{2\pi Q_O\, n} \quad (5)$$

Because there are two conditions an efficiency ratio appears, namely what is known as the relative-rotative efficiency (η_{RR}). Rotative efficiency is a comparison between propeller efficiency in conditions behind the ship with propeller efficiency in conditions in open water.

$$\eta_{RR} = \frac{\eta_B}{\eta_O} = \frac{\dfrac{T_B \times Va}{2\,\pi\, n\, Q}}{\dfrac{T_O \times Va}{2\,\pi\, n\, Q}} = \frac{T_O}{T_B} \quad (6)$$

So that the relative-rotative efficiency (η_{RR}) is actually not a characteristic of the actual efficiency quantity (not a power conversion). This efficiency is only a comparison of the different efficiency values [5]. So the magnitude of the relative-rotative efficiency can also be greater than one, but in general, the value is taken to be around one.

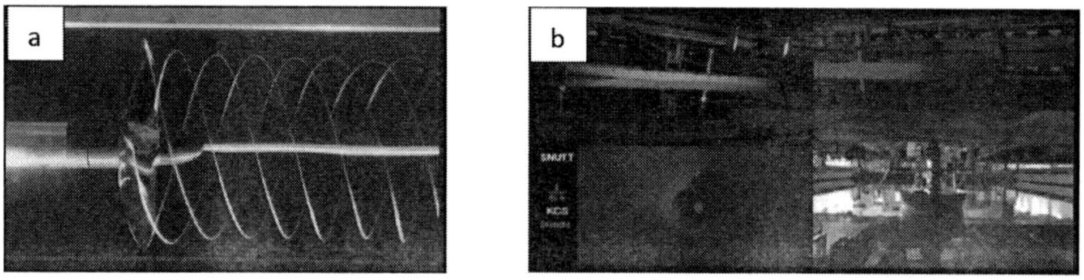

Figure 1. Cavitation and Self Propulsion Test : (a) Cavitation Test, (b) Self Propulsion Test

2.2 Procedures for Sample Work Instructions Calibration of Load Cells and Ship Models

Procedure for sample work instructions calibration of load cells can be applied to the calibration of a strain gauge load cell to measure strength and weight in either tension-compression or compression-tension direction serving a dual purpose (hereinafter referred to as load cell). Some of the instructions in this procedure include technical requirements (classification, basic characteristic, creep characteristic, drift), calibration condition (standard calibration condition, loading condition, temperature balance, preheating, atmosphere pressure, etc), items and method of calibration first-time calibration for a new load cell or after repair (preparation, calibration method, asymmetry, relative deviation, creep, the influence of temperature, random calibration, repeat calibration, etc).

Procedure for ship model is to ensure correct hull construction. Where the basic requirement of the model is that it must have geometrically similar to a ship, especially the part that is submerged in water. Models are usually cut from a re-drawing of the shipping plan. The hull model tolerances for width (Y) and depth (Z) should be within ± 1.0 mm. Tolerances for the model length should be within ± 0.05% LPP or ± 1.0 mm whichever is greater. The instructions in this procedure include : model manufacture (hull, propeller, appendages), turbulence stimulations (hull, propeller), preparation of model testing (ballasting & trimming, wax model), and documentation.

2.3 Standard Open Water Procedure and Computational Fluid Dynamics Methods Efficiency

Open water test on motorized container barge is used to determine the performance of the propeller that has been designed. The intended performance includes the thrust coefficient, torque coefficient, and hull efficiency. CFD (computational fluid dynamics) is a branch of fluid mechanics that uses numerical methods and algorithms to solve and analyze problems that occur in fluid flow. The purpose of CFD is to accurately predict fluid flow, heat transfer, and chemical reactions in complex systems, which involve one or all of the above phenomena. Computational fluid dynamics consists of three main elements, namely :

- The pre-processor preprocessor is the initial stage in computational fluid dynamic (CFD), which is the data input stage which includes determining the domain and boundary conditions. At this stage, meshing is also carried out, where the object being analyzed is divided into a certain number of grids.
- The next stage is the processor stage, where at this stage the process of calculating the data that has been entered is carried out using related equations iteratively so that the results obtained can reach the smallest error value.
- Postprocessor is the last stage, where the results of calculations at the processor stage will be displayed in images, graphics, and animation [7-9].

3. Result and Discussion

3.1 Modelling Hull

At this stage, a 3D model of the basic design of the hull and propeller will be created. Furthermore, the modification was made in the form of changing from a single screw to a twin screw. In the final stage, a twin-screw hull design modification is also made, especially in the bossing section to increase the value of $\boldsymbol{V_a}$ [10,11]. The first model is a type of hull model that has a shape according to the actual hull model. With the form of a standard stern bulb type twin-screw vessel. The second type of hull model is a type of hull model of the first model with the addition of a half-cylinder parabolic basin on the stern part of the hull near the stern bulb.

The purpose of making this basin shape is to increase the flow of water to the propeller so that its efficiency increases. The third type of hull model is a form of the first model of the hull with the addition of kell on the part that is parallel to the stern bulb. The purpose of adding this kell is to compress the flow so that the flow rate is higher and the efficiency increases. The four hull model is the first type of hull model by removing the stern bulb shape and replacing it with a hanging skeg. The purpose of modification to this form is so that the flow to the propeller is not obstructed by the stern bulb shape to increase the efficiency of the ship (Figure 2).

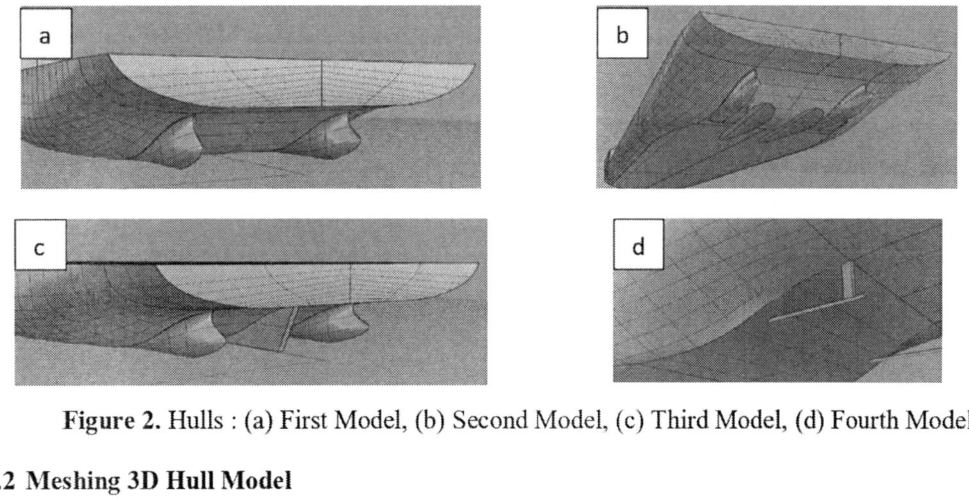

Figure 2. Hulls : (a) First Model, (b) Second Model, (c) Third Model, (d) Fourth Model

3.2 Meshing 3D Hull Model

Meshing is a stage in the planning of the division of the ship's parts to be analyzed discrete [12] (Figure 3). By dividing each part of the hull into a square shape it will be easier to analyze using heavier equations. The main principle of calculating using Navier stokes is to apply force to each rectangular part of the hull so that it will produce a response force on the x, y, and z axes. This force is later read as resistance. The meshing criteria and the shape of the hull mesh at the adapt to the geometry stage are as follows (Table 1).

Table 1. Mesh Criteria

Patch	Max. Nb of ref	Criteria	Diffusion
Deck	6	Target cell size (0,0,0)	Global
Hull, Transom	7	Target cell size (0,0,0)	Global
Bulb, Small, Bilge Keel	8	Target cell size (0,0,0)	Global
Bilgekeel fillet, Rudder, Shaft, Skeg	9	Target cell size (0,0,0)	Global
Cap, Hub, Strut	10	Target cell size (0,0,0)	Global
Hull shaft	10	Target cell size (0,0,0)	3
Gap, Iner	11	Target cell size (0,0,0)	Global

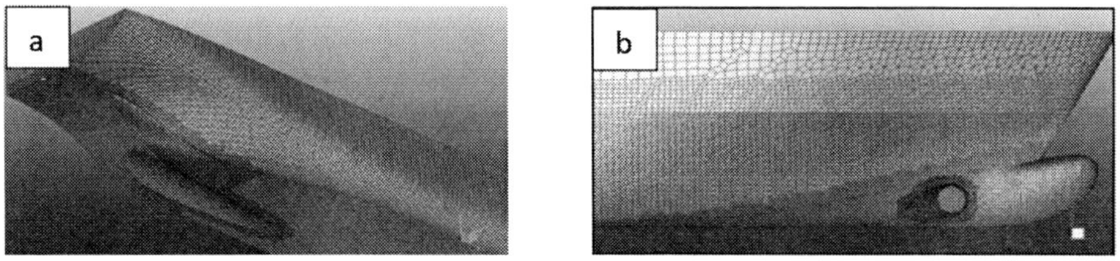

Figure 3. Final Meshing Results of : (a) The Stern Bulb Basin, (b) Bulbus Bow and Bow Thruster

Motorized container barge ship specification data includes the main size of the ship and the scale size of the hull model to be tested. As for the suitability of dimensions and suitability of shape, it determines the accuracy of data collection in the hydrodynamics laboratory (Figure 4). The following is a description of the specification data on the 1000 TEU container vessels (Table 2).

Table 2. Specifications of 1000 TEU Container Ship

No.	Items	Symbols	Scores	
			Models	Prototypes
1	Length on waterline	Lwl	6056 m	181.67 m
2	Breadth Moulded	B	1.033 m	31.00 m
3	Displacement (vol)	A	0.837 m^3	22603.21 m^2
4	Draught 1	T_1	0.160 m	4.8 m
	Draught 2	T_2	0.167 m	5.0 m

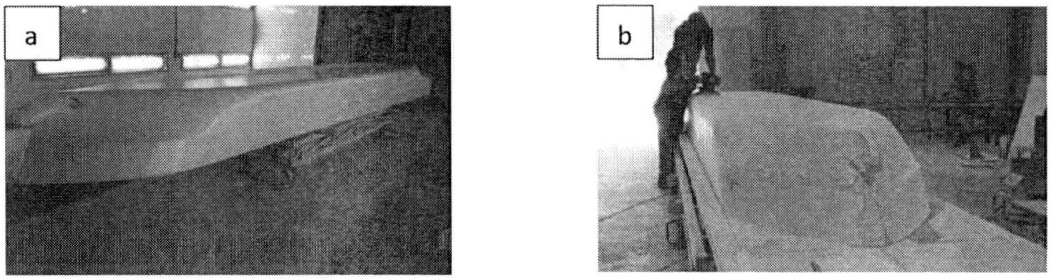

Figure 4. The Hull Shape Is Ready To Be Tested

IOP Conf. Series: Materials Science and Engineering 1052 (2021) 012057 doi:10.1088/1757-899X/1052/1/012057

Motorize container barge ship propeller specification data is used to analyze the performance of the open water test and self-propulsion test. As for the suitability of dimensions and suitability of shape, it determines the accuracy of data collection in the hydrodynamics laboratory. The following is the main size data for the motorized container barge propeller ship as follows (Table 3), and testing parameter on LHI (Table 4).

Table 3. Main Size of Container Ship Propeller Model and Prototype

No.	Items	Symbols	Scores	
			Models	Prototypes
1	Propeller Diameter	D	0.010 m	3.000 m
2	Blade Area Ratio	Ae/Ao	0.5	0.583
3	Pitch/Diameter Ratio	P/D	0.802	0.802
4	Pitch Propeller	$P_{0.7R}$	0.0802 m	2.406 m
5	Number of Propeller	Z	4 pcs	4 pcs

Table 4. Testing Parameter in LHI

Test Conditions	Draft		Speed	
	Models (m)	Prototypes (m)	Models (m/s)	Prototypes (m/s)
		Resistances		
First Draft	0.160/0.160	4.800/4.800	0.564-1.315	6-14
Second Draft	0.167/0.167	5.000/5.000	0.564-1.315	6-14
		Propulsions		
First Draft	0.160/0.160	4.800/4.800	0.939-1.315	10-14
Second Draft	0.167/0.167	5.000/5.000	0.939-1.315	10-14

Ship resistance testing is used to determine the value of ship resistance or ship resistance in calm water conditions. The results of this ship resistance value will be used as the basis for determining the value of engine power that must be installed. The resistance test in the hydrodynamics laboratory is set at a model speed of 0.56 m/s to 1.32 m/s. The test was carried out at a load of 4.8 m and 5 m (Figure 5).

The resistance test for the container ship model using the CFD method is carried out in the same conditions as when tested in the hydrodynamics laboratory. The type of meshing quality used in this test is medium. This aims to minimize the time during the process (Figure 6a). After getting the hull model test results using the CFD method, the next step is to compare the results of the hull resistance value (Figure 6b).

Figure 5. Resistance Test in LHI

IOP Conf. Series: Materials Science and Engineering 1052 (2021) 012057 doi:10.1088/1757-899X/1052/1/012057

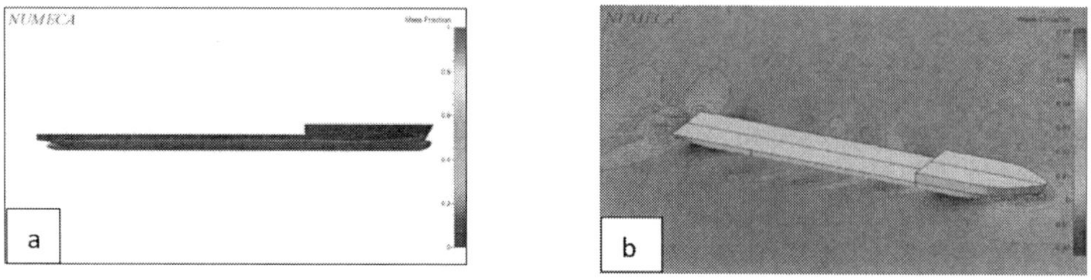

Figure 6. Resistance Test using CFD Method : (a) Water-Immersed Area, (b) Flow Form

Open water test on the motorized container barge is used to determine the performance of the propeller that has been designed. The intended performance includes the thrust coefficient, torque coefficient, and hull efficiency. Open water test using the CFD method (Figure 7b) and testing in LHI (Figure 7a).

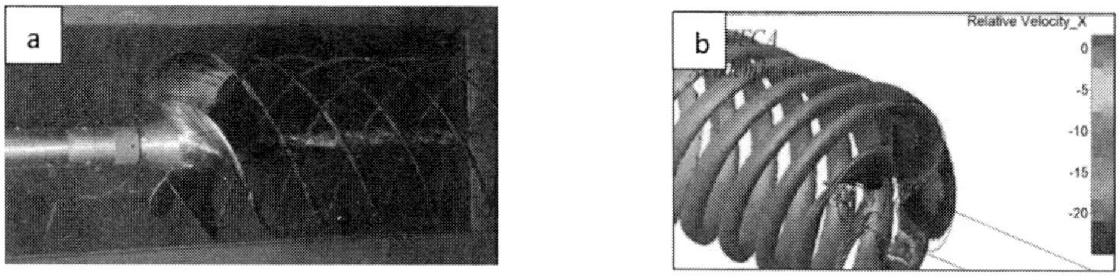

Figure 7. Open Water Test with : (a) Testing in LHI, (b) CFD Method

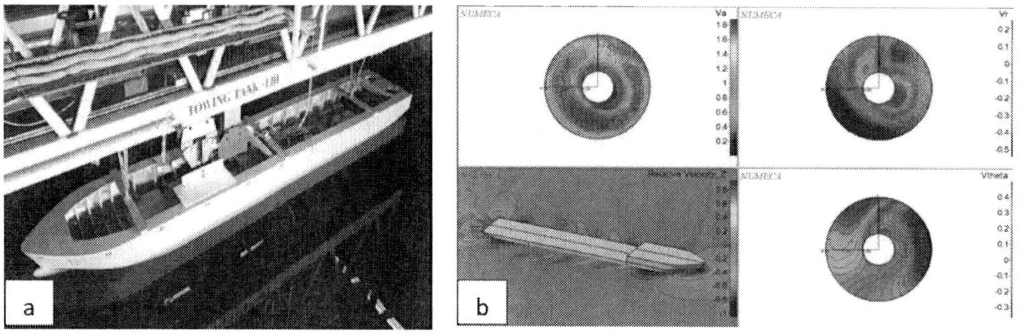

Figure 8. The Self-Propulsion Test with : (a) Testing in The Hydrodynamics Laboratory, (b) CFD Method

The self-propulsion test on a motorized container barge ship is used to determine the performance of the propeller that has been designed when it is operated right at the stern of the ship. The self-propulsion test in the hydrodynamics laboratory is carried out by adjusting the condition of the propeller installed on the hull, and varying the propeller rotation according to the thrust requirements of the hull model. Furthermore, note the value of the model ship torque which will be converted into channeled engine power (DHP) (Figure 8a).

The self-propulsion test using the CFD method was used to determine the value of the advance velocity ($\boldsymbol{V_a}$), $\boldsymbol{V_r}$, and $\boldsymbol{V_\alpha}$ of the container hull. As for the self-propulsion test, it is necessary to pay attention to the input of the thrust propeller value according to the results of the open water test.

3.3 Comparison and Analysis

3.3.1 Hull Efficiency

From the simulation results that have been carried out on 4 types of ships with testing in LHI and using CFD method, the data results are presented in the graph (Figure 9) and table (Table 5). This comparison aims to determine the performance of the value of the ship's resistance (resistance) which will have an impact on the need for power to overcome it so that the ship can go at the desired speed. Besides, the comparison of this resistance value is to see how significant the impact of changes in the shape of the hull is to changes in the resistance value so that researchers can produce a hull design that has a high propulsive coefficient value without having to increase the resistance value significantly.

Table 5. Hull Efficiency in LHI

VS	V_s (m/s)	w	R	T	t	V_a (knot)	η_H	$V_a = v - (w.v)$
6	3.0864	0.241	77	99.3	0.225	4.6	0.98	2.3
7	3.6008	0.241	102	131	0.225	5.3	0.98	2.7
8	4.1152	0.237	134	173	0.225	6.1	0.98	3.1
9	4.6296	0.232	173	222	0.224	6.9	0.99	3.6
10	5.144	0.23	210	271	0.224	7.7	0.99	4.0
11	5.6584	0.227	255	328	0.224	8.5	1.00	4.4
12	6.1728	0.221	311	400	0.223	9.3	1.00	4.8
13	6.6872	0.212	382	491	0.222	10.2	1.01	5.3
14	7.2016	0.202	471	605	0.221	11.2	1.02	5.7

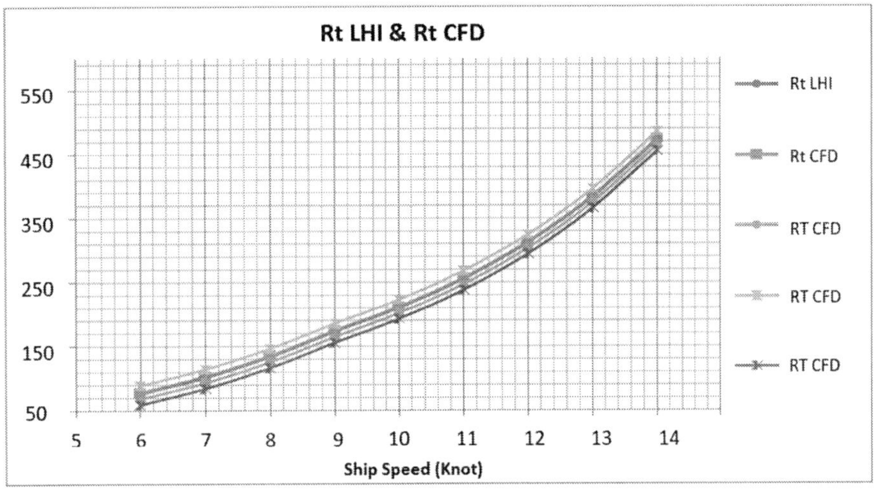

Figure 9. Graph of Resistance Value At Each Ship Speed

Based on the test results of the four hull models, it is known that the value of the resistance for the first hull model from the test at LHI has a value of 311 kN at a speed of 12 knots (service speed) while the comparison is that model 1 with the CFD method has a value of 313 kN. Comparison of the four hull models with the CFD method at service speed produces the following hull resistance value data: model one is 313 kN, model two has a hull resistance value. ship 303 model three has a hull resistance value of 324 kN and model four has a hull resistance value of 294 kN. Based on these data it can be concluded that the triple hull model has the highest hull resistance value.

3.3.2 Open Water and Relative Rotative Efficiency

The open water test efficiency value of the hull is a comparison between the output of the propeller in the form of thrust and the input power in the form of torque. In this research, only the value of the efficiency of the open water test from the LHI test was carried out with the open water test on the CFD software. As for the hull modification, using the same propeller as the model from the hydrodynamics laboratory (Table 6). Based on Figure 10, it is known that the comparison of tests in the hydrodynamics laboratory has a lower value when compared to testing on the computational fluid dynamic (CFD) method. Despite the lower value, the error limit for the two methods used is still tolerable, which is below 5%.

Table 6. Open Water and Relative Rotative Efficiency in LHI

VS	Vm (knot)	NM (m/s)	FN (Hz)	KT	KQ	KQ-Q	J	WT	ETA-O	ETA-R	N RPM	N RPS
6.0	0.564	9,605	0.073	0.1946	.02573	0.02627	0.445	0.241	0.525	1.021	105.2	1.753
7.0	0.657	11,099	0.085	0.1928	.02555	0.02609	0.449	0.241	0.529	1.021	121.6	2.027
8.0	0.751	12,738	0.097	0.1926	.02553	0.02606	0.450	0.237	0.529	1.021	139.5	2.325
9.0	0.845	14,437	0.110	0.1927	.02554	0.02608	0.450	0.232	0.529	1.021	158.1	2.635
10.0	0.939	15,986	0.122	0.1917	.02543	0.02597	0.452	0.230	0.531	1.021	175.1	2.918
11.0	1,033	17,605	0.134	0.1911	.02537	0.02591	0.454	0.227	0.533	1.021	192.9	3.215
12.0	1,127	19,426	0.146	0.1917	.02544	0.02597	0.452	0.221	0.531	1.021	212.8	3.547
13.0	1,221	21,425	0.158	0.1931	.02558	0.02611	0.449	0.212	0.528	1.021	234.7	3.912
14.0	1,315	23,663	0.171	0.1952	.02579	0.02633	0.444	0.202	0.523	1.021	259.2	4.320
										Mean	177.6 778	2.961

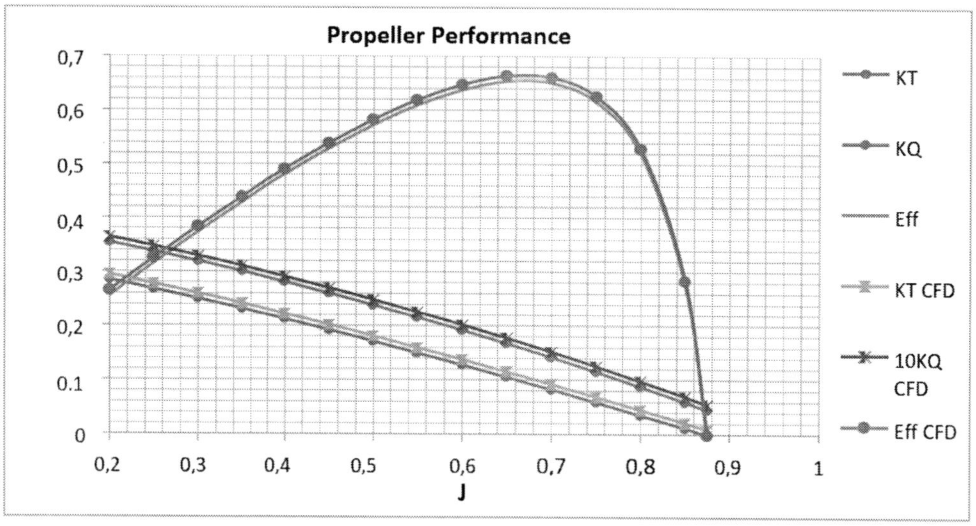

Figure 10. Comparison Chart of LHI and CFD Propeller Performance Testing

3.3.3 Quazy Propulsive Efficiency

The quazy propulsive coefficient (QPC) value is the total efficiency of the three efficiencies, namely hull efficiency, open water test efficiency and the last is relative rotative efficiency. This QPC value has a significant effect on how much delivery power will be converted into thrust to overcome ship resistance so that the ship can go according to planning (Table 7). Based on the test results of the four hull models (Figure 11), it is known that the value of the quazy propulsive coefficient for the first hull model from the test at LHI has an average value of 0.54; while the comparison is model 1 with the CFD method has an average value of 0.63.

The comparison of the four hull models with the same method, namely CFD (Table 8), produces an average data value of the quazy propulsive coefficient of model one, namely 0.63; model two has an average quazy propulsive coefficient value of 0.55; model three has an average quazy propulsive coefficient value of 0.63, and model four has an average quazy propulsive coefficient value of 0.51. Based on these data, it can be concluded that the triple hull model has the highest quazy propulsive coefficient value.

Delivery horsepower is the amount of power that is channeled through the shaft to the propeller to generate thrust to overcome the resistance value as a consequence of the movement of the ship moving according to speed. The DHP value is a comparison of the EHP (Effective Horse Power) divided by the Quazy Propulsive Coefficient (QPC). Based on the graph below, it can be concluded that even though the power transmitted to the ship's shaft is the same, the value of the thrust generated can vary, this depends on the value of propulsive (Figure 12).

Table 7. Quazy Propulsive Efficiency in LHI

η_H	η_O	η_{RR}	QPC
0.98	0.525	1.021	0.52
0.98	0.529	1.021	0.53
0.98	0.529	1.021	0.53
0.99	0.529	1.021	0.53
0.99	0.531	1.021	0.54
1.00	0.533	1.021	0.54
1.00	0.531	1.021	0.54
1.01	0.528	1.021	0.55
1.02	0.523	1.021	0.55
1.00	**0.528667**	**1.021**	**0.54**

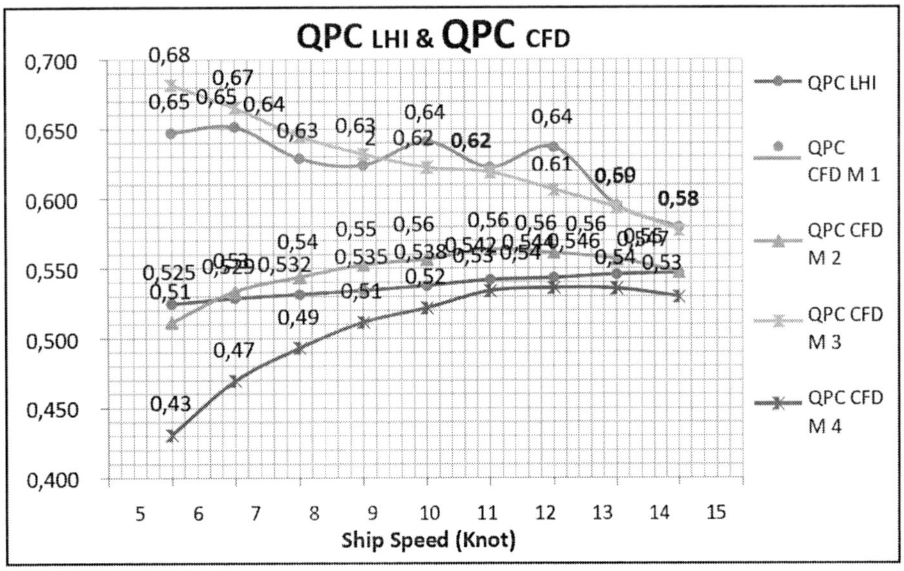

Figure 11. Quazy Propulsive Coefficient (QPC) Each Hull Model

Table 8. Comparison of QPC

Models	η_H	η_0	η_{RR}	QPC
M1	1.02	0.53	1.16	0.63
M2	0.97	0.53	1.07	0.55
M3	1.01	0.53	1.18	0.63
M4	0.94	0.53	1.02	0.51

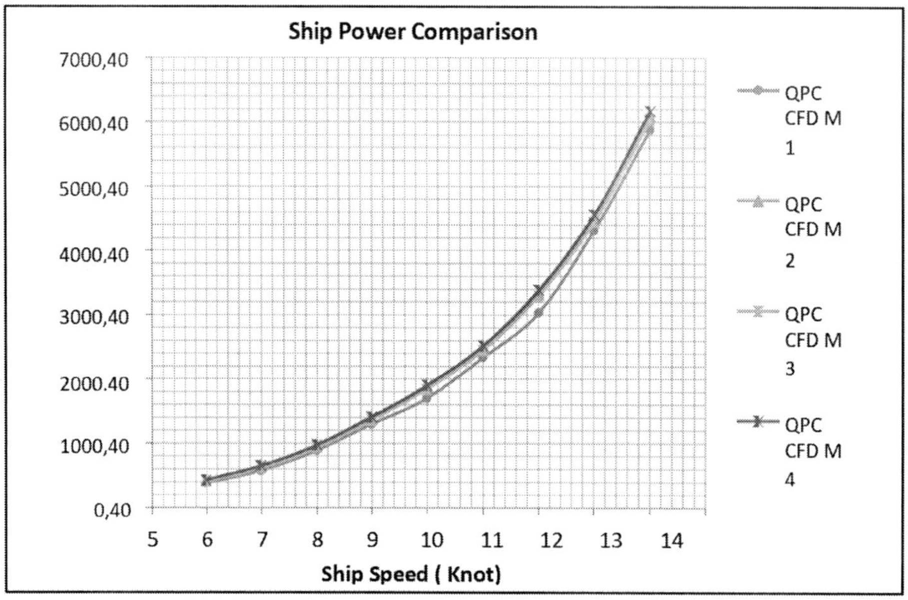

Figure 12. Ship Power Comparison

4. Conclusion

The accumulative values of the three efficiencies in this study are combined into a value commonly referred to as the quazy propulsive coefficient (QPC). This value represents how much power is available in the ship's propulsion system to become a thrust that can overcome the value of the ship's resistance so that the ship can go according to planning. The value of the QPC on the LHI test is 0.54; while testing using the CFD method on model one, model two, model three, and model four yields a value of 0.63; 0.55; 0.63; and 0.51. So it can be concluded that the triple hull model has the largest QPC value.

Acknowledgement

The authors are grateful to Institut Teknologi Sepuluh Nopember, Surabaya and our research partner Agency for the Assessment and Application of Technology - Hydrodynamics Laboratory Indonesia (BPPT-LHI).

References

[1] J. S. Carlton, *Marine Propellers and Propulsion. 2nd Edition.* 2012.
[2] D. A. Taylor, *Introduction to Marine Engineering. Revised 2nd Edition.* 1996.
[3] A. F. Molland, *Theoretical and Analytical Methods Relating to Propeller Action*. 2008.
[4] H. Schneekluth, and V. Bertram, *Ship Design for Efficiency and Economy. 2nd Edition.* 1998.
[5] D. G. M. Watson, *Practical Ship Design.* 1998.

IOP Conf. Series: Materials Science and Engineering 1052 (2021) 012057 doi:10.1088/1757-899X/1052/1/012057

[6] L. C. Burrill, "Propellers in Action Behind a Ship", *Transactions of The North East Coast Institution of Engineers And Shipbuilders,* vol. 76, pp. 25–34, 1996.
[7] C. Gallin,. , H. Hiersig, dan Heiderich,O. *Ships and Their Propulsion Systems – Developments in Power Transmission.* 1983.
[8] J.A. Holtrop, Statistical Re-Analysis Of Resistance And Propulsion Data .1984.
[9] Kim, J., Park, I.R., Van, S.H., dan Park, N.J., "Numerical Computation For The Comparison Of Stern flows Around Various Twin Skegs", *Journal of Ship and Ocean Technology*, Vol. 10, No. 2. 2006.
[10] W.P.A. Lammeran, , dan Van, N.D. *Resistance, Propulsion and Steering of Ship.* 1962.
[11] E.V. Lewis, *Principles of Naval Architecture. The Society of Naval Architects and Marine Engineers.* 1989.
[12] A.F. Molland, *The Maritime Engineering Reference Book.* 2008.

IOP Conf. Series: Materials Science and Engineering 1052 (2021) 012058 doi:10.1088/1757-899X/1052/1/012058

Improved PID Controller for DC Motor Control

Mohammed Ahmed[1], Nura Musa Tahir[2], Aminu Yahaya Zimit[3], Musa Idi[2], Kamal Abubakar Abubakar[4], Salihu Abdulmumini Jalo[5]

[1]Department of Electrical and Electronics Engineering, Abubakar Tafawa Balewa University, P. M. B. 0248 Bauchi, Nigeria.
[2]Department of Mechatronics and System Engineering, Abubakar Tafawa Balewa University, P. M. B. 0248 Bauchi, Nigeria.
[3]University of Hafr Albatin Housing, Saudi Arabia.
[4]Department of Electrical and Electronics Engineering, Nigerian Army University, Biu, Nigeria.
[5]Department of Electrical and Electronics Engineering, Adamawa State Polytechnic Yola, Nigeria.

inunugoloma@yahoo.com

Abstract. Presented is an improved version of the PID controller (PIDC) for DC motor control. The PIDC is a very cheap and easy controller to implement and is currently in use in many systems. There are instances whereby higher precision is required and an improved version of the PID can serve. The systems might not be necessary limited to marine systems. Hence, saving the need for more costly redesign. The PIDC proposed was compared with the PID. Results show that the PIDC has superior performance compared to the PID. Hence, it is an indication that the PID could still withstand future developments; that is maintaining its characteristics of simplicity and cheapness.

1. Introduction

Electric motors have a wide range of applications as results of remarkable improvements in technology. Some the areas include positioning in large telescopic antennas, tracking of radar systems, aircraft systems actuators, marine systems actuators and robotic systems [1-3]. Achieving desired objectives sometimes only require simple and cheap designs.

Rubaai and Kotaru [4] and Benard [5] explored the artificial neural networks for direct current (DC) motor control. In the works of Hameed and Mohamed [6], the fuzzy logic control technique was combined with the artificial neural network counterpart was applied. The fuzzy logic technique was investigated by Sadiq et al in their studies [7-9]. Rajesekhar et al. [10, 11], utilised the fractional-order proportional integral derivative (PID) along with the artificial bee colony algorithm in their works. In the studies conducted by Mishra et al., the proportional-integral (PI) control method was harnessed for motor control. The PID control methods have the advantage of simplicity, easy implantation and also a lot was known about it. Despite all these, some systems require better performance which cannot be attained with the ordinary PID control scheme. Hence, the need enhanced version of the PID [12]. The study was on exploring the reaching law for enhancing the PID controller performance which is referred to as the PIDC for the armature controlled direct current (DC) motors.

2. Methodology

2.1. Plant modelling

The system plant is an armature controlled DC motor. Equations (1) – (4) give the motor transfer function. The parameters; are the motor torque constant K_T, total moment of inertia for the motor system J, armature inductance of the motor L_a, armature resistance of the motor R_a, total viscous friction B and back e.m.f constant K_b. The values of the aforementioned parameters are: 0.085 Nm, 5.0 x 10^{-4} kgm^2/s^2, negligible, 5 ohms, 6.0 x 10^{-3} Nm/sec and 0.1 V/rad/sec respectively.

$$G(s) = \theta(s)/E_a(s) \quad (1)$$

$$G(s) = K_T/(As^3 + Bs^2 + Cs) \quad (2)$$

$$A = JL_a; B = JR_a + BL_a \quad (3)$$

$$C = BR_a + K_bK_T \quad (4)$$

Equation (5) resulted after substituting the values. Replacing the armature voltage E_a with the control signal, u made equation (6) emerges. The plant mathematical model in the canonical form is as given by Equation (7).

$$\theta(s)/E_a(s) = 34/(s^2 + 15.4s) \quad (5)$$

$$\theta/u = 34/(s^2 + 15.4s) \quad (6)$$

$$\ddot{\theta} = 34u - 15.4\dot{\theta} \quad (7)$$

2.2. The Improved PID Control (PIDC) System

Equation (8) is the reaching law and equation (9) is the sliding mode function. Replacing left-hand side of (8) with the control signal and f with s in the sgn function and negating the right-hand side, and inserting (9) after the operation resulted in (10). Equation (10) now has the property of the proportional and derivative of the PID controller. Equation (11) resulted after adding the integral component in equation (10). The improved PID control law was as given by (11). Where: e is the tracking error, ė its derivative, G is the proportional constant, H is the derivative constant, M is the integral constant, Sedt refers to the integral of the tracking error, f is chosen as the desired response of the system, D is a constant and has to be greater than unity.

$$\dot{s} = -D * sgn(s) - Ks \quad (8)$$

$$s(t) = ce(t) + \dot{e}(t) \quad (9)$$

$$u = D * sgn(f) + Kce + K\dot{e} \quad (10)$$

$$u = D * sgn(f) + Ge + H\dot{e} + MSedt \quad (11)$$

The system with the improved Proportional Integral Derivative controller (PIDC) compared with the system implemented with the normal Proportional Integral Derivative (PID) control method.

3. Results

The presentation of results was made in an orderly manner. The first presented result was without any disturbance effect. Next was with the positive pulse effect, followed by with the negative pulse effects. These try to simulate what happens when the armature voltage suddenly surges and in the second case sudden dip. It was followed with the dynamic friction reduced to 30% and then increase to 110%; in reality, the parameter is not fixed it changes. An effort was made to have a replica of real operating conditions. The next was on reduction of the system combined moment of inertia to 30% and then increased by 110%; which was an effort to see what happens to the system due to changes loading. Finally, the presentation of the results was closed with its summary.

3.1. Unit step response without disturbance

Figure 1a is the unit step performance of the system without disturbance effect. Table1 gives a tabulation of the response. Figure 1b was the error plots of the system during the transition.

Table 1. Unit step response without disturbance.

Parameters	**Controllers**	
	PID	***PIDC***
Overshoot (%)	8.0000	0.0000
Rise time (s)	0.0700	0.1000
Settling time (s)	0.2000	0.2000

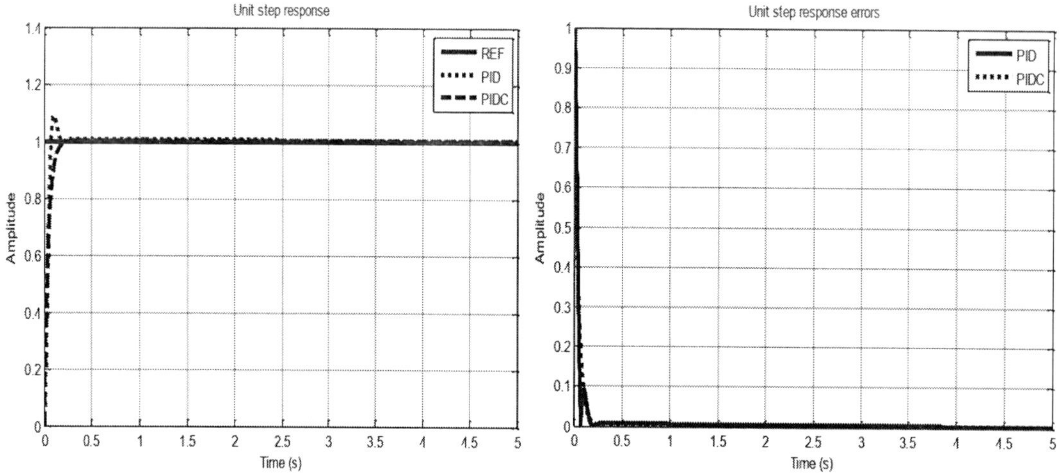

Figure 1. Unit step response and error graphs without disturbance.

The results from the Table 1 and Figure 1a show that the system response with the PIDC was the best. It has no overshoot, with a rise time of 0.1s and 0.2s settling time. The system having the PID was characterized with an overshoot of 8% overshoot, a rise time of 0.07s, settling time of 0.2 and also a steady-state error of about 1.5% lapsing up to 2.5s response duration.
It can be seen from Figure 1b the system with the PIDC has less error accumulation than with PID. The error is the difference between the reference and the response obtained. All the errors of the system with both control scheme diminish with time which implied high stability tendency of the system [13-15].

3.2. Unit step response with pulse disturbance (positive)

The response with positive pulse disturbance was as given by Figure 2a and Table 2. The disturbance (DIS) appeared between 2.5 to 3.5s of the system response. Figure 2b is the graphical presentation of the errors that occurred.

Table 2. Unit step response with pulse disturbance (positive).

Parameters	**Controllers**	
	PID	***PIDC***
Overshoot (%)	8.0000	0.0000
Rise time (s)	0.0700	0.1000
Settling time (s)	0.2000	0.2000
Compensation ability (%)	93.0000	100.0000

IOP Conf. Series: Materials Science and Engineering 1052 (2021) 012058 doi:10.1088/1757-899X/1052/1/012058

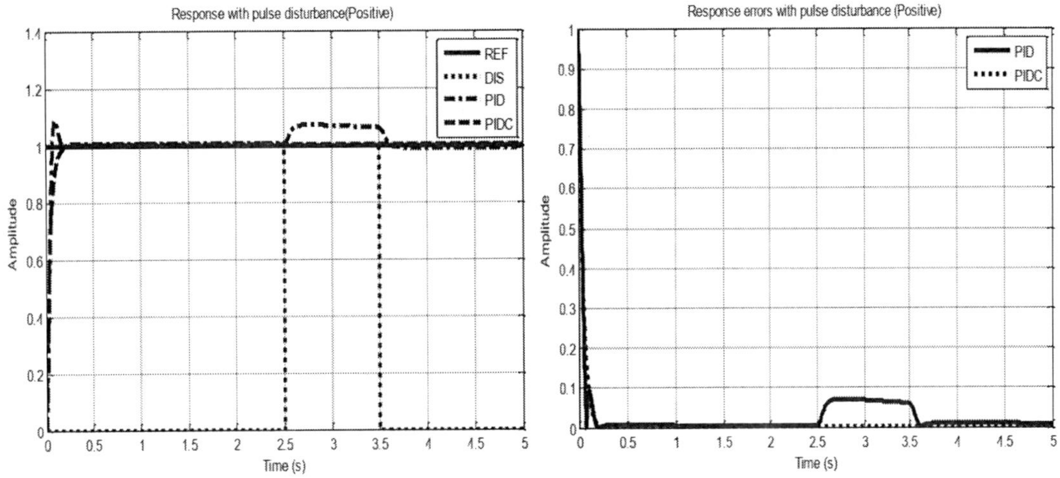

Figure 2. Unit step response and error plots with positive pulse disturbance.

Figure 2a give the system response and was tabulated in Table 2. The results are similar to without disturbance except that it can be seen that with the PIDC there was a total compensation of the disturbance whereas with the PID it was annulled up to 93%.

Figure 2b shows the errors of the DC motor system. The system designed with the PIDC has by far lesser amount of error compared to with the PID. The appearance of the pulse disturbance makes the system amount of error higher than without it as presented earlier, while it remains the same with the PIDC. This ability makes the PIDC system response better than with the PID. The errors all diminish with time, hence the chances of the systems being stable are high.

3.3. Unit step response with pulse disturbance (negative)

The system performance response with the negative pulse disturbance was as given by Figure 3a and Table 3. The disturbance (DIS) also lapses between 2.5 to 3.5s and Figure 3b are the errors plots of the system.

Table 3. Unit step response with pulse disturbance (negative)

Parameters	**Controllers**	
	PID	***PIDC***
Overshoot (%)	8.0000	0.0000
Rise time (s)	0.0700	0.1000
Settling time (s)	0.2000	0.2000
Compensation ability (%)	93.0000	100.0000

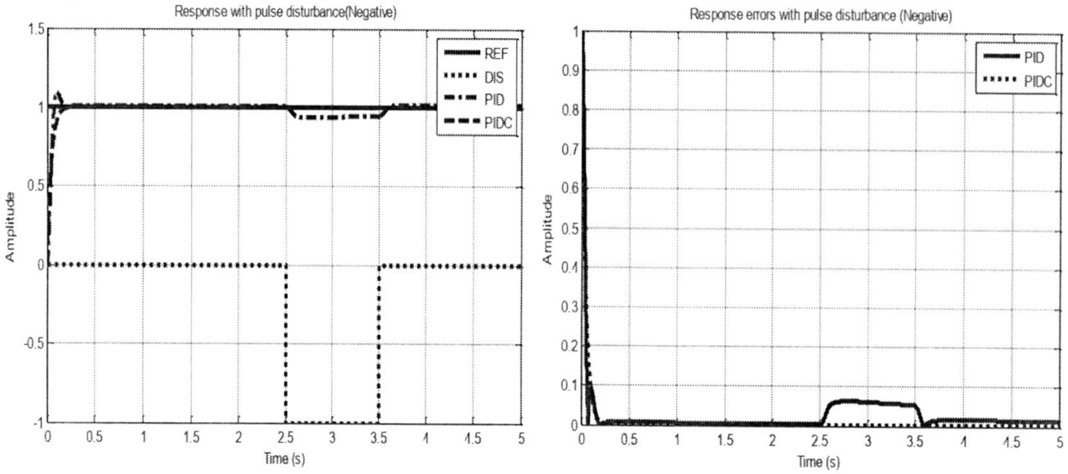

Figure 3. Unit step response and plot of errors with negative pulse disturbance.

Table 3 and Figure 3a showed the tabulated results obtained which are similar to, with the positive pulse disturbance. Except that the pulse direction is opposite.

The error plots are as given by Figure 3b. They are also similar to the ones obtained with the positive pulse disturbance.

3.4. Unit step response with B reduced to 30%

The effect of reduction of the system dynamic friction (*B*) to 30% on the performance were as given by Figure 4a which were then tabulated in Table 4. Figure 4b give the errors aroused in the process.

Table 4. Unit step response with *B* reduced to 30%.

Parameters	**Controllers**	
	PID	***PIDC***
Overshoot (%)	39.0000	0.0000
Rise time (s)	0.0500	0.1000
Settling time (s)	0.3500	0.2000

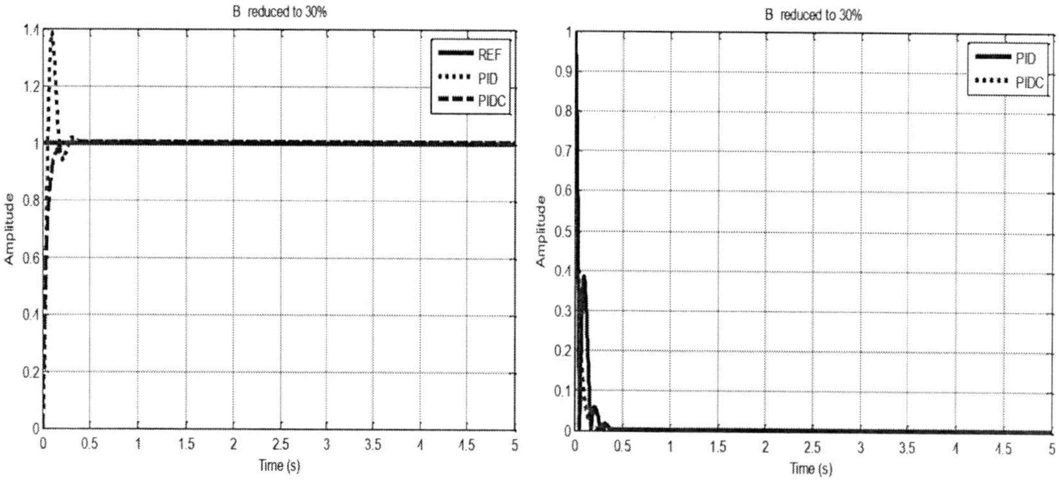

Figure 4. Unit step response and errors portrait with B reduced to 30%.

The results in Table 4 and Figure 4a showed that the system with the PIDC has a better result. It maintained its parameters as the case without disturbance effect. The system implemented with the PID has its performance worsens. It resulted in 39% overshoot, settling time of 0.35s and also has a steady-state error of about 1% throughout.

In terms of the errors resulted, the system with the PIDC has the least which was same as without any disturbance. It worsens for the system with the PID as illustrated by Figure 4b. The dying-out nature of the error graphs indicates the possibility of the systems to be stable.

3.5. Unit step response with B increased by 110%

The results achieved for the system with its dynamic friction (B) added by 100% as illustrated by Figure 5a. Table 5 shows the results in tabular form and Figure 5b portrays the resulted errors of the system.

Table 5. Unit step response with B increased to 110%.

Parameters	Controllers	
	PID	***PIDC***
Overshoot (%)	0.0000	0.0000
Rise time (s)	0.0500	0.1000
Settling time (s)	0.2500	0.2000

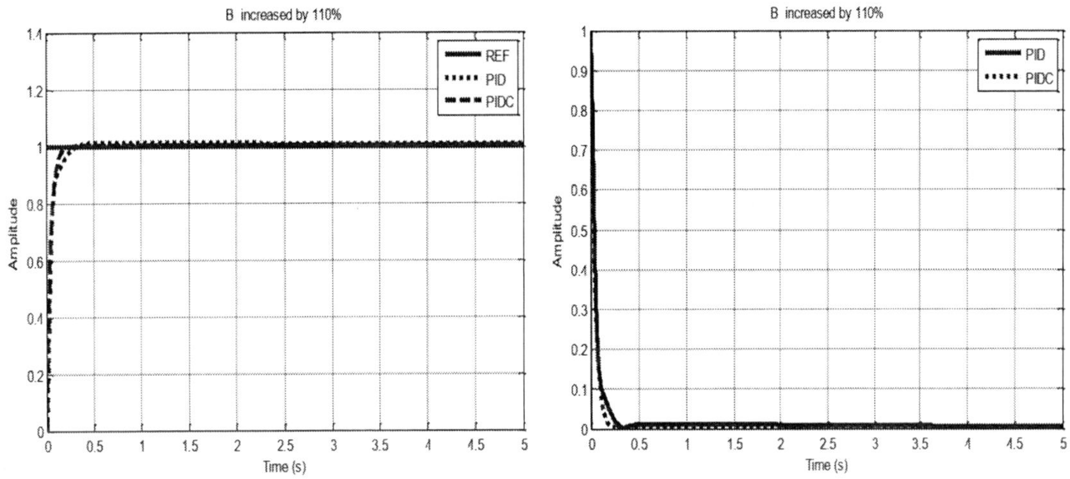

Figure 5. Unit step response and errors graph with B increased by 110%.

Figure 5a and Table 5 are illustrations of the system performance. The system response, with the PIDC, was the same as without any disturbance effect and with B reduced to 30%. Here as well it was better than with PID. Although there is a bit of improvement with PID compared to when the B was 30% in terms of the other parameters, the 1.5% steady-state error still exist.

The PIDC maintain the lead in terms of the amount of error resulted as shown in Figure 5b. The error of the system with the PID reduced compared to the previous case. The resulted errors also diminished with time. Hence, indicating the likelihood of system stability.

3.6. Unit step response with J reduced to 30%

Presented in Table 6 was the system performance obtained with the combined moment of inertia reduced by 50% which ties to perceive what happens when the loading effect suddenly decreases by close to 50% on the motor system. Figure 6a also shows the response while Figure 6b is the resulting system errors.

Table 6. Unit step response with *J* Reduced to 30%

Parameters	Controllers	
	PID	*PIDC*
Overshoot (%)	0.0000	0.0000
Rise time (s)	0.0500	0.1000
Settling time (s)	0.2000	0.2000

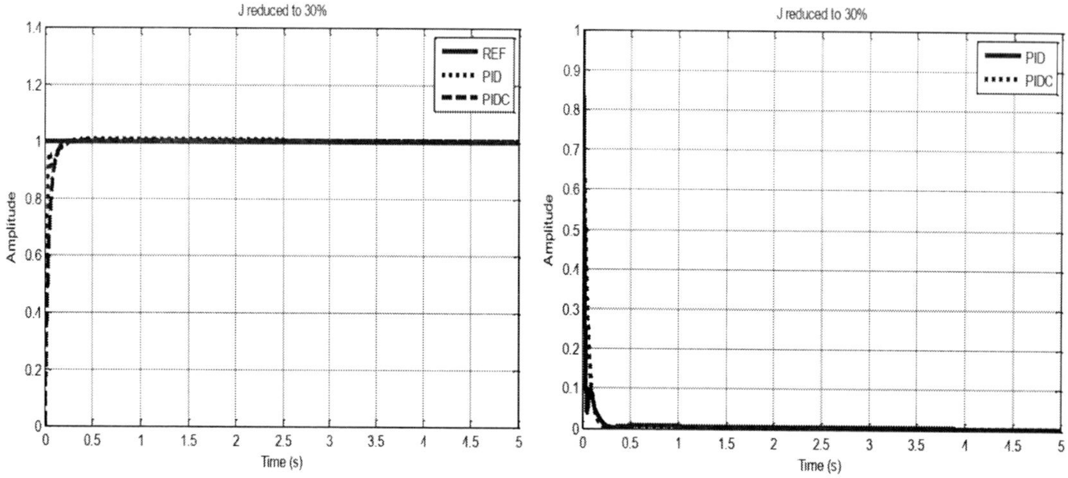

Figure 6. Unit step response and plot of errors with J decreased to 30%.

The system performance as given by Figure 6a and Table 6. The overshoot, rise time and settling time for the PIDC remained as 0%, 0.1s and 0.2s respectively, which are 0%, 0.05s and 0.2s with the PID. The system with the PID still has the 1.5% steady-state error which in this case diminishes to about 1% around 2.5s and continues to the end.

Regarding the errors as shown in Figure 6b. It was same for the PIDC and it was lesser with the reduced J and all subsided with time, hence, an indication of stability.

3.7. Unit step response with J increased by 110%

The response with the moment of inertia raised by 110% is as given by Table 7 and Figure 7a. Figure 7b was the graphs of the errors.

Table 7. Unit step response with *J* Increased by 110%

Parameters	Controllers	
	PID	*PIDC*
Overshoot (%)	20.0000	0.0000
Rise time (s)	0.1000	0.1000
Settling time (s)	0.3000	0.2000

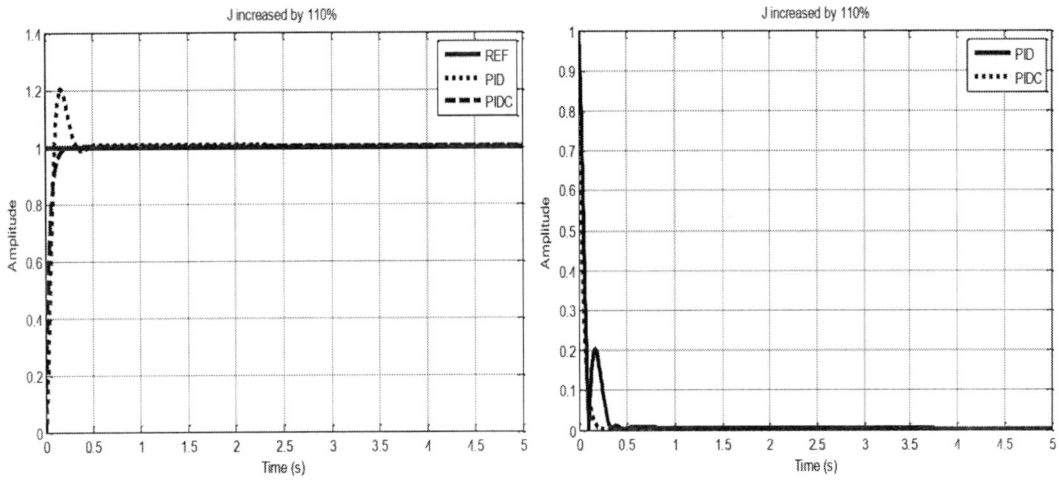

Figure 7. Unit step response and errors plot with J increased by 110%.

Figure 7a shows the system response and its summary in Table 7. The system with the PIDC was better. Performance of the system with the PID worsened in terms of overshoot and settling time compared with reduced *J* to 30%.

The error plots were as given in Figure 7b. The PIDC has the least amount of error. The errors here also all die-down with time portraying high probability of the system stability.

3.8. Summary of results

The system was subjected to different test and results revealed the high robustness of the PIDC. Despite the varying conditions the system implemented using the PIDC maintained the same results throughout which was the same as without any changes. Some level of robustness was shown by the system with PID which changes as the level of disturbance changes. An insight into the system stability was the dying-away of the error signals under all conditions. Hence, the proposed control scheme has portrayed the ability to remarkably improve the motor control system.

4. Conclusion

The improved PID controller PIDC was successfully applied for control of an armature controlled DC motor. It was evaluated most likely associated disturbances referred to as the positive pulse effect and the negative pulse effects; these were to simulate what happens when the armature voltage suddenly increases and in the latter case sudden decrease. Others are the dynamic friction reduction to 30% and increase to 110%; to simulate changes in the dynamic friction which is, in reality, is not fixed it changes. Next was the reduced system moment of inertia to 30% and then increased by 110%; which was an effort to simulate what happens to the system due to changes loading. Even though different conditions are imposed on the system, the system implemented with the PIDC maintained the same results throughout which was the same as without any disturbance injected. Thus, a high level of robustness of the PIDC was portrayed. A certain level of robustness was shown by the system with PID which changes depending on the introduced perturbation. Information on the system stability was revealed as the error die away with time under all examined conditions meaning that the system has a high chance of being stable under the conditions. Therefore, the presented control method has indicated the ability to remarkably improve the armature controlled motor control system. It may not be limited to such a system as further works could reveal other possibilities. Hence, it indicated that the PID could still withstand future developments while maintaining its characteristics of being simple and cheap. The work is novel in the sense that to best of the authors there no such proposed.

References

[1] J. Fu, J.-C. Maré, and Y. Fu, "Modelling and simulation of flight control electromechanical actuators with special focus on model architecting, multidisciplinary effects and power flows," *Chinese Journal of Aeronautics,* vol. 30, pp. 47-65, 2017.

[2] B. Drury, *Control techniques drives and controls handbook*: IET, 2001.

[3] H. Lendenmann, R. R. Moghaddam, A. Tammi, and L.-E. Thand, "Motoring ahead," *ABB review,* 2011.

[4] A. Rubaai and R. Kotaru, "Online identification and control of a DC motor using learning adaptation of neural networks," *IEEE Transactions on industry applications,* vol. 36, pp. 935-942, 2000.

[5] A. Bernard, "Speed control of separately excited DC motor using artificial intelligent approach," Masters, Universiti Tun Hussein Onn Malaysia, 2013.

[6] W. I. Hameed and K. A. Mohamad, "Speed control of separately excited dc motor using fuzzy neural model reference controller," *International Journal of Instrumentation and Control Systems (IJICS) Vol,* vol. 2, pp. 27-39, 2012.

[7] A. Sadiq, G. Bakare, E. Anene, and H. B. Mamman, "A fuzzy-based speed control of DC motor using combined armature voltage and field current," *IFAC Proceedings Volumes,* vol. 46, pp. 387-392, 2013.

[8] A. Sadiq, H. Mamman, and M. Ahmed, "Field current speed control of direct current motor using fuzzy logic technique," *International Journal of Information and Computation Technology,* vol. 3, pp. 751-756, 2013.

[9] A. Sadiq, H. Mamman, and I. Musa, "Fuzzy Speed Control of Separately Excited DC Motor Using Armature Voltage Control Method," in *The International Conference on Digital Information Processing, E-Business and Cloud Computing (DIPECC)*, 2013, p. 225.

[10] A. Rajasekhar, S. Das, and A. Abraham, "Fractional order PID controller design for speed control of chopper fed DC motor drive using artificial bee colony algorithm," in *2013 World Congress on Nature and Biologically Inspired Computing*, 2013, pp. 259-266.

[11] A. Rajasekhar, R. K. Jatoth, and A. Abraham, "Design of intelligent PID/PIλDμ speed controller for chopper fed DC motor drive using opposition based artificial bee colony algorithm," *Engineering Applications of Artificial Intelligence,* vol. 29, pp. 13-32, 2014.

[12] C. Lynch and M. Popovic, "Closed-loop control for FES: Past work and future directions," in *10th Annual Conference of the International FES Society*, 2005, pp. 2-4.

[13] N. S. Nise, *CONTROL SYSTEMS ENGINEERING, (With CD)*: John Wiley & Sons, 2015.

[14] J.-J. E. Slotine and W. Li, *Applied nonlinear control* vol. 199: Prentice hall Englewood Cliffs, NJ, 1991.

[15] H. K. Khalil, "Nonlinear systems, 3rd," *New Jewsey, Prentice Hall,* vol. 9, 2002.

Performance of Rescue Boat Operation when Operated in Waves

I K A P Utama[1], M H N Aliffrananda[1], A Sulisetyono[1], D Setyawan[1], I K Suastika[1], D Utama[1], W D Aryawan[1] and Y A Hermawan[1].

[1]Department of Naval Architecture, Institut Teknologi Sepuluh Nopember, Indonesia

Abstract. Search and rescue operations, especially in ship accidents, require speed and reliability from the rescue vessels in facing the challenges of extreme weather and sea waves. A rescue boat hydrodynamic performance study is very important to do to determine the ship's ability to withstand dangerous conditions while operating. Therefore, a rescue boat design is needed that is capable of high acceleration and can penetrate various extreme sea condition. Three rescue boat hull models have been developed to obtain the hull model with the best seakeeping performance. A numerical arrangement based on computational fluid dynamics (CFD) techniques is used to predict seakeeping behaviour from three rescue boat hull model in wave. Three-dimensional diffraction panel method uses as a CFD solver method to predict the seakeeping performance of rescue boat high-speed operation in wave. From the study results, it was found that the first rescue boat hull model had a good seakeeping performance compared to other hull models.

1. Introduction

The National SAR Agency as an institution that is responsible for search and rescue problems needs to complete the need for adequate marine facilities in the form of search and rescue vessels capable of operating in accordance with the Indonesia Search Rescue Region (Indonesia SRR). Rescue boats are also needed a high-speed craft that are available in each port that have function as rescue ships. A design for a rescue boat that has high acceleration and can penetrate various extreme sea conditions. This is important because the rate of ship accidents in Indonesia is quite high. The Minister of Transportation of the Republic of Indonesia, through the Instruction of the Minister of Transportation of the Republic of Indonesia Number IM 10 of 2018, is to carry out security and order in the Work Environment and Port Interests for users of transportation modes and to coordinate with related agencies. This instruction is to respond to the high level of ship accidents in Indonesia.

Rescue Boat is necessary so as not to cause a lot of loss and casualties on sea transportation. Rescue boat itself is a rescue ship which is usually used during MOB or Man Overboard events (people who fall from the boat and float in the sea, or during natural disasters such as water overflows due to flooding in certain areas, etc.). Rescue and handling of shipping accidents must be carried out effectively and efficiently so as not to cause significant losses.

The speed and weight of the ship greatly affect the work of the ship when it is on the waves [1]. Likewise, the wave characteristics consisting of height and period and direction of wave propagation determine the response of motion on the ship. Motion responses such as rolling, pitching, and heaving will ultimately affect the comfort and safety of the crew and the goods being transported. Based on the explanation above, this study was conducted to evaluate and analyze the response of high-speed vessel motion in various conditions in waves based on the conditions or the character of the voyage, especially in the response of the rolling, heaving, and pitching motion of the ship [2].

2. Rescue boat model

The simulation is done by modeling the hull of the rescue boat. Three prototype hull models are analyzed. The three hull models of the rescue boat have different hull characteristics from one another. **Table 1** shows the principal dimension of the hull model. It can be seen form the picture that the first model ship has a length of 4.56 meters and a displacement of 936.4 kg, the second model has a length of 5,296 meters with a displacement of 1414 kg, and the third model has a length of 3,758 meters with a displacement of 904 kg. The second model has the largest physical size, and the third model has the smallest physical size. The three hull models will operate at an operating speed of 20 knots.

Table 1 Principal Dimension of Rescue Boat Hull Model.

Parameter	Unit	Model 1	Model 2	Model 3
LOA	m	4.56	5.296	3.758
LWL	m	4.3	4.756	3.559
B	m	1.52	1.815	1.363
T	m	0.35	0.396	0.3
H	m	0.92	1.035	0.607
Volume Displacement	m^3	0.936	1.414	0.904
Displcement	kg	936.4	1414	904

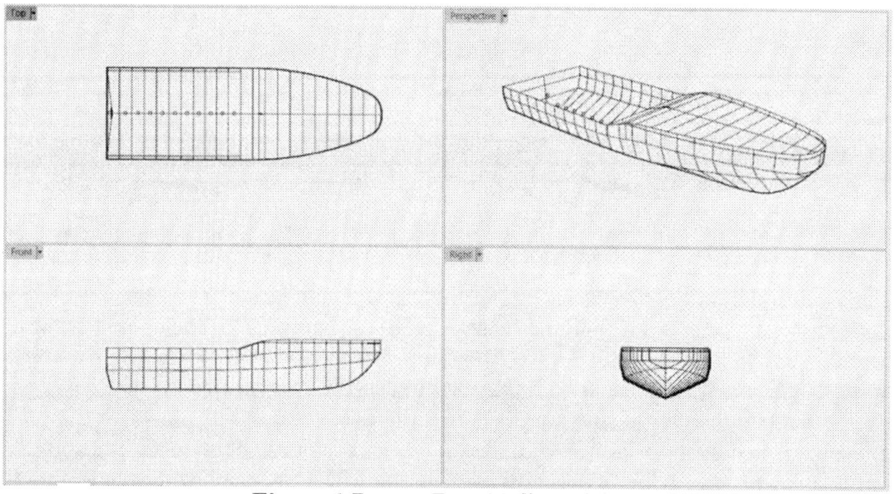

Figure 1 Rescue Boat Hull Model 1.

Figure 1 shows the hull of the first rescue ship model. It has a V-shaped hull with a deadrise angle of up to 30°. This hull model does not have a chine in its hull, so it is a full V-shaped hull. Deadrise with a large angle can reduce the impact load that occurs when slamming and make it easier for the hull to break the incoming waves [3].

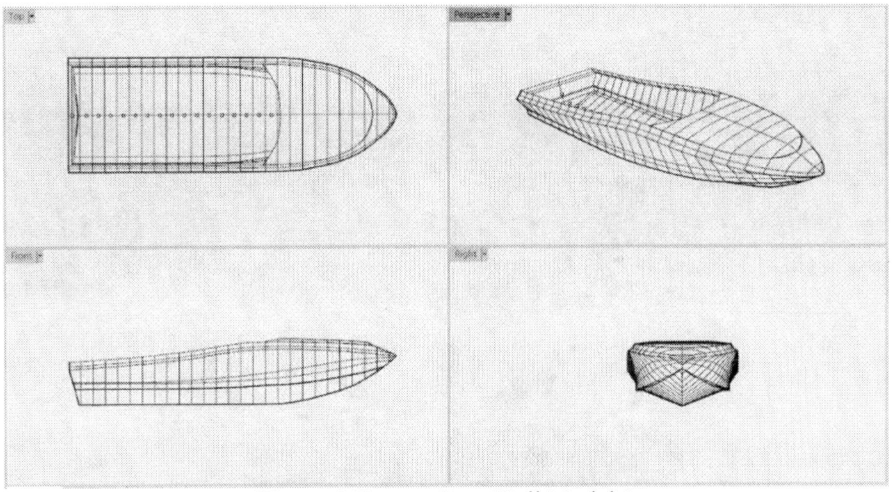

Figure 2 Rescue Boat Hull Model 2.

Figure 2 shows the hull of the second model rescue boat. This hull model has a V-shaped hull type with a deadrise angle of 20°. This hull model has a chine that serves to increase hydrodynamic lift when the ship is traveling at high speed [4]. The second hull model has a shape like a high-speed pleasure craft with a sharp bow that can increase difficulty in production. The shape of the hull is not simple because there are many curves in its hull when compared to the first hull model.

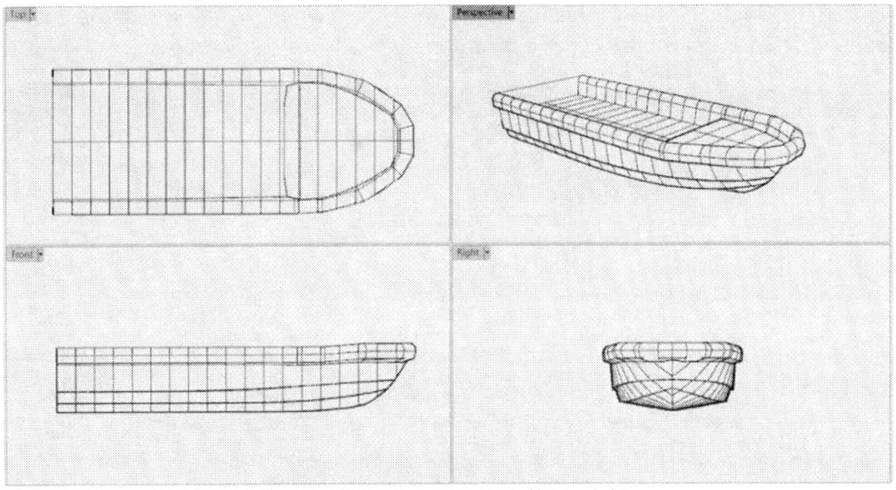

Figure 3 Rescue Boat Hull Model 3.

Figure 3 shows the hull of the third model rescue boat. The third hull model also has a V-shaped hull, but with a much smaller deadrise angle of 15°. The small deadrise angle provides a large enough impact load when slamming occurs and has the potential to accelerate the aging of the construction at the bottom area [3]. The third model is also equipped with a chine on the hull to increase hydrodynamic lift when operating. The shape of the hull is simple and easy to manufacture and has a less curve shape in its hull.

3. Mathematical model and computing method

CFD solver three-dimensional diffraction panel method uses for seakeeping analysis. The three hull models are discretized so that they are arranged into small panels that compose the shape of the hull.

IOP Conf. Series: Materials Science and Engineering 1052 (2021) 012059 doi:10.1088/1757-899X/1052/1/012059

The three-dimensional Panel method can predict motion and wave load that occurs in the hull of a ship with forwards speed on regular waves [5].

3.1. Computational Grid

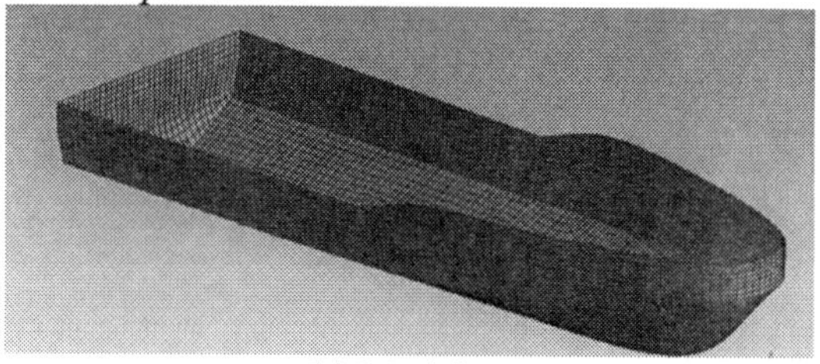

(a) Model 1

(b) Model 2

(c) Model 3

Figure 4 Computational Grid Uses for Motion Analysis.

Figure 4 shows the mesh used. A total of approximately five thousand panel elements are used to model the shape of the hull which is used as a model for ship seakeeping analysis. Estimates of the number of elements used are obtained from the results of the grid independence study which can be seen in Figure 5. There is a difference or error of 2.2% in the grid independence study process.

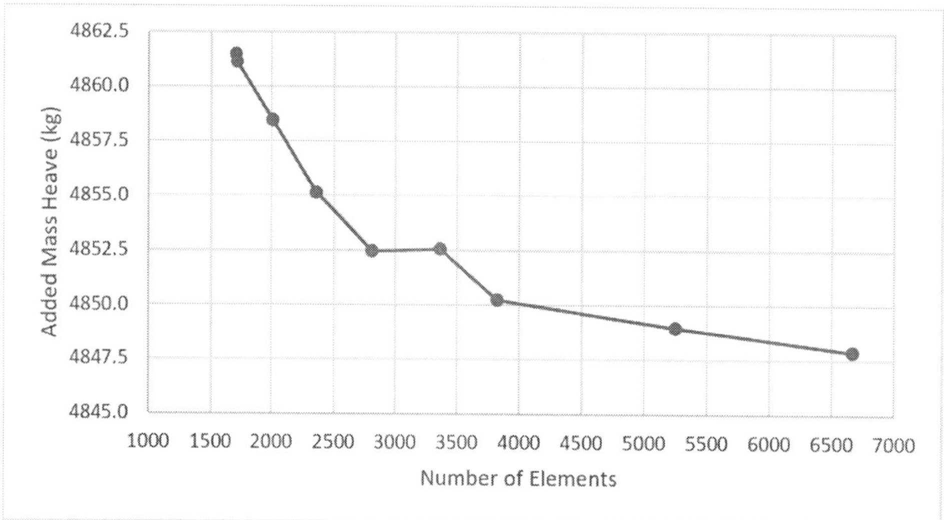

Figure 5 Grid Independence Study of Total Element Used.

3.2. Equations of motion on floating structures

The equation for the motion of a floating structure can be analyzed using Newton II's law:

$$\sum F = mi\xi i \tag{1}$$

Where i translational and rotational modes of motion in structures, $\sum F$ represent resultant of force, m is structural mass in kg and ξ is structure acceleration.

The equation above shows a relationship between the resultant force and mass and the acceleration of the structure. The resultant force in the floating building is the result of the reduction of the action force with the reaction force. The action force referred to in this case is the excitation force received by the structure due to the uneven pressure distribution of the wave. The impact of this uneven pressure is called the Froude-Krillof force [6]. Besides, the presence of large structures will also cause changes in the pressure distribution. The impact of a change in the pressure distribution due to a large structure is called the diffraction force. The amount of excitation force will vary with changes in time. While the reaction force referred to here is the force that opposes the action force. The following is an equation that describes the relationship between the resultant force and the action and reaction force:

$$\sum F = F_{eksitasi} - F_{reaksi} \tag{2}$$

The reaction force is influenced by added mass, damping, and stiffness. The following is an explanation of each of these components:

1. Added mass
   Floating buildings that move due to waves will cause their displacement to change and push the fluid particles around them in a certain mass. This is considered to increase the mass of the object so it is called added mass. The added mass depends on the shape of the object and its mode of motion. The hydrodynamic force that occurs due to added mass is an inertia force (Fa). Here is the formula for the inertia force:
   $$F_a = (m + a)i\xi i \tag{3}$$
   Where a is added mass in kg.
2. Damping
   Damping is the impact of the energy dissipation of the structure. 'Even lost due to rubbing against the surface of the medium. In other words, if the structure is vibrated in calm water, the kinetic energy of the structure (movement) will decrease over time, this is called damping. The damping value is the same as the added mass, depending on the shape of the object and the mode of motion of the object. The following is the equation for the damping force
   $$F_b = bi\xi i \tag{4}$$
   Where a is system damping in N m/s.
3. Stifness
   Floating structures that move on the surface of calm water without waves will have the rigidity to return to the initial position which is influenced by the hydrostatic characteristics of the structure. Where this is called a restoring force. This phenomenon only occurs in three modes of structural motion, namely heave, roll, and pitch [7]. This is because the other modes of motion do not have natural frequencies which technically do not have their stiffness mechanism manifesting as return force. Here is the return style equation:
   $$F_c = ci\xi i \tag{5}$$
   Where c is structure stifness in N/m.

Based on the explanation above, equations 3, 4, and 5 can be substituted into equation 2 to become:

$$\sum F = F_{eksitasi} - (F_a + F_b + F_c) \tag{6}$$

By substituting equation 1 into equation 6, then equation 7 is obtained as follows:

$$mi\xi i = F_{eksitasi} - (F_a + F_b + F_c) \tag{7}$$

$$F_w = (mi + ai)\ddot{\xi}i + bi\dot{\xi}i + ci\xi i \tag{8}$$

Where ξ is position response of structure motion in m, $\dot{\xi}$ is velocity of structure in m/s, and $\ddot{\xi}$ is acceleration of structure motion in m/s^2.

3.2.1. Response Amplitude Operator

Response Amplitude Operator (RAO) is called the transfer function. It is a response function that transfers the wave force into a dynamic response to the structure in the frequency range. The RAO graph contains the frequency parameter in its abscissa and the ratio between the amplitudes of the motion of a particular mode in its ordinate.

According to [7], RAO in translational motion in formulated as follows:

$$RAO\ (\omega) = \frac{\zeta_{k0}(\omega)}{\zeta_0\ (\omega)}\ (m/m) \tag{9}$$

Where $\zeta_{k0}(\omega)$ is structure amplitude in meter and $\zeta_0\ (\omega)$ is wave amplitude in meter.

Meanwhile, RAO on rotational motion which is the ratio between the amplitude of the rotational motion and the slope of the wave, namely the multiplication of the wave number and the wave amplitude ($k_w = w^2 / g$) is formulated as follows:

$$RAO\ (\omega) = \frac{\zeta_{k0}(\omega)}{\zeta_0\ (\omega)}\ (rad/m) \tag{10}$$

$$RAO\ (\omega) = \frac{\zeta_{k0}(\omega)}{\left(\frac{\omega^2}{g}\right)\zeta_0}\ (rad/m) \tag{11}$$

3.2.2. Wave Spectra

The wave spectrum used has 2 parameters, namely significant wave height (Hs) and zero-up crossing period (Tz). The spectrum can be represented in the JONSWAP spectrum. From data collected at the Joint North Sea Wave Observation Project or simply JONSWAP, Hasselmann found that there was an additional factor in the previously developed spectrum, namely Pierson-Moskowitz. Thus, the JONSWAP spectrum is the Pierson-Moskowitz spectrum multiplied by the peak enhancement factor γ^r [8]. The Pierson-Moskowitz spectrum can be seen in equation 12 below [9].

$$S_{(PM)} = \frac{Hs^2}{4\pi}\left(\frac{2\pi}{Tz}\right)\omega^{-5}\ exp\left[-\frac{1}{\pi}\left(\frac{2\pi}{Tz}\right)^4\omega^{-4}\right] \tag{12}$$

Where $S_{(PM)}$ is the Pierson-Moskowitz wave spectrum in m^2/(rad/s), Hs is the significant wave height in meters, Tz is the zero-up crossing period in seconds, and ω is the wave frequency in rad/s. Then Pierson-Moskowitz is given a curve in the wave frequency function.

Recently, JONSWAP spectral formulations are widely used in the design and analysis of offshore structures operated in Indonesia. However, from some studies, it is recommended that Indonesian waters use smaller parameters, around 2.0 to 2.5. In essence, this is to reduce the dominance of energy contributed by certain wave frequencies only [7]. This processing can be seen in equation 13.

$$S_{(JWP)} = S_{(PM)}\gamma^r \tag{13}$$

Where $S_{(JWP)}$ is a JONSWAP wave spectrum, γ is a peak enhancement factor where the value for Indonesian waters is 2 to 2.5, r is formulated with $r = exp\left[\frac{(\omega - w_0)^2}{2\ \tau\ \omega_0{}^2}\right]$, ω_0 is the wave peak frequency and τ is a shape parameter with 0.07 if $\omega \leq \omega_0$, and 0.09 if $\omega > \omega_0$.

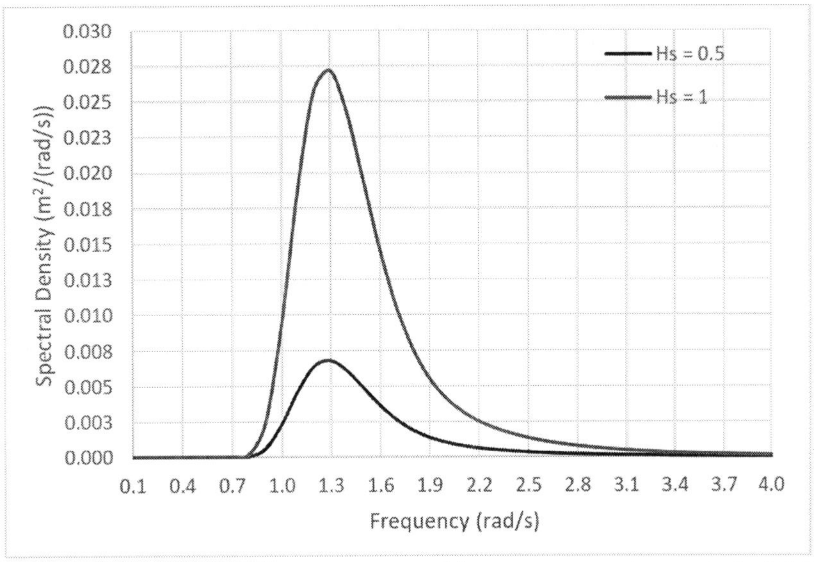

Figure 6 JONSWAP Wave spectrum with zero up crossing period of 3.5 seconds, with variations in wave height of 0.5 and 1.0 meters

From **Figure 6** JONSWAP Wave spectrum with zero up crossing period of 3.5 seconds, with variations in wave height of 0.5 and 1.0 meters. It can be seen from the graph that the spectral density for waves with a height of 1.0 meter is greater than the wave height of 0.5 meters, this makes sense because the energy generated by the waves will increase as the wave height gets higher. In the graph, it is also shown that a wave with zero up crossing 3.5 seconds has the greatest energy at a wave frequency between 1.0 rad/s to 1.6 rad/s so it is possible that if there is a ship operating at that wave frequency it has the potential to have sufficient motion response. big. And it can also be seen that with small wave frequencies and high wave frequencies the spectral density has a fairly small value.

3.2.3. Response Spectra

Effect of hydrodynamic interaction on the added mass, damping potential, and the external force in analyzing the response of buildings floating on a regular wave. The response of marine buildings to random waves is to transform the wave spectrum into a response spectrum. The response spectrum is the energy density response to the structure due to wave loads. it can be formulated as follows [7]:

$$S_{\zeta r}(\omega) = [RAO\ (\omega)]^2 x\ S_{\zeta}(\omega) \tag{14}$$

Where $S_{\zeta r}(\omega)$ is response spectra in m^2/s, $S_{\zeta}(\omega)$ is wave spectra in m^2/s and ω is wave frequency in rad/s.

4. Results and Discussion

Ship motion obtained from the CFD simulation results is represented in the transfer function or RAO. Ship motion which is analyzed is the heave translation motion and the rotational motion of pitch and roll. The simulation is carried out at a speed of 0 knots up to a speed of 20 knots which is the speed. operation of a rescue boat. When the ship is moving in wave sea conditions, the ship will get an external force from the waves which cause translational or rotational movements due to the response of external forces due to waves [10]. The response from the ship will vary depending on the magnitude of the wave frequency and wave height. RAO describes the response from the ship due to differences in wave frequency received by the ship.

4.1. Rescue Boat Ship Motion RAO

Figure 7 shows the response of the rescue boat first hull model heave motion in a function of wave frequency. It can be seen from the graph that the highest response to heave motion occurs at a vessel speed of 20 knots with a wave frequency of between 1.5 rad/s to 2.5 rad/s.

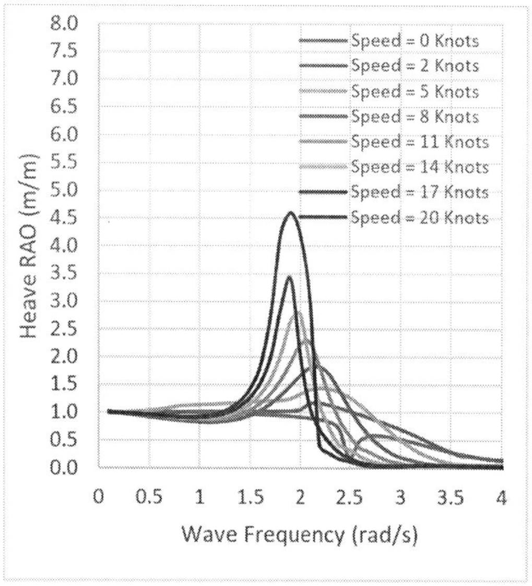

(a) Hull Model 1

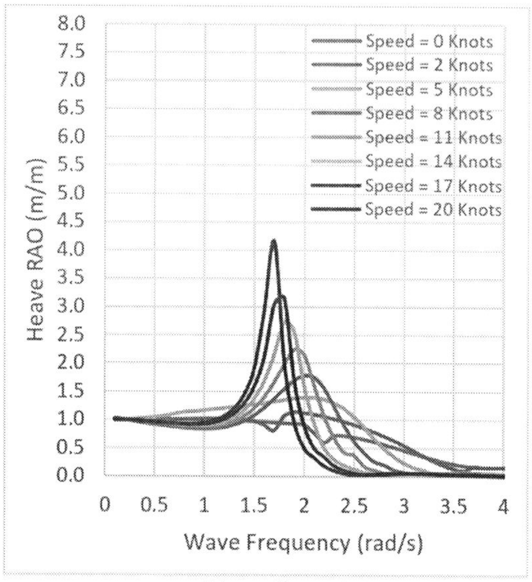

(b) Hull Model 2

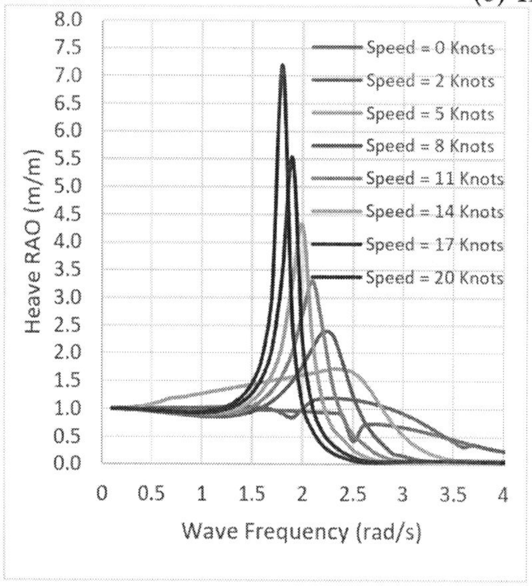

(c) Hull Model 3

Figure 7 Heave Motion RAO Rescue on Head Sea.

As the ship speed increases, the motion peak frequency of the ship's response will shift to a low wave frequency and the heave response value is also higher, this is because at high speed the hydrodynamic lift is the dominant force so that the higher the speed of the ship, the greater the ship's lifting force which causes heave response will be higher [11]. It can be seen at a higher frequency, increasing the speed of the vessel results in smaller motions.

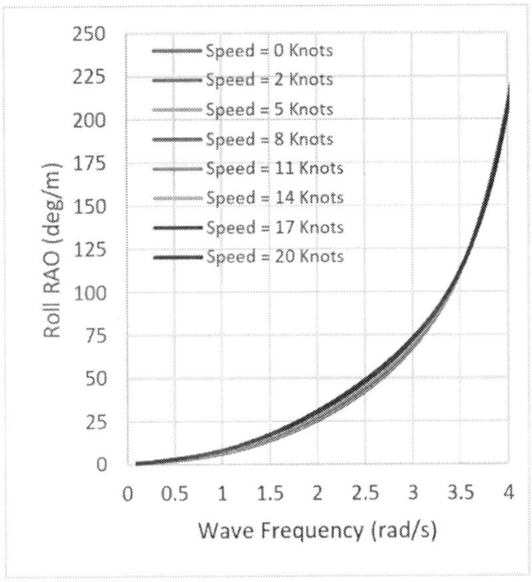

(a) Hull Model 1

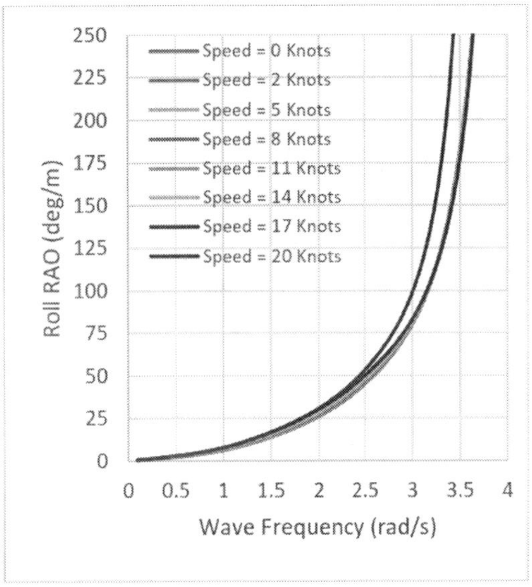

(b) Hull Model 2

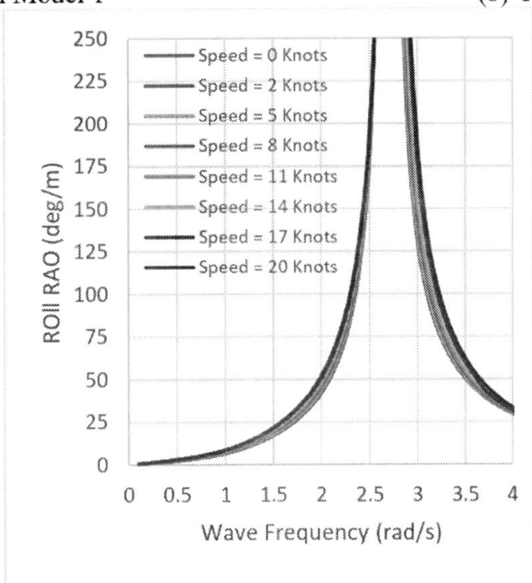

(c) Hull Model 3

Figure 8 Roll Motion RAO Rescue Boat on Beam Sea

Figure 8 show the roll response on the ship. It can be seen from the graph, that the response of the ship's roll will increase as the speed of the ship's operation increases, and the response will be very high at a wave frequency of 2 rad/s to 3.5 rad/s for hull model 3. In contrast to the case of heave motion which experiences a shift in peak frequency with increasing speed, heave motion does not experience a shift in peak frequency. Hull's third model has a very different response compared to other prototype models, with a very large roll response. Meanwhile, hull model 1 and hull model 2 have almost the same tendency for roll motion behaviour, but hull model 2 has a higher roll response, especially on high-frequency waves.

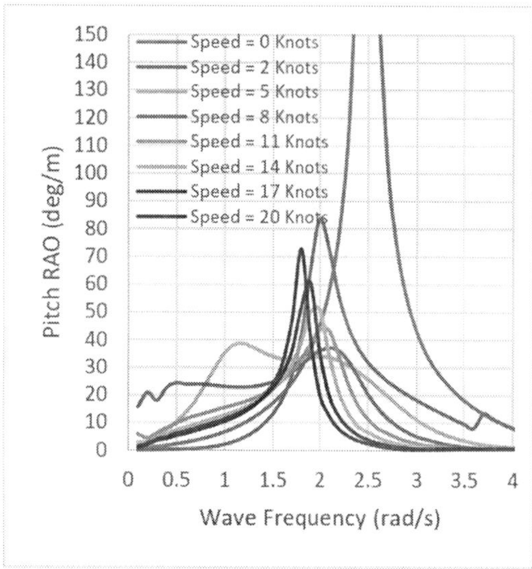

(a) Hull Model 1

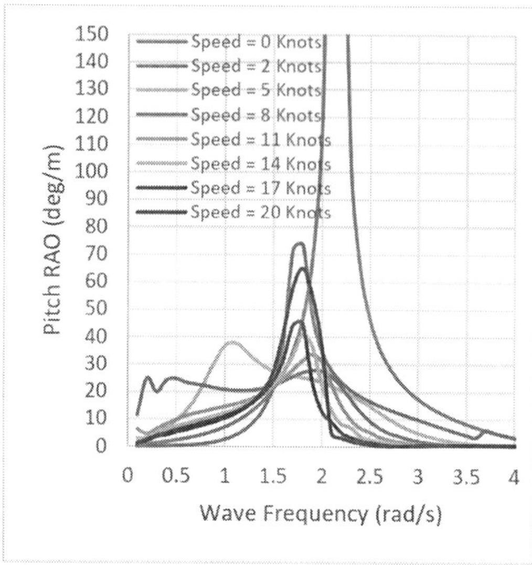

(b) Hull Model 2

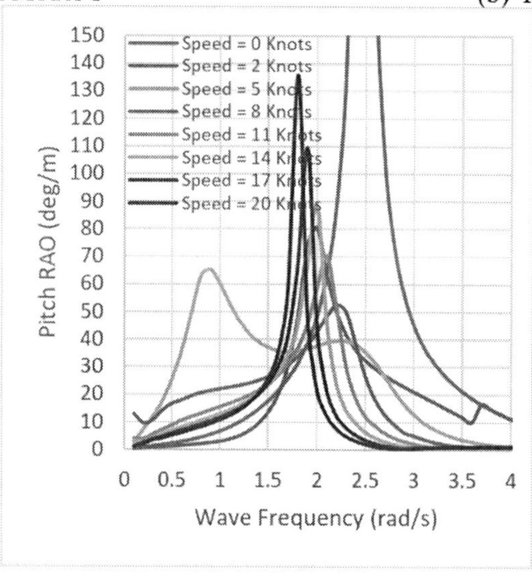

(c) Hull Model 3

Figure 9 Pitch Motion RAO Rescue Boat on Head Sea

Figure 9 shows the ship pitch response due to variations in wave frequency. It can be seen from the graph, that the pitch response has a trend similar to that of heave motion whereas the speed increases, the response will increase and there will be a shift in the peak frequency towards low frequency waves [11]. However, there is another thing that is different, namely in a stationary condition the ship experiences a high pitch response where the pitch response at rest exceeds the response of the ship when traveling at a speed of 20 knots and occurs at a fairly on high wave frequency. Also, there is a difference in the pitch attitude tendency at a speed of 5 knots where at this speed the ship has a large response at low frequencies, namely between 0.5 rad/s to 1.5 rad/s. Overall the ship will experience a high pitch response if it operates at a wave frequency of 1.5 rad/s to 2.5 rad/s.

4.2. Rescue Boat Motion Response

Motion response is obtained by using equation 12, namely by multiplying the RAO square with the wave spectrum to get the motion spectrum response from the ship. Then statistical analysis was

IOP Conf. Series: Materials Science and Engineering 1052 (2021) 012059 doi:10.1088/1757-899X/1052/1/012059

performed to obtain the Root Mean Square (RMS) motion of each ship's motion mode, the ship's heave, roll, and pitch. Ship motion RMS is obtained using equation 15 below [12]:

$$RMS = \sqrt{m_0} \tag{15}$$

Where m_0 is the area under the spectrum response curve obtained using numerical calculations. The area under the curve can be formulated in the following equation 16 below [12]:

$$m_0 = \int_0^\infty \omega_e{}^0 \ S_\zeta \ d\omega \tag{16}$$

Therefore, from the formula above, the ship's response is obtained for the three modes of motion for each operating speed, which is shown in **Figure 10**, **Figure 11** and **Figure 12**. Motion simulation is performed using regular waves with the characteristics of zero up crossing period of 3.5 seconds and a wave amplitude of 0.5 meters and 1.0 meters.

It can be seen in **Figure 10** that the heave motion response in head sea condition will increase as the ship's speed increases, this is because when the ship is traveling at high speed, the hydrodynamic lift will increase in proportion to the square of the ship's speed [3]. Besides, it can also be seen that at a speed of 5 knots there is a difference in trend where there is a heave response which is almost the same as the heave response at a speed of 14 knots so that it can be concluded that if the ship operates at low speed, especially the operating speed of 5 knots and the operation area is very wavy it is dangerous to the ship and needs to be avoided. It can also be seen that the first hull model has a fairly good heave performance where the heave response value at speeds of 8 to 17 knots is quite small and increases significantly at speeds of 20 knots. However, overall heave performance, the heave performance of the first hull model is better than the heave performance of other hull models.

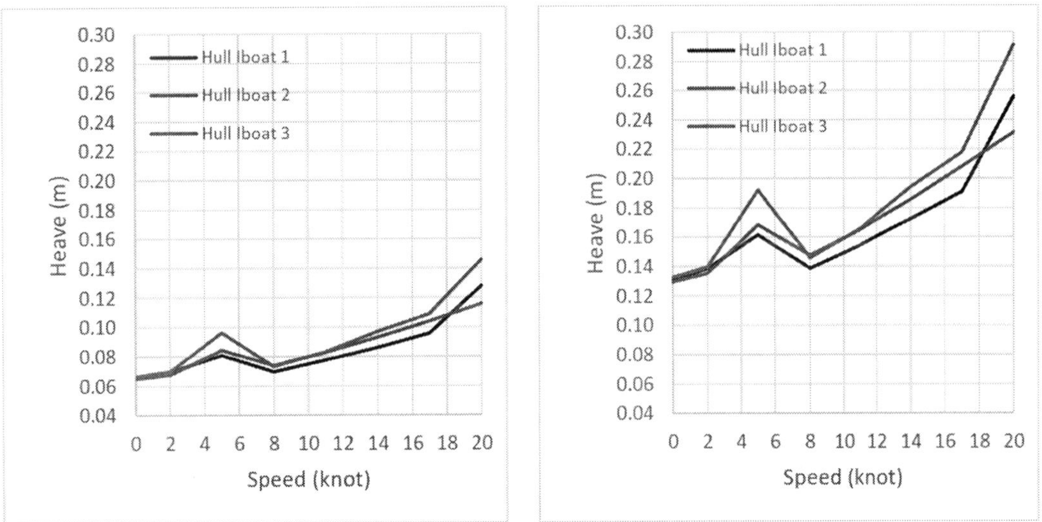

(a) Heave Response on Hs = 0.5 m Heave Response on Hs = 1.0 m

Figure 10 Rescue Boat Heave Response RMS on Head Sea (180°) with 0.5 meter and 1.0 meter Wave Amplitude.

Figure 11 shows the roll motion response for each model of the rescue boat hull in a function of operating speed with beam sea conditions. It can be seen from the graph the first hull model and the second hull model have a good roll response. Where the response to the 0.5-meter wave amplitude ranges an average of 1.6 degrees to 2.4 degrees. While the third hull model has a roll response that is very extreme and dangerous, it can be seen at rest condition, the ship experiences an extreme roll response and along with the increasing speed of the ship the roll response decreases, but the decrease in roll response is still classified as extreme conditions because the wave amplitude is 0.5 meters. with a 20 knots speed of the third hull, the model has a roll response of 7.4 degrees, and the ship's roll response will increase the higher the wave amplitude increases. Therefore, from the results of the analysis, it is concluded that the smallest roll response was experienced by the first hull model with an average roll value of 1.6 degrees at a wave amplitude of 0.5 meters and an average of 3.2 degrees at a wave amplitude of 1.0 meter.

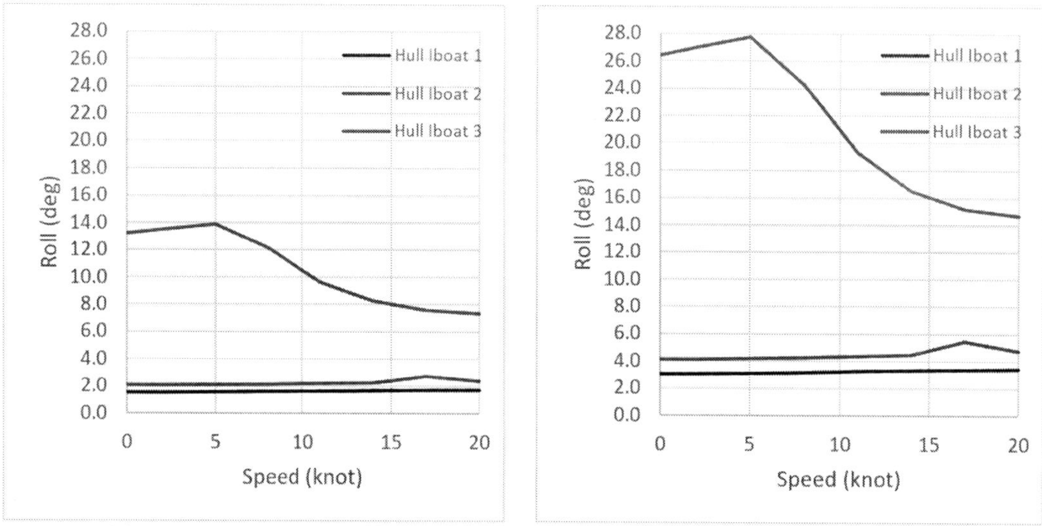

(a) Roll Response on Hs = 0.5 m (b) Roll Response on Hs = 1.0 m

Figure 11 Rescue Boat Roll Response RMS on Beam Sea (90°) with 0.5 meter and 1.0 meter Wave Amplitude.

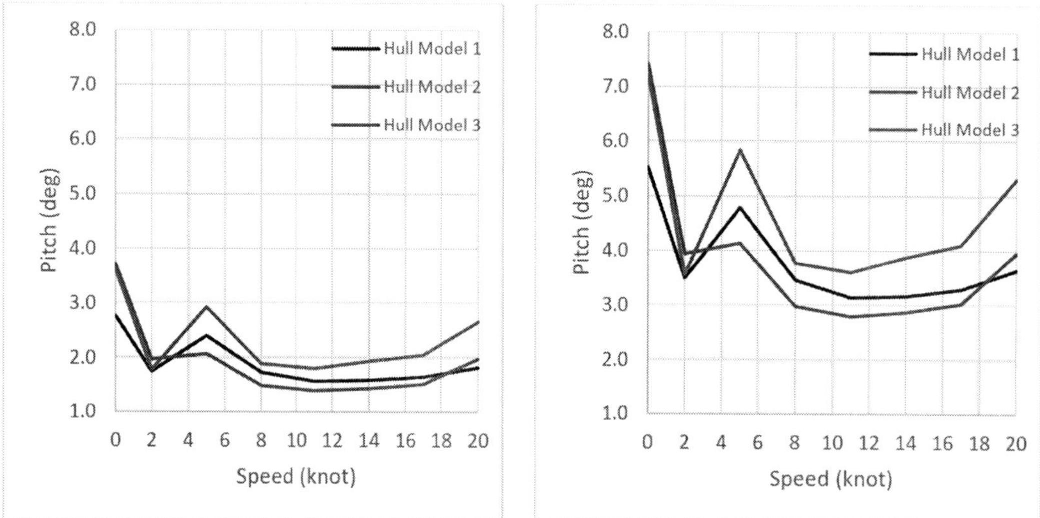

(a) Pitch Response on Hs = 0.5 m (b) Pitch Response on Hs = 1.0 m

Figure 12 Rescue Boat Pitch Response RMS on Head Sea (180°) with 0.5 meter and 1.0 meter Wave Amplitude.

Figure 12 shows the pitch motion response of a rescue boat model in the operating area with a wave amplitude of 0.5 and 1 meter in the function of the ship's operating speed. It can be seen from the graph, that the pitch motion has the same trend as the heave motion response whereas the ship's speed increases, the pitch motion will be higher. This is also due to the additional hydrodynamic lift force of the ship which causes an increase in the length of the ship. Extreme conditions also occur at speeds of 5 knots and speeds of 0 knots or rest condition. It can be seen the pitch motion at low speed is quite high and exceeds the pitch response at high speed so that it is very dangerous for the ship to operate at a low speed in rough sea conditions. And second hull model has the smallest pitch response where at a speed of 8 knots to 17 knots the pitch response is smaller than other ship models. So, it can be concluded that the best pitch response is experienced by ships with the second hull model.

5. Conclusion

Three high-speed craft hull models have been developed which will be used as hull models for rescue boats. The three-boat model has different hull shape characteristics so that they have different hydrodynamic performance, especially the seakeeping performance. So that the seakeeping performance analysis of the three hull models of the rescue boat was carried out using Computational Fluid Dynamics (CFD). The three-dimensional diffraction panel method is used to predict the motion behaviour of rescue boats operating at high speeds with operating speeds of up to 20 knots. The three hull models are simulated with two variations of the wave amplitude, the wave amplitude of 0.5 and 1.0 meter. From the simulation results, several things can be concluded regarding the motion behaviour of the three hull models.

1. As the speed of the ship increases, there will be a shift in the peak frequency of the heave to low-frequency waves as shown in **Figure 7**.
2. The three ship models have a very large pitch response at rest which can be seen in **Figure 9**. This is because when the ship is at rest the ship does not provide resistance to the load given by waves, it becomes different if the ship is in a moving condition the ship has kinetic energy when the ship is moving which is still able to absorb the load due to waves.
3. The heave motion response will increase with increasing ship speed. This occurs due to the increase in hydrodynamic lift, whereas the ship's speed increases, the hydrodynamic lift will increase which is directly proportional to the square of the ship's speed.
4. Roll motion will be very dominant in beam sea conditions. In beam sea conditions, the roll motion does not experience significant changes in the first and second hull models even though the ship's operating speed increases. In contrast to the third hull model where the roll motion will decrease as the speed of the ship increases, it is estimated that this is due to differences in the shape of the hull and the characteristics of the different hulls between the three-ship models.
5. The first ship model performed better for heave and roll motion than the other two models. Meanwhile, the second hull model has the best pitch response compared to other hull models. The best performance is chosen based on the response value of the three modes of motion with the smallest value. In this study, the first hull model was selected as the model with the best seakeeping performance.

The author realizes that there are still many deficiencies in research and in writing this paper, therefore to be able to provide development for future research, the author wants to provide some suggestions regarding research development. It would be better if the simulation is carried out using irregular short crest waves rather than using irregular long crest waves as the author has done, this is done to obtain simulations that are almost the same as the actual conditions. Also, the motion caused by the ship's maneuvering at high speed needs to be done because the ship's maneuvering at high speed is a high-speed craft operation that is very dangerous and has the potential to cause accidents.

Acknowledgment

The authors wished to thank the Institut Teknologi Sepuluh Nopember to finance the research under a scheme called "Penelitian Unggulan" with contract number: 816/PKS/ITS/2020.

IOP Conf. Series: Materials Science and Engineering 1052 (2021) 012059 doi:10.1088/1757-899X/1052/1/012059

References

[1] V. Bertram, Practical Ship Hydrodynamics, 2nd Edition., Butterworth-Heinemann, 2011.

[2] F. Sprenger, A. Maron, G. Delefortrie and T. V. Zwinjnsvoorde, "Experimental Studies on Seakeeping and Maneuverability of Ships in Adverse Weather Conditions," *Journal of Ship Research,* vol. 61, pp. 131-152, 2017.

[3] O. M. Faltinsen, Hydrodynamics of High-Speed Marine Vehicles, Cambridge: Cambridge University Press, 2005.

[4] D. Savitsky, "Hydrodynamics Design of Planing Hulls," *Marine Technolgy Vol 1,* pp. 71-95, 1964.

[5] A. Papanikolaou and T. E. Schellin, "Three-dimensional panel method for motions and loads of ships with forward speed," *Ship Technology Research,* vol. 39, pp. 147-156, 1992.

[6] O. M. Faltinsen, Sea Loads On Ships and Offshore Structures, United Kingdom: Cambridge University Press, 1990.

[7] E. B. Djatmiko, Perilaku dan Operabilitas Bangunan Laut di Atas Gelombang Acak, Surabaya: ITS Press, 2012.

[8] K. Hasselmann, T. P. Barnett, E. Bouws, H. Carlson, D. E. Cartwright, K. Enke, J. A. Ewing, H. Gienapp, D. E. Hasselmann, P. Kruseman, A. Meerburg, P. Muller, D. J. Olbers, K. Richter, W. Sell and H. Walden, "Measurements of WInd-Wave Growth and Swell Decay during the Joint North Sea Wave Project (JONSWAP)," Hearusgegeben vom Deutschen Hydrographiscen Institut, Hamburg, 1973.

[9] American Bureau of Shipping (ABS), Spectral-Based Fatigue Analysis for Vessel, Houston: American Bureau of Shipping, 2016.

[10] A. Sulisetyono, "Computing Porpoising Instability and Motions of a High Speed Boat in Waves," DalTech Dalhoisie Univeristy, Master Thesis, Nova Scotia, 1999.

[11] F. Prini, R. W. Birmingham, S. Benson and P. J. Sheppard, "Seakeeping Analysis of High-Speed Search and Rescue Craft by Linear Potential Theory," in *International Conference on Lightweight Design of Marine Structures (LIMAS 2015)*, 2015.

[12] R. Bhattacharya, Dynamic of Marine Vehicle, Marryland: U.S. Naval Academy, Annapolis, 1978.

Automated door with face recognition: using artificial neural network approach

A Bashir[1], N M Tahir[1], A Y Zimit[2], U I Bature[3,4], K A Abubakar[5] and A Y Nasir[3]

[1]Department of Mechatronics and System Engineering Abubakar Tafawa Balewa University Bauchi, Nigeria
[2]University of Hafr Albatin, 39524 Hafar Al Batin City, Saudi Arabia.
[3]Department of Computer and Communications Engineering, Abubakar Tafawa Balewa University Bauchi, Nigeria.
[4]Department of Electrical and Electronic Engineering, Universiti Teknologi Petronas, Malaysia.
[5]Department of Materials & Metallurgical Engineering, Nigerian Army University Biu, Borno, Nigeria.

nuratahir85@gmail.com, biusman@atbu.edu.ng

Abstract. The implementation of an accurate face recognition system in a hardware device is a very important aspect of various security applications. In a security system, any type of password is used to access private and confidential data, some are hard to remember and can be stolen or guessed, misplaced, forgotten, purloined or duplicated, or become corrupted and unreadable. While there have been significant improvements in the algorithms with increasing recognition accuracy, only a few types of research were conducted on implementing these face recognitions in hardware devices. Door access control systems based on face recognition is geared towards simplifying much-difficult face recognition problems in uncontrolled environments. This report comprised mainly of three sub-phases: face detection, face recognition, and automatic door access control. The face is detected by using the Viola-Jones method and face recognition is implemented by using the Principal Component Analysis (PCA). If a face is recognized, it is known, else it is unknown. The door will open automatically for the known person due to the command of the microcontroller (Atmega328P) received from the MATLAB program. On the other hand, the alarm will ring for an unknown person.

1. Introduction

Security is an aspect that is given top priority by organizations, educational institutions, political and government in personal identity and forgery issues [1], also the demand for going smart is continuously increasing [2]. Although the risk and security issues concern related to this category of systems may be enormous, the inclusion of a more general model of privacy and security in the design phase will reduce a lot of risks that are classified as moderate [3]. Monitoring door entrance is a vital issue, considering it as the only means of structure access. So, an efficient and precise door security system is vital to secure our assets and properties. Nowadays, manual or traditional methods of things are evolving to smart access control. Therefore, there is a need to change our traditional methods of accessing things to modernize methods of smart connectivity, this will provide enhance security and eliminate the existing drawbacks. Security separate access and threats, access control constitute three main components; identification, authentication, and authorization, thus there is a need to actualize this component on door system for better security and control [4-6]. Traditional door access requires keys, and existing ones requires a password using the keypad, pattern, card, or identification ID [7].

Interestingly, human face image has analogous features to fingerprint, they have uniqueness and they are nonexchangeable. Therefore, face image remains the potential means for substituting password, card, and PIN-based security systems [7]. Drawbacks in password, PIN, and card security systems are easily transferable, guess, or crack. Password or PIN are commonly created using names, special dates of the user or user's family, etc. Furthermore, keys can be misplaced, the password is easily leaked through phones, handbooks, or jotters. They can be hard to remember, if stored in a memory device, the device may be corrupted thus rendered unreadable. The traditional lock using keys or password system and the use of radio frequency identification (RFID) lock systems requires keys, PIN, and cards respectively, these systems do not verify the authenticity of the user [8, 9]. Therefore, the security level of these systems seems to be diminutive. But a system that generates a one-time password (OTP) or uses a near field communication tends to be more secured [10]. However, obtaining the OTP requires extra communication linkage between the device and a server which often seems to be unreliable. Using the encryption method is another good technique of establishing security in locking systems [11], only the users who know the key can have access to the system since the original information is being transformed to another format [12]. However, the setbacks with this are losing the key which renders the system vulnerable, and security level depends upon the surety of the encryption keys, it can be too slow, and energy-consuming and complex system [13]. The IoT based security lock system interacts with the server to provide some level of security [14], but the user normally transmits his confidential information over the network, such may offer his confidential data vulnerable to attackers. Thus, by developing a face recognition system, it will be more secure because the facial image had to be used as the ID. It will also help in avoiding any duplicated identification. Either for security or human-computer interaction, there is a wide application to face recognition. Recent developments in communication technology and the call for better security applications have turned face-recognition into focus. Face recognition systems have become the subject of increased interest [15]. Chances of unauthorized access to critical information and infrastructure have become a major concern [16]. Thus, a system must be implemented so that the security system becomes more robust and the probability of unauthorized access to the system becomes minimal.

The general idea of face recognition is to extract certain data from the region of interest of a human facial image and to compare them to stored data for identification. Face identification identifies an input image that belongs to a person in a database whereas face verification searches for the existence of a person's image in the database. The feature extraction stage represents the backbone of face recognition systems because of the direct dependency on the accuracy of the extracted facial features. In this system, Viola jones method of face detection is used, since it has been developed with a scale of the invariant detector [17]. Principal Component Analysis (PCA) is an effective face recognition method, it involves feature extraction, reduction, and classification [18]. Face recognition is realized using the PCA algorithm, and the Euclidean method of classification and recognition is used. This system is implemented using a microcontroller (Atmega 328P), it possesses good features and integrated peripherals, the microcontroller is used to receive the signal sent from the personal computer (PC) that process the software programming of the face recognition system that was developed using MATLAB© and send a command back either to the door motor circuit or to the alarm circuit. The connection between the microcontroller and the PC is achieved using USB (RS232) converter, facial image databases are used to study the behaviour of the proposed face recognition system under various types of images. There are two cases in this system. The first is an automatic door opening for the recognized person (known) and the second is ringing the alarm for the unauthenticated person (unknown). This proposed smart door automation system using the face recognition method, since it has become an important research area because of its usefulness in numerous applications such as access control, security systems, credit card verification, and criminal identification.

2. Research method

Artificial Neural Network (ANN) is an information processing model just like the biological nervous systems of the brain processes information.

In simpler terms, ANN is a mathematical model of the brain which is used to process nonlinear relationships between inputs and outputs data or information. In this work, the supervised learning of ANN is used with the K-nearest neighbour algorithm of the lazy learner's method to store all the instances that correspond to the training data in n-dimensional space. This algorithm is used because it does not focus on constructing a general internal model. Instead, it works on storing instances of training data. The overall block diagram of automatic door-access system using face recognition is shown in figure 1.

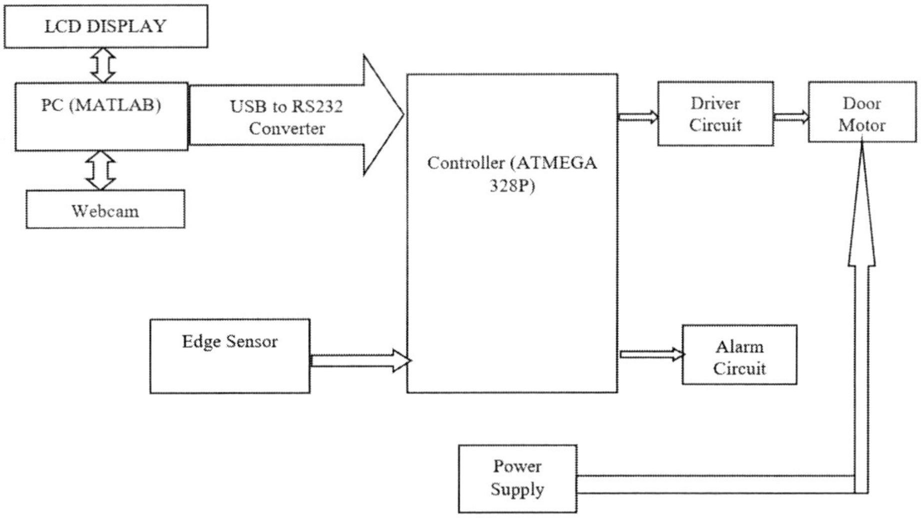

Figure 1. Overall block diagram of automatic door access system using face recognition

2.1. Arduino Uno/Microcontroller (ATMEGA 328P)
The selection of the appropriate hardware device for the proposed door access control system is important to achieve higher recognition accuracy and high performance. In this paper, the Arduino hardware platform is used due to its good features and integrated peripherals as shown in figure 2. The microcontroller is used to receive the signal sent from the PC and send a command back either to the door motor circuit or to the alarm circuit.

Figure 2. Arduino Uno Board

The ATmega328 chip found on the Uno has the following specifications:

- Flash 32k bytes (of which 5k is used for the boot loader)
- SRAM 2k bytes
- EEPROM 1k byte

2.2. Motor Driver (L293D)

The Device is a monolithic integrated high voltage, high current four channel driver designed to accept standard DTL or TTL logic levels and drive inductive loads (such as relays, solenoids, DC and stepping motors) and switching power transistors. A separate supply input is provided for the logic, allowing the operation of the DC motor to move backwards and forward. The flowchart for the automatic door closing and opening system is shown in figure 3.

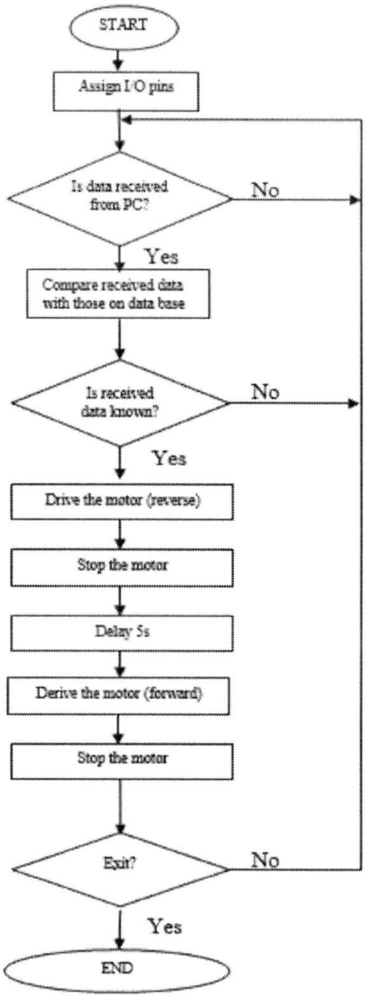

Figure 3. Flowchart for the automatic door opening and closing system

2.3. DC Motor

A motor converts electrical energy into mechanical energy. There are two types of motor: AC motor & DC motor. A simple DC motor uses electricity and magnetic field for producing torque which rotate the motor. The applied voltage describes the speed of motor while current in the armature windings shows the torque. If applied load increased in the motor shaft, then in order to sustain its speed, the motor draws more current from the supply. If the supply is unable to provide enough current, the motor speed will be affected. In this research, the motor is used to drive the door to and fro to give access into or out of the building.

2.4. Buzzer

A buzzer is a small yet efficient component to add sound features to the system. It is a very small and compact 2-pin structure. Hence it can be easily used on breadboard, perforated board and even on PCBs, this makes it a widely used component in most electronic devices. This buzzer can be powered using a DC power supply ranging from 4V to 9V. A simple 9V battery can also be used, but it is recommended to use a regulated +5V or +6V DC supply. The buzzer is normally associated with a switching circuit to either turn it ON or OFF at the required time and interval. In this research work, the buzzer is used as an Alarm Circuit, where the user is to be notified about an intruder trying to access the door.

2.5. Push Button

A push-button (also known as button) is a simple switch mechanism for controlling some aspect of a machine or a process. The surface is usually flat or shaped to accommodate the human finger or hand, so as to be easily pressed or pushed. Buttons are most often biased switches, although many un-biased buttons (due to their physical nature) still require a spring to return to their un-pushed state. In this research, the push button is placed inside the room for the authorized person to press, thereby giving him/her access to go out of the room since high level of security inside the room is not needed.

2.6. Circuit Diagram of the Proposed System

The hardware configuration of this system is composed of the microcontroller (ATMEGA 328P), L293D driver IC, USB to RS232 converter, push-button, DC motor, and buzzer. USB to RS232 converter is used as the interface between the personal computer and the microcontroller. The push-button is used as the inputs of the microcontroller and other components are used as outputs of the microcontroller. This system used ATMEGA 328P microcontroller because of its good features and integrated peripherals. The microcontroller is used to receive the signal sent from the PC and send a command back either to the door motor circuit or to the alarm circuit. The simulation diagram of the proposed system using proteus software is shown in figure 4 while the hardware circuit of the proposed system is shown in figure 5.

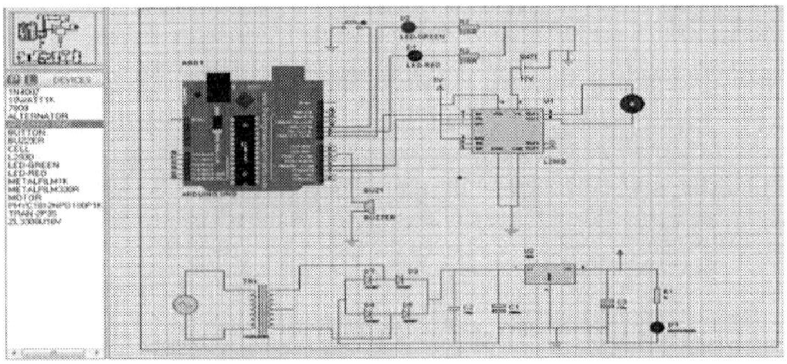

Figure 4. Simulation Diagram of the Proposed System in Proteus

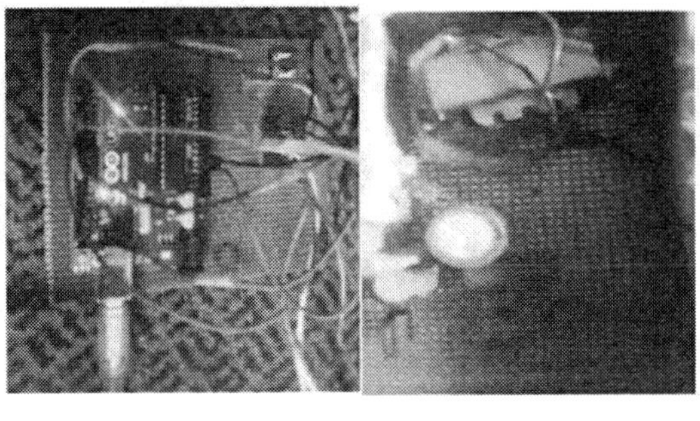

Figure 5. Hardware circuit of the proposed system

3. Results and analysis

3.1. Facial Database

In this work, facial image databases are used to study the behaviour of the proposed face recognition system under various types of images. There is a total of six images of three people; each person has two images with different poses and expressions. Only one image from each person is used for training and the remaining one image is used for the testing phase. This database is chosen because of the availability and simplicity of various instances of front views for each subject. Figure 6 depicts the facial databases of the proposed system.

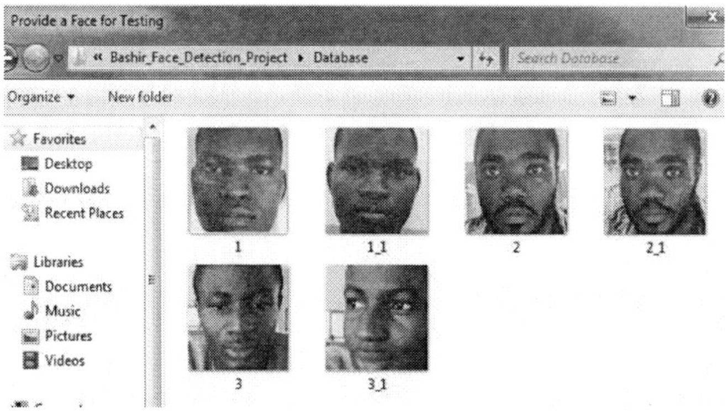

Figure 6. Facial databases of the system

3.2. Hardware simulation results

The overall automatic door access system using face recognition is simulated by using MATLAB. A MATLAB code is created to perform automatic face detection and recognition. There are two cases in this system. The first is an automatic door opening for the recognized person and the second is ringing the alarm for the unauthenticated person.

3.2.1. CASE I. When the test image is recognized as the authenticated person, the green LED will turn ON and the door motor is rotated with the reverse direction until the micro-switch one is activated. When micro-switch one is activated, the door motor is stopped. After five seconds, the red LED will turn ON while that of the green will go OFF and the door motor is rotated with forwarding direction until micro-switch two is activated. Figure 7 and 8 show the simulation results for the authenticated person.

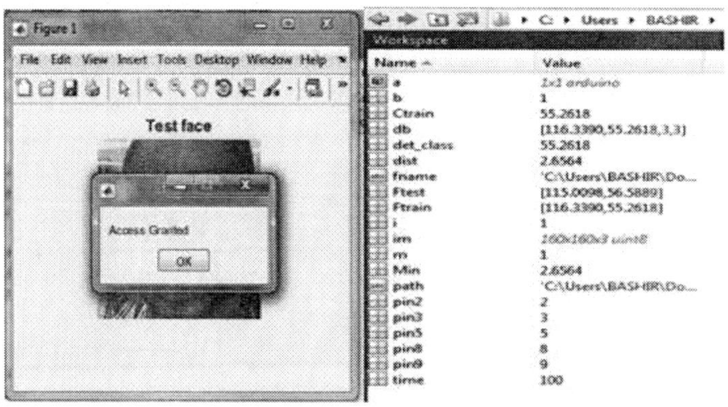

Figure 7. Result when test face is detected and recognized as an authenticated person

Figure 8. Hardware test for an authenticated person when door open

3.2.2. CASE II. When the test image is not recognized as the authenticated person, the red LED will turn ON blinking ten times, and the door motor will not rotate, thus ringing an alarm to the user that an unauthenticated person is about to enter the home. Figure 9 and 10 show the simulation results for the unauthenticated person.

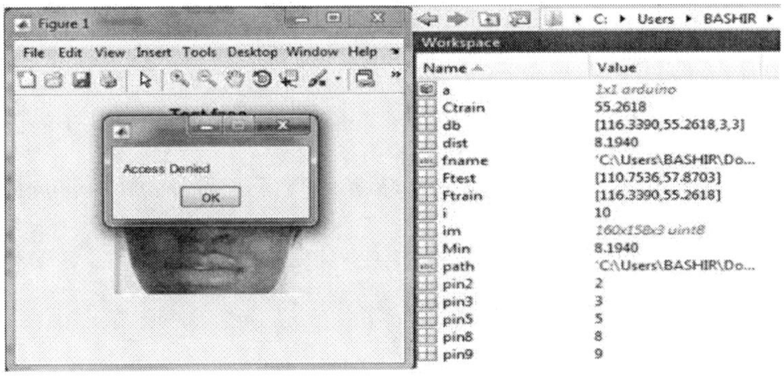

Figure 9. Result when test face is recognized as an unauthenticated person

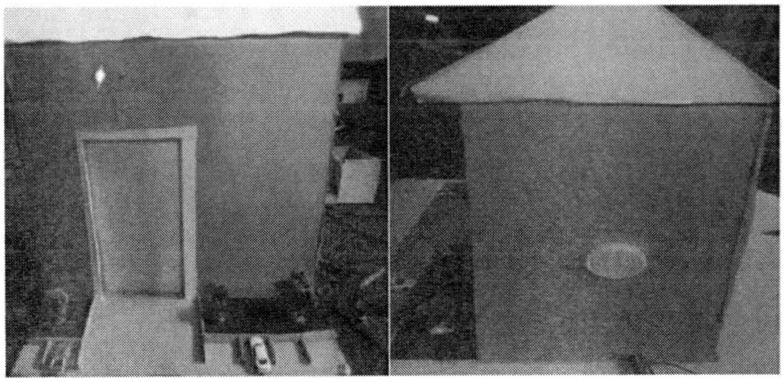

Figure 9. Hardware test for an unauthenticated person when the door does not open

4. Conclusion

In this paper, an automatic door access system using face recognition and detection is presented. The automatic face detection and recognition are done by the MATLAB© program on PC. The microcontroller is used to control the door access system depending on the incoming data sent from the personal computer (PC). The door is opened immediately after confirming that the person is authenticated. After five seconds, the door is closed automatically. The majority of previous research was focused on improving the accuracy of the face recognition algorithm in light of some problems such as illumination, pose, and facial expression changes. However, only a few of these studies focused on implementing these face recognition techniques with an embedded system, such as a door access control system. The problem of high computational time and high power consumption are also faced by current door access control systems either supervised or not fully automatic. Therefore, more research should be conducted in an embedded system taking into consideration problems such as illumination, pose, and facial expression changes.

Acknowledgments

The authors wish to thank the Department of Mechatronics and System Engineering, Abubakar Tafawa Balewa University Bauchi for the provision of laboratory's equipment.

References

[1] Nag A, Nikhilendra J N and Mrutyunjay K 2018 *Int. Conf. on Con. in Technol.* (Pune) vol 3 (Pune: India) 1
[2] Dutta J and Roy S 2017 *Int. Conf. on Cloud Computing, Data Sci. & Eng.-Confluence.* (Noida: India) pp 237-242
[3] Jacobsson A, Boldt M and Carlsson B 2016 Fut. Gen. Comp Sys. B **56** 719
[4] Wijaya I G P S, Husodo A Y and Jatmika A H 2016 *Proc. Int. Sem. On Intel. Technol. and its Appl.* (Lombok: Indonesia) pp 151-156
[5] Simon O A, Bature U I, Jahun K I and Tahir N M 2020 *Int. J. Infor. & Commun. Technol.* **9** 212-220
[6] Tshomo K, Tshering K, Gyeltshen D, Yeshi J and Muramatsu K 2019 *Proc. Int. Conf. on Converg. in Technol.* (Bombay: India) pp 1-5
[7] Divya R S and Mathew M 2017 *Proc. Int. Conf. on Cir., Power and Comput. Technol.* (Kollam: India) pp 1-3
[8] Tombeng M T and Laluyan H S 2017 *Proc. Int. Conf. on Cyb. and IT Serv. Managt.* (Denpasar) vol 5 (Indonesia: IEEE) pp 1-5
[9] Ahtsham M, Yan H Y and Ali U 2019 *Proc. Int. Conf. on Networ. Sen. and Contrl.* (Banff, AB: Canada: IEEE) pp 448-453
[10] Brown D R L 2017 (Certicom Corp,) U.S. Patent 9,734,322
[11] Chen H, Liu J and Yang C F 2016 *Advanc. in Mech. Eng.* **8** 1-7
[12] James A and Sarikak T 2020 *Stud. in Indian Plac. Nam.* **40** pp 9-15
[13] Sadkhan S B and Salman A O 2018 *Proc. Int. Conf. on Advanc. Of Sust. Eng. and its Appl.* (Wasit: Iraq) pp 105-108
[14] Ha I 2015 *Int. J. Sec. and its Appl.* **9** 45-54
[15] Tham K Y, Cheam T W, Wong H L and Fauzi M F A 2020 *Comput. Sci. and Technol.* vol 603 (Singapore: Springer) pp 227-236
[16] Petrakis E G, Antonopoulos F, Sotiriadis S and Bessis N 2020 *J. Amb. Intell. And Huma. Comput.* **11** 929-943
[17] Tavallali P, Yazdi M and Khosravi M R 2017 *Proc. Int. Conf. on Conput. Sci. and Comput. Intell.* (Las Vegas: USA) pp 828-831
[18] Barnouti N H, Al-Dabbagh S S M, Matti W E and Naser M A S 2016 *Int. J. Advan. Computer Sci. and Appl.* **7** 371-377

IOP Conf. Series: Materials Science and Engineering 1052 (2021) 012061 doi:10.1088/1757-899X/1052/1/012061

Theoretical Study for Predicting Response Water Column in a Moonpool

O Turbaningsih[1] and H I Nur[1]

[1]Department of Marine Transportation Engineering, Institut Teknologi Sepuluh Nopember, Surabaya, Indonesia.

Abstract. Moonpool commonly installed in production barge, drilling vessel, pipe laying vessel, survey vessel and offshore diving vessel. Moonpool is an opening vertical wall through the deck from top to bottom of the vessel. The main purpose of using moonpool is to perform lifting/lowering of subsea equipment. It considers as the most effective subsea handling method, but it has disadvantages with oscillation of water column inside moon pool. The purpose of this study is to understand the correlation between response water column inside the moonpool with environmental load and the motion of the floating structure during operational time or transit condition. During transit condition, the forward speed of the barge is zero, the excitation force only caused by incoming waves and barge heave motion. The study started by defining the barge vertical movement and all the forces acting in the body of moonpool to obtain simplified harmonic equation for water column motion and plotting the result using MATLAB software. The amplitude of heave barge and water column piston mode is strongly influenced by environmental conditions, in this case the significant wave height and also the dimension of the moonpool area since it is giving additional added mass into the motion. The large amplitude of the water column likely to occur if critical frequency is in the range of wave excitation. Thus, to prevent any over flowing condition during lowering/lifting operation via moonpool, the limiting sea-state condition should be identified for the designated location.

1. Introduction

In business practice, marine offshore operation can be categorized as offshore facilities operation and vessel operation. Many vessels have been specifically designed to meet the technical and functional to serve the operational purposes such as exploration, maintenance and construction works in the ocean. The offshore vessels mainly classified as offshore exploration and drilling vessel, offshore support vessels, construction/special purpose vessel and offshore productions vessel. The efficiency of marine offshore operation is depending to the ability for handling the project equipment. The handling methods that commonly use in offshore practice can be broadly group into side handling, stern handling, open handling, and moonpool [1]. Moonpool commonly installed in drilling vessel, pipe laying vessel, survey vessel and offshore diving vessel. Moonpool is an opening vertical wall through the deck from top to bottom of the vessel. The main purpose of using moonpool is to perform lifting and lowering of subsea equipment, riser and pipeline installation, ROV launching and diving bell handling. By using moonpool, the operator has full control over the load while its being lifting or lowering from the sea. The equipment such as diving bell or drilling hammer can be launched using the lifting arrangement system built in the vessels safer than side or stern handling without affecting much to the ship stability. Figure 1 shows the illustrations on how the lifting operation via moonpool.

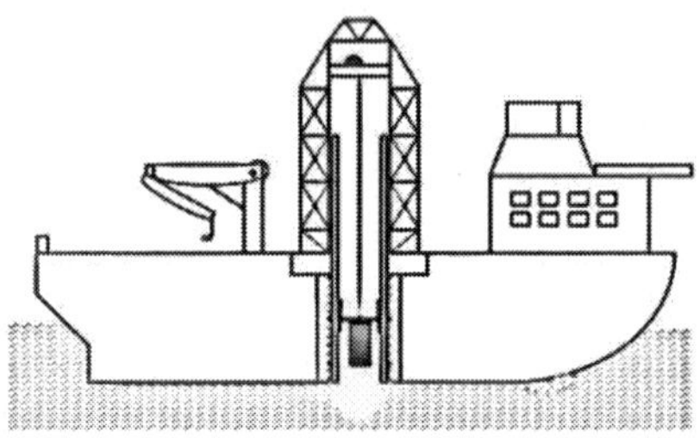

Figure 1. Illustration for vessel with lifting arrangement via moonpool [2].

Table 1 show the advantages and disadvantages for equipment handling towards moonpool [2][3].

Table 1. The advantages and disadvantages of handling equipment via moonpool.

Advantages	Disadvantages
Protection of equipment from environmental forces, thus the limiting sea-state can be increased during the operation.	Water plugs within moonpool can result in flooding on vessel deck and large loads on equipment in moonpool.
Minimise installation cost and intervention operation (during equipment installation)	The equipment has higher risk to hit the cursor frame during lowering/lifting operation
Moonpool close to vessel roll and pitch axis which minimise dynamic force	Size limitation of equipment

The offshore field operators required to keep the operability of subsea equipment as high as possible to reduce shut down cost. It is advisable to avoid any sea-state condition that may potentially damage subsea equipment. To date, the sea state limit has been increased by using module handling systems that enables launch and recovery through moonpool placed near the centre of roll and pitch axis of the vessel [4]. The water motions inside the moonpool have two type of natural modes [5] which are sloshing modes (back and forth movement) and piston modes (heave motion). The piston mode is very dominant. During resonant condition, the water elevation inside the moonpool is higher than usual, which able to cause flooding on the vessel deck. The lifting/lowering equipment operation much likely happened during transit condition. The vessel will be either perform dynamic positioning, anchored or using single point mooring system depend on the locations. During stationary waves, large water motion likely happened if vertical excitation resulting from the ship motions and waves is matched with natural frequency of the water column [6].

The excitation caused by the pressure difference in bottom of moonpool. The pressure variation is depending on the wave forces and cross section of the moonpool which called as F_{WH} (wave exciting force). This force is well known as Froude Krylov wave force. This force can be calculated from incident wave potential with consideration of interaction force between added mass and damping coefficient [7]. Heave motion is the most dominant motion among other the six type of motions acting in the floating structures. One of research investigated that the strongest excitation of the moonpool was caused by the forced heaving, depending on the oscillation amplitude and its damping[7]. Theoretical approach was performed to determine simplified equation to predict the heave response in moonpool during stationary waves. The study investigates the main parameter affecting the piston mode in the moonpool.

The purpose of this study is to understand the correlation of response water column inside the moonpool to the environmental condition and the floating structure movement. The main outcome of this study is to generate simplified equation to predict response of water column inside moonpool in relative to sea water elevation and vertical motion of the floating structure. This equation can be used to check the limiting environmental factor that considered still safe to perform lifting/lowering equipment through moonpool.

2. Work methodology

The theoretical approach to define the water column response was performed in the two stages, first calculating heave response in the floating structures and then include it as an input to determine the water column equation of motion. The floating structure modelled in this study based on barge shape. Figure 1 shows the flowchart for the work methodology in this study.

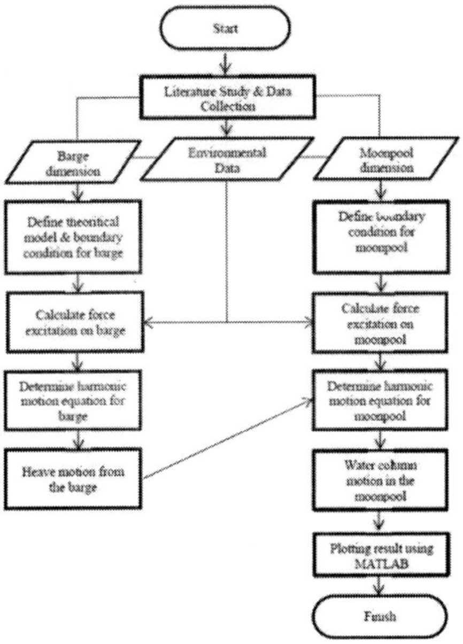

Figure 2. Work methodology flowchart.

The study will be based on the parameter shows at Table 2.

Table 2. Barge details and environmental data used in the study.

Parameter Description	Value
Barge Length (m)	227.0
Barge Width (m)	42.0
Water depth (m)	100
Moonpool dia. (m)	10; 20; 30
Draft (m)	8; 12; 18
Hs (m); Tp(s) (1-year)	2.9; 9.1
Hs (m); Tp(s) (10-year)	4.1; 10.3
Hs (m); Tp(s) (100-year)	5.3; 11.1

The simplified equation described in this study is based on the following assumption:

- The wave theory for deep water is applied.
- Ideal fluid, where fluid is inviscid, incompressible and irrotational.
- Flow of the fluid is uniform and constant over time.
- The long wave theory is applied.
- The floating structure being simulated is barge type without moonpool
- Only vertical motion of the barge taken into consideration
- The volume of the barge is calculated by the total volume minus the moonpool volume.
- Wave excitation force calculated is combination of Froude Krylov force and diffraction force.
- If length overall barge far less than wave length, the diffraction force is negligible, the additional forces to be counted for is radiation force, where radiation damping equal to zero and added mass m_{ij} is equal to m_{ji}.
- Added mass at the moonpool calculated using slender body theory, where the added mass is obtained from the 3-dimensional integration of 2D added mass cuts.

Understanding the six degree of freedom in the floating structure is important. Figure 3 shows the rotational and translational movement in every axis.

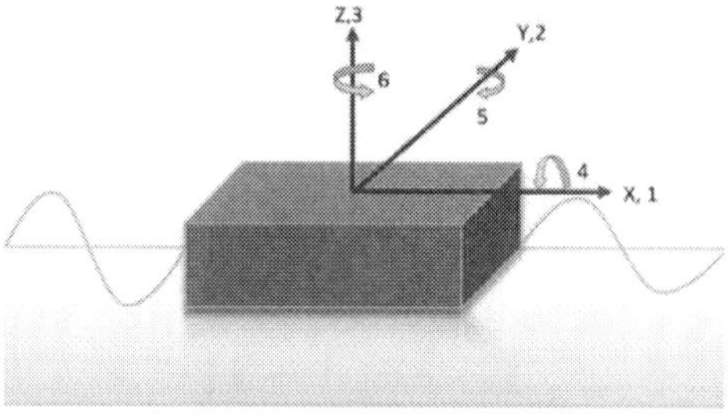

Note: 1=surge, 2=sway, 3=heave,4=pitch,5=roll, 6=yaw
Figure 3. Six degree of freedom in floating structure.

where, M_{ij} is matrix mass component, A_{ij} is added mass component, C_{ij} is restoring hydrostatic force component, F_j is excitation force amplitude.

$$M_{jk} = \begin{bmatrix} M_{11} & 0 & 0 & 0 & M.Z_{\epsilon G} & 0 \\ 0 & M_{22} & 0 & -M.Z_{\epsilon G} & 0 & 0 \\ 0 & 0 & M_{33} & 0 & 0 & 0 \\ 0 & -M.Z_{\epsilon G} & 0 & I_{44} & 0 & -I_{46} \\ M.Z_{\epsilon G} & 0 & 0 & 0 & I_{55} & 0 \\ 0 & 0 & 0 & -I_{46} & 0 & I_{66} \end{bmatrix} \quad (1)$$

$$A_{jk} = \begin{bmatrix} A_{11} & 0 & A_{13} & 0 & A_{15} & 0 \\ 0 & A_{22} & 0 & A_{24} & 0 & A_{26} \\ A_{31} & 0 & A_{33} & 0 & A_{35} & 0 \\ 0 & A_{42} & 0 & A_{44} & 0 & A_{46} \\ A_{51} & 0 & A_{53} & 0 & A_{55} & 0 \end{bmatrix} \quad (2)$$

General equation for floating structure motion stated in Equation (3).

$$\sum_{j=1}^{6}[(M_{ij} + A_{ij})X_j + B_{ij} X_j + C_{ij}X_j] = F_j e^{imt} \quad (3)$$

3. Theoretical study of response water column in a moonpool

This study of response water column is started by examining the forces acting on the barge influenced by the actual environmental data on site.

3.1. Heave motion equation for the barge

Basic assumption required to evaluate sea loads and motions acting on the ships. The governing equation applied for fluid incompressible and inviscid with constant density [8] use Laplace formula where $\nabla^2\emptyset = 0$ is:

$$\frac{\partial^2\emptyset}{\partial x^2} + \frac{\partial^2\emptyset}{\partial y^2} + \frac{\partial^2\emptyset}{\partial z^2} = 0 \quad (4)$$

Kinematic boundary condition for fixed body in a moving fluid [8] as follow:

$$\frac{6\emptyset}{6n} = 0;\ \text{on body surface} \quad (5)$$

The pressure in the water surface is equal to constant atmosphere pressure in the free surface, thus, the kinematic boundary condition applied as follows:

$$g\zeta + \frac{\partial\Phi}{\partial t} + \frac{1}{2}\left(\frac{\partial^2\emptyset}{\partial x^2} + \frac{\partial^2\emptyset}{\partial y^2} + \frac{\partial^2\emptyset}{\partial z^2}\right) = 0 \quad (6)$$

By linearizing Equation (6) and assume that the ship in transit condition, no forward speed and current is zero [8]. The dynamic boundary condition become:

$$g\zeta + \frac{6\Phi}{6t} = 0 \text{ on } z = 0 \quad (7)$$

Hydrodynamic forces acting on the barge can be calculated using pressure area method. Barge vertical motion on can be calculated using strip theory. The wave excitation force calculated using Froude Krylov theory. The potential velocity used in the analysis:

$$\Phi(x, z, t) = \frac{g\zeta_a}{\omega} e^{kz}\cos(\omega t - kx) \tag{8}$$

The excitation force on barge calculated using Froude Krylov:

$$F_I = \iint_S \left(-\rho\frac{\partial\phi}{\partial t}\right) n\, dS = \int_{-\frac{L_b}{2}}^{\frac{L_b}{2}} f_{13}(x)\, dx \tag{9}$$

$$f_{13}(x) = \int_{-\frac{B_w}{2}}^{\frac{B_w}{2}} P(x) n_z\, d = \int_{-\frac{B_w}{2}}^{\frac{B_w}{2}} \rho g\zeta_a e^{kz}\sin(\omega t - kx)\, dy = \rho g\zeta_a B_w e^{kz}\sin(\omega t - kx) \tag{10}$$

$$F_{13} = \int_{-\frac{L_b}{2}}^{\frac{L_b}{2}} \rho g\zeta_a B_w e^{kz}\sin(\omega t - kx)\, dx = \rho g\zeta_a B_w e^{-kD}\int_{-\frac{L_b}{2}}^{\frac{L_b}{2}} \sin(\omega t - kx)\, dx \tag{11}$$

$$F_{13} = \rho g\zeta_a B_w e^{-kD}\left[\frac{2}{k}\sin\frac{kL_b}{2}\sin\omega t\right] \tag{12}$$

where, F_{13} is Froude Krylov force, B_w is barge width, L_b is barge length, D is barge draft, P is pressure in the bottom of the barge, and ζ_a is sea water elevation.
Added mass of structure, A_{33} is depending on its shape, since barge is rectangular shape, the added mass A_{33}^{2D} value can be taken from Lewis graphics [8] as 0.75 $\rho B_w D$ which resulted following equation:

$$A_{33} = L_b A_{33}^{2D} = 0.75\rho B_w D L_b \tag{13}$$

Radiation force is affected by the added mass and the damping coefficient in the structures.
The radiation force can be formulated as follows:

$$F_R = -A_{33}\ddot{Z}_3 - B_{33}\dot{Z}_{33} = -L_b A_{33}^{2D}\ddot{Z}_{33} = -0.75\rho B_w D L_b\ddot{Z}_{33} \tag{14}$$

where B_{33} is heave damping, in this scenario the heave damping is considered as zero.

When a body is free floating, the restoring force will follow hydrostatic and mass considerations [8] as shows in the following equation:

$$F_H = -\rho g\int_S z\, n\, dS = -\rho g A_{wp} Z_{33} = -\rho g B_w L_b Z_{33} \tag{15}$$

where, F_K is restoring forces (hydrostatic force) and $C_{33} = \rho g B_w L_b$ is restoring moment coefficient.
The heave motion is following Newton II principle where $\sum F = m.a$ show as follows:

$$M\ddot{Z}_{33} = F_H + F_R + F_{13} \tag{16}$$

$$M\ddot{Z}_{33} = -\rho g A_{wp} Z_{33} - L_b A_{33}^{2D}\ddot{Z}_{33} + \rho g\zeta_a B_w e^{-kD}\left[\frac{2}{k}\sin\frac{kL_b}{2}\sin\omega t\right] \tag{17}$$

$$(M + L_b A_{33}^{2D})\ddot{Z}_{33} + \rho g A_{wp} Z_{33} = \rho g\zeta_a B_w e^{-kD}\left[\frac{2}{k}\sin\frac{kL_b}{2}\sin\omega t\right] \tag{18}$$

$$(M + A_{33})\ddot{Z}_{33} + C_{33}Z_{33} = F_{E3}(t) \tag{19}$$

where, M is total mass of the barge (kg), A_{33} is added mass, C_{33} is restoring moment coefficient, $F_{E3}(t)$ is barge excitation force, Z_{33} is heave motion barge.
Natural period calculated using this equation:

$$-\omega^2(M + A_{33}) + \rho g B_w L_b = 0 \tag{20}$$

$$\omega_n = \left(\frac{\rho g B_w L_b}{M + A_{33}}\right)^{\frac{1}{2}} = \left(\frac{g B_w}{B_w D + A_{33}^{2D}}\right)^{\frac{1}{2}} = \left(\frac{g}{D + 0.75BD}\right)^{\frac{1}{2}} \tag{21}$$

$$T_n = 2\pi\left(\frac{D + 0.75BD}{g}\right)^{\frac{1}{2}} \tag{22}$$

In heave motion, damping assumed to be 0, which is called as forced vibration system.

$$(M + A_{33})\ddot{Z}_{33} + C_{33}Z_{33} = F_{E3}(t) \tag{23}$$

$$F_{E3} = \rho g \zeta_a B_w L_b \tag{24}$$

$$Z_{33} = \frac{\frac{\rho g \zeta_a B_w L_b}{\rho g B_w L_b}}{1 - \left(\frac{\omega}{\omega_n}\right)^2} \sin \omega t = \zeta_a \sin \omega t \tag{25}$$

3.2. Piston mode in a moonpool

The response water column is studied during transit time where the barge forward speed is 0, so there is no flow separation and the excitation force acting is based on Froude Krylov theory. The symbol Ω in Figure 4 state the fluid domain, S_F is the free surface, S_B is the fluid body limitation and S_0 is the fluid limitation between wall tank and seabed.

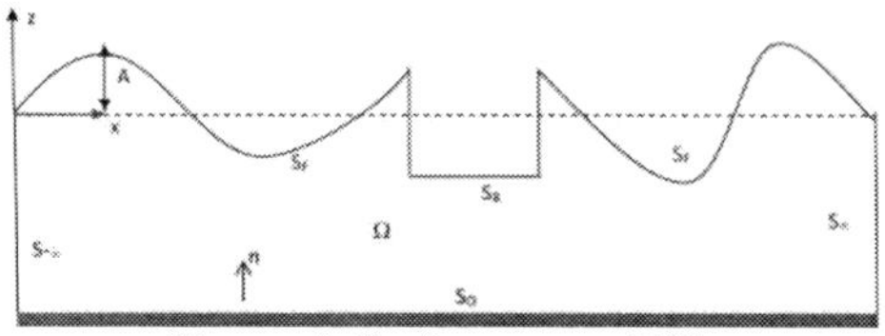

Figure 4. Surface control [8].

The governing equation applied in the fluid domain Ω is:

$$\frac{\partial^2 \emptyset}{\partial x^2} + \frac{\partial^2 \emptyset}{\partial y^2} + \frac{\partial^2 \emptyset}{\partial z^2} = 0 \tag{26}$$

The kinematic boundary condition at the free surface in the S_B plane body is:

$$\frac{\partial \zeta}{\partial t} = \frac{\partial \Phi}{\partial z} \tag{27}$$

The dynamic boundary condition, the pressure in the water surface is equal to the constant atmospheric pressure in the free surface shown as the following:

$$P + \rho g y + \frac{1}{2}\rho(u^2 + v^2 + w^2) = \text{Constant} \tag{28}$$

$$P = P_0 \text{ , on } S_F \tag{29}$$

The kinematic boundary condition in the Equation 42 become:

$$g\zeta + \frac{\partial \Phi}{\partial t} = 0 \text{ , for } z = 0 \tag{30}$$

The total energy E in the volume Ω, consist of potential energy and kinetic energy, shown as follow:

$$E(t) = \rho \int_{\Omega} \left(\frac{1}{2}V^2 + gz\right) d\tau \tag{31}$$

$$\frac{\partial E(t)}{\partial t} = \rho \int_S \left(\frac{\partial \emptyset}{\partial t}\frac{\partial \emptyset}{\partial n} - \left(\frac{(P - P_0)}{\rho} + \frac{\partial \emptyset}{\partial t}\right) U_n \right) ds \tag{32}$$

The fluid speed applied as follows:

$$u_n = \frac{\delta \Phi}{\delta n} \text{ , on } S_F \text{ and } S_B \tag{33}$$

The Equation (32) become

$$\frac{\partial E(t)}{\partial t} = \rho \int_S \left(\frac{(P - P_0)}{\rho} \right) U_n \, ds - \rho \int_{S\infty + S - \infty} \frac{\partial \emptyset}{\partial t}\frac{\partial \emptyset}{\partial n} ds \tag{34}$$

$$\rho \int_S \left(\frac{(P - P_0)}{\rho} \right) U_n \, ds = \frac{\partial z}{\partial t} \int_{S0} (P - P_0) Z \, ds \tag{35}$$

$$\rho \int_s \left(\frac{(P - P_0)}{\rho}\right) U_n \, ds = \frac{\partial z}{\partial t} \int_{s0} A_z \; Z + B_z z + CZ - \rho g V \quad (36)$$

The forces acting in a moonpool is Froude Krylov, radiation forces and hydrostatic forces. The Froude Krylov for inertia force in circular moonpool as per follows:

$$F_I = \iint_s \left(-\rho \frac{\partial \emptyset}{\partial t}\right) n dS = \iint_s \rho g \zeta_a \, e^{kz} \sin(\omega t - kx) \, n dS = \rho g \zeta_a \, A e^{-kD} \sin(\omega t - kx) \quad (37)$$

The radiation force caused by the added mass and damping of the structure. The radiation forces equation as per below:

$$\vec{F}_R(t) = \iint_{SB} P_R \, n dS \quad (38)$$

$$F_R(t) = f_3 \cos(\omega t - \theta) = f_3 \cos\theta \cos\omega t + f_3 \sin\theta \sin\omega t = \frac{f_3 \cos\theta}{m^2 Z} Z(t) + \frac{f_3 \sin\theta}{mZ} Z(t) \quad (39)$$

$$F_R(t) = -[A_z] Z(t) - [B_z] Z(t) \quad (40)$$

$$F_R = -A_z (Z_3 + Z) - [B_z](Z) \quad (41)$$

Added mass calculated using slender body theory [9]. The added mass for circular moonpool such as follows:

$$A_z = \int_x m_{33} \, dx = \int_{-\frac{l}{2}}^{\frac{R}{2}} \rho \pi R^2 \, dx = \rho \pi R^3 \text{ (R is moonpool radius)} \quad (42)$$

The damping coefficient calculated using slender body theory

$$B_{33} = \int_{-\frac{R}{2}}^{\frac{R}{2}} \lambda_{33} \, dx \quad (43)$$

The damping coefficient for the circular moonpool, equation is:

$$B_z = \frac{\rho g^2 R}{\omega^3} \left(\frac{\zeta_a}{Z_3}\right)^2 \quad (44)$$

The force acting in the body of moonpool in control volume system is gravitation force and inertia force caused by the heave motion z_3.

$$F_H = \int_V \rho g dV = -\rho g A \int_{-D}^{z} dz = -\rho g A (Z + Z_{33}) \quad (45)$$

The heave motion is following Newton II principle where $\sum F = m.a$, all the forces acting in the structures equal to mass times it accelerations.

$$\Sigma F_Z = F_I + F_H + F_R + F_{FR} \quad (46)$$

$$(\rho AD) Z = -A_z (Z_3 + Z) - B_z Z - \rho g A (Z + Z_3) + \rho g \zeta_a A e^{-kD} \sin(\omega t - kx) \quad (47)$$

$$(\rho AD) Z + A_z (Z_3 + Z) + B_z (Z) + \rho g A (Z + Z_3) = \rho g \zeta_a A e^{-kD} \sin(\omega t - kx) \quad (48)$$

$$(M + A_z) Z - B_z Z - CZ = F(t) \quad (49)$$

$$F(t) = \rho g A \zeta_a e^{-kD} \sin(\omega t - kx) - A_z Z_3 - \rho g A Z_3 \quad (50)$$

Where: M is mass of water column in moonpool (ρAD), B_z is damping calculated by $\frac{\rho g^2 R}{m^3} \left(\frac{\zeta_a}{Z_3}\right)^2$, C is hydrodynamic stiffness ($\rho g A$), Z is water column elevation in a moonpool, Z_3 is heave motion in barge, A is moonpool area. The natural frequency in a moonpool during no un-damping, formulated as:

$$\omega_n = \left(\frac{\rho g A}{\rho A D + A_z}\right)^{\frac{1}{2}} = \left(\frac{\rho g A}{\rho A D + R C_A \rho A}\right)^{\frac{1}{2}} = \left(\frac{g}{D+R}\right)^{\frac{1}{2}} \quad (51)$$

The water column response is depended on the force excitations. The force excitation is directly proportioned with wave amplitudes. The amplitude for response water column in a moonpool is:

$$X = \frac{\frac{F}{\rho g A_{wp}} \sin(\omega t - \alpha)}{\left(\left[1-\left(\frac{\omega_F}{\omega_n}\right)^2\right]^2 + \left[2\frac{\epsilon}{\epsilon_c}\left(\frac{\omega_F}{\omega_n}\right)\right]^2\right)^{\frac{1}{2}}}; \text{ where } \alpha = \tan^{-1}\left[\frac{2\frac{\epsilon}{\epsilon_c}\left(\frac{\omega_F}{\omega_n}\right)}{1-\left(\frac{\omega_F}{\omega_n}\right)^2}\right] \quad (52)$$

$$X = X_m \sin(\omega_f t - \alpha) \quad (53)$$

The water column response is depending on the wave excitation force acting on the body. This excitation force is directly proportionated with the wave amplitude. Table 3 shows the water column response relative to the sea water elevation and heave barge response.

Table 3. The response of water column based on different environmental data.

1-year return period	10-year return period	100-year return period
Heave amplitude: 0.57 m	Heave amplitude: 0.95 m	Heave amplitude: 1.22 m
Water column amplitude: 2.08m	Water column amplitude: 3.12 m	Water column amplitude: 4.03 m

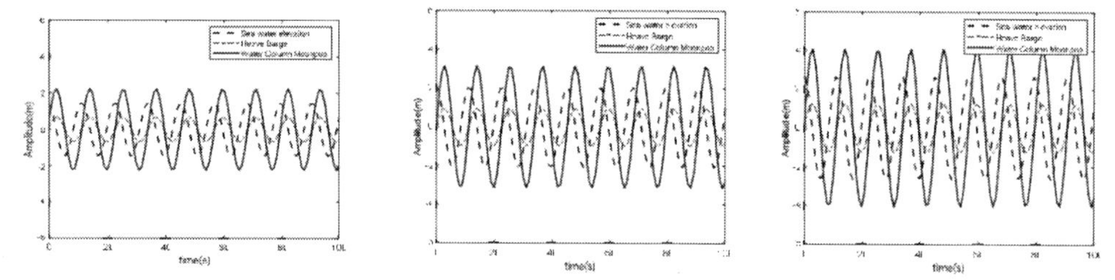

The response of water column in a moonpool in Table 3 is relative to heave barge amplitude and seawater elevation based on environmental data 1-year, 10-year and 100-year return period based on moonpool with diameter 10 meters. It clearly shows that elevation water column in moonpool will be getting higher if the seawater level outside the floating structure is increased. The maximum amplitude of piston mode occurred during 100-year storm environmental condition. For the further analysis, we will use the 100-year storm data. The study investigates the influence of vessel draft and moonpool sizing to the water column elevation inside the moonpool shows in Table 4 and Table 5.

Table 4. The response of water column with the variant of moonpool diameter.

Moonpool dia. = 10m	Moonpool dia. = 18m	Moonpool dia. = 25m
Water column amplitude: 4.03m	Water column amplitude: 5.5 m	Water column amplitude: 7.80 m

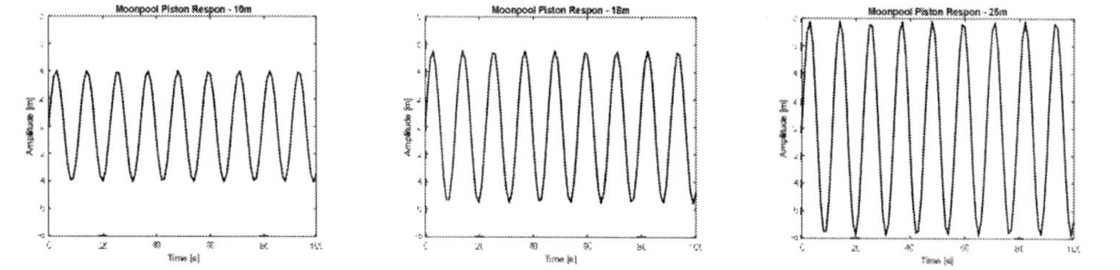

Table 5. The response of water column during several barge draft condition.

Draft = 8m	Draft = 12m	Draft = 18m
Heave amplitude: 2.40 m Water column amplitude: 3.85 m	Heave amplitude: 1.22 m Water column amplitude: 4.03 m	Barge heave amplitude: 0.71 m Water column amplitude: 5.28 m
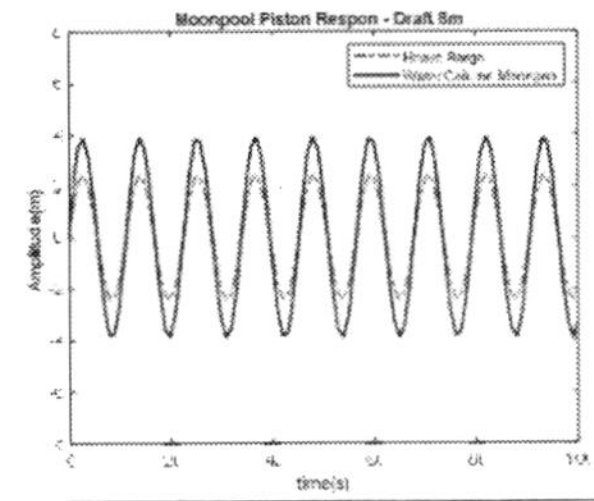	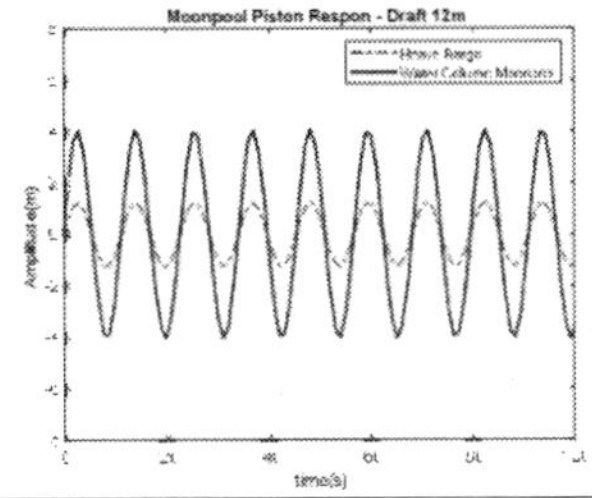	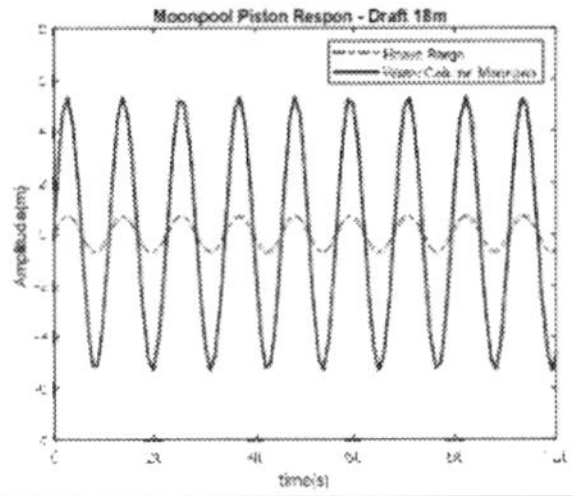

During loaded condition or deeper draft, the water column elevation will be higher. The draft of the barge is giving additional weight to the floating structure total masses. The natural frequency of moonpool is inversely proportional with the floating structure draft.

4. Verification of result and discussion

The harmonic motion equation in this study will be verified using studies performed by Matusiak [10] and DNV [3]. Further analysis to investigate the response water column will be using irregular waves theory. The motion response of the floating structure in irregular wave theory using wave spectrum Pierson Moskowitz as shown in the Equation below:

$$S_{PM}(\omega) = \frac{5}{16} H_s^2 \omega_p^4 \omega^{-5} \exp\left[-\frac{5}{4}\left(\frac{\omega}{\omega_p}\right)^{-4}\right] \quad (54)$$

Sea water elevation calculated based on consideration long wave theory and random waves formulated as follow [8]:

$$\zeta = \sum_{j=1}^{N} A_j \sin(\omega_j t - k_j x + \varepsilon_j) \text{ where } [A_j^2] = 2S(\omega_j)\Delta\omega_j \quad (55)$$

The steady state response of the floating structure in each wave direction shown as:

$$\sum_{j=1}^{N} A_j |H(\omega)| \sin(\omega_j t - k_N x + \varepsilon_j) \quad (56)$$

where, $H(\omega)$ is transfer function or amplitude response per unit of amplitude wave and ε_j is the random phase within 0 and 2π.

$$|H(\omega)| = K_s[(K_s - m\omega^2)^2 + c_1^2\omega^2]^{-\frac{1}{2}} \quad (57)$$

DNV [3] simplified the formula to calculate the fluid motion inside moonpool as per below:

$$MX + C_{s1}(X - X_s) + KX = F(t) \quad (58)$$

$$-\omega^2 MX + i\omega C_{s1}(X - G_s.X_w) + KX = G_w.X_w \quad (59)$$

$$-\omega^2\rho AJD + \kappa\sqrt{A}X + i\omega 2\zeta\sqrt{KM}(X - G_s.X_w) + \rho gAX = G_w.X_w \quad (60)$$

$$F(t) = \rho FK.A + A_{33}X_s = \rho AX_w(g.e^{-KD} - \omega^2\kappa\sqrt{A}G_s) \quad (61)$$

$$\frac{X}{X_w} = \frac{i\omega C_{s1}G_s + G_w}{-\omega^2 M + i\omega C_{s1} + K} = \frac{\frac{G_w}{\rho gA} + 2i\omega\zeta G_s\frac{\omega}{\omega_0}}{1 - \left(\frac{\omega}{\omega_0}\right)^2 + 2i\zeta\frac{\omega}{\omega_0}} \quad (62)$$

The other study is based on Jerzy Matusiak [10]. The force working in the bottom of moonpool consist of the reaction F_R due to water flux, force F_D due to dynamic pressure, hydrostatic force F_H, radiation force F_A, and Froude Krylov force F_{FK} stated as below:

$$F_{SB} = \int_{SB} f\, dS = F_R + F_D + F_H + F_A + F_{FK} \quad (63)$$

$$F_{SB} = \rho S_B \left[\dot{z}|\dot{z}| - \frac{1}{2}(\dot{z})^2 + g(T - z_3)\right] - A_z(\ddot{z} + \ddot{z}_3) + S_B\left[P_a + P_{FK}(z,t)\right] \quad (64)$$

The added mass calculated using: $A_{33} = \frac{2}{3}\rho\pi R^3$

The only force acting on body of moonpool is gravitational force and inertia force due to ship heave motion z_3

$$F_{SFS} = \int_{SFS} f\, dS = -S_{FS}P_a \int_V \rho\, g\, V = \rho A(g + \ddot{z}_3)\int_0^{T+z} dh = \rho A(g + \ddot{z}_3)(T + z) \quad (65)$$

Since the forwards speed is zero during transit period, the harmonic equation for vertical motion in the moonpool become:

$$\rho A g(T - z_3) - A_z(\ddot{z} + \ddot{z}_3) + \rho A\dot{z} + AP_a + P_{FK}(z,t).A - \rho A(g + z_3)(T + \ddot{z}) = 0 \quad (66)$$

$$\rho A(D + z + A_z)(\ddot{z} + \ddot{z}_3) + \rho A\dot{z} + \rho Ag(z + z_3) = P_{FK}(z,t) \quad (67)$$

$$\rho A(D + z)(\ddot{z}) + \rho A\dot{z} + \rho Agz = P_{FK}(z,t) - A_z\ddot{z}_3 - \rho Agz_3 \quad (68)$$

Table 6 shows the comparison of parameter in the calculation of response water column in moonpool.

Table 6. Comparison for water column motion inside moonpool.

Motion Equation: $MX + CX + KX = F(t)$			
Description	Moonpool Study	DNV RP H103	Jerzy Matusiak
Mass of barge (kg)	$9.78.10^8$	$9.78.10^8$	$9.78.10^8$
Stiffness of barge (kg/s²)	$9.58.10^7$	$9.58.10^7$	$9.58.10^7$
Natural frequency(rad/s)	0.16	0.16	0.16
Excitation force F(t)	$P_{FK}(z,t) - A_z\ddot{Z}_3$	$\rho FK.A + A_{33}\ddot{X}_s$	$P_{FK}(z,t) - A_z\ddot{z}_3 - \rho Agz_3$
Equivalent mass in the moonpool (kg)	$\rho AD + A_z$ $1.31.10^7$	$\rho A[D + \kappa\sqrt{A}]$ $1.31.10^6$	$\rho A(D + z + A_z)$ $1.17.10^6$
Stiffness (kg/s²)	ρgA $7.89.10^5$	ρgA $7.89.10^5$	ρgA $7.89.10^5$
Damping	$\frac{\rho g^2 R}{\omega^3}\left(\frac{\zeta_a}{Z_3}\right)^2$ $3.45.10^3$	$2\zeta\sqrt{KM}$ $1.62.10^5$	ρA $1.21.10^5$
Natural Frequency Moonpool(rad/s)	0.78	0.77	0.80
Natural Period (s)	8.10	8.15	7.88

Based on the above result, the highest damping factor resulted from studies by DNV. DNV using the linearized damping as critical damping ratio multiplied by the damping of structure-dependent damping in the moonpool, so that the amplitude of the water column in the moonpool based on DNV resulted the smallest amplitude for the response water column. Damping factor will reduce the water elevation trapped inside the moonpool. Figure 5 shows the comparison of response water column in a moonpool based on this study, DNV [3], and Matusiak [10] and Table 7 shows its statistical parameter. The highest water elevation inside the water column is based on the authors' study.

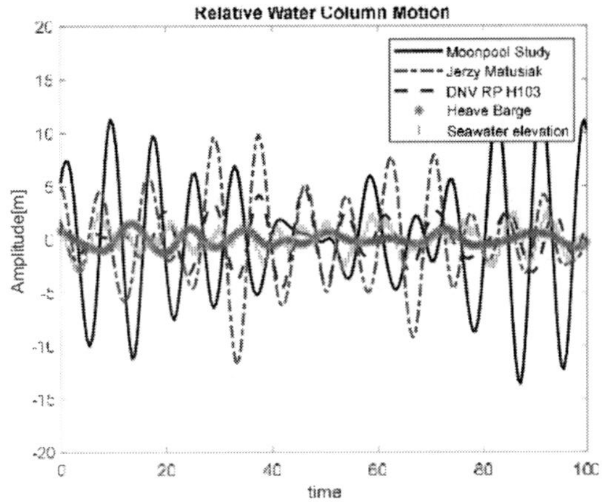

Figure 5. Relative response of water column in a moonpool.

Table 7. Statistical parameter extracted from response water column in a moonpool.

Reference	**Moonpool Study**		**DNV RPH103**		**Jerzy Matusiak**	
Parameter	Zero -Up Crossing	Zero Down Crossing	Zero -Up Crossing	Zero Down Crossing	Zero -Up Crossing	Zero - Down Crossing
H_{max} (m)	25.70	27.09	9.10	9.51	21.31	21.66
T_{max} (s)	16.51	13.02	12.85	12.67	12.41	12.04
H_{av} (m)	15.07	14.85	5.67	5.71	11.30	11.34
T_{av} (s)	8.97	9.00	9.36	9.49	8.59	8.32
H_s (m)	22.79	22.52	7.70	7.69	16.76	16.45
T_s (s)	8.26	8.49	8.76	8.58	8.38	8.37

The water column motion is almost symmetrical with respect to the still water level. Based on the statistical parameter analysis (zero up and down crossing) the max elevation is 25.70m and peak period 16.51 s. The water column elevation inside the moonpool able to present the overflowing water that may cause flooding on deck when the water oscillation in the moonpool occurred.

5. Conclusion

Based on the study, the parameters that influenced the amplitude of piston motion in the moonpool such as follow:
- Oscillation water column in a moonpool is relative motion to the ship movement.
- Amplitude of heave barge and water column piston mode is strongly influenced by environmental conditions, in this case the significant wave height. Thus, in the marine offshore operation, the limitation for sea state should be identified for the designated offshore location.
- The study on barge with 227m(length) x 42m(width) x 12m(draft) and moonpool diameter 10m, the highest water column amplitude obtained during 100-year return period with seawater amplitude 2.6m, barge heave motion 1.22m and response water column in a moonpool is 4.03m. This shown that the water column inside the moonpool always have higher elevation compare to the seawater elevation outside the barge.
- The moonpool area is in linear with the added mass, thus the bigger moonpool area will increase water column elevations inside moonpool.
- To reduce the amplitude of water column in a moonpool, damping device is required.

For the future works, the heave motion will be more accurate if the floating structure with moonpool is modelled to obtain more accurate response. The current analysis, it assumed that the fluid is inviscid, for further analysis should be done for viscous fluid to consider friction factor between fluid and the moonpool wall. For the more accurate marine lifting operation modelling, it is recommended to analyse the hydrodynamics forces acting on the object lifted through moonpool.

6. References

[1] Kuo C 1978 *A Controlled Handling Method for Effective Offshore Support Operations* OTC 3318 (Offshore Technology Reference)
[2] Revheim E 2012 *Moonpool Operations on Havila Subsea – improvement study* (Norway: University of Stavanger) p 14
[3] Det Norse Veritas 2011 *DNV RPH103 Modeling and Analysis of Marine Operations* (Det Norske Veritas) pp 50-57
[4] Stald TC 2011 *Assessment of critical factors when running and retrieving Framo pump modules through moonpool* (Norway: University of Stavanger) p 42
[5] Molin B 1999 On the piston mode in moonpools *International Workshop on Water Waves and Floating Bodies* p 1
[6] Gaillarde G, Cotteleer A. 2001 *Water motion in moonpools empirical and theoretical approach* (Netherland: Maritime Research Institute Netherlands MARIN)
[7] Aalbers AB 1984 *The Water Motion in Moonpool* Ocean Engineering 11(6) pp 557–79.
[8] Faltinsen, O 1990 *Sea Loads on ship and Offshore Structure* (United Kingdom: Cambridge University Press) pp 10-48
[9] Newman JN 1980 *Marine Hydrodynamics* (Cambridge: The MIT Press)pp152
[10] Matusiak J 1997 *Water Colum Motion in A Moonpool of a Ship* (Rakenteiden Mekanikka Vol.30)

Acknowledgments

Both authors would like to acknowledge for all the support given by Department of Marine Transportation Engineering, ITS. Apart from that, the first author wishes to thanks the lecturers in Ocean Engineering, ITB for the guidance during her studies.

IOP Conf. Series: Materials Science and Engineering 1052 (2021) 012062 doi:10.1088/1757-899X/1052/1/012062

Resistance Analysis of Rescue Boat in Calm Water Condition

I K A P Utama[1], Sutiyo[1], I K Suastika[1], A Sulisetyono[1], Hasanudin[1], Y A Hermawan[1], and W D Aryawan[1]

[1]Department of Naval Architecture, Institut Teknologi Sepuluh Nopember, Indonesia

Abstract. Search and rescue operations, especially in ship accidents, require speed and reliability from the rescue vessels in facing the challenges of extreme weather and sea waves. A rescue boat hydrodynamic performance study is very important to be carried out in order to determine the ship's ability to withstand dangerous conditions while in operation. A combination of using Maxsurf and CFD method are implemented to estimate the resistance of 3 (three) selected designs of crew boat, which demonstrated a good agreement between the two techniques. Selection of the final design also uses seakeeping criteria, i.e. having lower pitching and rolling responses. The final model is further tested in a towing tank belongs to the Institut Teknologi Sepuluh Nopember (ITS) and agrees well with Maxsurf and CFD calculation.

1. Introduction

Although the number of investigations has dramatically increased, there is an overwhelming need to study the design of an efficient high-speed craft. In recent years, motorboats, pilot boats, personnel carriers, search and rescue (SAR) boats and some passenger boats have all been considered for attaining high speeds. Accelerating to a high speed is directly related with resistance characteristics and hull form[1]. At high speeds, the hull is subjected to dynamic effects, because of the hydrodynamic characteristics of the geometry, as well as the heave and pitch motions, termed trim and sinkage. At high speeds, according to the dynamic conditions of the rigid body, the total resistance increases rapidly and leads to excessive fuel consumption[2]. Generally, with Froude numbers between 0.4 and 1.58, a typical high-speed craft enters the transient (hump) regime. As the vessel advances through higher speeds, the wave resistance also increases[3]. To estimate the total resistance of the full-scale ship, in addition to the drag values, all attitudes including trim and sinkage must be considered and predicted accurately[4].

The planing hulls moving at high speed also generate hydrodynamic lift, which reduces the wave-making resistance and may generate spray around the bow zone. The spray should also be considered and visualized in the early stages of ship design, because it is intricately tied to the geometry, trim and speed of the hull. For the most accurate results, high speed craft model tests have been the industry standard for decades[5]. Model tests estimate total resistance characteristics of a high-speed boat, but trim and sinkage values must also be measured correctly to be included in full-scale analyses[6].

Detailed sketches of ship resistance which consist of several ship resistance components, the interactions between the drag components and the quantity of their interactions, are some of the things that concern ship designers in designing fast ships[7]. Likewise, how to arrange the mechanism for the formation of ship resistance components including how to calculate or estimate the amount of resistance in the case of fast boats such as hovercraft, flying boat, and rescue boat. The results of the amount of resistance are especially indispensable in determining the amount of propulsion of a ship to meet the ship's speed as desired [8]. Ship resistance will be divided into several components which include resistance due to friction and pressure resistance due to viscosity, as well as resistance due to waves.

The interaction between the ship resistance components can be indicated by the amount of the ship drag coefficient and the Reynold and Froude numbers [9]. For fast ships, the mechanism for the formation of ship resistance components develops from the resistance formed due to the characteristics of the ship's motion. Fast ships get hydrostatic and hydrodynamic forces, the body of the ship that is immersed in the water is getting smaller, causing lower ship resistance.

Rescue boat have hard-chine planning hulls which the hard-chine planning hull is configured to develop positive dynamic bottom pressures at high speed. These positive pressures lift the hull and thereby reduce the buoyant component of hull support. Consequently, the wetted bottom area, when planning, is substantially smaller than the static wetted area. Unfortunately, the addition of induced drag associated with the development of dynamic lift results in a total resistance- weight ratio that is substantially greater than those for a displacement or semi-displacement vessel at their design speeds.

In this paper, a rescue boat analysis was carried out with a variation of 3 models with 3 approaches, among others: (a) Maxsurf Resistance, (b) CFD, and (c) Experiment. The Maxsurf software approach uses the Savitsky method where the ship uses a planning hull with Fr = 0.40 to 1.58 and a stern trim will occur. Furthermore, the CFD approach uses ANSYS CFX software with steady state simulations and multiphase domains [10]. The study objectives were to determine the total resistance of rescue boat will be validated with experimental model.

2. Rescue boat model

The high-speed craft referred to in this paper is a type of rescue boat. Rescue modelling was carried out with 3 variations of the principal dimensions as shown in Table 1

Table 1 Principle Particular of Rescue Boat.

Parameter	Units	Boat-1	Boat-2	Boat-3
LOA	m	4.56	5.296	3.758
LWL	m	4.3	4.756	3.559
B	m	1.52	1.815	1.363
T	m	0.35	0.396	0.30
H	m	0.92	1.035	0.607
Vol. Displacement	m^3	0.936	1.414	0.904
Displacement	kg	936.4	1414	904

The rescue boat model uses a v-hull with a variation of chine. The first variation is without using chine, as shown in Figure 1. The next model is using 1 chine as in Figure 2 and in the third variation using the double chine. the selection of this variation is expected to be able to obtain smaller resistance at high speeds.

IOP Conf. Series: Materials Science and Engineering 1052 (2021) 012062 doi:10.1088/1757-899X/1052/1/012062

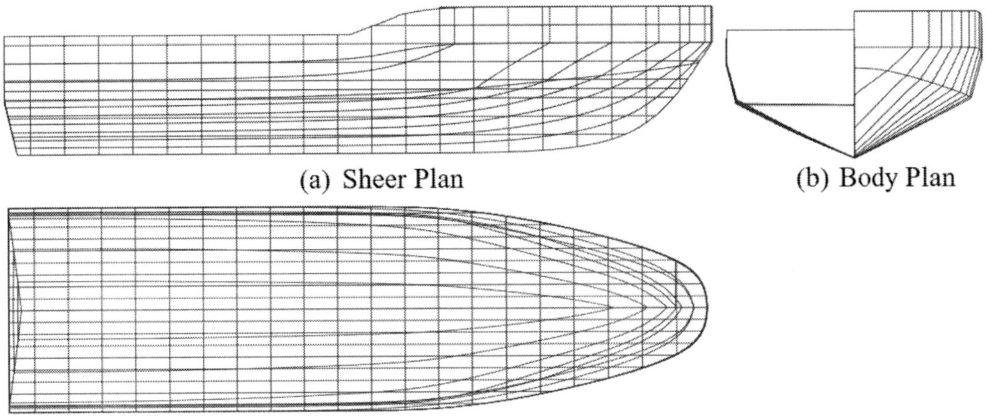

(a) Sheer Plan (b) Body Plan

(c) Breadth Plan

Figure 1 Lines and Body Plan (Boat-1).

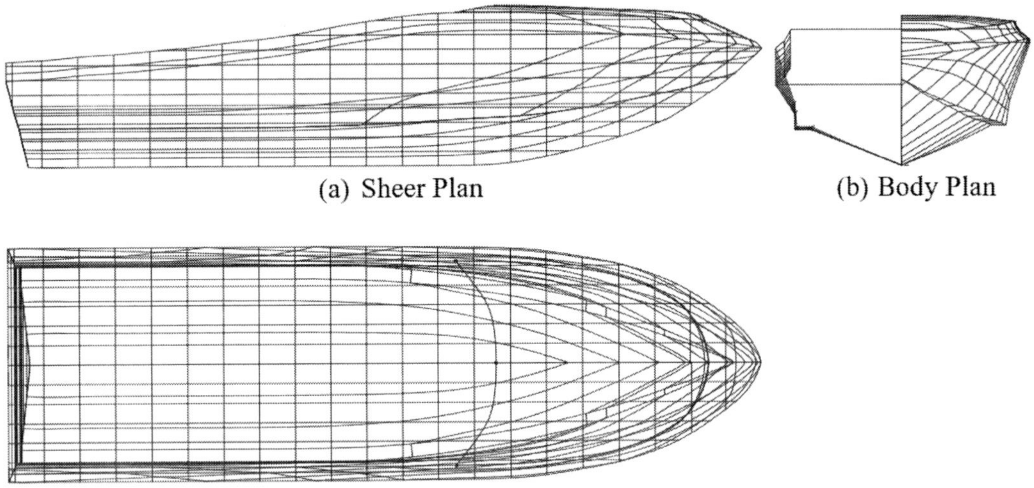

(a) Sheer Plan (b) Body Plan

(a) Breadth Plan

Figure 2 Lines and Body Plan (Boat-2).

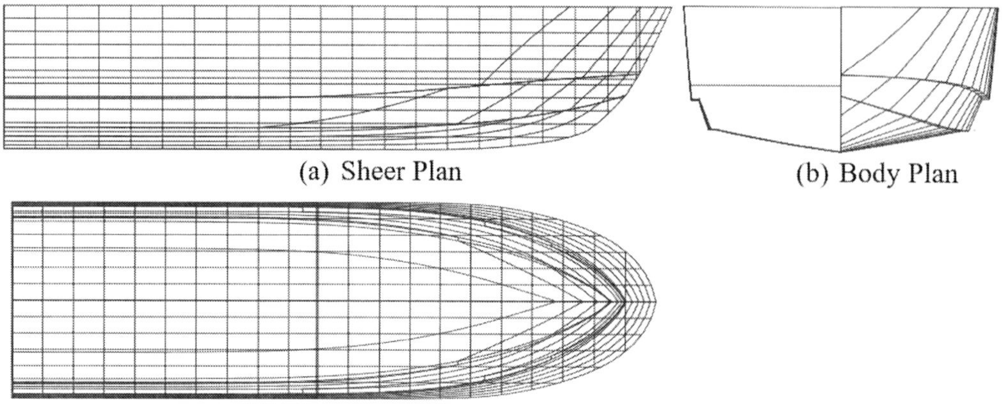

(a) Sheer Plan (b) Body Plan

(c) Breadth Plan

Figure 3 Lines and Body Plan (Boat-3).

3. Method

3.1. Maxsurf Resistance

Rescue Boat resistance calculations using the Maxsurf resistance with the Savitsky method. It was used for estimating the resistance of planing hulls when in the planing speed regime[11]. In 1964, Savitsky [12] proposed a semi-empirical technique to evaluate the hydrodynamic performance of planning hulls which is simple and fast. The results of [13] showed that his presented equation for resistance prediction is very accurate in comparison with experimental tests at higher Froude numbers Fr > 1. It was also shown in [14] that this semi-empirical method is possible to implement for different prismatic hulls and has an acceptable accuracy in comparison with measurements. Generally, the equations of Savitsky's method were presented for prismatic hulls. In a more realistic view, a planning hull has a variable deadrise angle along its length; therefore, the simple assumption of prismatic hull form is not applicable to warped planning hulls. According to [15] and [16] , the effect of variable deadrise could be implemented on the [12] equations by taking effective deadrise and beam at LCG section, as following formula (1), (2) and (3), as shown at Figure 4.

$$D = D_p + \frac{D_f}{\cos\tau} \tag{1}$$

Where:
D = Total drag
D_p = Pressure drag;

$$D_p = \Delta \tan\tau \tag{2}$$

D_f = Friction drag
;

$$D_f = \frac{C_f v^2(\lambda b^2)}{2cos\beta^4} \tag{3}$$

τ = trim angle
Δ = load
C_f = Friction Coeficient
v = velocity
λb^2 = Viscous Drag
β = deadrise angle

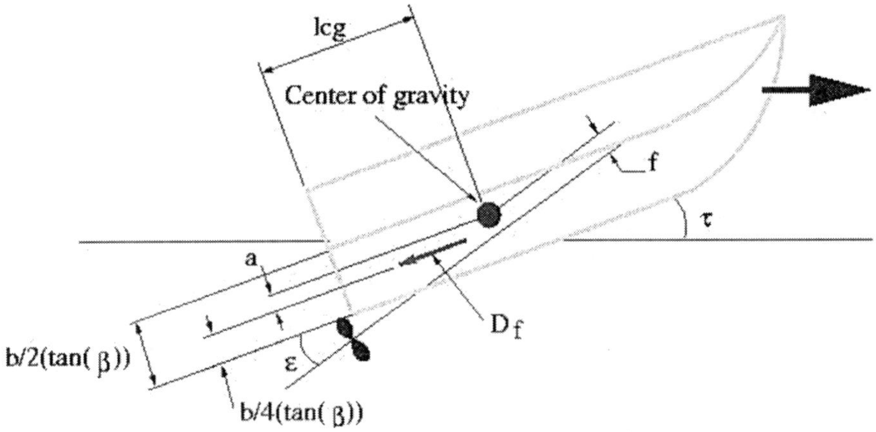

Figure 4 High Speed Planning Hull.

3.2. Numerical Simulation

Ship Resistance analysis also uses CFD approach. Analysis using CFD involves solving the governing equations of fluid flow numerically. The three governing equations of fluid flow are the continuity equation or the mass conservation equation, the momentum conservation equation or the Navier-Stokes equations, and the energy conservation equation. Every CFD code solves the mass and momentum conservation equations in the background as this form the basis of any fluid calculations.

3.2.1 Governing Equations

The averaged continuity and momentum equations for incompressible flows may be given as in the following two equations [17]. The instantaneous equations of mass, momentum can be written as follows in Equations (4) and (5):

$$\frac{\partial \rho}{\partial t} + \nabla.(\rho U) = 0 \tag{4}$$

$$\frac{\partial(\rho U)}{\partial t} + \nabla.(\rho U \; x \; U) = -\nabla p + \nabla.\tau + S_M \tag{5}$$

where the stress tensor, τ is related to the strain rate as follows:

$$\tau = \mu\left(\nabla U + (\nabla U)^T - \frac{2}{3}\delta\nabla.U\right) \tag{6}$$

The computation which is applied uses the SST model that provides boundary layer modelling with high accuracy because it uses k-ε and k-ω combined for numerical simulations. The SST model equation provides high accuracy by separating fluid flow in turbulent flow areas. By applying those two models, it includes the boundary layer area that is close to the model wall and which is far from the model wall can be appropriately covered. Bradshaw's relationship is also applied in SST model to perform predictive turbulence with good separation [18], as following Equations (4) and (5);

$$\frac{\partial}{\partial t}(\rho k) + \frac{\partial}{\partial x_i}(\rho k u_i) = \frac{\partial}{\partial x_j}\left(\Gamma\omega\frac{\partial\omega}{\partial x_j}\right) + G^{\sim}k + Yk + Sk \tag{7}$$

$$\frac{\partial}{\partial t}(\rho\omega) + \frac{\partial}{\partial x_i}(\rho\omega u_i) = \frac{\partial}{\partial x_j}\left(\Gamma\omega\frac{\partial\omega}{\partial x_j}\right) + G^{\sim}\omega - Y\omega + D\omega + S\omega \tag{8}$$

where $G^{\sim}k$ represents the generation of turbulence kinetic energy due to mean velocity gradients, Gω represents the generation of ω, Γk and Γω represent the effective diffusivity of k and ω, respectively. Sk and Sω are user-defined source terms. Yk and Yω represent the dissipation of k and ω due to turbulence. Dω represents the cross-diffusion term.

3.2.2 Numerical Simulation

Calculation have been performed with the commercial CFD code ANSYS CFX. The code is based on Finite Volume method. The parameters and solver settings used to define the numerical solution are provided in Table 2.

Table 2 Numerical Simulation Properties.

Properties	Fine Mesh
Grid	Unstructured with Inflation
Total Element (x10^3)	Approx. 568 to 2,876
Domain	Air + Water, SST Turbulence
Boundary physics:	
Inlet	Velocity Inlet Fr=0.40 to 1.58
Outlet	Velocity Outlet Fr=0.40 to 1.58
Top	Opening
Bottom	Wall with symmetry
Model Rescue Boat	Wall with no-slip
Solver settings:	
Advection scheme	High Resolution (ANSYS, 2017)
Convergence criteria	Residuary type: RMS, Target: 0.0001

The computations using free-surface model apply, velocity outlet, and symmetry conditions for the inlet, the outlet, and centre/side/top/bottom boundary planes, respectively as shown at Figure 5. The standard wall function is used to enhance computational efficiency [19, 20].

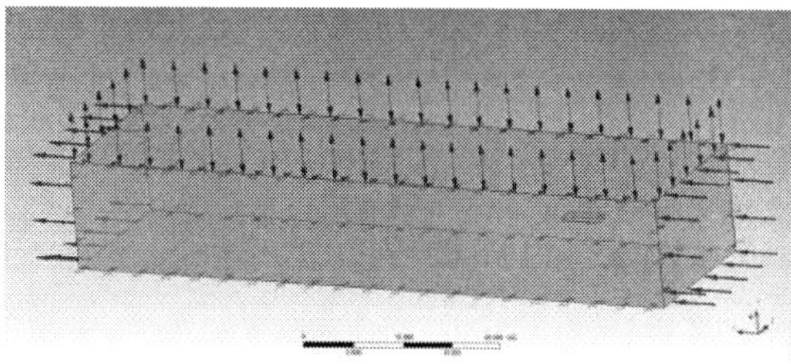

Figure 5 Computational Domain.

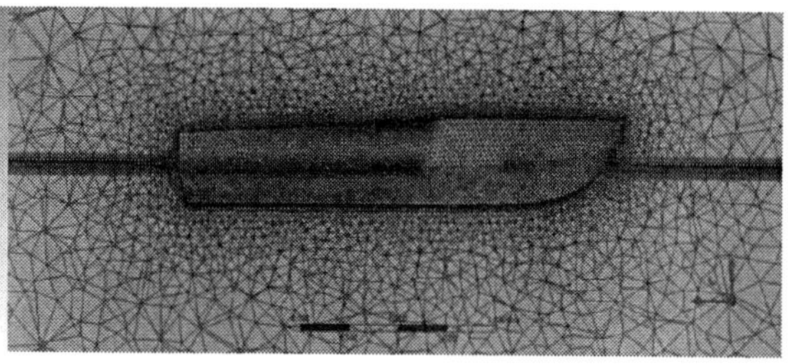

Figure 6 Unstructured mesh with inflation.

The grid generator of Design Modeler was used for meshing the computational domain with tetrahedral grid elements with approx. 568 to 2,876 (x10^3) with inflation. The mesh was refined over the hull surface and in the region of water free surface near the ship hull to correctly predict the value of hull pressure resistance and to get a sharp free surface. Figure 6 show the unstructured tetrahedral grid with inflation mesh of the computational domain and on the Rescue Boat hull surface used with ANSYS CFX solver.

Selection of the meshing amount through the independent grid study approach as shown in Table 3 and Figure 7. Independent Grid Simulation was carried out on the boat at velocity 17 knot. The independent grid study shows that the optimal model simulation is on a grid of 1,583 x10^3 with resistance of 2.265kN with an optimal difference of Cv of 1.63%.

Table 3 Grid Independence.

Element (x 1000)	568	1,103	1,583	2,876
Resistance (kN)	2.969	2.465	2.265	2.228
Difference (%)		17.00	8.11	1.63

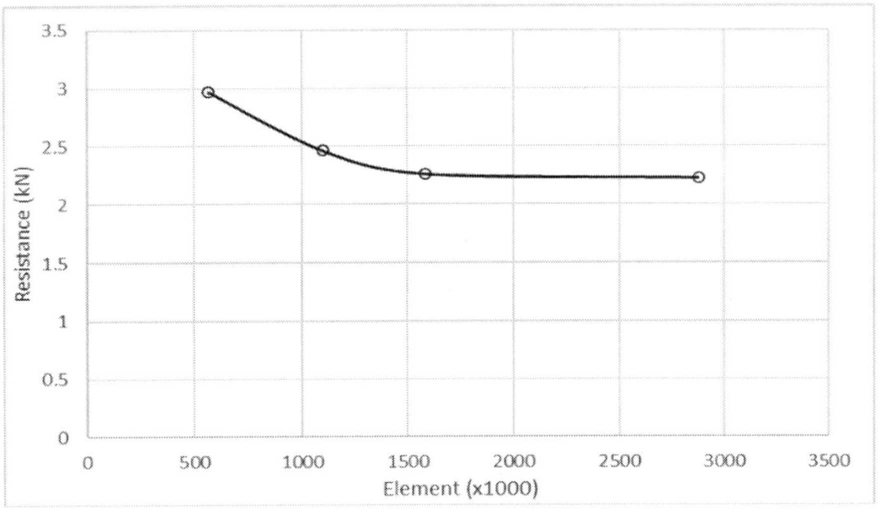

Figure 7 Grid Independence Study.

3.3. *Experiment*

The experiments were carried out in the Hydrodynamic Laboratory of Naval Architecture Department at the Institut Teknologi Sepuluh Nopember (ITS) at Surabaya, Indonesia. Rescue boats were tested using a scale of 1: 11.3 in order to obtain the model sizes as shown in Table 4

Table 4 Physical Model Test.

Parameter	Boat-1	Model	Unit
LOA	4.560	0.403	m
B	1.520	0.106	m
T	0.350	0.031	m
H	0.920	0.082	m
Displacement of model	936.400	0.645	kg

The test model creation procedure followed ITTC recommendations 7.5-02-05-01[21]. The size of the model is 1: 11.3 from the actual ship size, $\lambda = 1/7$ [22]. The size of this winged ship test model adjusts the size of the towing tank test facility at the Hydrodynamic Laboratory of Naval Architecture Department at the ITS with a length of 50 m of pulling pool; width 3 m; and a depth of 2 m. The maximum towing carriage speed is 4.5 m/s. The mass density of the air towing tank is 999.1 kg / m3, and the mass density of the air is + 1.164 kg/m3, (for temperatures of 30 ^{0}C), while the water temperature ranges from 16 - 18°C.

The test model is attached to the towing carriage through the towing guide (see Figure 8) with the setup that the test model can only move heave and pitch freely, cannot move yawing or swaying so that no rolling or heeling moments arise. The trim meter was installed in an upright position at the front and at the rear of the test model in a position that did not interfere with the towing guide.

The test model is installed with the center line of the model right in the center line of the draw pool and parallel to the side wall of the pull pool. The instrumentation cable is laid in such a way that at the time of measurement it does not interfere with the motion of the test model.

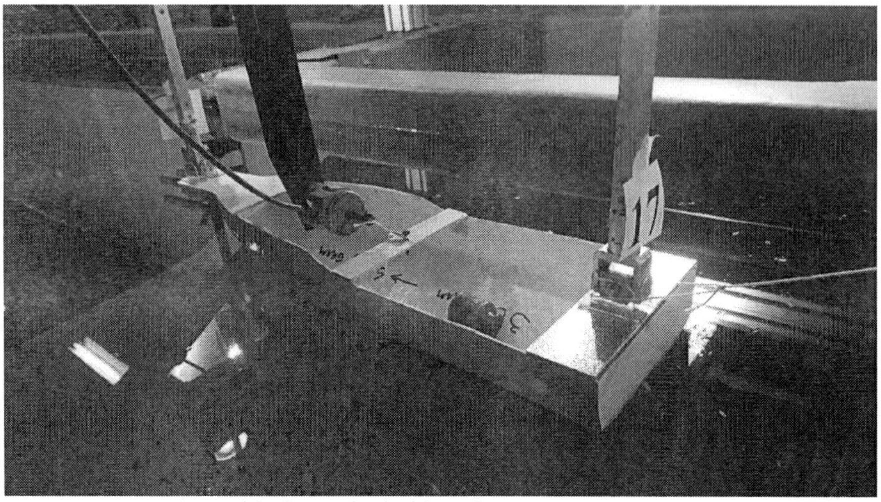

Figure 8 Towing guide frame for holder of Boat-1 test model in Towing Tank.

4. Results and Discussion

Rescue Boat resistance calculations with the Savitsky and CFD approaches show the same trend, as shown in table 5. In sequence, the total resistance coefficient of the rescue boat with Savitsky starting from the Boat-2 model produces a C_T of 2.9696, then the Boat-1 model with a C_T of 2.0453 and finally the Boat-3 model with a C_T of 1.6963. As well, the largest order of total resistance coefficient CFD rescue boat by starting from the model Boat-2 generate C_T of 3.2937, then the model Boat-1 with C_T at 2.5946 and the last with C_T models Boat-3 at 2.4330.

Table 5 Total Resistance Coefficient of Rescue Boat.

Fn	Savitsky			CFD		
	Boat-1	Boat-2	Boat-3	Boat-1	Boat-2	Boat-3
0.40	0.0217	0.0277	0.0270	0.0208	0.0269	0.0261
0.63	0.2567	0.4139	0.3303	0.1551	0.3454	0.2783
0.87	0.6068	0.9622	0.6759	0.5567	0.8283	0.6068
1.11	1.0288	1.5931	0.9933	1.1092	1.4807	1.0426
1.35	1.4858	2.2362	1.3126	1.8209	2.2564	1.6300
1.58	2.0453	2.9696	1.6963	2.5946	3.2937	2.4330

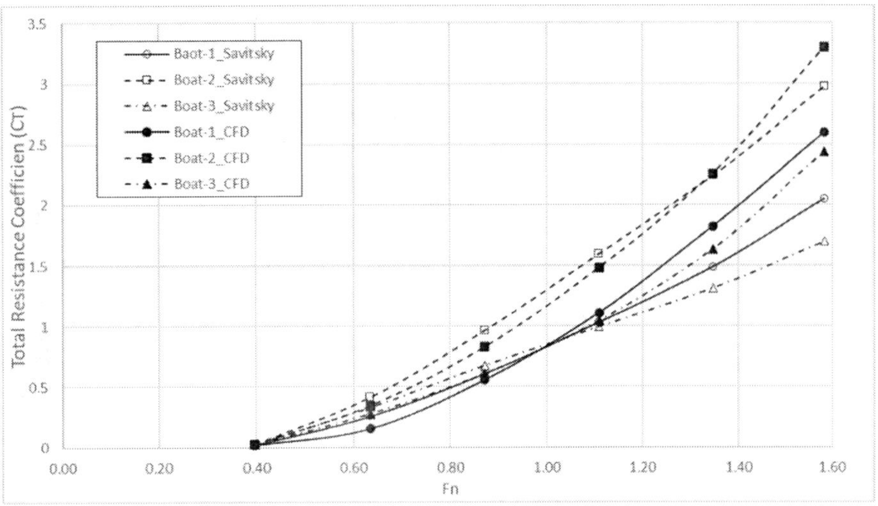

Figure 9 Difference of Rescue Boat Total Resistance Coefficient.

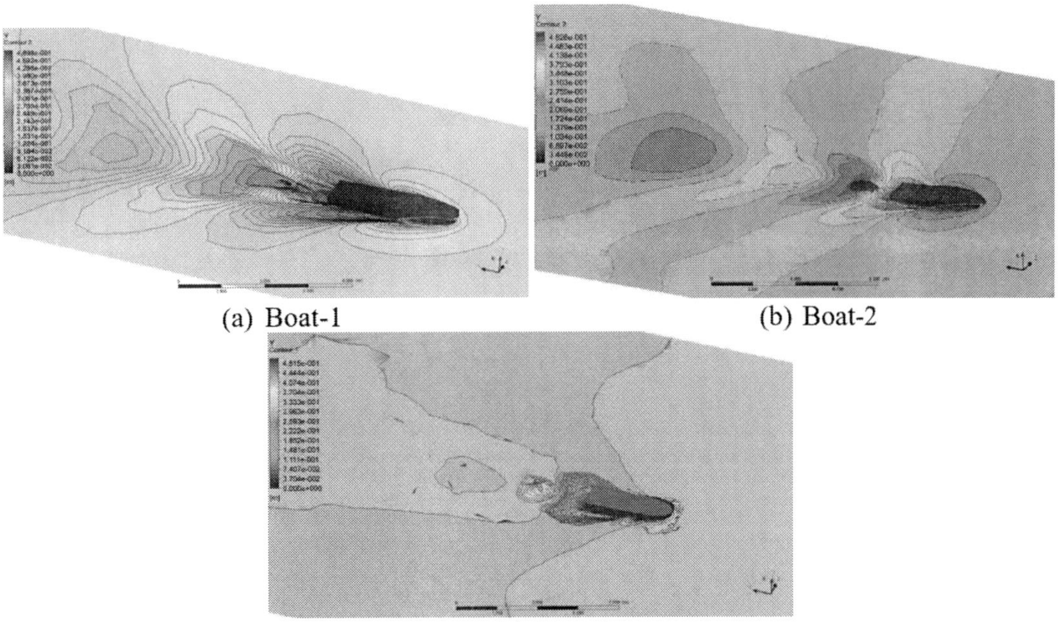

(a) Boat-1 (b) Boat-2

(c) Boat-3

Figure 10 Wave Height Contour of Rescue Boat.

Presented numerically computed resistance force (see Table 6 and Figure 11) shows that its pattern mostly coincides with the experimental data. The accuracy of the presented numerical approach is due to the generated mesh around the hull and the prism layer mesh, which have been used to create the computational domain. Also, the other important factor in resolving the generated wave is the capturing scheme which is introduced in the volume of fluid section of the current paper. In addition to the CFD results with experiment have 1.09% in average, this is due to the sum of the negative and positive differences so that the result is small. However, when viewed from the difference at each speed, the difference is quite large. This sizable difference is caused by the behaviour of fast boats that occurred in trim during the test but in numerical simulations it is not treated the same (fix body) as shown in Figure 10. Furthermore, the results of Savitsky's method for resistance are shown Table 6. The average resistance error for the Savitsky's method is 10.23% compared to the experimental results. Whisker spray and deadrise angle corrections had significant effects on the resultant magnitudes of the Savitsky's resistance force. It can be said that both CFD and Savitsky's method has an acceptable trend on the resistance force in comparison with experimental results.

Table 6 Total Resistance Coefficient of Boat-1.

Fr	Savitsky	CFD	Experiment	Difference (%)	
				Savitsky vs Experiment	CFD vs Experiment
0.40	0.0217	0.0208	0.023	-5.65	-9.57
0.63	0.2567	0.1551	0.1814	41.51	-14.50
0.87	0.6068	0.5567	0.4493	35.05	23.90
1.11	1.0288	1.1092	0.9936	3.54	11.63
1.35	1.4858	1.8209	1.9373	-23.31	-6.01
1.58	2.0453	2.5946	-		
		Average		10.23	1.09

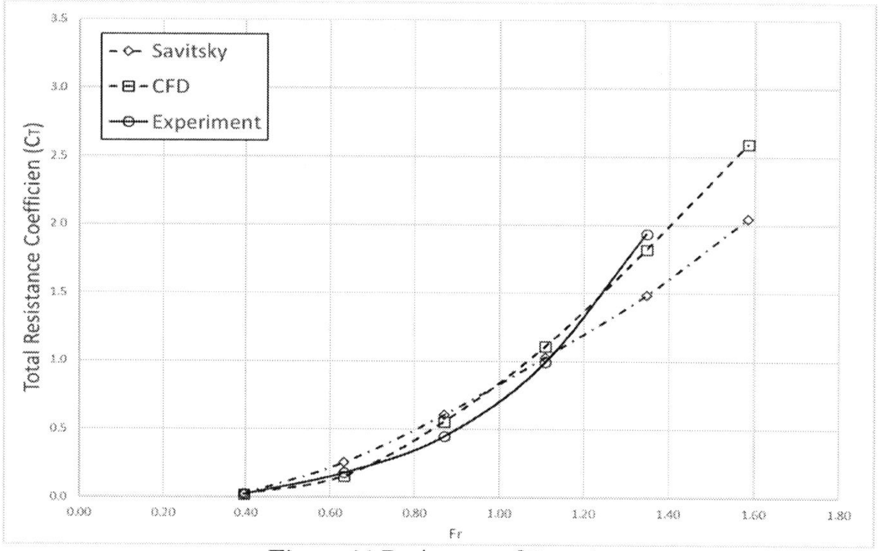

Figure 11 Resistance of Boat-1.

(a) Experiment (b) CFD

Figure 12 Wave elevation.

5. Conclusion

The calculation of the resistance of rescue boats with the 3 models using the Maxsurf (together with the Savitsky method for the calculation), CFD and experiment showed good agreement and provided several interesting conclusions: (1) each approach gives different results, but nevertheless has the same trend of resistance; (2) the difference between the calculation using the Maxsurf-Savitsky and CFD methods varied from 3.5 % to 41.5 %, thus further study is recommended; (3) however, CFD calculation is trusted to provide a more accurate picture than the Savitsky method, including in providing visualization of the flow around the rescue boat.

Since the experimental study was only carried out on Boat-1, the comparative study between Boat-2 and Boat-3 against the Maxsurf-Savitsky and CFD cannot me made. Further experimental test on the two models is found to be necessary to provide optimum results on the resistance analysis of the rescue boat.

Acknowledgment

The authors wished to thank the Institut Teknologi Sepuluh Nopember (ITS) for funding the current work under a scheme called "Research Excellence" (Penelitian Unggulan) with contract number: 816/PKS/ITS/2020.

References

[1] D. J. Kim, S. Y. Kim, Y. J. You, K. P. Rhee, S. H. Kim, and Y. G. Kim, "Design of high-speed planing hulls for the improvement of resistance and seakeeping performance," *Int. J. Nav. Archit. Ocean Eng.*, 2013, doi: 10.3744/JNAOE.2013.5.1.161.

[2] N. Bialystocki and D. Konovessis, "On the estimation of ship's fuel consumption and speed curve: A statistical approach," *J. Ocean Eng. Sci.*, 2016, doi: 10.1016/j.joes.2016.02.001.

[3] E. O. Tuck, D. C. Scullen, and L. Lazauskas, "Wave Patterns and Minimum Wave Resistance for High-Speed Vessels," *24th Symp. Nav. Hydrodyn.*, 2002.

[4] T. Tezdogan, Y. K. Demirel, P. Kellett, M. Khorasanchi, A. Incecik, and O. Turan, "Full-scale unsteady RANS CFD simulations of ship behaviour and performance in head seas due to slow steaming," *Ocean Eng.*, 2015, doi: 10.1016/j.oceaneng.2015.01.011.

[5] ITTC, "ITTC - Recommended Procedures and Guidelines - Resistance Test (7.5-02-02-01)," 2011.

[6] M. Fathi Kazerooni and M. S. Seif, "On the scale effects of resistance model tests of high-speed monohulls," *J. Brazilian Soc. Mech. Sci. Eng.*, vol. 41, no. 4, pp. 1–14, 2019, doi: 10.1007/s40430-019-1695-x.

[7] K. J. Paik, P. M. Carrica, D. Lee, and K. Maki, "Strongly coupled fluid-structure interaction method for structural loads on surface ships," *Ocean Eng.*, 2009, doi: 10.1016/j.oceaneng.2009.08.018.

[8] O. M. Faltinsen, *Hydrodynamics of high-speed marine vehicles*. 2006.

[9] L. Nikolopoulos and E. Boulougouris, "A study on the statistical calibration of the holtrop and Mennen approximate power prediction method for full hull form, low froude number vessels," *J. Sh. Prod. Des.*, 2019, doi: 10.5957/JSPD.170034.

[10] ITTC, "ITTC – Recommended Procedures and Guidelines - Practical guidelines for ship CFD applications. 7.5-03-02-03 (Revision 01)," *ITTC – Recomm. Proced. Guidel.*, 2014.

[11] BENTLEY, *User Manual MAXSURF Resistance*. 2018.

[12] D. Savitsky, "Hydrodynamic design of planing hulls," *Mar. Technol.*, 1964.

[13] D. Savitsky, "Procedures for Hydrodynamic Evaluation of Planing Hulls in Smooth and Rough Water," *Mar. Technol. SNAME News*, 1976.

[14] T. C. Fu, R. Akers, T. O'Shea, K. Brucker, D. G. Dommermuth, and E. Lee, "Measurements and computational predictions of a deep-V monohull planing hull," in *11th International Conference on Fast Sea Transportation, FAST 2011 - Proceedings*, 2011.

[15] D. Savitsky, "On the Subject of High-Speed Monohulls," *Greek Sect. Soc. Nav. Archit. Mar. Eng.*, 2003.
[16] N. Santoro, E. Begovic, C. Bertorello, A. Bove, S. De Rosa, and F. Franco, "Experimental study of the hydrodynamic loads on high speed planing craft," in *Procedia Engineering*, 2014, doi: 10.1016/j.proeng.2014.11.143.
[17] J. H. Ferziger and M. Perić, *Computational Methods for Fluid Dynamics*. 2002.
[18] F. R. Menter, M. Kuntz, and R. Langtry, "Ten Years of Industrial Experience with the SST Turbulence Model Turbulence heat and mass transfer," *Cfd.Spbstu.Ru*, 2003.
[19] H. A. Ahmad, A. F. Ayob, and P. History, "State of the Art Review of the Application of Computational Fluid Dynamics for High Speed Craft," *J. Ocean. Mech. Aerospace-Science Eng.*, 2017.
[20] Sutiyo, E. Yuliora, and I. K. A. P. Utama, "Numerical Investigation Into The Pressure Distribution And Form Factor Effect Of Slenderbody Catamaran," in *Proceeding of the International Conference on Ship and Technology (ICSOT) 2019, Developments Marine Design Architecture*, 2019, no. November, pp. 25–26.
[21] ITTC, "ITTC – Recommended Procedures Testing and Extrapolation Methods Resistance Test," *Int. Towing Tank Conf.*, 2002, doi: 10.1017/CBO9781107415324.004.
[22] V. Bertram, "3 rd International EuroConference on High-Performance Marine Vehicles," no. September, pp. 14–17, 2002.

Analysis of pontoon multi pendulum motion response trimaran model at ocean wave power plant based on pendulum system (PLTG-SB)

I K A P Utama[1], R Hantoro[2], E Septyaningrum[2], Q Khasanah[2], J Prananda[3] and I S Arief[3]

[1]Department of Naval Architecture, Institut Teknologi Sepuluh Nopember, Surabaya, Indonesia
[2]Department of Engineering Physics, Institut Teknologi Sepuluh Nopember, Surabaya, Indonesia
[3]Department of Marine Engineering, Institut Teknologi Sepuluh Nopember, Surabaya, Indonesia

Abstract. Ocean waves are one of the renewable energy potentials for an archipelago country like Indonesia. The development of Wave Energy Converter presented in the research carried out is a system that uses a pontoon with a pendulum for energy extraction, which is called the Pendulum System Wave Power Plant (PLTG-SB). The shape of the pontoon uses a trimaran model with a cylindrical shape as the main body and support. The simulation of the response of the pontoon with multi pendulums shows the maximum RAO pitch value of 1.15 ° / cm in the 6 second wave period and the response to the change in deviation is the maximum value in the 6 second wave period with a deviation of 198.07°. The variation in the length of the pendulum arm gave the largest deviation value of 106.7 cm, while in the variation of the number of pendulums the largest deviation occurred on the pontoon with the number of pendulums n = 2.

1. Introduction

Indonesia is an archipelago that has a wider water area than the land area, which is about 3,5 million km^2 [1]. Because the area of Indonesian waters is wider than land, there is a potential source of marine energy that it is be converted into electrical energy. According to the theory of global marine energy resources equivalent to 17,400 TWh/year [2]. Compared to other renewable energy sources such as solar and wind energy, ocean wave energy provides up to 90% availability with an unlimited potential area [3]. Until now, several types of WECs have been developed in the world, including Wave Activated Bodies (WABs), Oscillating Water Columns (OWCs) and Overtopping Devices [4].

In several countries, WECs have been developed as an alternative energy substitute for fossil energy. In 2015 China through the Guangzhou Institute of Energy Conversion (GIEC) developed a Sharp Eagle WEC prototype with a size of 36m x 24m x 16m with a power capacity of 100kW on Wanshan Island [5]. Yoshio Masuda developed a buoy that was later called a floating oscillating water column and was developed in 1965 [6]. Denmark has also developed a WECs type overtopping device called Wavedragon [7]. Norway is also developing WECs of a similar type called the Tapered Channel Wave Power Device (Tapchan) [8].

In this research, a tool for converting ocean wave energy into electric energy in the pendulum system (PLTG-SB) has been developed. This research was first conducted by Zamrisyaf in 2002[9]. The working principle of PLTG-SB is a system that has been designed in the form of a pontoon, placed to float above sea level. Then the system will follow the motion or wave flow according to the frequency of the ocean waves, causing the pendulum to move like a bell. This movement is then transmitted in a rotating motion to drive the dynamo which then becomes electrical energy. The power generated by the PLTG-SB is much influenced by the dimensions of the pontoon, the length of the pendulum, the mass of the pendulum, and the arrangement of the pontoon [9].

Research related to the multi-pendulum pontoon has been conducted [10]. This research was conducted using an octagonal pontoon with n = 3 pendulums placed horizontally to determine the amount of electrical energy and the resulting characteristics of the pendulum mass variation, pendulum arm length, wave frequency and amplitude. The results of this study indicate that the stress value will increase as the number of pendulums increases.

Based on the research that has been done, further research is needed regarding the multi-pendulum pontoon to the changes in the deviation of each pendulum due to ocean waves. This paper attempts to convey the results of a CFD simulation study related to the relationship between changes in pendulum length, pendulum number, and wave period to the response of the multi-pendulum pontoon motion to determine the characteristics of the pontoon.

2. Methodology

2.1. Pontoon and Pendulum Design

The design of the pontoon and the pendulum as in previous research, shown in Figure 1, varies the number of pendulums, the length of the pendulum arm, and the period of the wave hitting the pontoon. The ratio of the pontoon is 1: 1 with the dimensions as in Table 1.

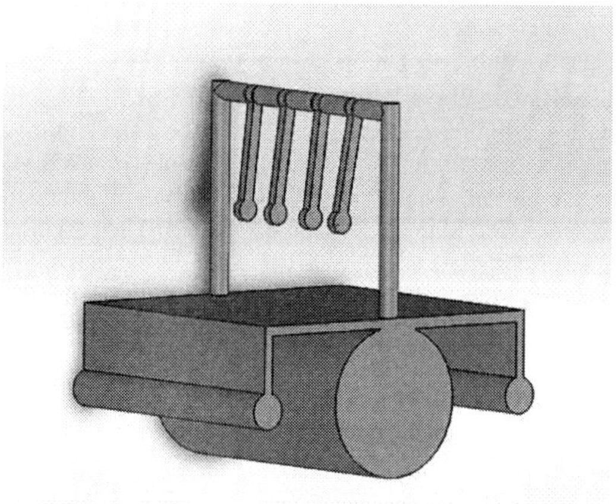

Figure 1. Design 3D Pontoon 1:1.

Table 1. Geometry of Pontoon.

Parameter	Unit (cm)
Diameter of Big Cylinder	165
Diameter of Small Cylinder	40
Length of Big Cylinder	320
Length of Small cylinder	320
Height of Jukung	83
Length of Plate	320
Wide of Plate	170

Parameter	Unit (cm)
Pendulum Pole Height (from plate)	250
Pendulum Pole Arm	168

2.2. Simulation

Simulations were carried out to determine the interaction of the waves with the pontoon using AQWA Hydrodynamics Diffractions and Hydrodynamic Responses, and to determine the response of the pontoon to the pendulum using Rigid Dynamics. In this simulation stage, computational fluid dynamics (CFD) are carried out through the pre-processing, processing, and post-processing stages.

In the pre-processing stage, a setup is carried out by inputting data in the form of:

- water size, on the x and y-axes in the geometry section.
- Water draft.
- Center of gravity (COG) at the point mass.
- Input of inertia at the point mass.

Then meshing is performed by inputting the defeaturing tolerance and maximum element size. In the smaller pendulum, mesh sizing is added.

Table 2. Boundary condition of Pontoon Modelling.

Parameter	Unit
Water size X	15000 cm
Water size Y	5000 cm
Water draft	1/2 D
Center of Gravity	57 cm
Input of Inertia	Ixx = 4095482247.7 kg.cm^2
	Iyy = 9219950155.6 kg.cm^2
	Izz = 7330776665.4 kg.cm^2

Table 3. Detail of Mesh.

Parameter	Unit
Defeaturing Tolerance	4 cm
Max Element Size	10 cm
Total Nodes	29915
Total Elements	29934

The processing stage is carried out by wave simulation by inputting the wave period and wave direction used. This stage is performed during the simulation on AQWA Hydrodynamics Diffractions. Meanwhile, the AQWA Hydrodynamics Response simulation includes data in the form of wave amplitude and wave period with regular wave types. The final stage of simulation (post-processing) is a solution or result in the form of a pontoon response due to ocean waves, then the RAO (Response Amplitude Operator) graph.

In the simulation using Rigid Dynamics, the data entered is in the form of the actual response value from AQWA to determine the value of the pendulum deviation due to the motion of the pontoon. In the pre-processing stage, the data entered is in the form of a point mass for each component. Then given the connections in the form of joint cylindrical to define the relationship between the pontoon with the ground and the pontoon with the pendulum. Furthermore, at the processing stage, the input is given in the form of actual response data in the Joint - Rotation section. Then the post-processing stage, the results of the simulation shown by the change in the deviation value due to the motion of the pontoon on each pendulum on the joint probe are displayed.

3. Result and Discussion

3.1. The response of the pontoon motion to the wave period

The response of the pontoon motion to the wave variations hitting the pontoon is the RAO (Response Amplitude Operator) value and the actual response. The RAO pitch value has the same characteristic tendency towards variations of the pontoon with different arm lengths. The pitching value decreases as the wave period is enlarged. This provides a confirmation with the greater the value of the wave period so that the pitching angle is getting smaller, but the pitching speed will get bigger. Under these conditions, the frequency of the pontoon oscillation will increase and make it easier for the pendulum to oscillate [11].

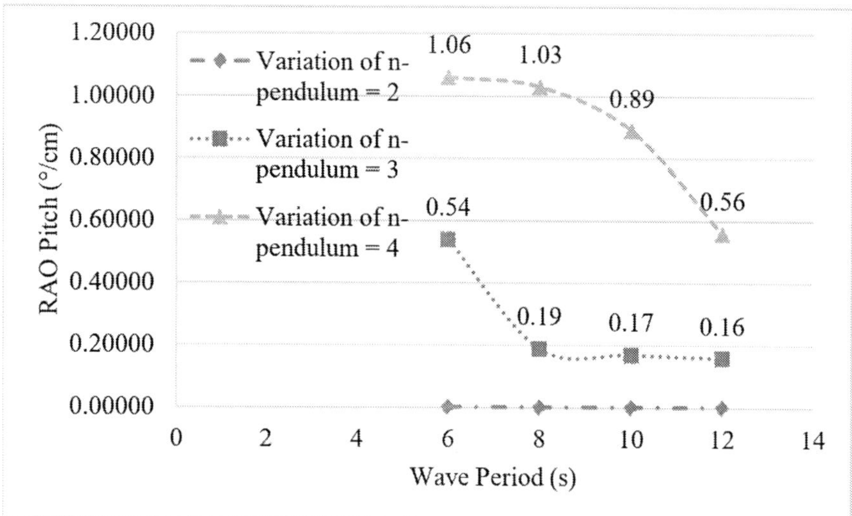

Figure 2. RAO pitching at arm's length of pendulum 106.7 cm.

RAO pitching peaks occur at the same variation in the 6 second period in all variations of the number of pendulums. The maximum RAO pitching value with the number of pendulums n = 4 (four) also experienced the same condition in the 6 second period. The maximum pitching values at variations in the length of the pendulum arm (106.7; 110; 130 and 165 cm) were 1.06° / cm, 1.15° / cm, 1.05° / cm and 1.14° / cm, respectively.

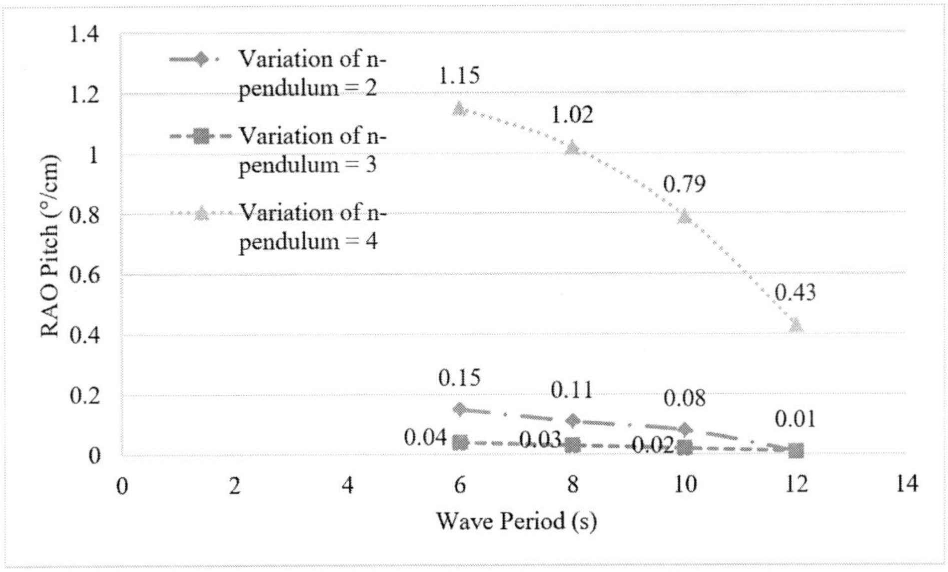

Figure 3. RAO pitching at arm's length of pendulum 110 cm.

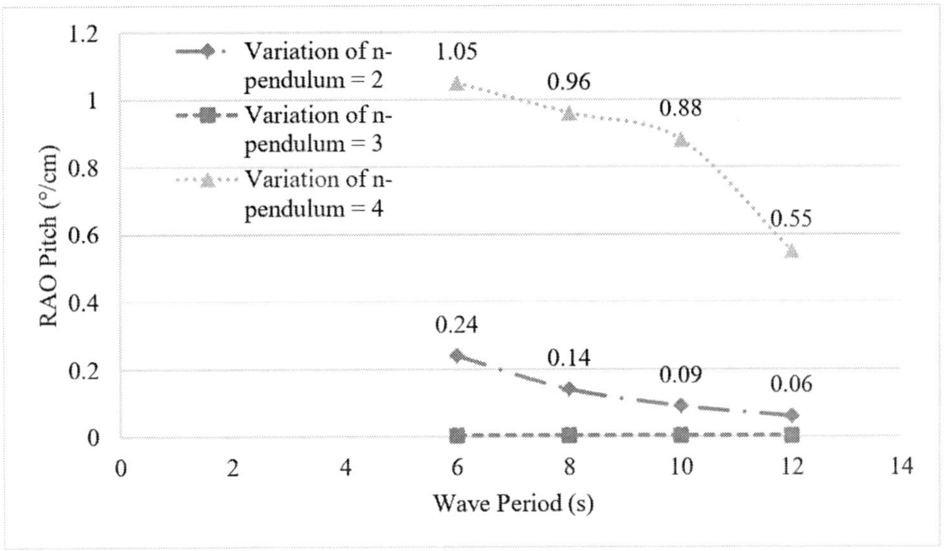

Figure 4. RAO pitching at arm's length 130 cm.

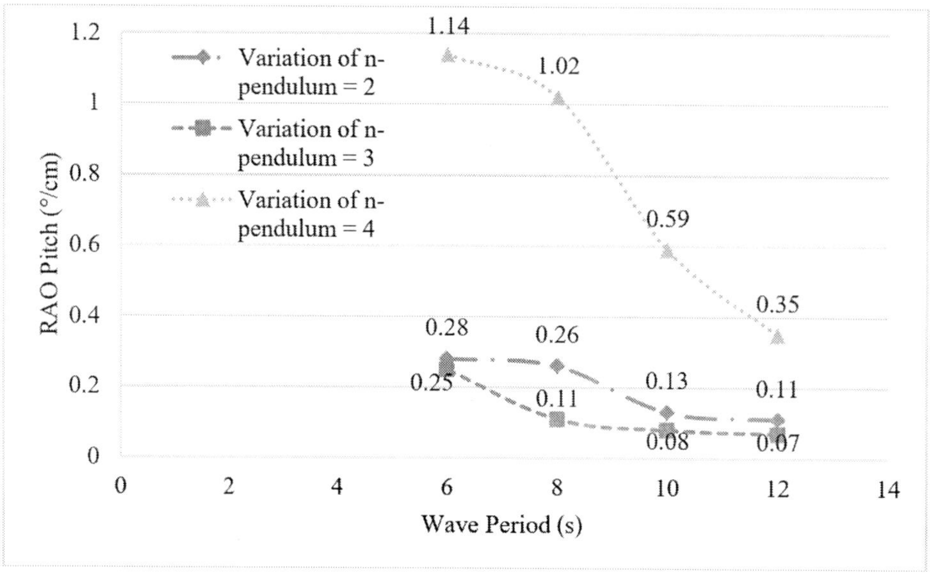

Figure 5. RAO pitching at arm's length 165 cm.

3.2. Effect of Wave Period, Pendulum Length, and n-Pendulum on Pendulum Oscillation

The response to the pontoon movement in the form of an actual response causes changes in the pendulum oscillation motion. Analysis of the pendulum motion is carried out by varying the wave period, the pendulum length, and the number of pendulums used on one pontoon.

3.2.1. Responses of Pendulum with Variation of Wave Period

Analysis of the pendulum motion due to variations in the wave period (6, 8, 10, and 12 seconds) was carried out at an amplitude of 20 cm and laden with water 1 / 2D. The effect of period changes on pendulum drift can be seen in Figures 6 - 9.

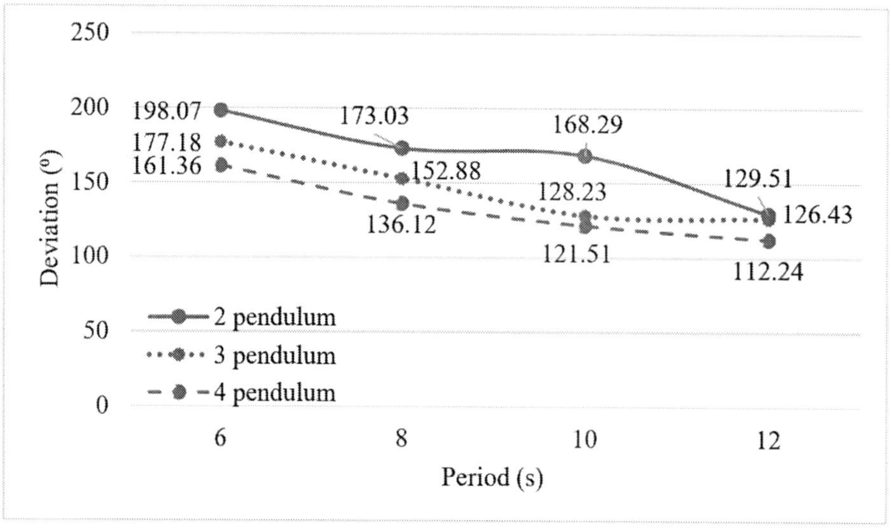

Figure 6. Deviation of pendulum with wave period of the pendulum length 106.7 cm.

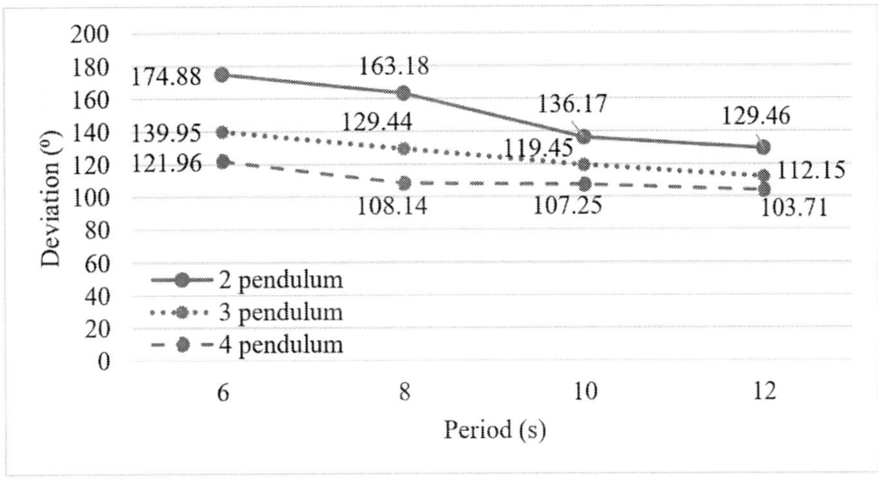

Figure 7. Deviation of pendulum with wave period of the pendulum length 110 cm.

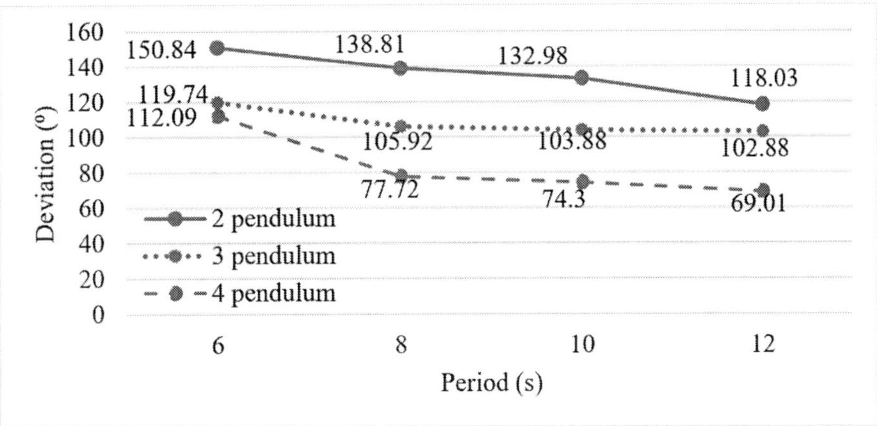

Figure 8. Deviation of pendulum with wave period of the pendulum length 130 cm.

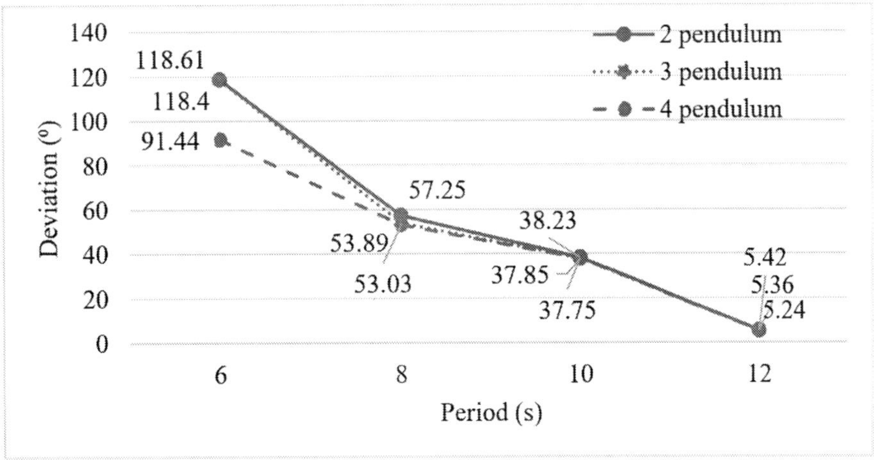

Figure 9. Deviation of pendulum with wave period of the pendulum length 165 cm.

In the variation of the pontoon with (n = 2, 3, and 4) the pendulum and the length of the arm (106,7; 110; 130; 165 cm), it shown that the change in pendulum deviation will decrease with increasing wave period. This is made possible by the magnitude of the wave period which is inversely proportional to the deviation value as described in equations (1) and (2). The pitching value that gets smaller with increasing wave period also affects the magnitude of the deviation change in the pendulum. When viewed from Figures 6 to 9 it has the same deviation change characteristics where the largest deviation occurs on the pontoon with n = 2 pendulums during 6 seconds of the period.

$$\theta = \frac{2\pi t}{T} \tag{1}$$

$$Y = A \sin\frac{2\pi t}{T} \tag{2}$$

3.2.2. The Motion of Pendulum with Variation of n – Pendulum

Pendulum motion analysis is also seen from the variation in the number of pendulums used, namely 2, 3, and 4 pendulums in one pontoon. Based on the results of this simulation, it is hoped that the effect of changes in pendulum deviation on the number of pendulums shown in Figures 10 to 13.

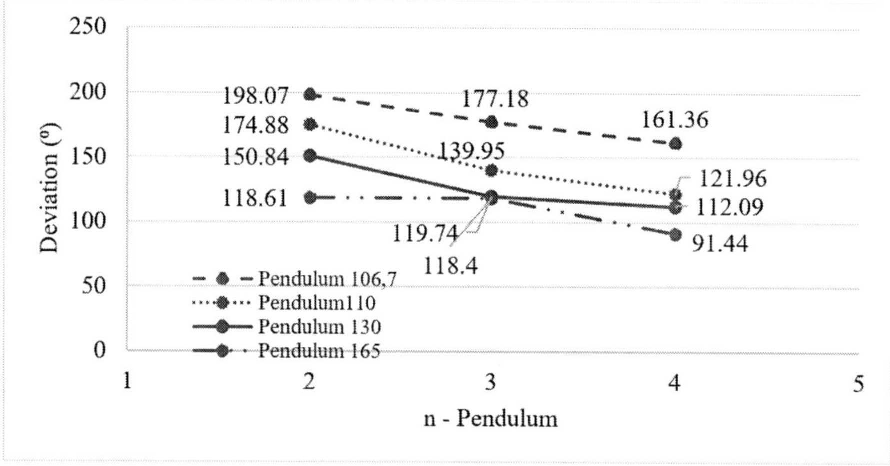

Figure 10. Relation of n-pendulum with wave period of 6 second.

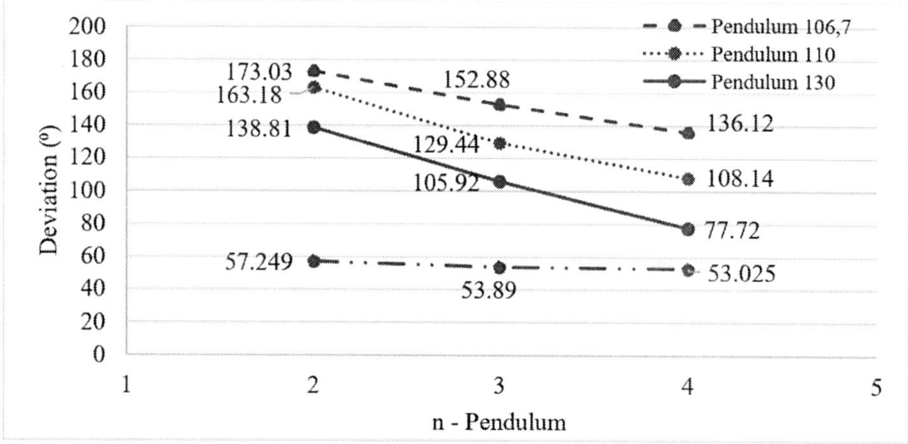

Figure 11. Relation of n-pendulum with wave period of 8 second.

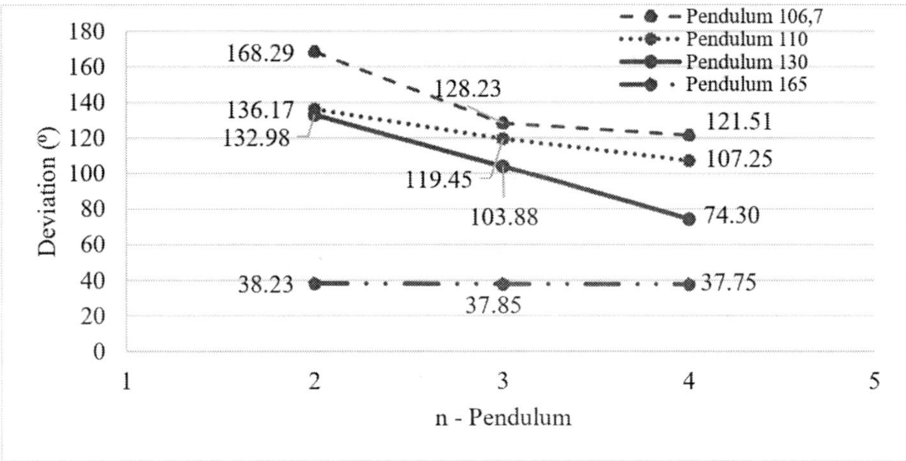

Figure 12. Relation of n-pendulum with wave period of 10 second.

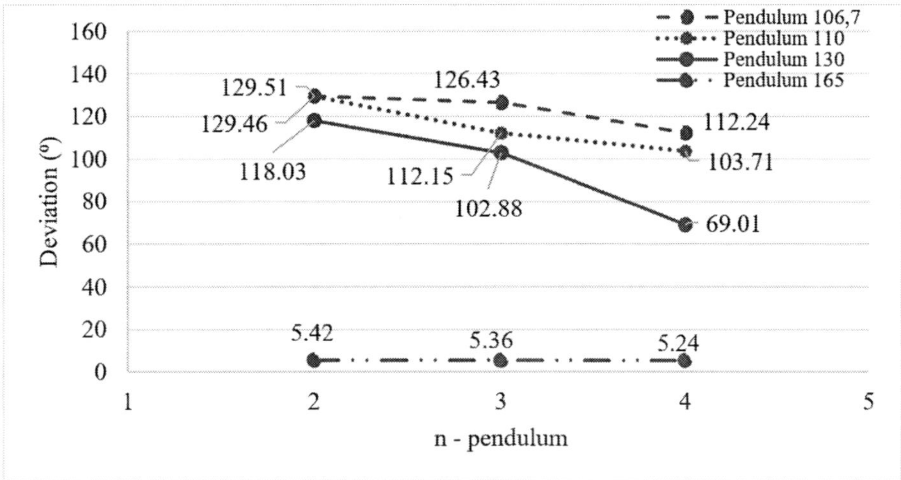

Figure 13. Relation of n-pendulum with wave period of 12 second.

Based on Figures 10 to 13, shown that the maximum deviation value occurs on the pontoon with the number of pendulums n = 2 during the 6 second period with a value of 198.07°. The alteration of pendulum deviation with the number of pendulums is caused by the mass of the pontoon. When the number of pendulums in 1 pontoon is increasing, mass of the pontoon increasing too.

3.2.3. Motion of Pendulum with Variation of Pendulum Length

The analysis of the motion of the pendulum deviation changes was also reviewed for the pendulum length variations, namely 106.7, 110, 130, and 165 cm. Based on the simulation results, the value of the change in pendulum deviation to the pendulum length variation is obtained as shown in the graphs in Figures 14 to 16.

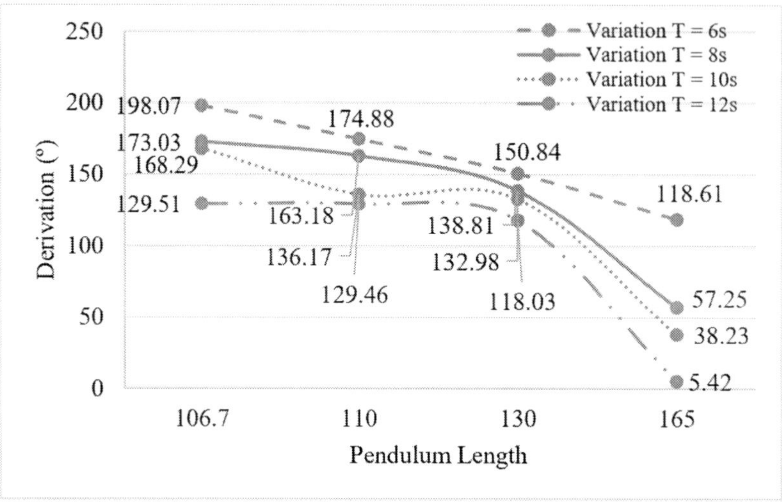

Figure 14. Relation of pendulum length with deviation of pendulum n = 2.

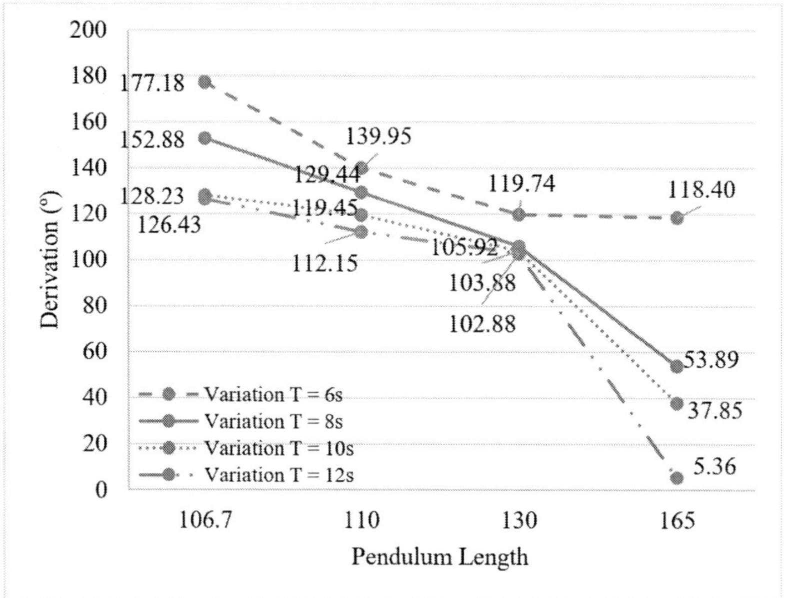

Figure 15. Relation of pendulum length with deviation of pendulum n = 3.

Based on the Figures 14-16 of the pendulum length relationship to the deviation change, shown the same change characteristics. The longer the pendulum arm has the smaller deviation change. This is the following equations (2) and (3) where the value of the pendulum arm length is directly proportional to the wave period and inversely proportional to the deviation so that it can result in a smaller deviation as the arm length increases. From the Figures above, the maximum deviation value occurs at the arm's length of 106.7 cm.

$$T = 2\pi\sqrt{\frac{l}{g}} \tag{3}$$

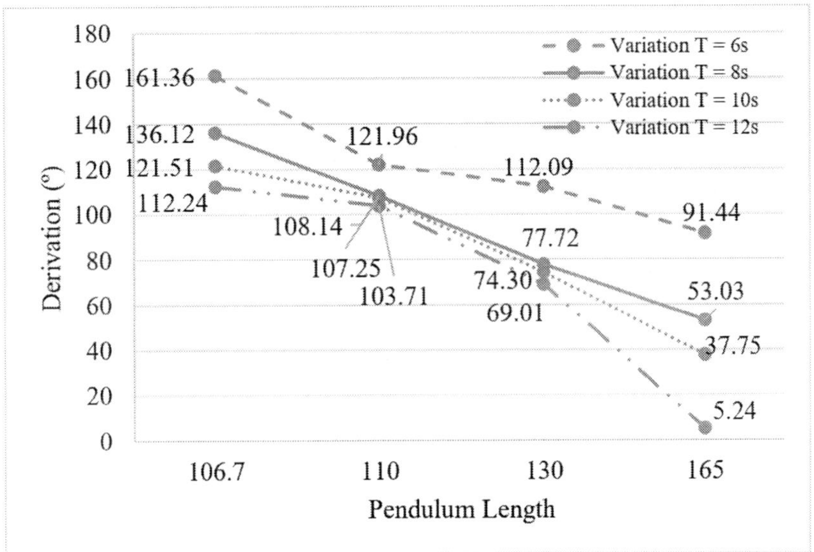

Figure 16. Relation of pendulum length with deviation of pendulum n = 4.

3.3. Pendulum Motion with Irregular Wave

Simulation in irregular waves aims to determine the response of the pontoon motion in real conditions. This simulation uses Jonswap Hs with a peak frequency of 0.17 Hz, gamma 2 [12], and a wave height of 40cm. The cross swell is set with a 45° wave direction parameter.

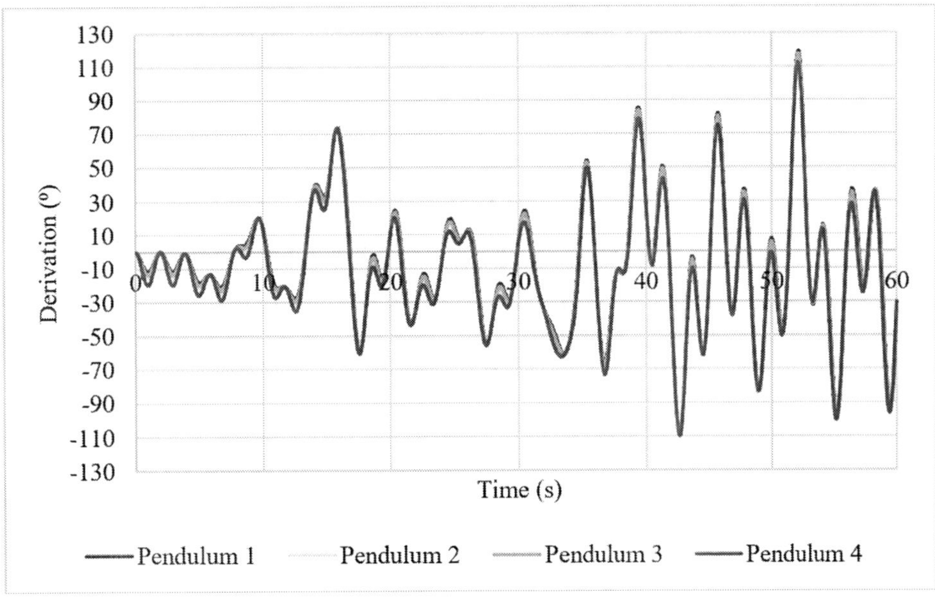

Figure 17. Deviation of pendulum in irregular wave.

Figure 17. shows the simulation results of irregular waves where there is no phase difference between the 1st pendulum to the 4th pendulum. The speed of motion is relatively the same, although the angle of deviation is different. However, the difference between them is minor. In the 1st pendulum, the maximum deviation formed is 119º. This is the biggest deviation by the 2nd, 3rd, and 4th pendulums. The difference that occurs due to the pontoon is caused by pitching motion.

3.4. Torque and Power

The pontoon motion that causes the pendulum to move from equilibrium results in torque and power values. The torque is obtained from mass, length of the pendulum arm, and the angle of inclination of the pendulum from its equilibrium state. Meanwhile, the power value is obtained from the multiplication of torque and angular speed. The torque and power in each n-pendulum variation can be seen in Figures 18 to 20.

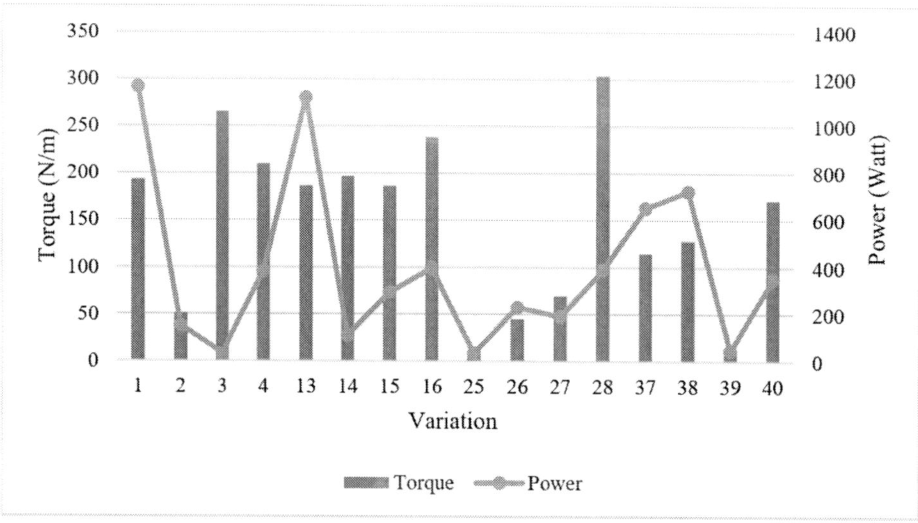

Figure 18. Torque and power of ponton with pendulum n = 2.

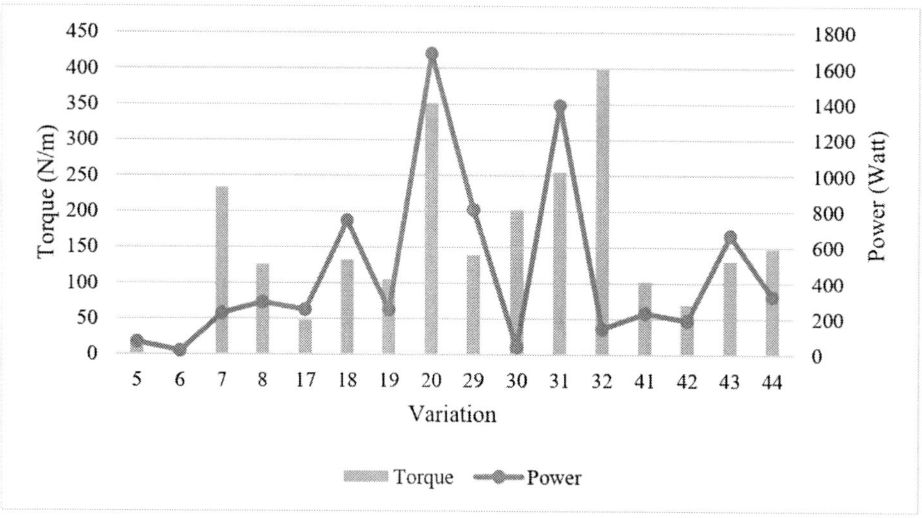

Figure 19. Torque and power of ponton with pendulum n = 3.

The maximum torque and power value on the pontoon with n = 2 pendulums in variation 1 has a torque value of 193 N / m and a power of 1167 Watt, this is different from variation 28 which has a large torque value of 303 N / m but gives a small power value of 387 Watt. The deviation that occurs in variation 1 is the maximum deviation in the pontoon variation with n = 2 pendulums of 198.07°, while in the variation of 28 the resulting deviation has a small value of 38.23°.

On the pontoon with n = 3, the pendulum also has the same phenomenon in the variation of 20, the maximum power generated is 1680.93 Watt with a torque value of 351 N / m. This value is quite small when compared to the maximum torque value at variation 32 which is 399 N / m, but the resulting power only 143.7 Watt. The deviation angles in variations 20 and 32 are 53.03° and 37.83, respectively. This value is very small when compared to the deviation value in variation 1.

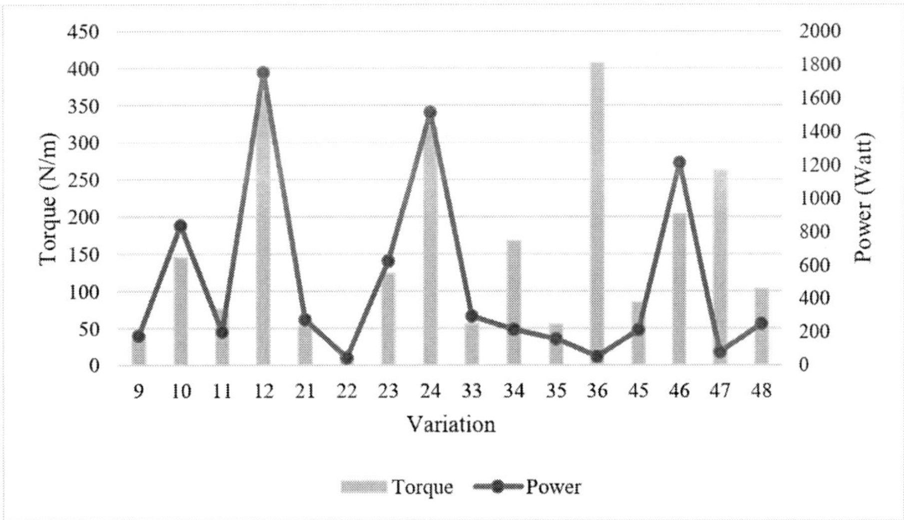

Figure 20. Torque and power of ponton with pendulum n = 4.

On a pontoon with n = 4 pendulums, the difference between the maximum torque and power values is not significant. Variation 12 has a maximum power value of 1750Watt with a torque of 364 N / m, compared to variation 36 which has a maximum torque value of 406 N / m. with a power of 50 Watts. The large difference between the torque and the power produced is not only due to the size of the resulting pendulum angle of deviation but also the mass and length of the pendulum arm as described in equations (4) (5) and (6).

$$\omega = \frac{v}{L} \tag{4}$$

$$\tau\ (torque) = m \times g \times Lcos^2\theta \tag{5}$$

$$P = \omega \times \tau \tag{6}$$

4. Conclusion

Based on the research objectives as well as the research that has been done on the ocean wave power plant pendulum system, it can be concluded as follows:

- Low wave periods can result in large pendulum deviations because the magnitude of the wave period is inverse of the deviation generated by the pendulum.
- The number of n-pendulums causes the pendulum deviation to get smaller because of the pendulum movements subtracting each other, so the deviation becomes smaller.
  The response of the pontoon motion at the Ocean Wave Power Plant Pendulum System (PLTG-SB) is influenced by changes in the wave period. When the wave period increases at an amplitude of 20 cm and 1 / 2D water-laden, the RAO pitching value tends to increase in 6 seconds.
- The change in pendulum deviation that occurs due to changes in the wave period shows the maximum deviation in 6 seconds with a value of 198.07°.

- In the variation of the arm's length of the pendulum, the largest deviation occurs at the arm's length of 106.7 cm. While in the variation of the n - pendulum the largest deviation occurred on the pontoon with the number of pendulums n = 2.
- The power generated due to the variation of the wave period, the arm's length of the pendulum, and the number of pendulums has a maximum value of 1167 Watt on the pontoon with the number of pendulums n = 2, 1680.93 Watt on the pontoon with the number of pendulums n = 3 and 1750.37 Watt on the pontoon with the number of pendulums n = 4.

This research still needs improvement, so for further research, the wave simulation is required to use a mooring system so that the pontoon does not move when it is hit by a wave. It is also necessary to simulate variations in the height and length of jukung to the response of the multi-pendulum pontoon motion and how it changes the deviation. Simulation of a multi-pendulum pontoon in real life by simulating using irregular wave types in more detail is needed too.

5. Acknowledgments

The author wished to thank the Ministry of Research and Technology (KEMENRISTEK-BRIN) for funding the research under a scheme of PTUPT with contract number: 1333/PKS/ITS/2020.

6. Reference

[1] I. R. H. Irfan, "Respon to Pontoon and Pendulum Motion at Wave Energy Converter Based on Pendulum System," *E3S Web of Conferences 43,01022 (2018),* 2017.

[2] J. K. a. G. Bhuyan, "Ocean Energy: Global Technology Development Status," *Report prepared by Powertech Labs for the IEA-OES (Implementing Agreement on Ocean Energy Systems - International Energy Agency,* 2009.

[3] I. T. M. M.I. Nafis, "Analisa Perbandingan Gerakan Ponton Model Tripod Sama Kaki dan Siku-Siku untuk Energi Gelombang Sistem Bandulan," *Jurnal teknik POMITS,* **vol. 3**, p. 1, 2014.

[4] A. Goldman, "Introduction to Wave Energy Converter," *Renewable Green Energy Power,* 2012.

[5] "An Overview of Ocean Energy Activities in 2018," *Ocean Energy Systems Annual Report,* p. 67, 2018.

[6] H. Titah, "An Up-to-Date Technologies Review and Evaluation of Wave Energy Converters," *International Review of Electrical Engineering ,* **vol. 10**, 2015.

[7] P. F. E. F.-M. a. H. S. J.P Kofoed, "Prototype Testing of the Wave Energy Converter Wave Dragon," *Renew Energy,* **Vols. 181-189**, p. 31, 2006.

[8] A. d. O. F. D.V. Evan, "Hydrodynamic of Ocean Wave Energy Utilization," *Springer,* pp. 51-55, 1986.

[9] J. Falnes, "A Review of wave Energy Extraction," *Science Direct Marine Structure,* vol. 20, pp. 185-201, 2007.

[10] Ibrahim, "Rancang Bangun Dan Studi Eksperimen Respon Dinamis Model Pembangkit Listrik Tenaga Gelombang Laut Sistem Ponton Dengan Tiga Bandul", 2015.

[11] Djatmiko, "Perilaku dan Operabilitas Bangunan Laut di Atas Gelombang Acak", Surabaya, ITS Press, 2012.

[12] Setiawan, "Spectral Representation In Sabang And Jepara Coast," *Eco Rekayasa,* **vol. 9**, pp. 34-41, 2013.

Field investigation on the state of marine safety of traditional fishery

S Nugroho and S A Nugraha

Department of Marine Transportation Engineering, Institut Teknologi Sepuluh Nopember Kampus ITS Sukolilo, Surabaya, Indonesia

Abstract. Gaining an understanding of the characteristics that affect maritime safety is not easy. It is necessary to have sufficient information about what can support safety. Problems in traditional fishing are visible from the low level of safety and the availability of information that supports the fishers' safety. Several accidents that occur due to negligence of work prove that the low quality of information is also the cause of failure to maintain safety. What commonly happens is that there are differences and discrepancies in recorded accident data with reality on the field, where this is rarely known even though accidents occur. It shows that there is no data integrity to support safety. When talking about safety, there must be a potential failure, which can be identified from the information obtained. However, in reality, the information used to support fishers' safety is not available, so traditional fishing boats' safety cannot run optimally due to data unavailability. When safety governance cannot be carried out using correct data, safety management becomes ineffective. By investigating the state of marine safety and examining the role of information on traditional fishers' safety, it is hoped that it can produce a better mechanism for traditional fishers' safety.

1. Introduction

The problems in traditional fishers are visible from the low level of safety and the flow of information that supports the safety management of traditional fishermen themselves. According to [1], information is data that has been processed into a meaningful form for the recipient and is useful for current or future decision making. Concerning safety, especially traditional fishers safety, one of the most critical pieces of information is the last position of a fishing boat when an accident occurs in the middle of the sea. The last coordinate part that provides information on the accident incident is often an obstacle to both the rescue and evacuation of victims. Law Number 17 of 2008 on Shipping, [2] it is stated that ship worthiness is the condition of the ship that meets the requirements for ship safety, prevention of water pollution from ships, manning, loading, health and welfare of the crew and passengers, and the legal status of the ship to sail in certain waters. In [3], we look at the importance of safety management and its maritime applications, where the choice of destination will depend on the area of interest. One of them is the importance of information when the goal is to make facts, figures, instructions, dates, and knowledge available to those needing them in forms that are relevant, up-to-date and user friendly. The resources involve social skills and facilities such as computer hardware and software. The activities are coordinated to ensure that the goal can be met. The scope of the arrangements in marine operation includes the area of the sea, marine development, marine management, marine development, marine space management and marine environmental protection, defense, security, law enforcement, safety at sea, governance and institutions, and public participation [4]. We see that the state must be present in the context of marine safety. Important point is also discussed in [5] where estimation and detection of dangerous situations is one of the key problems of maritime activities, monitoring, and decision support, which governments can easily do. We can define government as an entity with agency that can form a relationship with fishers to exchange significant information. To form a relationship to communicate between humans and agents with intentionality, autonomy, rationality, individuality, and interactivity, humans must understand the purpose behind an agent's behavior [6]. Operators, equipment, procedures and the entire radiocommunication system, as a whole, constitute a fundamental pillar for the safety of human life at sea, as well as for the safety of navigation [7]. Humans have a tendency to avoid changing their behaviour, also when it comes to safety critical tasks. People sometimes express their understanding of the need for change, but in practice use all means to sabotage new procedures, in [8] we call this unwillingness to change. This unwillingness to change does not only happen to traditional fishers but also agencies related to marine safety. Apart from safety culture issues, the weather is also an obstacle for traditional fishers carrying out their activities at sea. Where the availability of information also plays a vital role in helping traditional fishers to be able to carry out activities at sea safely. The question that may arise is whether the agency can provide this information. The frequency of occurrence of types of adverse or unfavorable weather conditions is extremely important for sea ports, for their effective work. No matter what cargo is handled in a given port, strong and very strong winds are to be treated as weather conditions which disturb or even do not allow working properly [9]. A fishing vessel can be a dangerous working environment. But the risk of accidents will be greatly reduced if proper precautions are taken [10]. This research was conducted at the Port of Pasuruan and involved several agencies related to the safety of traditional fishers in Pasuruan City. In this study, more investigations are carried out in the field to get a big and bare picture of how the interactions between agencies related to marine safety are, what data is available there, what is the role of each agency in helping marine safety, how technology plays a role in the process of recording data, how much is the ability of an agency to carry out search and rescue activities of traditional fishers. All these questions are fundamental in achieving marine safety in the regional scope, where the government is also obliged to play a role. The benefits of this research provide a big picture with the findings in the field and are the first step to build a marine safety mechanism and as a lever in improving the quality of the information in supporting the safety of traditional fishers. This research will present more qualitative data. However, findings and field investigations will provide a reality in addition to answering the existing questions and showing that the success of improving the safety of traditional fishers is determined by the integrity of the data and completeness of the information.

2. Identification of the Availability of Data for Traditional Fishery

In observations made either through direct observation or interviews with agencies related to the safety of traditional fishing boats in Pasuruan City, there are six agencies including Harbormaster (KSOP Class V of Pasuruan City) which also includes KPLP (Sea and Coast Guard), Lantamal V Indonesian Navy, Polair (Water Police Unit), Transportation Department of Pasuruan City, Marine and Fisheries Department of Pasuruan City, Disaster Management Agency of Pasuruan City (BPBD). From the six agencies then identified the availability of data on traditional fishers' accidents. As for what is meant by the availability of data on a traditional fishers' accidents is that the agency can provide data in the form of notes or reports that show the incidents of traditional fishers' accidents in areas that are its responsibility.

2.1. Availability of Data for Traditional Fishery at Harbormaster (KSOP Class V of Pasuruan City)

Harbormaster (KSOP Class V of Pasuruan City) is a Technical Implementing Unit (UPT) within the Directorate General of Sea Transportation (DJPL) which is responsible to the Director-General of Sea Transportation and has the task of carrying out supervision and law enforcement in the field of safety and shipping security, coordination of government activities at ports as well as regulating, controlling and supervising port activities at ports that are commercially operated. In its activities in the field of marine safety, the Harbormaster does not record any incidents of traditional fishers accidents. The main recording or data collection activities carried out by the Harbormaster are records related to ships call and data related to shipping and maritime activities, including the activity of issuing ship certificates, carried out for one year.

Figure 1. Harbormaster Sample Annual Report Book (KSOP Class V of Pasuruan City).

In its activities in the field of marine safety, the Harbormaster because it does not record traditional fishers accidents, it can be said that the Harbormaster digitally records traditional fishers accident data was also not carried out. However, it is known that in terms of annual recording and reporting carried out, computer technology has been used.

Table 1. Findings at the Harbormaster (KSOP Class V of Pasuruan City).

Issue	Findings
Accident Data Availability	X
Accident Data Recording Using Computer Technology (Digital)	X

2.2. Availability of Data for Traditional Fishery at Indonesian Navy (Lantamal V)

The Main Base of the Indonesian Navy V (Lantamal V) is a service element with one of the main tasks of carrying out maritime security patrols in the work area of the Indonesian Navy Main Base. Where it is known that Lantamal V Pasuruan City is included in the elements involved in the safety of traditional fishers, in its activities in the field of maritime safety and security, the Main Base of the Indonesian Navy V (Lantamal V) always records accidents that occur within its working area. The elements of information contained in the recording or data collection of accident events conducted by the Lantamal V Pasuruan City are as follows :

Table 2. The information elements contained in the records by Lantamal V.

Column	Contain
DAY	- The day of the incident - The date of the incident
TIME	- Time of incident
EVENTS *(different format for each incident)	- Chronological sequence of events - Name of ship - Size of the ship - Name of ship owner - List of crew (ABK) names - Address of crew members - The name of the ship's captain

IOP Conf. Series: Materials Science and Engineering 1052 (2021) 012064 doi:10.1088/1757-899X/1052/1/012064

In its activities in the field of marine safety and security, Lantamal V Pasuruan City is known to record traditional fishers accidents. However, data recording of traditional fishers accidents is done manually without using technology such as computers or similar technological devices that are capable of storing data in digital form. It is known that in recording the archives, it is appropriately stored using a folio-sized book with stripes.

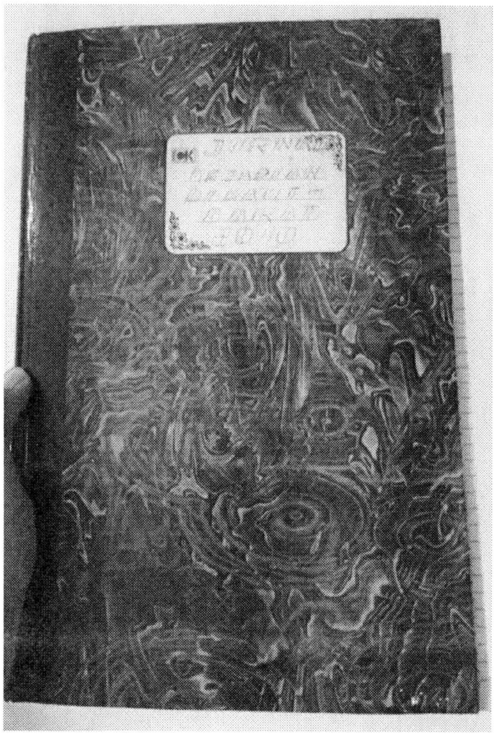

Figure 2. Folio-Sized Record Book by Lantamal V.

Table 3. Findings at the Lantamal V.

Issue	Findings
Accident Data Availability	✓
Accident Data Recording Using Computer Technology (Digital)	X

2.3. Availability of Data for Traditional Fishery at Polair (Water Police Unit)

The Water Police Unit or what is commonly called the Polair, is an element that carries out the main tasks of the city police (Polres) who are under the Police Chief. The Water Police Unit is in charge of carrying out police functions in the waters, which includes patrolling waters, law enforcement in the waters, fostering coastal and other marine communities, as well as searching and rescuing accidents in waters (SAR). Where the Water Police Unit (Polair) is one of the elements involved in the safety of traditional fishers, in its activities in the field of marine safety, the Water Police Unit of Pasuruan City always records accidents that occur in its working area. The elements of information contained in the recording or data collection of accident events conducted by the Water Police Unit (Polair) of Pasuruan City are as follows :

Table 4. The information elements contained in the records by Polair.

Column	Contain
NO	- Listing serial number (every change of month)
UNIT	- The unit that records the incident
POLICE REPORT	- Record report number
TYPE OF CASE	- Types of accident cases
EVENT DESCRIPTION	- A brief chronology of the sequence of events which also tells the time of the incident and the name of the reporter
IDENTITY OF SUSPECTED	- The name of the suspect
VICTIM	- The name of the victim
WITNESS	- Name of witness
BREAKING THE ARTICLE	- Types of articles being violated
PROOF GOODS	- Evidence of events
DESCRIPTION	- Status of the victim's condition

NO	KESATUAN	LAPORAN POLISI	JENIS KASUS	URAIN KEJADIAN
1	2	3	4	5
1	JANUARI			
	Sat polair Polres Pasuruan	Laporan Segera : Nomor :LAPGA/ 01/I / 2017 / Satpolair	Laka Laut	Pada hari Kamis tanggal 26 Januari 2017 sekira pukul 16.00 Wib, telah datang ke kantor Sat Polair Polres Pasuruan Sdr TANTOWI warga Kelurahan Pangungrejo Rt 01 Rw 01 Kec, Panggungrejo Kota Pasuruan melaporkan bahwa pada pukul 13.00 wib telah terjadi laka laut di perairan utara Dsn Semare Kec. Kraton di karenakan adanya angin kencang dan ombak sehingga perahu milik korban terbalik dan sampai saat ini nelayan tersebut belum di temukan

Figure 3. Sample Fishers' Accident Record by Polair.

In its activities in the field of marine safety, the Water Police Unit (Polair) records traditional fishing accidents, and data of traditional fishing accidents recorded by the Water Police Unit (Polair) is carried out using computer technology (digital).

Table 5. Findings at the Polair.

Issue	Findings
Accident Data Availability	✓
Accident Data Recording Using Computer Technology (Digital)	✓

IOP Conf. Series: Materials Science and Engineering 1052 (2021) 012064 doi:10.1088/1757-899X/1052/1/012064

2.4. Availability of Data for Traditional Fishery at Transportation Department of Pasuruan City

The Department of Transportation is an element implementing government affairs in the transportation sector, which is the regional authority. In its activities in the field of marine safety, the Transportation Department does not record any incidents of traditional fishers accidents. The data collection activity carried out by Transportation Department is data related to registration of ship documents with a size of below 7 GT, provided that this authority only applies in 2016 and has been returned in 2017 to the Harbormaster following the Regulation of the Minister of Transportation of the Republic of Indonesia Number PM 39 of 2017 on Registration and Nationality of Ships and is valid until now.

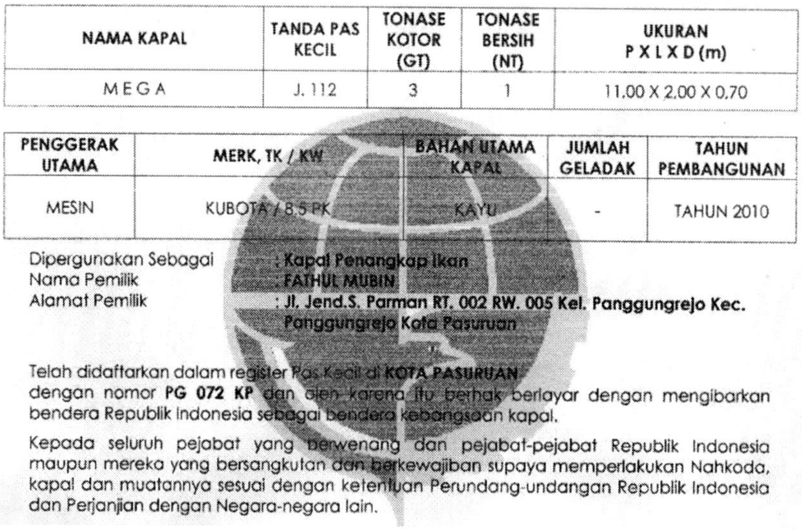

NAMA KAPAL	TANDA PAS KECIL	TONASE KOTOR (GT)	TONASE BERSIH (NT)	UKURAN P X L X D (m)
MEGA	J. 112	3	1	11,00 X 2,00 X 0,70

PENGGERAK UTAMA	MERK, TK / KW	BAHAN UTAMA KAPAL	JUMLAH GELADAK	TAHUN PEMBANGUNAN
MESIN	KUBOTA / 8,5 PK	KAYU	-	TAHUN 2010

Dipergunakan Sebagai : **Kapal Penangkap Ikan**
Nama Pemilik : **FATHUL MUBIN**
Alamat Pemilik : **Jl. Jend.S. Parman RT. 002 RW. 005 Kel. Panggungrejo Kec. Panggungrejo Kota Pasuruan**

Telah didaftarkan dalam register Pas Kecil di **KOTA PASURUAN** dengan nomor **PG 072 KP** dan oleh karena itu berhak berlayar dengan mengibarkan bendera Republik Indonesia sebagai bendera kebangsaan kapal.

Kepada seluruh pejabat yang berwenang dan pejabat-pejabat Republik Indonesia maupun mereka yang bersangkutan dan berkewajiban supaya memperlakukan Nahkoda, kapal dan muatannya sesuai dengan ketentuan Perundang-undangan Republik Indonesia dan Perjanjian dengan Negara-negara lain.

Figure 4. Sample Data Record at The Transportation Department of Pasuruan City.

In its activities in the field of marine safety, the Department of Transportation because it does not record traditional fishers accidents, it can be said that the digital recording of traditional fishers accident data by the Department of Transportation is also not carried out. However, in terms of recording others that have been implemented have used computer technology.

Table 6. Findings at the Transportation Department of Pasuruan City.

Issue	Findings
Accident Data Availability	X
Accident Data Recording Using Computer Technology (Digital)	X

2.5. Availability of Data for Traditional Fishery at Marine and Fisheries Department of Pasuruan City

Marine and Fisheries Department of Pasuruan City is an element implementing government affairs in the field of Fisheries and Marine, which is the regional authority. In its activities in the field of marine safety, Marine and Fisheries Department of Pasuruan City have not recorded any incidents of traditional fishers accidents. Although at the time of the accident involving fishers, the Marine and Fisheries Department received information on the incident from both the Pokwasmas Nelayan (Fishers Community Group), as well as Polair and BPBD. Pokwasmas Nelayan is a community chaired by a traditional fishing group that connects traditional fishers with agencies related to marine safety.

Table 7. Findings at the Marine and Fisheries Department of Pasuruan City.

Issue	Findings
Accident Data Availability	X
Accident Data Recording Using Computer Technology (Digital)	X

2.6. Availability of Data for Traditional Fishery at Disaster Management Agency of Pasuruan City (BPBD)

Disaster Management Agency (BPBD) of Pasuruan City is a non-departmental government institution that carries out disaster management tasks in the area of Pasuruan City based on policies set by the National Disaster Management Agency (BNPB). Disaster Management Agency of Pasuruan City has the main task of implementing disaster management. In its activities in the field of marine safety, the Disaster Management Agency (BPBD) of Pasuruan City is known not to have recorded incidents of traditional fishers accidents. Although at the time of the accident involving fishers, Disaster Management Agency (BPBD) of Pasuruan City sometimes gets information about the incident from both the Pokwasmas Nelayan (Fishers Community Group) and Polair.

Table 8. Findings at the Disaster Management Agency (BPBD) of Pasuruan City.

Issue	Findings
Accident Data Availability	X
Accident Data Recording Using Computer Technology (Digital)	X

2.7. Recapitulation of Traditional Fishers Accident Data Availability at Each Agency

From the six agencies observed and interviewed, only Lantamal V and Polair recorded the data. The data recording referred to in this case, is every fisher accident that occurs on purpose. Moreover, stored in the form of dates, the sequence of events, number of victims, types of accidents, and so on, in a medium such as books, reports, or digital data storage media, so that when someone asks for the data, it is useful data every quarter, yearly, or in the more time, the agency can provide the desired data. There is one case such as happened to the Marine and Fisheries Department, which got data when an accident occurred through an Android-based application, namely WhatsApp, but the data did not just become stored data. The agencies were observed and interviewed, and among the Lantamal V and Polair, which were previously known to have records of fishing accidents, only Polair used computer technology in the data recording process. As for other four agencies such as the Harbormaster, Transportation Department, Marine and Fisheries Department, and Disaster Management Agency (BPBD) of Pasuruan City are automatically excluded from the agency, which uses computer technology (digital) in the process of recording fisher accident data. The findings and field investigations regarding the availability of data on the six agencies related to marine safety show a lack of awareness to record every incident that is known to exist even though data exchange has occurred and may have occurred between the six agencies. Also, it can be seen that the poor management of information is shown by most of these agencies where only one out of six agencies keep records digitally and are neatly arranged in table form so that it can be said that only 17% of these agencies have the power to support marine safety in terms of data availability. Although there is one other agency that records events by hand, we know that non-digital recording has a high risk of data loss. The application of technology is an absolute must if marine safety goals are to be realized not only on the agency's side but also by traditional fishers.

IOP Conf. Series: Materials Science and Engineering 1052 (2021) 012064 doi:10.1088/1757-899X/1052/1/012064

Table 9. Findings at all the agencies.

Agency	Accident Data Availability	Accident Data Recording Using Computer Technology (Digital)
Harbormaster (KSOP Class V of Pasuruan City)	X	X
Indonesian Navy (Lantamal V)	✓	X
Water Police Unit (Polair)	✓	✓
Transportation Departement	X	X
Marine and Fisheries Department	X	X
Disaster Management Agency (BPBD)	X	X

3. Identification of the Role of each Agency in Traditional Fisheries Safety

From the six agencies that were observed and interviewed, all of them had a relationship to coordinate in terms of marine safety even though they were not in one task group or agency function. For example, the Harbormaster, followed by the Sea and Coast Guard (KPLP), Water Police Unit (Polair), Indonesian Navy (Lantamal V), and Disaster Management Agency (BPBD) coordinating with search and rescue operations (SAR) when the accident occurs involving a traditional fisher. The Water Police Unit (Polair) and Indonesian Navy (Lantamal V) coordinate on patrols to prevent accidents. The Transportation Department and Marine and Fisheries Department coordinate in terms of providing guidance and socialization related to marine safety. We can conclude that agency has a role both in terms of prevention and handling of marine safety even though their duties do not overlap with each other.

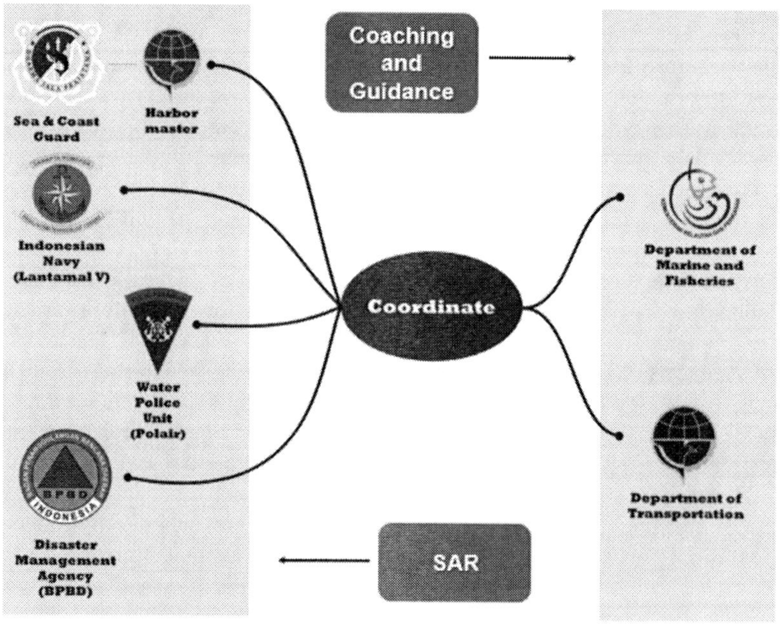

Figure 5. The role of local agencies in marine safety.

3.1. The Role of Traditional Fisheries Safety Development

The role of training in the safety of traditional fishers is carried out by two local agencies, namely Transportation Departement as the main driving force in the field of unique marine safety guidance for fishers who use traditional fishing boats under 7 GT as well as fisherwomen groups, then assisted by the Marine and Fisheries Department as the supervisor of the fishing community in Pasuruan City. Guiding and coaching activities are generally carried by dividing the guidance areas based on sub-districts which have fishers in the community. The sub-districts that are the location of socialization and guidance include Panggungrejo Village, Tamba'an Village, Mandaranrejo Village, Ngemplakrejo Village, Gadingrejo Village, and Bugul Lor Village. The Marine and Fisheries Department delivered the socialization materials in addition to guidance in increasing the productivity of fisheries, the socialization by the Transportation Department, among others, was conveying socialization related to marine safety which included a briefing on knowledge of safety equipment that fishers must carry when doing activities at sea, as well as providing direction on the importance of community participation in participating in monitoring and maintaining order in the implementation of marine activities, where the socialization activities are related to the role of local governments to participate in fostering communities around ports according to Law Number 17 of 2008 on Shipping.

Figure 6. Socialization and guidance activities.

3.2. The Role of Search and Rescue of Traditional Fisheries

From the six agencies that were observed and interviewed, there were four agencies that carried out SAR actions, namely the Harbormaster followed by the Sea and Coast Guard Unit (KPLP), the Water Police Unit (Polair), Indonesian Navy (Lantamal V), as well as the Disaster Management Agency of Pasuruan City (BPBD). SAR measures can be carried out by four of these agencies supported by the availability of patrol boats, except for BPBD which is only supported by owned rubber boat.

Table 10. Availability of patrol boats for SAR activities.

Agency	Availability	Amount
Harbormaster	✓	1 Patrol Boat
Water Police Unit (Polair)	✓	4 Patrol Boats
Indonesian Navy (Lantamal V)	✓	1 Patrol Boat
Disater Management Agency (BPBD)	X	1 Rubber Boat
Total		6 Patrol Boats

Figure 7. Patrol boat fleet at Port of Pasuruan.

Bad weather is one of the obstacles in the Search and Rescue (SAR) process where it is known that the maximum search time that can be carried out is seven days so that the process can be stopped when the search process has reached seven days even though without getting any results (no victims of shipping accidents were found missing). For the availability of SOP (Standard Operating Procedure), the following is a recapitulation of SOP availability at four agencies carrying out Search and Rescue (SAR) activities :

Table 11. Availability of Standard Operating Procedure for SAR activities.

Agency	Availability
Harbormaster	X
Water Police Unit (Polair)	✓
Indonesian Navy (Lantamal V)	X
Disater Management Agency (BPBD)	X

As for the results of interviews and observations, it is known that there is no integrated joint SOP between four SAR-related agencies carried out together to form standard coordination. It is also known that of the four agencies carrying out SAR, only the Water Police Unit (Polair) has an SOP for SAR activities. In contrast, the Harbormaster (KSOP), Indonesian Navy (Lantamal V), and the Disaster Management Agency of Pasuruan City (BPBD) does not have SOP (Standard Operating Procedure) for SAR activities.

4. The Reality that Occurs in the Marine Safety of Traditional Fisheries

After knowing the availability of data on traditional fishers accidents and the role of local agencies in supporting the safety of traditional fishers, it is necessary to identify the realities of what happens in the field regarding the safety of traditional fishers. The reality that occurs in the field is often far from expectations, where the best hope is that it becomes something ideal. One example of the reality that occurs in the safety of traditional fishers is the difference in the recording data of traditional fishing accidents between the Indonesian Navy (Lantamal V) and the Water Police Unit (Polair). Ideally, the two agencies have the same data and even complement each other, but the reality in the field shows that the data owned by one agency is different from that of other agencies.

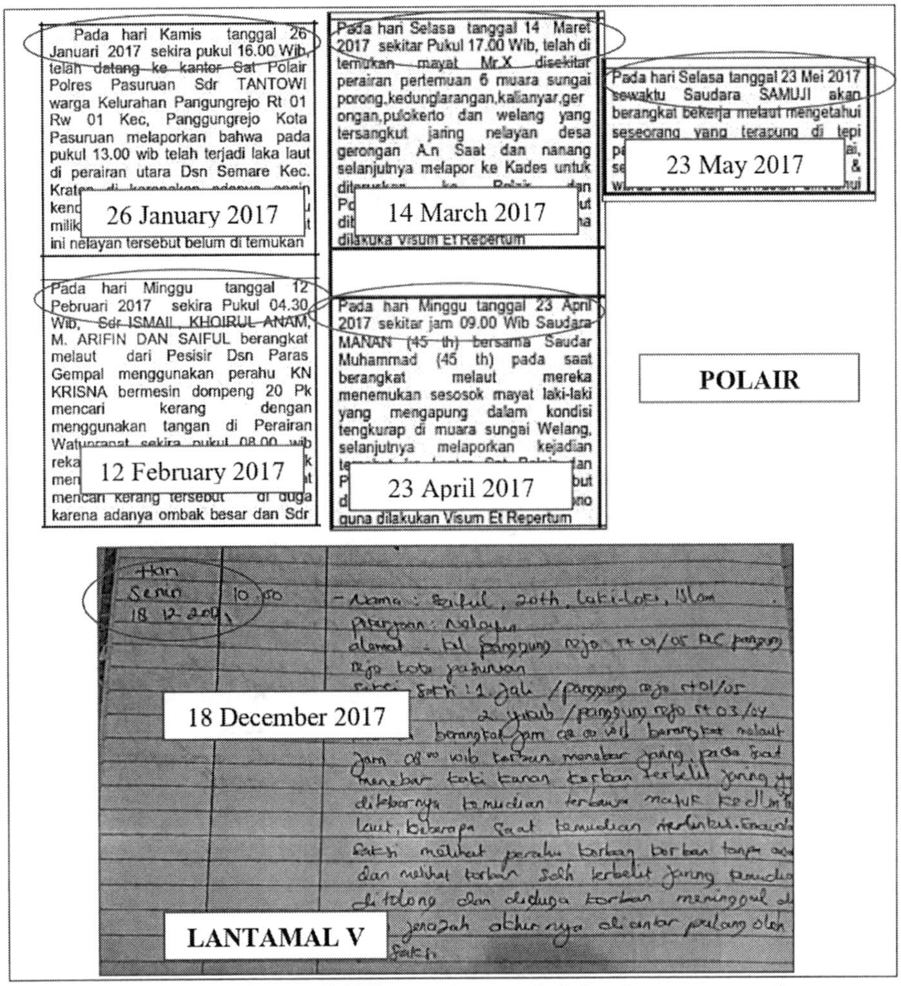

Figure 8. Evidence of difference in recorded data between agencies.

From the evidence above, it is known that the data recorded on Indonesian Navy (Lantamal V) and the Water Police Unit are different. For data from the Water Police Unit (Polair), in 2017 it started in January and lasted only until May so that in 2017 there were a total of 5 incidents recorded. Lantamal V data, the initial data for 2017 the incident was in December so that in 2017 there was a total of 1 incident. From the two records, it was also known that there was no record of the same accident. However, for data in 2018, both Lantamal V and the Water Police Unit have the same data, namely, for 2018 there were no incidents of traditional fishing accidents.

Table 12. Differences in recorded data between agencies.

Agency	Event Date	Number of Events
Water Police Unit (Polair)	26 January 2017 12 February 2017 14 March 2017 23 April 2017 23 May 2017	5
Indonesian Navy (Lantamal V)	18 December 2017	1

Accidents that occur to traditional fishers can occur on the fishing boat and the fishers themselves, whether in a docked condition or carrying out fishing activities in the middle of the sea, even though fishers have made efforts to avoid them. Accidents involving traditional fishers have many scenarios in reporting events to the evacuation process of victims. Various incident scenarios can be influenced by several factors, including the reporting tools used by fishermen, the domicile of the accident victim and the domicile of the accident reporter, and the first recipient of the accident report. The factors that influence the number of scenarios in the SAR (Search and Rescue) process are the realities that occur in the field and of course, of the many possible scenarios some obstacles make the scenario a reality that is not ideal.

5. Conclusion

The main conclusions drawn are :

- Lack of marine safety management due to poor management of information shows that we are further away from the goal of safety.
- Investigations and field findings indicate that the role of government agencies in supporting safety needs to be supported by technological advances.
- The discrepancy of data recorded in local government agencies is one finding that shows that the safety of our fishers is at a slightly unfavourable level.
- More methodological work needed for future study on how to strongly capture the impacts and outcomes of a lack of marine safety management in this study, including further cost-benefit analysis and exploration of implications when information to support safety is limited.

6. References

[1] Gordon B D 1985 Management Information Systems (McGraw-Hill Inc.,US)
[2] Republic of Indonesia 2008 Law Number 17 of 2008 on Shipping
[3] Chengi K 2007 Safety Management and its Maritime Application (The Nautical Institute, UK)
[4] Republic of Indonesia 2014 Law Number 32 of 2014 on Maritime Affairs
[5] Adam W and Tomasz N 2015 Information, Communication and Environment: Marine Navigation and Safety of Sea Transportation p 140 (The Nautical Institute, UK)
[6] Sakae Y 2013 Human Interface and the Management of Information (Springer Heidelberg Dordrecht London, NewYork, USA)
[7] Adam W and Tomasz N 2019 Advances in Marine Navigation and Safety of Sea Transportation Transportation p 94 (The Nautical Institute, UK)
[8] Svein K 2005 Maritime Transportation: Safety Management and Risk Analysis (Elsevier Butterworth-Heinemann, UK)
[9] Adam W and Tomasz N 2011 Miscellaneous Problems in Maritime Navigation, Transport and Shipping: Marine Navigation and Safety of Sea Transportation p 33 (The Nautical Institute, UK)
[10] Transport Canada 2003 Small Fishing Vessel Safety Manual (Department of Transport, CA)

Acknowledgments

This paper and the research behind it would not have been possible without the exceptional support of my supervisor, Dr.Ing Setyo Nugroho. His enthusiasm, knowledge, and exacting attention to detail have been an inspiration and kept my work on track from my first encounter to the final draft of this research. We are also grateful to all levels of regional agencies in the City of Pasuruan who participated in providing useful information for the completion of this research.

IOP Conf. Series: Materials Science and Engineering 1052 (2021) 012065 doi:10.1088/1757-899X/1052/1/012065

Regulatory framework of a computer-based stowage planning: safety and efficiency considerations

S Nugroho[1], E B Djatmiko[2], Murdjito[2], E W Ardhi[1], H Supomo[3] and I G N S Buana[1]

[1] Marine Transportation Engineering, Institut Teknologi Sepuluh Nopember Surabaya, Indonesia
[2] Ocean Engineering, Institut Teknologi Sepuluh Nopember Surabaya, Indonesia
[3] Naval Architecture and Shipbuilding Engineering, Institut Teknologi Sepuluh Nopember Surabaya, Indonesia

Abstract. Stowage planning is a tedious job. It involves technical and administrative tasks. Evolving data along the process and information completeness should be considered, when dealing with the updating of stowage plans. Recent IMO Regulations have brought forward a set of advanced stability criteria, far beyond the computation of metacentric height. The paper outlines the challenges on the implementation of proper stowage planning practices from the perspectives of safety and operational efficiency.

1. Introduction

A growing global trade is deeply connected to the ocean freight. Some ninety percent of the commodities traded globally are transported by ships. The ship size grows as as a result of economy of scale, in particular for container shipping. Efficiency improvement has always been a key in order to make every sector competitive and profitable. For port operation, it means a higher cargo handling productivity [1].

Information on the incoming cargoes must be available in a sufficient period of time beforehand. In order to maximize the certainty of the so-called cargo loading list (CLL), many container terminals apply the closing time rule. Containers must be ready at the container yard about two days before the ship's arrival. The above is strictly applied for all international cargoes. Unfortunately, it does not necessarily apply for domestic cargoes in Indonesia.

On the other hand, the portrait of marine safety still raises awareness. Number of accidents remains high. Apparently a huge portion of the source of accidents attributed to the loading process [2]. Some of this accidents could be viewed as a systematic in nature, instead of a human error. It is therefore imperative to emphasize the importance to improve the quality of load planning, also known as stowage planning. This paper outlines the typology and requirements of a stowage planning tool, its regulatory framework, efficiency and human error issues and how to overcome them.

2. Loading Sofware: Typology and Requirements

The above portrait of a modern container shipping operations becomes a challenge for the planning process of allocating containers on board the ship. This process is called stowage planning. Time available for allocating thousand of containers on board virtually, through six big LED monitors, becomes an increasing challenge. It has direct cost consequences. A restow is a process where a container must be forced to be discharged first, in order to give a way for discharging another container, then this container will be reloaded again on board the ship. This restow is costly. For domestic shipping in Indonesia, it is approximately 10% of the freight rate. For bigger ships, and for voyages with multiple ports of call, the chance of producing restows is significantly higher. An optimality strategy has always been the vision of a majority of algorithm development in order to automize the planning process, fully or partially [3,4].

Approved stability software is not a substitute for the approved stability information, and is used as a supplement to the approved stability information to facilitate stability calculations. The input/output information shall be easily comparable with approved stability information so as to avoid confusion and possible misinterpretation by the operator relative to the approved stability information.

The types stowage planning software also known as loading softaware or stability software are determined from the scope of calculations performed [5]:

1. Type 1 Software calculating intact stability only (for vessels not required to meet a damage stability criterion).
2. Type 2 Software calculating intact stability and checking damage stability on basis of a limit curve (e.g. for vessels applicable to SOLAS Part B -1 damage stability calculations, etc.) or checking all the stability requirements (intact and damage stability) on the basis of a limit curve.
3. Type 3 Software calculating intact stability and damage stability by direct application of pre-programmed damage cases based on the relevant Conventions or Codes for each loading condition (for some tankers etc.).
4. Type 4 Software calculating damage stability associated with an actual loading condition and actual flooding case, using direct application of user defined damage, for the purpose of providing operational information for safe return to port (SRtP).

The following regulations and guidelines serve as a reference:

1. MARPOL 73/78, Annex 1, Regulation 28 for Oil Tanker
   Maritime Pollution (MARPOL) 73/78 is an international convention signed in 1973 and amended by protocol in 1978. Its Annex 1, Regulation 28 stipulates the obligations of oil tanker using loading instrument.
2. IBC Code Chapter 2 for Chemical Tanker
   International Bulk Chemical Code (IBC Code) is an international code for the construction and equipment of ships carrying dangerous chemical in bulk. Its chapter 2 specifies that a bulkcarrier of the length of over 150 meter must be fitted with a loading instrument.
3. IGC Code Chapter 2 for Gas Carrier
   International Gas Carrier Code (IGC Code) is an international code for the construction and equipment of ships carrying liquefied gas in bulk. Its purpose is to provide an international standard for the safe carriage, by sea in bulk, of liquefied gases and certain other substances.
4. SOLAS 1974 Chapter XI I/11 for Bulkcarrier
   Safety of Life at Sea (SOLAS)1974 is the international convention that regulates shipping safety especially on three aspects namely ship construction, equipment and operation. Its Chapter XI I/11 specifies additional measures for bulkcarrier.
5. SOLAS 1974 Chapter I I-1/8-1.3 for Passenger Ship
   Safety of Life at Sea (SOLAS)1974 is an international convention that regulates shipping safety in the areas of ship construction, equipment and operation. Chapter I I-1/8-1.3 stipulates that Passenger vessel constructed after 1 January 2014 must be installed with loading instrument.
6. IS Codes
   Intact Stability Code (IS Code) is an international code that presents both mandatory and recommendatory stability criteria and other measures for ensuring the safe operation of ships, in order to minimize the risk to the ship, to the personnel on board and to the environment.
7. MSC.1/Circ.1229 Guidelines for the Approval of Stability Instruments
   Maritime Safety Committee (MSC).1/Circ. 1229 presents guidelines for the approval of stability instruments, and specifies the acceptable tolerance for the approval.
8. IACS UR L5
   International Association of Classification Societies (IACS) UR L5 is present Onboard computers for stability calculation.

9. IS Code Part B Chapter 4
   Intact Stability Code (IS Code) Part B chapter 4 is present Stability calculation performed by stability instruments.

Table 1. Acceptable tolerance for loading software examination

Stability Instrument	Acceptable Tolerance
Hull Form Dependent	
Displacement	2%
Longitudinal center of buoyancy, from AP	1% / 50 cm max
Vertical center of buoyancy	1% / 5 cm max
Transverse center of bouyancy	0.5% of B / 5 cm max
Longitudinal center of floatation, from AP	1% / 50 cm max
Moment to trim 1 cm	2%
Transverse metacentric height	1% / 5 cm max
Longitudinal metacentric height	1% / 50 cm max
Cross curve of stability	50 mm
Compartment dependent	
Volume or deadweight	2%
Longitudinal center of gravity, from AP	1% / 50 cm max
Vertical center of gravity	1% / 5 cm max
Transverse center of gravity	0.5% of B / 5 cm max
Free surface moment	2%
Shifting moment	5%
Level of contents	2%
Trim and Stability	
Draught (forward, aft, mean)	1% / 5 cm max
GMt	1% / 5 cm max
GZ values	5% / 5 cm max
FS Correction	2%
Downfoading angle	2%
Equilibrium angles	1
Distance to unprotected opening or margin line from WL, if applicable	± 5% / 50 mm
Areas under righting arm curve	5% or 0.0012 mrad

A stability booklet contains in particular the geometry, hydrostatic properties, tank tables of a ship. A validity of a stability booklet is shown by the class approval. A loading software refers to the stability booklet, which it represents. The classification society or a recognized organization examines the accuracy of the calculation in line with the IACS Guidelines, see Table 1 [5].

3. Regulatory Framework

The mandatory installation of loading software is scheduled in phases, see Table 2. Bulkcarrier of the length of over 150 meter must be fitter with a loading instrument, starting from 1 July 2016 Gas carrier, liquified gas, oil tanker and chemical carrier constructed before 1/1/2016 must be installed before 1/1/2016 [6]. Thus applies also for tankers with a renewal survey between 1 January 2016 and 1 January 2021 [7]. Passenger vessel constructed 1 January 2014 must be installed with it [8].

Standards of Training, Certification and Watchkeeping for Seafarers (STCW) – Manila Amendments 2010 emphasizes the importance of adopting ICT in the seafarers' education and training [9]. A computer-based education for stowage planning must be deployed in their education.

After decades of waiting, Indonesia finally ratified the International Maritime Organization (IMO) regulations on Safety of Life At Sea (SOLAS) in December 2017, under them Presidential Act 57/2017

[10]. It demonstrates a firm effort to improve the marine safety of the Indonesian maritime industry. This is a major step the government undertakes to improve the marine safety portrait of Indonesia.

National regulations have paid the stability issues as a measure to ensure marine safety. Ministry of Transportation Regulation no. 39/2016 on Loadline and Ship Loading asserts accurate stability calculations as a pre-requisite for the Port Clearance (Surat Pemberitahuan Berlayar) [11]. Considering the increasing role of the information technology, the Ministry of Transportation Regulation no. 154/ 2015 introduces an Online Harbour Master Declaration (Surat Pemberitahuan Syahbandar), also known as SPS Online [12]. SPS Online contains three main services, namely: a. Clearance-in; b. Maneuvering and Clearance-out approval documents.

Table 2. Schedule stability instrument IMO

Type of Ship	IMO Regulation	Constructed	Application
Tanker	MARPOL 73/78 Annex 1.Reg 28(6)	01/01/2016	Survey of the ship on or after 1 January 2016 but not later than 1 January 2021
Bulk Carrier	IBC Code Chapter 2	01/01/2016	Bulk Carriers of 150 m in length and upwards shall be fitted with a loading instrument
Gas Carrier	IBC Code Chapter 2.2.7	01/01/2016	Survey of the ship on or after 1 January 2016 but not later than 1 January 2021
Passenger ships	SOLAS 1974 Chapter II-1/8-1.3	01/01/2020	Passenger ships constructed on or after 1 January 2014 shall have: 1. Onboard stability computer 2. Shore-based support

In spite of the above progress, a firm regulation mandating the inspection on ships on their compliance towards the deployment of a loading computer still does not exist yet.

4. Efficiency Issues

Efficiency is a key success factor for well-functioning of all industries. Shorter port time serves as an important instrument to boot efficiency in marine transport. This instrument affects the execution of stowage planning considerably. Preparing a stowage plan for a container ship of about 200 teus capacity takes some two hours by using a spreadsheet program. A closing time rule is a time windows provided to the shipper to deliver their containers. The purpose is to ensure that all containers are already available lates one day before the arrival of the ship. By doing so, the stowage plan could be done in time and accurately.

Time pressure is obvious, especially at ports where no closing time rule is applied. Cancellation of booked containers or arrival of new containers may still take place even two hours before the departure of the ship. This practice puts the chief officer in time pressure to update the list of cargoes, the bayplan and the stability calculation. This portrait repeats almost every time when a ship calls a port.

Other types of vessel have a different operation profile at port. A ferry must complete both discharging and loading operations in 45 minutes. A tanker and bulkcarrier with usually one or only a few types of cargoes has been in a better position. The loading list is usually fixed. The stowage planner has sufficient time to prepare a stowage plan. Their stowage plan is usually prepared in phases, the so-called loading sequence. This loading sequence is usually in a sequence of 0%, 25%, 50% and 100% loading conditions.

Another important issue in loading operations of a ship is the information accuracy of cargoes. Ministry of Transportation Regulation no. 53/2018 on the Feasibility of Containers and Verified Container Gross Mass obliges shipper to inform the verified gross mass to the ship and port prior to delivering them to the terminal. This measure is established in order to ensure the input correctness for stability calculation.

In spite of the above measures, the compliance on fulfilling the above regulations raises some concern. The result, the quality of stability calculation could be less trustworthy.

5. Human Error Issues

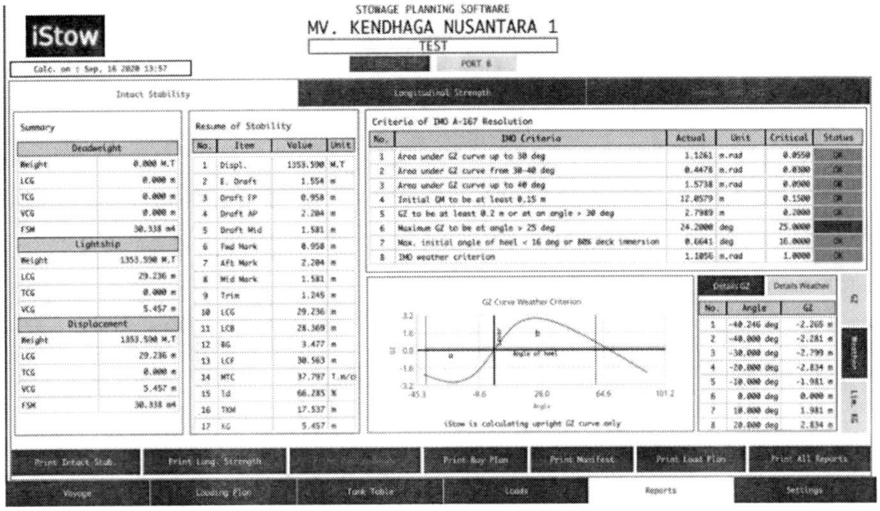

Figure 1. Stability criteria compliance in iStow: visual

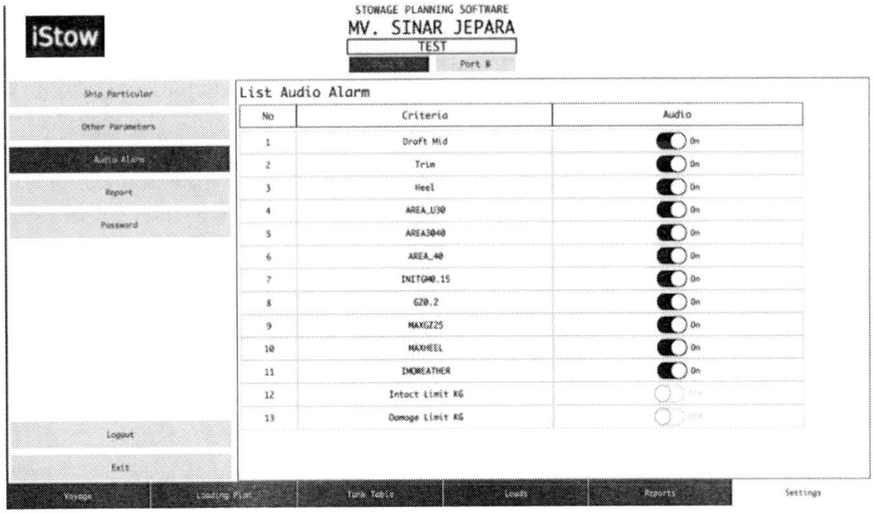

Figure 2. Stability criteria compliance of iStow: audio setting

The portion of accidents involving human cannot simply be neglected. Human error is a dominating source of ship accidents. For clearance-out purposes, the metacenter height (GM) suffices. This operational requirement in practice is much softer than what IMO obliges to all operating ships.

iStow stowage planning software has been equipped with an visual warning system by showing the compliance of the stability criteria for the ship type in concern with green colour, other wise red, see Figure 1. Additionally, an audio warning system is also embedded, in order to improve the stability criteria compliance, see Figure 2.

Having those facilities in place, unnecessary human error could be minimized.

6. Concluding Remarks

1. At present, a stability calculation summary is often prepared by using a computer spreadsheet or manually. An accurate stability calculation summary is a pre-requisite for the issuance of a clearance-out document.
2. This stability calculation summary regards the metacentric height (GM) as the only stability criterion, which is far less than the full list of IMO stability criteria. Depending up on its type, a ship must fulfill at least seven stability criteria.
3. The usage of a loading software enables us to automate the process of stability calculations and the checking of stability criteria compliance. By doing so, unnecessary human error could be minimized, operational efficiency could be accomplished.
4. It is recommended to implement usage of loading computer strictly, in order to make ship operations satisfy both the operational efficiency and IMO safety requirements.

Acknowledgement

The authors would like to express a sincere gratitude to RISPRO Program of LPDP of the Ministry of Finance for providing a research grant in the period of 2020-2022 under the contract no. 35/LPDP/2020. We thank to the Ministry of Research, Technology and Higher Education for supporting us conduct the research and diffusion of the stowage planning tool for various types of ships since 2007. We are grateful for a very nice partnership with Samudera Indonesia Group, Pertamina Shipping, Edricko Shipping, Dok Perkapalan Surabaya (DPS), Biro Klasifikasi Indonesia, ClassNK Indonesia, RINA Indonesia, Indian Register of Shipping and Lloyd' Register (LR). A big thank goes also to our great teams at the Department of Marine Transportation Engineering of Institut Teknologi Sepuluh Nopember (ITS) and PT Pranala Digital Transmaritim to make this innovation meaningful for the society.

References

[1] Nugroho S. (2007). "Reoccurence in planning: Foundational principles and lessons learned from it", Proc. of Operations and Supply Chain Management (OSCM), Thailand, May 18-20.
[2] Komite Nasional Keselamatan Transportasi. (2014). Investigasi Kecelakaan Pelayaran: Tergulingnya KM BJL 1 di Dermaga 107 Pelabuhan Tanjung Priok, Jakarta. Laporan Final KNKT - 1 4 - 0 1 - 0 2 - 0 3, Komite Nasional Keselamatan Transportasi, 14 January.
[3] Nugroho S. (2009). Penerapan Piranti Lunak Perencanaan Stowage Kapal Peti Kemas. Program insentif percepatan difusi dan pemanfaatan iptek, Institut Teknologi Sepuluh Nopember (ITS).
[4] Nugroho S, Abidin A Z, and Zulkarnaen F. (2017) "Stowage planning for container vessels: Methodology development and implementation issues in the light of intelligent transportation systems implementation". Int. J. of Business Economics, 2(6):119–134.
[5] IACS. (2017). UR L5 Onboard Computers for Stability Calculations. Technical report.
[6] International Maritime Organization. (2020). IBC CODE, 2020 ed. International Maritime Organization.
[7] International Maritime Organization. (2017). MARPOL 73/78 Annex 1 Rev. 3 Reg. 28(6) Subdivision and Damage Stability, 2017 ed. International Maritime Organization.
[8] International Maritime Organization. (1974). SOLAS, 2018 Consolidated ed. International Maritime Organization.

[9] International Transport Workers' Federation. (2018). STCW a Guide for Seafares, 2018 ed. International Transport Workers' Federation.
[10] Presiden Republik Indonesia. (2017). Peraturan Presiden No.57 tahun 2017 tentang Pengesahan Protocol of 1988 Relating to the International Convention for the Safety of life at Sea. Technical report.
[11] Menteri Perhubungan Republik Indonesia. (2016). Peraturan Menteri Perhubungan No.39 tahun 2016 tentang Garis Muat Kapal dan Pemuatan. Technical report.
[12] Menteri Perhubungan Republik Indonesia. (2018). Peraturan Menteri Perhubungan No.154 tahun 2015 tentang Surat Pemberitahuan Berlayar Online. Technical report.

IOP Conf. Series: Materials Science and Engineering 1052 (2021) 012066 doi:10.1088/1757-899X/1052/1/012066

User Aspiration and Design of Functional Architecture in Seafarer's Certification Management

S Nugroho[1], A T Putro[1]
[1]*Institute Teknologi Sepuluh Nopember, Kampus ITS, Sukolilo, Surabaya*
Department of Marine Transportation Engineering, FTK - ITS
snugroho@seatrans.its.ac.id

Abstract. Maritime (kepelautan) is matters related to manning, education, certification, authority, rights and seafarer's obligations. In marine transportation activities, there are several main instruments that must be in place, namely transportation facilities (ships), transportation facilities providers (shipping companies / government), ship operator service providers (seafarer), certified seafarers / producers of seafarers (training institutions) and infrastructure, such as ports, channels, warehouses, and so on. The relationship between seafarers and seafarer certification (education and training institutions / government) and shipping companies is a concern in this study, especially in relation to the impact of many falsification of seafarer certificates, despite the implementation of an online database certification system that is centralized in Directorate General of Sea Transportation (DJPL). There are 3 (three) major problems that are still popular at this time, for instance unoriginal of seafarer certificates, unpaid seafarers or late receiving wages for more than 1 (one) wage period, and seafarers have a difficult time to find jobs on board. One of the available instruments to solve this problem is the intelligent transport system (ITS) architecture which is liable to be adopted in mapping the aspirations of marine transportation stakeholders, blockchain technology, such as decentralizing (one of the pillars / characteristics of blockchain technology) seafarer certificate database. A cost benefit analysis will be used in this study to obtain how much does it cost to improve the system and to see the benefit of the product.

1. Introduction

Table 1 Supply and Demand for Global Seafarers

Years	Supply of Seafarer			Demand of Seafarer			Balance (Supply-Demand)		
	Officer	Rating	Total	Officer	Rating	Total	Officer	Rating	Total
2005	466,000	721,000	1,187,000	476,000	586,000	1,062,000	-10,000	135,000	125,000
2010	624,000	747,000	1,371,000	637,000	747,000	1,384,000	-13,000	0	-13,000
2015	774000	873000	1647000	790500	754500	1,545,000	-16,500	118,500	102,000
2020	789,500		789,500	881,500		881,500	-92,000	0	-92,000
2025	805,000		805,000	952,500		952,500	-147,500	0	-147,500

Source : BIMCO Manpower report 2015 [3]

IOP Conf. Series: Materials Science and Engineering 1052 (2021) 012066 doi:10.1088/1757-899X/1052/1/012066

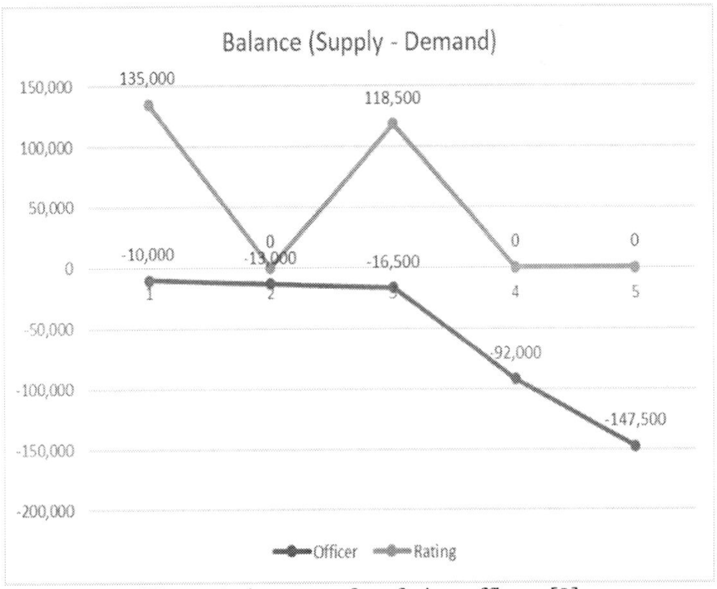

Figure 1 shortage of seafaring officers [3]

As in the Table 1 and Figure 1, it can be seen that in 2015 the world had a shortage of 16,500 officers and in 2020 it is estimated that the world has a shortage of 92,000 officers, until 2025 it is estimated that the shortage of marine officers is 147,500 officers [3].
According to the Baltic and International Maritime Council (BIMCO) report [3], the shortage of seafarers and the high population growth which increase job competition becomes more selective and competitive, furthermore it can be seen as an opportunity by individuals who want to take advantage of it by looking for shortcuts to make seafarers' certificates that can be traded and owned without a process of education and training [9].
Due to the high demanding, the person sellis the illegal certificates with a price range approximately Rp. 700,000 to Rp. 20,000,000 depending on the level of the certificate.it doesn't stop there, the illegal certificate seller also hacked the Indonesian seafarer's website to enter the certificate data they sold so that it gave a genuine impression. and currently there are more than 5041 illlegal seafarers certificates that have entered the Indonesian seafarers' database [9]. in detail we can see in the following table:

Table 2 The Level of Seafarers' Certificates that are Traded

Certificate Level	Price Per Certificate	
Sertifikat keterampilan pelaut	Rp	700,000
Electrical Technical Officer (ETO)	Rp	1,500,000
Competency Certificate Class IV	Rp	7,000,000
Competency Certificate Class IIII	Rp	10,000,000
Competency Certificate Class II	Rp	15,000,000
Competency Certificate Class I	Rp	20,000,000

Source : Tribunnews.com 2020 [9]

those cases seriously threatens the existence of Indonesian seafarers on international shipping, where currently there are 300,000 Indonesian seafarers working on foreign ships [19] and if it becomes a serious problem in international law then the worst consequence is that all Indonesian seafarers' certificates will be banned and unable to sail in international seas.
The next problem is related to unpaid seafarers by the company that employs them or the company that employs them is late providing wages for more than 1 (one) period of wages [11]. the reports are quite

often received by the International Transport Workers' Federation (ITF). Every year, more than 1500 ships crew reported that they are not paid and on the other hand, the company is not able to pay [11]. The third problem is a contradicts report by BIMCO that Seafarers have a hard time find jobs on board. when the world comes up short of seafarers, however it is difficult for seafarers to get a job. it can be concluded that the third problem is job vacancies are not conveyed to seafarers, or it could be that Indonesian seafarers are just looking for and applying for jobs at local companies [12].

2. Literature Review

The International Convention on the Standards of Training, Certification and Watchkeeping for Seafarers (or STCW), 1978 establishes the standard qualifications for captains, officers and guards on board merchant ships sailing. STCW was born in 1978 from the International Maritime Organization (IMO) conference in London [6].

1. A certificate of competence for both deck and engine seafarers;
2. Latest training, updated requirements;
3. Assignment of responsibility to members, including those relating to licensing, and flags of states employing foreign countries, to ensure seafarers meet the standard requirements of competence.

International Labor Organization (ILO), IMO does not have the capacity to make a comprehensive legal instrument on the protection of seafarers, so it is appropriate if the ILO makes the maritime labor convention (MLC) 2006 as an international legal instrument [5]. The MLC provides a legal basis relating to the fulfillment of several rights of seafarers [7], [8], namely:

1. A safe and secure workplace according to proper safety standards;
2. Fair terms of employment;
3. Work and workplace conditions on board are decent; and
4. Work protection, health care, welfare and other forms of social protection (Health protection, medical care, welfare measures and other forms of social protection).

Intelligent transport system (ITS) architecture, there are several basic things that are needed [1], [2], [4] [18], namely :

1. Change Stakeholder Aspirations into formal User Needs, whose functions can be expanded;
2. Divide functionality into components that can be produced;
3. Make an outline specification of these components;
4. Submit these specifications to the review team;
5. Compare 'where you are' with 'where you want to be;
6. Develop a placement plan.

3. Methodology

3.1 Research Location

This study was conducted in several places in Surabaya - Politeknik Pelayaran Surabaya and Port of Tanjung Perak, Surabaya. This study was also conducted in Jakarta - Directorate General of Sea Transportation (DJPL) Jakarta.

3.2 Data Collection Methods

Data were collected by using 2 (two) methods:

1. Field observations to find out how much does it costs and how long does it takes for seafarers to take education and training until they get a seafarer certificate and work on a ship.
2. Interviews with seafarers to find the information towards the duration in finding a job on board and how they apply for it.

All data will be analysed by using the cost benefit analysis.

3.3 Case Study

Various problems that arise when falsification of seafarers' certificates continues, are:

1. Indonesian seafarers are no longer trusted in the world shipping company, therefore, it is increasingly difficult to work in international companies;

2. If it is found out that the seafarers working on their ships are using illegal certificates, the company will suffer material losses because the ship is not allowed to sail until there is a replacement seafarer and makes another work contract with the replacement sailor and pays additional mooring fees at the port;
3. All seafarers' certificates issued by the government in Indonesia are doubted in the world shipping company, so that seafarers cannot work, reducing foreign exchange and increasing unemployment.

4. Result and Discussion

4.1 Observation

In the observation on falsification of seafarers' certificates, it is found that 5041 illegal certificates were successfully sold and the data entered into the database of Indonesian seafarers [9]. Some data obtained from observations, they are:

1. training rates at the politeknik pelayaran surabaya
2. costs and time spent on education and training seafarers
3. the supply data of the number of seafarer each year from the politeknik pelayaran surabaya
4. Minimum standards of wages for seafarers

data obtained from observation can be seen as follows Table 3

Table 3 Training Rate at The Politeknik Pelayaran Surabaya

No	Types of Training Programs	Average Fare
1	Basic education and training program	Rp. 54.060.167 /person
2	Upgrading training program	Rp. 16.130.000 /person/ grade
3	Updating training program	Rp. 1.083.125 /certificate
4	Proficiency training program	Rp. 1.550.815 /certificate
5	Revalidation of certificate profeciency	Rp. 490.000 /certificate

(Source: Poltekpel Surabaya 2020, author)

The tariff description from the table above is the average cost for education and training at politeknik pelayaran Surabaya.

Furthermore, it is shown the amount of money needed for seafarers and the time consumed to obtain a seafarer certificate at the highest level.

Table 4 Seafarer Training Costs

Seafarer Education and Training Certification	Cost	Other expenses (living cost and transport)	Time
Diklat pembentukan keahlian (Certificate Of Competance/COC)	Rp 54,060,167	Rp 98,550,000	3 years
Diklat peningkatan keahlian (Certificate Of Competance/COC)	Rp 48,390,000	Rp 40,500,000	7 years 3 month
Diklat Keterampilan (certificate of profeciency)	Rp 26,121,407	Rp 540,000	6 day per certificate
Buku Pelaut	Rp 100,000	Rp 450,000	5 day
Total	Rp 128,671,574	Rp 140,040,000	10.5 years

(Source: Poltekpel Surabaya 2020, author)

The table 4, shows that in order to obtain seafarer certification up to the highest level it requires training fees of Rp. 128,671,574,- and other costs Rp. 140,041,000,- in the fastest period of 10.5 years.

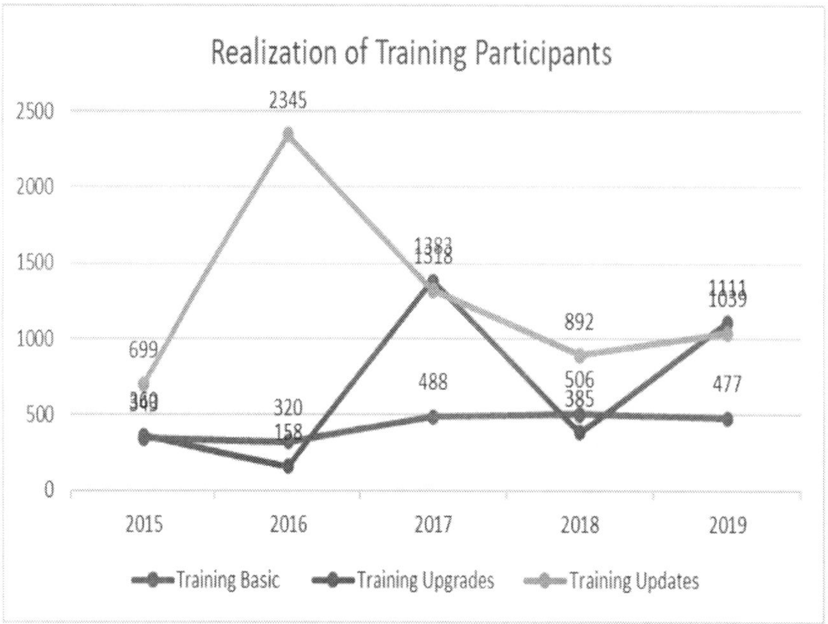

Figure 2 the number of seafarers' graduates per year at Poltekpel Surabaya.

Follow the Figure 2, the number of training participants is varied and tends not to be maintained in annual numbers.
from the Tabel 5, it can be seen the data related to the minimum wage standard for seafarers :

Table 5 Minimum Wage Standards For Seafarers

Position Level	Dry Ship (Bulk/ Ro-Ro/ Container Etc) GT 3000 - 20000		Tanker Ship (Oil/Cemical/Gas Etc) GT 3000 - 20000	
	USD	RUPIAH	USD	RUPIAH
Engineer cadet Deck cadet	100	1,487,800	200	2,975,600
Able Deck Able Engine	400	5,951,200	500	7,439,000
Bosun	500	7,439,000	600	8,926,800
Electrician Radio officer chief. Cook	800	11,902,400	1,000	14,878,000
third engineer third officer	1,200	17,853,600	1,500	22,317,000
second engineer second officer	1,600	23,804,800	1,900	28,268,200
first engineer chief officer	2,000	29,756,000	2,200	32,731,600
chief engineer Captain	2,500	37,195,000	2,800	41,658,400

(Source: ITF ILO Minimum Wage 2020, author)

The minimum wage in the Tabel 5 is the international standard recommended by the ITF ILO minimum wage scale [5].

4.2 Interview

Interview was conducted to obtain information from seafarers:

1. How do Indonesian seafarers apply for work?
2. How is the process for Indonesian seafarers to obtain the job vacancy information?
3. How do Indonesian seafarers choose the targeted job?

From those questions, the following business process diagram is obtained:

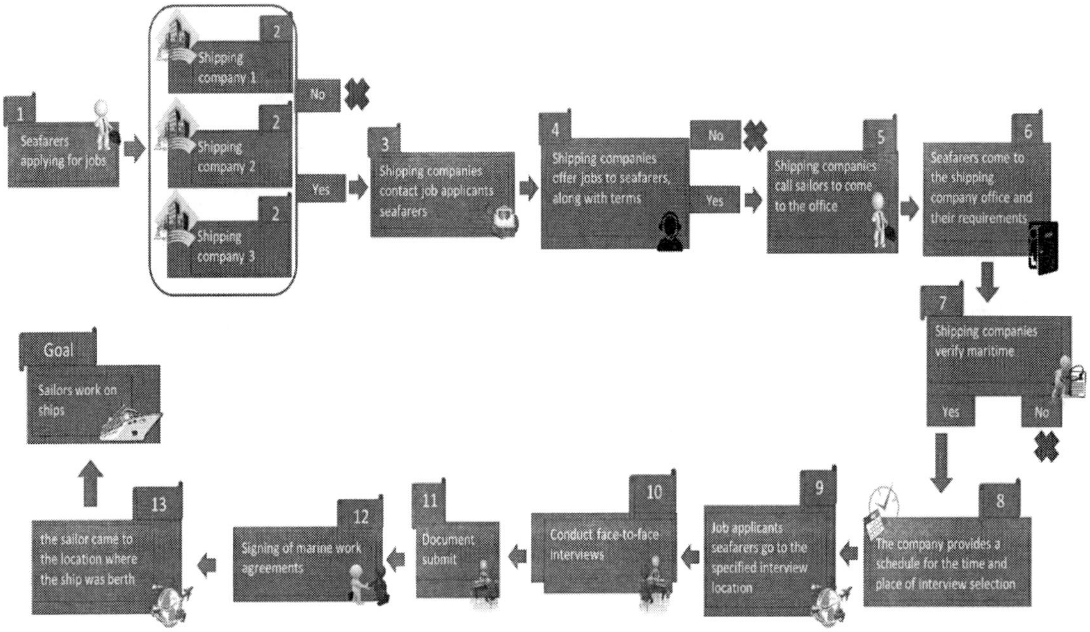

Figure 3 General processes for seafarers to work on board

Following Figure 3 there are 14 steps that a sailor must take to be able to work on the ship, each stage in Figure 3 will be explained in table 6.

Table 6 Process of Seafarers' Activities of Finding and Applying for Jobs

No.	Description of activities	Time
1	Seafarers are looking for and applying for jobs by entering a curriculum vitae (CV) to companies that they know both digitally and manually and of course seafarers are already in a city where many shipping companies are domiciled so that seafarers also live temporarily in that city.	3 month
2 & 3 & 4	Shipping companies receive application files from seamen and when shipping companies need seafarers then The company contacted seafarers Offer jobs and requirements that must be met	10 minutes
5	The shipping company calls the sailor to come to the company office by telephone	10 minutes
6	Seafarers come to the shipping company office and bring the required maritime documents	2 day
7 & 8	Shipping companies verify seafarers' documents and when the documents match the company sets a schedule for interviews regarding time and place	30 minutes

9 & 10	Face to face interview Seafarers come to the specified interview location Conducting interviews	1 day
11 & 12	Seafarers come to the shipping company office and sign the sea work agreement (PKL) and submit maritime documents	2 hours
13	Seafarer go to the location where the ship is berth a. Transportation to the airport b. Air Transport c. Transportation to the inn d. Lodging e. Transportation to the port where the ship is berth	2 day

(Source : observation and interviews)

Follow the Figure 3 and table 6, seafarers carry out the process of looking for work and applying for jobs by visiting the companies they know one by one. So it takes extra cost and time to do it. Likewise seafarers do not know whether the company requires seafarers who meet their specifications or not. And by using this process the average seafarer costs more and takes more than 3 months to get a job on board.

5. Conclusion

Related to the problems of concern in this study, namely

1. The number of illegal seafarer certificates registered in the Indonesian seafarers database.
2. In Indonesia there is no official facility or institution that can receive complaints from seafarers regarding seafarers not receiving a salary.
3. Seafarers find it difficult to find work onboard.

from those problems finding, then the authors conclude that the problem is possible to solve by adopting a work system from the intelligence transport system (ITS framework) and the concept of data server decentralization (blockchain technology).

The following can be seen an overview of the intelligent transport system (ITS) architecture so that it is possible to be able to solve the problems in this study.

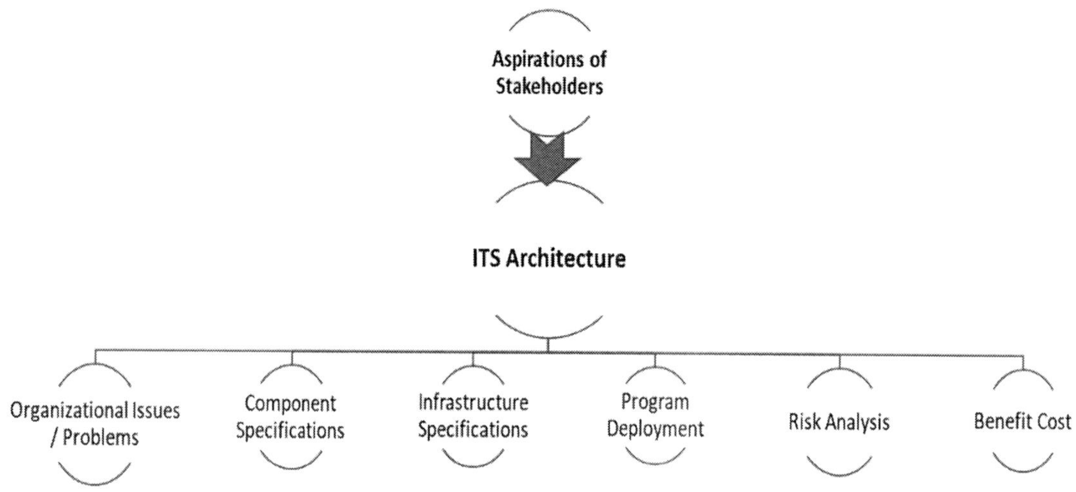

Figure 4 ITS Framework Diagram [1]

Follow the Figure 4 where in the ITS architecture it accommodates user complaints or user needs so that it can bring together the needs of service providers and users, In data processing, data analysis is needed and one of the components in the ITS architectural framework is the data analysis method, it is called cost benefit analysis [1], [2], [4], [18].

Where the inter-stakeholder needs linkages can be seen in the following figure 5:

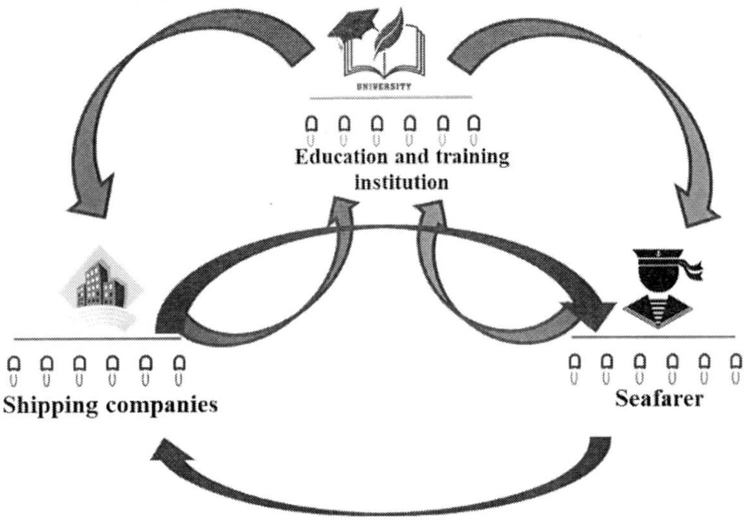

Figure 5 diagram of the relationship between stakeholder needs

Following Figure 5 regarding the linkage of needs between stakeholders will be explained in the following table :

Table 7 Stakeholder needs

Stakeholder who need it	Providers		
	Education and training institution	Shipping companies	Seafarer
Education and training institution	1. Share information related to the needs of each training institution. 2. Share information regarding what matters are needed by international shipping companies so that sailors produced by training institutions can be accepted by shipping companies with pleasure.	1. Training institutions require graduates to be accepted to work in companies. 2. Training institutions require information related to the specifics of expertise and skills of seafarers expected by shipping companies	1. Training institutions require training participants 2. Education and training institutions need information related to data on seafarers who have met the requirements for education and training which are then offered to take education and training in their place
Shipping companies	Shipping companies need seafarers whose competencies are legally certified and have a good educational record	1. share information related to the needs of each company. 2. share information related to services obtained from seafarers and training institutions.	Shipping companies require experienced seafarers
Seafarer	1. Seafarers need education and training to improve their competence 2. Share information with seafarers regarding the latest knowledge	Seafarers need a good and decent job	Share experiences and knowledge

(Source : observation and interviews)

Follow the Figure 5 and Table 7 In the diagram of stakeholder needs, the linkage of needs between them can be accurately informed and communicated using the ITS architecture so that every stakeholder's needs can be easily and quickly met.

Furthermore, related to the data obtained from this research will be used as follows :

1. Follow the data in Tables 3 and 4 are used for:
   a. Calculating times and costs required by seafarers to obtain certification to the highest level;
   b. Calculating seafarers' losses with widespread falsification of seafarer certificates.
2. Follow the Figure 2 shows that it is difficult to maintain the number of training participants so that the difference in numbers is much different from the previous year. in addition, the using of ITS architecture will make it possible for education and training institutions to be able to find sailors who have met the requirements to take training.
3. Follow the data in Table 5 is used to calculate how much currency exchange in one year is contributed by seafarers who receive US dollar wages so that it can be seen how much the state loses through foreign exchange earnings from the export of seafarers.
4. Follow the Figure 3 and the Table 6, the process of seafarers looking for and applying for jobs is very long, so it takes extra cost and time to do it. Likewise seafarers do not know whether the company requires seafarers who meet their specifications or not. It is also discovered that by using this process the average seafarer costs more and takes more than 3 months to get a job on board. Then using the telematics architecture is expected to shorten and make it easier for seafarers to obtain job vacancies and submit the applications

Acknowledgment
The first author is thanks to the Ministry of Transportation, Human Resources Development of Sea Transportation Center.

References

[1] www.frame-online.net European ITS Fram ework Architecture, and the FRAME projects.
[2] Mikulski, Jerzy. 2010. *Using Telematics in Transport.*
[3] The Baltic and International Maritime Council (BIMCO). *Manpower report 2015 executive summary.*
[4] Nugroho, Setyo. 2018. *Intelligent Transportation Architecture to Address Challenges of Traditional Shipping Operations (PELRA).*
[5] International Labour Organization (2020). *Recommended minimum wage for an AB – extra rates applicable from* 1st January 2020.
[6] International maritime organization (IMO). *Standard training certification of seafarer (STCW) 2010 as amendment.*
[7] Kusuma Candra Ade. 2018. *Manajemen Pengawakan Pelaut Kapal Niaga dan Tanggung Jawab Perusahaan Keagenan Awak Kapa*l.
[8] International Labour Organization (ILO). *Maritime labour convention* 2006.
[9] https://wartakota.tribunnews.com Retrieved 25 June 2020.
[10] https://pdppsdmpl.bpsdm.dephub.go.id Retrieved 4 January 2020.
[11] https://www.itfseafarers.org Retrieved 17 August 2020.
[12] https://www.pelaut.club Retrieved 25 August 2020.
[13] https://pelaut.dephub.go.id Retrieved 8 February 2020.
[14] Badan Pengembangan Sumber Daya Manusia Perhubungan, PK.07/BPSDM-2016 Tentang Kurikulum Pendidikan Dan Pelatihan Pembentukan Dan Peningkatan Kompetensi dibidang Pelayaran.
[15] Kementerian Perhubungan, PM 70 Tahun 2013 Tentang Pendidikan dan Pelatihan, Sertifikasi Serta Dinas Jaga Pelaut.
[16] Kementerian Perhubungan, PM 140 Tahun 2016 Tentang Perubahan Atas Peraturan Menteri Perhubungan Nomor PM 70 Tahun 2013 Tentang Pendidikan dan Pelatihan, Sertifikasi Serta Dinas Jaga Pelaut.
[17] Kementerian Perhubungan, PP No. 07 Tahun 2000 Tentang Kepelautan.
[18] Williams B 2008 *Intelligent Transport Systems Standards* (ARTECH HOUSE, INC: Norwood).
[19] Sugihardjo. 2020. *Pemberdayaan SDM dan Peluang Karir Pelaut di Sector Kemaritiman*. Jakarta: webinar Indonesia maritime club discussion series #18.

Trip Planning Support Ship Mockup Design: A Case Study of Pioneer Ship Services in the Maluku Islands

S Nugroho[1], M G Makatita[1].
[1]Department of Sea Transportation Engineering, Institut Teknologi Sepuluh Nopember Surabaya.

Abstract. The process of mobilizing passengers and goods on more than five hundred islands in Maluku Province uses pioneer shipping where the route is regulated by the government to serve areas that are not reached by modern commercial ships, including: outermost, remote, border, and disadvantaged areas. The condition in which one port serves ten islands makes the impact of information availability very large, it was known in previous research by Kelwulan, 2019 [5]. Explain that the impact of information uncertainty causes additional costs that are 2-3 times the cost of transportation itself. This research was conducted to analyze and design support for the planning of pioneering ships. ITS Architecture or also known as Telematics Architecture is an intelligent transportation system where one of its basic objectives is information management by utilizing information and communication technology as a tool for making transportation systems that can manage data from pioneering shipping.

1. Introduction.

Maluku Province is an archipelago in the eastern region of Indonesia which is included in the criteria for pioneer shipping services. The total archipelago in Maluku Province is 993 islands consisting of 10 large islands and 983 small islands [9] and there are 559 inhabited islands. Maluku province also has 55 ports which are divided into 3 types of ports, namely public ports, collectors, and local ports. Thus, one port serves 10 islands or 10.16 islands / ports [5].

The impact of uncertain information causes service users to travel long before the ship arrives. Due to the relatively long distance, such as: inter-island trips to the port and having to wait at the port because there is no certainty of information about the arrival of ships.

There are four factors that have a significant impact on improving the performance of marine transportation in the Maluku region, namely the availability of port infrastructure, ship services, the availability of information and communication infrastructure, and the availability of other modes of infrastructure. The factor index of the availability of information and communication infrastructure has a higher impact on transportation performance than the other three factors, namely 58.4%, while other supporting factors are: port infrastructure by 34.3%, ship services by 16.3%, and other modes of 24.1% [9].

In today's world, the development of information and communication technology (ICT) is getting wider so that it includes the land, sea and air transportation sector. In the development of ICT for the transportation sector, it is inseparable with the term Intelligent Transport System (ITS) which is also referred to as transport telematic systems where an Intelligent Transportation System. Recently, the mode of water transportation (sea / river / lake), especially sea transportation, has received special attention from ITS, related to the development of a maritime intelligent transport system, where the main focus will be marine navigation with a telematics approach.

2. Literature Review.

ITS first appeared in the late 1960s as a computerized traffic signal control system with the aim of making urban traffic flow smooth, the ITS concept was developed again in the 1980s, to be precise in 1988, OECD (Organitation for Economic Corporation and Development) in Paris.

Development continued in 1996 in America, namely National ITS Architecture, in 2000 in Europe with the KAREN Project. Recently, water transportation modes (sea, river, and lake), especially sea transportation, have received special attention from ITS, related to the development of a maritime intelligent transport system in the future. The main focus is marine navigation with a telematics approach.

Although ITS is often seen in the context of road traffic, because travelers and other modes of transportation share goods, ITS also includes rail, water and air transportation systems (Williams, 2008).

ITS service types, according to ISO 14813-1 include several domains. Among them:

- Traveler information.
- Traffic management and operations.
- Vehicle services.
- Freight transport.
- Public transport.
- Emergency.
- Electronic payments related to transportation, etc.

According to a group of organizations that are responsible for the maintenance and development of the European ITS Framework Architecture (FRAME Architecture) methodology to create an ITS Architecture from the FRAME Architecture which is illustrated in Figure 2.1. This is important for two reasons: firstly an ITS architecture created using a methodology will not become obsolete through technological advances or the development of certain products, and the second opens up the possibility for the development of new technologies to allow certain functionality to be provided.

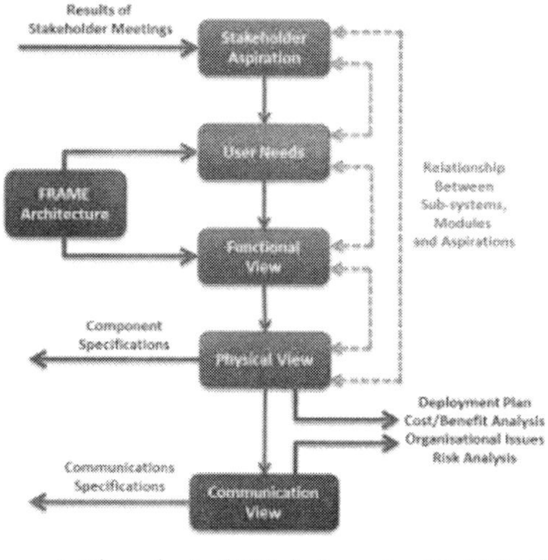

Figure 1. Flow chart of ITS design using FRAME [3].

3. Methodology.

3.1 Research Location

This research was conducted in the Maluku archipelago and several ports on the R-49 with the KM Belt Nusantara 103 ship as a pioneer ship serving the R-49 route.

3.2 Data collection methods

The data collection process is carried out by three methods, including:

- Study literature to identify stakeholders by studying documents related to pilot ship activities. Literature studies are also carried out at the completion of the observation process to maintain what possibilities if unidentified stakeholders.
- Field observations to find out directly the obstacles faced by stakeholders and observe the process of data and information flow to and from the pilot ship.

- Interview with stakeholders to get initial aspirations related to their needs, problems, issues and goals.

4. Result and Discussion.

4.1 Stakeholders

Identified stakeholders:

- Main Port Authority Office. Main Port Authority Office is a government institution at the port as the authority that carries out regulatory, controlling and supervisory functions of port activities that are commercialized.
- Main Harbormaster Office. Main Harbormaster Office is a government official who is appointed by the minister and has the highest authority to carry out and supervise compliance with the provisions of laws and regulations to ensure the safety and security of shipping.
- Port Authority and Harbormaster Office. The Harbormaster and Port Authority Office is the Technical Implementation Unit within the Ministry of Transportation under and responsible to the Director General of Sea Transportation.
- Port Health Office. Port Health Office is a technical implementation unit within the Ministry of Health which is responsible technically and administratively to the Director General of Disease Prevention and Health.
- Agricultural Quarantine Office. The Agricultural Quarantine Office is to carry out the operational activities of animal and plant quarantine, as well as supervision of biosafety, animal and vegetable.
- Fish Quarantine and Fish Quality Control Office. Who carry out operational activities in the context of preventing the entry and spread of quarantine fish pests and diseases from one area to another in the country as well as controlling the quality and safety of fishery products.
- Port Business Entity. Business entities whose business activities are specifically in the field of operating terminals and port facilities.
- National Sea Transportation Company. National Sea Transportation Company is a sea transportation company with an Indonesian legal entity that carries out sea transportation activities in Indonesian territorial waters and / or to foreign ports. The Sea Transportation Company also functions as a pioneer ship operator.
- Loading and Unloading Companies. Companies that contribute to smooth port operations in the form of loading and unloading goods from and to ships.
- Service User. Prospective passengers or passengers who use Pioneer Ship services as a mode of transportation for the mobilization of people and goods.
- The captain Crew of Pioneer Ship. The captain is the crew of the ship who is the supreme leader on the ship and has certain powers and responsibilities in accordance with the provisions of laws and regulations. The crew includes Chief Commander I, Chief Second Officer, Chief of the Engine Room who has the authority to transfer information and data.

4.2 Identification of Stakeholder Aspirations.

The process of identifying the aspirations of stakeholders is carried out by means of interviews with stakeholders. The aspirations of stakeholders are divided into two parts, where the first is the aspirations of pioneer ship operators as well as related agencies and service users.

4.2.1 Ship operators and related agencies.

The interview process was carried out to PT. Pelni as the operator of the domestic ship, the captain and the ship's crew as well as related agencies related to data transfer activities and ship information.

Table 1. Aspirations of Ship Operators and Related Agencies.

No	Description
1	Information needs, data reporting is not properly monitored
2	Reports relating to easily lost and damaged documents
3	Data recapitulation is still manual

4	The customer database could not be fully known
5	ships of difficulty sending information and data in Real time
6	Information provided is often late
7	The arrival of the ship cannot be properly monitored
8	The departure of the ship cannot be properly monitored
9	Cargo loading is not well integrated
10	SOP not working properly
11	Poor telephone and internet networks in remote areas

Source: Results of Data Processing.

4.2.2 Pioneer ship service users.

The interview process is carried out to users of pioneer ship services either using pioneer ship services to send goods or as prospective passengers.

Table 2. Needs, Problems, Issues, and Purpose of Pioneer Ships

No	Description
1	Passengers have difficulty obtaining information on special goods delivery procedures.
2	Gaps in ship delay information
3	Gap in port schedule information
4	The cost of seeking information is relatively high
5	Port waiting fees are expensive
6	The cost of living on board is expensive
7	Alternative boats are expensive
8	There are no alternative ships in some ports
9	Information gaps related to the mechanism of delivery of goods
10	Passengers want to get information easily
11	Leave the goods at ABK Deck for the delivery process
12	Not knowing the delivery mechanism
13	Want to know the position of the ship in real time
14	Want to know the position of the goods (can track the shipment)
15	More security in shipping goods
16	There was a queue at the manual ticket reception

Source: Results of Data Processing.

4.3 Data and Information Identification.

Field observations of the pioneer ship serving route R-49. Following are the ship and route specifications for the R-49.

ship specifications:

Name of Ship: KM Sabuk Nusantara 103

LOA : 62.80 M

LPP : 57.36 M

B : 12,00 M

H : 4,00 M

T : 2,70 M

V : 12 Knot

Engine Power: 2 x 1000 HP

Capacity:

ABK : 36 people

Class I Passengers : 8 People

Class II Passengers :18 People

Economy Passengers :372 People

R-49 Voyage Route 17,2019:

Ambon – Marsela – Kroing – Dawelor – Seira – Adaut – Saumlaki – Tutukembong – Romean – Larat – Molu – Elat – Tual – Ambon.

Rute R-49 Voyage 18,2019:
Ambon – Tual – Elat – Molu – Larat – Romean – Tutukembong – Saumlaki – Tutukembong – Romean – Larat – Molu – Elat – Tual – Ambon.

4.3.1 Data and information to and from the ship.
Identification of data and information from and to the ship is carried out by studying the documents relating to the ship's activities at the port of origin, while sailing, and the ship arriving at the port of destination. Data and information can be seen in table 3.

4.3.2 Data and information on service users.
There are 3 data and information for service users where:

- Information Search Process for Ticket Purchases.
- Goods Delivery Mechanism.
- Search for Ship Arrival Information and Ship Schedule

Table 3. Identified data and information to and from the ship

Data and Information Services	Types of Data and Information	Description
Pioneer Ship Routes Network	Emplooi	Emplooi Trip report Ship Travel Monitor
	Employee's Correction	Proposed Change of Schedule Employee's Correction
	Omission	Omission Report Emlooi Correction for Omisi
	Deviation	Emplooi Correction for Deviation
Service Standards	Eating of the Crew	Submitting for the ABK Food Fee Giving Meal Costs for ABK
Fuel and Lubricants	BBM	Request for filling fuel Approval Letter for BBM Request
	Lubricant	Lubricating Oil Demand Lubricant Request Approval Letter Oil Supply and use reports
Ship Maintenance	Emergency Docking	Breakdown News Event (emergency docking) Emergency docking approval letter
	Maintenance of the Ship When Operated	submission of SPB (Ship Request Letter) SPB / FPP approval letter
Updates	Updating of Ship Documents and Certificates	Submission of endorsements Endorsement Approval Letter
Procurement	Insurance	Incident Minutes
Financial Management on Pioneer Ships	Recording of Cash Receipts and Disbursements	Ship Cash Report
Ticket Management	Use of Boat Tickets	Minutes of Use of Boat Tickets
	Use of Partner Tickets	Partner Ticket Usage Report
	Ticket Distribution	Receipt of tickets on board
The ship entered	Incoming Ship	PKK (Notification of Entry Vessels) RPK / PKKA / PPKM / PK Ship Health Documents Manifest List of Dangerous Payments Quarantine Documents SPM (Ship Entry Approval Letter)
Issuance and Receipt of Goods	Unloading and Loading	Goods Transportation Report (LAB)
Ship Out Service	Ship Out	Request for Ship Out Issuance of Sailing Health Certificate SPB (Sailing Approval Letter)

Extend Mooring Period	SPOG (Movement Order) Application for Extension of Mooring Period
Moving Ship Services	PPK issuance Request for moving ships / shifting
Cancellation of Services	Publishing Request Cancellation of Service Issuance of Ship Service Determination (PPK) for Service Cancellation

Source: Results of Data Processing.

4.4 Interpreted Ship Data and Information.
Integrated data and information is data and information that has been developed in the form of apps or web, both from ship operators and related agencies related to pioneer ship information transfer activities.

4.4.1 Pioneer Ship Application.
The function of the pioneer ship application is management of marketing and ticket sales of pioneer ships and this application has been implemented in 45 pioneer ships and 38 branches / terminal points (until the end of 2018).

Table 4. Pioneer Ship Application Module.

No	Module Application	Description	User
1	Central Administration Module	This application module is inside the PELNI Portal application that serves for managing data master such as ship, port, route, schedule / emplooi, and tariff.	Head Office: •Passenger Transportation Marketing Division of Passenger & Perintis Ships.. •Passenger & Perintis Ship Operations Division.
2	Ships Administration Module	This application module is installed on the ship server that serves for managing the upload data of revenues / passenger, download master data, and reporting passenger list and ship income.	Ships
3	Mobile Module	This application module is mobile and installed on android smartphone. This module serves for conducingt sales and check-in / checking passenger tickets.	Ships

Source: PT. Pelni [7].

4.4.2 Inaportnet.
Inaportnet is an internet / web-based electronic single service system to integrate a standard port information system in serving ships and goods physically from all agencies and stakeholders at the port.

4.4.3 SPS Online.
Online Syahbandar Approval Letter (SPS Online) is a mail service system harbormaster approval or syahbandar approval letter service system or licensing in the field of harbormaster at the online office of the technical unit of the Directorate General of Sea Transportation. SPS Online has been

implemented in several ports of the Technical Implementation Unit of the Directorate General of Sea Transportation which includes the Ambon Class I Port Authority and Harbormaster Office.

Table 5. Integrated ship data and information mapping.

Identified data and information from and to the ship			Ship Application Pioneer	Reported	Sps Online
Data and Information Services	Types of Data and Information	Description			
Pioneer Ship Routes Network	Emplooi	Emplooi	•		
		Trip report			
		Ship Travel Monitor			
Updates	Updating of Ship Documents and Certificates	Submission of endorsements		•	
		Endorsement Approval Letter		•	
Ticket Management	Use of Boat Tickets	Minutes of Use of Boat Tickets	•		
	Use of Partner Tickets	Partner Ticket Usage Report	•		
	Ticket Distribution	Receipt of tickets on board	•		
The ship entered	Incoming Ship	PKK (Notification of Entry Vessels)		•	•
		RPK/PKKA/PPKM/PK		•	•
		Ship Health Documents		•	•
		Manifest		•	•
		List of Dangerous Payments		•	•
		Quarantine Documents		•	•
		SPM (Ship Entry Approval Letter)		•	•
Issuance and Receipt of Goods	Unloading and Loading	Freight Report (LAB)		•	
Ship Out Service	Ship Out	Outboard Ship Application		•	•
		Issuance of Sailing Health Letters		•	•
		SPB (Sailing Approval Letter)		•	•
Extend Mooring Period		SPOG (Letter of Exercise Order)		•	
		Application for Extension of Time Off		•	
Moving Ship Services		Publication of PPK		•	
		Ship / Shifting Application		•	
		Publication		•	
Cancellation of Services		Service Cancellation Application		•	
		Issuance of Ship Service Determination (PPK) for Service Cancellation		•	

Source: Results of Data Processing.

4.5 Impact of non-availability of information.

In this research, the reason is that the unavailability of information related to cargo and passengers at the port of destination has resulted in ships still visiting the port. In table 6 it can be seen that the unavailability of information on goods and passengers at Romean Port, Tutukembong, and returning to Romean

Table 6. Manifest, Voyage 17.2019

Port Name	Cargo		Passengers	
	Loading M^3	Unloading M^3	On Board	Off Board
Ambon	97		66	
Tual	4	22	73	3
Elat		4		3
Molu		9		15
Larat		14	21	30
Romean				
Tutukembong				
Saumlaki	4	52	14	109
Tutukembong			2	
Rumean				
Larat		2	21	2
Molu			31	9
Elat				
Tual				39
Ambon		2		18
Total	105	105	228	228

Source: PT. Pelni.

In this session, researchers will calculate the cost of ship fuel or marine fuel oil (MFO) on Voyage 17.2019 to see the big impact on the 3 port visits.

Table 7. Fuel Consumption, Voyage 17.2019.

Port Name		LENGTH OF JOURNEY (hour)	MFO (Ton)
from	**to**		
Ambon	Star		
Ambon	Tual	36.54	35,000
Tual	Elat	03.12	26,533
Elat	Molu	12.30	25,878
Molu	Larat	04.36	22,706
Larat	Rumean	01.30	21,858
Rumean	Tutukembong	05.36	
Tutukembong	Saumlaki	06.18	19,561
Saumlaki	Tutukembong	07.12	17,028
Tutukembong	Rumean	03.42	15,234
Rumean	Larat	01.48	14,174
Larat	Molu	04.00	13,226
Molu	Elat	12.54	12,378
Elat	Tual	03.12	9,462
Tual	Ambon	34.42	8,826
Ambon	Finish		0,600

Source: PT. Pelni.

In table 7 it is known that the total distance covered by the ship following the initial route is 1,199.8 ml, the fuel used is 34,400 tonnes, and the fuel consumption per ml is 0.029 tonnes with an average speed of 8.6 knots. If the ship has information about the unavailability of cargo at the three ports, the distance for the ship becomes 1,129.9 ml and the fuel used is 32,394 tons, then there is a difference in fuel of 2,006 tons with the MFO base price at PT. Pertamina for region 4 (Maluku, NTT, Papua), the ship can save budget for fuel as much as Rp 19,308,993.10.

4.6 Recommendations for the Implementation of ITS Architecture.

Based on the constraints faced, the aspirations of stakeholders and the conditions of existing applications, the authors propose an ITS Architecture design and involve technology related to data management and information on pioneer ship services so that they can process and send data when needed. The ITS Architecture design process includes:

- Needs and Aspirations Analysis.
  Analysis of needs and aspirations to determine priority needs and aspirations of users.
- Functional Architecture.
  The functional architectural design or implementation package is in accordance with the aspirations and needs of the current stakeholders and impactors.
- Physical Architecture.
  Physical Architecture Design is in accordance with the functional architecture that was designed earlier.
- Communication Architecture.
  This architecture describes the way of transferring information at each stage with certain conditions of the information.
- Mockup Design.
  The purpose of the mockup design is to be able to see the visual appearance of the previous design.
- Performance evaluation.
  This process is carried out using the method of cost benefit analysis and investment analysis.

5. Conclusion

1. The implementation of ITS in the marine transportation system, especially in the Maluku islands, is one of the recommendations to improve pioneer ship services to be more productive.
2. Availability of good information can save ship operating costs.
3. The availability of good information can reduce the waiting time for pioneer ship service users.
4. The availability of easily accessible information can reduce the cost of searching for information related to ship routes and schedules.
5. Users of pioneer ship services find it easier to make decisions to mobilize people and goods.

6. References

[1] Andersen J and S SutciifJe S 2000 *Intelugent Transport Systems (ITS) An Overview* (IFAC Technology Transfer in Developing Countries: Pretoria, South Africa)
[2] Arizal J F 2000 *Perancangan Arsitektur ITS (Intelligent Transport System) Pelayanan Barang Pelabuhan Pelayaran Rakyat: Studi Kasus Pelabuhan Pelayaran Rakyat Kalimas Surabaya* (Tugas Akhir: Institut Teknologi Sepuluh Nopember Surabaya)
[3] Bossom R Jesty P Spence A Frötscher A and Ebner R 2011 *Extend FRAMEwork architecture for cooperative systems FRAME Architecture: Overview* (Version 4.1: PJCL E-FRAME Project the European Commission, UK)
[4] Jesty P H and Spence A 2011 *Extend FRAMEwork architecture for cooperative systems* (Version 1.0: PJCL E-FRAME Project., the European Commission, UK)
[5] Kelwulan L M 2019 *Peran Ketersediaan Informasi Terhadap Kinerja Transportasi Laut Wilayah Kepulauan : Studi Kasus Propinsi Maluku* (Disertasi: Institut Teknologi Sepuluh Nopember Surabaya)
[6] McQuenn B and McQueen J 1999 *Intelligent Transportation Systems Architectures* (3rd edition: ARTECH HOUSE, INC., Norwood)
[7] PT. Pelayaran Nasional Indonesia 2018 *Laporan Tahunan* (DKI Jakarta)
[8] PT. Pelayaran Nasional Indonesia 2019 *Laporan Tahunan* (DKI Jakarta)
[9] Prasetia A 2016 *Ekomoni Maritim Indonesia* (Edisi Pertama: Diandra Kreatif & Anggota IKAPI (062/DIY/08)., Yogyakarta)
[10]Williams B 2008 *Intelligent Transport Systems Standards* (ARTECH HOUSE, INC: Norwood)

Logistics System of Exported Fishery Product : a Case Study of Fisheries Centre in Natuna

A. Mustakim[1], **S. D. Lazuardi**[1] **and A. K. Caterina**[2]

[1] Department of Marine Transportation Engineering, Faculty of Marine Technology, Institut Teknologi Sepuluh Nopember, Surabaya
[2] PT. Salim Pacific Indonesia Lines, Surabaya

Abstract. As one of potential fishery areas in Natuna, the Integrated Marine and Fisheries Center (SKPT) has been producing the fishery commodities of 350,000 kg per month. Currently, SKPT Natuna is cooperating with Japan to export one of fishery products, which is frozen octopus. However, the existing export logistics system is considered ineffective and inefficient due to the higher logistics cost and higher lead time. This study aims to develop logistics system model with respect to obtain the optimum travel duration and the schedule of ship arrival at each transit point by considering minimum unit cost. The method used is the Shortest Path Problem [1] to optimize 24 cargo delivery schedules for one year so that the optimum routes defined are different for each delivery schedule depending on the availability of ship. The result shows the total duration of the delivery required is 32 days and the total shipping cost is Rp. 31,3 million/TEU. Thus, the optimum route obtained is able to reduce the shipping cost of frozen octopus by 45% and shorten the lead time by 15% compared to the current logistics system.

Key words: Cold Chain Logistics, Frozen Octopus, Logistics Cost, Shortest-Path Problem

1. Introduction

Natuna Regency as one of the largest potential fishery areas in Indonesia produced 2.687,8 tons of fishery products in 2017. There were nine types of fishery products with the largest number during 2017, namely grouper fish (301.9 tons), kurisi bali (233.1 tons), red snapper (208.9 tons), kurisi (252 tons), sea catfish (318.1 tons), octopus (138.5 tons), squids (237.9 tons)), tuna (311.2 tons), and flying fish (451.7 tons) [2]. Due to this area has the highest potential fishery resources, the Ministry of Marine Affairs and Fisheries in Natuna Regency had established the Integrated Marine and Fisheries Center (SKPT) in this area. The establishment of SKPT Natuna has purpose to increase the economy of fishermen in Natuna regency after facing economic problem[3].

The main export commodity at SKPT Natuna is frozen octopus, where Japan is the biggest importing country, which imports it regularly in every month [2]. During the distribution process of frozen product, the cold chain logistics should take into account to keep the product in cold temperature until the end customer received [4] [5]. Thus, it effects on the higher logistics cost and the higher lead time from Natuna to Japan. In order to solve this problems, optimization of logistics system is required to create the optimum route under several constraints, such as: ship schedule, ship capacity, product quality in term of handling the cargo, etc dan to obtain the minimum of total logistics cost[6] .

2. Problem definition and datasets

SKPT Natuna is collaborating with the Indonesian Public Fisheries Company (Perum Perindo) for exporting product to Nagoya, Japan. This collaboration has been established since 2020 and will be continuing for the next few years. Japan is currently the main importer of frozen octopus product from

SKPT Natuna. This shipment is scheduled in twice a month, while the export volume of frozen octopus in one shipment is 15 tons. Moreover, reefer container trucks and container ships are used during this shipments because frozen octopus products can only be transported in reefer container to keep the standard temperature of frozen octopus in -18°C. Those trucks are used in the delivery of cargo from SKPT Natuna to the Selat Lampa Maritime Highway Port, while container ships are used in the delivery of cargo from Natuna to Japan which passes through several transit points. The existing route of fishery product shipments from Natuna to Japan is SKPT Natuna - Lampa Strait Maritime Highway Port – Tanjung Priok - Tanjung Pelepas - Hong Kong - Shanghai - Kobe – Nagoya. Transportation cost from each port affects on the total logistics cost significantly, whilst storage cost and cargo handling cost are the parameter of port competitiveness[7]. Hereby, the datasets used in this study as below:

Table 1. Locations (Nodes).

Code	Location	Code	Location
T1	SKPT Natuna	T11	Nansha
T2	Pelb. Tol Laut	T12	Yantian
T3	Pontianak	T13	Gwangyang
T4	Belawan	T14	Tokyo
T5	Jakarta	T15	Shanghai
T6	Tj. Pelepas	T16	Port Klang
T7	PSA	T17	Xiamen
T8	Laem Chabang	T18	Kobe
T9	Hongkong	T19	Yokohama
T10	Kaohsiung	T20	Nagoya

Table 2. Cost Component Belawan – PSA.

Cost Component	Value (IDR/trip)
Capital Cost	37,882,340
Operating Cost	307,625,678
Voyage Cost	1,103,792,019
Cargo Handling Cost	2,090,400,000
Total Cost	**3.539.700.038**
Unit Cost	**3.403.557**

Table 1 and Table 2 present the locations used in the analysis including the code of each location and cost components in the route of Belawan – PSA in IDR per trip, respectively. In addition, Table 3 depicts the name of vessel used in every route, which will use in the model to calculate the unit cost by considering the load factor of 70% for each vessel [8].

Table 3. Vessels and the routes.

Route	Vessel	Route	Vessel	Route	Vessel	Route	Vessel
T2 - T3	KM. Logistik Nusantara 4	T4 - T7	MV. Meratus Medan 5	T9 - T11	MV. Nordmaas	T14 - T19	MV. TS Tokyo
T2 - T4	KM. Logistik Nusantara 4	T4 - T16	MV. Newark	T9 - T12	MV. MCC Qingdao	T15 - T18	MV. Kuo Chang
T2 - T5	KM. Logistik Nusantara 4	T6 - T7	MV. Tove Maersk	T9 - T15	MV. JPO Pisces	T16 - T20	MV. Bomar Renaissance
T3 - T7	MV. Lintas Bengkulu	T6 - T9	MV. PL Germany	T10 - T11	MV. Filotimo	T17 - T20	MV. Areopolis
T5 - T6	MV. CMA CGM Bellini	T7 - T8	MV. Nordlion	T11 - T9	MV. Maersk Songkhla	T18 - T20	MV. One Olympus
T5 - T17	MV.Bellavia	T7 - T10	MV. Kmarin Azur	T11 - T14	MV. MCC Cebu	T19 - T20	MV. APL Pusan
T4 - T6	MV. Kmarin Atlantica	T7 - T16	MV. Lucky Merry	T12 - T13	MV. Chespeake Bay	T13 - T19	MV. Merkur Horizon
		T8 - T9	MV. Seaspan Hannover				

3. Research Methodology

3.1. Stages of study

There are six stages needed to solve the problems in this study as follows:

IOP Conf. Series: Materials Science and Engineering 1052 (2021) 012068 doi:10.1088/1757-899X/1052/1/012068

- Identify the main problem
  The first step is to identify the main problems occured in the distribution of fishery export products from Natuna to Japan. Based on the data showed the problems are high logistics cost and long lead time during the distribution process.
- Analyze the current situation
  This stage should be conducted in order to find the current logistic system in Natuna, the existing route, the number of fishery products exported, the type of modes used, the initial logistic cost and the current lead time.
- Define the alternative routes
  Due to inefficiency occurred in the current route, thus alternative routes should be defined to obtain the minimum unit cost and the shorten lead time.
- Determine the optimum route
  There are three factors considered in the analysis, such as: logistic costs, delivery time and quality of product. The type of fishery export product is frozen octopus, so that the theory of cold chain logistic should take into account when builds up the model. Optimum route will be determined by minimizing the unit cost of distribution. Afterwards, total cost can be calculated by considering the number of export products and the amount of unit cost.
- Compare the results
  After determining the optimum route, the results should be compared by the current situation with respect to total logistic cost and lead time.
- Provide conclusion and future research
  The conclusion will provide the optimum route selected and the comparison between current situation and the model result. Moreover, the suggestion regarding future research will be add in this stage.

3.2. Mathematical model

This study focuses to obtain optimum solution for octopus product shipment. Mathematical model for the optimization is as below:

Objective Function:

$$Min\ Z = \sum_{i=1}^{I}\sum_{j=1}^{J}\sum_{k=1}^{K} X_{ijk} \cdot C_{ijk} + X_{ijk} \cdot (D_{ijk} - A_{ijk}) \cdot AC_{ijk} \tag{1}$$

Subject to:

$$\sum_{k=1}^{K}\sum_{j=1}^{J} X_{1jk} = 1 \tag{2}$$

$$\sum_{k=1}^{K}\sum_{i=1}^{J} X_{ijk} = 1 \qquad j=J \tag{3}$$

$$\sum_{k=1}^{K}\sum_{i=1}^{I} X_{ijk} - \sum_{k=1}^{K}\sum_{i=1}^{I} X_{ijk} = 0,\ j = 1,2, \dots \dots J \tag{4}$$

$$\sum_{k=1}^{K}\sum_{j=1}^{J}\sum_{i=1}^{I} D_{ijk} - A_{ijk} \geq 0 \tag{5}$$

$$C_{ijk} = \frac{CC_{ijk} + OC_{ijk} + VC_{ijk} + CHC_{ijk}}{Q_{ijk}} \tag{6}$$

Where:

X_{ijk} : Path of origin i destination j transportation mode k (1 = path is chosen, 0 = path is not chosen)
C_{ijk} : Cost of origin i destination j transportation mode k (IDR/TEU)
AC_{ijk} : Container stacking cost (IDR/day)
D_{ijk} : Departure time transportation mode k from origin i destination j
A_{ijk} : Arrival time transportation mode k from origin i destination j
CC_{ijk} : Capital Cost of origin i destination j transportation mode k (IDR)
OC_{ijk} : Operating Cost (IDR)
VC_{ijk} : Voyage Cost (IDR)
CHC_{ijk} : Cargo Handling Cost (IDR)
Q_{ijk} : Vessel Capacity x load factor (TEU)

Equation 1 represents the objective function to minimize the unit cost. Equation 2 and 3 depicts the cargo must be shipped from the origin and must be delivered to the destination, respectively. Equation 4 presents for each destination selected must be chosen for the origin in the next trip. Equation 5 means the arrival of vessel must be defined before the departure of the next vessel used. Equation 6 shows the formula to determine the unit cost, which consists of capital cost, operating cost, voyage cost and cargo handling cost[9] [10].

4. Results and discussion

The optimization model solves the problem in SKPT Natuna, when the first shipment of frozen octopus is shipped on January, 1st 2020. The optimum routes is SKPT Natuna - Maritime Highway Port – Belawan - Port Klang – Nagoya. The details of this solution are as below:

Table 4. Optimum path of frozen octopus product.

No.	Location	ETA	ETD	Transportation mode
1	SKPT Natuna	1/1/2020	1/1/2020	Container truck
2	Maritime Highway Port	1/1/2020	1/11/2020	KM. Logistik Nusantara 4
3	Belawan	1/13/2020	1/15/2020	MV. Newark
4	Port Klang	1/19/2020	1/26/2020	MV. Bomar Renaissance
5	Nagoya	4/2/2020	-	-

The total cost of this optimal solution is 32,605,806 IDR/TEU and lead time is 34.6. The details can be seen in Table 5 as below:

Table 5. Total cost.

Route	ETA cargo	ETD cargo	Waiting time (day)	Transport TIme (day)	Total Waiting Cost (IDR)	Logistics Cost (IDR)
SKPT Natuna - Pelabuhan Tol Laut	1/1/2020	1/1/2020	-	0.3		4.380.000
Pelabuhan Tol Laut - Belawan	1/1/2020	1/11/2020	9.7	2.6	483,183	6.491.764
Belawan - Port Klang	1/13/2020	1/15/2020	1.5	3.9	97,350	4.068.135
Port Klang - Nagoya	1/19/2020	1/26/2020	6.9	9.7	840,451	16.244.922
Total:			**18.1**	**16.5**	**1,420,985**	**31.184.821**
			34.6		**32.605.806**	

Therefore, this model is conducted repeatedly for every shipment in 2020. The results for every shipment can be seen in Table 6 as below:

Table 6. Total cost and lead time.

Shipment Schedule		Lead Time (day)	Total Cost (Mio - IDR)
January	01/01/2020	34.74	32.61
	15/01/2020	31.3	25.88
February	01/02/2020	30.74	32.38
	14/02/2020	31.74	32.47
Maret	01/03/2020	26.3	25.57
	15/03/2020	30.74	32.47
April	01/04/2020	40.74	32.88
	15/04/2020	26.74	32.47
May	01/05/2020	35.3	26.09
	15/05/2020	24.74	32.02
June	01/06/2020	35.74	32.63
	15/06/2020	35.74	32.6
July	01/07/2020	33.74	32.47
	15/07/2020	25.98	25.98
August	01/08/2020	35.3	26.14
	15/08/2020	38.74	32.99
September	01/09/2020	28.74	31.95
	15/09/2020	34.74	32.47
October	01/10/2020	32.74	32.29
	15/10/2020	32.74	32.29
November	01/11/2020	37.74	32.79
	15/11/2020	36.74	32.53
Desember	01/12/2020	34.74	32.47
	15/12/2020	20.74	35.58
Mean		**32,38**	**31.25**

Table 6 shows the mean of lead time of proposed model is 32,38 and total cost of proposed model is 31.25 Mio-IDR. Compare to the current condition in shipment frozen octopus product from SKPT Natuna, the proposed model has better results than the current condition. The details of the comparison results can be seen as below:

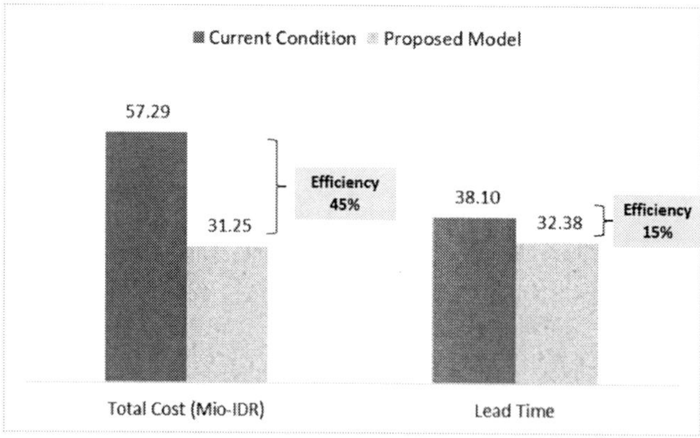

Figure 1. Comparison results of current condition and proposed model.

Figure 1 illustrates the comparison of the total cost and lead time for current condition and proposed model. Total cost of current condition is 57.29 Mio – IDR, while the total cost of proposed model is significantly more efficiency (45%) than the current condition cost. On the other hand, the current lead time is 38.10 days, while the lead time of proposed model obtains 15% more efficiency than current condition.

5. Conclusion

Logistics system of fishery product is important especially for the Natuna fishermen. The proposed model can obtain significant efficiency in term of total cost and lead time. This study provides the mathematical model based on the shortest-path problem. The current total cost is 57.29 Mio-IDR and total lead time is 38.10 days whereas the average of total cost obtained by the model is 31.25 Mio-IDR and lead time is 32.28 days. Overall, this model gains the efficiency on total cost by 45% and shorten lead time by 15%. For the future research is important to conduct the stochastic problem optimization because the schedule of the vessel shows the stochastic properties.

Acknowledgments

Authors wishing to acknowledge assistance or encouragement from colleagues at Department of Marine Transportation Engineering and special work from team of Integrated Marine and Fisheries Center (SKPT) Natuna for providing the data required in this study.

References

[1] Yao. G.-C. & M. 2011, "Shortest path problem with uncertain arc lengths," *Elsevier*, Accessed: Nov. 21, 2020. [Online].
[2] K. K. dan Perikanan, *Rekomendasi Pengembangan Perikanan Tangkap di Natuna dan Sekitarnya*. Depok: PT. Raja Grafindo, 2017.
[3] F. Hadi *et al.*, "UTILIZATION OF SHELL WASTE AS THE ACCELERATE ADMIXTURE ON PAVING IN ORDER TO IMPROVE THE QUALITY OF THE LIVING ENVIRONMENT OF THE KENJERAN COASTAL COMMUNITY," *Int. J. Civ. Eng. Technol. (IJCIET*, vol. 9, no. 9, pp. 9–09, 2018,
[4] P. Wuryaningrum, T. Achmadi, A. Mustakim, H. Iqbal Nur, S. Dwi Lazuardi, and M. Kusumadewi, "Optimization of Product Oil Shipment System for Archipelegic Region," *J. Sci. Technol. Rekayasa*, vol. 12, no. 2, pp. 120–125, 2019, doi: 10.21107/rekayasa.v12i2.5929.
[5] K. Likar and M. Jevšnik, "Cold chain maintaining in food trade,"
[6] S. D. Kumalasari, A. M. A. Rasyid, S. Nugroho, and A. Mustakim, "Perencanaan Transportasi Laut Batubara: Studi Kasus Taboneo-Surabaya-Tuban-Semarang," *J. Tek. Transp.*, vol. 1, no. 1, pp. 25–39, 2020.
[7] A. Mustakim, T. Achmadi, … I. B.-P. of the, and undefined 2018, "Model of Container Terminal Competition with Game Theory Approach," *search.proquest.com*,
[8] S. D. Lazuardi, B. van Riessen, T. Achmadi, I. Hadi, and A. Mustakim, "Analyzing the National Logistics System through Integrated and Efficient Logistics Networks: A Case Study of Container Shipping Connectivity in Indonesia," *Appl. Mech. Mater.*, vol. 862, pp. 238–243, Jan. 2017.
[9] A. Mustakim and F. Hadi, "Innovated Conceptual Design of Loading Unloading Tool for Livestock at the Port," in *Journal of Physics: Conference Series*, Mar. 2018, vol. 979, no. 1, p. 12061.
[10] H. I. Nur, C. Boyke, S. Permana, T. Achmadi, E. W. Ardhi, and A. Mustakim, "Study of Container Dwelling Time in Indonesia: Current Condition & Challenge," *pdfs.semanticscholar.org*, 2019.

Analysis of Ferry Fleet Operation Planning: A Case Study of Ujung - Kamal Ports

P. Wuryaningrum[1], A. Mustakim[1], A. M. Sodik[2] and C. B. S. Permana[1]

[1] Department of Marine Transportation Engineering, Faculty of Marine Technology, Institut Teknologi Sepuluh Nopember, Surabaya
[2] PT. PAL Indonesia, Surabaya

Abstract. Ujung - Kamal Ports is a ferry port that connects the island of East Java with Madura Island. These ports have a distance of about 2.5 nautical miles with shipping times of approximately 30-45 minutes. Since the operation of the Suramadu Bridge in 2009, the flow of passenger and vehicle visits at the Ujung-Kamal ports has decreased. this causes shipping operators for the Ujung-Kamal ferry port to reduce fleet size and operating time. Not only that, the number of fleets operated is uncertain, this has resulted in an inconsistent schedule for ship departures. besides that, it also resulted in many passengers having to wait or be late with ship departures. This study aims to plan the optimal Ujung-Kamal ferry operation. The method used in this research is the optimization method with scedulling problem. After optimizing the number of operating vessels, 2 ships were selected. The vessels are KMP Jokotole and KMP Tongkol. the ship sailing schedule in one day is 10 and 8 trips. The minimum cost for KMP Jokotole is IDR 6.357.855 per day and for KMP Tongkol is IDR 5.615.357 per day and the total cost per day is IDR 11,973,213.

Key words: ferry fleet, optimization method, schedulling problem

1. Introduction

Before the Suramadu Bridge, Ujung - Kamal Port was one of the ferry ports that was visited by many passengers. The Ujung - Kamal ferry port is an inter-island ferry port that connects Madura Island with Java Island which crosses the Madura Strait. This port connects Madura Island, namely the Port of Kamal with Ujung Port, which is located in Surabaya City [1], with a distance of about 2.5 nautical miles and a shipping time of approximately 25-30 minutes. Kamal port has been the main gateway to the island of Madura since the Dutch era. Since that time, the Port of Kamal Madura has played a very important role in terms of transportation and also the economy[2]. Over time, from year to year there has been a very significant development in this Kamal Port. When a few years ago Kamal Port was very busy, the bustle at this port almost never rested, every day many people went back and forth to enter Kamal Port, this port has facilities, namely 3 docks, garbata and several ships. Several ships that have operated, namely KMP Jokotole, KMP Trunojoyo, KMP Potre Koneng and KMP Madura 1 and 2 straits.

Since the operation of the Suramadu Bridge, which is expected to grow the economy in Madura [3], users of this port have experienced a drastic decline (up to 40%) [4]. Due to the tendency of vehicles to pass through the Suramadu Bridge compared to the port of Kamal, then the shipping operator reduced its fleet from initially having 19 fleets with an operational time of 24 hours to 3 fleets in just 16 hours. Not only that, the number of fleets operated is uncertain, this has resulted in an inconsistent schedule for ship departures. besides that, it also resulted in many passengers having to wait or be late with ship departures.

IOP Conf. Series: Materials Science and Engineering 1052 (2021) 012069 doi:10.1088/1757-899X/1052/1/012069

2. Literatur Review

This chapter describes a literature review of methods for optimizing the ferry route system. Passenger ferries have a route system that has the same characteristics as other transit systems in terms of objectives and constrain.

Ding at all developed a ferry operational simulation method using micro traffic simulation modeling. Modeling is carried out using the VISSIM program in 2D and 3D analysis[5]. Mankowska conduct detailed market studies to predict passenger traffic that will be used as the basis for ferry route optimization. They identified the passengers' motivations for using the ferry and factors that made them choose ferries over other transportation modes[6]. Maiorov at all made a simulation of the route network and ferry traffic intensity based on discretization and circos plot intensity diagram. With this method, they can perform an analysis by considering the number of passengers, ferry terminals, and the number of ferry operators[7].

3. Problem definition

The problem that becomes the focus of the object of research in this paper is to determine the optimal ferry operation at the Ujung-Kamal ports. The distance between the port of Ujung and the port of Kamal is 3 nautical miles with the route as in the Figure 1.

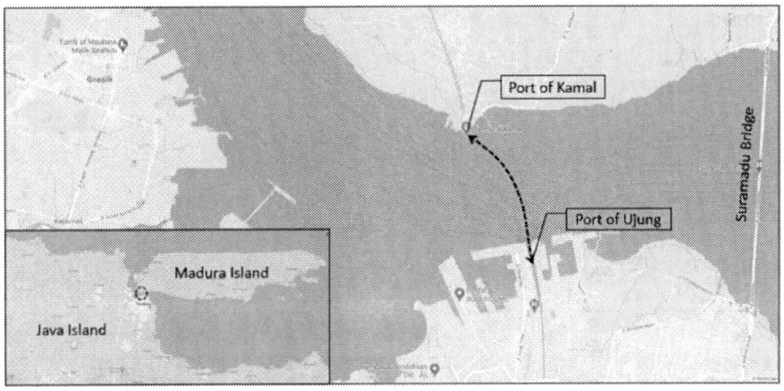

Figure 1. Ujung – Kamal Ferry Shipping Route.

Figure 2 shows that Kamal Port has 3 piers which are of the Mobile Bridge type or commonly called MB pier, each MB pier has different facilities. The MB 1 pier is located at the westernmost point of the port. The facility at this pier is a passenger bridge access that commonly called Garbarata that makes it easy for passengers to get on the ship without having to be obstructed by vehicles.[8] The MB II pier is located in the middle between the MB I and MB III pier, this MB II pier does not have facilities like other piers. This pier only has a road that is given a building to check tickets and fences for passengers who do not use vehicles, so that passengers on foot are avoided from vehicles that will enter or exit the ship. The MB III pier is located east of the port where this pier does not have the same facilities as the MB I pier, Garbarata, but the MB III pier still has a special lane for passengers who want to get on or off the ship so that it does not disturb motorized vehicle users or vice versa. The special lane is given a protective fence so that passengers on foot are not hit by a motorized vehicle besides that the special lane has a canopy or roof so that passengers are not exposed to rain during the rainy season.

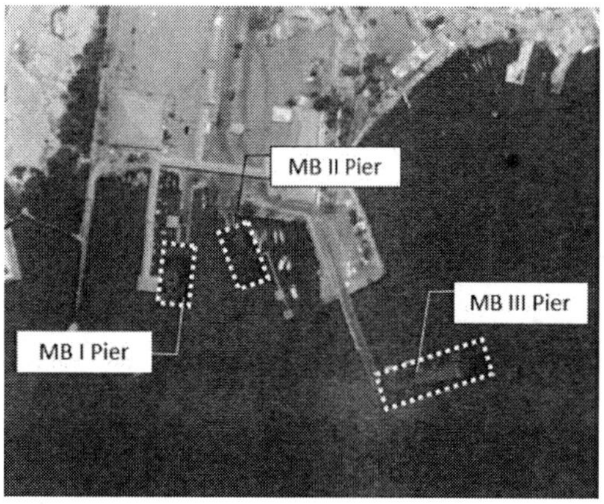

Figure 2. Port of Kamal, Madura.

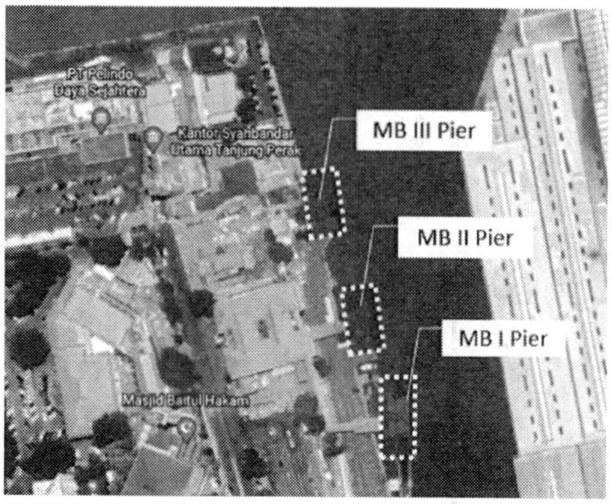

Figure 3. Port of Ujung, Surabaya.

Figure 4 shows Ujung Port has 3 Mobile Bridge Docks or commonly called MB Docks. Each pier at the end port has different facilities, the MB I Pier, which is located in the south of the port, has a jet bridge which makes it easier for passengers to walk. Similar to the MB I pier, the MB II pier, which is located in the middle between the MB I pier and the MB III jetty, has a jet bridge facility and the MB III jetty is not equipped with a jet bridge, so that passengers on foot mix with motorized vehicles. however, this MB III pier has not been in operation for a long time considering the small number of fleets so that only MB I pier and MB II pier are operated.

Furthermore, for passenger data, sample data is taken for passenger visits for one month. The data is obtained from shipping companies serving the Ujung-Kamal route. The data is taken from the number of passengers on the ships that serve the route, namely KMP Gajahmada, KMP Tongkol and KMP Jokotole. Specifications of each ship can be seen at table 1.

Table 1. Ship Specification.

No.	Ship Specifications	Ship Name		
		KMP. Gajah Mada	KMP. Tongkol	KMP. Jokotole
1.	Ship Type	Ferry/Ro ro	Ferry/Ro ro	Ferry/Ro ro
2.	LOA (meter)	37,50	36,74	31,50
3.	B (meter)	13,20	8,90	9,00
4.	T (meter)	3,65	2,90	1,30
5.	H (meter)	4,20	3,00	2,40
6.	Vs (knot)	13	5	8
7.	DWT (ton)	155	110	150
8.	GT	512	259	192
9.	Capacity			
	Passenger (pax)	290	134	220
	Vehicle (unit)	18	18	
10.	Power (HP)	1100	600	870

The number of passenger traffic for KMP Gajah Mada for one month is 37.916 passengers, with an average per day of 1.223 passengers. where the largest number of traffic reached 2184 in one day. Meanwhile, for KMP. Tongkol the number of passengers traffic for one month is 27.474 passengers, with an average per day of 886 passengers. where the largest number of traffic reached 1296 in one day. then for KMP. Jokotole has 32.253 passenger traffic per month, with an average of 1.040 passengers per day and where the largest number of traffic reached 1558 in one day. For more details about the passenger traffic on these vessels, see figure 4, figure 5 and figure 6.

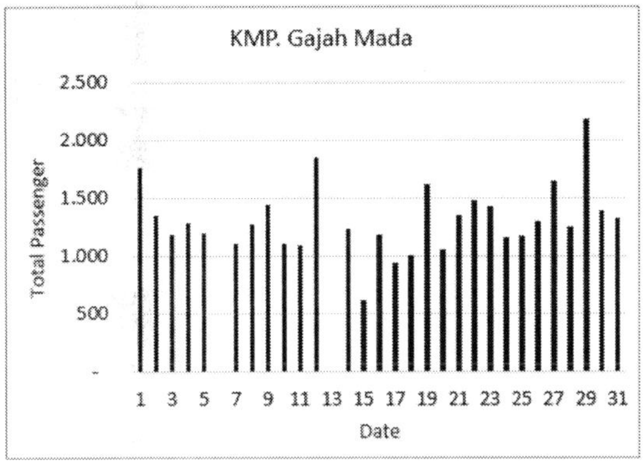

Figure 4. Passenger Traffic KMP. Gajah Mada.

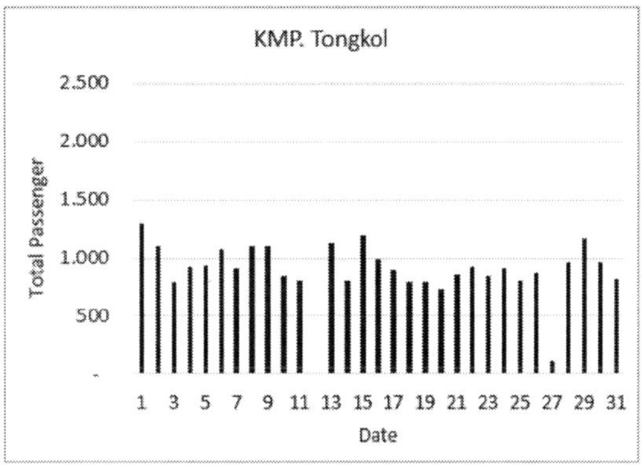

Figure 5. Passenger Traffic KMP. Tongkol.

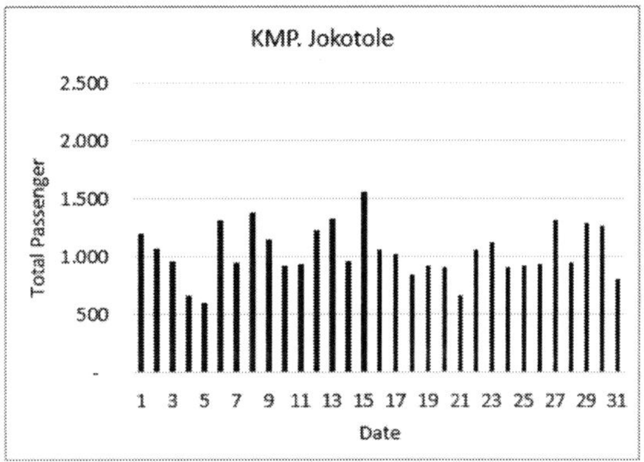

Figure 6. Passenger Traffic KMP. Jokotole.

4. Research Methodology

4.1 Stages of Study

There are six stages needed to solve the problems in this study as follows:

- Identify the main problem
  The first step is to identify the main problems that occur in the crossing in Ujung - Kamal which is affected by the operation of the Suramadu Bridge. The flow of passenger and vehicle visits at the Ujung-Kamal port has decreased. this causes shipping operators to reduce fleet size and operating time. So that the number of operated fleets is uncertain and the ship's departure schedule is not appropriate
- Analyze the current situation
  This stage is carried out to determine the existhing conditions regarding ferry shipping in Ujung - Kamal. The methodology to determine existhing conditions with primary and secondary surveys and questionnaires to ferry service users.
- Analyze the Shipping Cost
  In this stage, calculate the shipping cost of all alternative ships that can operate in Ujung – Kamal. components in shipping cost are capital cost, voyage cost, operating cost and cargo handling cost. However, in this study, because the types of cargo are vehicles and passengers, cargo handling costs can be ignored.

- Determine the optimum schedule
  This stage is carried out to find out which ship will be selected to serve Ujung - Kamal and how the schedule is. The optimal scheduling will be determined by minimizing the total shipping cost. in choosing a ship, it will consider passenger demand and the maximum ship capacity. For the optimization model can be seen in the sub-chapter 4.2.
- Provide conclusions and future research
  This conclusion will provide the optimal scheduling and selected vessels to serve Ujung - Kamal. And the shipping cost that must be borne. In addition, suggestions regarding further research will be added at this stage.

4.2 Mathematical Model
This paper focuses to obtain optimum solution for ship scheduling for Ujung – Kamal Ferry Transportation. Mathematical model for the optimization process is as below:
Objective Function:

$$Min\ Z = \sum_{t=1}^{T} \sum_{v=1}^{V} C_{tv} X_{tv} \tag{1}$$

Subject to:

$$\sum_{v=1}^{V} X_{tv}.P_v \geq D_t \qquad t = 1,2,3,4, \dots T \tag{2}$$

$$X_{tv} = \begin{cases} 1, & Vessel\ v\ at\ time\ t\ is\ assigned \\ 0, & Vessel\ v\ at\ time\ t\ is\ not\ assigned \end{cases} \tag{3}$$

Where:

X_{tv} : Decision assigned for vessel v time t (1 = is assigned, 0 = is not assigned)
C_{tv} : Cost of assigned for vessel v time t (IDR)
P_v : Transport Capacity of vessel v (pax)
D_t : Demand at time t (pax)
t : time in scheduling

Equation 1 represents the objective function to minimize total shipping cost. Where in the shipping cost calculation the fix cost and variable cost are calculated. the fixed cost component is the capital cost and operating cost, then the variable cost component is the voyage cost[9]. Equation 2 represents the Constraint of this optimization model. Constraint of this model is the ship capacity must be greater than passenger demand at a certain time. Time is the time according to ship scheduling in one day. Equation 3 represents for decision assigned for vessel at each time. And for the optimization model, the decision variable is a binary form.

5. Results and discusssion

In the previous section, it was mentioned that one of the methods to determine the existing conditions of the Ujung-Kamal ferry shipping was done by using a questionnaire to passengers.

5.1 Ferry service conditions from the response of users
Figure 8 shows that the majority of Ujung - Kamal ferry service users are motorbike users, which is 80% and the others are trucks and pedestrians. Then for the type of work, it is almost equal for students and workers, namely 41% and 54% as can be seen in Figure 9. the largest use of ferry service time is 2-3 times per week at 40%, then 22% for 1-2 times per month, after that 20% for 1-2 times per week and 18% for daily use.

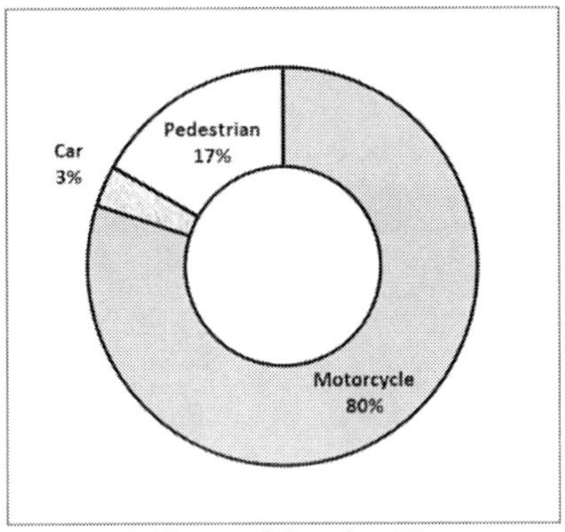

Figure 7. Type of Service Users.

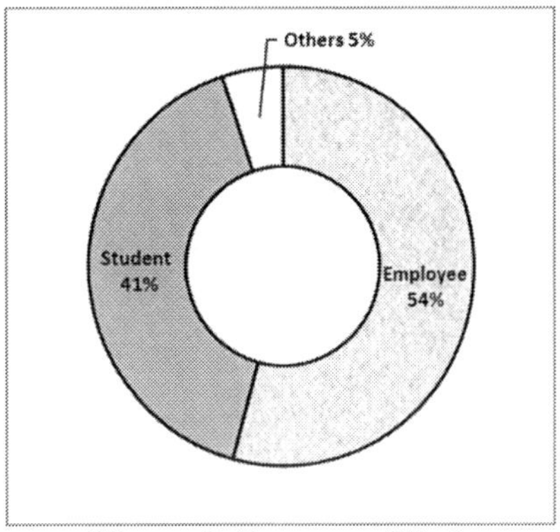

Figure 8. Passenger Work.

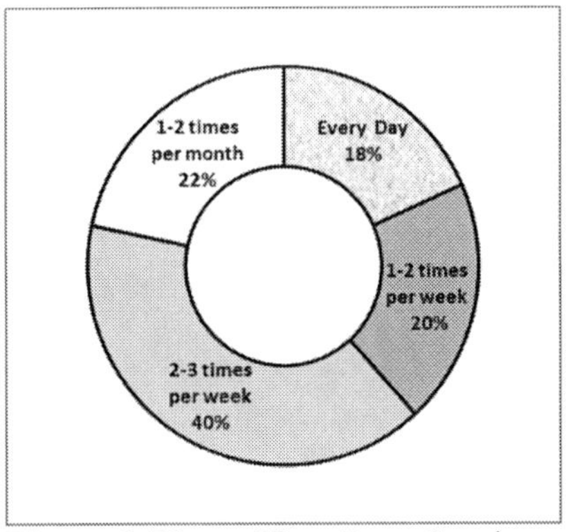

Figure 9. Usage Time Ferry Services.

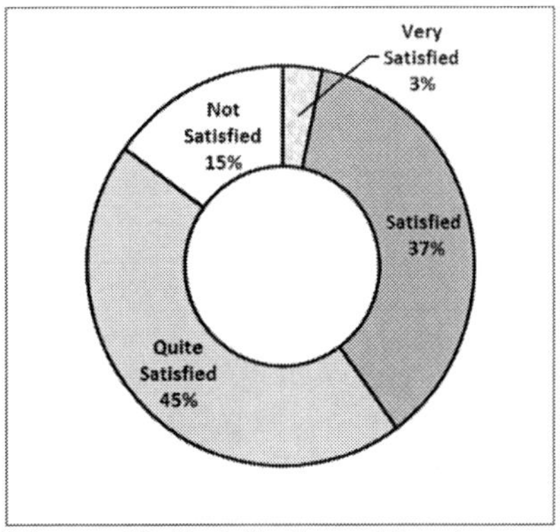

Figure 10. Satisfaction of Services.

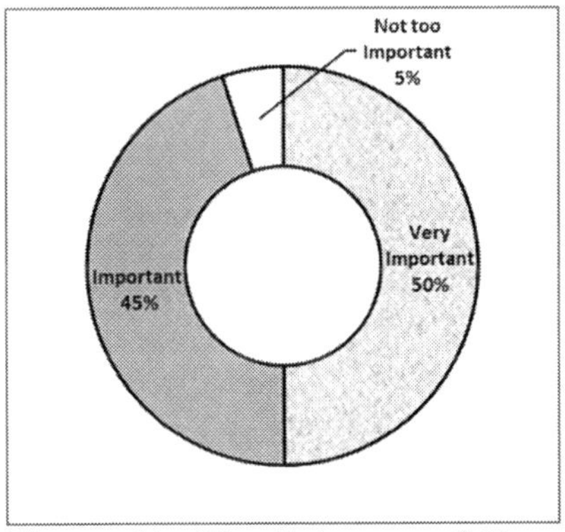

Figure 11. The Existence of Ferry Ujung – Kamal.

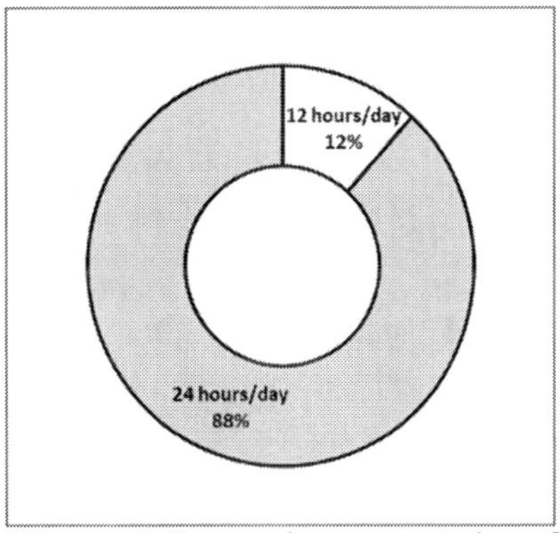

Figure 12. Respondents expectation of ship operation schedule.

Furthermore, for the biggest customer satisfaction stated quite satisfied, namely 45% and 37% satisfied. For the existence of the Ujung - Kamal ferry service, 50% is very important and 45% important.

One of the most important things is the expectations of service users for the Ujung - Kamal ferry service time. From the survey results, it was found that 88% expected there was a ship schedule 24 hours per day and the remaining 12% was 12 hours per day as can be seen in Figure 13.

5.2 Analysis of Shipping Cost

Shipping costs include fixed costs and variable costs. To get the value of fixed costs and variable costs, several assumptions such as data on shipping distances, specifications and variations of fleets, ship procurement costs, fuel, and port costs are used as the basis for calculations in the optimization model. the component of fixed cost is capital cost and Operational Cost, while the component of variable cost is Voyage Cost[10]. With the number of shipcalls in one year. The following will explain the proportions of each cost component in Table 2.

Table 2. Shipping Cost.

Cost	KMP Tongkol	KMP Gajahmada	KMP Jokotole	
Capital Cost	2.581.674	5.812.592	2.790.682	IDR/Day
	78.525.915	176.799.659	84.883.238	IDR/ Month
	942.310.981	2.121.595.911	1.018.598.860	IDR/ Year
Operational Cost	9.225.008	10.566.978	9.284.936	IDR/Day
	224.354.000	262.052.250	226.176.800	IDR/ Month
	3.367.128.000	3.856.947.000	3.389.001.600	IDR/ Year
Voyage Cost	19.957.932	22.488.797	22.174.389	IDR/Day
	598.737.958	670.683.191	665.231.684	IDR/ Month
	7.184.855.499	8.048.198.287	7.982.780.205	IDR/ Year
Total Cost	31.764.614	38.868.367	34.250.007	IDR/Day
	901.617.873	1.109.535.100	976.291.722	IDR/ Month
	11.494.294.480	14.026.741.198	12.390.380.665	IDR/ Year

5.3 Optimization of Schedulling

Regarding the operation of 2 ships and 3 ships in 19 hours on the schedule, not every trip gets a lot of passengers, resulting in a loss in the voyage cost of each ship. Therefore, in this section scheduling optimization is carried out, the constraint of the scheduling is the time where the time is divided into 3 parts, namely Morning between 05.30 AM - 08.00 AM, then Afternoon between 12.00 PM- 1:00 PM and Evening between 4:00 PM - 7:30 PM. Besides that, the constraint is the capacity of the ship, where this capacity will be compared with the number of passengers at the 3 times. If the number of passengers is more than the capacity of the ship, the schedule will be added at that time. Scheduling optimization aims to reduce voyage costs.

From the calculation and optimization results, it is obtained the optimal ship assignment schedule as shown in Table 3. From the table, the value 1 is if the ship is assigned, and the value 0 is if not assigned. For the total schedule in one day, KMP Jokotole is assigned to each schedule and for KMP Tongkol is scheduled for the whole day except for the day. Meanwhile, for optimal results, only two ships were assigned and KMP Gajah Mada was not assigned.

Table 3. Scheduling Optimization Results.

Name of Ship	Time Schedule				
	Morning	Morning	Afternoon	Evening	Evening
KMP Jokotole	1	1	1	1	1
KMP Tongkol	1	1	0	1	1
KMP Gajah Mada	0	0	0	0	0

Table 4. Ship Departure Schedule.

Port	Ship	Morning	Morning	Afternoon	Evening	Evening
Kamal	KMP Jokotole	5:30 AM	7:10 AM	12:00 PM	4:00 PM	7:20 PM
	KMP Tongkol	6:20 AM	8:00 AM		5:00 PM	9:00 PM
Ujung	KMP Tongkol	5:30 AM	7:10 AM		4:00 PM	7:20 PM
	KMP Jokotole	6:20 AM	8:00 AM	1:00 PM	5:00 PM	9:00 PM

Table 5. Total Trip and Total Cost.

Ship	Total Trip/Day	Cost/Day (IDR)
KMP Jokotole	10	6.357.855
KMP Tongkol	8	5.615.357
	Total Cost/Day	11.973.213

table 4 displays the schedule of ship departures. this schedule is less than the previous schedule. This is because the existing schedule and the number of passengers have decreased, resulting in higher costs while the load factor is getting smaller. From the analysis, it is found that the optimal number of operating vessels is 2, namely KMP Jokotole and KMP Tongkol. where the number of trips for each ship is 10 trips for KMP Jokotole and 8 trips for KMP Tongkol. The total cost per day is IDR 11,973,213.

6. Conclusion

In this study, primary and secondary surveys were carried out, primary surveys of users of ferry transportation services. and analysis of sea transportation cost calculations and optimization for ship scheduling and assignment. from these results, the following conclusions were obtained:

- The interest of users of ferry transportation services at the port of Ujung - Kamal is still high. This is obtained from the results of a service user questionnaire, where 95% of users think this transportation is important and very important, besides the expectations of service users that transportation services in Ujung - Kamal can operate 24 hours per day.
- Optimization results for ferry fleet scheduler obtained 2 ships assigned, namely KMP Jokotole and KMP Tongkol. for KMP Jokotole it operates 10 trips per day, while for KMP Tongkol it operates 8 trips per day.
- Ship scheduling is carried out 5 times per day, 2 times for the morning, 1 time for the afternoon and 2 times for the evening. where the number of vessels in operation for the entire time is 2 ship, except on the afternoon only one ship that is KMP Jokotole
- From the calculation of shipping cost, where cost is an objectivity function in the scheduling optimization model. The minimum cost for KMP Jokotole is IDR 6.357.855 per day and for KMP Tongkol is IDR 5.615.357 per day and the total cost per day is IDR 11,973,213.

For the future research is important to conduct the stochastic problem optimization because the schedule of the vessel shows the stochastic properties.

References

[1] E. Apriliani, D. Irhamah, and J. Statistika, "Peramalan Pengguna Kapal Ferry Ujung-Kamal dengan Metode Intervensi," Dec. 2016. doi: 10.12962/J23373520.V5I2.17624.

[2] S. Hotijah and H. Cahyono, "Perkembangan Industri Dan Pendapatan Daerah Kabupaten Bangkalan Sebelum Dan Sesudah Pembangunan Jembatan Suramadu," 2013.

[3] E. Akbarwati, E. Akbarwati, and P. G. Ariastita, "Revitalisasi Kawasan Pelabuhan Kamal Di Madura," *J. Tek. ITS*, vol. 2, no. 2, pp. C104–C108, Sep. 2013, Accessed: Nov. 21, 2020. [Online]. Available: http://ejurnal.its.ac.id/index.php/teknik/article/view/4324.

[4] S. (Sri) Amiranti and K. E. (Katherine) Permanasari, "928X Print)," Sepuluh Nopember Institute of Technology, 2013. doi: 10.12962/J23373520.V2I2.3807.

[5] L. Ding and S. Venglar, "Analysis for the Port Ferry Operation and Control Alternatives by Using Traffic Micro Simulation Modeling," *Procedia - Soc. Behav. Sci.*, vol. 43, pp. 805–812, 2012, doi: 10.1016/j.sbspro.2012.04.155.

[6] M. Mańkowska, "The Concept Of Development Of Passenger Ferry Services In The Baltic Sea Region In Terms Of The Growing Interbranch Competition," no. May 2015, pp. 285–298, 2016.

[7] N. Maiorov, V. Fetisov, S. Krile, and D. Miskovic, "Simulation of the route network and ferry traffic intensity based on the process of discretization and circos plot intensity diagram," *Transp. Probl.*, vol. 14, no. 4, pp. 23–30, 2019, doi: 10.20858/tp.2019.14.4.2.

[8] A. Mustakim and F. Hadi, "Innovated Conceptual Design of Loading Unloading Tool for

Livestock at the Port," in *Journal of Physics: Conference Series*, Mar. 2018, vol. 979, no. 1, p. 12061, doi: 10.1088/1742-6596/979/1/012061.

[9] P. Wuryaningrum, T. Achmadi, A. Mustakim, H. I. Nur, S. D. Lazuardi, and M. Kusumadewi, "Optimization of Product Oil Shipment System for Archipelegic Region," *Rekayasa*, vol. 12, no. 2, pp. 120–125, Oct. 2019, doi: 10.21107/rekayasa.v12i2.5929.

[10] M. Stopford, *Maritime Economics Second Edition*. 1997.

Supplier Selection in Shipbuilding Industry – A Review

S R W Pribadi [1], T W Pribadi [1], and S I Wahidi [1]
[1]Department of Naval Architecture, Faculty of Marine Technology, Institut Teknologi Sepuluh Nopember, Surabaya, Indonesia

Abstract. There are many criteria and methods that can be used to select the right supplier. Selection of a suitable supplier is needed, especially in the shipbuilding industry. Selection of suppliers is to reduce risky purchases, maximize price suppression with the best quality goods for buyers and establish long-term relationships between suppliers and buyers to facilitate future cooperation with today's increasingly competitive industry. This study was made to critically analyze the factors and appropriate methods in selecting suppliers in the shipbuilding industry. With many methods and criteria that are of particular concern to today's industrial developments, in this paper the literature has been thoroughly reviewed to help provide solutions to these problems. The results of the reviews that have been carried out can be seen that the right criteria for selecting suppliers include quality, price, delivery, service, technical ability, financial strength, geographical location, reputation, reciprocal arrangements, etc. Likewise, the methods used, namely ANP-PROMETHEE and LARG, can be taken into consideration for a good supplier selection method.

1. Introduction

In most industries including shipbuilding, one of the proportions that has a large enough value in development financing is the cost of components and raw materials. Therefore, it is necessary to be careful in choosing the right supplier of raw materials in the procurement process as an opportunity for companies to reduce production costs, so that shipyard companies can offer excellent products at competitive price levels, each company must try to press or reduce all costs without reducing product quality or the standards that have been set [1]. However, choosing the appropriate supplier is not easy because it requires criteria that must be checked carefully and the fact is that in the field there are still many companies choosing vendors based on their experience and intuition, where this approach is subjective [2, 3]. Over the years the criteria for selecting suppliers have been based on the vendors who are able to provide the lowest prices. However, making price the main criterion for the current era has begun to be inefficient and has transformed into a more comprehensive multi-criteria approach. Apart from price, new criteria that need to be considered include environmental, social, political and customer satisfaction factors in addition to paying attention to quality, delivery, cost and service [4].

Currently, it is commonly used a distributed manufacturing system consisting of manufacturing components at different physical locations and processed by a supply chain management system, combining all components and assembling them into more complex products [5]. The way companies work with their supply chain partners creates a sustainable competitive advantage because the company has claimed suppliers to form a competitive advantage for the buying company [6]. This pattern can also be applied when ship was built. Supply chain itself, if described is a concept that has the aim of being able to connect and manage resources, flow and control of raw materials from various perspectives based on various functions and various levels of suppliers [7]. The shipyard company's ability to set strategies in choosing good quality suppliers will continuously increase customer satisfaction which is expected to increase company profits and also in a sustainable manner can be a more effective way to be able to select suppliers for future projects. This selection is done by selecting several suppliers to make it easier to compare so that one can choose the one that best fits the company's needs. Noting the strategic importance of the role of suppliers in the supply chain function, researchers have developed a number of criteria, methods, and models for supplier selection. Therefore, this research was conducted to provide a reference in selecting the right supplier, especially in the shipbuilding industry using the ANP-PROMETHEE and LARG methods and based on certain criteria such as quality, price, delivery, service, technical ability, financial strength, geographical location. , reputation, reciprocal arrangements, etc. The relevant literature has been thoroughly reviewed and is presented below.

2. Literature Review

A.N. Sariyun [8] wrote an article about "Manajemen Rantai Pasokan (Supply Chain Management): Konsep Dan Hakikat " providing an explanation of supply chain management which has been widely discussed and has become a common thing that is often found in print media, books and discussion forums. However, because it is a relatively new discipline, there are still many who do not quite understand the term as many think that supply chain is a software and can only be owned by manufacturing companies. In the supply chain there are several main players who are companies that have the same interests, namely:

1. Supplies
2. Manufactures
3. Distribution
4. Retail Outlet
5. Customers

From this article it can also be concluded that the supply chain management implemented in the company is expected to increase the effectiveness and efficiency in the process of purchasing raw goods, fulfillment of customer orders and the process of distributing goods. The application of supply chain management today is suitable, because this system has the advantage of being able to regulate the flow of goods or products in a supply chain.

P. Om, K.G. Amit [4] made a review on "Supplier Selection Criteria and Methods in Supply Chains: A Review" concluded that there has been a lot of research on how to select suppliers since 1960. Where is the importance of selecting suppliers that will impact the entire supply chain starting from material procurement raw material until the process of sending the finished product to the customer. Researchers have provided reviews and methods of appropriate methods in making decisions in choosing suppliers. The reviews are price, quality, delivery, track record, policy, technical capability, industry reputation and position, attitude, geographic location, financial position, repair services, quantity. previous business, passion for business, procedural compliance and packaging skills, employee relations and reciprocal arrangement. Based on the review, it is not rational to suggest that the problem of supplier selection needs further attention to align a combination of qualitative and quantitative criteria to develop the best criteria and methods for selecting the best supplier. While the suggested methods to be able to select many suppliers and reduce them to smaller groups with acceptable criteria are categorical methods, Data Envelopment Analysis (DEA) and Cluster Analysis (CA).

M. Wicaksono, S. Bayu, etc. [9] wrote about "Implementasi Metode ANP-PROMETHEE untuk Pemilihan Supplier (Studi Kasus PT. Lamongan Marine Industry)". Analityc Network Process (ANP) has 2 controls, namely the control hierarchy showing the relationship between the criteria and sub-criteria interests and the control of the relationship between criteria and clusters. As a general theory

IOP Conf. Series: Materials Science and Engineering 1052 (2021) 012070 doi:10.1088/1757-899X/1052/1/012070

the relative measurement of ANP functions as a derivative of the compost priority ratio from the ratio reflecting the relative measurement of the effects of mutually adjustable elements due to control criteria [10]. PROMETHEE (Preference Ranking Organization Method for Enrichment Evaluation) is a simple and flexible method for analyzing multi-criteria problems using the value in the outranking relationship as an estimate of the dominaso criteria [11].

R.P. Magdalena, J.A.F. Francisco, etc [12] wrote a journal on "Shipbuilding 4.0 Index Approaching Supply Chain". Using the LARG paradigm: Lean, Agile, Resilient and Green. The conclusion that can be drawn from the journal is an evaluation that can be carried out in shipbuilding adapting Industry 4.0, developing factors that make it possible to find out, current supply chain developments and also how to evaluate technology that allows it to adapt more quickly and efficiently to Industry 4.0. This development has been carried out in collaboration by professional experts from the shipbuilding sector and academics who have links to this sector. The shipbuilding industry shows a special interest in the supply chain system being a major factor in various changes, as well as a control tool in carrying out repairs. In addition, the adaptation of supply chain use in industry 4.0 also shows an interest and some disorientation, which can be used by managers to decide which technology has priority over others to achieve their goals. It also highlights the need for future research to establish different criteria to improve the criteria studied.

3. The Criteria for Selecting Suppliers

The process of identifying and analyzing the criteria for selecting suppliers of components and materials used has been widely reviewed by professional experts and related academics. Dickson (1966) in his work sent 300 questionnaires to companies, mostly in the manufacturing industry. The procurement divisions of these companies were asked to identify what factors were most important for selecting suppliers. This questionnaire is divided into two categories, namely vendor selection by companies and individual selection. Then we get what factors influence vendor selection, namely [13]:

1. Quality
2. Price
3. Delivery
4. Service
5. Technical Ability
6. Financial strength
7. Geographical location
8. Reputation
9. Reciprocal Arrangements
10. Another factor

If discussed further, these factors are indeed important criteria in selecting suppliers of materials to be used in the construction process, one of which is for the construction of vessels.

1. Quality

Seeing whether the supplier is able to provide raw materials that are of high quality and at least have the minimum standards required by local organizations and governments to maintain the safety of the final product and the durability of the final product when used for a long time.

2. Price

Suppliers can provide competitive and varied prices. The lowest and competitive price is still the company's priority.

3. Delivery

The company will see suppliers who are able to provide flexible shipping facilities with less complicated rules and are able to deliver the materials on time according to a predetermined schedule.

4. Service

Suppliers provide good service to customers, especially when communicating during the procurement process. A supplier who is nimble and friendly and able to provide good explanations has a plus point.

5. Technical Ability

Suppliers already know what to do on the field from the initial negotiations to product delivery. Can solve technical problems quickly and be able to adapt by utilizing existing technologies that are in accordance with the times so as to make work more efficient.

6. Financial strength

Assessment of financial stability and the physical prospects of suppliers are increasingly important factors in line with the growing trend between suppliers and prospective buyers of goods. Suppliers who do not have a solid and stable financial foundation will find it difficult to contribute to the partnership business, because suppliers will focus more on efforts to improve their financial stability. Both buyers and suppliers will look for partners who can work with in the long and sustainable way who will contribute to good relations so that both suppliers and buyers will pay attention to the financial condition of their partners before making cooperation decisions [13].

7. Geographical location

The position where the vendor is located greatly influences the supplier selection process by the company. The thing that is most considered is the distance between the project that will be running and the vendor because this will affect the delivery process, especially the ones that will be considered are the means of transportation used, the terrain to be traveled and the length of the delivery process which will affect the costs that must be incurred.

8. Reputation

The reputation of the vendor that will be selected must not go unnoticed. It is necessary to investigate the track record of the vendor to minimize the bad possibilities that may occur in the future. Looking for positive feedback from previous customers so that long term trust can be established between companies that will supply materials or materials from the vendor.

9. Reciprocal Arrangements

There is an agreement or regulation in writing between the vendor and the company that contains monitoring, supervision and regulation of trade and settlement of transactions.

10. Another Factor

The use of the internet to centrally integrate supply chains can greatly contribute to a company's profit and competitiveness. It is possible that the widespread use of e-procurement will allow companies to move closer, saving time and effort. Suppliers will also minimize inventory costs and data can be accessed and updated anywhere without fear of loss or damage [13].

4. Various Supplier Selection Methods (Prequalification)

Initial qualification is a process where multiple vendors are selected and eliminated into smaller scopes until suppliers who meet the criteria are found. The methods that can be used are:

1. Categorical Methods

Existing or known suppliers are evaluated based on the required criteria, based on the buyer's history and experience. The advantage of using this method is that it helps to make assessments in a clear and systematic manner.

2. Data Development Analysis (DEA)

A grouping system which divides suppliers into two categories, namely efficient and inefficient and is assessed based on input and output criteria.

3. Cluster Analysis (CA)

CA is a basic statistical method that uses an algorithmic system to classify objects based on clusters that have similar characteristics. This grouping is used to reduce the larger suppliers into smaller subsets so that they are easy to manage.

5. Multi Attribute Decision Making (MADM) Techniques

The MADM technique is used to help select suppliers which usually involve more than one criterion where these criteria contradict one another. Some of the MADM techniques are:

1. Analytical Hierarchical Process (AHP)

AHP is a method that is quite easy to apply where this method provides an alternative ranking or sequence when there are several criteria and sub-criteria in decision making [14]. These criteria are complete, operational, not excessive and minimum [15].

2. Analytic Network Process (ANP)

ANP is a comprehensive decision-making method that captures the results of dependencies and feedback within and between groups of elements. AHP is the starting point of ANP and ANP is the more common form of AHP network. ANP is a combination of two criteria, namely, the first part

IOP Conf. Series: Materials Science and Engineering 1052 (2021) 012070 doi:10.1088/1757-899X/1052/1/012070

consists of a hierarchy of controls or a network of criteria and subcriteria that controls interaction, while the second part is a network of influence between elements and clusters [16].

3. Total Cost of Ownership (TCO) Models

The TCO-based model in supplier selection basically consists of simplifying and calculating all or some of the costs associated with the selected vendor and adjusting the price agreed upon by the supplier [17].

4. Multiple Attribute Utility Theory (MAUT)

MAUT is a linear weighting technique. Enable professionals to formulate a viable sourcing strategy capable of addressing the problem of conflicting attributes. However, this method is only used to select international suppliers which are more complicated and risky [18].

5. Outranking Methods

Used to solve multi criteria problems. This method only partially solves the problem and handles situations where there is uncertainty [19].

6. Technique for the Order Performance by Similarity to Ideal Solution (TOPSIS)

The closest coefficient is defined to determine the supplier ranking and the linguistic value is used to assess the rating and factor weights. TOPSIS based on the optimal alternative must have the shortest distance from the positive ideal solution (PIS) and the farthest distance from the negative ideal solution (NIS) [20].

6. Mathematical Programming (MP) Models

MP is a model that only considers quantitative criteria and allows considering various constraints in choosing the best supplier. This model is ideal because it uses a single objective model and a multiple objective model with maximum results in supplier selection. There are 2 MP programs namely Multi-Objective Models and Goal Programming Models [21], [22], and [23].

7. Artificial Intelligence [AI] Method

Systems that use computers as a basis for processing based on historical data and experience. This system has the advantage of overcoming complex and uncertain supplier selection.

1. Case-Based-Reasoning (CBR) Systems

CBR is a relatively new system. Used to access useful information and experience from previous decisions that are similar to making decisions.

2. Artificial Neural Network (ANN)

Models that require qualified experts but the ANN model can save time and money in processing.

8. Fuzzy Logic Approach

Ranks and weights are assessed using linguistic values and expressed in trapezoidal or triangular fuzzy numbers [4].

9. Hybrid Methods

Hybrid method is a method that combines several methods in the process of supplier selection. Such as MAUT and LP as solving problems in supplier selection, ANP and TOPSIS to identify interdependent criteria, and Fuzzy and TOPSIS as decision makers when facing uncertainty [4].

10. Using the ANP-PROMETHEE Method to Select Suppliers in Shipyard Companies

The ANP and PROMETHEE methods have their respective functions in solving supplier selection problems at shipyards. The ANP method can solve the problem in terms of the complexity of the large number of supplier choices available and the sub-criteria variable so that it remains objective in making an assessment or it can be called ANP tends to solve problems in terms of MODM. While the PROMETHEE method is able to handle problems in terms of MADM because it is used to deal with problems in terms of the complexity of the attributes of each variable in order to be precise in determining the level of importance of each predetermined attribute. A significant impact is seen on the values that exist in PROMETHEE when compared with the ANP values from weighting results in determining the final decision for supplier ratings. By combining the ANP-PROMETHEE method, the best alternative supplier output results are more rational when compared to doing manual and subjective assessments [10].For example, research conducted by Ma'had, Bayu and Farikhin (2020) at the shipyard PT. Lamongan Marine Industry, for the supplier of plate 6 feed 14 mm. Selected data from 8 different suppliers (MG, ATT, ESA, YSA, INT, GC, SUT, BIZ). The data is then classified into criteria, namely bid price, discount, delivery accuracy, shipping costs, packing, document completeness, product quality, ability to provide consistent quality, distance between locations, management role, information technology, reward systems, organizational structure and cooperation. Research is carried out manually and based on the values made by experts (P). The results of the experts' net flow calculations were made from P1-P6 with a consistency ratio (CR) value that should not be more than 0.1 (their opinion is considered consistent). The output from P1-P6 is then recapitulated to be the final result of the alternative ranking. The supplier with the largest Net flow value from the recapitulation is the best supplier. According to the calculation results that have been carried out, the best supplier for the 6 feed 14 mm plate is the supplier of SUT.

11. Paradigma LARG on the Supply Chain

The survey results show that Lean gets the highest score for the LARG paradigm as a shipbuilding supply chain followed by Agile and Resilient with the lowest weight. Lean is the most widespread paradigm and can be used for a long time. Lean is considered optimal because it is able to produce on a large scale in small batches, focuses on customers producing a lot of production with less waste, fast customer response, reduces inventory, increases work cycle efficiency, helps achieve more concise processes and provides instructions and advanced delivery times. The Agile Paradigm is not completely in line with the volume and variation of production required in the shipbuilding supply chain, while the Resilient Paradigm is the paradigm that most requires supply chain visibility. It is also known that Resilient is quite difficult to practice, this is due to the high cost required to practice this paradigm and the possible results will be close to the results of the lean paradigm. As for the Green paradigm, it shows surprising results because it has a lower value, where this paradigm is actually considered a key factor in shipbuilding to increase company competitiveness and improve energy efficiency [12].

12. Method Comparison

After conducting a literature study on the method of selecting suppliers, then a comparison of the methods that have been studied is made, it is intended to select how the suitable supplier selection method to be applied in shipyards. Table 1 below give resume about characteristic of each method.

Table 1. Comparison for Supplier Selection Method

No	Method	Characteristic
1	Categorical Methods	helps to make assessments in a clear and systematic manner
2	Data Envelompent Analysis	assessed based on input and output criteria
3	Cluster Analysis (CA)	reduce the larger suppliers into smaller subsets so that they are easy to manage
4	Analytical Hierarchical Process	provides an alternative ranking or sequence when there are several criteria and sub-criteria in decision making
5	Analytic Network Process (ANP)	combination of two criteria, namely, the first part consists of a hierarchy of controls or a network of criteria and subcriteria that controls interaction, while the second part is a network of influence between elements and clusters
6	Total Cost of Ownership (TCO) Models	simplifying and calculating all or some of the costs associated with the selected vendor
7	Multiple Attribute Utility Theory (MAUT)	formulate a viable sourcing strategy capable of addressing the problem of conflicting attributes
8	Outranking Methods	Used to solve multi criteria problems
9	Technique for the Order Performance by Similarity to Ideal Solution (TOPSIS)	determine the supplier ranking and the linguistic value is used to assess the rating and factor weights
10	Mathematical Programming (MP) Models	uses a single objective model and a multiple objective model with maximum results in supplier selection
11	Case-Based-Reasoning (CBR) Systems	access useful information and experience from previous decisions
12	Artificial Neural Network (ANN)	save time and money in processing
13	Fuzzy Logic Approach	Ranks and weights are assessed using linguistic values
14	Hybrid Methods	combines several methods in the process of supplier selection
15	ANP-PROMETHEE Method	best alternative supplier output results are more rational when compared to doing manual and subjective assessments
16	Paradigma LARG	focuses on customers producing a lot of production with less waste, fast customer response, reduces inventory, increases work cycle efficiency, helps achieve more concise processes and provides instructions and advanced delivery times

Through the comparison table that has been made, it can be seen that not all supplier selection methods are suitable for use in the shipbuilding industry. This is because shipbuilding industry is one of industry manufacture type that based on demand or projects and their characteristic are not same with mass industry. From the review, it can be conclude that ANP-Promethee Method and Paradigma LARG is the method of supplier selection that suitable to use in shipbuilding industry. It is because ANP-Promethee method is supplier selection method that can solve the problem in terms of the complexity of the large number of supplier choices available. This method accommodate MADM method in making decision that need in supplier selection process for shipbuilding because number of supplier are large and variated. Paradigma LARG method focuses on customers staisfaction, efficiency in production and delivery times, so this method give a lot of parameters in supplier selection process. Therefore this method is suitable in shipbuilding industry.

13. Conclusion

The conclusion that can be drawn based on the review that has been made is that the selection of suppliers for shipyards can be done by taking into account criteria such as quality, price, delivery, service, technical capability, financial strength, geographical location, reputation, reciprocal arrangements and other factors . As well as the ANP-PROMETHEE and LARG methods can be taken into consideration in determining a good supplier.

14. References

[1] Lestari F. and Dinata R.S.,*Green Supply Chain Management untuk Evaluasi Manajemen Lingkungan Berdasarkan Sertifikasi ISO 14001*, 2019, vol. 8, pp. 209–217.

[2] Hwang C. L. and Yoon K., Multiple Attribute Decision Making, Springer-Verlag, Berlin, 1981.

[3] Kontio J., A case study in applying a systematic method for COTS selection, in: Proceedings of the 18th International Conference on Software Engineering, 25–30 March 1996, 1996, pp. 201–209.

[4] Om P. and mit K.G., Supplier Selection Criteria and Methods in Supply Chains: A Review, in: International Journal of Social, Behavioral, Educational, Economic, Business and Industrial Engineering, 2013, Vol:7, No:10, pp. 2667.

[5] AguayoF., Marcos M., Sánchez M., Lama J.,Sistemas Avanzados de Fabricación, Distribuida;RA-MA Editorialy Publicaciones S.A.: Madrid, Spain, 2007.

[6] Koufteros X., Vickery S.K., Dröge C., The effects of strategic supplier selection on buyer competitive advantage in matched domains: does supplier integration mediate the relationships?, Journalof Supply ChainManagement, 2012, 93–115 (April).

[7] John T. M., DeWitt W, etc, Defining Supply Chain Management, in: Journal Of Business Logistics, 2001, Vol.22, No. 2, pp. 6.

[8] Anwar S. N., Manajemen Rantai Pasokan (Supply Chain Management): Konsep Dan Hakikat, in: Jurnal Dinamika Informatika, 2011, Vol:3, No:2, pp. 1.

[9] Wicaksono M., Surarso B., etc. Implementasi Metode ANP-PROMETHEE Untuk Pemilihan Supplier (Studi Kasus PT. Lamongan Marine Industry), In: Jurnal Sistem Informasi Bisnis, 2020, vol10iss1, pp. 36-45.

[10] Saaty T.L., Decision Making with Dependence and Feedback The Analytic Network Process, RWS Publications, 1996, Pittsburgh.

[11]Brans, J.P. and Vinckle, P.H., A Preference Ranking Organisation Method: (The PROMETHEE Method for Multiple Criteria Decision-Making, 1986, Managemenet Science (31), 647-656.

[12] Ramirez-Pena M., Fraga F. J. AF., etc, Shipbuilding 4.0 Index Approaching Supply Chain, in: www.mdpi.com/journal/materials, 2019, pp. 1-18.

[13] Dadashzadeh M., Subramanian M., etc, Critical Success Factors For Supplier Selection: An Update, in: Journal of Applied Business Research, 2004, Vol: 20, No: 2, pp. 1-18.

[14] Chen, Y. M. and Wang, W. C., An adaptive rescheduling scheme based heuristic algorithm for cloud services applications, In: International Conference on Machine Learning and Cybernetics, 2011, Vol:3, pp. 961-966. IEEE.

[15]TahririF., OsmanM. R., AliA., YusuffR. M., and EsfandiaryA.,AHP approach for supplier evaluation and selection in a steel. Journal of Industrial Engineering and Management,2008, Vol:1, No:2, pp 54-76.

[16] Saaty T. L., Fundamentals of Analytical Process, ISAHP 1999, Kobe, Japan, Aug 12 – 14, 1999.

[17] Ellram, Lisa M., Total Cost of Ownership: An analysis approach for purchasing, International Journal of Physical Distribution & Logistics Management, 1995, 25(8), pp 4-23.

[18] Min H. International Supplier Selection: a Multi-attribute Utility Approach, International Journal of Physical Distribution & Logistics Management, 24(5) (1994) 24-33.

[19] De Boer, L., Van der Wegen, L., Telgen, J., Outranking methods in support of supplier selection, European Journal of Purchasing and Supply Management, 1998, Vol:4, pp 109-118.

[20] Elanchezhian C., Vijaya R. B., and Kesavan R., Vendor Evaluation Using Multi Criteria Decision Making, International Journal of Computer Applications, 2010,Vol:5, No:9, pp 0975 – 8887

[21] Aissaoui, N., M. Haouari and E. Hassini, Supplier selection and order lot sizing modeling: A review, Computers & Operations Research, 2007, vol:34, pp3516-3540

[22] Monczka, R. M. and Trent, R. J., Cross-Functional Sourcing Team Effectiveness, Tempe, AZ: Center for Advanced Purchasing Studies, (1993).

[23] Hinkle, C.L., Robinson, P. J., Green, P. E., Vendor evaluation using cluster analysis, Journal of Purchasing, 1969, Vol:5, No:3, pp 49-58.

IOP Conf. Series: Materials Science and Engineering 1052 (2021) 012071 doi:10.1088/1757-899X/1052/1/012071

Comparison Between Fish-Catching Techniques with Purse Seine Fishing Equipment in Japan and Indonesia

D. Faturachman[1], Z. Fitri[2], S. Aslan[3], K.A. Sulaeman[4]

[1] Marine Engineering Department, Faculty of Marine Technology, Darma Persada University, Jakarta 13450, Indonesia
[2] Japanese Language and Culture Department, Faculty of Letters, Darma Persada University, Jakarta 13450, Indonesia
[3] P.T. Latinusa Tbk, Jakarta, Indonesia
[4] English Language and Culture Department, Faculty of Letters, Darma Persada University, Jakarta 13450, Indonesia

Abstract. Japan is a country that has highly developed marine fishery resources. In the field of fisheries, Indonesia is still very behind compared to Japan. One of the factors related to Indonesia being lagging behind Japan in the fisheries sector lies in the fishing technique using the purse seine fishing gear. There are several conditions related to fishing gear in Indonesia, such as: simple navigation and fishing gear, almost 80% of fishing gear operation still uses human power with manual steering systems and fishing operations in Indonesia which can take some time weeks or even months. This has an impact on the less than optimal fish catch. Indonesia is still very far from being able to catch up with the technology owned by Japan, especially in the operation of purse seine fishing gear. The purpose of this study was to determine and explain the purse seine fishing gear and to know the similarities and differences between fishing techniques and purse seine fishing gear in Japan and Indonesia. This study used a qualitative approach with literature review to collect data and make observations directly into the field, namely in Indonesia, especially on the Tri Mina Barokah Ship, Juwana area, Pati Regency and in Japan especially the Yamasen Ship, Choshi area, Chiba Prefecture. The results of the study that show the most significant difference between fishing techniques in Indonesia and in Japan are the number of vessels and fishing aids used. In Indonesia, fishing is still using 1 unit of ship which has a function as fish finder, net carrier and fish catcher, besides that the fishing aid used is still very simple. In Japan, 4 vessels are engaged in fishing (*honsen, tansakusen, unpansen*, recko boat) and are equipped with sophisticated and modern fishing tools.

1. Introduction

Indonesia is an archipelago country that has the biggest natural resources in marine especially in fisheries. As one of the pillars that support economy of nation, national fishery has other aim such as to: accomplish targeted production, provide national fish need, export fish to other countries and provide workforce/labour. Adequate available means and infrastructures in catching fish are totally required in order to increase fast of: production growth, fishermen and units of fish-catching vessel year over year. There are many ways to get fast and excellent caught-fish result. One of them is to amplify speed of the fish-catching vessel by upgrading the power, adding number of vessel engine or augmenting purse seine net size. In Asia, Japan is very advanced in fish-catching technology.

According to FAO (Food and Agriculture Organization), in 1988 total caught-fish quantity of Japan reached 12 tons or 13% of total world caught-fish number. Not only are the Japanese fish-catching armadas operating in their own water territory but they are also operating even until Pacific Ocean, Indian Ocean and other international water areas. Indonesian fishermen only operate by using one unit of vessel while in Japan they at least operate 2 units of vessels, even more, most fishermen there use 4 units of vessels altogether to catch fish, therefore it is very likely that fish- catching result in Japan is more dominant than in Indonesia. In addition to using fish-catching equipment, geographical location of Japanese fishing area is exceedingly supporting because fish- catching area is only 20 miles from the shore.

In Japan, most of the operations of purse seine use 4 units of vessels consisting of fishnet vessel (*honsen*), fish-searching vessel (*tansakusen*), caught-fish storing vessel (*unpansen*) and assisting vessel (*recko boat*). Every vessel is equipped with navigation tool based on type of the vessel, while purse seine fishing vessels in Indonesia mostly just use one vessel in addition to functioning as net vessel, also as caught-fish storing vessel. In operating fish-catching purse seine in Indonesia, almost 80% of them use human labour; this is totally opposite with operation in Japan. In Indonesia purse seine fishing vessel facility along with navigation tool used is not as complete as in Japan. Japanese vessels have complete navigation tool like navigating ability to monitor: wind direction, sea water currents, sear water depth, vessel position; and every crew has each bed, whereas in Indonesia purse seine fishing vessel all beds are put together, navigation tool is limited and steering gear is still manual. During its operation to catch fish, purse seine fishing vessel in Japan can always returns to harbour every day, in contrast in Indonesia it will take weeks or even months to do so. This condition can affect fish-catching process speed and also the caught fish quality. If the fishing vessel is able to return to harbour everyday the caught fish will be fresher compared to those put in the vessel storage by weeks or months because the fishing time takes weekly or monthly.

2. Literature Review

2.1 Catching Device of Purse Seine

Purse seine is fish-catching device that is categorized into group of surrounding net equipped with wrinkle cord and ring to draw together the bottom during operation. The ring has double function as moving area of ring cord and as the sinker. Until now purse seine is still the most productive fish-catching device to get small pelagic fish. Role of net to the caught fish is to contain the fish so they cannot escape the capturing net when it is encircling the fish. Initially purse seine was used in Rhode Island water area to catch menhaden (brevootia tyrannus). Then purse seine was patented under the name of *Berent Velder* from Bergen in Norway on March 12, 1859. [10].

In Indonesia purse seine or 'pukat cincin' in Indonesia language initially was introduced in northern java coast by Sea Fishery Research Institute (Lembaga Penelitian Perikanan Laut-LPPL) in 1970 in cooperation with fishery businessman by the name of Djajuri in Batang and it worked very well. Soon afterward in 1973-1974 this device was implemented in Mucar, Banyuwangi Regency, Central Java and since then it has been developing fast till now. At the beginning during its growing process in Muncar there was social conflict between traditional fishermen and fishermen using purse seine, but in the long run this fish-catching method was accepted by traditional fishermen [5].

2.2 Type of Purse Seine

Various purse seines are developed in order to be in line with the fishermen need and use [1]. Generally, purse seines are categorized based on: mesh basic design, species of fish to catch, number of vessel used during the operation, and fishing time operated.

a. Based on mesh basic design, purse seine size is divided into: square, trapezium and two symmetrical indentations.
b. Based on species of fish captured by purse seine, the fish are divided into: purse seine sardines, purse seine tuna, purse seine seed mackerel, purse seine Indian mackerel, etc.
c. Based on number of vessels used during operation, they are divided into: two-boat system purse seine and one-boat system purse seine.
d. Based on operating time, they are divided into: daylight purse seine and night time purse seine.

Based on the dimension, purse seines are divided into:

a. Mini purse seine: length is not more than 300 meters, developed in shallow seas (Java Sea, Malacca strait, and East Aceh water area) or throughout fishery coastal areas.
b. Medium size purse seine: length more than 300 meters up to 600 meters operated in water area further away or in off shore fishery area.
c. Big size purse seine: length is more than 600 meters up to 1000 meters, operated in deep sea in Exclusive Economic Zone of Indonesia.
d. Super Purse Seine: length is more than 1000 meters, operated in high sea fishery area [7].

3. Result and Discussion

3.1 Fish-Catching Technique with Purse Seine catching device in Indonesia and Japan

a. Area and Time of Catching

In Indonesia, catching fish using purse equipment is carried out in the night time (between sunset/dusk to sunrise) that is between 9 pm to 3 am in the morning; however, there is also purse seine operated in daytime. By and large, in Indonesia the process setting or net deploying is done twice a day that is during dusk and morning to dawn, but on certain condition volume of catching can be reduced or increased, yet it can also be no setting or no net deploying due to weather condition such as high sea wave or bad weather. Most of purse seine vessels in Indonesia during the fish-catching process take 1 month to 2 months to do so depending on condition of fish, season and weather. Areas of catching fish cover Java Sea, Natuna islands, Karimata strait, Makassar, Kangean islands, Masalembo and Bawean Island. In determining fishing ground, fishermen refer to moving cycle of little pelagic fish in water area of Natuna islands, Karimata strait, Java sea and Makassar strait which naturally can alter and swift based on seasonal change [6].

In Japan, areas of catching fish in Japan are divided into Sea of Japan (Nihonkai) and Pacific Ocean (Taiheiyo). In most cases, fishermen in Pasifc Ocean (Taiheiyo) tend to use purse seine. Within a year, the fishing ground has been moving based on season and captured fish. Season will determine the location of the fish because pelagic fish tend to be in between cold current (oyashio) and warm current (kuroshio). Catching zone can change based on natural condition but in general it will not be so significantly different than information about catching situation in water area explained above. Fish-catching time in Japan also depends on captured fish species and catching duration. Local government has decided the duration and frequency of catching. Concerning the duration, frequency of net setting can be one time to twice. Most fish- catching activity using purse seine in Japan is carried out in one day. For example, if fishermen and the vessels leave in the early evening, they will come back to the harbour around 2 o'clock in the morning, and if they leave midnight, they will come back around 5 o'clock in the morning, that's why caught fish using purse seine in Japan are very fresh [4].

b. Dimension of Purse Seine Catching Device

Dimension of purse seine in Indonesia is determined by its length that is from one wing to the other wing and its width that is distance between upper rope and lower rope (in meter). Width or height of purse seine in backward trapezium is measured at center part or the one that makes the pouch/bag [7].

In Japan, minimum of purse seine net depends on vessel condition that will bring or operate that net. Several factors that affect dimension of purse seine catching device are: purse seine net minimum length is 15 meters of vessel length; net minimum depth depends on length of purse seine net; length and bunt minimum depth is equal to length of vessel; while selection of net mesh size depends on size and type of captured fish.

c. Type of Purse Seine

In Indonesia, generally, mini purse seine consists of vessel, catching device that is net and its part, also special lifeboat for purse seine with light supporting device and supporting device combination of light and Fishing Aggregating Device (FAD). Vessel used during purse seine operation is teak (tectona grandis) based motor boat with length of 12-20 m, breadth of 3.5 m, Depth of 1.2 – 1.5 m and Gross Tonnage (GT) is 10 – 20 GT. Net of purse seine has material of polyamide with average length of 500 – 700 m and width of 45 – 50 m, net mesh is 1 – 2 inch. Lifeboat with length of 4 – 6 m, breadth of 0.5 – 0.9 m and Depth of 0.5 m is specifically dedicated to purse seine with light supporting device and combination between light and FAD. Fish-catching operation using purse seine usually takes one day. It is equipped with fish- catching supporting device such as fishing aggregating device for daylight catching activity and light for nigh time catching activity. Vessel used is equipped with attached motor with 25-36 horsepower engine. Usually 12 to 16 people as vessel crews are needed. Other than mini purse seine, there is also big purse seine which the catching zone is water area much further away that goes from Java Sea to Malacca strait. In order to do this, bigger vessel with 160-250 horsepower engine is required. For Large purse seine is purse seine that essentially has length more than 600 m. One of the examples of large purse seine is Cakalang purse seine. This seine has length approximately 1.000 meters and width 60-70 meters, its mesh net is variably from 1 inch for net that can function as bunt and 4 inches at the outermost part. Method of operation of both large purse seine and small/mini purse seine principally is similar [4].

Type of purse seine used in Japan mostly is purse seine one-boat system so called in Japanese isomaki and two-boat system is nisomaki. Isomaki is fish-catching vessel with purse seine catching device that during its operation is operated by 1 unit of net vessel while nisomaki is operated by 2 units of net vessels. In Japan, purse seine can be categorized into: One-Boat Horse Sardine Purse Seine, Two-Boat Sardine Purse Seine, One-Boat Horse Mackerel and Mackerel Purse Seine, Two-Boat Horse Mackerel and Mackerel Purse Seine, One-Boat Skipjack and Tuna Purse Seine, Two- Boat skipjack and Tuna Purse Seine [4].

d. Supporting Catching Device that is used

As for supporting devices used to operate purse seine in Indonesia catching device are: fishing aggregating device to attract school of fish to gather around it so they can be easily captured; roller to function as device to drag purse line during catching fish placed at left or right hull of the vessel where hauling is being processed; lifeboat is a small vessel without independent moving engine but will be dragged by the main vessel to and from fishing ground. It is equipped with a paddle/oar and a pressurized paraffin lamp; Spade as helping device to move captured fish from capturing device to vessel deck, this spade is made from nylon material whereas the grip and diameter are made of wood or rattan material; and light as helping device to lure or entice fish interest to gather underneath the light. Types of light used are pressurized paraffin lamp, neon lamp and mercury lamp [9].

Supporting catching device of Japan purse seine are power block, net hauler, pass winch, side roller, light, spade, supporting vessel. Main supporting devices used in Japan are: power block that is main part that functions to drag the net to the vessel which the working process has to be simultaneously working with net hauler. This net hauler has same function with power block to drag the net; pass winch that is main part to function to drag purse line and side roller that is a supporting device used to drag the net when it makes the pouch designed to facilitate the fishermen to spend less energy when dragging quickly the net into the vessel [4].

e. Catching Method and Catching Technique

Purse seine vessels in Indonesia mostly use method of chasing after school of fish, therefore it can be done in night time and daytime by encircling the aggregate of fish with big net. Catching technique in Indonesia water area even though has used purse seine vessel but the supporting device used is still very rudimentary and during the setting and hauling, it tends to use human labour.

Purse Seine vessel in Japan during its operation to catch fish mostly has implemented method of chasing school of fish, it can be done both in the night and daylight. Technically what has been done in Japan is same with what has been done in Indonesia. What is more superior in Japan is only supporting devices that are used which are net setting, net hauling and handling the captured fish [4].

f. *Type of Captured Fish*
Types of fish caught in Indonesia water area by using purse seine catching device are: seed mackerel, yellow tail scads, horse/jack mackerel, Sardine/Phil card, narrow-barred Spanish mackerel. The fish that are the destination for catching purse seine nets are pelagic fish that are clustered and close to sea level [3]. If the fish have not been collected in a catchable area or beyond the ability to catch the net, efforts must be made to ensure that the fish come together by using light, FADs, floating raft, and others [8].

Types of fish captured using purse seine supporting device in Japan water areas are: Mackerel (Masaba), Dotted Mackerel (Goma saba), Sardine (Iwashi), Dotted Sardine (Ma iwashi), Pointy mouth Sardine (Katakuchi iwashi), Horse Mackerel (Aji), Squids (Ika), Pikes (Sanma) [4].

g. *Operating Vessel*
Purse seine fishing armada generally is operated by individual business using wood-made hull. Machines that are used vary with 20-360 horsepower engine depending on the vessel size and fish-catching area. Most of purse seine vessels in Indonesia use one vessel that has various functions as fish-searching vessel, fish-catching vessel and caught fish-storing vessel. Small purse seine vessel (10-30 GT) has 20 horsepower engine which operates on one-day fishing time. Medium size purse seine vessel (30-50 GT) has 120 horsepower engine which uses operating time based on how many days per trip like 7-20 days per trip. Big size purse seine vessel (30-50 GT and 100-130 GT) has 120-360 horsepower engine which uses catching time operation based on days like 10-40 days per trip [5].

Based on function, purse seine vessels in Japan consist of:

a. Net vessel (Honsen) is a main vessel to carry purse seine net and to function as fish-searching vessel equipped with navigation tool more advanced than other vessel and also functions as fish-storing vessel. This is carried out if unpansen cannot store fish anymore due to overcapacity.
b. Vessel for searching fish (Tansakusen) is a vessel that functions as fish-searching vessel where a school of fish is located, also to function to drag left hull of net vessel (honsen) during dragging the net and lifting up the fish out of the sea so the vessel is still balanced and has stable distance with fish-storing vessel (unpansen). This vessel is designed to have speed faster than other vessel that is around 15-20 knot and has storing power of 70 GT.
c. Vessel for storing fish (unpansen) is vessel to store and bring captured fish. Vessel for storing fish tends to have bigger size than other vessels. Unpansen used in Japan has storing capacity of 300-400 GT with speed of 15-18 knot. Maximum capacity of unpassen to store fish is 300 tons.
d. Helping vessel (Recko boat) is a vessel to function to set net from honsen during catching operation and also to drag unpansen from left hull during dragging up captured fish.

3.2 Comparison between fish-catching technique in Japan and Indonesia

Similarity between fish catching technique in Japan and Indonesia is both catch pelagic fish by using technique of encircling school of fish with big net. Type and material of purse seine catching device used has similar net mesh because it has been fitted with targeted pelagic fish. The smaller fish to capture the smaller the net mesh is. Fish-catching time for pelagic fish in Indonesia and Japan is carried out mostly in the night and early morning to dawn [2].

Actually there is not much different between Japanese and Indonesia regarding the process of handling fish. In Japan in handling captured fish, the fish mixed with ice will be put in the vessel and then they will be brought to the harbour unloading and be sent to the next consumer, meaning the fish are very fresh. Meanwhile in Indonesia the captured fish will be put in storing room for 1

IOP Conf. Series: Materials Science and Engineering 1052 (2021) 012071 doi:10.1088/1757-899X/1052/1/012071

week to 4 weeks because the fishing ground is far away so the fish will be returning to the harbour after the ship returns to the original harbour.

On the whole, the difference between fish-catching technique using purse seine in Japan and Indonesia is on vessel type and supporting device that is used. In Indonesia, material of vessel is wood while in Japan mostly is made of stainless steel which has been equipped with sophisticated tools. In Japan, fish catching is carried out by several vessels that have each individual function in one team and are equipped with modern supporting device while in Indonesia it is carried out by only one vessel and only equipped with simple supporting device. During full moon, Purse seine vessel in Indonesia cannot do catching because it is still uses light supporting device and fish aggregating device to gather fish. Light that is used to gather fish cannot function during full moon because the light will be biased by moonlight. In Japan full moon doesn't affect the fish-catching activity, because the catching system is carried out by chasing and searching a school of fish [4].

Table 1. Similarities and differences between fish-catching technique in Japan and Indonesia.

	Item	Purse Seine in Japan	Purse Seine in Indonesia
Purse seine vessel	Net Vessel	O	O
	Fish-Searching Vessel	O	X
	Fish-Storing Vessel	O	X
	Helping Vessel	O	X
	Recko Boat	O	X
Supporting catching device	Fishing Aggregating Device (FAD)	X	O
	Roller	O	O
	Supporting Vessel	O	O
	Spade	O	O
	Light	O	O
	Power Block	O	X
	Net Hauler	O	X
	Pass winch	O	X
	Side Roller	O	X

Conclusion

Fish-catching method using purse seine carried out in Japan and Indonesia has similar method order but different in fish-catching technique due to number of vessels and supporting device used. From above table we can see that number of vessel and supporting catching device used have significant differences. Those differences affect the fish-catching process. Number of armadas consisting of 4 vessels will be faster to locate school of fish and sophisticated supporting catching device will facilitate process of net deployment, net dragging and handling captured fish.

References

[1] Amin. 1984. *Teknik Penangkapan Ikan*. Singaraja: Angkasa.
[2] Ardidja, Supardi. 2007. Alat Penangkapan Ikan. Jakarta: Sekolah Tinggi Perikanan.
[3] Ayodhyoa, A.U. 1981. *Metode Penangkap Ikan*. Bogor: Yayasan Dewi Sri.
[4] Chichiro Miyazaki, Minoru Inoue, Yoshinori Shimozaki, Yunosuke Idaka. 1977. *Iwashi Aji Saba Makiami Gyogyou*. Showa: Koseisha Seikaku.

[5] H. Sudirman. 2013. *Mengenal Alat dan Metode Penangkapan Ikan.* Jakarta: Rineka Cipta.
[6] Karman, Amirul. 2008. Spesifikasi Perikanan Pukat Cincin (Purse Seine) yang Berpangkalan di PPI Bajomulyo Kecamatan Juwana Kabupaten Pati Jawa Tengah. Ternate: Universitas Khairun
[7] Nuraeni, Indri. 2014. *Alat Tangkap Purse Seine.* https://academia.edu/9479009 /alat _tangkap_purse_seine.
[8] Sismadi. 2006. Analisis Efisiensi Penggunaan Alat Tangkap Purse Seine di Pekalongan. Semarang: Undip
[9] Tea, Ikaupea. 2013. Penggunaan Alat Bantu pada Teknologi Penangkapan Alat Tangkap Purse Seine Surface Lamp. http://yhoyoji. blogspot.co.id/2013 /09/alat-bantu-pada-purse-seine.html.
[10] Wanaf. 2012. *Purse Seine.* https://wandaaf.wordpress.com/2012/11/03/purse-seine/.

Bedroom Module Design on Offshore Hotel as a Support for Maritime Based Tourism Accommodation
(Study Case in Sendang Biru, Malang, Jawa Timur, Indonesia)

A A Rucitra[1], K Sambodho[2], K K Wardani [3], Y Mulyadi[4]

[1] Interior Design Department, Institut Teknologi Sepuluh Nopember, Indonesia
[2] Marine Engineering Department, Institut Teknologi Sepuluh Nopember, Indonesia
[3] Visual Communication Design Department, Institut Teknologi Sepuluh Nopember, Indonesia
[3] Marine Engineering Department, Institut Teknologi Sepuluh Nopember, Indonesia

Abstract. Aaquaculture in Indonesia has only been used as a place for aquatic biota cultivation. At the same time, the structure of the building above the grove has not been used to the fullest. Therefore, with the development of the existing floating net design, support facilities for the offshore tourism sector are being developed, one of which is the Hotel. Hotel on aquaculture is an interesting development concept because it utilizes space on offshore sea buildings with facilities that educate about aquaculture and maritime potential in Indonesia for tourists staying at hotels. In designing the hotel on top of the aquaculture building, it is necessary to analyze the space requirements for hotel users to be comfortable on top of the building in the middle of the sea. Apart from that, the hotel design aesthetic is also needed, in order to attract the attention of tourists. Based on the design by considering the macro and micro concepts, as well as the arrangement of the hotel space, a design with a natural minimalist concept is obtained but still can meet the needs of the user space.
Keywords: *Aquaculture, Hotel, Maritime, Tourism*

1. Introduction

The marine and tourism sector is the strategy sector which is expected to realize Indonesia's development mission. Some of Indonesia's missions in maritime-based development are: maintaining regional sovereignty, maintaining economic independence by securing marine resources, and reflecting Indonesia's personality as an island nation, realizing high quality, advanced and prosperous Indonesian human life, realizing a competitive country, realizing Indonesia become an independent, developed, strong, and maritime nation based on national interests and realize a society that has a personality in culture.

According to RPJMN 2015-2019 in the field of maritime affairs and fisheries, to prepare Indonesia to be a maritime nation, there are several things that need to be prepared to encourage, strengthen and develop various development potentials in the maritime sector, one of which is marine buildings and marine tourism. One of the points of the development mission in the marine tourism sector is to develop competitive industries, sustainable goals and use responsible marketing (responsible marketing).

The marine tourism sector has the aim of developing tourism which is able to make a significant contribution to the national economy and the welfare of the people which are described as the following targets, increased expenditure and length of stay of tourists, the realization of the management of tourism destinations, increasing the number of foreign tourist arrivals to Indonesia and the movement of domestic tourists, supports increasing the contribution of tourism to the national economy on GDP, employment and investment (Peraturan Menteri Pariwisata dan Ekonomi Kreatif Republik Indonesia Nomor 1 Tahun 2014 Penyelenggaraan Sertifikasi Usaha Pariwisata, 2014).

Tourism makes a significant contribution to Indonesia's finances. Creation of GDP in the tourism sector through domestic tourists, the tourism budget, foreign tourists and investment in businesses. The tourism sector has a strategic role in creating added value for the nation and creating high foreign exchange-earners. In 2015 the tourism sector generated US $ 11.9 billion or equivalent to Rp 163 trillion (up 113% compared to 2014, which reached the foreign exchange rate of US $ 11.17 billion). One that contributed to the increase in Tourism GDP in 2016 was accommodation from Rp18,832 billion up to Rp 61,118.26 billion (Lembaga Penyelidikan Ekonomi dan Masyarakat Fakultas Ekonomi dan Bisnis - Universitas Indonesia, 2018).

There are many ways to exploit the maritime potential in Indonesia, such as Aquaculture. Aquaculture is a fish farming technology that is environmentally friendly and sustainable. Indonesia has a variety of aquaculture building models developed, one of which is a floating net. So far, aquaculture buildings in Indonesia have only been used as a place for aquatic biota cultivation. At the same time, the structure of the building above the aquaculture building has not been used optimally. With the development of the existing aquaculture building design, the concept of supporting facilities for the needs of the offshore tourism sector was made, which is a Hotel by utilizing a structure chart for aquaculture buildings. Hotel on aquaculture is an interesting development concept because it utilizes empty space in offshore sea buildings with facilities that can educate about aquaculture and the potential of the sea in Indonesia for tourists staying at hotels. Therefore, this research will discuss the study of hotel interiors used for offshore aquaculture tourism consisting of: 1) Study of standard furniture & interior materials for offshore buildings; 2) Study of standard acoustic materials in the interior of offshore buildings; 3) Hotel Review of Activities and User Needs; 4) Determination of the bedroom module concept.

2. Method

In general, this research develops the design of offshore aquaculture structures that are easy to move and operate, for this reason, the following steps are required:

2.1 Literature Study

The data collected is used as a reference for the interior design to be made. Several aspects are used as a reference in making designs, such as: wall materials, floor materials, furniture, and so forth.

2.2 Analysis of Room Users Room Space Needs Analysis

In terms of building use, data that need to be used as a reference are Ergonomics and Anthropometry and minimum building area. Ergonomics represents all the calculations / aspects of humans that are needed to determine the size of construction or pleasure that will be used in marine buildings in order to create a sense of comfort. Anthropometry is a contribution of the human body.

2.3 Determination of the Room Module Design Concept

Results and data collector used as a basis for creating interior design concepts for offshore hotel bedrooms. The concept used is expected to improve the quality of safety, comfort, and meet offshore building standards. Some things to consider are function, material, design, and cost (Pile, 1994). In the process of determining the macro concept of interior design of offshore hotel rooms, the grouping process is obtained from the main variables discussed in the initial chapter and the results of data analysis. Where this interest consists of three variables including:

a. Problems from the company side, namely the main problem of the design object in terms of the identity of the aquaculture building image
b. The problem of applicable offshore building standards, namely the problem of design objects related to the material aspects that are available specifically for buildings on the sea / ship.
c. The problem of the object side which is a hotel, while the hotel is identical to the place used for vacation, temporary rest, and for relaxation.

Next is to determine the micro concept, which is to determine the concept for each material that builds the sleep module for the offshore hotel building module

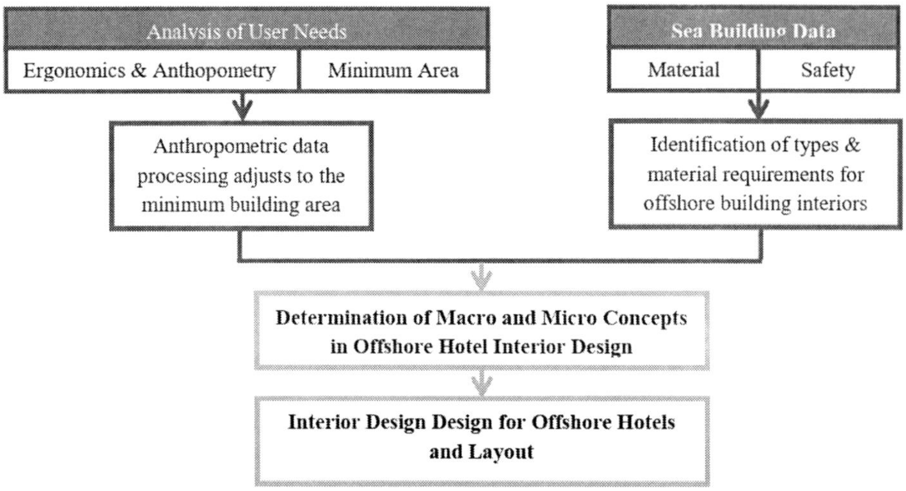

Figure 1. Research methodology

3. Results and Discussion

At this stage, it will discuss the bedroom needs and the development of bedroom design concepts based on identified micro and macro concepts

3.1 Analysis of User Needs

In the analysis phase of hotel room user needs in an aquaculture building, the assessment method used is ergonomics. Ergonomics is a method that involves people, tools, and work systems. In interior design, ergonomic aspects of applied safety, health, security, and work safety aspects are reviewed both the safety and psychological aspects. The following are the results of the analysis for needs

Table 1. Standard Variations of Single Bed
Source: Human Dimension and Interior - Space A Source Book of Design Reference Standards (Panero & Zelnik, 1979)

No.	Aspect	in	cm
1	Standard Widths of Bed	36	91.4
2	Standard Lengths of Bed	78	198.1
3	Standard Widths of Mattress	39	99.1
4	Standard Lengths of Mattress	84	213.4
5	Dimension Varies with Furniture Plan	60	152.4
6	Varies	16-17	40.6-43.2

Figure 2 shows the space requirements for the use of a standard bedroom with a single bed. Taking into account the condition of hotel rooms in aquaculture buildings that have extensive limitations, the minimum distance required by the user agrees in Figure 2.

3.2 Macro Concept

Based on the main variables, translated into a number of points about the interior design concept of a floating grove space building, then the following are problems and solutions which can then be used to get the design concept.

The design concept developed for hotel rooms over the sea leads to the Natural Minimalist style. The description of the Natural Minimalist style relates to the sub-themes obtained by the processing of each variable's solution, namely the identity's sub-theme as a representative of the problem of the corporate side; Natural and Relaxation sub themes as representatives of Resort problem variables used for vacation and Minimalist sub themes as representatives of applicable offshore building standard problem variables. Natural Minimalism is the style / style / main style that underlies this design. Characteristics and application of Minimalist Natural elements in the turbidity of the space will be combined with the preparation of each sub-theme.

IOP Conf. Series: Materials Science and Engineering 1052 (2021) 012072 doi:10.1088/1757-899X/1052/1/012072

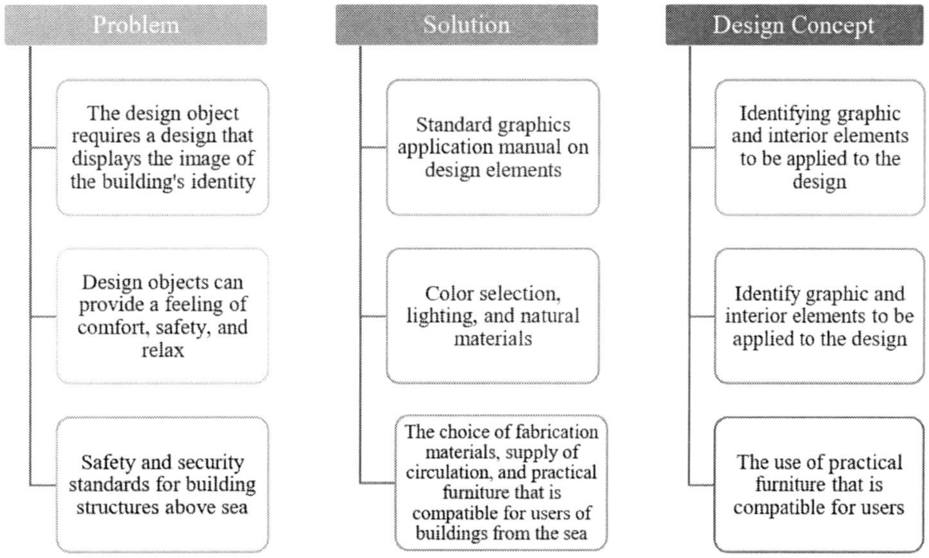

Figure 2. Micro Concept Design

This style is basically a combination of minimalist style Natural style. Form follow function is one solution for the utilization of a fairly limited space with the use of an appropriate design, especially on the side of formation. The simplicity of the formation of a minimalist style and prioritizing the function will later be combined with natural elements that tend to impress "earthy" through form, original texture of the material by combining the various fabrication material choices available by considering aspects of safety and security standards for Buildings on water that apply according to the Journal Byun.LS (Byun, 2006). Therefor the design will be made functional but can also offset the preferences of visitors' tastes through the texture or appearance of the material

3.3 Micro Concept

Development of micro concepts based on macro concepts aimed at the Minimalist Natural style, by that, for the application of Natural Minimalist design concepts, important aspects that need to be considered such as furniture details are used. Furniture that is used specifically is made on request. Furniture must use materials that are lightweight, waterproof, fireproof, and easy to install. In addition, furniture must have a space saving feature considering the area of the room in the building over the sea. The following materials must be considered in the design

3.3.1Ceiling

The ceiling concept in the building above uses standard materials which have aesthetic element requirements. The use of continuous ceilings is the standard of use for buildings above with dimensions, Width: 100, 200, 300 mm and Length: 600-5000 mm with a ceiling buffer frame Suspension Line Replacement Aluminum: square pipe, (20 × 20 or 20 × 40 mm). Here is one example of ceiling installation using these materials:

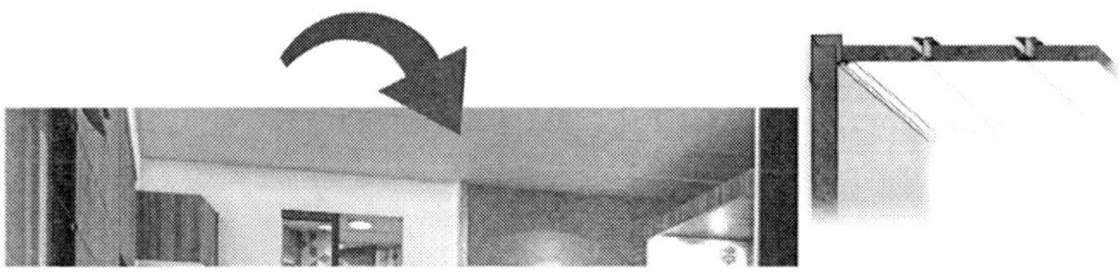

Figure 3. Application of Ceiling Design

This type of continuous ceiling is also a non-flammable material because it has insulation from mineral wool, so that safety protection for users can be obtained.

3.3.2Wall

For aquaculture buildings, the wall material used must have sufficient strength and have good soundproofing capabilities, requiring the required structure consisting of three parts namely: mineral wool (soft core), vermiculite (hardcore panel), and aluminum (metal sandwich panel).

Soft-core panels are materials that have good resistance and are non-flammable (non-flammable materials are substances that will not burn, burn, support combustion, or release flammable vapor when exposed to fire or heat, in the form where it used and under anticipated conditions), but does not have its own panel structure. Upper and lower structural profiles on soft-core panel panels must be added vinyl sheets which are useful for holding air and scratch resistance. In general, the thickness of a soft-core panel structure is around 25mm or 50mm, depending on the criteria of the fire resistance structure.

For the Hardcore Panel structure, the material used is Vermiculite. Vermiculite is a non-flammable material and has a weak resistance to air. The hardcore panel is connected with aluminum profiles and steel connections. Aluminum can be applied to wall panels, panel boards and boards for furniture. Meanwhile, metal sandwich panels are manufactured using a patented production process by printing 2 pieces of metal which are added using adhesives on both sides

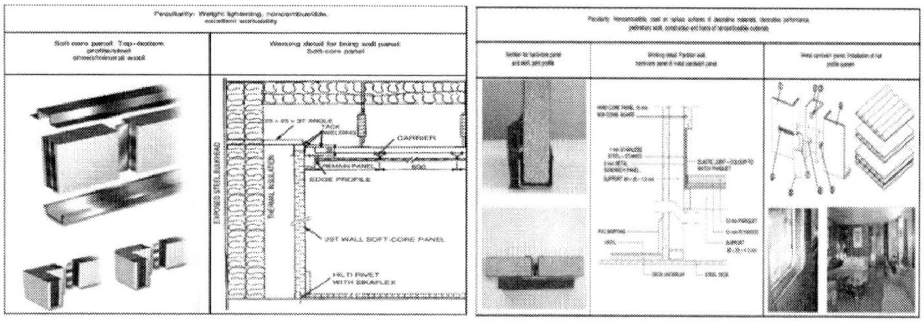

a) Soft-core Panels b) Hard-core Panels

Figure 4. Hotel room wall structure

The metal used is aluminum, because it has the characteristics of a lightweight material. Metal sandwiches have the advantages of lighter material, good acoustics and very dynamic. Many industries use these materials and methods for building walls above the sea.

3.3.3Floor

Floor finishing can use ceramic tiles, vinyl, carpet, wood, artificial stone, and cats. In addition to meeting the criteria determined by SOLAS Convention and IMO, only material with specific criteria can be used. The carpet can be used with specifications: 70-80% wool, 20-30% nylon with a pole weight of 1600 h / m2, and a mast height of 7mm to minimize gas released during a fire.

3.3.4Furniture

The use of furniture is divided into 2 criteria, custom-made and ready-made. Custom-made furniture that has collections must be installed permanently which cannot be moved. Ready-made furniture is furniture that can be moved. And must follow the criteria of the SOLAS Convention on furniture (2000).

- Furniture such as tables, cabinets, dressers, credenza, cabinets must use non-flammable construction materials of not less than 2mm;
- Free standing furniture such as chairs, sofas and small tables must use a non-flammable frame;
- Upholstered furniture must have materials that are resistant to fire.

In addition, the furniture used must be multifunctional. Multifunctional furniture is used as an alternative solution for minimal space. Furthermore, the shape and finishing must be neutral and simple so that the impression of the resort to recover can be conveyed properly. With PVC foam board material, also known as Chevron boards or Andy boards are widely used for indoor and outdoor applications.

PVC foam board made from a chemical composition, Poly Vinyl Chloride, which is used in the furniture, building and advertising industries. PVC is lightweight and foamy, withstands moisture, chemicals and corrosion. The thickness of the material will increase from 6 mm to 45 mm. Can be used to carve, emboss, paint, laminate and grind board surfaces as needed.

3.3.5Lighting

The lighting concept for bedrooms uses artificial lighting used for cottages (warm white lights). Warm white lights can give the impression of warmth and relaxation. For general lighting use lamps that depend on the exposed ceiling and down lighting on the upper and lower ceilings.

3.3.6Aesthetic Element

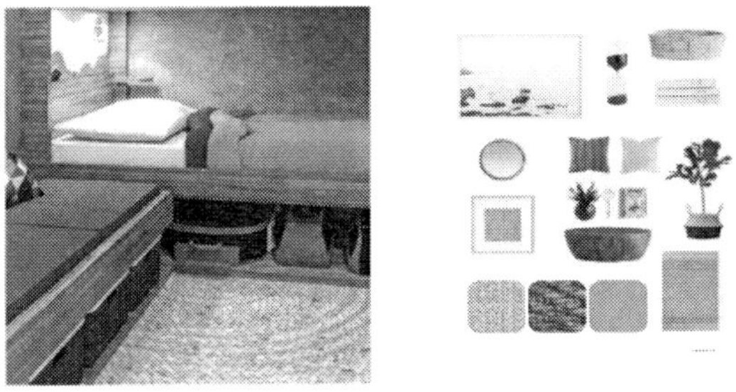

One application of the concept of Natural to aesthetic elements is interior pot accessories, pillows, cushions, and others. This can be applied to public areas or private areas, but must still use and visual appearance in order to characterize the minimalist look.

Figure 5. Application of Aesthetic Element

3.4 Development of Aquaculture Hotel Room Design

The development of hotel room design for aquaculture buildings based on macro and micro concepts will use the Tree Concept method with an aquaculture building concept that emphasizes the physical and technical structure of the building, as well as the 'Natural Minimalist' concept which has a philosophy of closeness to the environment and is easily created through designs that provide opportunities for humans to live and work in healthy places, at least stress levels, and provide a prosperous life by integrating design with nature.

In the macro concept and application, a decent percentage will be applied, taking into account the minimum space requirements with the many activities that will occur in the room. The function tree method of the aquaculture sub-theme will be more dominant in terms of function and aesthetics that will be applied to the design. Then the second is a Minimalist sub-theme which is represented by the blue color, which includes the use of Furniture and Materials, which will largely determine the use of the 35% design. And the rest use colors and accent shapes that will be applied to the design amounting to 15%.

Figure 6. Result Design Concept of Bedroom

Figure 8 shows the design for a hotel room with a minimalist natural concept. The bedroom concept consists of bed facilities consisting of a bunk-bed with a capacity of 2 people, a seating facility that is compatible for beds and bags, then also provided space for reading tables and chairs. The use of colors for walls and furniture that uses space uses monochrome colors synonymous with minimalist styles, such as gray and navy blue. Both colors can give the impression of clean, peaceful, light, which can affect the psychological aspects of users who want to declare themselves from the bustle of the city.

3.5 Alternative Layout Design

Bedroom is used as a resting place for guests who will visit Aquaculture. This is the one alternative layout designs as follows:

3.5.1Alternative 1

Alternative 1 has a bathroom design in the center / center position, as illustrated in Figured 9

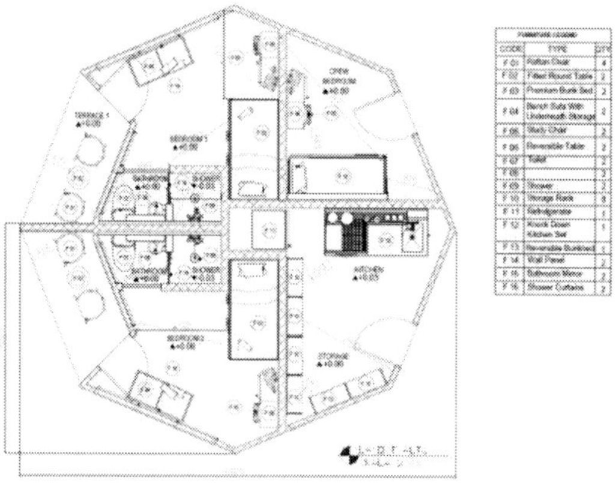

Figure 7. Layout Bedroom (Alternative 1)

Weakness

- Let the bed slightly uncomfortable due to being partially covered by the bathroom wall. Laying the bed is also in place also adjust the placement of furniture so that there is enough area for the crew's room.
- The kitchen is less ergonomic in terms of the cooking sequence (area and shape of the kitchen space that is less comfortable)

Advantages

- Laying a sofa bench in front of a window can give an exclusive and comfortable impression
- There is enough area for hot air circulation from the refrigerator.

3.5.2Alternative 2

The second alternative layout still uses the concept of a bathroom in the middle but is different and the bathroom is equipped with the facilities provided in Figure 10

Weakness

- Lay the sofa bench and study the table area that is directly opposite the entrance that causes discomfort and as if the room is closed.

Advantages

- The privacy of the bathroom consists of a shower and toilet which is not visible directly from the entrance
- Circulation that does not lead to the bathroom.
- Service area for guest servants is very important, so area alley and the bed crew get guests that are not too big. In this design both of these rooms feel spacious with different arrangements with alternative 1.

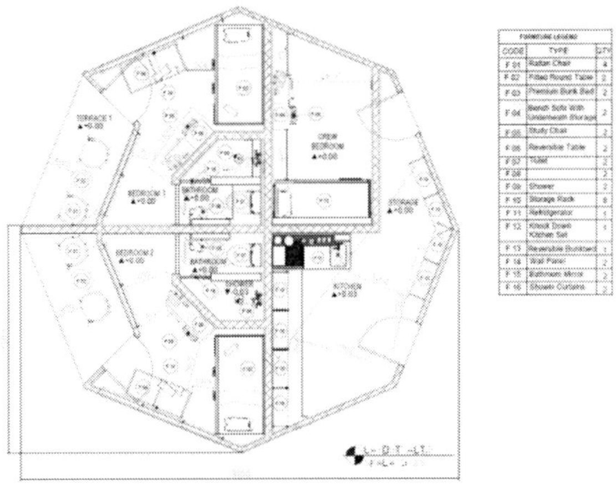

Figure 8. Layout Bedroom (Alternative 2)

3.5.3Alternative 3

From the previous alternatives, this layout design focuses more on the needs of the space and furniture chosen.

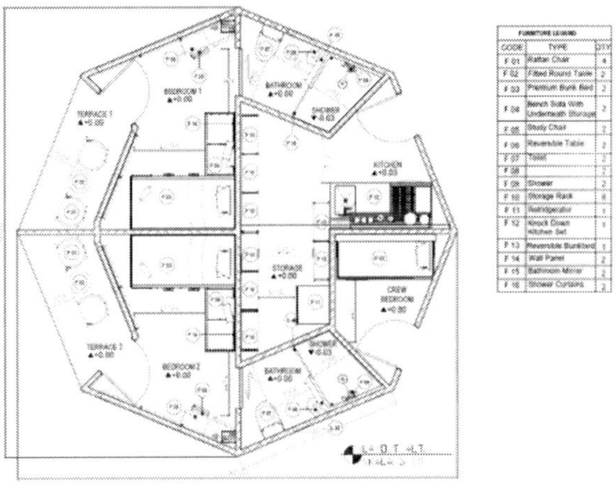

Figure 9. Layout Bedroom (Alternative 3)

Weakness

- There are still areas that are not used, for example in front of the bunkbed which is a dead corner

Advantages

- Is the most relevant alternative in terms of space requirements and available facilities. Configure the bed and study table in line to determine the type of activity for the space through the laying of furniture.
- Storage is placed on a horizontal line which will be done as a weighting point on the structure.

3.5.4Alternative 4

In this alternative concentrated on the distribution of the burden on the arrangement of laying furniture in Figs.

Weakness

- Storage space which becomes one side that causes poor loading makes the load point behind the floating net sinks.

Advantages

- Good configuration in laying furniture causes a fairly spacious room in the middle area, causing a comfortable effect.
- Centralized piping arrangements make bathroom facilities such as showers and washrooms comfortable.

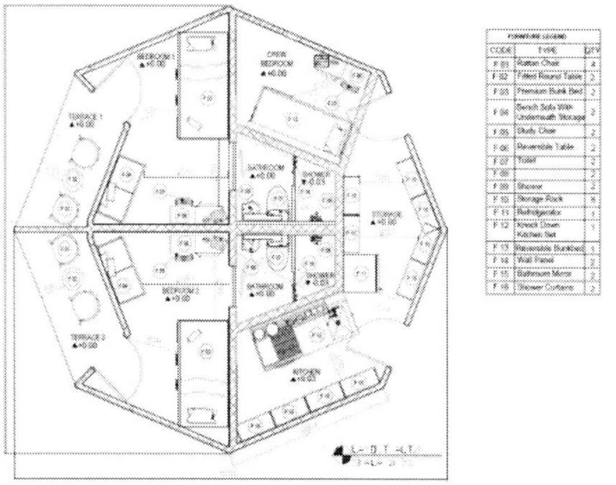

Figure 10. Layout Bedroom (Alternative 4)

3.5.5 Alternative 5

This design is almost similar to the 4th alternative, but there are differences in the laying of the bunkbed Figure

Weakness

- As in alternative 4, storage space depends on one side of the back of the floating net sinks, so it can change the point of load (unbalanced).

Advantages

- Laying a horizontal bunkbed has a roomy effect and doesn't take up too much space.
- Bathroom facilities consist of showers and washrooms which are quite comfortable due to centralized piping

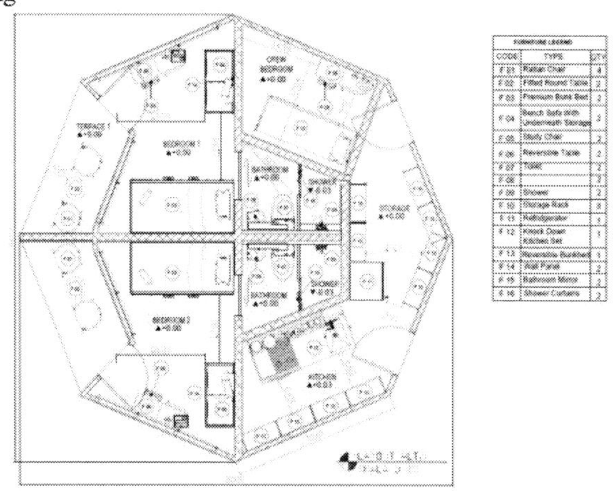

Figure 11. Layout Bedroom (Alternative 1)

Based on the five alternatives, analysis of layout selection using a weighted method based on the process of designing the layout, suitability of the facilities to the needs of users is the most important thing in designing the hotel. It is followed by the area of space for the activities and transportation paths that are important to be transferred, and the last is the suitability of alternative layouts with resort design and branding concepts inspired by the octagonal shape. Then the following alternative designs 3 are selected.

Table 2. Weighting Method Alternative Layout

Aspect	Weight	Alternative				
		1	2	3	4	5
Flow of Circulation	0.14	45	45	49	51	51
Suitability of Space to Activities	0.29	47	44	50	50	50
Suitability of Facilities Against User Requirement	0.36	49	48	50	39	37
Alternative Layout Suitability with the design concept	0.21	22	21	26	25	25
Total		40.72	39.28	43.72	42.64	41.92

Based on the calculation of the weighting method for the five alternative layouts, the third layout design was chosen because it has the highest score. For suitability of facilities against user requirement, the third alternative has the highest score, where this aspect is the main aspect in choosing an alternative design for the aquaculture building layout

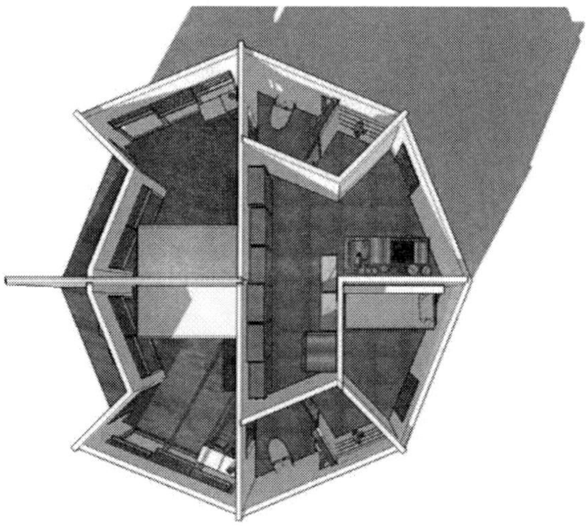

Figure 12. Selected Design Layout

So, based on the design of macro and micro concepts, as well as the selection of alternative designs for the layout of aquaculture buildings, the layout designs are as follows. The following designs describe a natural, minimalist room atmosphere and match the user's needs.

4. Conclusions

The results of data analysis and discussion, resulting in the conclusion of the concept that will be raised as the interior design concept of Aquaculture is Natural Minimalism. Characteristics and application of Minimalist Natural style elements in the space to be applied and integrated into each sub-theme. To adjust the macro concept, then choose components to fill the hotel building with floating jarring (micro concept) towards a minimalist natural concept, such as the example of choosing the floor concept using Bamboo Plastic Composites. This concept will give a natural impression on space because it has a structure like bamboo.

The alternative layout used in the design of aquaculture building is the third alternative layout, the selection is based on the weighting method of four influential aspects to meet user comfort, namely circulation, the suitability of space for activities, the suitability of facilities to user requirements, as well as the suitability of layout alternatives with the design concept.

5. Acknowledgments
We would like to express our special thanks of gratitude to The Center for Marine Studies, a research center under the Institute of Research and Community Service (LPPM) ITS Surabaya for providing the research opportunity and develop studies. Secondly, we would also like to thanks to Miss Nazia, Miss Irbah, Mrs Titis, and Mr Reno as an assistant, designer, reporting, and supporting the course of this research.

6. Reference
[1] Byun, L. S. (2006). Peculiarity of Interior Design Materials for accommodation areas of cruise ships: A state-of-the-art review. *Ships and Offshore Finish Aquaculture, 1:3*, 171-183.
[2] Lembaga Penyelidikan Ekonomi dan Masyarakat Fakultas Ekonomi dan Bisnis - Universitas Indonesia. (2018). *Laporan Akhir: Kajian Dampak Sektor Pariwisata Terhadap Perekonomian Indonesia.* Indonesia.
[3] Panero, J., & Zelnik, M. (1979). *Human Dimension and Interior Space: A Source Book of Design Reference Standards.* New York: Whitney Library of Design.
[4] (2014). *Peraturan Menteri Pariwisata dan Ekonomi Kreatif Republik Indonesia Nomor 1 Tahun 2014 Penyelenggaraan Sertifikasi Usaha Pariwisata.* Jakarta: Kementerian Pariwisata dan Eonomi Kreatif Republik Indonesia.
[5] Pile, J. F. (1994). *Interior design second edition.* Harry N. Abrams, inc.
[6] SOLAS. (2000). International Convention for the Safety of Life at Sea 1999/2000 Amandements. *International Convention for the Safety of Life at Sea, Consolidated Edition* , (pp. pp. Chapter II-2, Part B, C, D). London, UK.

IOP Conf. Series: Materials Science and Engineering 1052 (2021) 012073 doi:10.1088/1757-899X/1052/1/012073

Distribution Pattern of Capture Fisheries Industry in Southern of East Java Analysed from Operational Aspect and Interconnection Between Ports

M R Mubarok[1,3], O S Gurning[1] and N I Arvitrieda[2]

[1]Department of Marine Engineering, Faculty of Marine Technology, Institut Teknologi Sepuluh Nopember (ITS), Surabaya, East Java 60111, Indonesia

[2]Department of Industrial Engineering, Faculty of Industrial Technology and System Engineering, Institut Teknologi Sepuluh Nopember (ITS), Surabaya, East Java 60111, Indonesia

[3]Corresponding author : mubarokrizqi19@gmail.com

Abstract. East Java is one of the provinces that has the largest capture fisheries resources in Indonesia. It is recorded that the sustainable potential of East Java fisheries is more than 500.000 tons / year [1]. Capture fisheries production in the southern part of East Java Province is dominated by 3 Regencies, there are Trenggalek Regency, Pacitan Regency and Banyuwangi Regency [2]. As one of the Provinces that greatly contributed to the National income, East Java province is very potential if the government wants to make strategic plans related to the capture fisheries industry in East Java province. This strategic plan will determine a suitable development area for the capture fisheries industry on the southern part of East Java by considering operational aspects and interconnection between ports. By considering these two factors, it is hoped that the center for the development of the capture fisheries industry in southern of East Java will become more structured so can supports the government program, National Fisheries Logistics System (SLIN).

Keywords. *capture fisheries, industry, operational, interconnection ports, logistics*

1. Introduction

Indonesia has large sea areas, coastal areas and small islands which have strategic significance as pillars of national economic development. Apart from having economic value, marine resources also have ecological value, in addition, Indonesia's geographical condition lies in a strategic geopolitical position, namely between the Pacific and Indian oceans, which are the most dynamic regions in the flow of world politics, defence and security. These geo-economic and geo-political conditions make the marine sector an important sector in national development. One of the areas that has the largest capture fisheries products in Indonesia is East Java Province. East Java is surrounded by WPP-RI 573 and WPP-RI 712 which have the potential for sustainable fisheries, respectively 491.7 thousand tons / year and 836.6 thousand tons / year [3]. WPP-RI 573 has a wealth of resources that are abundant and have a variety of renewable and non-renewable natural resources. In terms of renewable natural resources, the potential that can be developed in WPP-RI 573 includes the capture fisheries sector, tourism, seaweed cultivation, salt ponds, ecosystem protection, development of ecosystem areas and the fishery industry. For the capture fisheries sector, WPP-RI 573 is the management area for kind of tuna, skipjack and mackerel in Indian Ocean. As the high seas, the Indian Ocean is connected to the Indian Ocean Rim countries so that their fisheries management is part of the Indian Ocean Tuna Commission / IOTC management area. Data on the potential and utilization of tuna, skipjack and mackerel are determined based on the best data that can be obtained from the Regional Fisheries Management Organization (RFMO) and / or the Ministry of Marine Affairs and Fisheries [4]. Based on the results of a study by the National Commission for the Study of Fish Resources which was carried out in 2016, the estimation of the potential of fish resource groups in WPP-RI 573 was dominated by large pelagic fish (tuna, skipjack and mackerel) where the production of fishery catch in the 2005-2014 period ranged from 127,815 - 218,359 tons / year with an average of 182,034 tons / year [5]. Another indicator that is no less important regarding fish resource stocks is the presence of larvae. A comprehensive and holistic larval study in the study of fish stocks is essential for sustainable fisheries resource management. Larvae study is needed to determine the optimum conservation and exploitation area so as to protect and enrich the environment [6]. UNESCO illustrates that larvae studies can provide information on spawning areas and seasons, absolute stock abundance and subsequent interactions that can affect stocks. Larval studies are also useful in determining the recruitment patterns of fish populations. Based on this, increasing the understanding and utilization of fish resources in WPPN-RI 573 is very important to be studied in an effective, efficient, and sustainable capture fisheries management book.

2. Materials and methods

The methodology type of this research is descriptive statistics and the approach method is a qualitative approaches. Descriptive statistics were chosen because the form of the Research Method serves to present a complete picture of the Social Settings or for Exploration and Clarification, which is formed by describing a phenomenon of events that occur in the basis of the statistical results that have been recorded [7]. Meanwhile, qualitative approaches was chosen so that this research would broaden the point of view in determining the location for determining the capture fisheries distribution centre in the southern part of East Java by considering institutional and operational aspects. The methods for this research can be seen in figure 1 below.

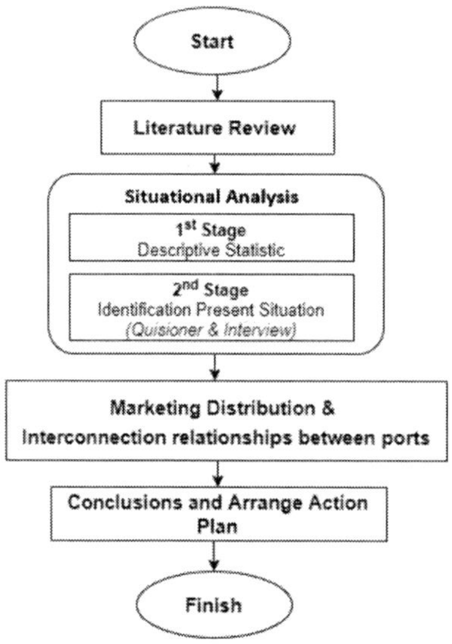

Figure 1. Methods.

Designing and managing the capture fisheries industry supply chain is a complex matter, and it must take into social issues involved in it. This complexity is not considered in the process of designing a supply chain management system using quantitative methods. Quantitative methods are able to improve coordination and efficiency in the supply chain, but do not consider key aspects such as ethics, sustainability and human values in the supply chain [8]. Therefore this study aims to design a model of the capture fisheries industry supply chain system.

2.1. Literature review

Literature review is carried out by create a critical analysis of the relationship between journal articles from the work of previous researchers, and their relationship with the research to be proposed. bibliography includes all secondary sources of research, for example films, audios, presentations, lectures, handwritten diaries, archives, laws and regulations, artifacts, and others [9].

2.2. Situational analysis

The situational analysis stage is intended to obtain a portrait of the problem situation in the capture fisheries industry supply chain at the research location [10]. The portrait of the situational analysis is described as the profile of the fishery industry, the profile of the capture fisheries industry in the East Java region, especially in the southern coast. The profile image is made based on statistical data and policies and regulations related to the capture fisheries industry obtained through literature studies in the form of books, journals and other information obtained from related agencies.

2.2.1. Descriptive statistic

Data related to situational analysis is in the form of statistical data from third parties who provide credible data, such as Ministry of Marine Fisheries and Central Bureau of Statistics of Indonesia. Furthermore, primary data is obtained from questionnaires and in-depth interviews and / or expert / expert surveys are carried out to obtain opinions / thoughts and knowledge possessed by experts related to the capture fisheries industry supply chain management system [11]. The appointment of an expert as a source of knowledge or a respondent is based on the following considerations and criteria :

a. Existence, ease and willingness to be interviewed.
b. Reputation, position and credibility as an expert.
c. Expert expertise and experience that demonstrates the ability to give correct advice and can assist with problem solving.

The types of respondents used as primary data sources can be seen in table 1 below :

Table 1. Types of primary data sources.

No	Data source	Types of Data Required	Data Collection Techniques
1	**Respondents From Agencies Government :** - Experts, Decision Makers, and / or Experts, Associated with the Supply Chain Management System - Marine Capture Fisheries Industry	**Related matters Directly or not straight away, especially associated with :** A. Economics, B. Social, C. Environment, D. Technology, E. Resources, F. Institutional, G. Supply Chain, H. Operations, I. Interconnection between ports	**Questionnaire, expert survey;** - Interview (in-depth interview); - Field observations
2	**Respondents from business actors:** Business actors who are directly or indirectly related to the value chain of marine capture fisheries business activities.		
3	**Respondents from stakeholders other related interests:** Other stakeholders who are directly or indirectly related to the marine capture fisheries industry supply chain.		

IOP Conf. Series: Materials Science and Engineering 1052 (2021) 012073 doi:10.1088/1757-899X/1052/1/012073

2.2.2. Identification present situation
Identify a picture of the problem situation obtained at the situational analysis stage used as a reference in the process of making the rich picture analysis. In addition, in-depth interviews were conducted with experts and supply chain actors to explore and identify human activity systems related to the capture fisheries industry supply chain. The results of the analysis obtained are then poured out in order to obtain a rich picture, root definition and a conceptual model (purposive activity model) [12]. The results of this stage then become a reference in designing a supply chain management system for the capture fisheries industry in the Southern part of East Java.

2.3. Marketing distribution & interconnection relationship between ports
The next step is to analyze how marketing distribution and the effect of interconnection between ports in supporting the marketing distribution process. A fishing port is a support for the management and utilization of fish resources and the environment from pre-production, production, processing and marketing [13].

2.4. Conclusions and arrange action plans
The stages of planning an action plan are carried out after get the assumptions used to build a supply chain management system so as to realize a structured capture fisheries supply chain management system in accordance with the existing conditions of East Java. The identification of the modeling design structure is carried out based on the assumptions obtained by the elaboration process from the results of the Situational Analysis and identifying the effects of marketing distribution and interconnection between ports.

3. Results and discussion

3.1. Distribuion Pattern of Catch Fishery in Southern of East Java
The first method to know the distribution pattern of catch fishery in Southern of East Java is describe the situational analysis using statistical data. The statistical data using credible data based on a survey issued by Central Bureau of Statistics of Indonesia [14], it concluded that the fishery production of East Java Province in the year was 414,644 tons during 2017. With dominated by 3 regency, there are Trenggalek Regency with its development center at PPN Prigi, Pacitan Regency with PPP Tamperan as its base and PP Muncar in Banyuwangi Regency. Production being in PP Muncar, which is in Banyuwangi district of 47,433 tons, followed by Trenggalek Regency with 18,472 tons and district Pacitan amounted to 11,054 tons. The volume map of the three districts can be seen in Figure 2 below.

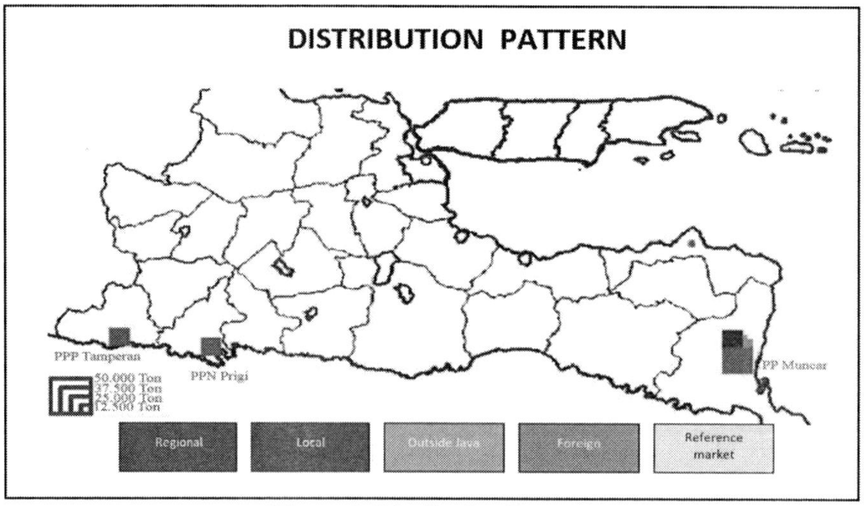

Figure 2. Distribution Pattern.

Based on production data from PP Muncar, it is stated that the total capture fisheries production from PP Muncar during 2017 amounted to 47,433 tonnes with money in circulation amounting to 958,929 billion. As for the amount of this production, the fish is distributed to local markets as well as international markets.

3.2 Distribution pattern of catch fishery analysed from connection between ports

After know distribution pattern of catch fishery in Southern of East Java, to know the connection between port is using descriptive statistic method. The descriptive statistic method based on situational analysis compared with primary data obtained from questionnaires and in-depth interviews from three parties, there are from government, business actor and stakeholders. From descriptive statistic know that based on the inter-port connectivity, the distribution of capture fisheries products that amounting to more than 32 thousand tons of capture fisheries products from PP Muncar are marketed in local markets which include East Java itself. As for the Bali market itself, PP Muncar supplied around 3 tons annually in 2018. As for the West Java, Jakarta, Central Java and surrounding areas, PP Muncar supplied approximately 11 thousand tons during 2018. As for the distribution of capture fisheries products and interconnection between ports, it can be seen in Figure 3 below.

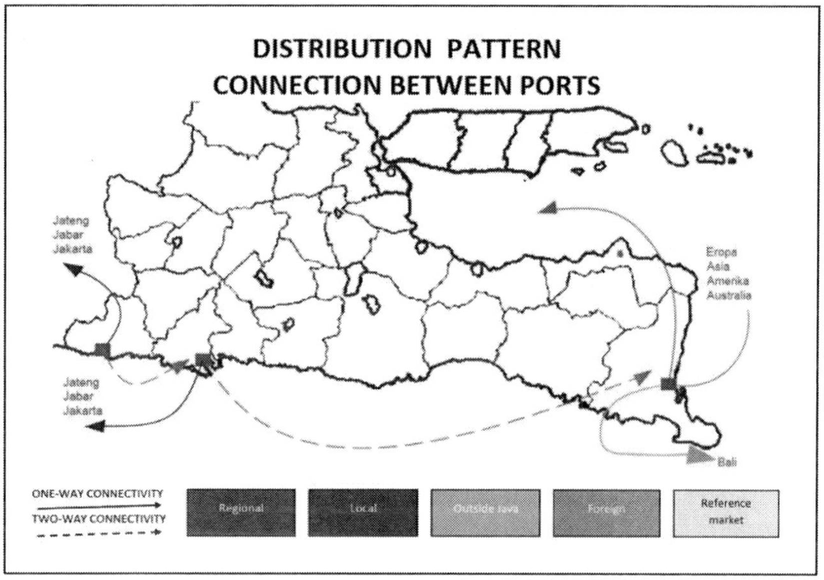

Figure 3. Distribution Pattern Connection Between Ports.

Based on Figure 3 above, the effect of interconnection between ports affects the marketing distribution around the southern part of East Java. From this figure, it can be explained that PP Muncar as a distribution center for capture fisheries products to export, local and long-distance, this is indicated by one-way connectivity means that PP Muncar only supplies market needs. Meanwhile, Pacitan Regency and Trenggalek Regency support a little need from PP Muncar which is shown by two-way connectivity. This means that capture fisheries products from PPN Prigi and PPI Tamperan go to PP Muncar to meet market demand in PP Muncar, but the capture fisheries production from PP Muncar also goes to both places.

4. Conclusions

The distribution of capture fisheries products is more than 73 thousand tons of captured fishery products from PPI Muncar are marketed in local markets covering East Java and Bali. PPI Muncar as a distribution center for capture fisheries products to export, local, long-distance and indicated by one-way connectivity means that PPI Muncar only supplies market needs. Meanwhile, Pacitan and

Trenggalek districts support a little need from PPI Muncar, which is shown by two-way connectivity. This means that capture fisheries products from PPI Tamperan and PPN Prigi in addition to PPI Muncar to meet market demand at PPI Muncar, but in marketing their capture fisheries production from PPI Muncar also goes to these two places. It can be concluded that the interconnection between ports is very influential on the marketing distribution of catch fisheries in Southeast of East Java, it is because the interconnection between ports will support the market needs of ports or other markets, this is indicated by two-way connectivity between ports.

Acknowledgements
It is hoped that the design of the capture fisheries industry supply chain management system can be used as input for policy makers, especially in building a capture fisheries industry supply chain management system based on regional situational conditions.

References

[1] Dinas Kelautan dan Perikanan Provinsi Jawa Timur. 2019. *Pemantapan Pelaksanaan Program Kegiatan Tahun 2019*. Jakarta.
[2] KKP. 2015. *Kelautan dan Perikanan dalam Angka Tahun 2015*. Pusat Data Statistik dan Informasi Kementerian Kelautan dan Perikanan. Jakarta.
[3] KKP. 2014. *Kelautan dan Perikanan dalam Angka Tahun 2014*. Pusat Data Statistik dan Informasi Kementerian Kelautan dan Perikanan. Jakarta.
[4] KKP. 2016. *Kelautan dan Perikanan dalam Angka Tahun 2016*. Pusat Data Statistik dan Informasi Kementerian Kelautan dan Perikanan. Jakarta.
[5] KKP. 2014. *Kelautan dan Perikanan dalam Angka Tahun 2014*. Pusat Data Statistik dan Informasi Kementerian Kelautan dan Perikanan. Jakarta.
[6] Rogers M. 1993. *Exercising Responsibility in the Seafood Supply Chain: A Case Study on How a Retailer Implements its Commitment to Sustainable Seafood & Experiences of Other Seafood Buyers*. Thesis Master of Science in Environmental Management and Policy, Lund, Sweden.
[7] Ronald E.Walpole. *Pengantar Statistika*. Jakarta.
[8] Minister of Marine and Fisheries. 2018. *Produktivitas Perikanan Indonesia*. Jakarta. Evaluasi Pelaksanaan Anggaran Tahun 2017.
[9] Angela Thody. 2006. *Writing and Presenting Research*. University of Lincoln, UK.
[10] Cutlip, Center, & Broom. 2000. *Effective Public Relations*. Prentice-Hall. New Jersey.
[11] Dergibson Siagian & Sugiarto. 2002. *Metode Statistika untuk Bisnis dan Ekonomi*. Jakarta: PT Gramedia Pustaka Utama.
[12] Checkland P, Poulter J. 2010. *Soft System Methodology*. Published In Association with Springer-Verlag. London.
[13] Regulation Minister of Marine and Fisheries of The Republic of Indonesia. Number PER.08 / MEN About Fishing Ports. Jakarta. 2012.
[14] Milestad RR, Bartel-Kratochvil H, Leitner P, Axmann. Being close : *The quality of social relationship in a local organic cereal and bread network in lower Austria*. Journal of Rural Studies 26 (3) : 228-240. Austria. 2010.

AUTHOR INDEX